The ***Streeter/Hutchison*** Series in Mathematics

**Baratto**

# ***basic mathematical skills***

***seventh*** edition

## with ***geometry***

 The Streeter/Hutchison Series in Mathematics Basic Mathematical Skills with Geometry

The *Streeter/Hutchison* Series in Mathematics

Baratto | Bergman

# *basic mathematical skills*

***seventh** edition*

## with ***geometry***


**Stefan Baratto**

Clackamas Community College

**Barry Bergman**

Clackamas Community College

**Higher Education**

Boston Burr Ridge, IL Dubuque, IA New York San Francisco St. Louis
Bangkok Bogotá Caracas Kuala Lumpur Lisbon London Madrid Mexico City
Milan Montreal New Delhi Santiago Seoul Singapore Sydney Taipei Toronto

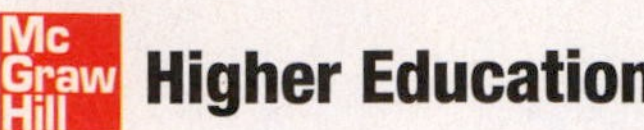

**Higher Education**

BASIC MATHEMATICAL SKILLS WITH GEOMETRY, SEVENTH EDITION

Published by McGraw-Hill, a business unit of The McGraw-Hill Companies, Inc., 1221 Avenue of the Americas, New York, NY 10020. 

This book is printed on recycled, acid-free paper containing 10% postconsumer waste.

2 3 4 5 6 7 8 9 0 QPD/QPD 0 9 8 7
1 2 3 4 5 6 7 8 9 0 QPD/QPD 0 9 8 7 6

ISBN 978–0–07–304833–8
MHID 0–07–304833–X

ISBN 978–0–07–325777–8
MHID 0–07–325777–X (Annotated Instructor's Edition)

Publisher: *Elizabeth J. Haefele*
Sponsoring Editor: *Richard Kolasa*
Senior Developmental Editor: *Michelle L. Flomenhoft*
Marketing Manager: *Barbara Owca*
Project Manager: *Lora Kalb*
Senior Production Supervisor: *Kara Kudronowicz*
Senior Media Project Manager: *Sandra M. Schnee*
Media Producer: *Amber M. Huebner*
Senior Designer: *David W. Hash*
Cover/Interior Designer: *Asylum Studios; Mona Grigaliunas, Tom J. Nemoda*
Senior Photo Research Coordinator: *Lori Hancock*
Photo Research: *Connie Mueller*
Supplement Producer: *Melissa M. Leick*
Compositor: *Interactive Composition Corporation*
Typeface: *10/12 New Times Roman*
Printer: *Quebecor World Dubuque, IA*

**Chapter 1:** Opener: © Greg Baker/AP Wide World; p. 25: © National Motor Museum/HIP/The Image Works; p. 27: © Dynamic Graphics/Jupiter Images; p. 44: © Greg Baker/AP Wide World; p. 53: © Alistair Berg/PhotoDisc/Getty Images; p. 87: © Business & Industry 3/Corbis; p. 92: © Susan VanEtten/Photo Edit; p. 104: © Mary Kate Danny/PhotoEdit.

**Chapter 2:** Opener: © BananaStock/PunchStock; p. 124: © Education/PhotoDisc; p. 134: © PhotoLink/Getty Images; p. 145: © The McGraw Hill Companies, Inc./Connie Mueller, photographer; p. 160: © The McGraw Hill Companies, Inc./Christopher Kerrigan, photographer; p. 177: © Vol. 16/PhotoDisc/Getty Images; p. 193: © BananaStock/PunchStock; p. 205: © Business & Industries/PhotoDisc.

**Chapter 3:** Opener: © Royalty-Free/Corbis; p. 248: © Royalty-Free/Corbis; p. 261: © APFile Photo/ Kevin German; p. 276: © PhotoLink/Getty Images; p. 284: © Mature Lifestyles/PhotoDisc/Getty Images.

**Chapter 4:** Opener: © Donovan Reese/Getty Images; p. 308: © Royalty-Free/Corbis; p. 312: © Comstock Images/JupiterImages; p. 338: John A. Rizzo/Getty Images; p. 339: © Brand X Pictures/Getty Images; p. 354: © Donovan Reese/Getty Images.

**Chapter 5:** Opener: © Ryan McVay/Getty Images: p. 369: © Nick Koudis/Getty Images; p. 376: © Getty Images/SW Productions; p. 389: © Getty Images; p. 414: © Ryan McVay/Getty Images.

**Chapter 6:** Opener: © John Lamb/Getty Images; p. 432: © Education/PhotoDisc/Getty Images; p. 433: © Digital Vision/Punch Stock; p. 450: © Getty Images/ Digital Vision; p. 456: © Education/PhotoDisc/Getty Images; p. 460: © Javier Pierini/Getty Images; p. 463: © The McGraw Hill Companies, Inc./John Flournoy, photographer; p. 465: © Royalty-Free/Corbis; p. 467: © The McGraw Hill Companies, Inc./Connie Mueller, photographer; p. 469: © Ingram Publishing/ SuperStock; p. 480: © Alan Schein Photography/Corbis.

**Chapter 7:** Opener: © Royalty-Free/Corbis; p. 511: © Steve Dunwell/The Image Bank/Getty Images; p. 522: © Antonia Reeve/Photo Researchers, Inc.; p. 532: © Digital Vision/PunchStock; p. 538: © Royalty-Free/Corbis.

**Chapter 8:** Opener: © Vol. 146/Corbis; p. 565: © Eat, Drink, Dine/PhotoDisc/Getty Images; p. 567: © PhotoLink/Getty Images; p. 569: © Business & Industry 2/Corbis; p. 584: © Hisham F. Ibrahim/Getty Images; p. 600: © Vol. 146/Corbis; p. 612: © Andrew Ward/Life File/Getty Images.

**Chapter 9:** Opener: © NOAA Office of Marine and Aircraft Operations; p. 636: © Brand X Pictures/PunchStock; p. 645: © Jack Star/PhotoLink/Getty Images; p. 657: © Recreational Sports/Corbis; p. 684: © Jack Star/PhotoLink/Getty Images; p. 700: © NOAA Office of Marine and Aircraft Operations; p. 708: © Business On The Go/PhotoDisc/Getty Image.

**Chapter 10:** Opener: © Norbert Schaefer/Corbis; p. 717: © New York City/Corbis CD; p. 740: © NOAA Photo Library, NOAA Central Library; OAR/ERL/ National Severe Storms Laboratory (NSSL); p. 743: © Vol. 74/PhotoDisc/Getty Images; p. 750: © Reuters/ Corbis; p. 775: © John A. Rizzo/Getty Images.

**Chapter 11:** Page 806: © Steve Mason/Getty Images; p. 818: © Brand X/SuperStock RF; p. 832: © Digital Vision/Getty Images.

www.mhhe.com

# dedications

Don Hutchison has literally changed our lives. Don provided us with the opportunity of a lifetime by inviting us to write with him on the Streeter/Hutchison texts. Just as Don learned a great deal from James Streeter, we have learned and grown in many ways thanks to Don. Not only does Don have the wonderful ability to present mathematical ideas in simple, clear terms, he is also generous, has a wonderful family, and is a pleasure to spend time with. Don serves as an excellent role model for both of us, but most importantly, we are proud to call Don a friend.

**Stefan Baratto and Barry Bergman**

 

# about the authors

Stefan Baratto

**Stefan** currently enjoys teaching math at Clackamas Community College in Oregon. He has also taught at the University of Oregon, Southeast Missouri State University, York County Technical College (Maine), and in New York City middle schools.

Stefan is involved in many professional organizations, including AMATYC and ORMATYC, where he is a frequent presenter. He has also used statistics, technology, and Web design to apply math to various fields.

Stefan, and his wife Peggy, make their home in Portland and spend their time enjoying the wonders of Oregon and the Pacific Northwest.

Barry Bergman

**Barry** has enjoyed teaching mathematics to a wide variety of students over the years. He began in the field of adult basic education and moved into the teaching of high school mathematics in 1977. He taught at that level for 11 years, at which point he served as a K–12 mathematics specialist for his county. This work allowed him the opportunity to help promote the emerging NCTM Standards in his region.

In 1990 Barry began the present portion of his career, having been hired to teach at Clackamas Community College. He maintains a strong interest in the appropriate use of technology and visual models in the learning of mathematics.

Throughout the past 29 years, Barry has played an active role in professional organizations. As a member of OCTM, he contributed several articles and activities to the group's journal. He has made presentations at OCTM, NCTM, ORMATYC, and ICTCM conferences. Barry also served as an officer of ORMATYC for four years and participated on an AMATYC committee to provide feedback to revisions of NCTM's Standards.

Donald Hutchison

**Don** began teaching in a preschool while he was an undergraduate. He subsequently taught children with disabilities, adults with disabilities, high school mathematics, and college mathematics. Although all of these positions were challenging and satisfying, it was breaking a challenging lesson into teachable components that he most enjoyed.

It was at Clackamas Community College that he found his professional niche. The community college allowed him to focus on teaching within a department that constantly challenged faculty and students to expect more. Under the guidance of Jim Streeter, Don learned to present his approach to teaching in the form of a textbook.

Don has also been an active member of many professional organizations. He has been president of ORMATYC, AMATYC committee chair, and ACM curriculum committee member. He has presented at AMATYC, ORMATYC, AACC, MAA, ICTCM, and numerous other conferences.

# contents

## Chapter 9

# Data Analysis and Statistics



## Chapter 10

# The Real Number System



## Chapter 11

# An Introduction to Algebra

# to the student

**You** are about to begin a course in mathematics. We made every attempt to provide a text that will help you understand what mathematics is about and how to use it effectively. We made no assumptions about your previous experience with mathematics. Your progress through the course will depend on the amount of time and effort you devote to the course and your previous background in math. There are some specific features in this book that will aid you in your studies. Here are some suggestions about how to use this book. (Keep in mind that a review of *all* the chapter and summary material will further enhance your ability to grasp later topics and to move more effectively through the text.)

1. If you are in a lecture class, make sure that you take the time to read the appropriate text section *before* your instructor's lecture on the subject. Then take careful notes on the examples that your instructor presents during class.
2. After class, work through similar examples in the text, making sure that you understand each of the steps shown. Examples are followed in the text by *Check Yourself* exercises. You can best learn mathematics by being involved in the process, and that is the purpose of these exercises. Always have a pencil and paper at hand, work out the problems presented, and check your results immediately. If you have difficulty, go back and carefully review the previous exercises. Make sure you understand what you are doing and why. The best test of whether you do understand a concept lies in your ability to explain that concept to one of your classmates. Try working together.
3. At the end of each chapter section you will find a set of exercises. Work these carefully to check your progress on the section you have just finished. You will find the solutions for the odd-numbered exercises following the problem set. If you have difficulties with any of the exercises, review the appropriate parts of the chapter section. If your questions are not completely cleared up, by all means do not become discouraged. Ask your instructor or an available tutor for further assistance. A word of caution: Work the exercises on a regular (preferably daily) basis. Again, learning mathematics requires becoming involved. As is the case with learning any skill, the main ingredient is practice.
4. When you complete a chapter, review by using the *Summary.* You will find all the important terms and definitions in this section, along with examples illustrating all the techniques developed in the chapter. Following the *Summary* are *Summary Exercises* for further practice. The exercises are keyed to chapter sections, so you will know where to turn if you are still having problems.
5. When you finish with the *Summary Exercises,* try the *Self-Test* that appears at the end of each chapter. It is an actual practice test you can work on as you review for in-class testing. Again, answers with section references are provided.
6. Finally, an important element of success in studying mathematics is the process of regular review. We provide a series of *Cumulative Reviews* throughout the textbook, beginning at the end of Chapter 2. These tests will help you review not only the concepts of the chapter that you have just completed but also those of previous chapters. Use these tests in preparation for any midterm or final exams. If it appears that you have forgotten some concepts that are being tested, don't worry. Go back and review the sections where the idea was initially explained or the appropriate chapter summary. That is the purpose of the Cumulative Review.

We hope that you will find our suggestions helpful as you work through this material, and we wish you the best of luck in the course.

**Stefan Baratto**
**Barry Bergman**
**Donald Hutchison**

# preface

## Message from the Authors

We believe the key to learning mathematics, at any level, is active participation—**MASTERING MATH THROUGH PRACTICE.** Students who are active participants in the learning process have the opportunity to construct their own mathematical ideas and make connections to previously studied material. Such participation leads to understanding, retention, success, and confidence. We developed this text with that philosophy in mind and integrated many features throughout the book to reflect that philosophy. Our goal is to provide *content in context.* The opening vignette for each chapter, one activity in that chapter, and several section exercises all relate to the same topic in order to engage students and allow them to see the relevance of mathematics. The Check Yourself exercises are designed to keep the students active and involved with every page of exposition. The optional calculator references involve students actively in the development of mathematical ideas.

The exercise sets are organized to showcase the different types of exercises available to the students: Basic Skills, Advanced Skills, Vocational-Technical Applications, Calculator/Computer, and Above and Beyond (which encompasses challenging exercises, writing exercises, and collaborative exercises). Answer blanks for these exercises appear in the margin allowing the student to actively use the text, making it more than just a reference tool. Many of these exercises are designed to awaken interest and insight in students: all are meant to provide continual practice and reinforcement of the topics being learned. For those seeking additional understanding, some exercises include icons next to them to indicate that a video stepping through that exercise is available. Not all the exercises will be appropriate for every student, but each one gives another opportunity for both the instructor and the student. Our hope is that every student who uses this text will become a better mathematical thinker as a result.

 

# Overview

When preparing to write this text, the authors solicited feedback from the market about what types of pedagogical tools would better help students to understand and retain the key basic mathematical concepts, in addition to what content should be updated or expanded. Increasing the number of applications and integrating them throughout the sections was one of the most prevalent responses. Modifying the geometry coverage and incorporating some geometric material earlier was another common request. Based on this feedback, the authors focused on these themes as they revised the text.

In terms of pedagogical tools, this text seeks to provide carefully detailed explanations and accessible pedagogy to introduce basic mathematic skills to the students. The authors use a three-pronged approach to present the material and encourage critical thinking skills. The areas used to create the framework are communication, pattern recognition, and problem solving. Items such as Math Anxiety boxes, Check Yourself exercises, and activities represent this approach and the underlying philosophy of mastering math through practice. A new feature is Reading Your Text—these quick exercises presented at the end of each section quiz students' vocabulary knowledge and help strengthen their communication skills.

Market research has reinforced the importance of the exercise sets to the student's ability to process the content. To that end, the exercise sets in this edition have been expanded, organized, and clearly labeled (as previously outlined). Vocational and professional-technical exercises have been added throughout. In addition, exercises with fractions, decimals, and negative numbers have been added as appropriate. Repeated exposure to this consistent structure should help advance the student's skills in relating to mathematics.

# Features

A number of features have contributed to the previous success of the Hutchison texts. All of these features are also included in this text and help reinforce the authors' philosophy of *mastering math through practice.* Each feature was thoroughly discussed by the authors and review panels during the development of the text. More than ever, we are confident that the entire learning package is of value to your students and to you as an instructor. We will describe each of the key features of our package.

### > Make the Connection—Chapter-Opening Vignettes

The chapter-opening vignettes were substantially revised to provide students with interesting, relevant scenarios that will capture their attention and engage them in the upcoming material. Furthermore, exercises and activities related to the opening vignette were added or updated in each chapter. These exercises are marked with a special icon next to them.

CHAPTER 4

Decimals

INTRODUCTION

When Barry is not at his job, he loves to be outside partaking in many outdoor activities. Some of these activities include bicycling, hiking, and tennis. Lately, Barry has been spending a lot of time on his bike. He bought a new road bike because of the long distances he was riding. Two years ago, Barry noticed a sign in his neighborhood advertising a 150-mile bike ride to raise money for muscular dystrophy. He thought it was a good cause and started training for the ride. Barry completed the ride and experienced a great feeling of accomplishment in doing so.

The MD ride inspired Barry to look for more rides that raised money for important causes. He has since ridden for cerebral palsy and cancer research. Barry likes the idea that he can raise money for a good cause while getting exercise doing something he enjoys.

Lance Armstrong is one of Barry's inspirational heroes. Armstrong is a cancer survivor who battled the disease and came back to win the Tour de France seven times. He founded the Lance Armstrong Foundation for Cancer Research and the yellow Livestrong wristbands.

You can learn more about the Tour de France by doing Activity 12 on page 354. For more information on the Tour de France or Lance Armstrong you can also go to www.letour.fr or lancewins.com.

CHAPTER 4 OUTLINE



289

## > Activities

Activities are included in each chapter. They promote active learning by requiring students to find, interpret, and manipulate real world data. One activity in each chapter relates to the chapter-opening vignette, providing cohesiveness to the chapter. Students can complete the activities on their own, but these are best solved in small groups.

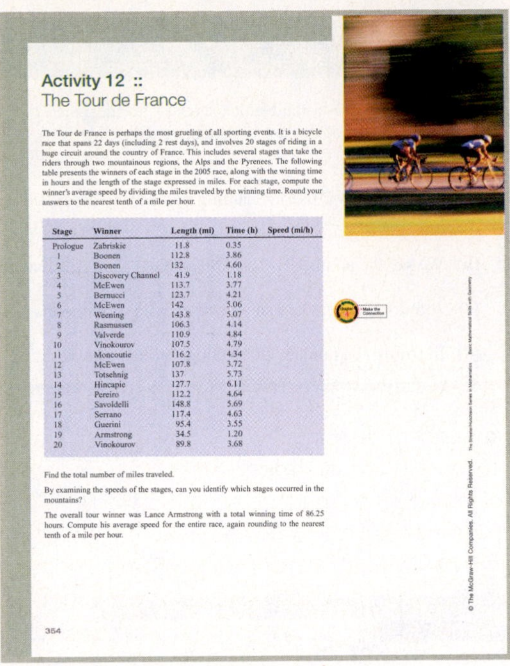

Activity 12 ::
The Tour de France

The Tour de France is perhaps the most grueling of all sporting events. It is a bicycle race that spans 22 days (including 2 rest days), and involves 20 stages of riding in a huge circuit around the country of France. This includes several stages that take the riders through two mountainous regions, the Alps and the Pyrenees. The following table presents the winners of each stage in the 2005 race, along with the winning time in hours and the length of the stage expressed in miles. For each stage, compute the winner's average speed by dividing the miles traveled by the winning time. Round your answers to the nearest tenth of a mile per hour.

| Stage | Winner | Length (mi) | Time (h) | Speed (mi/h) |
|---|---|---|---|---|
| Prologue | Zabriskie | 11.8 | 0.35 | |
| 1 | Boonen | 112.8 | 3.86 | |
| 2 | Boonen | 132 | 4.60 | |
| 3 | Discovery Channel | 41.9 | 1.18 | |
| 4 | McEwen | 113.7 | 3.77 | |
| 5 | Bernucci | 123.7 | 4.21 | |
| 6 | McEwen | 142 | 5.06 | |
| 7 | Weening | 143.8 | 5.07 | |
| 8 | Rasmussen | 106.3 | 4.14 | |
| 9 | Valverde | 110.9 | 4.84 | |
| 10 | Vinokourov | 107.5 | 4.79 | |
| 11 | Moncoutie | 116.2 | 4.34 | |
| 12 | McEwen | 107.8 | 3.72 | |
| 13 | Totschnig | 137 | 5.73 | |
| 14 | Hincapie | 127.7 | 6.11 | |
| 15 | Pereiro | 112.2 | 4.64 | |
| 16 | Savoldelli | 148.8 | 5.69 | |
| 17 | Serrano | 117.4 | 4.63 | |
| 18 | Guerini | 95.4 | 3.55 | |
| 19 | Armstrong | 34.5 | 1.20 | |
| 20 | Vinokourov | 89.8 | 3.68 | |

Find the total number of miles traveled.

By examining the speeds of the stages, can you identify which stages occurred in the mountains?

The overall tour winner was Lance Armstrong with a total winning time of 86.25 hours. Compute his average speed for the entire race, again rounding to the nearest tenth of a mile per hour.

354

## > Check Yourself Exercises

The Check-Yourself exercises have been the hallmark of the Hutchison series; they are designed to actively involve students throughout the learning process. Every example is followed by an exercise that encourages students to solve a problem similar to the one just presented and to check and practice what they have just learned. Answers are provided at the end of the section for immediate feedback.

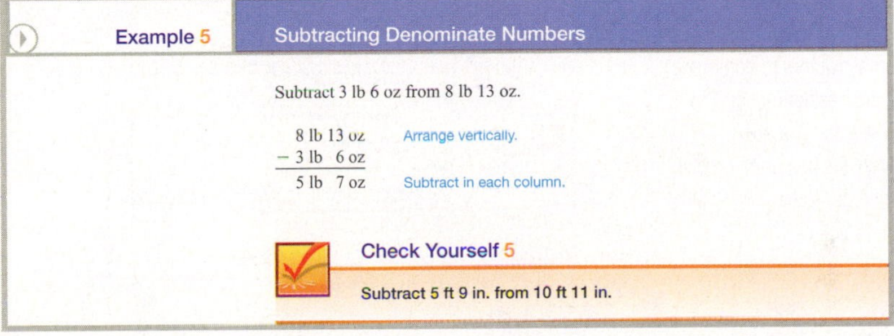

Example 5 — Subtracting Denominate Numbers

Subtract 3 lb 6 oz from 8 lb 13 oz.

8 lb 13 oz — Arrange vertically.
− 3 lb 6 oz
5 lb 7 oz — Subtract in each column.

Check Yourself 5

Subtract 5 ft 9 in. from 10 ft 11 in.

## > Overcoming Math Anxiety Boxes

The Overcoming Math Anxiety boxes are located within the first few chapters. These suggestions are designed to be timely and useful. They are similar to the same suggestions most instructors make in class, but having them in print in the text provides another opportunity to impact the student.

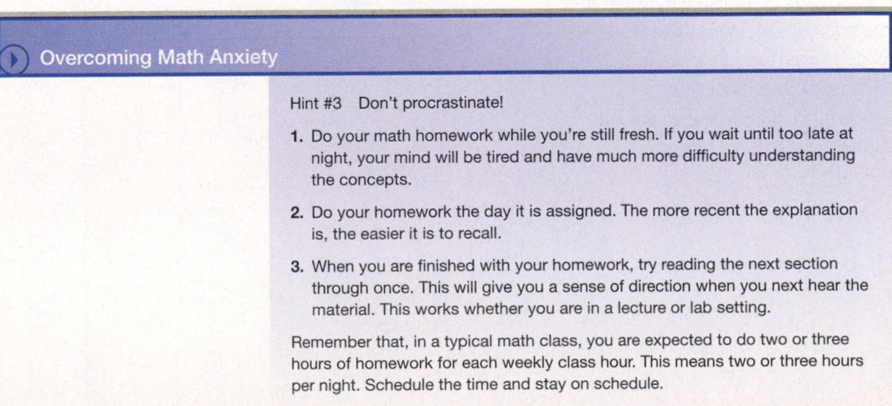

Overcoming Math Anxiety

Hint #3 Don't procrastinate!

1. Do your math homework while you're still fresh. If you wait until too late at night, your mind will be tired and have much more difficulty understanding the concepts.
2. Do your homework the day it is assigned. The more recent the explanation is, the easier it is to recall.
3. When you are finished with your homework, try reading the next section through once. This will give you a sense of direction when you next hear the material. This works whether you are in a lecture or lab setting.

Remember that, in a typical math class, you are expected to do two or three hours of homework for each weekly class hour. This means two or three hours per night. Schedule the time and stay on schedule.

## > Reading Your Text

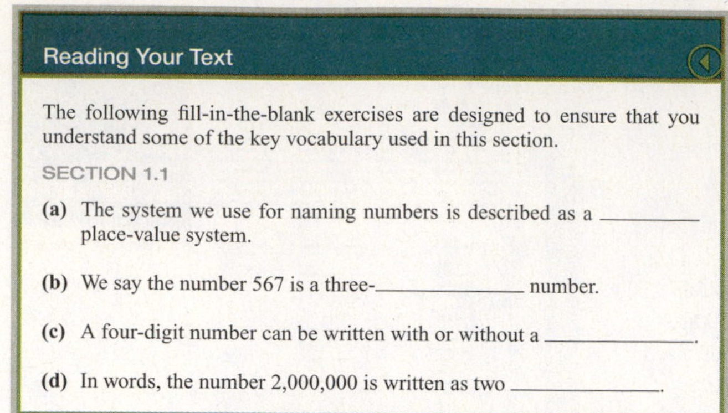

Reading Your Text

The following fill-in-the-blank exercises are designed to ensure that you understand some of the key vocabulary used in this section.

SECTION 1.1

(a) The system we use for naming numbers is described as a ________ place-value system.

(b) We say the number 567 is a three-____________ number.

(c) A four-digit number can be written with or without a ____________.

(d) In words, the number 2,000,000 is written as two ____________.

The new Reading Your Text feature is a set of quick exercises presented at the end of each section meant to quiz students' vocabulary knowledge. These exercises are designed to encourage careful reading of the text. If students do not understand the vocabulary, they cannot communicate effectively. The Reading Your Text exercises address the vocabulary issue that many students struggle with in learning and understanding mathematics. Answers to these exercises are provided at the end of the book.

## > End-of-Section Exercises

The comprehensive end-of-section exercises have been reorganized to more clearly identify the different types of exercises being presented. This structure highlights the progression in level and type of exercise for each section. This will not only provide clarity for students, but will also make it easier for instructors to determine the exercises for their assignments. The application exercises that are now integrated into every section are a crucial component of this organization.

## > Summary and Summary Exercises

The comprehensive Summaries at the end of each chapter enable students to review important concepts. The Summary Exercises provide an opportunity for students to practice these important concepts. The answers to odd-numbered exercises are provided in the Answers Appendix.

## > Icon Key

**Make the Connection.** Indicates the exercises, activity, and chapter opener that relate to each other.

**Check Yourself.** Indicates an exercise that is tied to the preceding example.

**Video.** Indicates an exercise that has a video walking through how to solve it.

**Caution.** Points out potential trouble spots.

## > Margin Notes and Recall Notes

Margin notes and Recall notes are provided throughout the text. Margin notes are designed to help students focus on important topics and techniques, while Recall notes give students references to previously learned material.

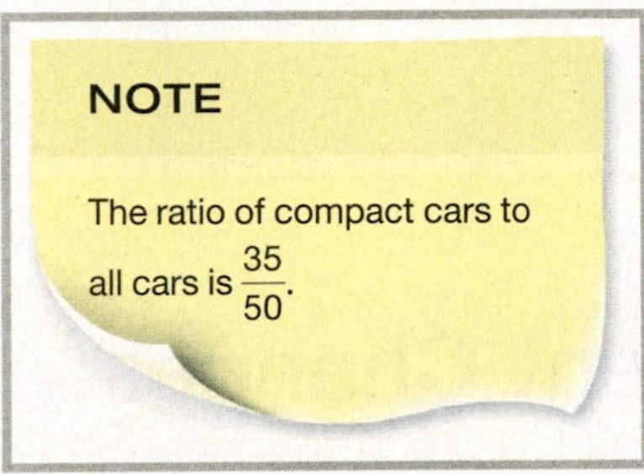


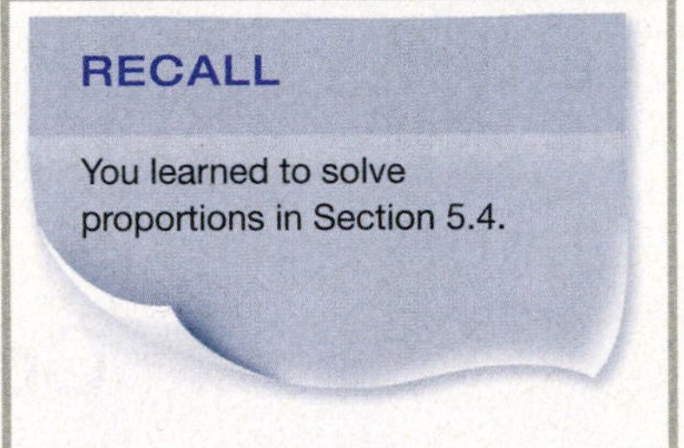


## > Learning Objectives

Learning objectives are clearly identified for each section. Annotations for the objectives appear next to examples, showing when a particular objective is about to be developed. References are also included within the exercise sets to help students quickly identify examples related to topics where they need more practice.

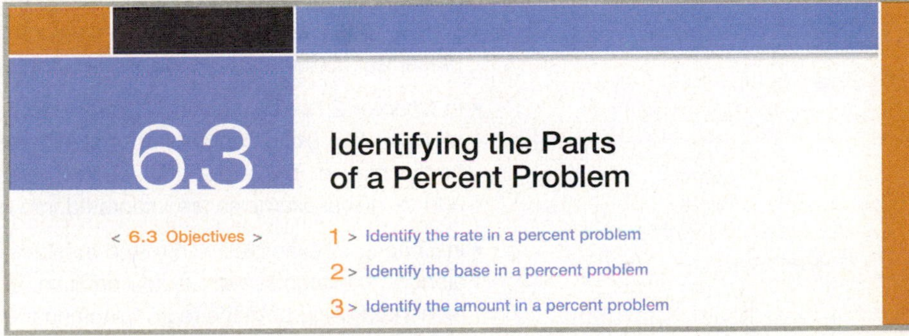


## > Self-Tests

Self-tests appear in each chapter to provide students with an opportunity to check their progress and to review important concepts, as well as provide confidence and guidance in preparing for in-class tests or exams. The answers to the Self-Test exercises are given at the end of the book. Section references are given with the answers to help the student.

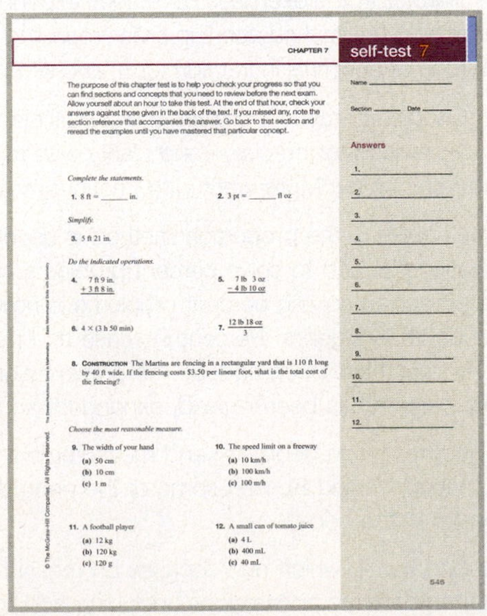

CHAPTER 7

**self-test 7**

The purpose of this chapter test is to help you check your progress so that you can find sections and concepts that you need to review before the next exam. Allow yourself about an hour to take this test. At the end of that hour, check your answers against those given in the back of the text. If you missed any, note the section reference that accompanies the answer. Go back to that section and reread the examples until you have mastered that particular concept.

Name ______

Section ______ Date ______

*Complete the statements.*

1. 8 ft = ______ in.
2. 3 pt = ______ fl oz

*Simplify.*

3. 5 ft 21 in.

*Do the indicated operations.*

4. 7 ft 9 in. + 3 ft 8 in.
5. 7 lb 3 oz − 4 lb 10 oz
6. 4 × (3 h 50 min)
7. $\frac{12 \text{ lb } 18 \text{ oz}}{3}$
8. **Construction** The Martins are fencing in a rectangular yard that is 110 ft long by 40 ft wide. If the fencing costs $3.50 per linear foot, what is the total cost of the fencing?

*Choose the most reasonable measure.*

9. The width of your hand
   (a) 50 cm
   (b) 10 cm
   (c) 1 m
10. The speed limit on a freeway
   (a) 10 km/h
   (b) 100 km/h
   (c) 100 m/h
11. A football player
   (a) 12 kg
   (b) 120 kg
   (c) 120 g
12. A small can of tomato juice
   (a) 4 L
   (b) 400 mL
   (c) 40 mL

Answers

1. ______
2. ______
3. ______
4. ______
5. ______
6. ______
7. ______
8. ______
9. ______
10. ______
11. ______
12. ______

545

## > Cumulative Reviews

Cumulative Reviews are included starting with Chapter 2, following the Self-Tests. These reviews help students build on previously covered material and give them an opportunity to reinforce the skills necessary in preparing for midterm and final exams. These reviews assist students with the retention of knowledge throughout the course. The answers to these exercises are also given at the end of the book, along with section references.

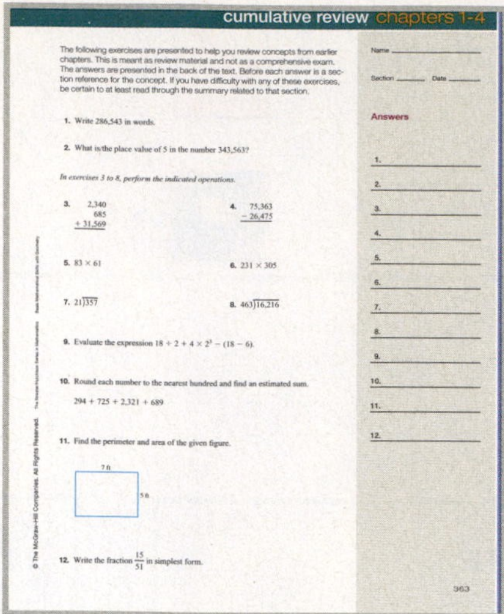

cumulative review chapters 1-4

The following exercises are presented to help you review concepts from earlier chapters. This is meant as review material and not as a comprehensive exam. The answers are presented in the back of the text. Before each answer is a section reference for the concept. If you have difficulty with any of these exercises, be certain to at least read through the summary related to that section.

Name ______

Section ______ Date ______

1. Write 286,543 in words.

2. What is the place value of 5 in the number 343,563?

*In exercises 3 to 8, perform the indicated operations.*

3. 2,340 + 685 + 31,569

4. 75,363 − 26,475

5. $83 \times 61$

6. $231 \times 305$

7. $21\overline{)357}$

8. $463\overline{)16{,}216}$

9. Evaluate the expression $18 \div 2 + 4 \times 2^3 - (18 - 6)$.

10. Round each number to the nearest hundred and find an estimated sum.

$294 + 725 + 2{,}321 + 689$

11. Find the perimeter and area of the given figure.



12. Write the fraction $\frac{15}{51}$ in simplest form.

Answers

1. ______
2. ______
3. ______
4. ______
5. ______
6. ______
7. ______
8. ______
9. ______
10. ______
11. ______
12. ______

363

# Content Changes

- In *Chapter 1,* the concepts of "borrowing" and "carrying" are now presented using more mathematically precise language, such as "regrouping." Geometry is now integrated into appropriate sections. Other forms for multiplication are presented, including the centered dot and parentheses.
- In *Chapter 2,* the number of examples and exercises with exponents, multiplication using centered dot notation, and multiplication using parentheses has increased. The explanation of finding factors has been clarified. Geometric and technical examples are integrated into appropriate sections.
- In *Chapter 3,* examples with mixed numbers have been added throughout, along with examples with vertical addition. Subtracting mixed numbers is now presented using both the regrouping and the improper-fractions approaches. A new section on the order of operations and fractions has been added, while more order-of-operations examples and exercises have been added throughout.
- In *Chapter 4,* the exercises have been expanded to include exercises involving arranging decimal numbers in order, rounding to the nearest cent or dollar, division of decimals in fraction form, and simpler order-of-operations calculations.
- In *Chapter 5,* ratios are given using multiple notations. The exposition has been revised for precision and clarity with regards to rates, ratios, unit rates, and unit ratios. More work with fractions is also included.
- In *Chapter 6,* the proportion method is used to describe the percent relationship and to solve percent problems and applications. The equation approach to solving percent problems is now in Chapter 11, after introducing students to algebra. Percent increase and percent decrease examples and exercises have been added. The presentation of the table with percent-equivalents has been moved, expanded, and clarified.
- The chapter on geometry and measurement has been split into two chapters (Chapters 7 and 8), with some of the geometry material now being presented earlier in the book.
- In *Chapter 7,* which now focuses on measurements, material has been added on temperature conversions, including additional examples.

- *Chapter 8* now focuses on geometric topics and includes more precise vocabulary terms.
- *Chapter 9* now contains technical examples and applications where appropriate, as do the other chapters.
- In *Chapter 10,* more explanation and examples for multiple negative signs are included, while the discussion of order of operations has been revised.
- In *Chapter 11,* the coverage of solving linear equations is more consistent.

# Supplements

## Multimedia Supplements

www.mathzone.com

### MathZone

McGraw-Hill's *MathZone* is a complete, online tutorial and course management system for mathematics and statistics, designed for greater ease of use than any other system available. Attainable with selected McGraw-Hill texts, the system allows instructors to *create and share courses and assignments* with colleagues and adjuncts with only a few clicks of the mouse. All assignments, questions, e-Professors, online tutoring, and video lectures are directly tied to text-specific materials.

MathZone courses are customized to your textbook, but you can edit questions and algorithms, import your own content, and create announcements and due dates for assignments.

MathZone has automatic grading and reporting of easy-to-assign algorithmically generated homework, quizzing, and testing. All student activity within MathZone is automatically recorded and available to you through a fully integrated grade book that can be downloaded to Excel.

MathZone offers:

- *Practice exercises* based on the text and generated in an unlimited number for as much practice as needed to master any topic you study.
- *Videos* of classroom instructors giving lectures and showing you how to solve exercises from the text.
- *e-Professors* to take you through animated, step-by-step instructions (delivered via on-screen text and synchronized audio) for solving exercises in the book, allowing you to digest each step at your own pace.
- *NetTutor,* which offers live, personalized tutoring via the Internet.

### Instructor's Testing and Resource CD (Instructors Only)

This cross-platform CD-ROM includes a computerized test bank utilizing Brownstone Diploma algorithm-based testing software that enables users to create customized exams quickly. This user-friendly program enables instructors to search for questions by topic, format, or difficulty level; to edit existing questions or to add new ones; and to scramble questions and answer keys for multiple versions of the same test. Hundreds of text-specific open-ended and multiple-choice questions are included in the question bank. Sample chapter tests and final exams in Microsoft Word and PDF formats are also provided.

### NetTutor

Available through MathZone, NetTutor is a revolutionary system that enables students to interact with a live tutor over the World Wide Web. NetTutor's Web-based, graphical chat capabilities enable students and tutors to use mathematical notation and even to draw graphs as they work through a problem together. Students can also submit questions and receive answers, browse previously answered questions, and view previous live-chat sessions. Tutors are familiar with the textbook's objectives and problem-solving styles.

### Video Lectures on Digital Video Disk (DVD)

In the videos, qualified teachers work through selected exercises from the textbook, following the solution methodology employed in the text. The video series is available on DVD or online as an assignable element of MathZone. The DVDs are closed-captioned for the hearing impaired, subtitled in Spanish, and meet the Americans with Disabilities Act Standards for Accessible Design. Instructors may use them as resources in a learning center, for online courses, and/or to provide extra help for students who require extra practice.

### ALEKS

www.ALEKS.com

ALEKS (**A**ssessment and **LE**arning in **K**nowledge **S**paces) is an artificial-intelligence–based system for individualized mathematics learning, available over the Web 24/7. ALEKS's unique adaptive questioning continually assesses each student's math knowledge. ALEKS then provides an individualized learning path, guiding the student in the selection of appropriate new study material. ALEKS 3.0 now links to text-specific video, multimedia tutorials, and textbook pages in PDF format. The system records each student's progress toward mastery of curricular goals in a robust classroom management system.

ALEKS improves students' performance by assessing what they know, guiding them to what they are ready to learn, and helping them master key mathematical concepts.

## Printed Supplements

### Annotated Instructor's Edition (Instructors Only)

This ancillary contains answers to exercises in the text, including answers to all section exercises, all Summary Exercises, Self-Tests, and Cumulative Reviews. These answers are printed in a special color for ease of use by the instructor and are located on the appropriate pages throughout the text. Exercises, Self-Tests, Summary Exercises, and Cumulative Reviews are annotated with section references to aid the instructor who may have omitted certain sections from study.

### Instructor's Solutions Manual (Instructors Only)

The Instructor's Solutions Manual provides comprehensive, worked-out solutions to all exercises in the text. The methods used to solve the problems in the manual are the same as those used to solve the examples in the textbook.

### Student's Solutions Manual

The Student's Solutions Manual provides comprehensive, worked-out solutions to all the odd-numbered exercises. The steps shown in the solutions match the style of solved examples in the textbook.

# Acknowledgments

Putting together a text such as this one requires much more than an author team. There are three primary groups that make it possible to write and assemble the material. The most important of those groups is the set of end users, the students. Thanks to the many students who have contributed to the development of this text by telling us what worked (and didn't work) in helping them to learn the material.

At the other end of the process we have the many people who participated in the production of the text. The McGraw-Hill team includes our publisher Elizabeth Haefele who assembled this fine team, our editor Rich Kolasa who made certain that we had the resources and support that we needed to put the project together, and the project manager Lora Kalb. Other important contributors at or through McGraw-Hill include Barb Owca, David Hash, Kara Kudronowicz, Amber Huebner, Lori Hancock, Sandra Schnee, Harold (Hal) Whipple, Kelly Jackson, A. Elena Bogardus, Dorothy Brown, and Melissa Leick. Special thanks go to our developmental editor Michelle Flomenhoft. Michelle kept us on track and on time. She also ensured that we would never suffer the loneliness of waking up to an empty e-mail inbox.

The third group that made this edition possible was the set of reviewers who took the time to make observations and suggestions that helped improve this text. We specifically want to thank the Board of Advisors who contributed feedback throughout the process. We also thank the sixth edition users for their willingness to respond to surveys as we refined our revision plan.

## Board of Advisors

Marcia Kleinz, *Atlantic Cape Community College*

Patricia Lazzarino, *Northern Virginia Community College*

Jordan Neus, *Suffolk County Community College—Brentwood*

Sandra Tannen, *Camden County College*—Lead member

## Manuscript Reviewers

Cedric E. Atkins, *Mott Community College*

Laurel L. Berry, *Bryant & Stratton College*

A. Elena Bogardus, *Camden County College*

Jerome Brown, *Harford Community College*

Alina Coronel, *Miami Dade College—Kendall*

Mike Everett, *Santa Ana College*

Jeremiah A. Gilbert, *San Bernardino Valley College*

Catherine Griffin, *Lansing Community College*

Elizabeth Hamman, *Cypress College*

Kelly Jackson, *Camden County College*

Joe Jordan, *John Tyler Community College*

Carolyn T. Krause, *Delaware Tech & Community College*

Shawn Krest, *Genesee Community College*

Anna P. Leder, *Gloucester County College*

Joan H. Macneil, *Quinebaug Valley Community College*

Lois Martin, *Massasoit Community College*

Darrell (Stan) Mattoon, *Merced College*

Shai Neumann, *Brevard Community College*

Rodney Oberdick, *Delaware Tech & Community College*

Carolyn Pabian, *Delaware Tech & Community College—Wilmington Campus*

Maria Consuelo (Suzie) C. Pickle, *St. Petersburg College*

Frank Pizza, *Heald College*

Marilyn G. Platt, *Gaston College*

Susan Poss, *Spartanburg Community College*

Yolanda Roddy, *Albany Technical College*

Jeannine Stein, *Los Medanos College*

Brad Sullivan, *Community College of Denver*

Lori Wall, *University of New England*

Kenneth Williams, *Albany Technical College*

Debra Wolfson, *Suffolk County Community College*

# A commitment to accuracy

You have a right to expect an accurate textbook, and McGraw-Hill invests considerable time and effort to make sure that we deliver one. Listed below are the many steps we take to make sure this happens.

# Our accuracy verification process

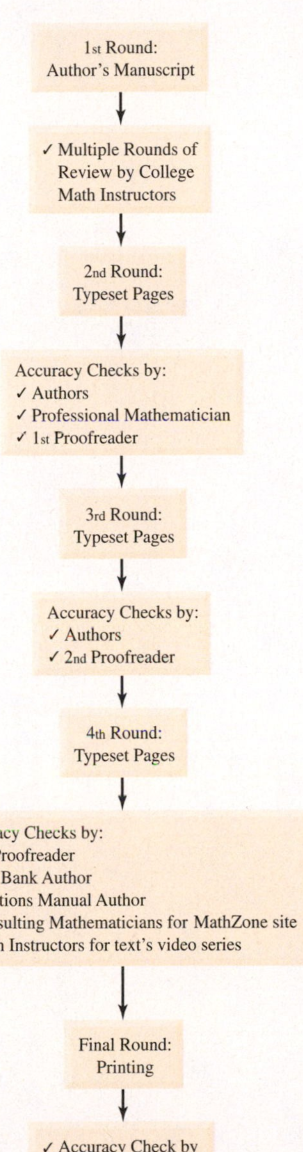


## *First Round*

Step 1: Numerous **college math instructors** review the manuscript and report on any errors that they may find, and the authors make these corrections in their final manuscript.

## *Second Round*

Step 2: Once the manuscript has been typeset, the **authors** check their manuscript against the first page proofs to ensure that all illustrations, graphs, examples, exercises, solutions, and answers have been correctly laid out on the pages, and that all notation is correctly used.

Step 3: An outside, **professional mathematician** works through every example and exercise in the page proofs to verify the accuracy of the answers.

Step 4: A **proofreader** adds a triple layer of accuracy assurance in the first pages by hunting for errors; then a second, corrected round of page proofs is produced.

## *Third Round*

Step 5: The **author team** reviews the second round of page proofs for two reasons: (1) to make certain that any previous corrections were properly made, and (2) to look for any errors they might have missed on the first round.

Step 6: A **second proofreader** is added to the project to examine the new round of page proofs to double check the author team's work and to lend a fresh, critical eye to the book before the third round of paging.

## *Fourth Round*

Step 7: A **third proofreader** inspects the third round of page proofs to verify that all previous corrections have been properly made and that there are no new or remaining errors.

Step 8: Meanwhile, in partnership with **independent mathematicians,** the text accuracy is verified from a variety of fresh perspectives:

- The **test bank author** checks for consistency and accuracy as they prepare the computerized test item file.
- The **solutions manual author** works every single exercise and verifies their answers, reporting any errors to the publisher.
- A **consulting group of mathematicians,** who write material for the text's MathZone site, notifies the publisher of any errors they encounter in the page proofs.
- A video production company employing **expert math instructors** for the text's videos will alert the publisher of any errors they might find in the page proofs.

## *Final Round*

Step 9: The **project manager,** who has overseen the book from the beginning, performs a **fourth proofread** of the textbook during the printing process, providing a final accuracy review.

⇒ What results is a mathematics textbook that is as accurate and error-free as is humanly possible, and our authors and publishing staff are confident that our many layers of quality assurance have produced textbooks that are the leaders of the industry for their integrity and correctness.

# applications index

CHAPTER

1

# Operations on Whole Numbers

## INTRODUCTION

To supplement her income while attending college, Nadia answered an ad in a local paper to become a census worker. She got the job at a census bureau in her community. Because she was juggling school, studying, and raising a 4-year-old daughter, she was able to work only part-time. Nadia thought this would be a good way to earn some extra money.

Eventually, Nadia began to realize how important her work was. She spent time studying data pertaining to people and businesses in her county. She saw that by analyzing facts such as age, education, income, and family size, she could formulate certain ideas about her environment. She began working with city planners to help make projections pertaining to population growth. This enabled them to sketch out a blueprint to accommodate new housing, office buildings, roads, and a new school, based on certain predictions. Until this new job, Nadia never realized the impact math has on the lives of ordinary people. She decided to change her major to something that would allow her to continue her work with strategic planning. She also realized that she would need to take more mathematics and statistics classes to continue on this track.

To learn more about the U.S. Census Bureau data, do Activity 1 on page 44.

## CHAPTER 1 OUTLINE

# 1 pretest



Name ____________

Section ________ Date ________

This pretest provides a preview of the types of exercises you will encounter in each section of this chapter. The answers for these exercises can be found in the back of the text. If you are working on your own, or ahead of the class, this pretest can help you identify the sections in which you should focus more of your time.

**Answers**

1. ____________
2. ____________
3. ____________
4. ____________
5. ____________
6. ____________
7. ____________
8. ____________
9. ____________
10. ____________
11. ____________
12. ____________
13. ____________
14. ____________

1.1 **1.** Write 107,945 in words.

1.2 **2.** The statement $2 + (3 + 5) = (2 + 3) + 5$ illustrates which property of addition?

1.3 **3.** $35{,}147 + 2{,}873 - 7{,}305 - 3{,}101 = ?$

1.5 **4.** The statement $5 \times 7 = 7 \times 5$ is an illustration of which property of multiplication?

**5.** $392 \times 51 = ?$

**6.** $187 \times 300 = ?$

1.6 **7.** $3{,}234 \div 7 = ?$

Use long division to find each quotient.

**8.** $7\overline{)8{,}431}$

**9.** $267\overline{)21{,}758}$

1.7 **10.** **(a)** $5 + 3 \times 7 = ?$
**(b)** $7 \times (3 + 5) = ?$

**11.** **Statistics** Suppose that you need a total of 360 points on four tests during the semester to receive an A for the course. Your scores on the first three tests were 84, 91, and 92. What is the lowest score you can get on the fourth test and still receive an A?

**12.** **Business and Finance** A refrigerator is advertised as follows: "Pay $50 down and $30 a month for 24 months." If the cash price of the refrigerator is $619, how much more will you pay if you buy on the installment plan?

**13.** Evaluate $2^3 \div 2 \times 3 - (5 - 2 + 3)$.

**14.** Find the perimeter and area of the following figure.

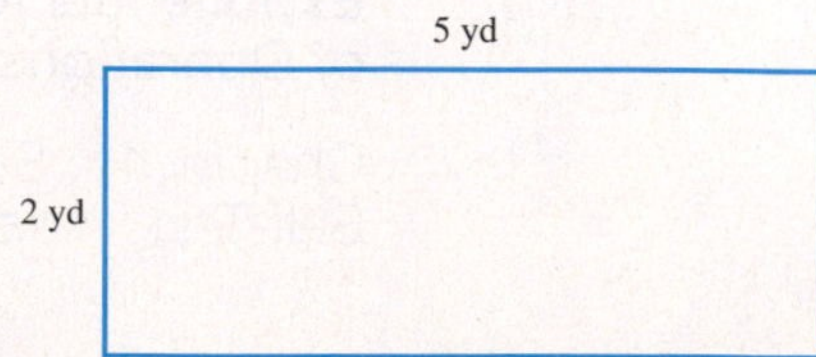

# 1.1 The Decimal Place-Value System

< **1.1 Objectives** >

1 > Write numbers in expanded form

2 > Determine the place value of a digit

3 > Write a number in words

4 > Given its word name, write a number

## Overcoming Math Anxiety

Throughout this text, we present you with a series of class-tested techniques that are designed to improve your performance in this math class.

Hint #1 Become familiar with your textbook.

Perform each of the following tasks.

1. Use the Table of Contents to find the title of Section 5.1.
2. Use the Index to find the earliest reference to the term *mean*. (By the way, this term has nothing to do with the personality of either your instructor or the textbook authors!)
3. Find the answer to the first Check Yourself exercise in Section 1.1.
4. Find the answers to the pretest for Chapter 1.
5. Find the answers to the odd-numbered exercises in Section 1.1.
6. In the margin notes for Section 1.1, find the origin of the term *digit*.

Now you know where some of the most important features of the text are. When you have a moment of confusion, think about using one of these features to help you clear up that confusion. You should also find out whether a solutions manual is available for your text. Many students find these to be helpful.

Number systems have been developed throughout human history. Starting with simple tally systems used to count and keep track of possessions, more and more complex systems were developed. The Egyptians used a set of picturelike symbols called **hieroglyphics** to represent numbers. The Romans and Greeks had their own systems of numeration. We see the Roman system today in the form of Roman numerals. Some examples of these systems are shown in the following table.

| Numerals | Egyptian | Greek | Roman |
|---|---|---|---|
| 1 | I | I | I |
| 10 | ∩ | Δ | X |
| 100 | ϙ | H | C |

**NOTES**

The prefix *deci* means 10. Our word *digit* comes from the Latin word *digitus,* which means finger.

Any number, no matter how large, can be represented using the 10 digits of our system.

Any number system provides a way of naming numbers. The system we use is described as a **decimal place-value system.** This system is based on the number 10 and uses symbols called **digits.** (Other numbers have also been used as bases. The Mayans used 20, and the Babylonians used 60.)

The basic symbols of our system are the digits 0, 1, 2, 3, 4, 5, 6, 7, 8, 9.

These basic symbols, or digits, were first used in India and then adopted by the Arabs. For this reason, our system is called the Hindu-Arabic numeration system.

Numbers may consist of one or more *digits*.

The numbers 3, 45, 567, and 2,359 are examples of the **standard form** for numbers. We say that 45 is a two-digit number, 567 is a three-digit number, and so on.

As we said, our decimal system uses a *place-value* concept based on the number 10. Understanding how this system works will help you see the reasons for the rules and methods of arithmetic that we will be introducing.

## Example 1 Writing a Number in Expanded Form

< Objective 1 >

**NOTES**

Each digit in a number has its own place value.

Here the parentheses are used for emphasis. (4 × 100) means 4 is multiplied by 100. (3 × 10) means 3 is multiplied by 10. (8 × 1) means 8 is multiplied by 1.

Look at the number 438.

We call 8 the *ones digit.* As we move to the left, the digit 3 is the *tens digit.* Again as we move to the left, 4 is the *hundreds digit.*

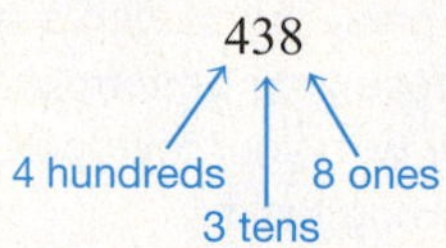


If we rewrite a number such that each digit is written with its units, we have used the **expanded form** for the number.

In expanded form, we write 438 as

$400 + 30 + 8$ or

$(4 \times 100) + (3 \times 10) + (8 \times 1)$

### Check Yourself 1

Write 593 in expanded form.

The following place-value diagram shows the place value of digits as we write larger numbers. For the number 3,156,024,798, we have

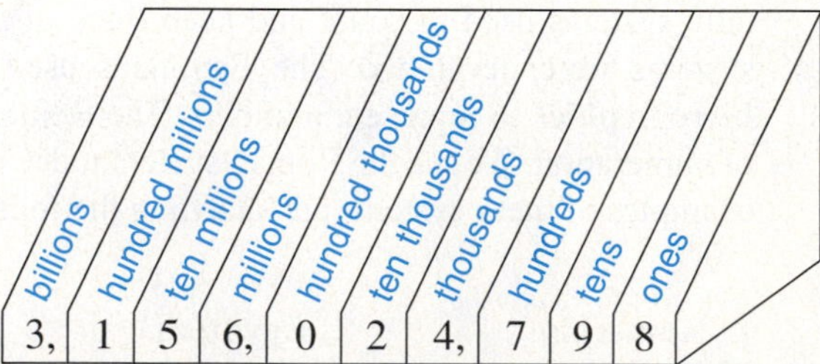


Of course, the naming of place values continues for larger numbers beyond the chart.

For the number 3,156,024,798, the place value of the digit 4 is thousands. As we move to the left, each place value is 10 times the value of the previous place. The place value of 2 is ten thousands, the place value of 0 is hundred thousands, and so on.

## Example 2 Identifying Place Value

< Objective 2 >

Identify the place value of each digit in the number 418,295.

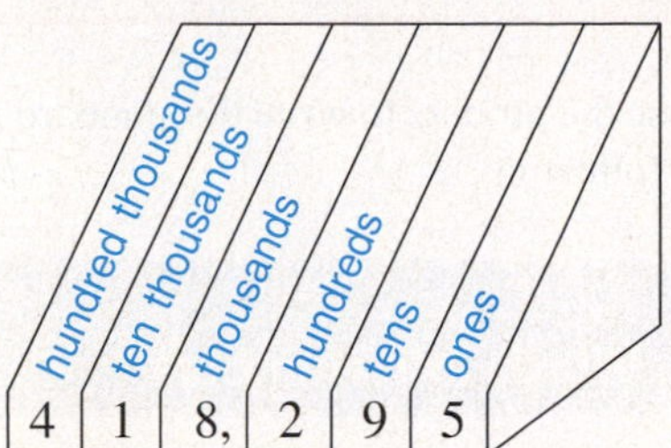

### Check Yourself 2

Use a place-value diagram to answer the following questions for the number 6,831,425,097.

**(a)** What is the place value of 2?
**(b)** What is the place value of 4?
**(c)** What is the place value of 3?
**(d)** What is the place value of 6?

Understanding place value will help you read or write numbers in word form. Look at the number

7 2, 3 5 8, 6 9 4

Millions Thousands Ones

> **NOTE**
>
> A four-digit number, such as 3,456, can be written with or without a comma. We have chosen to write them with a comma in this text.

Commas are used to set off groups of three digits in the number. The name of each group—millions, thousands, ones, and so on—is then used as we write the number in words. To write a word name for a number, we work from left to right, writing the numbers in each group, followed by the group name. The following chart summarizes the group names.

| Billions Group | | | Millions Group | | | Thousands Group | | | Ones Group | | |
|---|---|---|---|---|---|---|---|---|---|---|---|
| Hundreds | Tens | Ones | Hundreds | Tens | Ones | Hundreds | Tens | Ones | Hundreds | Tens | Ones |

## Example 3 Writing Numbers in Words

< Objective 3 >

> **NOTE**
>
> The commas in the word statements are in the same place as the commas in the number.

Write the word name for each of the following.

27,345 is written in words as twenty-seven *thousand,* three hundred forty-five.

2,305,273 is two *million,* three hundred five *thousand,* two hundred seventy-three.

**Note:** We do *not* write the name of the ones group. Also, the word *and* is not used when a number is written in words. It will have a special meaning later.

### Check Yourself 3

Write the word name for each of the following numbers.

**(a)** 658,942 **(b)** 2,305

We reverse the process to write the standard form for numbers given in word form. Consider the following.

### Example 4 Translating Words into Numbers

< Objective 4 >

Forty-eight thousand, five hundred seventy-nine in standard form is

48,579

Five hundred three thousand, two hundred thirty-eight in standard form is

503,238

Note the use of 0 as a placeholder in writing the number.

### Check Yourself 4

Write twenty-three thousand, seven hundred nine in standard form.

### Check Yourself ANSWERS

**1.** $(5 \times 100) + (9 \times 10) + (3 \times 1)$ **2. (a)** Ten thousands; **(b)** hundred thousands; **(c)** ten millions; **(d)** billions **3. (a)** Six hundred fifty-eight thousand, nine hundred forty-two; **(b)** two thousand, three hundred five **4.** 23,709

### Reading Your Text

The following fill-in-the-blank exercises are designed to ensure that you understand some of the key vocabulary used in this section.

SECTION 1.1

**(a)** The system we use for naming numbers is described as a ________ place-value system.

**(b)** We say the number 567 is a three-____________ number.

**(c)** A four-digit number can be written with or without a ____________.

**(d)** In words, the number 2,000,000 is written as two ____________.

# 1.1 exercises

< Objective 1 >

*Write each number in expanded form.*

**1.** 456

**2.** 637

**3.** 5,073

**4.** 20,721

< Objective 2 >

*Give the place values for the indicated digits.*

**5.** 4 in the number 416

**6.** 3 in the number 38,615

**7.** 6 in the number 56,489

**8.** 4 in the number 427,083

**9.** In the number 43,729

**(a)** what digit tells the number of thousands?

**(b)** what digit tells the number of tens?

**10.** In the number 456,719

**(a)** what digit tells the number of ten thousands?

**(b)** what digit tells the number of hundreds?

**11.** In the number 1,403,602

**(a)** what digit tells the number of hundred thousands?

**(b)** what digit tells the number of ones?

**12.** In the number 324,678,903

**(a)** what digit tells the number of millions?

**(b)** what digit tells the number of ten thousands?

< Objective 3 >

*Write the word name for each of the following.*

**13.** 5,618

**14.** 21,812

**15.** 200,304

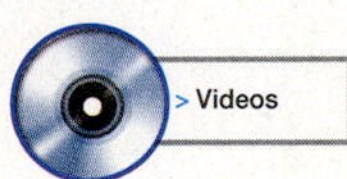


**16.** 103,900

< Objective 4 >

*Write each of the following in the standard form of a number.*

**17.** Two hundred fifty-three thousand, four hundred eighty-three

**18.** Three hundred fifty thousand, three hundred fifty-nine

**19.** Five hundred two million, seventy-eight thousand

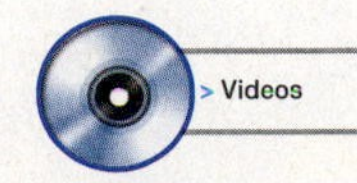


**20.** Four billion, two hundred thirty million

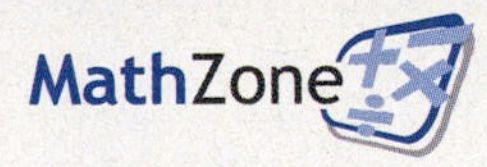


Boost your grade at mathzone.com!

- > Practice Problems
- > NetTutor
- > Self-Tests
- > e-Professors
- > Videos

Name ____________

Section ______ Date ______

## Answers

1. ____
2. ____
3. ____
4. ____
5. ____
6. ____
7. ____
8. ____
9. ____ 10. ____
11. ____ 12. ____
13. ____
14. ____
15. ____
16. ____
17. ____ 18. ____
19. ____
20. ____

## Answers

21. ____________

22. ____________

23. ____________

24. ____________

*Write the whole number in each sentence in standard form.*

21. **Statistics** The first-place finisher in the 2004 U.S. Open won one million one hundred twenty-five dollars.

22. **Science and Medicine** Scientific speculation is that the universe originated in the explosion of a primordial fireball approximately fourteen billion years ago.

23. **Social Science** The population of Kansas City, Missouri, in 2000 was approximately four hundred forty-one thousand, five hundred.

24. **Social Science** The Nile river in Egypt is about four thousand, one hundred forty-five miles long.

Basic Skills | Advanced Skills | Vocational-Technical Applications | Calculator/Computer | Above and Beyond

*Sometimes numbers found in charts and tables are abbreviated. The following table represents the population of the 10 largest cities in the 2000 U.S. census. Note that the numbers represent thousands. Thus Detroit had a population of 951 thousand or 951,000.*

chapter 1 > Make the Connection

| Name | Rank | Population (thousands) |
|---|---|---|
| New York City, NY | 1 | 8,008 |
| Los Angeles, CA | 2 | 3,695 |
| Chicago, IL | 3 | 2,896 |
| Houston, TX | 4 | 1,954 |
| Philadelphia, PA | 5 | 1,518 |
| Phoenix, AZ | 6 | 1,321 |
| San Diego, CA | 7 | 1,223 |
| Dallas, TX | 8 | 1,189 |
| San Antonio, TX | 9 | 1,145 |
| Detroit, MI | 10 | 951 |

*Source:* U.S. Census Bureau

**Social Science** In exercises 25 to 28, write your answers in standard form, using the preceding table.

**25.** What was the population of San Diego in 2000?

**26.** What was the population of Chicago in 2000?

**27.** What was the population of Philadelphia in 2000?

**28.** What was the population of Dallas in 2000?

*Determine the number represented by the scrambled place values.*

**29.** 4 thousands
1 tens
3 ten thousands
5 ones
2 hundreds

**30.** 7 hundreds
4 ten thousands
9 ones
8 tens
6 thousands

*Assume that you have alphabetized the word names for every number from one to one thousand.*

**31.** Which number would appear first in the list?

**32.** Which number would appear last?

## Answers

25. ____

26. ____

27. ____

28. ____

29. ____

30. ____

31. ____

32. ____

## Answers

33. ______

34. ______

35. ______

36. ______

37. ______

**33. Business and Finance** Inci had to write a check for $2,565. There is a space on the check to write out the amount of the check in words. What should she write in this space?

**34. Business and Finance** In a rental agreement, the amount of the initial deposit required is two thousand, five hundred forty-five dollars. Write this amount as a number.

**35. Allied Health** Doctor Edwards prescribes four hundred eighty thousand units of penicillin G benzathine to treat a 3-year-old child with a streptococcal infection. Write this amount as a number.

**36. Allied Health** Doctor Hill prescribes one thousand, one hundred eighty-three milligrams (mg) of amifostine to be administered together with an adult patient's chemotherapy to reduce the adverse effects of the treatment. Write this amount as a number.

| Child's Weight (lb) | Dose of Penicillin G Potassium (thousands of units) |
|---|---|
| 30 | 680 |
| 35 | 795 |
| 40 | 910 |
| 45 | 1,020 |
| 50 | 1,135 |

**37. Allied Health** Write your answers in standard form using the preceding table.

**(a)** Carla weighs 35 pounds (lb). What dose of penicillin G potassium should her doctor prescribe?

**(b)** Nelson weighs 50 lb. What dose of penicillin G potassium should his doctor prescribe?

**(c)** What dose of penicillin G potassium is required for a child weighing 40 lb?

| Appliance | Estimated Power Consumption [thousands of watts/hour (W/h)] |
|---|---|
| Drip coffeemaker | 301 |
| Electric blanket | 120 |
| Laser printer | 466 |
| Personal computer | 25 |
| Video game system | 49 |

**38.** **Electronics** Write your answers in standard form using the preceding table.

**(a)** What is the estimated power consumption of a laser printer?

**(b)** What is the estimated power consumption of a video game system?

**(c)** What is the estimated power consumption of an electric blanket?

**39.** **Number Problem** Write the largest five-digit number that can be made using the digits 6, 3, and 9 if each digit is to be used at least once.

Basic Skills | Advanced Skills | Vocational-Technical Applications | Calculator/Computer | **Above and Beyond**

**40.** What are the advantages of a place-value system of numeration?

**41.** **Social Science** The number 0 was not used initially by the Hindus in our number system (about 250 B.C.E.). Go to your library (or "surf the net"), and determine when a symbol for zero was introduced. What do you think is the importance of the role of 0 in a numeration system?

**42.** A *googol* is a very large number. Do some research to find out how big it is. Also try to find out where the name of this number comes from.

## Answers

38. ______

39. ______

40. ______

41. ______

42. ______

## Answers

We provide the answers for the odd-numbered exercises at the end of each exercise set. The answers for the even-numbered exercises are provided in the instructor's resource manual.

**1.** $(4 \times 100) + (5 \times 10) + (6 \times 1)$ **3.** $(5 \times 1{,}000) + (7 \times 10) + (3 \times 1)$
**5.** Hundreds **7.** Thousands **9.** **(a)** 3; **(b)** 2 **11.** **(a)** 4; **(b)** 2
**13.** Five thousand, six hundred eighteen
**15.** Two hundred thousand, three hundred four
**17.** 253,483 **19.** 502,078,000 **21.** $1,125,000 **23.** 441,500
**25.** 1,223,000 **27.** 1,518,000 **29.** 34,215 **31.** Eight
**33.** Two thousand, five hundred sixty-five **35.** 480,000
**37.** **(a)** 795,000; **(b)** 1,135,000; **(c)** 910,000 **39.** 99,963
**41.** Above and Beyond

1.2

# Addition

< 1.2 Objectives >

1 > Add single-digit numbers

2 > Identify the properties of addition

3 > Add groups of numbers with no carrying

4 > Use the language of addition

5 > Solve simple applications

6 > Add any group of numbers

7 > Find the perimeter of a figure

8 > Solve applications that involve perimeter

 Overcoming Math Anxiety

Hint #2 Become familiar with your syllabus.

In the first class meeting, your instructor probably handed out a class syllabus. If you haven't done so already, you need to incorporate important information into your calendar and address book.

1. Write all important dates in your calendar. This includes homework due dates, quiz dates, test dates, and the date and time of the final exam. Never allow yourself to be surprised by any deadline!
2. Write your instructor's name, contact number, and office number in your address book. Also include the office hours and e-mail address. Make it a point to see your instructor early in the term. Although this is not the only person who can help clear up your confusion, he or she is the most important person.
3. Make note of other resources that are made available to you. These include CDs, videotapes, Web pages, and tutoring.

Given all these resources, it is important that you never let confusion or frustration mount. If you can't "get it" from the text, try another resource. All the resources are there specifically for you, so take advantage of them!

**NOTE**

The three dots (. . .) are called an **ellipsis;** they mean that the indicated pattern continues.

The *natural* or *counting numbers* are the numbers we use to count objects.

The natural numbers are 1, 2, 3, . . .

When we include the number 0, we have the set of *whole numbers.*

The whole numbers are 0, 1, 2, 3, . . .

Let's look at the operation of *addition* on the whole numbers.

**Definition**

**Addition** — **Addition** is the combining of two or more groups of the same kind of objects.

This concept is extremely important, as we will see in our later work with fractions. We can only combine or add numbers that represent the same kind of objects.

From your first encounter with arithmetic, you were taught to add "3 apples plus 2 apples."

On the other hand, you have probably encountered a phrase such as "that's like combining apples and oranges." That is to say, what do you get when you add 3 apples and 2 oranges?

You could answer "5 fruits," or "5 objects," but you cannot combine the apples and the oranges.

What if you walked 3 miles and then walked 2 more miles? Clearly, you have now walked 3 miles + 2 miles = 5 miles. The addition is possible because the groups are of the same kind.

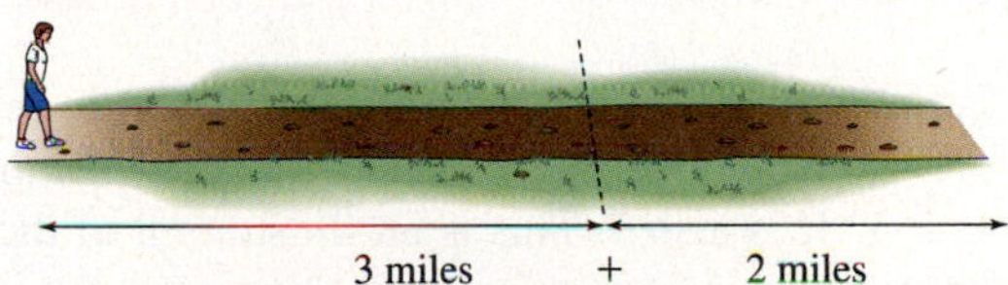


**NOTES**

The first printed use of the symbol + dates back to 1526 (Smith, *History of Math*, Vol. II.)

The point labeled 0 is called the **origin** of the number line.

Each operation of arithmetic has its own special terms and symbols. The addition symbol + is read **plus.** When we write 3 + 4, 3 and 4 are called the **addends.**

We can use a number line to illustrate the addition process. To construct a number line, we pick a point on the line and label it 0. We then mark off evenly-spaced units to the right, naming each marked point with a successively larger whole number.

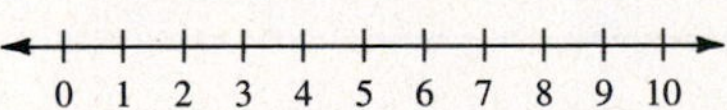


We use arrowheads to show the number line continues.

**Example 1** — **Representing Addition on a Number Line**

< Objective 1 >

Represent 3 + 4 on the number line.

To represent an addition, such as 3 + 4, on the number line, start by moving 3 spaces to the right of the origin. Then move 4 more spaces to the right to arrive at 7. The number 7 is called the *sum* of the addends.

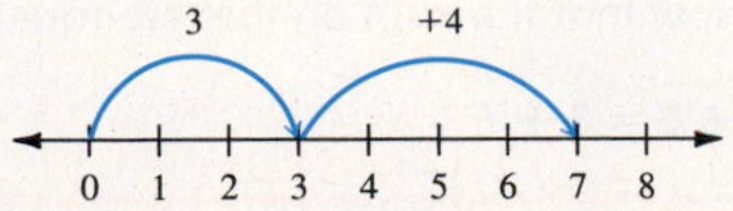

**NOTE**

Again, addition corresponds to combining groups of the same kind of objects.

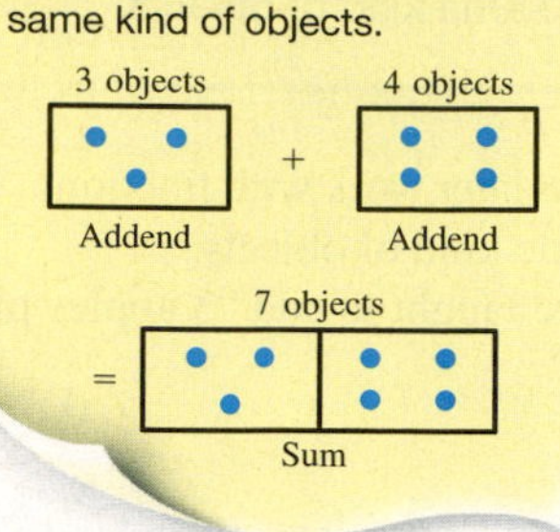


We can write 3 + 4 = 7

Addend Addend Sum

### Check Yourself 1

Represent 5 + 6 on the number line.

A statement such as 3 + 4 = 7 is one of the **basic addition facts.** These facts include the sum of every possible pair of digits. Before you can add larger numbers correctly and quickly, you must memorize these basic facts.

**Basic Addition Facts**

| + | 0 | 1 | 2 | 3 | 4 | 5 | 6 | 7 | 8 | 9 |
|---|---|---|---|---|---|---|---|---|---|---|
| **0** | 0 | 1 | 2 | 3 | 4 | 5 | 6 | 7 | 8 | 9 |
| **1** | 1 | 2 | 3 | 4 | 5 | 6 | 7 | 8 | 9 | 10 |
| **2** | 2 | 3 | 4 | 5 | 6 | 7 | 8 | 9 | 10 | 11 |
| **3** | 3 | 4 | 5 | 6 | 7 | 8 | 9 | 10 | 11 | 12 |
| **4** | 4 | 5 | 6 | 7 | 8 | 9 | 10 | 11 | 12 | 13 |
| **5** | 5 | 6 | 7 | 8 | 9 | 10 | 11 | 12 | 13 | 14 |
| **6** | 6 | 7 | 8 | 9 | 10 | 11 | 12 | 13 | 14 | 15 |
| **7** | 7 | 8 | 9 | 10 | 11 | 12 | 13 | 14 | 15 | 16 |
| **8** | 8 | 9 | 10 | 11 | 12 | 13 | 14 | 15 | 16 | 17 |
| **9** | 9 | 10 | 11 | 12 | 13 | 14 | 15 | 16 | 17 | 18 |

**NOTES**

To find the sum 5 + 8, start with the row labeled 5. Move along that row to the column headed 8 to find the sum, 13.

*Commute* means to move back and forth, as to school or work.

Examining the basic addition facts leads us to several important properties of addition on whole numbers. For instance, we know that the sum 3 + 4 is 7. What about the sum 4 + 3? It is also 7. This is an illustration of the fact that addition is a **commutative** operation.

**Property**

**The Commutative Property of Addition**

The order of two numbers around an addition sign *does not* affect the sum.

### Example 2 Using the Commutative Property

< Objective 2 >

8 + 5 = 13 = 5 + 8

6 + 9 = 15 = 9 + 6

**NOTE**

The *order* does not affect the sum.

### Check Yourself 2

Show that the sum on the left equals the sum on the right.

7 + 8 = 8 + 7

If we wish to add *more* than two numbers, we can group them and then add. In mathematics this grouping is indicated by a set of parentheses ( ). This symbol tells us to perform the operation inside the parentheses first.

## Example 3 Using the Associative Property

**NOTES**

We add 3 and 4 as the first step and then add 5.

Here we add 4 and 5 as the first step and then add 3. Again the final sum is 12.

$(3 + 4) + 5 = 7 + 5 = 12$

We also have

$3 + (4 + 5) = 3 + 9 = 12$

This example suggests the following property of whole numbers.

**Property**

**The Associative Property of Addition**

The way in which several whole numbers are grouped *does not* affect the final sum when they are added.

**NOTE**

In Example 3, the addend 4 could have been "associated" with the 3 or the 5.

### Check Yourself 3

Find

$(4 + 8) + 3$ and $4 + (8 + 3)$

The number 0 has a special property in addition.

**Property**

**The Additive Identity Property**

The sum of 0 and any whole number is just that whole number.

Because of this property, we call 0 the **identity** for the addition operation.

## Example 4 Adding Zero

Find the sum of **(a)** $3 + 0$ and **(b)** $0 + 8$.

**(a)** $3 + 0 = 3$

**(b)** $0 + 8 = 8$

### Check Yourself 4

Find each sum.

**(a)** $4 + 0$ **(b)** $0 + 7$

Next, we turn to the process of adding larger numbers. We apply the following rule.

**Property**

**Adding Digits of the Same Place Value**

We can add the digits of the same place value because they represent the same types of quantities.

**NOTE**

Remember that 25 means 2 tens and 5 ones; 34 means 3 tens and 4 ones.

Adding two numbers, such as $25 + 34$, can be done in expanded form. Here we write out the place value for each digit.

$$\begin{aligned} 25 &= 2 \text{ tens} + 5 \text{ ones} \\ +\ 34 &= 3 \text{ tens} + 4 \text{ ones} \\ \hline &= 5 \text{ tens} + 9 \text{ ones} \\ &= 59 \end{aligned}$$

↓ Add down.

In actual practice, we use a more convenient short form to perform the addition.

### Example 5 Adding Two Numbers

< Objective 3 >

**NOTE**

In using the short form, be very careful to line up the numbers correctly so that each column contains digits of the same place value.

Add $352 + 546$.

**Step 1** Add in the ones column.

$$\begin{array}{r} 352 \\ +\ 546 \\ \hline 8 \end{array}$$

**Step 2** Add in the tens column.

$$\begin{array}{r} 352 \\ +\ 546 \\ \hline 98 \end{array}$$

**Step 3** Add in the hundreds column.

$$\begin{array}{r} 352 \\ +\ 546 \\ \hline 898 \end{array}$$

### Check Yourself 5

Add.

$$\begin{array}{r} 245 \\ +\ 632 \\ \hline \end{array}$$

You have already seen that the word *sum* indicates addition. There are other words that also tell you to use the addition operation.

The *total* of 12 and 5 is written as

12 + 5 or 17

8 *more than* 10 is written as

10 + 8 or 18

12 *increased by* 3 is written as

12 + 3 or 15

## Example 6 Translating Words That Indicate Addition

< Objective 4 >

Find each of the following.

**(a)** 36 increased by 12.

36 increased by 12 is written as 36 + 12 = 48.

**(b)** The total of 18 and 31.

The total of 18 and 31 is written as 18 + 31 = 49.

### Check Yourself 6

Find each of the following.

**(a)** 43 increased by 25 **(b)** The total of 22 and 73

**NOTE**

Get into the habit of writing down *all* your work, rather than just an answer.

Now we consider applications, or word problems, that will use the operation of addition. An organized approach is the key to successful problem solving, and we suggest the following strategy.

**Step by Step**

**Solving Addition Applications**

**Step 1** Read the problem carefully to determine the given information and what you are being asked to find.

**Step 2** Decide upon the operation (in this case, addition) to be used.

**Step 3** Write down the complete statement necessary to solve the problem and do the calculations.

**Step 4** Write your answer as a complete sentence. Check to make sure you have answered the question of the problem and that your answer seems reasonable.

We work through an example, using these steps.

## Example 7 Setting Up a Word Problem

< Objective 5 >

**NOTE**

Remember to attach the proper unit (here "students") to your answer.

Four sections of algebra were offered in the fall quarter, with enrollments of 33, 24, 20, and 22 students. What was the total number of students taking algebra?

**Step 1** The given information is the number of students in each section. We want the total number.

**Step 2** Since we are looking for a total, we use addition.

**Step 3** Write $33 + 24 + 20 + 22 = 99$ students.

**Step 4** There were 99 students taking algebra.

### Check Yourself 7

**Elva Ramos won an election for city council with 3,110 votes. Her two opponents had 1,022 and 1,211 votes. How many votes were cast in that election?**

In the previous examples and exercises, the digits in each column added to 9 or less. We now look at the situation in which a column has a two-digit sum. This will involve the process of **carrying.** Look at the process in expanded form.

## Example 8 Adding in Expanded Form When Regrouping Is Needed

< Objective 6 >

**NOTE**

Regrouping in addition is also called **carrying.** Of course, the name makes no difference as long as you understand the process.

$$\begin{aligned} 67 &= 60 + 7 \\ + 28 &= 20 + 8 \\ \hline &\quad 80 + 15 \end{aligned}$$

We have written 15 ones as 1 ten and 5 ones.

or $\underbrace{80 + 10} + 5$

The 1 ten is then combined with the 8 tens.

or $90 + 5$

or $95$

The more convenient short form carries the excess units from one column to the next column to the left. Recall that the place value of the next column to the left is 10 times the value of the original column. It is this property of our decimal place-value system that makes carrying work. We work this problem again, this time using the short, or "carrying," form.

**NOTE**

Of course this is true for any sized number. The place value thousands is 10 times the place value hundreds, and so on.

**Step 1** **Step 2**

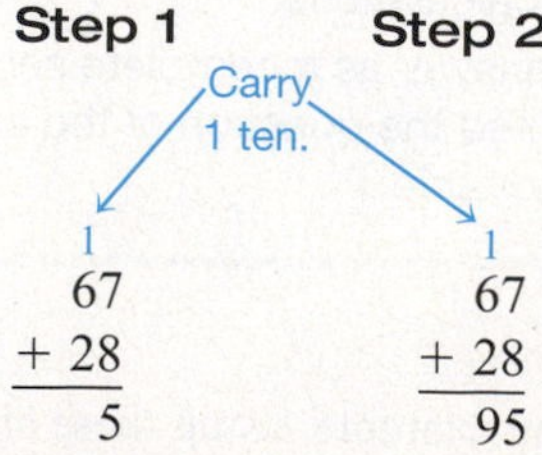


$$\begin{array}{r} ^1 \\ 67 \\ +\ 28 \\ \hline 5 \end{array} \qquad \begin{array}{r} ^1 \\ 67 \\ +\ 28 \\ \hline 95 \end{array}$$

**Step 1:** The sum of the digits in the ones column is 15, so write 5 and make the 10 ones **a 1** in the tens column. **Step 2:** Now add in the tens column, being sure to include the carried 1.

## Check Yourself 8

Add.

(a) $\begin{array}{r} 58 \\ +\ 36 \\ \hline \end{array}$ (b) $\begin{array}{r} 73 \\ +\ 18 \\ \hline \end{array}$ (c) $\begin{array}{r} 68 \\ +\ 25 \\ \hline \end{array}$

The addition process often requires more than one regrouping step, as is shown in Example 9.

## Example 9 Adding in Short Form When Regrouping Is Needed

Add 285 and 378.

$\begin{array}{r} \scriptstyle 1 \\ 285 \\ +\ 378 \\ \hline 3 \end{array}$ ← Write the 10 as 1 ten.

The sum of the digits in the ones column is 13, so write 3 and carry 1 to the tens column.

Carry 1 hundred. → $\begin{array}{r} \scriptstyle 1\,1\ \ \\ 285 \\ +\ 378 \\ \hline 63 \end{array}$

Now add in the tens column, being sure to include the carry. We have 16 tens, so write 6 in the tens place and carry 1 to the hundreds column.

$\begin{array}{r} \scriptstyle 1\,1\ \ \\ 285 \\ +\ 378 \\ \hline 663 \end{array}$

Finally, add in the hundreds column.

## Check Yourself 9

Add.

(a) $\begin{array}{r} 479 \\ +\ 287 \\ \hline \end{array}$ (b) $\begin{array}{r} 585 \\ +\ 368 \\ \hline \end{array}$

The regrouping process is the same if we want to add more than two numbers.

## Example 10 Adding in Short Form with Multiple Regrouping Steps

Add 53, 2,678, 587, and 27,009.

$\begin{array}{r} \scriptstyle 1\,1\,2\,2\ \ \\ 53 \\ 2{,}678 \\ 587 \\ +\ 27{,}009 \\ \hline 30{,}327 \end{array}$ ← Carries

Add in the ones column: 3 + 8 + 7 + 9 = 27. Write 7 in the sum and carry 2 to the tens column.

Now add in the tens column, being sure to include the carry. The sum is 22. Write 2 tens and carry 2 to the hundreds column. Complete the addition by adding in the hundreds column, the thousands column, and the ten thousands column.

### Check Yourself 10

**Add 46, 365, 7,254, and 24,006.**

Finding the *perimeter* of a figure is one application of addition.

**Definition**

**Perimeter** — **Perimeter** is the distance around a closed figure.

If the figure has straight sides, the perimeter is the sum of the lengths of its sides.

### Example 11 Finding the Perimeter

< Objective 7 >

We wish to fence in the field shown in the figure. How much fencing, in feet (ft), will be needed?

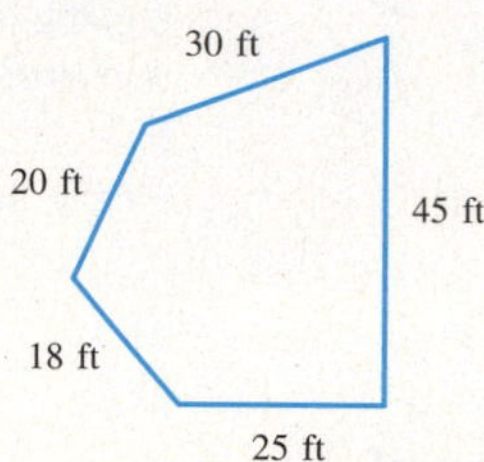


**NOTE**

Make sure to include the unit with each number.

The fencing needed is the perimeter of (or the distance around) the field. We must add the lengths of the five sides.

20 ft + 30 ft + 45 ft + 25 ft + 18 ft = 138 ft

So the perimeter is 138 ft.

### Check Yourself 11

**What is the perimeter of the region shown?**

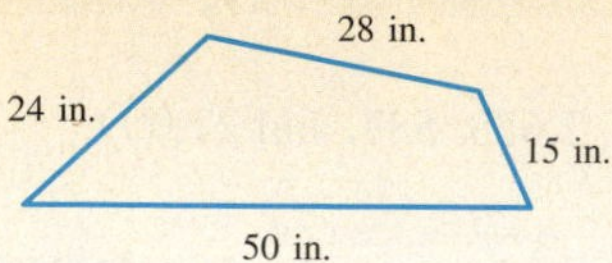


A **rectangle** is a figure, like a sheet of paper, with four equal corners. The perimeter of a rectangle is found by adding the lengths of the four sides.

## Example 12 Finding the Perimeter of a Rectangle

Find the perimeter in inches (in.) of the rectangle pictured here.

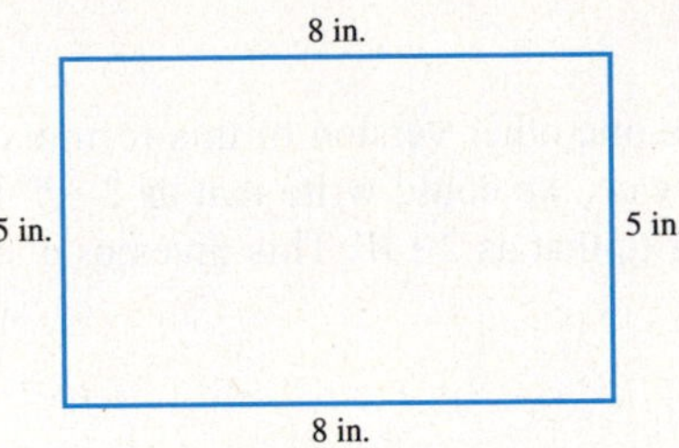


The perimeter is the sum of the lengths 8 in., 5 in., 8 in., and 5 in.

8 in. + 5 in. + 8 in. + 5 in. = 26 in.

The perimeter of the rectangle is 26 in.

### Check Yourself 12

**Find the perimeter of the rectangle pictured here.**

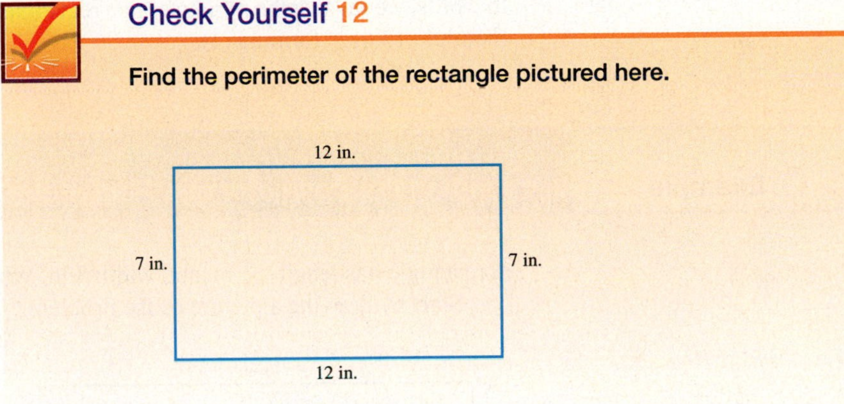


In general, we can find the perimeter of a rectangle by using a *formula*. A **formula** is a set of symbols that describe a general solution to a problem.

Look at a picture of a rectangle.

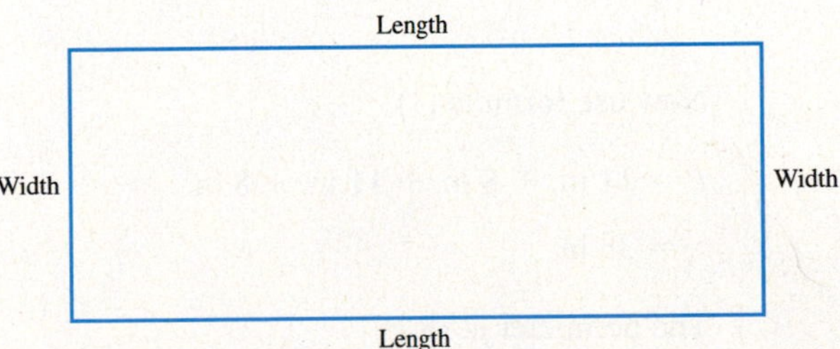


The perimeter can be found by adding the distances, so

Perimeter = length + width + length + width

To make this formula a little more readable, we abbreviate each of the words, using just the first letter.

**Property**

**Formula for the Perimeter of a Rectangle**

$$P = L + W + L + W \qquad (1)$$

There is one other version of this formula that we can use. Because we add the length ($L$) twice, we could write that as $2 \cdot L$. Because we add the width ($W$) twice, we could write that as $2 \cdot W$. This gives us another version of the formula.

**Property**

**Formula for the Perimeter of a Rectangle**

$$P = 2 \cdot L + 2 \cdot W \qquad (2)$$

In words, we say that the perimeter of a rectangle is twice its length plus twice its width. Example 13 uses formula (1).

### Example 13 Finding the Perimeter of a Rectangle

< Objective 8 >

A rectangle has length 11 in. and width 8 in. What is its perimeter?

Start by drawing a picture of the problem.

**NOTE**

We say the rectangle is 8 in. by 11 in.

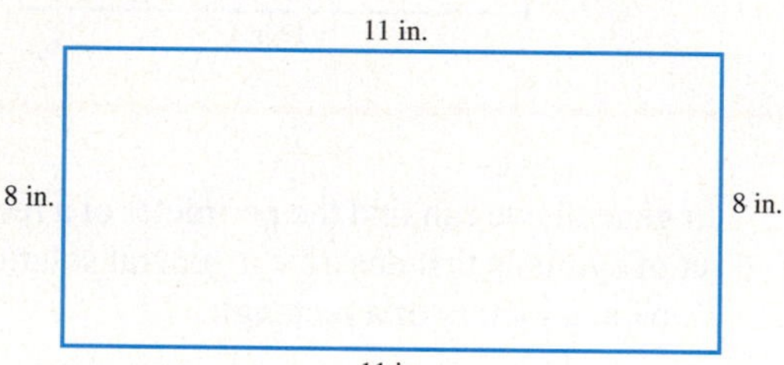


Now use formula (1).

$P = 11 \text{ in.} + 8 \text{ in.} + 11 \text{ in.} + 8 \text{ in.}$

$= 38 \text{ in.}$

The perimeter is 38 in.

**Check Yourself 13**

A bedroom is 9 ft by 12 ft. What is its perimeter?

## Check Yourself ANSWERS

**1.** $5 + 6 = 11$

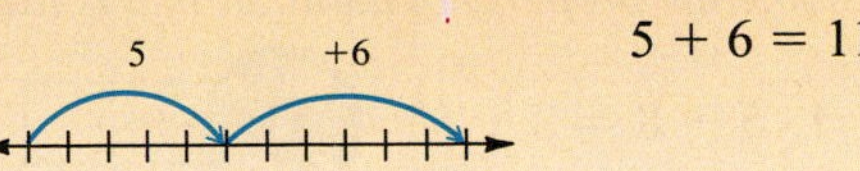


**2.** $7 + 8 = 15$ and $8 + 7 = 15$
**3.** $(4 + 8) + 3 = 12 + 3 = 15; 4 + (8 + 3) = 4 + 11 = 15$
**4.** **(a)** 4; **(b)** 7 **5.** 877 **6.** **(a)** 68; **(b)** 95 **7.** 5,343 votes
**8.** **(a)** 94; **(b)** 91; **(c)** 93 **9.** **(a)** 766; **(b)** 953 **10.** 31,671
**11.** 117 in. **12.** 38 in. **13.** 42 ft

## Reading Your Text

The following fill-in-the-blank exercises are designed to ensure that you understand some of the key vocabulary used in this section.

### SECTION 1.2

**(a)** The __________ or counting numbers are the numbers used to count objects.

**(b)** A statement such as $3 + 4 = 7$ is one of the basic __________ facts.

**(c)** The __________ of two numbers around an addition sign does not affect the sum.

**(d)** The first step in solving an addition application is to __________ the problem carefully.

Basic Skills | Advanced Skills | Vocational-Technical Applications | Calculator/Computer | Above and Beyond

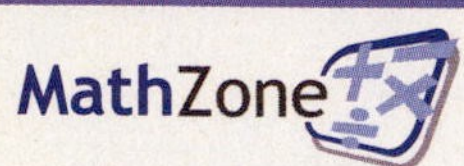


Boost your grade at mathzone.com!
> Practice Problems
> NetTutor
> Self-Tests
> e-Professors
> Videos

Name ____________

Section ________ Date ________

< Objective 2 >

*Name the property of addition that is illustrated. Explain your choice of property.*

**1.** $5 + 8 = 8 + 5$

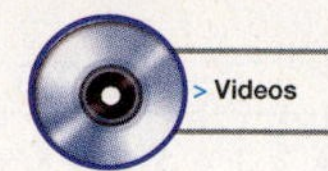


**2.** $2 + (7 + 9) = (2 + 7) + 9$

**3.** $(4 + 5) + 8 = 4 + (5 + 8)$

**4.** $9 + 7 = 7 + 9$

**5.** $4 + (7 + 6) = 4 + (6 + 7)$

> Videos

**6.** $5 + 0 = 5$

**7.** $5 + (2 + 3) = (2 + 3) + 5$

**8.** $3 + (0 + 6) = (3 + 0) + 6$

< Objectives 3, 6 >

*Perform the indicated addition.*

**9.** $\begin{array}{r} 2{,}792 \\ +\quad 205 \\ \hline \end{array}$

**10.** $\begin{array}{r} 5{,}463 \\ +\quad 435 \\ \hline \end{array}$

**11.** $\begin{array}{r} 2{,}345 \\ +\ 6{,}053 \\ \hline \end{array}$

**12.** $\begin{array}{r} 3{,}271 \\ +\ 4{,}715 \\ \hline \end{array}$

**13.** $\begin{array}{r} 2{,}531 \\ +\ 5{,}354 \\ \hline \end{array}$

**14.** $\begin{array}{r} 5{,}003 \\ +\ 4{,}205 \\ \hline \end{array}$

**15.** $\begin{array}{r} 21{,}314 \\ +\ 43{,}042 \\ \hline \end{array}$

**16.** $\begin{array}{r} 12{,}325 \\ +\ 35{,}403 \\ \hline \end{array}$

**17.** $\begin{array}{r} 3{,}490 \\ 548 \\ +\quad 25 \\ \hline \end{array}$

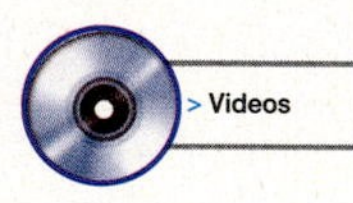


**18.** $\begin{array}{r} 678 \\ 4{,}533 \\ +\quad 70 \\ \hline \end{array}$

**19.** $\begin{array}{r} 2{,}289 \\ 38 \\ 578 \\ +\ 3{,}489 \\ \hline \end{array}$

**20.** $\begin{array}{r} 3{,}678 \\ 259 \\ 27 \\ +\ 2{,}356 \\ \hline \end{array}$

**21.** $\begin{array}{r} 23{,}458 \\ +\ 32{,}623 \\ \hline \end{array}$

**22.** $\begin{array}{r} 52{,}591 \\ +\ 59{,}739 \\ \hline \end{array}$

< Objective 4 >

**23.** In the statement $5 + 4 = 9$
5 is called an ________.
4 is called an ________.
9 is called the ________.

**24.** In the statement $7 + 8 = 15$
7 is called an ________.
8 is called an ________.
15 is called the ________.

## Answers

1. ____________
2. ____________
3. ____________
4. ____________
5. ____________
6. ____________
7. ____________
8. ____________
9. ______ 10. ______
11. ______ 12. ______
13. ______ 14. ______
15. ______ 16. ______
17. ______ 18. ______
19. ______ 20. ______
21. ______ 22. ______
23. ____________
24. ____________

< Objectives 7, 8 >

*Find the perimeter of each figure.*

**25.**

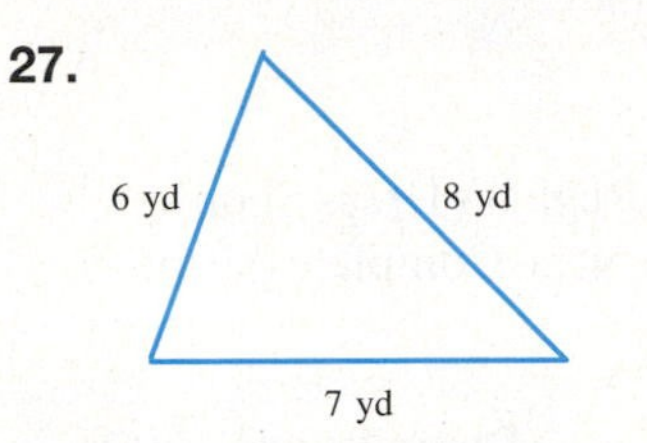


**26.**

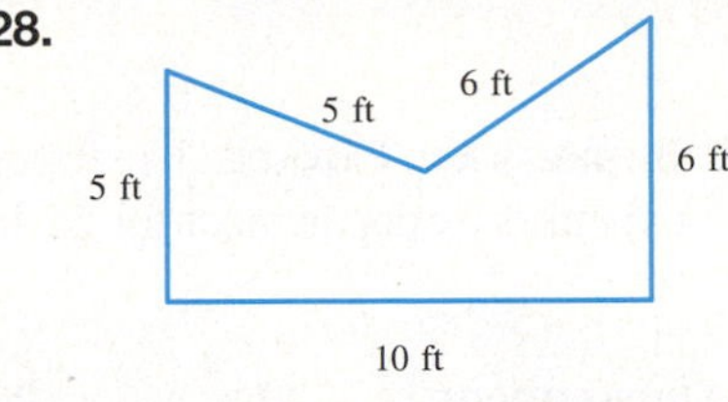


**27.**

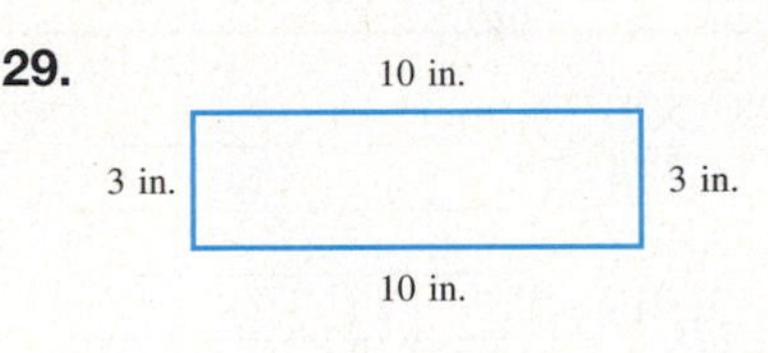


**28.**

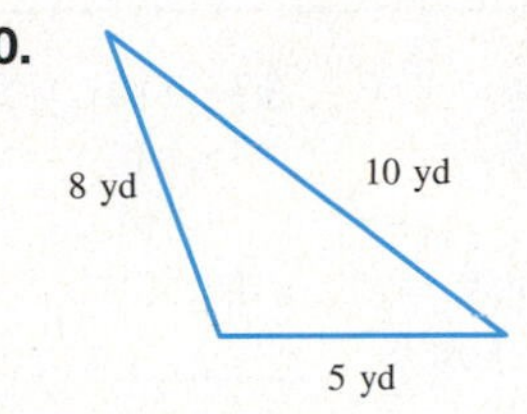


**29.**

10 in.
3 in.
3 in.
10 in.

**30.**

8 yd
10 yd
5 yd

**31.** In each of the following exercises, find the appropriate sum.

**(a)** Find the number that is 356 more than 1,213.

**(b)** Add 23, 2,845, 5, and 589.

**(c)** What is the total of the five numbers 2,195, 348, 640, 59, and 23,785?

**(d)** Find the number that is 34 more than 125.

**(e)** What is 457 increased by 96?

**32.** In each of the following exercises, find the appropriate sum.

**(a)** Find the number that is 567 more than 2,322.

**(b)** Add 5,637, 78, 690, 28, and 35,589.

**(c)** What is the total of the five numbers 3,295, 9, 427, 56, and 11,100?

**(d)** Find the number that is 124 more than 2,351.

**(e)** What is 926 increased by 86?

**33.** **Business and Finance** Tral bought a 1931 Model A for $15,200, a 1964 Thunderbird convertible for $17,100, and a 1959 Austin Healy Mark I for $17,450. How much did he invest in the three cars?

## Answers

25. ____________
26. ____________
27. ____________
28. ____________
29. ____________
30. ____________
31. ____________
32. ____________
33. ____________

Answers

34. ______

35. ______

36. ______

34. **Business and Finance** The following chart shows Family Video's monthly rentals for the first 3 months of 2006 by category of film. Complete the totals.

| Category of Film | Jan. | Feb. | Mar. | Category Totals |
|---|---|---|---|---|
| Comedy | 4,568 | 3,269 | 2,189 | ______ |
| Drama | 5,612 | 4,129 | 3,879 | ______ |
| Action/Adventure | 2,654 | 3,178 | 1,984 | ______ |
| Musical | 897 | 623 | 528 | ______ |
| Monthly Totals | ____ | ____ | ____ | ______ |

35. **Business and Finance** The following chart shows Regina's Dress Shop's expenses by department for the last 3 months of the year. Complete the totals.

| Department | Oct. | Nov. | Dec. | Department Totals |
|---|---|---|---|---|
| Office | $31,714 | $32,512 | $30,826 | ______ |
| Production | 85,146 | 87,479 | 81,234 | ______ |
| Sales | 34,568 | 37,612 | 33,455 | ______ |
| Warehouse | 16,588 | 11,368 | 13,567 | ______ |
| Monthly Totals | ____ | ____ | ____ | ______ |

36. **Social Science** The following table ranks the top 10 areas for women-owned firms in the United States.

| Metro Area | Number of Firms | Employment | Sales (in millions) |
|---|---|---|---|
| Los Angeles–Long Beach, Calif. | 360,300 | 1,056,600 | $181,455,900 |
| New York | 282,000 | 1,077,900 | 193,572,200 |
| Chicago | 260,200 | 1,108,800 | 161,200,900 |
| Washington, D.C. | 193,600 | 440,000 | 56,644,000 |
| Philadelphia | 144,600 | 695,900 | 90,231,000 |
| Atlanta | 138,700 | 331,800 | 50,206,800 |
| Houston | 136,400 | 560,100 | 78,180,300 |
| Dallas | 123,900 | 431,900 | 63,114,900 |
| Detroit | 123,600 | 371,400 | 50,060,700 |
| Minneapolis–St. Paul, Minn. | 119,600 | 337,400 | 51,063,400 |

**(a)** How many firms in total are located in Washington, Philadelphia, and New York?

**(b)** What is the total number of employees in all 10 of the areas listed?

**(c)** What are the total sales for firms in Houston and Dallas?

**(d)** How many firms in total are located in Chicago and Detroit?

37. **Number Problem** The following sequences are called *arithmetic sequences*. Determine the pattern and write the next four numbers in each sequence.

**(a)** 5, 12, 19, 26, _____, _____, _____, _____

**(b)** 8, 14, 20, 26, _____, _____, _____, _____

**(c)** 7, 13, 19, 25, _____, _____, _____, _____

**(d)** 9, 17, 25, 33, _____, _____, _____, _____

38. **Number Problem** Fibonacci numbers occur in the sequence

1, 1, 2, 3, 5, 8, 13, 21, 34, 55, . . .

This sequence begins with the numbers 1 and 1 again, and each subsequent number is obtained by adding the two preceding numbers.

Find the next four numbers in the sequence.

Basic Skills | Advanced Skills | **Vocational-Technical Applications** | Calculator/Computer | Above and Beyond

39. **Manufacturing Technology** An inventory of steel round stock shows 248 feet (ft) of $\frac{1}{4}$ inch (in.), 124 ft of $\frac{3}{8}$ in., 428 ft of $\frac{1}{2}$ in., and 162 ft of $\frac{5}{8}$ in. How many total feet of steel round stock are in inventory?

40. **Manufacturing Technology** B & L Industries produces three different products. Orders for today are for 351 of product A, 187 of product B, and 94 of product C. How many total products need to be produced today to fill the orders?

41. **Allied Health** The source-image receptor distance (SID) for radiographic images is the sum of the object-film distance (OFD) and the focus-object distance (FOD). Determine the SID if the distance from the object to the film is 8 inches (in.), and the distance from the object to the focus is 48 in.

42. **Allied Health** Total lung capacity, measured in milliliters (mL), is the sum of the vital capacity and the residual volume. Determine the total lung capacity for a patient whose vital capacity is 4,500 mL and whose residual volume is 1,800 mL.

**Answers**

37. ____________________

38. ____________________

39. ____________________

40. ____________________

41. ____________________

42. ____________________

## Answers

43. ______

44. ______

45. ______

46. ______

47. ______

**43. Information Technology** Computers can store data on different kinds of media such as a floppy disk, hard disk, data CD, or zip disk. The storage capacity of a floppy disk is one million, four hundred forty thousand bytes, of a hard disk is forty million bytes, of a data CD is six hundred fifty million bytes, and of a zip disk is one hundred million bytes. Which storage medium stores the largest amount of data, and which stores the least amount of data?

**44. Information Technology** On your PC, you create a document that is one million, four hundred forty-five thousand bytes. Will a floppy disk be able to hold the document (see Exercise 43)? Why or why not?

Basic Skills | Advanced Skills | Vocational-Technical Applications | **Calculator/Computer** | Above and Beyond

Although this text is designed to help you master the basic skills of arithmetic, it is occasionally preferable to perform complex calculations on a calculator. To that end, many of the exercise sets conclude with a short explanation of how to use a calculator to do an operation described in the section. This explanation will be followed by a set of exercises for which the calculator might be the preferred tool. As indicated by the placement of the explanation, you should refrain from using a calculator on the exercises that precede it.

To perform the addition

2,473 + 258 + 35 + 5,823

| | | |
|---|---|---|
| **Step 1** | Press the clear key. | [C] |
| **Step 2** | Enter the first number. | 2473 |
| **Step 3** | Enter the plus key followed by the next number. | [+] 258 |
| **Step 4** | Continue with the addition until the last number is entered. | [+] 35 |
| | | [+] 5823 |
| **Step 5** | Press the equal-sign key. | [=] |
| The desired sum should now be in the display. | | 8589 |

*Use your calculator to find the following sums.*

**45.** 3,295,153 + 573,128 + 21,257 + 2,586,241 + 5,291

**46.** 23,563 + 5,638,487 + 385,005 + 27,345

*Use your calculator to solve the following applications.*

**47. Business and Finance** The following table shows the number of customers using three branches of a bank during one week. Complete the table by finding the daily, weekly, and grand totals.

| Branch | Mon. | Tues. | Wed. | Thurs. | Fri. | Weekly Totals |
|---|---|---|---|---|---|---|
| Downtown | 487 | 356 | 429 | 278 | 834 | ______ |
| Suburban | 236 | 255 | 254 | 198 | 423 | ______ |
| Westside | 345 | 278 | 323 | 257 | 563 | ______ |
| Daily Totals | | | | | | ______ |

**48. STATISTICS** The following table lists the number of possible types of poker hand. What is the total number of hands possible?

| | |
|---|---:|
| Royal flush | 4 |
| Straight flush | 36 |
| Four of a kind | 624 |
| Full house | 3,744 |
| Flush | 5,108 |
| Straight | 10,200 |
| Three of a kind | 54,912 |
| Two pairs | 123,552 |
| One pair | 1,098,240 |
| Nothing | 1,302,540 |
| Total Possible Hands | ______ |

Basic Skills | Advanced Skills | Vocational-Technical Applications | Calculator/Computer | **Above and Beyond**

**49. NUMBER PROBLEM** A magic square is a square in which the sum along any row, column, or diagonal is the same. For example

| | | |
|---|---|---|
| 35 | 10 | 15 |
| 0 | 20 | 40 |
| 25 | 30 | 5 |

Use the numbers 1 to 9 to form a magic square.

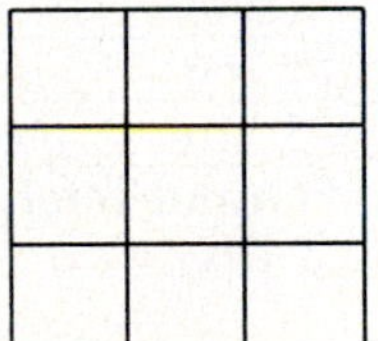

**50.** The following puzzle will give you a chance to practice some of your addition skills.

**Across**

1. 23 + 22
3. 103 + 42
6. 29 + 58 + 19
8. 3 + 3 + 4
9. 1,480 + 1,624
11. 568 + 730
13. 25 + 25
14. 131 + 132
16. The total of 121, 146, 119, and 132
17. The perimeter of a 4 × 6 rug

**Down**

1. The sum of 224,000, 155, and 186,000
2. 20 + 30
4. 210 + 200
5. 500,000 + 4,730
7. 130 + 509
10. 90 + 92
12. 100 + 101
15. The perimeter of a 15 × 16 room

| | | | | | |
|---|---|---|---|---|---|
| 1 | 2 | ■ | 3 | 4 | 5 |
| 6 | | 7 | ■ | 8 | |
| | ■ | 9 | 10 | | |
| 11 | 12 | | | ■ | |
| 13 | | ■ | 14 | 15 | |
| 16 | | | ■ | 17 | |

**Answers**

48. ______

49. ______

50. ______

## Answers

**1.** Commutative property of addition
**3.** Associative property of addition
**5.** Commutative property of addition
**7.** Commutative property of addition **9.** 2,997 **11.** 8,398
**13.** 7,885 **15.** 64,356 **17.** 4,063 **19.** 6,394 **21.** 56,081
**23.** 5 is an addend, 4 is an addend, 9 is the sum **25.** 22 ft **27.** 21 yd
**29.** 26 in. **31.** **(a)** 1,569; **(b)** 3,462; **(c)** 27,027; **(d)** 159; **(e)** 553
**33.** $49,750
**35.**

| Department | Oct. | Nov. | Dec. | Department Totals |
|---|---|---|---|---|
| Office | $31,714 | $32,512 | $30,826 | $95,052 |
| Production | 85,146 | 87,479 | 81,234 | $253,859 |
| Sales | 34,568 | 37,612 | 33,455 | $105,635 |
| Warehouse | 16,588 | 11,368 | 13,567 | $41,523 |
| Monthly Totals | $168,016 | $168,971 | $159,082 | $496,069 |

**37.** **(a)** 33, 40, 47, 54; **(b)** 32, 38, 44, 50; **(c)** 31, 37, 43, 49; **(d)** 41, 49, 57, 65
**39.** 962 ft **41.** 56 in. **43.** Data CD: 650,000,000; floppy disk: 1,440,000
**45.** 6,481,070
**47.**

| Branch | Mon. | Tues. | Wed. | Thurs. | Fri. | Weekly Totals |
|---|---|---|---|---|---|---|
| Downtown | 487 | 356 | 429 | 278 | 834 | 2,384 |
| Suburban | 236 | 255 | 254 | 198 | 423 | 1,366 |
| Westside | 345 | 278 | 323 | 257 | 563 | 1,766 |
| **Daily Totals** | 1,068 | 889 | 1,006 | 733 | 1,820 | 5,516 |
| | | | | | | **Grand Total** |

**49.**

| | | |
|---|---|---|
| 8 | 3 | 4 |
| 1 | 5 | 9 |
| 6 | 7 | 2 |

# 1.3 Subtraction

< 1.3 Objectives >

1 > Subtract whole numbers without borrowing

2 > Use the language of subtraction

3 > Solve applications of simple subtraction

4 > Use borrowing in subtracting whole numbers

5 > Solve applications that require borrowing

## Overcoming Math Anxiety

Hint #3 Don't procrastinate!

1. Do your math homework while you're still fresh. If you wait until too late at night, your mind will be tired and have much more difficulty understanding the concepts.
2. Do your homework the day it is assigned. The more recent the explanation is, the easier it is to recall.
3. When you are finished with your homework, try reading the next section through once. This will give you a sense of direction when you next hear the material. This works whether you are in a lecture or lab setting.

Remember that, in a typical math class, you are expected to do two or three hours of homework for each weekly class hour. This means two or three hours per night. Schedule the time and stay on schedule.

**NOTE**

By *opposite operation* we mean that subtracting a number "undoes" an addition of that same number. Start with 1. Add 5 and then subtract 5. Where are you?

We are now ready to consider a second operation of arithmetic—subtraction. In Section 1.2, we described addition as the process of combining two or more groups of the same kind of objects. Subtraction can be thought of as the *opposite operation* to addition. Every arithmetic operation has its own notation. The symbol for subtraction, −, is called a **minus sign.**

When we write $8 - 5$, we wish to subtract 5 from 8. We call 5 the **subtrahend.** This is the number being subtracted. And 8 is the **minuend.** This is the number we are subtracting from. The **difference** is the result of the subtraction.

To find the *difference* of two numbers, we will assume that we wish to subtract the smaller number from the larger. Then we look for a number which, when added to the smaller number, will give us the larger number. For example,

$8 - 5 = 3$ because $3 + 5 = 8$

This special relationship between addition and subtraction provides a method of checking subtraction.

**Property**

| **Relationship Between Addition and Subtraction** | The sum of the difference and the subtrahend must be equal to the minuend. |
|---|---|

### Example 1 — Subtracting a Single-Digit Number

< Objective 1 >

$12 - 5 = 7$

Check: $7 + 5 = 12$

Difference  Subtrahend  Minuend

Our check works because 12 − 5 asks for the number that must be added to 5 to get 12.

**Check Yourself 1**

**Subtract and check your work.**

**13 − 9**

The procedure for subtracting larger whole numbers is similar to the procedure for addition. We subtract digits of the same place value.

### Example 2 — Subtracting a Larger Number

| **Step 1** | **Step 2** | **Step 3** |
|---|---|---|
| 789 | 789 | 789 |
| − 246 | − 246 | − 246 |
| 3 | 43 | 543 |

We subtract in the ones column, then in the tens column, and finally in the hundreds column.

To check:

$$\begin{array}{r} 789 \\ -\ 246 \\ \hline 543 \end{array}$$

Add 543 + 246 = 789

The sum of the difference and the subtrahend must be the minuend.

**Check Yourself 2**

**Subtract and check your work.**

**(a)** $\begin{array}{r} 3{,}468 \\ -\ 2{,}248 \\ \hline \end{array}$  **(b)** $\begin{array}{r} 4{,}984 \\ -\ 1{,}081 \\ \hline \end{array}$

You know that the word *difference* indicates subtraction. There are other words that also tell you to use the subtraction operation. For instance, 5 *less than* 12 is written as

$12 - 5$ or $7$

20 *decreased* by 8 is written as

20 − 8 or 12

**Example 3** Translating Words That Indicate Subtraction

< Objective 2 >

Find each of the following.

**(a)** 4 less than 11

4 less than 11 is written 11 − 4 = 7.

**(b)** 27 decreased by 6

27 decreased by 6 is written 27 − 6 = 21.

**Check Yourself 3**

Find each of the following.

**(a)** 6 less than 19 **(b)** 18 decreased by 3

**Units ANALYSIS**

This is the first in a series of essays that are designed to help you solve applications using mathematics. Questions in the exercise sets will require the skills that you build by reading these essays.

A number with a unit attached (such as 7 **feet** or 26 **mi/gal**) is called a denominate number. Any genuine application of mathematics involves denominate numbers.

When adding or subtracting denominate numbers, the units must be identical for both numbers. The sum or difference will have those same units.

EXAMPLES:

\$4 + \$9 = \$13 Notice that, although we write the dollar sign first, we read it after the quantity, as in "four dollars."

7 feet + 9 feet = 16 feet

39 degrees − 12 degrees = 27 degrees

7 feet + 12 degrees yields no meaningful answer!

3 feet + 9 inches yields a meaningful result if 3 feet is converted into 36 inches. We will discuss conversion of units in later essays.

Now we consider subtraction word problems. The strategy is the same one presented in Section 1.2 for addition word problems. It is summarized with the following four basic steps.

Step by Step

**Solving Subtraction Applications**

**Step 1** Read the problem carefully to determine the given information and what you are asked to find.

**Step 2** Decide upon the operation (in this case, subtraction) to be used.

**Step 3** Write down the complete statement necessary to solve the problem and do the calculations.

**Step 4** Check to make sure you have answered the question of the problem and that your answer seems reasonable.

Here is an example using these steps.

## Example 4 Setting Up a Subtraction Word Problem

< Objective 3 >

Tory has \$37 in his wallet. He is thinking about buying a \$24 pair of pants and a \$10 shirt. If he buys them both, how much money will he have remaining?

First we must add the cost of the pants and the shirt.

\$24 + \$10 = \$34

Now, that amount must be subtracted from the \$37.

\$37 − \$34 = \$3

He will have \$3 left.

### Check Yourself 4

**Sonya has \$97 left in her checking account. If she writes checks for \$12, \$32, and \$21, how much will she have in the account?**

Difficulties can arise in subtraction if one or more of the digits of the subtrahend are larger than the corresponding digits in the minuend. We will solve this problem by using another version of the regrouping process called **borrowing.**

First, we look at an example in expanded form.

## Example 5 Subtracting When Regrouping Is Needed

< Objective 4 >

$$\begin{array}{r} 52 = 50 + 2 \\ -\ 27 = 20 + 7 \\ \hline \end{array}$$

Do you see that we cannot subtract in the ones column?

We regroup by borrowing 1 ten in the minuend and writing that ten as 10 ones:

$$\begin{array}{ll} & \overbrace{\quad 50 \quad} \quad + 2 \\ \text{becomes} & 40 + \underbrace{10 + 2} \\ \text{or} & 40 + \quad 12 \end{array}$$

We now have

$$\begin{array}{r} 52 = 40 + 12 \\ -\ 27 = 20 + \ \ 7 \\ \hline 20 + \ \ 5 \end{array}$$

We can now subtract as before.

or 25

In practice, we use a more convenient short form for the subtraction.

$$\begin{array}{r} 52 \\ -\ 27 \\ \hline \end{array} \qquad \begin{array}{r} {}^{4}\!\!\not{5}{}^{1}2 \\ -\ 27 \\ \hline 25 \end{array}$$

We indicate the fact that we have borrowed 1 ten by putting a slash through the 5 and then writing 4 tens. Add 10 ones to the original 2 ones to get 12 ones. We can then subtract.

Check: 25 + 27 = 52

## Check Yourself 5

Subtract and check your work.

$$\begin{array}{r} 64 \\ -\ 38 \\ \hline \end{array}$$

In Example 6, we work through a subtraction example that requires a number of regrouping steps. Here, zero appears as a digit in the minuend.

## Example 6 Subtracting When Regrouping Is Needed

**Step 1**

$$\begin{array}{r} 4{,}0\not{5}{}^{1}3 \\ -\ 2{,}365 \\ \hline 8 \end{array}$$

In this first step we regroup by borrowing 1 ten. This is written as 10 ones and combined with the original 3 ones. We can then subtract in the ones column.

**Step 2**

$$\begin{array}{r} \not{4}{,}\not{0}\not{5}{}^{1}3 \\ -\ 2{,}365 \\ \hline 8 \end{array}$$

We must regroup again to subtract in the tens column. There are no hundreds, and so we move to the thousands column.

**NOTE**

Here we borrow 1 thousand; this is written as 10 hundreds.

**Step 3**

$$\begin{array}{r} \not{4}{,}\not{0}\not{5}{}^{1}3 \\ -\ 2{,}365 \\ \hline 8 \end{array}$$

The minuend is now renamed as 3 thousands, 9 hundreds, 14 tens, and 13 ones.

**NOTE**

We now borrow 1 hundred; this is written as 10 tens and combined with the remaining 4 tens.

**Step 4**

$$\begin{array}{r} \not{4}{,}\not{0}\not{5}3 \\ -\ 2{,}365 \\ \hline 1{,}688 \end{array}$$

The subtraction can now be completed.

To check our subtraction: 1,688 + 2,365 = 4,053

## Check Yourself 6

Subtract and check your work.

$$\begin{array}{r} 5{,}024 \\ -\ 1{,}656 \\ \hline \end{array}$$

You need to use both addition and subtraction to solve some problems, as Example 7 illustrates.

## Example 7 Solving a Subtraction Application

< Objective 5 >

Bernard wants to buy a new piece of stereo equipment. He has \$142 and can trade in his old amplifier for \$135. How much more does he need if the new equipment costs \$449?

First we must add to find out how much money Bernard has available. Then we subtract to find out how much more money he needs.

\$142 + \$135 = \$277 The money available to Bernard

\$449 − \$277 = \$172 The money Bernard still needs

Bernard will need \$172.

### Check Yourself 7

**Martina spent \$239 in airfare, \$174 for lodging, and \$108 for food on a business trip. Her company allowed her \$375 for the expenses. How much of these expenses will she have to pay herself?**

### Check Yourself ANSWERS

**1.** 13 − 9 = 4 Check: 4 + 9 = 13 **2. (a)** 1,220; **(b)** 3,903

**3. (a)** 13; **(b)** 15 **4.** Sonya will have \$32 left.

**5.** $\begin{array}{r} \overset{5}{\not6}\,{}^{1}4 \\ -\ 38 \\ \hline 26 \end{array}$ To check: 26 + 38 = 64 **6.** 3,368 Check: 3,368 + 1,656 = 5,024

**7.** $\begin{array}{r} \$239 \\ 174 \\ +\ 108 \\ \hline \$521 \end{array}$ ← Total expenses

$\begin{array}{r} \$521 \\ -\ 375 \\ \hline \$146 \end{array}$ \$521 ← Total expenses; 375 ← Amount allowed; Martina will have to pay \$146.

## Reading Your Text

The following fill-in-the-blank exercises are designed to ensure that you understand some of the key vocabulary used in this section.

SECTION 1.3

**(a)** The ________ is the result of subtraction.

**(b)** 5 ________ than 12 is written as 12 − 5.

**(c)** The first step in solving a subtraction application is to ________ the problem carefully.

**(d)** The regrouping process used in subtraction is called ________.

Basic Skills | Advanced Skills | Vocational-Technical Applications | Calculator/Computer | Above and Beyond

# 1.3 exercises

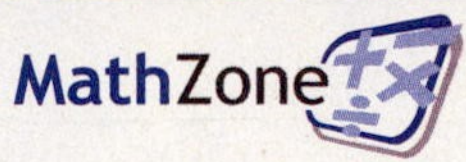


Boost your grade at mathzone.com!
- > Practice Problems
- > NetTutor
- > Self-Tests
- > e-Professors
- > Videos

Name ________________

Section ________ Date ________

< Objectives 1, 4 >

*In exercises 1 to 20, do the indicated subtraction and check your results by addition.*

**1.** 347 − 201

**2.** 575 − 302

**3.** 689 − 245

**4.** 598 − 278

**5.** 3,446 − 2,326 > Videos

**6.** 5,896 − 3,862

**7.** 64 − 27

**8.** 73 − 36

**9.** 627 − 358

**10.** 642 − 367

**11.** 6,423 − 3,678

**12.** 5,352 − 2,577

**13.** 6,034 − 2,569

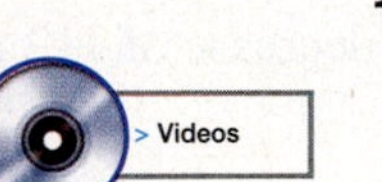


**14.** 5,206 − 1,748

**15.** 4,000 − 2,345

**16.** 6,000 − 4,349

**17.** 33,486 − 14,047

**18.** 53,487 − 25,649

**19.** 29,400 − 17,900

**20.** 53,500 − 28,700

< Objective 2 >

**21.** In the statement $9 - 6 = 3$

9 is called the ____________.
6 is called the ____________.
3 is called the ____________.
Write the related addition statement.

**22.** In the statement $7 - 5 = 2$

5 is called the ____________.
2 is called the ____________.
7 is called the ____________.
Write the related addition statement.

< Objective 3 >

**23.** Find the number that is 25 less than 76. > Videos

**24.** Find the number that results when 58 is decreased by 23.

**25.** Find the number that is the difference between 97 and 43.

**26.** Find the number that is 125 less than 265.

**27.** Find the number that results when 298 is decreased by 47.

**28.** Find the number that is the difference between 167 and 57.

## Answers

1. ________ 2. ________
3. ________ 4. ________
5. ________ 6. ________
7. ________ 8. ________
9. ________ 10. ________
11. ________ 12. ________
13. ________ 14. ________
15. ________ 16. ________
17. ________ 18. ________
19. ________ 20. ________
21. ________________
22. ________________
23. ________ 24. ________
25. ________ 26. ________
27. ________ 28. ________

Basic Skills | **Advanced Skills** | Vocational-Technical Applications | Calculator/Computer | Above and Beyond

## Answers

29. ______

30. ______

31. ______

32. ______

33. ______

34. ______

35. ______

36. ______

37. ______

38. ______

39. ______

40. ______

41. ______

*Based on units, determine if the following operations produce a meaningful result.*

**29.** 8 miles − 4 miles

**30.** \$560 + \$314

**31.** 7 feet + 11 meters

**32.** 18°F − 6°C

**33.** 17 yards − 10 yards

**34.** 4 mi/h + 6 ft/s

*In exercises 35 to 38, for various treks by a hiker in a mountainous region, the starting elevations and various changes are given. Determine the final elevation of the hiker in each case.*

**35.** Starting elevation 1,053 feet (ft), increase of 123 ft, decrease of 98 ft, increase of 63 ft.

**36.** Starting elevation 1,231 ft, increase of 213 ft, decrease of 112 ft, increase of 78 ft.

**37.** Starting elevation 7,302 ft, decrease of 623 ft, decrease of 123 ft, increase of 307 ft.

> Videos

**38.** Starting elevation 6,907 ft, decrease of 511 ft, decrease of 203 ft, increase of 419 ft.

< Objective 5 >

*Solve the following applications.*

**39.** **Social Science** Shaka's score on a math test was 87, and Tony's score was 23 points less than Shaka's. What was Tony's score on the test?

**40.** **Business and Finance** Duardo's monthly pay of \$879 was decreased by \$175 for withholding. What amount of pay did he receive?

> Videos

**41.** **Construction** The Sears Tower in Chicago is 1,454 ft tall. The Empire State Building is 1,250 ft tall. How much taller is the Sears Tower than the Empire State Building?

**42. Business and Finance** In one week, Margaret earned \$278 in regular pay and \$53 for overtime work, and \$49 was deducted from her paycheck for income taxes and \$18 for Social Security. What was her take-home pay?

**43. Business and Finance** Rafael opened a checking account and made deposits of \$85 and \$272. He wrote checks during the month for \$35, \$27, \$89, and \$178. What was his balance at the end of the month?

**44. Science and Medicine** Dalila is trying to limit herself to 1,500 calories per day (cal/day). Her breakfast was 270 cal, her lunch was 450 cal, and her dinner was 820 cal. By how much was she *under* or *over* her diet?

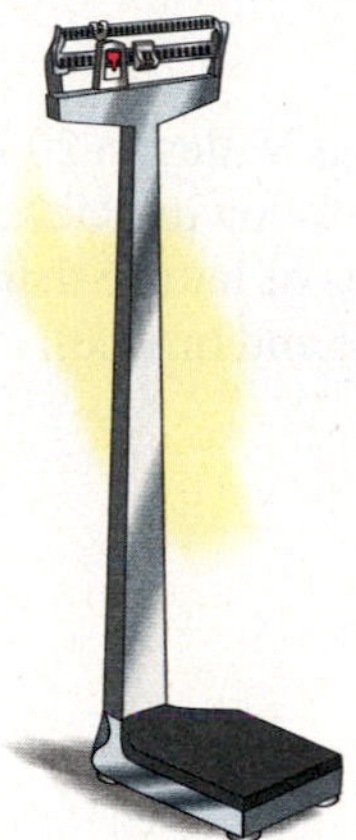

**45. Business and Finance** Complete the following record of a monthly expense account.

| | |
|---|---|
| Monthly income | \$1,620 |
| Share of rent | 343 |
| Balance | |
| Car payment | 183 |
| Balance | |
| Food | 312 |
| Balance | |
| Clothing | 89 |
| Amount remaining | |

**46. Business and Finance** To keep track of a checking account, you must subtract the amount of each check from the current balance. Complete the following statement.

| | |
|---|---|
| Beginning balance | \$351 |
| Check #1 | 29 |
| Balance | |
| Check #2 | 139 |
| Balance | |
| Check #3 | 75 |
| Ending balance | |

## Answers

42. ______

43. ______

44. ______

45. ______

46. ______

Answers

47. ________

48. ________

49. ________

50. ________

51. ________

52. ________

**47.** **Business and Finance** Carmen's frequent-flyer program requires 30,000 miles (mi) for a free flight. During 2004 she accumulated 13,850 mi. In 2005 she took three more flights of 2,800, 1,475, and 4,280 mi. How much farther must she fly for her free trip?

**48.** **Business and Finance** The value of all crops in the Salinas Valley in 2003 was nearly $3 billion. The top four crops are listed in the following table. **(a)** How much greater is the combined value of both types of lettuce than broccoli? **(b)** How much greater is the value of the lettuce and broccoli combined than that of the strawberries?

| Crop | Crop value, in millions |
|---|---|
| Head lettuce | $361 |
| Leaf lettuce | $277 |
| Strawberries | $298 |
| Broccoli | $259 |

Basic Skills | Advanced Skills | **Vocational-Technical Applications** | Calculator/Computer | Above and Beyond

**49.** **Allied Health** A patient's anion gap is used to help evaluate her overall electrolyte balance. The anion gap is equal to the difference between the serum concentration [measured in milliequivalents per liter (mEq/L)] of sodium and the sum of the serum concentrations of chloride and bicarbonate. Determine the patient's anion gap if the concentration of sodium is 140 mEq/L, chloride is 93 mEq/L, and bicarbonate is 24 mEq/L.

**50.** **Allied Health** To increase the geometric sharpness of a radiographic image, it is easiest to set the focus-object distance (FOD), which is the difference between the source-image receptor distance (SID) and the object-film distance (OFD), to its maximum value. What is the maximum FOD possible if the distance between the object and the film is fixed at 8 inches and the maximum distance possible between the source-image and the receptor is 72 inches?

**51.** **Information Technology** Sally's department needs a new printer. The old printer just died. Sally bought an ink-jet printer at a cost of $150. For a growing department, the ink-jet printer is not appropriate. The problem is she really needs a laser printer at a cost of $500. If she returns the ink-jet printer, how much extra money will she need to buy the laser printer?

**52.** **Information Technology** Max has a 20-foot roll of cable, and he needs to run the cable from a wiring closet to an outlet in a room that is adjacent to the closet. The distance from the wiring closet to the outlet is about 25 feet. How much cable will Max need to buy to be able to run the cable from the wiring closet to the outlet in the adjacent room?

**53.** **Electronics** Solder looks like flexible wire and typically comes wrapped on spools. When heated with a soldering iron or any other heat source, solder melts. It is used to connect an electronic component to wires, other components, or conductive traces. If a certain spool holds 10 pounds (lb) of solder, yet the shipping weight for the spool is 14 lb, how much does the empty spool and shipping materials weigh in pounds?

**54.** **Manufacturing Technology** Kinetics, Inc., sells an engine block for $168. Production of the block is $72 for materials, $58 for labor, and $19 for shipping and packaging. How much is the profit on the engine block?

Basic Skills | Advanced Skills | Vocational-Technical Applications | **Calculator/Computer** | Above and Beyond

Now that you have reviewed the process of subtracting by hand, look at the use of the calculator in performing this operation.

Find

23 − 13 + 56 − 29

Enter the numbers and the operation signs exactly as they appear in the expression.

23 [−] 13 [+] 56 [−] 29 [ENTER]

**Display** [37]

An alternative approach would be to add 23 and 56 first and then subtract 13 and 29. The result is the same in either case.

*Do the indicated operations.*

**55.** 5,830 − 3,987

**56.** 15,280 − 7,595

**57.** Subtract 235 from the sum of 534 and 678.

**58.** Subtract 476 from the sum of 306 and 572.

*Solve the following applications.*

**59.** **Business and Finance** Readings from Fast Service Station's storage tanks were taken at the beginning and end of a month. How much of each type of gas was sold? What was the total sold?

| | Diesel | Unleaded | Super Unleaded | Total |
|---|---|---|---|---|
| Beginning reading | 73,255 | 82,349 | 81,258 | |
| End reading | 28,387 | 19,653 | 8,654 | |
| Gallons used | ____ | ____ | ____ | ____ |

## Answers

53. ________

54. ________

55. ________

56. ________

57. ________

58. ________

59. ________

Answers

60. ______

61. ______

62. ______

63. ______

64. ______

65. ______

66. ______

67. ______

68. ______

*The land areas, in square miles ($mi^2$), of three Pacific coast states are California, 155,959 $mi^2$; Oregon, 95,997 $mi^2$; and Washington, 66,544 $mi^2$.*

**60.** **Social Science** How much larger is California than Oregon?

**61.** **Social Science** How much larger is California than Washington?

**62.** **Social Science** How much larger is Oregon than Washington?

Basic Skills | Advanced Skills | Vocational-Technical Applications | Calculator/Computer | **Above and Beyond**

**Number Problem** Complete the magic squares.

**63.**

| | | |
|---|---|---|
| | 7 | 2 |
| | 5 | |
| 8 | | |

**64.**

| | | |
|---|---|---|
| 4 | 3 | |
| | 5 | |
| | | 6 |

**65.**

| | | | |
|---|---|---|---|
| 16 | 3 | | 13 |
| | 10 | 11 | |
| 9 | 6 | 7 | |
| | | | 1 |

**66.**

| | | | |
|---|---|---|---|
| 7 | | | 14 |
| 2 | 13 | 8 | 11 |
| 16 | | | |
| | 6 | 15 | |

**67.** **Social Science** Use the Internet to find the population of Arizona, California, Oregon, and Pennsylvania in each of the last three censuses.

chapter 1 > Make the Connection

(a) Find the total change in each state's population over this period.

(b) Which state shows the greatest change over the past three censuses?

(c) Write a brief essay describing the changes and any trends you see in these data. List any implications that they might have for future planning.

**68.** Think of any whole number.

Add 5.
Subtract 3.
Subtract 2 less than the original number.
What number do you end up with?
Check with other people. Does everyone have the same answer? Can you explain the results?

## Answers

**1.** 146 **3.** 444 **5.** 1,120 **7.** 37 **9.** 269 **11.** 2,745
**13.** 3,465 **15.** 1,655 **17.** 19,439 **19.** 11,500
**21.** 9 is the minuend, 6 is the subtrahend, and 3 is the difference; 3 + 6 = 9
**23.** 51 **25.** 54 **27.** 251 **29.** Yes **31.** No **33.** Yes
**35.** 1,141 ft **37.** 6,863 ft **39.** 64 **41.** 204 ft **43.** $28
**45.** $1,277; $1,094; $782; $693 **47.** 7,595 mi **49.** 23 mEq/L **51.** $350
**53.** 4 lb **55.** 1,843 **57.** 977 **59.** Regular, 44,868 gal; unleaded, 62,696 gal; super unleaded, 72,604 gal; total, 180,168 gal **61.** 89,415 $mi^2$

**63.**

| | | |
|---|---|---|
| 6 | 7 | 2 |
| 1 | 5 | 9 |
| 8 | 3 | 4 |

**65.**

| | | | |
|---|---|---|---|
| 16 | 3 | 2 | 13 |
| 5 | 10 | 11 | 8 |
| 9 | 6 | 7 | 12 |
| 4 | 15 | 14 | 1 |

**67.** Above and Beyond

# Activity 1 :: Population Changes

The following table gives the population for the United States and each of the six largest states from both the 1990 census and the 2000 census. Use this table to answer the questions that follow.

| | 1990 Population | 2000 Population |
|---|---|---|
| United States | 248,709,873 | 281,421,906 |
| California | 29,760,021 | 33,871,648 |
| Texas | 16,986,510 | 20,851,820 |
| New York | 17,990,455 | 18,976,457 |
| Florida | 12,937,926 | 15,982,378 |
| Illinois | 11,430,602 | 12,419,293 |
| Pennsylvania | 11,881,643 | 12,281,054 |

1. Which two pairs of states switched population ranking between the two censuses?
2. By how much did the population of the United States increase between 1990 and 2000?
3. Which state had the greatest increase in population from 1990 to 2000? What was that difference?
4. Which state had the smallest increase in population from 1990 to 2000? What was that difference?
5. What was the total population of the six largest states in 1990?
6. How many people living in the United States did not live in one of the six largest states in 1990?
7. What was the total population of the six largest states in 2000?
8. How many people living in the United States did not live in one of the six largest states in 2000?
9. What regional trends might be true based on what you see in this table?
10. Go to the Website for this text to access regional data so that you can compare the data to what you speculated in question 9.

# 1.4 Rounding, Estimation, and Order

< 1.4 Objectives >

1 > Round a whole number to a given place value

2 > Estimate sums and differences by rounding

3 > Use the inequality symbols

It is a common practice to express numbers to the nearest hundred, thousand, and so on. For instance, the distance from Los Angeles to New York along one route is 2,833 miles (mi). We might say that the distance is 2,800 mi. This is called **rounding,** because we have rounded the distance to the nearest hundred miles.

One way to picture this rounding process is with the use of a number line.

## Example 1 Rounding to the Nearest Hundred

< Objective 1 >

To round 2,833 to the nearest hundred:

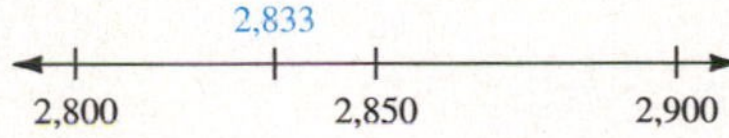


Because 2,833 is closer to 2,800 than it is to 2,900, we round *down* to 2,800.

### Check Yourself 1

**Round 587 to the nearest hundred.**

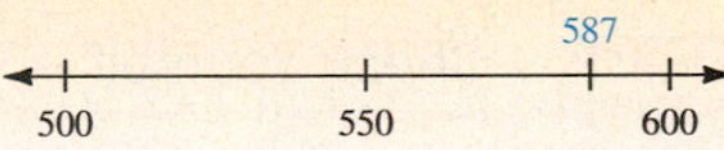


## Example 2 Rounding to the Nearest Thousand

To round 28,734 to the nearest thousand:

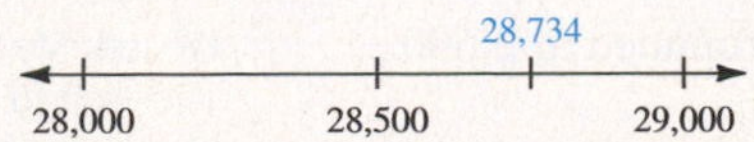


Because 28,734 is closer to 29,000 than it is to 28,000, we round *up* to 29,000.

**NOTES**

By a certain *place* in rounding, we mean tens, hundreds, thousands, and so on.

Step 3(a) is called **rounding up.**

Step 3(b) is called **rounding down.**

### Check Yourself 2

**Locate 1,375 and round to the nearest hundred.**

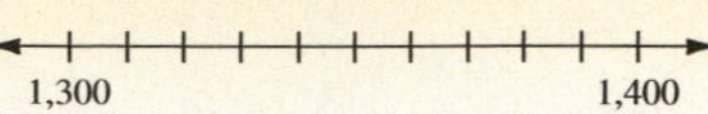


Instead of using a number line, we can apply the following rule.

**Step by Step**

**Rounding Whole Numbers**

Step 1 Identify the place of the digit to be rounded.

Step 2 Look at the digit to the right of that place.

Step 3 a. If that digit is 5 or more, that digit and all digits to the right become 0. The digit in the place you are rounding to is increased by 1.

b. If that digit is less than 5, that digit and all digits to the right become 0. The digit in the place you are rounding to remains the same.

### Example 3 Rounding to the Nearest Ten

Round 587 to the nearest ten:

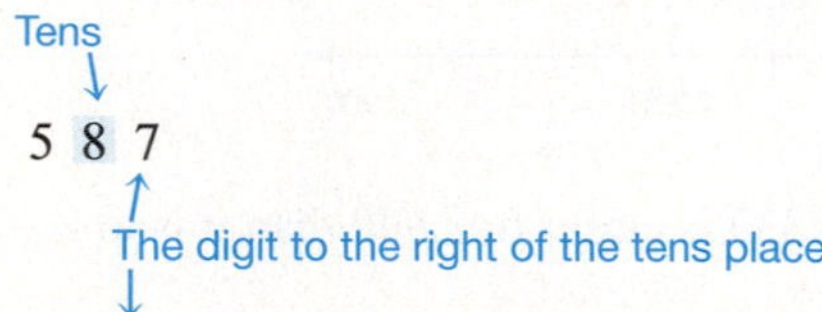


5 8 7 is rounded to 590

We identify the tens digit. The digit to the right of the tens place, 7, is 5 or more. So round up.

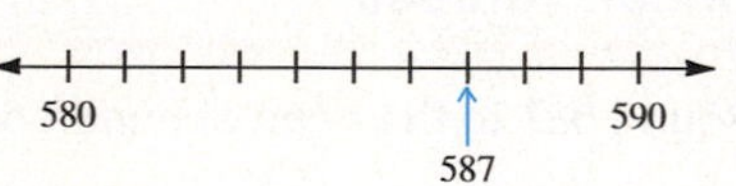


**NOTE**

587 is between 580 and 590. It is closer to 590, so it makes sense to round up.

### Check Yourself 3

**Round 847 to the nearest ten.**

### Example 4 Rounding to the Nearest Hundred

**NOTE**

2,638 is closer to 2,600 than to 2,700. So it makes sense to round down.

Round 2,638 to the nearest hundred:

2, 6 38 is rounded to 2,600

We identify the hundreds digit. The digit to the right, 3, is less than 5. So round down.

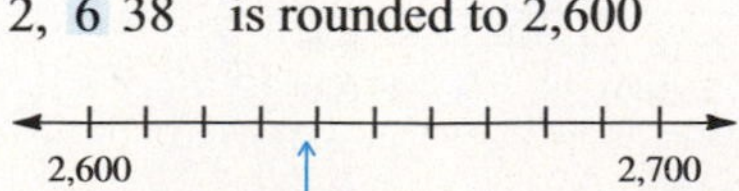

### Check Yourself 4

**Round 3,482 to the nearest hundred.**

Here are some further examples of using the rounding rule.

## Example 5 Rounding Whole Numbers

**(a)** Round 2,378 to the nearest hundred:

↓
2, 3 78 is rounded to 2,400

We identified the hundreds digit. The digit to the right is 7. Because this is 5 or more, the 7 and all digits to the right become 0. The hundreds digit is increased by 1.

**(b)** Round 53,258 to the nearest thousand:

↓
5 3 ,258 is rounded to 53,000

We identified the thousands digit. Because the digit to the right is less than 5, it and all digits to the right become 0, and the thousands digit remains the same.

**(c)** Round 685 to the nearest ten:

↓
6 8 5 is rounded to 690

The digit to the right of the tens place is 5 or more. Round up by our rule.

**(d)** Round 52,813,212 to the nearest million:

↓
5 2 ,813,212 is rounded to 53,000,000

### Check Yourself 5

**(a)** **Round 568 to the nearest ten.**
**(b)** **Round 5,446 to the nearest hundred.**

Now, look at a case in which we round up a 9.

## Example 6 Rounding to the Nearest Ten

**NOTE**

Which number is 397 closer to?

390 397 400

Suppose we want to round 397 to the nearest ten. We identify the tens digit and look at the next digit to the right.

↓
3 9 7

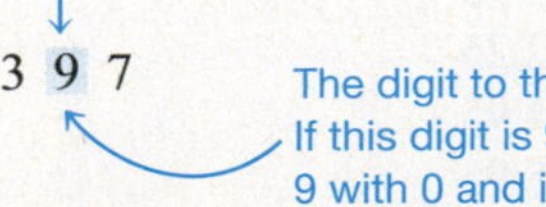

The digit to the right is 5 or more.
If this digit is 9, and it must be increased by 1, replace the 9 with 0 and increase the next digit to the *left* by 1.

So 397 is rounded to 400.

### Check Yourself 6

**Round 4,961 to the nearest hundred.**

**NOTE**

An estimate is basically a good guess. If your answer is close to your estimate, then your answer is reasonable.

Whether you are doing an addition problem by hand or using a calculator, rounding numbers gives you a handy way of deciding whether an answer seems reasonable. The process is called **estimating,** which we illustrate with an example.

## Example 7 Estimating a Sum

< Objective 2 >

**NOTE**

Placing an arrow above the column to be rounded can be helpful.

Begin by rounding to the nearest hundred.

$$\begin{array}{r} 456 \\ 235 \\ 976 \\ +\ 344 \\ \hline 2{,}011 \end{array} \qquad \begin{array}{r} 500 \\ 200 \\ 1{,}000 \\ +\quad 300 \\ \hline 2{,}000 \end{array} \leftarrow \text{Estimate}$$

By rounding to the nearest hundred and adding quickly, we get an estimate or guess of 2,000. Because this is close to the sum calculated, 2,011, our answer seems reasonable.

### Check Yourself 7

**Round each addend to the nearest hundred and estimate the sum. Then find the actual sum.**

**287 + 526 + 311 + 378**

Estimation is a wonderful tool to use while you're shopping. Every time you go to the store, you should try to estimate the total bill by rounding the price of each item. If you do this regularly, both your addition skills and your rounding skills will improve. The same holds true when you eat in a restaurant. It is always a good idea to know approximately how much you are spending.

## Example 8 Estimating a Sum in a Word Problem

Samantha has taken the family out to dinner, and she's now ready to pay the bill. The dinner check has no total, only the individual entries, as given below:

| | |
|---|---|
| Soup | $2.95 |
| Soup | 2.95 |
| Salad | 1.95 |
| Salad | 1.95 |
| Salad | 1.95 |
| Lasagna | 7.25 |
| Spaghetti | 4.95 |
| Ravioli | 5.95 |

What is the approximate cost of the dinner?

Rounding each entry to the nearest whole dollar, we can estimate the total by finding the sum

$3 + 3 + 2 + 2 + 2 + 7 + 5 + 6 = \$30$

### Check Yourself 8

**Jason is doing the weekly food shopping at FoodWay. So far his basket has items that cost \$3.99, \$7.98, \$2.95, \$1.15, \$2.99, and \$1.95. Approximate the total cost of these items.**

Earlier in this section, we used the number line to illustrate the idea of rounding numbers. The number line also gives us an excellent way to picture the concept of **order** for whole numbers, which means that numbers become larger as we move from left to right on the number line.

For instance, we know that 3 is less than 5. On the number line

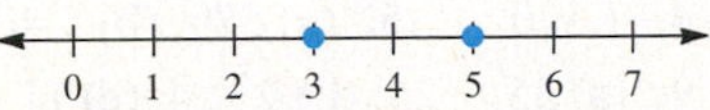


we see that 3 lies *to the left* of 5.

We also know that 4 is greater than 2. On the number line

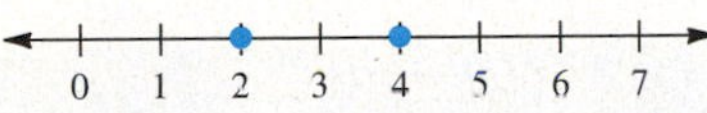


we see that 4 lies *to the right* of 2.

Two symbols, $<$ for "less than" and $>$ for "greater than," are used to indicate these relationships.

**NOTES**

3 is less than or smaller than 5.

4 is greater than or larger than 2.

The inequality symbol always "points at" the smaller number.

**Definition**

**Inequalities**

For whole numbers, we can write

1. $2 < 5$ (read "2 is less than 5") because 2 is *to the left* of 5 on the number line.
2. $8 > 3$ (read "8 is greater than 3") because 8 is *to the right* of 3 on the number line.

Example 9 illustrates the use of this notation.

### Example 9 Indicating Order with < or >

< Objective 3 >

Use the symbol $<$ or $>$ to complete each statement.

**(a)** 7 _____ 10
**(b)** 25 _____ 20
**(c)** 200 _____ 300
**(d)** 8 _____ 0

**(a)** $7 < 10$ 7 lies to the left of 10 on the number line.
**(b)** $25 > 20$ 25 lies to the right of 20 on the number line.
**(c)** $200 < 300$
**(d)** $8 > 0$

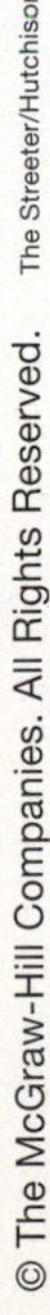

## Check Yourself 9

Use one of the symbols < and > to complete each of the following statements.

(a) 35 ___ 25
(b) 0 ___ 4
(c) 12 ___ 18
(d) 1,000 ___ 100

## Check Yourself ANSWERS

**1.** 600 **2.** 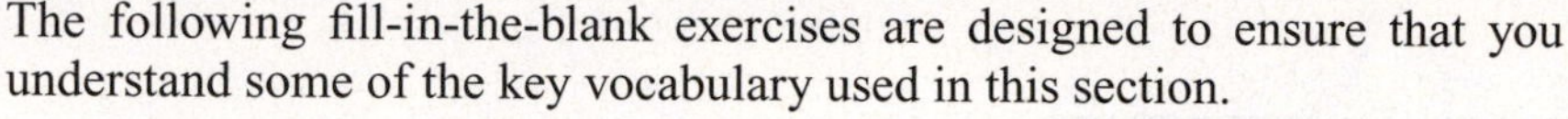 Round 1,375 *up* to 1,400.



**3.** 850 **4.** 3,500 **5.** **(a)** 570; **(b)** 5,400 **6.** 5,000 **7.** 1,500; 1,502
**8.** $21 **9.** **(a)** 35 > 25; **(b)** 0 < 4; **(c)** 12 < 18; **(d)** 1,000 > 100

## Reading Your Text

The following fill-in-the-blank exercises are designed to ensure that you understand some of the key vocabulary used in this section.

### SECTION 1.4

**(a)** The practice of expressing numbers to the nearest hundred, thousand, and so on is called __________.

**(b)** The first step in rounding is to identify the ____________ ____________ of the digit to be rounded.

**(c)** The number line gives us an excellent way to picture the concept of ______________ for whole numbers.

**(d)** The symbol < is read as "______________ than."

Basic Skills | Advanced Skills | Vocational-Technical Applications | Calculator/Computer | Above and Beyond

# 1.4 exercises

MathZone

Boost your grade at mathzone.com!
> Practice Problems
> NetTutor
> Self-Tests
> e-Professors
> Videos

Name ______________

Section ________ Date ________

< Objective 1 >

*Round each of the following numbers to the indicated place.*

**1.** 38, the nearest ten

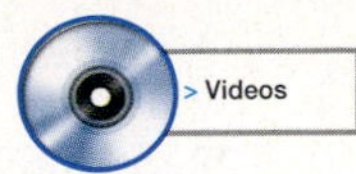


**2.** 72, the nearest ten

**3.** 253, the nearest ten

**4.** 578, the nearest ten

**5.** 696, the nearest ten

**6.** 683, the nearest hundred

**7.** 3,482, the nearest hundred

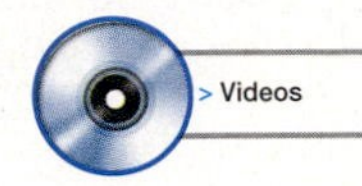


**8.** 6,741, the nearest hundred

**9.** 5,962, the nearest hundred

**10.** 4,352, the nearest thousand

**11.** 4,927, the nearest thousand

**12.** 39,621, the nearest thousand

**13.** 23,429, the nearest thousand

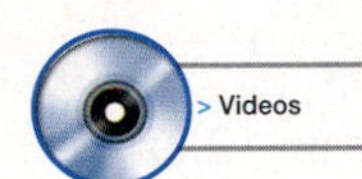


**14.** 38,589, the nearest thousand

**15.** 787,000, the nearest ten thousand

**16.** 582,000, the nearest hundred thousand

**17.** 21,800,000, the nearest million

**18.** 931,000, the nearest ten thousand

< Objective 2 >

*In exercises 19 to 30, estimate each of the sums or differences by rounding to the indicated place. Then do the addition or subtraction and use your estimate to see if your actual sum or difference seems reasonable.*

*Round to the nearest ten.*

**19.**
$$\begin{array}{r} 58 \\ 27 \\ +\ 33 \\ \hline \end{array}$$

**20.**
$$\begin{array}{r} 92 \\ 37 \\ 85 \\ +\ 64 \\ \hline \end{array}$$

**21.**
$$\begin{array}{r} 83 \\ -\ 27 \\ \hline \end{array}$$

**22.**
$$\begin{array}{r} 97 \\ -\ 31 \\ \hline \end{array}$$

*Round to the nearest hundred.*

**23.**
$$\begin{array}{r} 379 \\ 1{,}215 \\ +\ 528 \\ \hline \end{array}$$

**24.**
$$\begin{array}{r} 967 \\ 2{,}365 \\ 544 \\ +\ 738 \\ \hline \end{array}$$

**25.**
$$\begin{array}{r} 915 \\ -\ 411 \\ \hline \end{array}$$

**26.**
$$\begin{array}{r} 697 \\ -\ 539 \\ \hline \end{array}$$

## Answers

1. ________ 2. ________
3. ________ 4. ________
5. ________ 6. ________
7. ________ 8. ________
9. ________ 10. ________
11. ________ 12. ________
13. ________ 14. ________
15. ________ 16. ________
17. ________ 18. ________
19. ________
20. ________
21. ________
22. ________
23. ________
24. ________
25. ________
26. ________

Answers

27. ______

28. ______

29. ______

30. ______

31. ______

32. ______

33. ______

34. ______

35. ______

36. ______

Basic Skills | **Advanced Skills** | Vocational-Technical Applications | Calculator/Computer | Above and Beyond

*Round to the nearest thousand.*

**27.**
```
  2,238
  3,925
+ 5,217
```

**28.**
```
  3,678
  4,215
+ 2,032
```

**29.**
```
  4,822
− 2,134
```

**30.**
```
  6,120
− 4,890
```

*Solve the following applications.*

**31. Business and Finance** Ed and Sharon go to lunch. The lunch check has no total but only lists individual items:

| | |
|---|---|
| Soup $1.95 | Soup $1.95 |
| Salad $1.80 | Salad $1.80 |
| Salmon $8.95 | Flounder $6.95 |
| Pecan pie $3.25 | Vanilla ice cream $2.25 |

Estimate the total amount of the lunch check.

**32. Business and Finance** Olivia will purchase several items at the stationery store. Thus far, the items she has collected cost $2.99, $6.97, $3.90, $2.15, $9.95, and $1.10. Approximate the total cost of these items.

**33. Statistics** Oscar scored 78, 91, 79, 67, and 100 on his arithmetic tests. Round each score to the nearest ten to estimate his total score.

> Videos

**34. Business and Finance** Luigi's pizza parlor makes 293 pizzas on an average day. Estimate (to the nearest hundred) how many pizzas were made on a three-day holiday weekend.

**35. Business and Finance** Mrs. Gonzalez went shopping for clothes. She bought a sweater for $32.95, a scarf for $9.99, boots for $68.29, a coat for $125.90, and socks for $18.15. Estimate the total amount of Mrs. Gonzalez's purchases.

**36. Business and Finance** Amir bought several items at the hardware store: hammer, $8.95; screwdriver, $3.15; pliers, $6.90; wire cutters, $4.25; and sandpaper; $1.89. Estimate the total cost of Amir's bill.

< Objective 3 >

*Use the symbol < or > to complete each statement.*

**37.** 500 _____ 400 

**38.** 20 _____ 15

**39.** 100 _____ 1,000

**40.** 3,000 _____ 2,000

Basic Skills | Advanced Skills | **Vocational-Technical Applications** | Calculator/Computer | Above and Beyond

*Use the following chart for exercises 41 and 42.*

| Appliance | Power Required (in Watts/hour [W/h]) |
|---|---|
| Clock radio | 10 |
| Electric blanket | 100 |
| Clothes washer | 500 |
| Toaster oven | 1,225 |
| Laptop | 50 |
| Hair dryer | 1,875 |
| DVD player | 25 |

**41.** **Electronics** Assuming all the appliances listed in the table are "on," estimate the total power required to the nearest hundred watts.

**42.** **Electronics** Which combination uses more power, the toaster oven and clothes washer or the hair dryer and DVD player?

**43.** **Manufacturing Technology** An inventory of machine screws shows that bin 1 contains 378 screws, bin 2 contains 192 screws, and bin 3 contains 267 screws. Estimate the total number of screws in the bins.

**44.** **Manufacturing Technology** A delivery truck must be loaded with the heaviest crates starting in the front to the lightest crates in the back. On Monday, crates weighing 378 pounds (lb), 221 lb, 413 lb, 231 lb, 208 lb, 911 lb, 97 lb, 188 lb, and 109 lb need to be shipped. In what order should the crates be loaded?

## Answers

37. ______

38. ______

39. ______

40. ______

41. ______

42. ______

43. ______

44. ______

## Answers

45. ______

46. ______

47. ______

48. ______

49. ______

50. ______

45. **Number Problem** A whole number rounded to the nearest ten is 60. **(a)** What is the smallest possible number? **(b)** What is the largest possible number?

46. **Number Problem** A whole number rounded to the nearest hundred is 7,700. **(a)** What is the smallest possible number? **(b)** What is the largest possible number?

Basic Skills | Advanced Skills | Vocational-Technical Applications | Calculator/Computer | **Above and Beyond**

47. **Statistics** A bag contains 60 marbles. The number of blue marbles, rounded to the nearest 10, is 40, and the number of green marbles in the bag, rounded to the nearest 10, is 20. How many blue marbles are in the bag? (List all answers that satisfy the conditions of the problem.)

48. **Social Science** Describe some situations in which estimating and rounding would not produce a result that would be suitable or acceptable. Review the instructions for filing your federal income tax. What rounding rules are used in the preparation of your tax returns? Do the same rules apply to the filing of your state tax returns? If not, what are these rules?

49. The listed population of the United States on July 8, 2005, at 9:37 A.M. eastern standard time (EST) was **296,562,576 people.** Round this number to the nearest ten millions.

chapter 1 > Make the Connection

50. According to the U.S. Census Bureau, the population of the world was believed to be 6,457,380,056 on August 1, 2005. Round this number to the nearest million.

chapter 1 > Make the Connection

## Answers

**1.** 40 **3.** 250 **5.** 700 **7.** 3,500 **9.** 6,000
**11.** 5,000 **13.** 23,000 **15.** 790,000 **17.** 22,000,000
**19.** Estimate: 120, actual sum: 118 **21.** Estimate: 50, actual difference: 56
**23.** Estimate: 2,100, actual sum: 2,122
**25.** Estimate: 500; actual difference: 504
**27.** Estimate: 11,000, actual sum: 11,380
**29.** Estimate: 3,000, actual difference: 2,688 **31.** \$29 **33.** 420
**35.** \$255 **37.** > **39.** < **41.** 3,800 W **43.** 900 screws
**45.** **(a)** 55; **(b)** 64 **47.** 36, 37, 38, 39, 40, 41, 42, 43, 44
**49.** 300,000,000 people

# 1.5 Multiplication

< 1.5 Objectives >

1 > Multiply whole numbers

2 > Use the properties of multiplication

3 > Solve applications of multiplication

4 > Estimate products

5 > Find area using multiplication

**NOTES**

The use of the symbol × dates back to the 1600s.

The centered dot is the same as the times sign (×). We use the centered dot when we are using letters to represent numbers, as we have done with *a* and *b* here. We do that so the times sign will not be confused with the letter *x*.

Our work in this section deals with multiplication, another of the basic operations of arithmetic. Multiplication is closely related to addition. In fact, we can think of multiplication as a shorthand method for repeated addition. The symbol × is used to indicate multiplication.

$3 \times 4$ can be interpreted as 3 rows of 4 objects. By counting we see that $3 \times 4 = 12$. Similarly, 4 rows of 3 means $4 \times 3 = 12$.

| | 4 | | | |
|---|---|---|---|---|
| 3 | ★ | ★ | ★ | ★ |
| | ★ | ★ | ★ | ★ |
| | ★ | ★ | ★ | ★ |

| | 3 | | |
|---|---|---|---|
| 4 | ★ | ★ | ★ |
| | ★ | ★ | ★ |
| | ★ | ★ | ★ |
| | ★ | ★ | ★ |

The fact that $3 \times 4 = 4 \times 3$ is an example of the **commutative property of multiplication,** which is given here.

**Property**

**The Commutative Property of Multiplication**

Given any two numbers, we can multiply them in either order and we get the same result.

In symbols, we say $a \cdot b = b \cdot a$.

**Example 1** — **Multiplying Single-Digit Numbers**

< Objective 1 >

$3 \times 5$ means 5 multiplied by 3. It is read 3 *times* 5. To find $3 \times 5$, we can add 5 three times.

$3 \times 5 = 5 + 5 + 5 = 15$

In a multiplication problem such as $3 \times 5 = 15$, we call 3 and 5 the **factors.** The answer, 15, is the **product** of the factors, 3 and 5.

$$3 \times 5 = 15$$

Factor Factor Product

### Check Yourself 1

**Name the factors and the product in the following statement.**

**$2 \times 9 = 18$**

Statements such as $3 \times 4 = 12$ and $3 \times 5 = 15$ are called the **basic multiplication facts.** If you have difficulty with multiplication, it may be that you do not know some of these facts. The following table will help you review before you go on. Notice that, because of the commutative property, you need memorize only half of these facts!

**Basic Multiplication Facts Table**

| × | 0 | 1 | 2 | 3 | 4 | 5 | 6 | 7 | 8 | 9 |
|---|---|---|---|---|---|---|---|---|---|---|
| **0** | 0 | 0 | 0 | 0 | 0 | 0 | 0 | 0 | 0 | 0 |
| **1** | 0 | 1 | 2 | 3 | 4 | 5 | 6 | 7 | 8 | 9 |
| **2** | 0 | 2 | 4 | 6 | 8 | 10 | 12 | 14 | 16 | 18 |
| **3** | 0 | 3 | 6 | 9 | 12 | 15 | 18 | 21 | 24 | 27 |
| **4** | 0 | 4 | 8 | 12 | 16 | 20 | 24 | 28 | 32 | 36 |
| **5** | 0 | 5 | 10 | 15 | 20 | 25 | 30 | 35 | 40 | 45 |
| **6** | 0 | 6 | 12 | 18 | 24 | 30 | 36 | 42 | 48 | 54 |
| **7** | 0 | 7 | 14 | 21 | 28 | 35 | 42 | 49 | 56 | 63 |
| **8** | 0 | 8 | 16 | 24 | 32 | 40 | 48 | 56 | 64 | 72 |
| **9** | 0 | 9 | 18 | 27 | 36 | 45 | 54 | 63 | 72 | 81 |

**NOTE**

To use the table to find the product of $7 \times 6$: Find the row labeled 7, and then move to the right in this row until you are in the column labeled 6 at the top. We see that $7 \times 6$ is 42.

Armed with these facts, you can become a better, and faster, problem solver. Take a look at Example 2.

## Example 2 Multiplying Instead of Counting

< Objective 3 >

Find the total number of squares on the following checkerboard.

**NOTE**

This checkerboard is an example of a rectangular array, a series of rows or columns that form a rectangle. When you see such an arrangement, be prepared to multiply to find the total number of units.

You could find the number of squares by counting them. If you counted one per second, it would take you just over a minute. You could make the job a little easier by simply counting the squares in one row (8), and then adding 8 + 8 + 8 + 8 + 8 + 8 + 8 + 8. Multiplication, which is simply repeated addition, allows you to find the total number of squares by multiplying 8 × 8. How long that takes depends on how well you know the basic multiplication facts! By now, you know that there are 64 squares on the checkerboard.

## Check Yourself 2

**Find the number of windows on the displayed side of the building.**

The next property involves *both* multiplication and addition.

## Example 3 Using the Distributive Property

< Objective 4 >

**NOTE**

Multiplication can also be indicated by using parentheses. A number followed by parentheses or back-to-back parentheses represent multiplication.

$2 \times (3 + 4)$ could be written as $2(3 + 4)$ or $(2)(3 + 4)$.

$2 \times (3 + 4) = 2 \times 7 = 14$ We have added 3 + 4 and then multiplied.

Also,

$$\begin{aligned} 2 \times (3 + 4) &= (2 \times 3) + (2 \times 4) \\ &= 6 + 8 \\ &= 14 \end{aligned}$$

We have multiplied 2 × 3 and 2 × 4 as the first step.

The result is the same.

We see that $2 \times (3 + 4) = (2 \times 3) + (2 \times 4)$. This is an example of the **distributive property of multiplication over addition** because we distributed the multiplication (in this case by 2) over the "plus" sign.

**Property**

**The Distributive Property of Multiplication over Addition**

To multiply a factor by a sum of numbers, multiply the factor by each number inside the parentheses. Then add the products. (The result will be the same if we find the sum and then multiply.)

In symbols, we say $a \cdot (b + c) = a \cdot b + a \cdot c$

### Check Yourself 3

Show that

$3 \times (5 + 2) = (3 \times 5) + (3 \times 2)$

Regrouping must often be used to multiply larger numbers. We see how regrouping works in multiplication by looking at an example in the expanded form. When regrouping results in changing a digit to the left we sometimes say we "carry" the units.

## Example 4 Multiplying by a Single-Digit Number

**NOTE**

$3 \times 25$
$3 \cdot 25$
$(3)(25)$
all mean the same thing.

$$\begin{aligned} 3 \times 25 &= 3 \times (20 + 5) \\ &= 3 \times 20 + 3 \times 5 \\ &= 60 \qquad + 15 \\ &= 60 + \overbrace{10 + 5} \\ &= \underbrace{60 + 10} + 5 \\ &= 70 + 5 \\ &= 75 \end{aligned}$$

We use the distributive property again.

Write the 15 as 10 + 5.

Carry 10 ones or 1 ten to the tens place.

Here is the same multiplication problem using the short form.

**Step 1**

$$\begin{array}{r} \overset{1}{2}5 \\ \times\ 3 \\ \hline 5 \end{array}$$

1 ← Carry

Multiplying $3 \times 5$ gives us 15 ones. Write 5 ones and carry 1 ten.

**Step 2**

$$\begin{array}{r} \overset{1}{2}5 \\ \times\ 3 \\ \hline 75 \end{array}$$

Now multiply $3 \times 2$ tens and add the carry to get 7, the tens digit of the product.

### Check Yourself 4

Multiply.

(a) $\begin{array}{r} 34 \\ \times\ 6 \\ \hline \end{array}$

(b) $\begin{array}{r} 43 \\ \times\ 7 \\ \hline \end{array}$

### Units ANALYSIS

When you multiply a denominate number, such as 6 feet (ft), by an abstract number, such as 5, the result has the same units as the denominate number. Some examples are

$5 \times 6$ ft = 30 ft
$3 \times \$7 = \$21$
$9 \times 4$ A's = 36 A's

When you multiply two different denominate numbers, the units must also be multiplied. We will discuss this when we look at the area of geometric figures.

**RECALL**

It is best to write down the complete statement necessary for the solution of an application.

We briefly review our discussion of applications, or word problems.

As you will see, the process of solving applications is the same no matter which operation is required for the solution. In fact, the four-step procedure we suggested in Section 1.2 can be effectively applied here.

**Step by Step**

**Solving Applications**

**Step 1** Read the problem carefully to determine the given information and what you are asked to find.

**Step 2** Decide upon the operation or operations to be used.

**Step 3** Write down the complete statement necessary to solve the problem and do the calculations.

**Step 4** Check to make sure you have answered the question of the problem and that your answer seems reasonable.

## Example 5 Solving an Application Involving Multiplication

< Objective 4 >

A car rental agency orders a fleet of 7 new subcompact cars at a cost of $14,258 per automobile. What will the company pay for the entire order?

**Step 1** We know the number of cars and the price per car. We want to find the total cost.

**Step 2** Multiplication is the best approach to the solution.

**Step 3** Write

$$7 \times \$14{,}258 = \$99{,}806$$

We could, of course, *add* $14,258, the cost, 7 times, but multiplication is certainly more efficient.

**Step 4** The total cost of the order is $99,806.

### Check Yourself 5

Tires sell for $47 apiece. What is the total cost for 5 tires?

To multiply by numbers with more than one digit, we must multiply each digit of the first factor by each digit of the second. To do this, we form a series of partial products and then add them to arrive at the final product.

## Example 6 Multiplying by a Two-Digit Number

Multiply 56 × 47.

**Step 1**

```
   4
  56
× 47
 392
```

The first partial product is 7 × 56, or 392. Note that we had to carry 4 to the tens column.

**Step 2**

```
   2
   4
  56
× 47
 392
2240
```

The second partial product is 40 × 56, or 2,240. We must carry 2 during the process.

**Step 3**

```
    2
    4
   56
 × 47
  392
 2240
2,632
```

We add the partial products for our final result.

### Check Yourself 6

Multiply.

```
  38
× 76
```

**NOTE**

The three partial products are formed when we multiply by the ones, tens, and then the hundreds digits.

If multiplication involves two three-digit numbers, another step is necessary. In this case we form three partial products. This will ensure that each digit of the first factor is multiplied by each digit of the second.

## Example 7 Multiplying Two Three-Digit Numbers

< Objective 2 >

Multiply.

```
   22
   33
   22
  278
× 343
  834
11120
83400
95,354
```

In forming the third partial product, we must multiply by 300.

**Check Yourself 7**

Multiply.

$$\begin{array}{r} 352 \\ \times\ 249 \\ \hline \end{array}$$

Next, look at an example of multiplying by a number involving 0 as a digit. There are several ways to arrange the work, as our example shows.

**Example 8** **Multiplying Larger Numbers**

Multiply $573 \times 205$.

**Method 1**

$$\begin{array}{r} {\scriptstyle 1} \\ {\scriptstyle 31} \\ 573 \\ \times\ 205 \\ \hline 2865 \\ 0000 \\ 114600 \\ \hline 117{,}465 \end{array}$$

We can write the second partial product as 0000 to indicate the multiplication by 0 in the tens place.

We can look at a second approach to the problem.

**Method 2**

$$\begin{array}{r} {\scriptstyle 1} \\ {\scriptstyle 31} \\ 573 \\ \times\ 205 \\ \hline 2865 \\ 114600 \\ \hline 117{,}465 \end{array}$$

We can write a double 0 as our second step. If we place the third partial product on the same line, that product will be shifted *two* places left, indicating that we are multiplying by 200.

Because this second method is more compact, it is usually used.

**Check Yourself 8**

Multiply.

$$\begin{array}{r} 489 \\ \times\ 304 \\ \hline \end{array}$$

Example 9 will lead us to another property of multiplication.

## Example 9 Using the Associative Property

< Objective 3 >

$(2 \times 3) \times 4 = 6 \times 4 = 24$

We do the multiplication in the parentheses first, $2 \times 3 = 6$. Then we multiply $6 \times 4$.

Also,

$2 \times (3 \times 4) = 2 \times 12 = 24$

Here we multiply $3 \times 4$ as the first step. Then we multiply $2 \times 12$.

We see that

$(2 \times 3) \times 4 = 2 \times (3 \times 4)$

The product is the same no matter which way we *group* the factors. This is called the **associative property** of multiplication.

**Property**

**The Associative Property of Multiplication**

Multiplication is an *associative* operation. The way in which you group numbers in multiplication does not affect the final product.

### Check Yourself 9

**Find the products.**

**(a)** (5 × 3) × 6 **(b)** 5 × (3 × 6)

There are some shortcuts that let you simplify your work when multiplying by a number that ends in 0. Let's see what we can discover by looking at some examples.

## Example 10 Multiplying by 10

**RECALL**

Just as zero was the Additive Identity, one is the Multiplicative Identity.

First we multiply by 10.

$$\begin{array}{r} 67 \\ \times\ 10 \\ \hline 670 \end{array}$$

$10 \times 67 = 670$

Next we multiply by 100.

$$\begin{array}{r} 537 \\ \times\ 100 \\ \hline 53{,}700 \end{array}$$

$100 \times 537 = 53{,}700$

Finally, we multiply by 1,000.

$$\begin{array}{r} 489 \\ \times\ 1{,}000 \\ \hline 489{,}000 \end{array}$$

$1{,}000 \times 489 = 489{,}000$

## Check Yourself 10

**Multiply.**

**(a)** 257 × 100

**(b)** 2,436 × 1,000

**NOTE**

We talk about powers of 10 in greater detail in Section 1.7.

Do you see a pattern? Rather than writing out the multiplication, there is an easier way! We call the numbers 10, 100, 1,000, and so on **powers of 10.**

**Property**

**Multiplying by Powers of 10**

When a natural number is multiplied by a power of 10, the product is just that number followed by as many zeros as there are in the power of 10.

## Example 11 Multiplying by Numbers That End in Zero

Multiply 400 × 678.

Write

$$\begin{array}{r} 678 \\ \times\ \ 400 \\ \hline \end{array}$$

Shift 400 so that the two zeros are *to the right* of the digits above.

$$\begin{array}{r} {\scriptstyle 33} \\ 678 \\ \times\ \ 400 \\ \hline 271{,}200 \end{array}$$

Bring down the two zeros, then multiply 4 × 678 to find the product.

There is no mystery about why this works. We know that 400 is 4 × 100. In this method, we are multiplying 678 by 4 and then by 100, adding two zeros to the product by our earlier rule.

## Check Yourself 11

**Multiply.**

300 × 574

Your work in this section, together with our earlier rounding techniques, provides a convenient means of using estimation to check the reasonableness of our results in multiplication, as Example 12 illustrates.

## Example 12 Estimating a Product by Rounding

< Objective 5 >

Estimate the following product by rounding each factor to the nearest hundred.

$$\begin{array}{rcr} & & \text{Rounded} \\ 512 & \longrightarrow & 500 \\ \times\ 289 & \longrightarrow & \times\ 300 \\ \hline & & 150{,}000 \end{array}$$

You might want to find the *actual* product and use our estimate to see if your result seems reasonable.

### Check Yourself 12

**Estimate the product by rounding each factor to the nearest hundred.**

$$\begin{array}{r} 689 \\ \times\ 425 \\ \hline \end{array}$$

Rounding the factors can be a very useful way of estimating the solution to an application problem.

### Example 13 Estimating the Solution to a Multiplication Application

Bart is thinking of running an ad in the local newspaper for an entire year. The ad costs \$19.95 per week. Approximate the annual cost of the ad.

Rounding the charge to \$20 and rounding the number of weeks in a year to 50, we get

$50 \times 20 = 1{,}000$

The ad would cost approximately \$1,000.

### Check Yourself 13

**Phyllis is debating whether to join the health club for \$400 per year or just pay \$9 per visit. If she goes about once a week, approximately how much would she spend at \$9 per visit?**

### Units ANALYSIS

What happens when we multiply two denominate numbers? The units of the result turn out to be the product of the units. This makes sense when we look at an example from geometry.

The area of a square is the square of one side. As a formula, we write that as

$A = s^2$

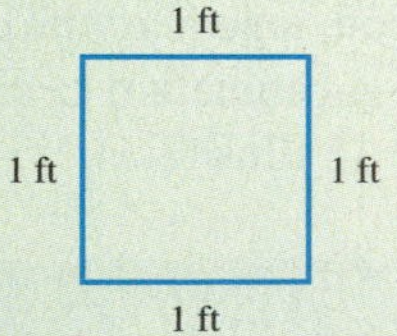


This tile is 1 ft by 1 ft.

$A = s^2 = (1\text{ ft})^2 = 1\text{ ft} \times 1\text{ ft} = 1\text{ (ft)} \times \text{(ft)} = 1\text{ ft}^2$

In other words, its area is one square foot (1 ft$^2$).

If we want to find the area of a room, we are actually finding how many of these square feet can be placed in the room.

Now look at the idea of **area.** Area is a measure that we give to a surface. It is measured in terms of **square units.** The area is the number of square units that are needed to cover the surface.

One standard unit of area measure is the **square inch** (written in.$^2$). This is the measure of the surface contained in a square with sides of 1 in.

**NOTE**

The unit inch (in.) can be treated as though it were a number. So in. × in. can be written in.$^2$. It is read "square inches."

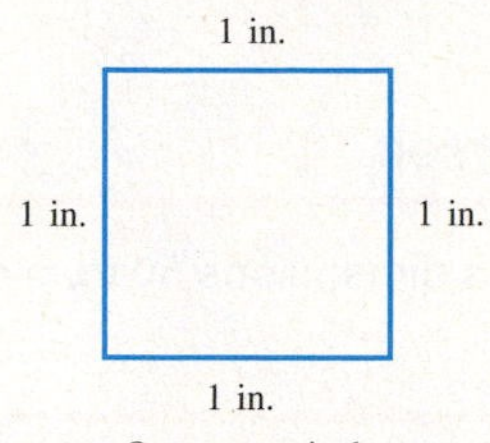

One square inch

Other units of area measure are the square foot (ft$^2$), the square yard (yd$^2$), the square centimeter (cm$^2$), and the square meter (m$^2$).

Finding the area of a figure means finding the number of square units it contains. One simple case is a rectangle.

The following figure shows a rectangle. The length of the rectangle is 4 in., and the width is 3 in. The area of the rectangle is measured in terms of square inches. We can simply count to find the area, 12 square inches (in.$^2$). However, because each of the four vertical strips contains 3 in.$^2$, we can multiply:

Area = 4 in. × 3 in. = 12 in.$^2$

**NOTE**

The length and width must be in terms of the same unit.

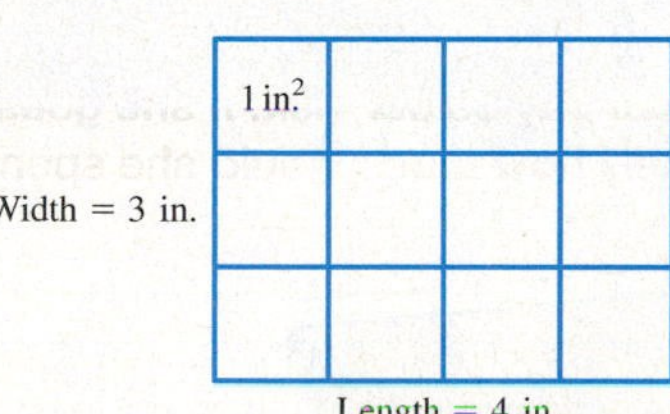


**Property**

**Formula for the Area of a Rectangle**

In general, we can write the formula for the **area of a rectangle:** If the length of a rectangle is $L$ units and the width is $W$ units, then the formula for the area $A$ of the rectangle can be written as

$A = L \cdot W$ (square units) (3)

**Example 14** **Find the Area of a Rectangle**

< Objective 6 >

A room has dimensions 12 ft by 15 ft. Find its area.

Use formula (3), with $L = 15$ ft and $W = 12$ ft.

$$A = L \cdot W$$
$$= 15 \text{ ft} \times 12 \text{ ft} = 180 \text{ ft}^2$$

The area of the room is 180 ft$^2$.

### Check Yourself 14

A desktop has dimensions 50 in. by 25 in. What is the area of its surface?

**NOTE**

$s^2$ is read "s squared."

We can also write a convenient formula for the area of a square. If the sides of the square have length $s$, we can write

**Property**

**Formula for the Area of a Square** $A = s \cdot s = s^2$ (4)

### Example 15 Finding the Area

You wish to cover a square table with a plastic laminate that costs 60¢ a square foot. If each side of the table measures 3 ft, what will it cost to cover the table?

We first must find the **area** of the table. Use the area formula with $s = 3$ ft.

$$A = s^2$$
$$= (3 \text{ ft})^2 = 3 \text{ ft} \times 3 \text{ ft} = 9 \text{ ft}^2$$

Now, multiply by the cost per square foot.

Cost = 9 × 60¢ = \$5.40

### Check Yourself 15

You wish to carpet a room that is a square, 4 yd by 4 yd, with carpet that costs \$12 per square yard. What will be the total cost of the carpeting?

Sometimes the total area of an oddly-shaped figure is found by adding the smaller areas. Example 16 shows how this is done.

## Example 16 Finding the Area of an Oddly Shaped Figure

Find the area of the figure.

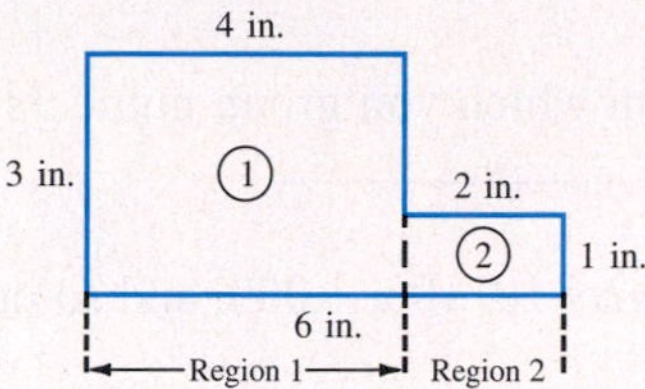


The area of the figure is found by adding the areas of regions 1 and 2. Region 1 is a 4 in. by 3 in. rectangle; the area of region 1 = 4 in. × 3 in. = 12 in.$^2$. Region 2 is a 2 in. by 1 in. rectangle; the area of region 2 = 2 in. × 1 in. = 2 in.$^2$.

The total area is the sum of the two areas:

Total area = 12 in.$^2$ + 2 in.$^2$ = 14 in.$^2$

### Check Yourself 16

Find the area of the figure.

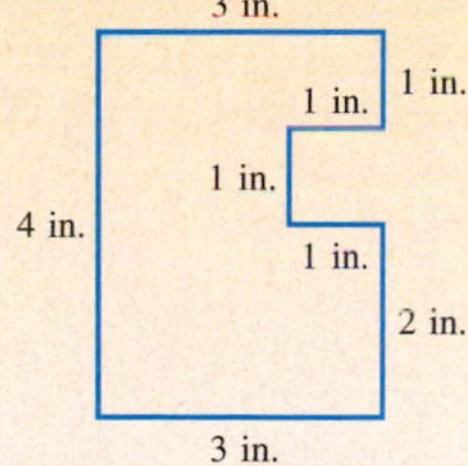


Hint: You can find the area by adding the areas of three rectangles, or by subtracting the area of the "missing" rectangle from the area of the "completed" larger rectangle.

### Check Yourself ANSWERS

**1.** Factors 2, 9; product 18 **2.** 24 **3.** 3 × 7 = 21 and 15 + 6 = 21 **4.** **(a)** 204; **(b)** 301 **5.** $235 **6.** 2,888 **7.** 87,648 **8.** 148,656 **9.** **(a)** 90; **(b)** 90 **10.** **(a)** 25,700; **(b)** 2,436,000 **11.** 172,200 **12.** 280,000 **13.** $500 **14.** 1,250 in.$^2$ **15.** $192 **16.** 11 in.$^2$

## Reading Your Text

The following fill-in-the-blank exercises are designed to ensure that you understand some of the key vocabulary used in this section.

### SECTION 1.5

**(a)** The final step in solving an application is to make certain that the answer is _________.

**(b)** The way in which you group numbers in multiplication does not affect the final ______________.

**(c)** The numbers 10, 100, 1,000, and so on are called the ______________ of 10.

**(d)** In general, we can write the equation for the area of a ______________ as $A = L \cdot W$.

Basic Skills | Advanced Skills | Vocational-Technical Applications | Calculator/Computer | Above and Beyond

# 1.5 exercises

MathZone

Boost your grade at mathzone.com!
> Practice Problems
> NetTutor
> Self-Tests
> e-Professors
> Videos

Name ______ Section ______ Date ______

< Objective 2 >

*Multiply.*

1. $\begin{array}{r} 5 \\ \times 3 \\ \hline \end{array}$

2. $\begin{array}{r} 7 \\ \times 4 \\ \hline \end{array}$

3. $6 \cdot 0$

4. $6 \cdot 6$

5. $4 \times 48$

6. $5 \times 53$

7. $\begin{array}{r} 508 \\ \times \;\; 6 \\ \hline \end{array}$

8. $\begin{array}{r} 903 \\ \times \;\; 9 \\ \hline \end{array}$

9. $\begin{array}{r} 75 \\ \times 68 \\ \hline \end{array}$

10. $235 \times 49$

11. $327 \times 59$ > Videos

12. $2{,}364 \cdot 67$

13. $4{,}075 \cdot 84$

14. $\begin{array}{r} 315 \\ \times 243 \\ \hline \end{array}$

15. $\begin{array}{r} 124 \\ \times 225 \\ \hline \end{array}$

16. $345 \times 267$

17. $639 \times 358$

18. $547 \cdot 203$

19. $668 \cdot 305$

20. $\begin{array}{r} 2{,}458 \\ \times 135 \\ \hline \end{array}$

21. $\begin{array}{r} 3{,}219 \\ \times 207 \\ \hline \end{array}$

22. $\begin{array}{r} 2{,}534 \\ \times 3{,}106 \\ \hline \end{array}$

23. $\begin{array}{r} 3{,}158 \\ \times 2{,}034 \\ \hline \end{array}$

24. $\begin{array}{r} 43 \\ \times 70 \\ \hline \end{array}$

25. $\begin{array}{r} 58 \\ \times 40 \\ \hline \end{array}$

26. $\begin{array}{r} 562 \\ \times 400 \\ \hline \end{array}$

27. $\begin{array}{r} 907 \\ \times 900 \\ \hline \end{array}$

28. $\begin{array}{r} 345 \\ \times 230 \\ \hline \end{array}$

29. $\begin{array}{r} 362 \\ \times 310 \\ \hline \end{array}$

30. $\begin{array}{r} 157 \\ \times 3{,}200 \\ \hline \end{array}$

31. Find the product of 304 and 7.

32. Find the product of 8 and 5,679.

> Videos

33. What is 21 multiplied by 551?

34. What is 135 multiplied by 507?

## Answers

| | |
|---|---|
| 1. | 2. |
| 3. | 4. |
| 5. | 6. |
| 7. | 8. |
| 9. | 10. |
| 11. | 12. |
| 13. | 14. |
| 15. | 16. |
| 17. | 18. |
| 19. | 20. |
| 21. | 22. |
| 23. | 24. |
| 25. | 26. |
| 27. | 28. |
| 29. | 30. |
| 31. | 32. |
| 33. | 34. |

## Answers

35. ________

36. ________

37. ________

38. ________

39. ________

40. ________

41. ________

42. ________

43. ________

44. ________

45. ________

46. ________

< Objective 4 >

*Name the property of addition and/or multiplication that is illustrated.*

**35.** $5 \times 8 = 8 \times 5$

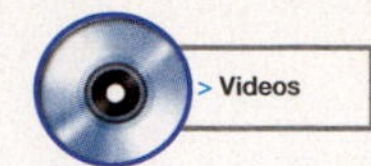


**36.** $3 \times (4 + 9) = (3 \times 4) + (3 \times 9)$

**37.** $2 \times (3 \times 5) = (2 \times 3) \times 5$

**38.** $5 \times (6 + 2) = 5 \times (2 + 6)$

Basic Skills | **Advanced Skills** | Vocational-Technical Applications | Calculator/Computer | Above and Beyond

*In exercises 39 and 40, complete the statement, using the given property.*

**39.** $7 + (3 \times 8) =$ Commutative property of multiplication

**40.** $3 \times (2 + 7) =$ Distributive property

< Objective 5 >

*Solve the following applications.*

**41.** **Business and Finance** A convoy company can transport 8 new cars on one of its trucks. If 34 truck shipments were made in one week, how many cars were shipped?

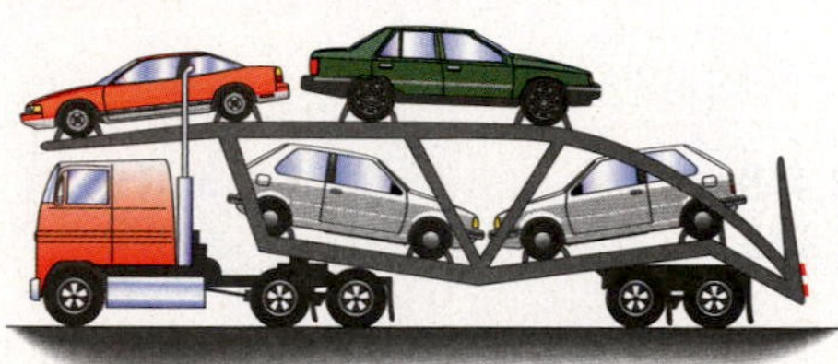

**42.** **Business and Finance** A computer printer can print 40 mailing labels per minute. How many labels can be printed in 1 hour (h)?

**43.** **Social Science** A rectangular parking lot has 14 rows of parking spaces, and each row contains 24 spaces. How many cars can be parked in the lot?

**44.** **Statistics** A petition to get Tom on the ballot for treasurer of student council has 28 signatures on each of 43 pages. How many signatures were collected?

**45.** **Business and Finance** The manufacturer of wood-burning stoves can make 15 stoves in 1 day. How many stoves can be made in 28 days?

**46.** **Business and Finance** Each bundle of newspapers contains 25 papers. If 43 bundles are delivered to Jose's house, how many papers are delivered?

## < Objective 6 >

*Find the area of each figure.*

**47.**

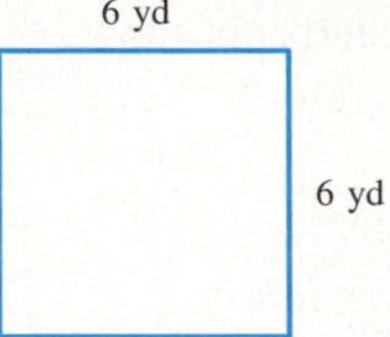


**48.**

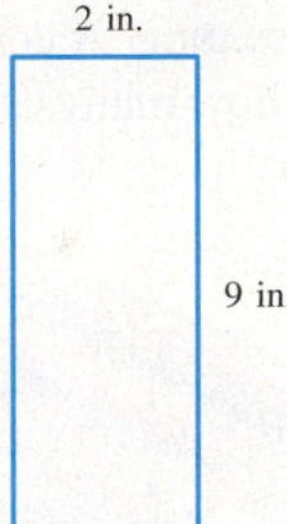


**49.**

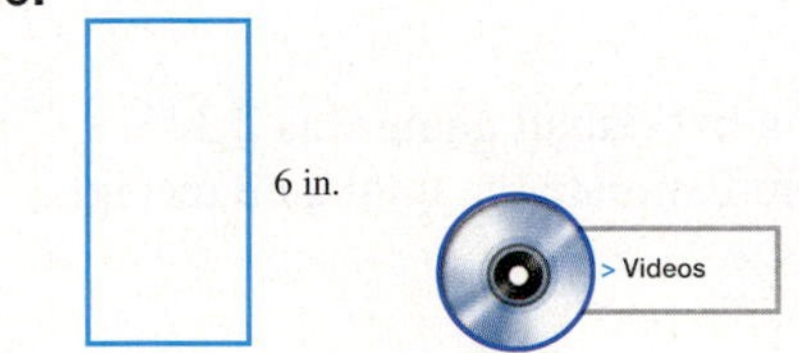


**50.**

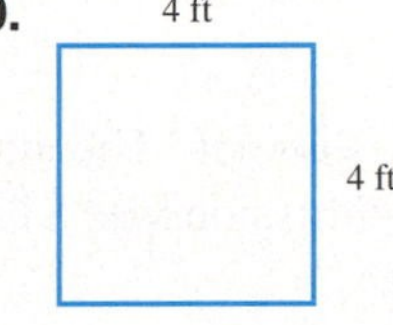


**51.**

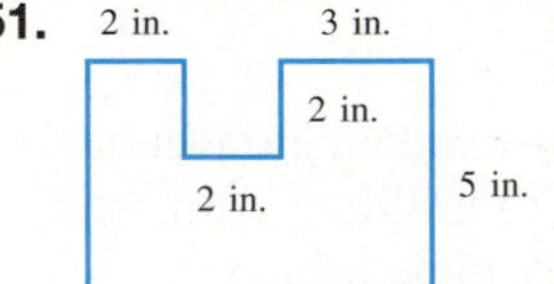


**52.**

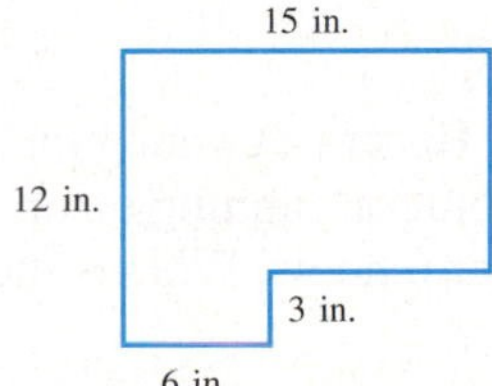


## < Objective 3 >

*Estimate each of the following products by rounding each factor to the nearest ten.*

**53.** $\begin{array}{r} 36 \\ \times\ 23 \\ \hline \end{array}$

**54.** $\begin{array}{r} 27 \\ \times\ 34 \\ \hline \end{array}$

*Estimate each of the following products by rounding each factor to the nearest hundred.*

**55.** $\begin{array}{r} 391 \\ \times\ 531 \\ \hline \end{array}$

**56.** $\begin{array}{r} 729 \\ \times\ 481 \\ \hline \end{array}$

## < Objective 4 >

*Solve the following applications.*

**57.** **Social Science** A movie theater has its seats arranged so that there are 42 seats per row. The theater has 48 rows. Estimate the number of seats in the theater.

## Answers

47. ______

48. ______

49. ______

50. ______

51. ______

52. ______

53. ______

54. ______

55. ______

56. ______

57. ______

Answers

58. ______

59. ______

60. ______

61. ______

62. ______

63. ______

64. ______

65. ______

**58.** **Statistics** There are 52 mathematics classes with 28 students in each class. Estimate the total number of students in the mathematics classes.

**59.** **Business and Finance** A company can manufacture 45 sleds per day. Approximately how many can this company make in 128 days?

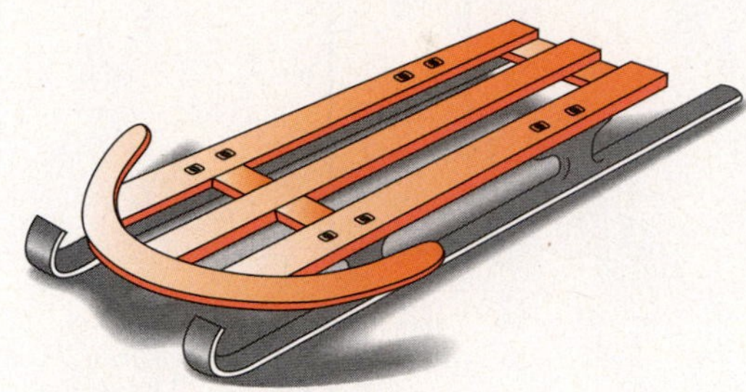

**60.** **Business and Finance** The attendance at a basketball game was 2,345. The cost of admission was $12 per person. Estimate the total gate receipts for the game.

Basic Skills | Advanced Skills | **Vocational-Technical Applications** | Calculator/Computer | Above and Beyond

**61.** **Allied Health** A young male patient is to be administered an intravenous (IV) solution running on an infusion pump set for 125 milliliters (mL) per hour for 6 hours. What is the total volume of solution to be infused?

**62.** **Allied Health** To help assess breathing efficiency, respiratory therapists calculate the patient's alveolar minute ventilation, in milliliters per minute (mL/min), by taking the product of the patient's respiratory rate, in breaths per minute, and the difference between the patient's tidal volume and dead-space volume, both in milliliters. Calculate the alveolar minute ventilation for a male patient with lung disease given that his respiratory rate is 10 breaths per minute, his tidal volume is 575 mL, and his dead-space volume is 200 mL.

**63.** **Information Technology** Jack needs a thousand feet of twisted pair cable for a network installation project. He goes to a local electronics store. The store sells cable at 10¢ a foot. How much (in cents) will one thousand feet of cable cost? How much is that in dollars?

**64.** **Information Technology** Amber needs to buy 100 new computers for new employees that have been hired by ABC consulting. She finds the cost of a decent computer to be $655 from Dell Computers. How much will she spend to buy 100 new computers?

**65.** **Electronics** An electronics-component distributor sells resistors, small components that "resist" the flow of electric current, in presealed bags. Each bag contains 50 resistors. If you purchased 25 bags of resistors, how many resistors would you have?

**66. ELECTRONICS** Assume that is takes 4 hours to solder all the components on a given printed circuit board. If you are given 36 boards to solder, how many hours will the project take? If you worked nonstop, how many hours would the project take?

**67. MANUFACTURING TECHNOLOGY** A small shop has 6 machinists each earning \$14 per hour, 3 assembly workers earning \$8 per hour, and one supervisor-maintenance person earning \$18 per hour. What is the shop's payroll for a 40-hour week?

**68. MANUFACTURING TECHNOLOGY** What is the distance from the center of hole A to the center of hole B in the following diagram?

A ○ ○ ○ ○ ○ ○ ○ ○ ○ ○ ○ ○ B

18 mm

Basic Skills | Advanced Skills | Vocational-Technical Applications | Calculator/Computer | **Above and Beyond**

**69. SOCIAL SCIENCE** We have seen that addition and multiplication are commutative operations. Decide which of the following activities are commutative.

**(a)** Taking a shower and eating breakfast
**(b)** Getting dressed and taking a shower
**(c)** Putting on your shoes and your socks
**(d)** Brushing your teeth and combing your hair
**(e)** Putting your key in the ignition and starting your car

**70. SOCIAL SCIENCE** The associative properties of addition and multiplication indicate that the result of the operation is the same regardless of where the grouping symbol is placed. This is not always the case in the use of the English language. Many phrases can have different meanings based on how the words are grouped. In each of the following, explain why the associative property would not hold.

**(a)** Cat fearing dog
**(b)** Hard test question
**(c)** Defective parts department
**(d)** Man eating animal

Write some phrases in which the associative property is satisfied.

**71. CONSTRUCTION** Suppose you wish to build a small, rectangular pen, and you have enough fencing for the pen's perimeter to be 36 ft. Assuming that the length and width are to be whole numbers, answer the following.

**(a)** List the possible dimensions that the pen could have.
(*Note:* A square is a type of rectangle.)

## Answers

66. __________

67. __________

68. __________

69. __________

70. __________

71. __________

(b) For each set of dimensions (length and width), find the area that the pen would enclose.

(c) Which dimensions give the greatest area?

(d) What is the greatest area?

## Answers

72.

73.

74.

**72. Construction** Suppose you wish to build a rectangular kennel that encloses 100 $ft^2$. Assuming that the length and width are to be whole numbers, answer the following.

(a) List the possible dimensions that the kennel could have. (*Note:* A square is a type of rectangle.)

(b) For each set of dimensions (length and width), find the perimeter that would surround the kennel.

(c) Which dimensions give the least perimeter?

(d) What is the least perimeter?

**73. Social Science** Most maps contain legends that allow you to convert the distance between two points on the map to actual miles. For instance, if a map uses a legend that equates 1 inch (in.) to 5 miles (mi) and the distance between two towns is 4 in. on the map, then the towns are actually 20 mi apart.

(a) Obtain a map of your state and determine the shortest distance between any two major cities.

(b) Could you actually travel the route you measured in part **(a)**?

(c) Plan a trip between the two cities you selected in part **(a)** over established roads. Determine the distance that you actually travel using this route.

**74. Number Problem** Complete the following number cross.

| Across | Down |
|---|---|
| **1.** $6 \times 551$ | **1.** $5 \times 7$ |
| **5.** $7 \times 8$ | **2.** $9 \times 41$ |
| **6.** $27 \times 27$ | **3.** $67 \times 100$ |
| **7.** $19 \times 50$ | **4.** $2 \times (49 + 100)$ |
| **10.** $3 \times 67$ | **8.** $4 \times 1{,}301$ |
| **12.** $6 \times 25$ | **9.** $100 + 10 + 1$ |
| **13.** $9 \times 8$ | **11.** $2 \times 87$ |
| **15.** $16 \times 303$ | **14.** $25 + 3$ |

## Answers

**1.** 15 **3.** 0 **5.** 192 **7.** 3,048 **9.** 5,100 **11.** 19,293
**13.** 342,300 **15.** 27,900 **17.** 228,762 **19.** 203,740
**21.** 666,333 **23.** 6,423,372 **25.** 2,320 **27.** 816,300

**29.** 112,220 **31.** 2,128 **33.** 11,571
**35.** Commutative property of multiplication
**37.** Associative property of multiplication
**39.** $7 + (8 \times 3)$ **41.** 272 cars **43.** 336 cars **45.** 420 stoves
**47.** 36 $yd^2$ **49.** 18 $in.^2$ **51.** 31 $in.^2$ **53.** 800
**55.** 200,000 **57.** 2,000 seats **59.** 6,500 sleds **61.** 750 mL
**63.** 10,000¢; \$100 **65.** 1,250 resistors **67.** \$5,040
**69.** Above and Beyond **71.** Above and Beyond **73.** Above and Beyond

1.6

# Division

< 1.6 Objectives >

1 > Write a division problem as repeated subtraction

2 > Use the language of division

3 > Divide whole numbers

4 > Estimate quotients

5 > Solve applications of division

Now we examine a fourth arithmetic operation, division. Just as multiplication is repeated addition, division is repeated subtraction. Division asks *how many times* one number is contained in another.

## Example 1 Dividing by Using Subtraction

< Objective 1 >

**NOTE**

The 8 is subtracted six times.

Joel needs to set up 48 chairs in the student union for a concert. If there is room for 8 chairs per row, how many rows will it take to set up all 48 chairs?

This problem can be solved by subtraction. Each row subtracts another 8 chairs.

| 48 | 40 | 32 | 24 | 16 | 8 |
|---|---|---|---|---|---|
| −8 | −8 | −8 | −8 | −8 | −8 |
| 40 | 32 | 24 | 16 | 8 | 0 |

Because 8 can be subtracted from 48 six times, there will be 6 rows.

This can also be seen as a division problem.

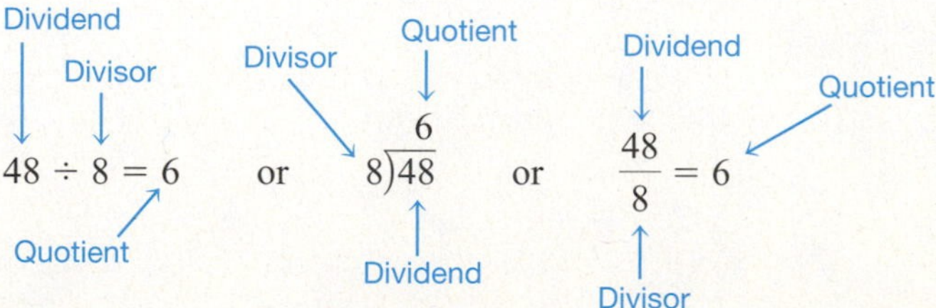


 The Streeter/Hutchison Series in Mathematics Basic Mathematical Skills with Geometry

No matter which method we use, we call 48 the **dividend,** 8 the **divisor,** and 6 the **quotient.**

### Check Yourself 1

Carlotta is creating a garden path made of bricks. She has 72 bricks. Each row will have 6 bricks in it. How many rows can she make?

### Units ANALYSIS

When you divide a denominate number by an abstract number, the result will have the units of the denominate number. Here are a couple of examples:

76 trombones ÷ 4 = 19 trombones

\$55 ÷ 11 = \$5

When one denominate number is divided by another, the result will get the units of the dividend over the units of the divisor.

144 miles ÷ 6 gallons = 24 miles/gallon (which we read as "miles per gallon")

\$120 ÷ 8 hours = 15 dollars/hour ("dollars per hour")

To solve a problem requiring division, first set up the problem as a division statement. Example 2 illustrates this idea.

### Example 2 Writing a Division Statement

< Objective 2 >

Write a division statement that corresponds to the following situation. You need not do the division.

The staff at the Wok Inn Restaurant splits all tips at the end of each shift. Yesterday's evening shift collected a total of \$224. How much should each of the 7 employees get in tips?

\$224 ÷ 7 employees (Note that the units for the answer will be "dollars per employee.")

## Check Yourself 2

Write a division statement that corresponds to the following situation. You need not do the division.

All nine sections of basic math skills at SCC (Sum Community College) are full. There are a total of 315 students in the classes. How many students are in each class? What are the units for the answer?

In Section 1.5, we used a rectangular array of stars to represent multiplication. These same arrays can represent division. Just as $3 \times 4 = 12$ and $4 \times 3 = 12$, so is it true that $12 \div 3 = 4$ and $12 \div 4 = 3$.

```
* * *
* * *   } 4 × 3 = 12        * * * *
* * *                      * * * *  } 3 × 4 = 12
* * *                      * * * *
  or                          or
12 ÷ 3 = 4                 12 ÷ 4 = 3
```

This relationship allows us to check our division results by doing multiplication.

## Example 3 Checking Division by Using Multiplication

**NOTE**

For a division problem to check, the *product* of the divisor and the quotient *must equal the dividend.*

**(a)** $7\overline{)21}$ with quotient 3 Check: $7 \times 3 = 21$

**(b)** $48 \div 6 = 8$ Check: $6 \times 8 = 48$

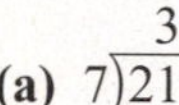

## Check Yourself 3

Complete the division statements and check your results.

**(a)** $9\overline{)45}$ **(b)** $28 \div 7$

**NOTE**

Because $36 \div 9 = 4$, we say that 36 is *exactly divisible* by 9.

In our examples so far, the product of the divisor and the quotient has been equal to the dividend. This means that the dividend is *exactly divisible* by the divisor. That is not always the case. Look at another example that uses repeated subtraction.

## Example 4 Dividing by Using Subtraction, Leaving a Remainder

**NOTE**

The remainder must be smaller than the divisor or we could subtract again.

How many times is 5 contained in 23?

$$\begin{array}{r} 23 \\ -\ 5 \\ \hline 18 \end{array} \quad \begin{array}{r} 18 \\ -\ 5 \\ \hline 13 \end{array} \quad \begin{array}{r} 13 \\ -\ 5 \\ \hline 8 \end{array} \quad \begin{array}{r} 8 \\ -\ 5 \\ \hline 3 \end{array}$$

We see that 5 is contained 4 times in 23, but 3 is "leftover."

The number 23 is not exactly divisible by 5. The "leftover" 3 is called the **remainder** in the division. To check the division operation when a remainder is involved, we have the following rule:

**Definition**

**Remainder** Dividend = divisor × quotient + remainder

### Check Yourself 4

How many times is 7 contained in 38?

## Example 5 Checking Division by a Single-Digit Number

> **NOTE**
> Another way to write the result is
> $5\overline{)23}$ with quotient 4 r3. The "r" stands for remainder.

Using the work of Example 4, we can write

$$5\overline{)23} \text{ with quotient } 4 \quad \text{with remainder } 3$$

To apply our previous rule, we have

> **NOTE**
> The multiplication is done before the 3 is added.

Divisor, Quotient

Dividend ⟶ $23 = 5 \times 4 + 3$ ⟵ Remainder

$23 = 20 + 3$

$23 = 23$ The division checks.

### Check Yourself 5

Evaluate $7\overline{)38}$. Check your answer.

We must be careful when 0 is involved in a division problem. There are two special cases.

**Property**

**Division and Zero**

1. Zero divided by any whole number (except 0) is 0.
2. Division by 0 is undefined.

The first case involving zero occurs when we are dividing into zero.

## Example 6 Dividing into Zero

$0 \div 5 = 0$ because $0 = 5 \times 0$.

**RECALL**

There are three forms that are equivalent:

$0 \div 5 = 0$

$5\overline{)0} = 0$

$\frac{0}{5} = 0$

### Check Yourself 6

(a) $0 \div 7$ (b) $9\overline{)0}$ (c) $\frac{0}{12}$

Our second case illustrates what happens when 0 is the *divisor.* Here we have a special problem.

## Example 7 Dividing by Zero

$8 \div 0 = ?$ This means that $8 = 0 \times ?$

Can 0 times some number ever be 8? From our multiplication facts, the answer is *no!* There is no answer to this problem, so we say that $8 \div 0$ is undefined.

### Check Yourself 7

Decide whether each problem results in 0 or is undefined.

(a) $\frac{9}{0}$ (b) $\frac{0}{9}$ (c) $15\overline{)0}$ (d) $0\overline{)15}$

It is easy to divide when small whole numbers are involved, because much of the work can be done mentally. In working with larger numbers, we turn to a process called **long division.** This is a shorthand method for performing the steps of repeated subtraction.

To start, look at an example in which we subtract multiples of the divisor.

## Example 8 Dividing by a Single-Digit Number

< Objective 3 >

**NOTE**

With larger numbers, repeated subtraction is just too time-consuming to be practical.

Divide 176 by 8.

Because 20 eights are 160, we know that there are at least 20 eights in 176.

**Step 1** Write

$$\begin{array}{r} 20 \\ 8\overline{)176} \\ \text{20 eights} \longrightarrow \underline{160} \\ 16 \end{array}$$

Subtracting 160 is just a shortcut for subtracting eight 20 times.

After subtracting the 20 eights, or 160, we are left with 16. There are 2 eights in 16, and so we continue.

**Step 2**

$$\begin{array}{r} 2 \\ 20 \\ 8\overline{)176} \\ \underline{160} \\ 16 \\ \underline{16} \\ 0 \end{array}$$

22 Adding 20 and 2 gives us the quotient, 22.

2 eights → 16

Subtracting the 2 eights, we have a 0 remainder. So $176 \div 8 = 22$.

**Check Yourself 8**

Verify the result of Example 8, using multiplication.

The next step is to simplify this repeated-subtraction process one step further. The result will be the long-division method.

## Example 9 Dividing by a Single-Digit Number

Divide 358 by 6.

The dividend is 358. We look at the first digit, 3. We cannot divide 6 into 3, and so we look at the *first two digits,* 35. There are 5 sixes in 35, and so we write 5 above the tens digit of the dividend.

$$\begin{array}{r} 5\phantom{8} \\ 6\overline{)358} \end{array}$$

When we place 5 as the tens digit, we really mean 5 tens, or 50.

Now multiply $5 \times 6$, place the product below 35, and subtract.

$$\begin{array}{r} 5\phantom{8} \\ 6\overline{)358} \\ \underline{30}\phantom{8} \\ 5\phantom{8} \end{array}$$

We have actually subtracted 50 sixes (300) from 358.

Because the remainder, 5, is smaller than the divisor, 6, we bring down 8, the ones digit of the dividend.

$$\begin{array}{r} 5\phantom{8} \\ 6\overline{)358} \\ \underline{30\downarrow} \\ 58 \end{array}$$

Now divide 6 into 58. There are 9 sixes in 58, and so 9 is the ones digit of the quotient. Multiply $9 \times 6$ and subtract to complete the process.

$$\begin{array}{r} 59 \\ 6\overline{)358} \\ \underline{30\downarrow} \\ 58 \\ \underline{54} \\ 4 \end{array}$$

We now have:
$358 \div 6 = 59$ r4

To check: $358 = 6 \times 59 + 4$

**NOTES**

Because 4 is smaller than the divisor, we have a remainder of 4.

Verify that this is true and that the division checks.

**Check Yourself 9**

Divide $7\overline{)453}$.

Long division becomes a bit more complicated when we have a two-digit divisor. It is now a matter of trial and error. We round the divisor and dividend to form a *trial divisor and a trial dividend.* We then estimate the proper quotient and must determine whether our estimate was correct.

## Example 10 Dividing by a Two-Digit Number

**NOTE**

Think: $4\overline{)29}$ with quotient 7

Divide.

$38\overline{)293}$

Round the divisor and dividend to the nearest ten. So 38 is rounded to 40, and 293 is rounded to 290. The trial divisor is then 40, and the trial dividend is 290.

Now look at the nonzero digits in the trial divisor and dividend. They are 4 and 29. We know that there are 7 fours in 29, and so 7 is our first estimate of the quotient. Now let's see if 7 works.

$$\begin{array}{r} 7 \\ 38\overline{)293} \\ \underline{266} \\ 27 \end{array}$$

7 ⟵ Your estimate

Multiply $7 \times 38$. The product, 266, is less than 293, so we can subtract.

The remainder, 27, is less than the divisor, 38, and so the process is complete.

$293 \div 38 = 7\text{ r}27$

Check: $293 = 38 \times 7 + 27$ You should verify that this statement is true.

**Check Yourself 10**

Divide.

$57\overline{)482}$

Because this process is based on estimation, our first guess is often incorrect.

## Example 11 Dividing by a Two-Digit Number

**NOTE**

Think: $5\overline{)43}$ with quotient 8

Divide.

$54\overline{)428}$

Rounding to the nearest ten, we have a trial divisor of 50 and a trial dividend of 430.

If you look at the nonzero digits, how many fives are in 43? There are 8. This is our first estimate.

```
    8
54)428
   432  ←—— Too large
```

We multiply 8 × 54. Do you see what's wrong? The product, 432, is too large. We can't subtract. Our estimate of the quotient must be adjusted *downward*.

We adjust the quotient downward to 7. We can now complete the division.

```
    7
54)428
   378
    50
```

We have

$428 \div 54 = 7 \text{ r}50$

Check: $428 = 54 \times 7 + 50$

**Check Yourself 11**

Divide.

63)557

We must be careful when a 0 appears as a digit in the quotient. There is an example in which this happens with a two-digit divisor.

## Example 12 Dividing with Large Dividends

**NOTE**

Our divisor, 32, divides into 98, the first two digits of the dividend.

Divide.

32)9,871

Rounding to the nearest ten, we have a trial divisor of 30 and a trial dividend of 100. Think, "How many threes are in 10?" There are 3, and this is our first estimate of the quotient.

```
     3
32)9,871
   9 6
     2
```

Everything seems fine so far!

Bring down 7, the next digit of the dividend.

```
     30
32)9,871
   9 6↓
     27
```

Now do you see the difficulty? We cannot divide 32 into 27, and so we place 0 in the tens place of the quotient to indicate this fact.

We continue by multiplying by 0. After subtraction, we bring down 1, the last digit of the dividend.

$$\begin{array}{r}
30 \\
32\overline{)9{,}871} \\
\underline{9\,6\phantom{71}} \\
27\phantom{1} \\
\underline{00}\downarrow \\
271
\end{array}$$

Another problem develops here. We round 32 to 30 for our trial divisor, and we round 271 to 270, which is the trial dividend at this point. Our estimate of the last digit of the quotient must be 9.

$$\begin{array}{r}
309 \\
32\overline{)9{,}871} \\
\underline{9\,6\phantom{71}} \\
27\phantom{1} \\
\underline{00\phantom{1}} \\
271 \\
\underline{288} \leftarrow \text{Too large}
\end{array}$$

We cannot subtract because 288 is larger than 271. The trial quotient must be adjusted downward to 8. We can now complete the division.

$$\begin{array}{r}
308 \\
32\overline{)9{,}871} \\
\underline{9\,6\phantom{71}} \\
27\phantom{1} \\
\underline{00\phantom{1}} \\
271 \\
\underline{256} \\
15
\end{array}$$

$9{,}871 \div 32 = 308 \text{ r}15$

Check: $9{,}871 = 32 \times 308 + 15$

**Check Yourself 12**

**Divide.**

$43\overline{)8{,}857}$

Because of the availability of calculators, it is rarely necessary that people find the exact answer when performing long division. On the other hand, it is frequently important that one be able to either estimate the result of long division or confirm that a given answer (particularly from a calculator) is reasonable. As a result, the emphasis in this section will be on improving your estimation skills in division.

Let's divide a four-digit number by a two-digit number. Generally, we will round the divisor to the nearest ten and the dividend to the nearest hundred.

## Example 13 Estimating the Result of a Division Application

< Objective 4 >

The Ramirez family took a trip of 2,394 miles (mi) in their new car, using 77 gallons (gal) of gas. Estimate their gas mileage (mi/gal).

Our estimate will be based on dividing 2,400 by 80.

$$80\overline{)2,400}\quad 30$$

They got approximately 30 mi/gal.

### Check Yourself 13

Troy flew a light plane on a trip of 2,844 mi that took 21 hours (h). What was his approximate speed in miles per hour (mi/h)?

As before, we may have to combine operations to solve an application of the mathematics you have learned.

## Example 14 Estimating the Result of a Division Application

Charles purchases a new car for \$8,574. Interest charges will be \$978. He agrees to make payments for 4 years. Approximately what should his monthly payments be?

First, we find the amount that Charles owes:

$$\$8,574 + \$978 = \$9,552$$

Now, to find the monthly payment, we divide that amount by 48 (months). To estimate the payment, we'll divide \$9,600 by 50 months.

$$50\overline{)9,600}\quad 192$$

The payments will be approximately \$192 per month.

### Check Yourself 14

One \$10 bag of fertilizer will cover 310 square feet ($ft^2$). Approximately what would it cost to cover 2,145 $ft^2$?

## Check Yourself ANSWERS

**1.** 12 rows **2.** 315 students ÷ 9 classes; students per class
**3.** **(a)** 5; $9 \times 5 = 45$; **(b)** 4; $7 \times 4 = 28$ **4.** 5 **5.** 5 r3; $38 = 5 \times 7 + 3$
**6.** **(a)** 0; **(b)** 0; **(c)** 0 **7.** **(a)** Undefined; **(b)** 0; **(c)** 0; **(d)** undefined
**8.** $8 \times 22 = 176$ **9.** 64 r5 **10.** 8 r26 **11.** 8 r53 **12.** 205 r42
**13.** 140 mi/h **14.** $70

## Reading Your Text

The following fill-in-the-blank exercises are designed to ensure that you understand some of the key vocabulary used in this section.

### SECTION 1.6

**(a)** The result from division is called the ______________.

**(b)** The dividend is equal to the divisor times the quotient plus the ______________.

**(c)** Zero divided by any whole number (except ______________) is 0.

**(d)** Division is a shortened form for repeated ______________.

Basic Skills | Advanced Skills | Vocational-Technical Applications | Calculator/Computer | Above and Beyond

# 1.6 exercises

MathZone

Boost your grade at mathzone.com!
> Practice Problems
> NetTutor
> Self-Tests
> e-Professors
> Videos

Name ________

Section ________ Date ________

< Objective 2 >

1. If 48 ÷ 8 = 6, 8 is the ________, 48 is the ________, and 6 is the ________.

2. In the statement $5\overline{)45}$ with quotient 9, 9 is the ________, 5 is the ________, and 45 is the ________.

3. Find 36 ÷ 9 by repeated subtraction.

4. Find 40 ÷ 8 by repeated subtraction.

5. **Crafts** Stefanie is planting rows of tomato plants. She wants to plant 63 plants with 9 plants per row. How many rows will she have?

> Videos

6. **Construction** Nick is designing a parking lot for a small office building. He must make room for 42 cars with 7 cars per row. How many rows should he plan for?

*Divide the following. Identify the correct units for the quotient.*

7. 36 pages ÷ 4

8. \$96 ÷ 8

9. 4,900 kilometers (km) ÷ 7

10. 360 gal ÷ 18

11. 160 miles ÷ 4 hours

12. 264 ft ÷ 3 s

13. 3,720 hours ÷ 5 months

14. 560 calories ÷ 7 grams

< Objective 3 >

*Divide using long division, and check your work.*

15. 54 ÷ 9

16. 21 ÷ 3

17. $6\overline{)42}$

18. $7\overline{)63}$

## Answers

1. ________
2. ________
3. ________ 4. ________
5. ________ 6. ________
7. ________ 8. ________
9. ________ 10. ________
11. ________
12. ________
13. ________
14. ________
15. ________
16. ________
17. ________
18. ________

## Answers

19. ______ 20. ______

21. ______ 22. ______

23. ______ 24. ______

25. ______ 26. ______

27. ______

28. ______

29. ______

30. ______ 31. ______

32. ______

33. ______ 34. ______

35. ______ 36. ______

37. ______ 38. ______

39. ______ 40. ______

41. ______ 42. ______

43. ______ 44. ______

45. ______ 46. ______

47. ______ 48. ______

49. ______ 50. ______

51. ______ 52. ______

53. ______

**19.** $4\overline{)32}$ **20.** $56 \div 8$

**21.** $5\overline{)43}$ **22.** $40 \div 9$

**23.** $9\overline{)65}$ **24.** $6\overline{)51}$

**25.** $57 \div 8$ **26.** $74 \div 8$

**27.** $0 \div 5$ **28.** $5 \div 0$

**29.** $4 \div 0$ **30.** $0 \div 12$

**31.** $0 \div 6$ **32.** $18 \div 0$

*Divide.*

**33.** $5\overline{)83}$ > Videos **34.** $9\overline{)78}$

**35.** $162 \div 3$ **36.** $232 \div 4$

**37.** $\frac{293}{8}$ **38.** $\frac{346}{7}$

**39.** $8\overline{)3{,}136}$ **40.** $5\overline{)4{,}938}$

**41.** $5{,}438 \div 8$ **42.** $3{,}527 \div 9$

**43.** $\frac{22{,}153}{8}$ **44.** $\frac{43{,}287}{5}$

Basic Skills | **Advanced Skills** | Vocational-Technical Applications | Calculator/Computer | Above and Beyond

**45.** $45\overline{)2{,}367}$ **46.** $53\overline{)3{,}480}$

**47.** $8{,}748 \div 34$ 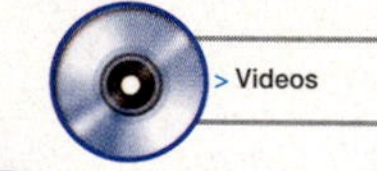

**48.** $9{,}335 \div 27$

**49.** $\frac{7{,}902}{42}$ > Videos **50.** $\frac{8{,}729}{53}$

< Objective 5 >

*Solve the following applications.*

**51.** **Business and Finance** There are 63 candy bars in 7 boxes. How many candy bars are in each box? > Videos

**52.** **Business and Finance** A total of 54 printers were shipped to 9 stores. How many printers were shipped to each store?

**53.** **Social Science** Joaquin is putting pictures in an album. He can fit 8 pictures on each page. If he has 77 pictures, how many will be left over after he has filled the last 8-picture page?

54. **Statistics** Kathy is separating a deck of 52 cards into 6 equal piles. How many cards will be left over?

55. **Social Science** The records of an office show that 1,702 calls were made in 1 day. If there are 37 phones in the office, how many calls were placed per phone?

56. **Business and Finance** A television dealer purchased 23 sets, each the same model, for $5,267. What was the cost of each set?

57. **Business and Finance** A computer printer can print 340 lines per minute (min). How long will it take to complete a report of 10,880 lines?

58. **Statistics** A train traveled 1,364 mi in 22 h. What was the speed of the train? (*Hint:* Speed is the distance traveled divided by the time.)

*Estimate the result in the following division problems. (Remember to round divisors to the nearest ten and dividends to the nearest hundred.)*

59. 810 divided by 38

60. 458 divided by 18

61. 4,967 divided by 96

62. 3,971 divided by 39

63. 8,971 divided by 91

64. 3,981 divided by 78

65. 3,879 divided by 126

66. 8,986 divided by 178

67. 3,812 divided by 188

68. 5,245 divided by 255

Basic Skills | Advanced Skills | **Vocational-Technical Applications** | Calculator/Computer | Above and Beyond

69. **Allied Health** The doctor prescribes 525 milligrams (mg) of bexarotene to be given once daily to an adult female patient with lung cancer. How many pills should the nurse give her every day if each pill contains 75 mg of bexarotene?

70. **Allied Health** Determine the flow rate, in milliliters per hour (mL/h), needed for an electronic infusion pump to administer 180 mL of a saline solution via intravenous (IV) infusion over the course of 12 hours (h).

**Answers**

54. ______
55. ______
56. ______
57. ______
58. ______
59. ______
60. ______
61. ______
62. ______
63. ______
64. ______
65. ______
66. ______
67. ______
68. ______
69. ______
70. ______

Answers

71. ______

72. ______

73. ______

74. ______

75. ______

76. ______

77. ______

78. ______

79. ______

80. ______

81. ______

82. ______

83. ______

84. ______

85. ______

86. ______

87. ______

**71.** **Information Technology** Marcela is in the process of building a new testing computer lab for ABC software. This lab has five rows of computers and nine computers per row. The distance from each computer to a switch is about 4 feet (ft). She has a 200-ft roll of cable for the job. How many 4-ft cables can she make with a 200-ft roll? Does she have enough cable on the roll for the job?

**72.** **Information Technology** Your modem on your computer typically transmits at 56,000 bits per second. How long will it take to transmit five hundred sixty thousand bits?

**73.** **Electronics** An electronics component distributor sells resistors in two package sizes. A small package contains 500 resistors. A large package contains 1,250 resistors. If 10,000 resistors are needed to make a batch of parts, how many packages would you need to buy if you bought all small packages? All large packages?

**74.** **Electronics** A vendor that makes small-quantity batches of printed circuit boards will sell 25 boards for \$400. What is the cost per board?

**75.** **Manufacturing Technology** An order of 24 parts weighs 1,752 lb. Assuming that the parts are identical, how much does each part weigh?

**76.** **Manufacturing Technology** Triplet Precision Machining has a 3,000-gallon (gal) liquid petroleum (LP) tank. If the cutting line consumes 125 gal of LP each day, how many days will the LP supply last?

Basic Skills | Advanced Skills | Vocational-Technical Applications | **Calculator/Computer** | Above and Beyond

*Use your calculator to perform the indicated operations.*

**77.** $583{,}467 \div 129$

**78.** $464{,}184 \div 189$

**79.** $6 + 9 \div 3$

**80.** $18 - 6 \div 3$

**81.** $24 \div 6 \times 4$

**82.** $32 \div 8 \times 4$

**83.** $4{,}368 \div 56 + 726 \div 33$

**84.** $1{,}176 \div 42 - 1{,}572 \div 524$

**85.** $3 \times 8 \times 8 \times 8 \div 12$

**86.** $5 \times 6 \times 6 \div 18$

Basic Skills | Advanced Skills | Vocational-Technical Applications | Calculator/Computer | **Above and Beyond**

**87.** **Construction** You are going to recarpet your living room. You have budgeted \$1,500 for the carpet and installation.

**(a)** Determine how much carpet you will need to do the job. Draw a sketch to support your measurements.

**(b)** What is the highest price per square yard you can pay and still stay within budget?

**(c)** Go to a local store and determine the total cost of doing the job for three different grades of carpet. Be sure to include padding, labor costs, and any other expenses.

(d) What considerations (other than cost) would affect your decision about what type of carpet to install?

(e) Write a brief paragraph indicating your final decision and give supporting reasons.

**88.** **Social Science** Division is the inverse operation of multiplication. Many daily activities have inverses. For each of the following activities, state the inverse activity.

(a) Spending money (b) Going to sleep

(c) Turning down the volume on your CD player (d) Getting dressed

**89.** If you have no money in your pocket and want to divide it equally among your four friends, how much does each person get? Use this situation to explain division of zero by a nonzero number.

**90.** **Number Problem** Complete the following number cross.

**Across**

1. 48 ÷ 4
3. 1,296 ÷ 8
6. 2,025 ÷ 5
8. 4 × 5
9. 11 × 11
12. 15 ÷ 3 × 111
14. 144 ÷ (2 × 6)
16. 1,404 ÷ 6
18. 2,500 ÷ 5
19. 3 × 5

**Down**

1. (12 + 16) ÷ 2
2. 67 × 3
4. 744 ÷ 12
5. 2,600 ÷ 13
7. 6,300 ÷ 12
10. 304 ÷ 2
11. 5 × (161 ÷ 7)
13. 9,027 ÷ 17
15. 400 ÷ 20
17. 9 × 5

## Answers

88. ____

89. ____

90. ____

## Answers

**1.** Divisor, dividend, quotient **3.** 4 **5.** 7 **7.** 9 pages **9.** 700 km **11.** 40 mi/h **13.** 744 h/month **15.** 6 **17.** 7 **19.** 8 **21.** 8 r3 **23.** 7 r2 **25.** 7 r1 **27.** 0 **29.** Undefined **31.** 0 **33.** 16 r3 **35.** 54 **37.** 36 r5 **39.** 392 **41.** 679 r6 **43.** 2,769 r1 **45.** 52 r27 **47.** 257 r10 **49.** 188 r6 **51.** 9 bars **53.** 5 pictures **55.** 46 calls **57.** 32 min **59.** 20 **61.** 50 **63.** 100 **65.** 30 **67.** 20 **69.** 7 pills **71.** Yes, 50 4-ft cables **73.** 20 packages; 8 packages **75.** 73 lb **77.** 4,523 **79.** 9 **81.** 16 **83.** 100 **85.** 128 **87.** Above and Beyond **89.** Above and Beyond

# Activity 2 :: Restaurant Management

In 2002 there were more than 30,000 McDonald's restaurants in the world. Of these approximately 14,000 were in the United States. In Great Britain there were 1,116 McDonald's franchises. The following data are taken from those 1,116 McDonald's.

| | |
|---|---|
| Total employees: | 49,726 |
| Office staff: | 545 |
| Management: | 2,974 |
| The rest are restaurant workers. | |
| Total employees 20 years old and under: | 34,241 |
| Total employees between 21 and 29 years old: | 10,607 |
| Total male employees: | 27,546 |

Use these data to answer the questions that follow.

1. What is the total number of restaurant workers in McDonald's franchises in Great Britain?
2. How many of the employees mentioned are of age 30 or older?
3. How many of the employees were females?
4. Divide the total number of employees by the number of franchises to determine approximately how many employees are in each McDonald's in Great Britain.
5. Assume that each franchise decided to give each employee a $125 bonus. If you use your result from question 4, how much would that cost each franchise?
6. What would the total cost of the $125 bonus be for all the employees?
7. Assume that there are 45 employees in each of the U.S. McDonald's. Approximately how many people would be employed by McDonald's in the United States?
8. If health care costs $400 per month for one employee, what would the annual cost for health care be?
9. Use the results of questions 7 and 8 to approximate the cost to McDonald's of providing health care for all its workers.
10. Go to the website for this text to find the total profit made by McDonald's in 2002.

# 1.7 Exponential Notation and the Order of Operations

< **1.7 Objectives** >

1 > Use exponent notation

2 > Evaluate expressions containing powers of whole numbers

3 > Evaluate expressions that contain several operations

## Overcoming Math Anxiety

**Preparing for a Test**

Preparation for a test really begins on the first day of class. Everything you do in class and at home is part of that preparation. However, there are a few things that you should focus on in the last few days before a scheduled test.

1. Plan your test preparation to end at least 24 hours before the test. The last 24 hours are too late, and besides, you need some rest before the test.
2. Go over your homework and class notes with pencil and paper in hand. Write down all the problem types, formulas, and definitions that you think might give you trouble on the test.
3. The day before the test, take the page(s) of notes from step 2, and transfer the most important ideas to a $3 \times 5$ card.
4. Just before the test, review the information on the card. You will be surprised at how much you remember about each concept.
5. Understand that, if you have been successful at completing your homework assignments, you can be successful on the test. This is an obstacle for many students, but it is an obstacle that can be overcome. Truly anxious students are often surprised that they scored as well as they did on a test. They tend to attribute this to blind luck. It is not. It is the first sign that you really do "get it." Enjoy the success.

Earlier we described multiplication as a shorthand for repeated addition. There is also a shorthand for repeated multiplication. It uses **powers of a whole number.**

## Example 1 Writing Repeated Multiplication as a Power

< Objective 1 >

$3 \times 3 \times 3 \times 3$ can be written as $3^4$.

This is read as "3 to the fourth power."

In this case, repeated multiplication is written as the power of a number.

In this example, 3 is the **base** of the expression, and the raised number, 4, is the **exponent,** or **power.**

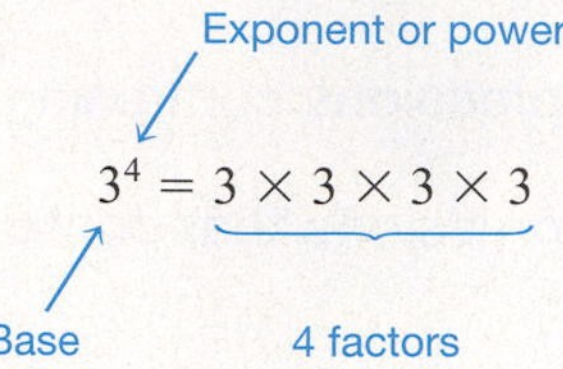

Exponent or power

$$3^4 = \underbrace{3 \times 3 \times 3 \times 3}_{\text{4 factors}}$$

Base

We count the factors and make this the power (or exponent) of the base.

**NOTE**

Recall that

$3 + 3 + 3 + 3 = 4 \times 3$

Repeated addition was written as multiplication.

René Descartes, a French philosopher and mathematician, is generally credited with first introducing our modern exponent notation in about 1637.

### Check Yourself 1

Write $2 \times 2 \times 2 \times 2 \times 2 \times 2$ as a power of 2.

**Definition**

**Exponents**

The *exponent* tells us the number of times the base is to be used as a factor.

## Example 2 Evaluating a Number Raised to a Power

< Objective 2 >

$2^5$ is read "2 to the fifth power."

$$2^5 = \underbrace{2 \times 2 \times 2 \times 2 \times 2}_{\text{5 times}} = 32$$

Here 2 is the base, and 5 is the exponent.

$2^5$ tells us to use 2 as a factor 5 times. The result is 32.

### Check Yourself 2

Read and evaluate $3^4$.

## Example 3 Evaluating a Number Raised to a Power

> CAUTION

$5^3$ is *entirely different* from $5 \times 3$.
$5^3 = 125$ whereas $5 \times 3 = 15$.

Evaluate $5^3$ and $8^2$.

$5^3 = 5 \times 5 \times 5 = 125$ Use 3 factors of 5.

$5^3$ is read "5 to the third power" or "5 cubed."

$8^2 = 8 \times 8 = 64$ Use 2 factors of 8.

And $8^2$ is read "8 to the second power" or "8 squared."

### Check Yourself 3

Evaluate.

(a) $6^2$ (b) $2^4$

We need two special definitions for powers of whole numbers.

**Definition**

**Raising a Number to the First Power**

A number raised to the first power is just that number.

For example, $9^1 = 9$.

**Definition**

**Raising a Number to the Zero Power**

A number, other than 0, raised to the zero power is 1.

For example, $7^0 = 1$.

## Example 4 Evaluating Numbers Raised to the Power of 0 or 1

**(a)** $8^0 = 1$ **(b)** $4^0 = 1$ **(c)** $5^1 = 5$ **(d)** $3^1 = 3$

### Check Yourself 4

Evaluate.

(a) $7^0$ (b) $7^1$

**NOTE**

Notice that $10^3$ is just a 1 followed by *three zeros*.

**NOTE**

$10^5$ is a 1 followed by *five zeros*.

We talked about *powers of 10* earlier when we multiplied by numbers that end in 0. Because the powers of 10 have a special importance, we list some of them here.

$10^0 = 1$

$10^1 = 10$

$10^2 = 10 \times 10 = 100$

$10^3 = 10 \times 10 \times 10 = 1{,}000$

$10^4 = 10 \times 10 \times 10 \times 10 = 10{,}000$

$10^5 = 10 \times 10 \times 10 \times 10 \times 10 = 100{,}000$

Do you see why the powers of 10 are so important?

**Definition**

**Powers of 10**

The powers of 10 correspond to the place values of our number system—ones, tens, hundreds, thousands, and so on.

**NOTE**

Archimedes (about 250 B.C.E.) reportedly estimated the number of grains of sand in the universe to be $10^{63}$. This would be a 1 followed by 63 zeros!

This is what we meant earlier when we said that our number system was based on the number 10.

If multiplication is combined with addition or subtraction, you must know which operation to do first in finding the expression's value. We can easily illustrate this problem. How should we simplify the following statement?

$3 + 4 \times 5 = ?$

Both multiplication and addition are involved in this expression, and we must decide which to do first to find the answer.

1. Multiplying first gives us

   $3 + 20 = 23$

2. Adding first gives us

   $7 \times 5 = 35$

> CAUTION

The answers differ depending on which operation is done first!

Only one of these results can be correct, which is why mathematicians developed a rule to tell us the order in which the operations should be performed. The rules are as follows.

**Step by Step**

**The Order of Operations**

If multiplication, division, addition, and subtraction are involved in the same expression, do the operations in the following order:

**Step 1** Do all multiplication and division in order from left to right.

**Step 2** Do all addition and subtraction in order from left to right.

## Example 5 Using the Order of Operations

< Objective 3 >

**NOTE**

By this rule, we see that strategy 1 from before was correct.

(a) $3 \times 4 + 5 = 12 + 5 = 17$ Multiply *first,* then add or subtract.

(b) $5 + 3 \times 6 = 5 + 18 = 23$

(c) $16 - 2 \times 3 = 16 - 6 = 10$

(d) $7 \times 8 - 20 = 56 - 20 = 36$

(e) $5 \times 6 + 4 \times 3 = 30 + 12 = 42$

### Check Yourself 5

Evaluate.

(a) $8 + 3 \times 5$ (b) $15 \times 5 - 3$ (c) $4 \times 3 + 2 \times 6$

**NOTE**

When learning the order of operations, students sometimes remember this order by relating each step to part of the phrase

"Please Excuse My Dear Aunt Sally."

P
E
MD
AS

We now want to extend our rule for the order of operations to see what happens when parentheses or exponents are involved in an expression.

**Step by Step**

### The Order of Operations

Mixed operations in an expression should be done in the following order:

**Step 1** Do any operations inside *p*arentheses.

**Step 2** Apply any *e*xponents.

**Step 3** Do all *m*ultiplication and *d*ivision in order from left to right.

**Step 4** Do all *a*ddition and *s*ubtraction in order from left to right.

## Example 6 Evaluating an Expression

Evaluate $4 \times 2^3$.

**Step 1** There are no parentheses.

**Step 2** Apply exponents.

$4 \times 2^3 = 4 \times 8$

**Step 3** Multiply or divide.

$4 \times 8 = 32$

### Check Yourself 6

Evaluate.

$3 \times 3^2$

## Example 7 Evaluating an Expression

Evaluate $(2 + 3)^2 + 4 \times 3$.

**Step 1** Do operations inside parentheses.

$(2 + 3)^2 + 4 \times 3 = 5^2 + 4 \times 3$

**Step 2** Apply exponents.

$5^2 + 4 \times 3 = 25 + 4 \times 3$

**Step 3** Multiply or divide.

$25 + 4 \times 3 = 25 + 12$

**Step 4** Add or subtract.

$25 + 12 = 37$

### Check Yourself 7

Evaluate.

(a) $4 + (8 - 5)^2$

(b) $(6 - 4)^3 + 3 \cdot 2$

## Example 8 Using the Order of Operations

**(a)** Evaluate $20 \div 2 \times 5$.

$$\begin{aligned} &\underbrace{20 \div 2} \times 5 \\ =\ & 10 \times 5 \\ =\ & 50 \end{aligned}$$

Because the multiplication and division appear next to each other, work in order from left to right. Try it the other way and see what happens!

So $20 \div 2 \times 5 = 50$.

**(b)** Evaluate $(5 + 13) \div 6$.

$$\begin{aligned} &\underbrace{(5 + 13)} \div 6 \\ =\ & 18 \div 6 \\ =\ & 3 \end{aligned}$$

Do the addition in the parentheses as the first step.

So $(5 + 13) \div 6 = 3$.

**(c)** Evaluate $(3 + 4)^2 \div (2^3 - 1)$.

$$\begin{aligned} &\underbrace{(3 + 4)^2} \div \underbrace{(2^3 - 1)} \\ =\ & (7)^2 \div (8 - 1) \\ =\ & 7^2 \div (7) \\ =\ & 49 \div 7 \\ =\ & 7 \end{aligned}$$

Perform operations inside parentheses first.

So $(3 + 4)^2 \div (2^3 - 1) = 7$.

### Check Yourself 8

Evaluate.

(a) $36 \div 4 \times 2$ (b) $(2 + 4)^2 \div (3^2 - 3)$ (c) $15 \div 3 + (3 - 2)^2 \times 4 - 2$

## Overcoming Math Anxiety

**Taking a Test**

Earlier in this section, we discussed test preparation. Now that you are thoroughly prepared for the test, you must learn how to take it.

There is much to the psychology of anxiety that we can't readily address. There is, however, a physical aspect to anxiety that can be addressed rather easily. When people are in a stressful situation, they frequently start to panic. One symptom of the panic is shallow breathing. In a test situation, this starts a vicious cycle. If you breathe too shallowly, then not enough oxygen reaches your brain. When that happens, you are unable to think clearly. In a test situation, being unable to think clearly can cause you to panic. Hence we have a vicious cycle.

How do you break that cycle? It's pretty simple. Take a few deep breaths. We have seen students whose performance on math tests improved markedly after they got in the habit of writing "remember to breathe!" at the bottom of every test page. Try breathing; it will almost certainly improve your math test scores!

### Check Yourself ANSWERS

**1.** $2^6$ **2.** "Three to the fourth power" is 81 **3. (a)** 36; **(b)** 16
**4. (a)** 1; **(b)** 7 **5. (a)** 23; **(b)** 72; **(c)** 24 **6.** 27 **7. (a)** 13; **(b)** 14
**8. (a)** 18; **(b)** 6; **(c)** 7

## Reading Your Text

The following fill-in-the-blank exercises are designed to ensure that you understand some of the key vocabulary used in this section.

SECTION 1.7

**(a)** Preparation for a test really begins on the ______________ day of class.

**(b)** Another name for a power is an ______________.

**(c)** The first step in the order of operations involves doing operations ______________ parentheses.

**(d)** A whole number (other than zero) raised to the zero power is always equal to ______________.

# 1.7 exercises

Basic Skills | Advanced Skills | Vocational-Technical Applications | Calculator/Computer | Above and Beyond

MathZone

Boost your grade at mathzone.com!

> Practice Problems
> NetTutor
> Self-Tests
> e-Professors
> Videos

Name ____________

Section ________ Date ________

## Answers

| | |
|---|---|
| 1. | 2. |
| 3. | 4. |
| 5. | 6. |
| 7. | 8. |
| 9. | 10. |
| 11. | 12. |
| 13. | 14. |
| 15. | 16. |
| 17. | 18. |
| 19. | 20. |
| 21. | 22. |
| 23. | 24. |
| 25. | 26. |
| 27. | 28. |
| 29. | 30. |
| 31. | 32. |

< Objectives 1, 2, 3 >

*Evaluate.*

1. $3^2$

2. $2^3$

3. $5^1$

4. $6^0$

5. $10^3$

6. $10^6$

7. $2 \times 4^3$

8. $(2 \times 4)^3$

9. $5 + 2^2$

10. $(5 + 2)^2$

11. $(3 + 2)^3 - 20$

12. $5 + (9 - 5)^2$

13. $(7 - 4)^4 - 30$

14. $(5 + 2)^2 + 20$

15. $8^2 \div 4^2 + 2$

16. $3 \times 5^2 + 2^2$

17. $24 - 6 \div 3$

> Videos

18. $3 + 9 \div 3$

19. $(24 - 6) \div 3$

20. $(3 + 9) \div 3$

21. $12 + 3 \div (3^2 - 2 \cdot 3)$

22. $(8^2 - 2^4) \div 2$

23. $8^2 - 2^4 \div 2$

24. $(5 - 3)^3 + (8 - 6)^2$

25. $30 \div 6 - 12 \div 3$

26. $5 + 8 \div 4 - 3$

Basic Skills | Advanced Skills | Vocational-Technical Applications | Calculator/Computer | Above and Beyond

27. $16 - 12 \div 3 \cdot 2 + (16 - 12)^2 \cdot 3$

28. $3 \cdot 5 + 3 \cdot 4^2 \div (6 - 4)^2$

29. $6 + 3 \cdot 2^4 - (12 - 7)(10 - 7)$

30. $27 \div (2^2 + 5) - (35 - 33)(24 - 23)$

31. $3 \times [(7 - 5)^3 - 8] + 5 \times 2$

32. $[(3 + 1) + (7 - 2)] \times 4 - 5 \times 7$

Basic Skills | Advanced Skills | **Vocational-Technical Applications** | Calculator/Computer | Above and Beyond

## Answers

33. ______

34. ______

35. ______

36. ______

**33. Electronics** Resistors are commonly identified by colored bands to indicate their approximate resistance, measured in ohms. Each band's color and position corresponds to a specific component of the overall value. In resistors with four colored bands, the third band is typically considered to be the exponent for a base 10. If the first two bands are decoded as 43 and the third band is decoded as 5, what is the total resistance in ohms?

$43 \times 10^5 = ?$

Which of the three bands is most important to read correctly? Why?

**34. Manufacturing Technology** The kinetic energy (KE) of an object (in Joules) is given by the formula:

$$\text{KE} = \frac{mv^2}{2}$$

Find the kinetic energy of an object that has a mass ($m$) of 46 kg and is moving at a velocity ($v$) of 16 meters per second.

**35. Manufacturing Technology** The power ($P$) of a circuit (in Watts) can be given by any of the following formulas:

$$P = IV$$

$$P = \frac{V^2}{R}$$

$$P = I^2 R$$

Find the power for each of the following circuits:

**(a)** Voltage ($V$) = 110 volts (V) and current ($I$) = 13 amperes (A).

**(b)** Voltage = 220 V and resistance ($R$) = 22 ohms.

**(c)** Current = 25 A and resistance = 9 ohms.

**36. Manufacturing Technology** A belt is used to connect two pulleys.

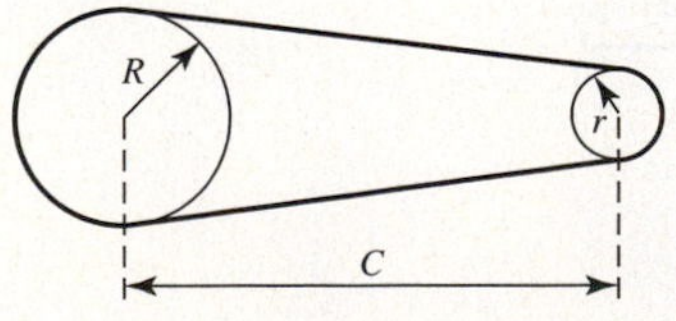


The length of the belt required is given by the formula:

$$\text{Belt length} = 2C + 3(R + r) + \frac{(2R + 2r)^2}{4C}$$

where $C$ is the distance between the centers of the two pulleys.

Find the approximate belt length required to go around a 4-in. radius ($r$) pulley and a 6-in. radius ($R$) pulley that are 20 in. apart.

Basic Skills | Advanced Skills | Vocational-Technical Applications | **Calculator/Computer** | Above and Beyond

## Answers

37. ______
38. ______
39. ______
40. ______
41. ______
42. ______
43. ______
44. ______
45. ______
46. ______
47. ______
48. ______
49. ______
50. ______
51. ______
52. ______

*Multiply, using your calculator.*

**37.** $78 \times 145 \times 36$

**38.** $358 \times 39 \times 928$

**39.** $24 \times 35 \times 48 \times 36$

**40.** $37 \times 15 \times 42 \times 29$

*Use your calculator to evaluate each of the following expressions.*

**41.** $4 \times 5 - 7$

**42.** $3 \times 7 + 8$

**43.** $9 + 3 \times 7$

**44.** $6 \times 0 + 3$

**45.** $4 + 5 \times 0$

**46.** $23 - 4 \times 5$

**47.** $5 \times (4 + 7)$

**48.** $8 \times (6 + 5)$

**49.** $5 \times 4 + 5 \times 7$

**50.** $8 \times 6 + 8 \times 5$

*Solve the following applications, using a calculator.*

**51.** **Business and Finance** A car dealer kept the following record of a month's sales. Complete the table.

| Model | Number Sold | Profit per Sale | Monthly Profit |
|---|---|---|---|
| Subcompact | 38 | $528 | ______ |
| Compact | 33 | 647 | ______ |
| Standard | 19 | 912 | ______ |
| | | **Monthly Total Profit** | ______ |

**52.** **Business and Finance** You take a job paying $1 the first day. On each following day your pay doubles. That is, on day 2 your pay is $2, on day 3 the pay is $4, and so on. Complete the table.

| Day | Daily Pay | Total Pay |
|---|---|---|
| 1 | $1 | $1 |
| 2 | 2 | 3 |
| 3 | 4 | 7 |
| 4 | ______ | ______ |
| 5 | ______ | ______ |
| 6 | ______ | ______ |
| 7 | ______ | ______ |
| 8 | ______ | ______ |
| 9 | ______ | ______ |
| 10 | ______ | ______ |

Basic Skills | Advanced Skills | Vocational-Technical Applications | Calculator/Computer | **Above and Beyond**

*Numbers such as 3, 4, and 5 are called* ***Pythagorean triples,*** *after the Greek mathematician Pythagoras (sixth century B.C.), because*

$3^2 + 4^2 = 5^2$

*Which of the following sets of numbers are Pythagorean triples?*

**53.** 6, 8, 10 **54.** 6, 11, 12 **55.** 5, 12, 13

**56.** 7, 24, 25 **57.** 8, 16, 18 **58.** 8, 15, 17

**59.** Is $(a + b)^P$ equal to $a^P + b^P$?

Try a few numbers and decide if you think this is true for all whole numbers, for some whole numbers, or never true. Write an explanation of your findings and give examples.

**60.** Does $(a \cdot b)^P = a^P \cdot b^P$?

Try a few numbers and decide if you think this is true for all whole numbers, for some whole numbers, or never true. Write an explanation of your findings and give examples.

## Answers

53. ________
54. ________
55. ________
56. ________
57. ________
58. ________
59. ________
60. ________

## Answers

**1.** 9 **3.** 5 **5.** 1,000 **7.** 128 **9.** 9 **11.** 105 **13.** 51 **15.** 6 **17.** 22 **19.** 6 **21.** 13 **23.** 56 **25.** 1 **27.** 56 **29.** 39 **31.** 10 **33.** 4,300,000 ohms; the third band; misreading it would lead to errors of powers of ten **35. (a)** 1,430 W; **(b)** 2,200 W; **(c)** 5,625 W **37.** 407,160 **39.** 1,451,520 **41.** 13 **43.** 30 **45.** 4 **47.** 55 **49.** 55 **51.** $20,064
21,351
17,328
$58,743
**53.** Yes **55.** Yes **57.** No

**59.** Above and Beyond

# Activity 3 :: Package Delivery

Read the following article and use the information you find there to answer the questions that follow.

> The 5-day workweek before Christmas is the busiest week of the year for package deliveries. In the year 2000, the U.S. Postal Service (USPS) averaged 750 million deliveries per day that week. In the same week, United Parcel Service (UPS) delivered about 18 million packages per day. Federal Express (Fed Ex) delivered nearly 5 million packages per day. Of the 5 million Federal Express deliveries, over 3 million of the orders were originated on the Internet.

1. Write out the number that represents the number of deliveries made by the USPS in the 5-day week before Christmas.
2. Write the word form for the answer to question 1.
3. Over the same 5 days, how many deliveries were made by UPS?
4. Over the same 5 days, how many deliveries were made by Fed Ex?
5. How many more deliveries were made by UPS than by Fed Ex that week?

| Definition/Procedure | Example | Reference |
|---|---|---|
| **The Decimal Place-Value System** | | **Section 1.1** |
| **Digits** Digits are the basic symbols of the system. | 0, 1, 2, 3, 4, 5, 6, 7, 8, and 9 are digits. | *p. 4* |
| **Place Value** The value of a digit in a number depends on its position or place. | 52,589 — 9: Ones; 8: Tens; 5: Hundreds; 2: Thousands; 5: Ten thousands | *p. 4* |
| The value of a number is the sum of each digit multiplied by its place value. | $2{,}345 = (2 \times 1{,}000) + (3 \times 100) + (4 \times 10) + (5 \times 1)$ | *p. 4* |
| **Addition** | | **Section 1.2** |
| ***The Properties of Addition*** | | |
| **The Commutative Property** The order in which you add two whole numbers does not affect the sum. | $5 + 4 = 4 + 5$ | *p. 14* |
| **The Associative Property** The way in which you group whole numbers in addition does not affect the final sum. | $(2 + 7) + 8 = 2 + (7 + 8)$ | *p. 15* |
| **The Additive Identity** The sum of 0 and any whole number is just that whole number. | $6 + 0 = 0 + 6 = 6$ | *p. 15* |
| ***Measuring Perimeter*** | | |
| The perimeter is the total distance around the outside edge of a shape. The perimeter of a rectangle is $P = 2 \cdot L + 2 \cdot W$. | Rectangle: 6 ft, 2 ft, 2 ft, 6 ft<br>$P = 2 \times 6 \text{ ft} + 2 \times 2 \text{ ft}$<br>$= 12 \text{ ft} + 4 \text{ ft} = 16 \text{ ft}$ | *p. 20* |
| **Subtraction** | | **Section 1.3** |
| **Minuend** The number we are subtracting from.<br>**Subtrahend** The number that is being subtracted.<br>**Difference** The result of the subtraction. | 15 ⟵ Minuend<br>− 9 ⟵ Subtrahend<br>6 ⟵ Difference | *p. 31* |
| **Rounding, Estimation, and Order** | | **Section 1.4** |
| **Step 1** To round a whole number to a certain decimal place, look at the digit to the right of that place. | | *p. 46* |
| **Step 2** **a.** If that digit is 5 or more, that digit and all digits to the right become 0. The digit in the place you are rounding to is increased by 1. | To the nearest hundred, 43,578 is rounded to 43,600. | |
| **b.** If that digit is less than 5, that digit and all digits to the right become 0. The digit in the place you are rounding to remains the same. | To the nearest thousand, 273,212 is rounded to 273,000. | |
| **Multiplication** | | **Section 1.5** |
| **Factors** The numbers being multiplied.<br>**Product** The result of the multiplication. | $7 \times 9 = 63$ ⟵ Product<br>(7 and 9: Factors) | *p. 56* |

| Definition/Procedure | Example | Reference |
|---|---|---|
| ***The Properties of Multiplication*** | | |
| **The Commutative Property** The order in which you multiply two whole numbers does not affect the product. | $7 \times 9 = 9 \times 7$ | *p.* 56 |
| **The Distributive Property** To multiply a factor by a sum of numbers, multiply the factor by each number inside the parentheses. Then add the products. | $2 \times (3 + 7) = (2 \times 3) + (2 \times 7)$ | *p.* 57 |
| **The Associative Property** The way in which you group numbers in multiplication does not affect the final product. | $(3 \times 5) \times 6 = 3 \times (5 \times 6)$ | *p.* 62 |
| ***Finding the Area of a Rectangle*** | | |
| The area of a rectangle is found using the formula $A = L \cdot W$. | 6 ft; 2 ft<br>$A = L \times W = 6\text{ ft} \times 2\text{ ft} = 12\text{ ft}^2$ | *p.* 65 |
| **Division** | | **Section 1.6** |
| **Divisor** The number we are dividing by.<br>**Dividend** The number being divided.<br>**Quotient** The result of the division.<br>**Remainder** The number "left over" after the division. | Divisor, Quotient<br>$7\overline{)38}$ ⟵ Dividend, quotient 5<br>35<br>3 ⟵ Remainder | *pp.* 77–78 |
| Dividend = divisor × quotient + remainder | $38 = 7 \times 5 + 3$ | *p.* 79 |
| Division by 0 is undefined. | $7 \div 0$ is undefined. | *p.* 79 |
| **Exponential Notation and the Order of Operations** | | **Section 1.7** |
| ***Using Exponents*** | | |
| **Base** The number that is raised to a power.<br>**Exponent** The exponent is written to the right and above the base. The exponent tells the number of times the base is to be used as a factor. | Exponent<br>$5^3 = \underbrace{5 \times 5 \times 5}_{\text{Three factors}} = 125$<br>Base | *p.* 94 |
| ***The Order of Operations*** | | |
| Mixed operations in an expression should be done in the following order:<br>**Step 1** Do any operations inside parentheses.<br>**Step 2** Evaluate any exponents.<br>**Step 3** Do all multiplication and division in order from left to right.<br>**Step 4** Do all addition and subtraction in order from left to right.<br>Remember *P*lease *E*xcuse *M*y *D*ear *A*unt *S*ally | $4 \times (2 + 3)^2 - 7$<br>$= 4 \times 5^2 - 7$<br>$= 4 \times 25 - 7$<br>$= 100 - 7$<br>$= 93$ | *p.* 97 |

# summary exercises :: chapter 1

This summary exercise set is provided to give you practice with each of the objectives of this chapter. Each exercise is keyed to the appropriate chapter section. When you are finished, you can check your answers to the odd-numbered exercises against those presented in the back of the text. If you have difficulty with any of these questions, go back and reread the examples from that section. The answers to the even-numbered exercises appear in the *Instructor's Solutions Manual*. Your instructor will give you guidelines on how to best use these exercises in your instructional setting.

**1.1** *In exercises 1 and 2, give the place value of each of the indicated digits.*

**1.** 6 in the number 5,674

**2.** 5 in the number 543,400

*In exercises 3 and 4, give word names for each of the following numerals.*

**3.** 27,428

**4.** 200,305

*Write each of the following in standard form.*

**5.** Thirty-seven thousand, five hundred eighty-three

**6.** Three hundred thousand, four hundred

**1.2** *In exercises 7 and 8, name the property of addition illustrated.*

**7.** $4 + 9 = 9 + 4$

**8.** $(4 + 5) + 9 = 4 + (5 + 9)$

*In exercises 9 to 13, perform the indicated operations.*

**9.**
$$\begin{array}{r} 784 \\ 385 \\ +\ 247 \\ \hline \end{array}$$

**10.**
$$\begin{array}{r} 2{,}570 \\ 498 \\ 21{,}456 \\ +\quad 28 \\ \hline \end{array}$$

**11.**
$$\begin{array}{r} 367 \\ 289 \\ 1{,}463 \\ +\ 2{,}682 \\ \hline \end{array}$$

**12.**
$$\begin{array}{r} 6{,}389 \\ 1{,}567 \\ 315 \\ +\ 113{,}602 \\ \hline \end{array}$$

**1.3** **13.** Find the value for the following:

**(a)** 34 decreased by 7

**(b)** 7 more than 4

**(c)** The product of 9 and 5, divided by 3

*Solve the following applications.*

**14.** **Statistics** An airline had 173, 212, 185, 197, and 202 passengers on five morning flights between Washington, D.C. and New York. What was the total number of passengers?

15. **Business and Finance** Future Stars summer camp employs five junior counselors. Their weekly salaries last week were $108, $135, $81, $135, and $81. What was the total salary for the junior counselors?

**1.3** *In exercises 16 to 20, perform the indicated operations.*

16. $\begin{array}{r} 5{,}325 \\ -\ 847 \\ \hline \end{array}$

17. $\begin{array}{r} 38{,}400 \\ -\ 19{,}600 \\ \hline \end{array}$

18. $\begin{array}{r} 86{,}000 \\ -\ 2{,}169 \\ \hline \end{array}$

19. $\begin{array}{r} 2{,}682 \\ -\ 108 \\ \hline \end{array}$

20. Find the difference of 7,342 and 5,579.

*Solve the following applications.*

21. **Business and Finance** Chuck owes $795 on a credit card after a trip. He makes payments of $75, $125, and $90. Interest of $31 is charged. How much remains to be paid on the account?

22. **Business and Finance** Juan bought a new car for $16,785. The manufacturer offers a cash rebate of $987. What was the cost after rebate?

**1.4** *Round the numbers to the indicated place.*

23. 6,975 to the nearest hundred

24. 15,897 to the nearest thousand

25. 548,239 to the nearest ten thousand

*Complete the statements by using the symbol < or >.*

26. 60 ______ 70

27. 38 ______ 35

*Find the perimeter of the following figures.*

28.

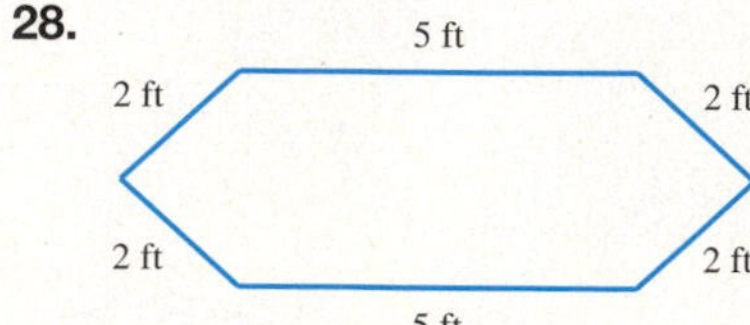


29.

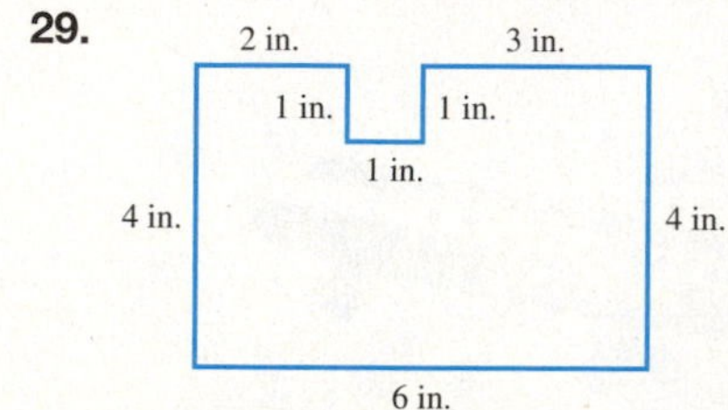

**1.5** *In exercises 30 to 32, name the property of multiplication that is illustrated.*

**30.** $7 \times 8 = 8 \times 7$

**31.** $3 \times (4 + 7) = 3 \times 4 + 3 \times 7$

**32.** $(8 \times 9) \times 4 = 8 \times (9 \times 4)$

*In exercises 33 to 35, perform the indicated operations.*

**33.** $\begin{array}{r} 58 \\ \times\ 32 \\ \hline \end{array}$

**34.** $\begin{array}{r} 25 \\ \times\ 43 \\ \hline \end{array}$

**35.** $\begin{array}{r} 378 \\ \times\ 409 \\ \hline \end{array}$

*Find the area of the given figures.*

**36.**

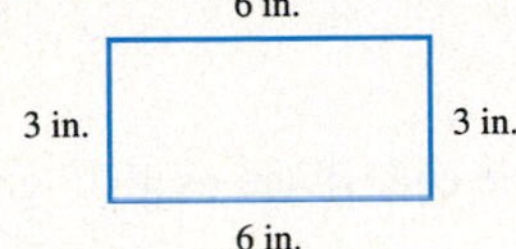


**37.**

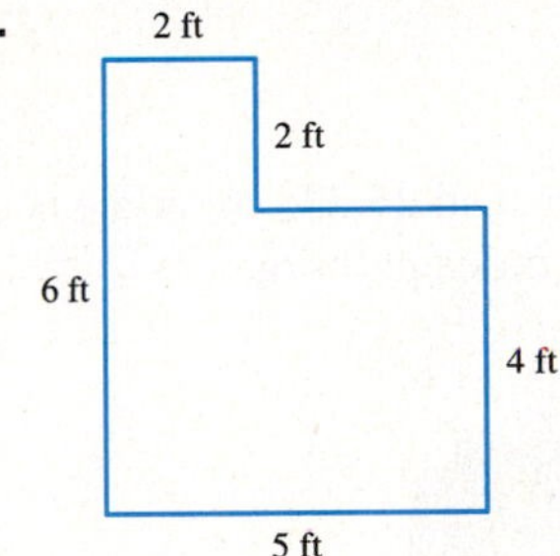


*Solve the following application.*

**38. CRAFTS** You wish to carpet a room that is 5 yd by 7 yd. The carpet costs $18 per square yard. What will be the total cost of the materials?

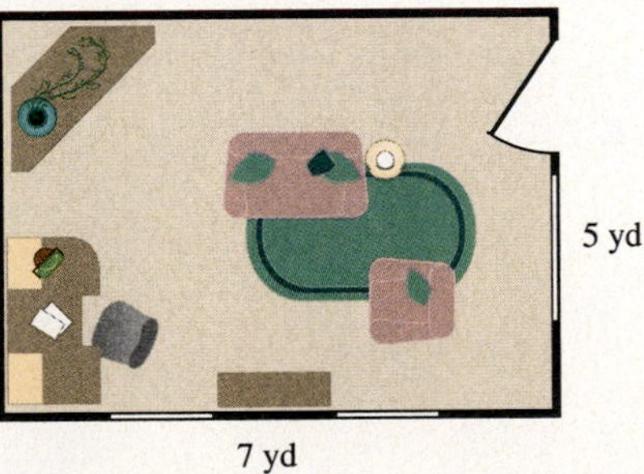


*Perform the indicated operation.*

**39.** $\begin{array}{r} 129 \\ \times\ 240 \\ \hline \end{array}$

*Estimate the product by rounding each factor to the nearest hundred.*

**40.** $\begin{array}{r} 1{,}217 \\ \times\ 494 \\ \hline \end{array}$

**1.6** *Divide if possible.*

**41.** $0 \div 8$

**42.** $5 \div 0$

*In exercises 43 to 46, divide.*

**43.** $8\overline{)2{,}469}$

**44.** $39\overline{)2{,}157}$

**45.** $64\overline{)31{,}809}$

**46.** $362\overline{)86{,}915}$

*Solve the following application.*

**47. STATISTICS** Hasina's odometer read 25,235 mi at the beginning of a trip and 26,215 mi at the end. If she used 35 gal of gas for the trip, what was her mileage (mi/gal)?

*Estimate the following.*

**48.** 356 divided by 37

**49.** 2,125 divided by 123

**1.7** *In exercises 50 to 59, evaluate the expressions.*

**50.** $5 \times 2^3$

**51.** $(5 \times 2)^3$

**52.** $4 + 8 \times 3$

**53.** $48 \div (2^3 + 4)$

**54.** $(4 + 8) \times 3$

**55.** $4 \times 3 + 8 \times 3$

**56.** $8 \div 4 \times 2 - 2 + 1$

**57.** $63 \times 2 \div 3 - 54 \div (12 \times 2 \div 4)$

**58.** $(3 \times 4)^2 - 100 \div 5 \times 6$

**59.** $(16 \times 2) \div 8 - (6 \div 3 \times 2)$

**1.5** *In exercises 30 to 32, name the property of multiplication that is illustrated.*

**30.** $7 \times 8 = 8 \times 7$

**31.** $3 \times (4 + 7) = 3 \times 4 + 3 \times 7$

**32.** $(8 \times 9) \times 4 = 8 \times (9 \times 4)$

*In exercises 33 to 35, perform the indicated operations.*

**33.** $\begin{array}{r} 58 \\ \times\ 32 \\ \hline \end{array}$

**34.** $\begin{array}{r} 25 \\ \times\ 43 \\ \hline \end{array}$

**35.** $\begin{array}{r} 378 \\ \times\ 409 \\ \hline \end{array}$

*Find the area of the given figures.*

**36.**

6 in.
3 in.
3 in.
6 in.

**37.**

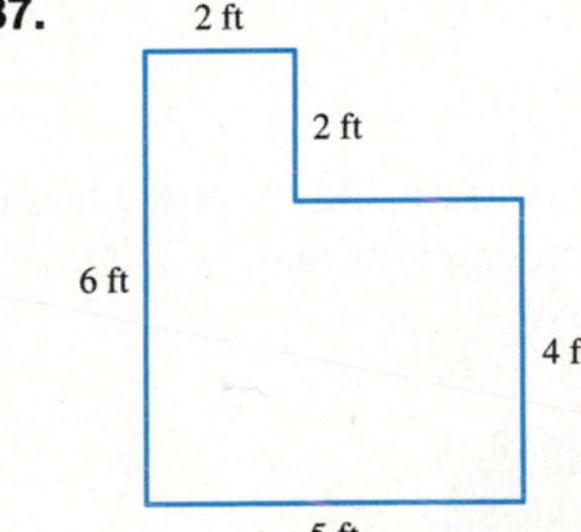


*Solve the following application.*

**38. Crafts** You wish to carpet a room that is 5 yd by 7 yd. The carpet costs $18 per square yard. What will be the total cost of the materials?

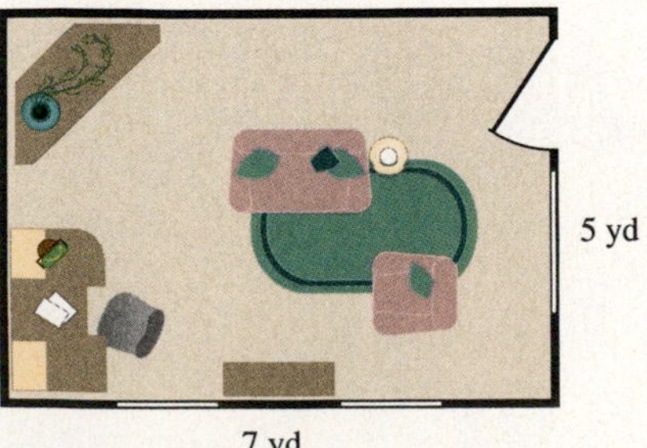


*Perform the indicated operation.*

**39.** $\begin{array}{r} 129 \\ \times\ 240 \\ \hline \end{array}$

*Estimate the product by rounding each factor to the nearest hundred.*

**40.** $\begin{array}{r} 1{,}217 \\ \times\ 494 \\ \hline \end{array}$

**1.6** *Divide if possible.*

**41.** $0 \div 8$

**42.** $5 \div 0$

*In exercises 43 to 46, divide.*

**43.** $8\overline{)2{,}469}$

**44.** $39\overline{)2{,}157}$

**45.** $64\overline{)31{,}809}$

**46.** $362\overline{)86{,}915}$

*Solve the following application.*

**47. STATISTICS** Hasina's odometer read 25,235 mi at the beginning of a trip and 26,215 mi at the end. If she used 35 gal of gas for the trip, what was her mileage (mi/gal)?

*Estimate the following.*

**48.** 356 divided by 37

**49.** 2,125 divided by 123

**1.7** *In exercises 50 to 59, evaluate the expressions.*

**50.** $5 \times 2^3$

**51.** $(5 \times 2)^3$

**52.** $4 + 8 \times 3$

**53.** $48 \div (2^3 + 4)$

**54.** $(4 + 8) \times 3$

**55.** $4 \times 3 + 8 \times 3$

**56.** $8 \div 4 \times 2 - 2 + 1$

**57.** $63 \times 2 \div 3 - 54 \div (12 \times 2 \div 4)$

**58.** $(3 \times 4)^2 - 100 \div 5 \times 6$

**59.** $(16 \times 2) \div 8 - (6 \div 3 \times 2)$

# self-test 1

Name ____________

Section ________ Date ________

The purpose of this self-test is to help you check your progress so that you can find sections and concepts that you need to review before the next exam. Allow yourself about an hour to take this test. At the end of that hour, check your answers against those given in the back of the text. If you missed any, note the section reference that accompanies the answer. Go back to that section and reread the examples until you have mastered that concept.

**1.** Give the place value of 7 in 3,738,500.

**2.** Give the word name for 302,525.

**3.** Write two million, four hundred thirty thousand as a number.

*Name the property of addition that is illustrated.*

**4.** $5 + 12 = 12 + 5$

**5.** $(7 + 3) + 8 = 7 + (3 + 8)$

*In exercises 6 and 7, perform the indicated operations.*

**6.**
$$\begin{array}{r} 489 \\ 562 \\ 613 \\ +\ 254 \\ \hline \end{array}$$

**7.**
$$\begin{array}{r} 13 \\ 2,543 \\ +\ 10,547 \\ \hline \end{array}$$

**8.** What is the total of 392, 95, 9,237, and 11,972?

*Solve the following application.*

**9.** **Statistics** The attendance for the games of a playoff series in basketball was 12,438, 14,325, 14,581, and 14,634. What was the total attendance for the series?

*In exercises 10 to 13, subtract.*

**10.** $289 - 54$

**11.** $53,294 - 41,074$

**12.** $32,345 - 1,575$

**13.** $55,342 - 14,787$

**14.** **Social Science** The maximum load for a light plane with full gas tanks is 500 pounds (lb). Mr. Whitney weighs 215 lb; his wife, 135 lb; and their daughter, 78 lb. How much luggage can they take on a trip without exceeding the load limit?

*Estimate the sum by rounding each addend to the nearest hundred.*

**15.**
$$\begin{array}{r} 943 \\ 3,281 \\ 778 \\ 2,112 \\ +\ 570 \\ \hline \end{array}$$

## Answers

1. ____________
2. ____________
3. ____________
4. ____________
5. ____________
6. ____________
7. ____________
8. ____________
9. ____________
10. ____________
11. ____________
12. ____________
13. ____________
14. ____________
15. ____________

## Answers

16. ______

17. ______

18. ______

19. ______

20. ______

21. ______

22. ______

23. ______

24. ______

25. ______

26. ______

27. ______

28. ______

29. ______

30. ______

*Complete the statements by using the symbol $<$ or $>$.*

**16.** 49 ______ 47

**17.** 80 ______ 90

*Find the perimeter of the figure shown.*

**18.**

2 in.
2 in.
2 in.
2 in.
2 in.
2 in.

*Name the property that is illustrated.*

**19.** $3 \times (2 \times 7) = (3 \times 2) \times 7$

**20.** $4 \times (3 + 6) = (4 \times 3) + (4 \times 6)$

*Find the products.*

**21.** $\begin{array}{r} 89 \\ \times\ 56 \\ \hline \end{array}$

**22.** $\begin{array}{r} 538 \\ \times\ 103 \\ \hline \end{array}$

**23. Business and Finance** A truck rental firm has ordered 25 new vans at a cost of $12,350 per van. What will be the total cost of the order?

*Find the area of the given figure.*

**24.**

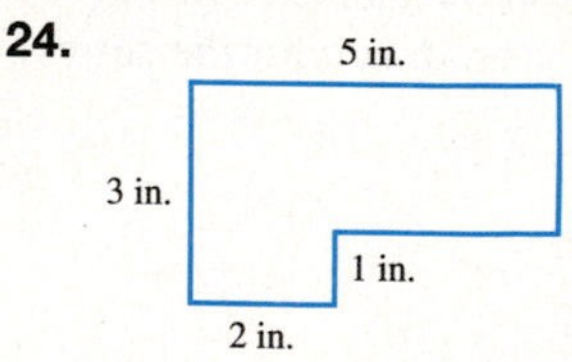


*Divide, using long division.*

**25.** $9\overline{)27{,}371}$

**26.** $28\overline{)2{,}135}$

**27.** $293\overline{)61{,}382}$

**28. Business and Finance** Eight people estimate that the total expenses for a trip they are planning to take together will be $1,784. If each person pays an equal amount, what will be each person's share?

**29.** Evaluate the expression $15 - 12 \div 2^2 \times 3 + (12 \div 4 \times 3)$.

**30.** Evaluate the expression $(3 + 4)^2 - (2 + 3^2 - 1)$.

CHAPTER 2

# Multiplying and Dividing Fractions

## INTRODUCTION

Rebecca got tired of working nine to five as a paralegal in a busy law office in Philadelphia. On many occasions, she had to stay at work until 6:00 or 6:30, and the commute back and forth was exhausting. She wanted a more flexible schedule and started looking for an opportunity to be her own boss.

Rebecca loved to cook and often made delicious meals when she had the time. After investigating certain job opportunities, Rebecca decided to become a personal chef. She read somewhere that personal chefs can make between $50,000 and $70,000 a year. She knew of many busy families in her area that did not have the time to cook meals at night, and she thought this might be a good opportunity. She quit her job and started making meals for some of the people in the law firm where she had previously worked. She planned menus, did the grocery shopping, and cooked the meals right in her clients' kitchens. Sometimes, she cooked enough meals for an entire week and left the meals in containers in the freezer with instructions on how to reheat. She had to adjust her recipes for larger or smaller amounts, depending on whom she was cooking for.

Before long, Rebecca's clients started referring her to friends and relatives and her business began to grow. She was very happy with her new schedule and the free time that she had.

## CHAPTER 2 OUTLINE

# 2 pretest



Name ______________________

Section __________ Date __________

This pretest provides a preview of the types of exercises you will encounter in each section of this chapter. The answers for these exercises can be found in the back of the text. If you are working on your own, or ahead of the class, this pretest can help you identify the sections in which you should focus more of your time.

**Answers**

1. ______________________
2. ______________________
3. ______________________
4. ______________________
5. ______________________
6. ______________________
7. ______________________
8. ______________________
9. ______________________
10. ______________________
11. ______________________
12. ______________________
13. __________ 14. __________
15. __________ 16. __________
17. __________ 18. __________
19. ______________________

**2.1**

**1.** List all the factors of 42.

**2.** For the group of numbers 2, 3, 6, 7, 9, 17, 18, 21, and 23, list the prime and composite numbers.

**3.** Use divisibility tests to determine whether 2, 3, or 5 is a factor of 5,130.

**2.2**

**4.** Write the prime factorization of 350.

**5.** Find the greatest common factor (GCF) of 24, 36, and 52.

**2.3**

**6.** Convert $\frac{37}{4}$ to a mixed number.

**7.** Convert $6\frac{4}{7}$ to an improper fraction.

**2.4**

**8.** Are $\frac{8}{12}$ and $\frac{20}{30}$ equivalent fractions?

**9.** Reduce $\frac{18}{30}$ to lowest terms.

In exercises 10 to 12, use the following group of numbers.

$$\frac{5}{6}, \frac{8}{7}, \frac{13}{9}, 2\frac{3}{5}, \frac{3}{8}, 7\frac{2}{9}, \frac{15}{8}, \frac{9}{9}, \frac{20}{21}, 3\frac{2}{7}, \frac{16}{5}, \frac{5}{11}$$

**10.** Identify the proper fractions.

**11.** Identify the improper fractions.

**12.** Identify the mixed numbers.

Carry out the indicated operations.

**2.5**

**13.** $2\frac{2}{5} \times 1\frac{3}{4}$

**14.** $\frac{18}{21} \times \frac{7}{9}$

**15.** $2\frac{2}{5} \times 5\frac{1}{4}$

**2.6**

**16.** $\frac{6}{25} \div \frac{3}{10}$

**17.** $\dfrac{\frac{1}{2}}{\frac{3}{4}}$

**18.** $1\frac{5}{6} \div 2\frac{4}{9}$

**19.** CRAFTS A piece of wood that is $15\frac{3}{4}$ in. long is to be cut into blocks $1\frac{1}{8}$ in. long. How many blocks can be cut?



The Streeter/Hutchison Series in Mathematics Basic Mathematical Skills with Geometry

# 2.1 Prime Numbers and Divisibility

< 2.1 Objectives >

1 > Find the factors of a number

2 > Determine whether a number is prime, composite, or neither

3 > Determine whether a number is divisible by 2, 3, 4, 5, 6, or 9

## Overcoming Math Anxiety

**Working Together**

How many of your classmates do you know? Whether you are by nature gregarious or shy, you have much to gain by getting to know your classmates.

1. It is important to have someone to call when you miss class or if you are unclear on an assignment.
2. Working with another person is almost always beneficial to both people. If you don't understand something, it helps to have someone to ask about it. If you do understand something, nothing will cement that understanding more than explaining the idea to another person.
3. Sometimes we need to commiserate. If an assignment is particularly frustrating, it is reassuring to find that it is also frustrating for other students.
4. Have you ever thought you had the right answer, but it didn't match the answer in the text? Frequently the answers are equivalent, but that's not always easy to see. A different perspective can help you see that. Occasionally there is an error in a textbook (here, we are talking about *other* textbooks). In such cases it is wonderfully reassuring to find that someone else has the same answer as you do.

**NOTE**

Also, 2 and 5 can be called *divisors* of 10. They divide 10 exactly.

In Section 1.5 we said that because $2 \times 5 = 10$, we call 2 and 5 **factors** of 10.

**Definition**

**Factor** A **factor** of a whole number is another whole number that *divides exactly* into the original number. This means that the division has a remainder of 0.

## Example 1 Finding Factors

< Objective 1 >

List all the factors of 18.

When finding all the factors for any number, we always start with division by 1 because 1 is a factor of every number.

$1 \times 18 = 18$ Our list of factors starts with 1 and 18.

We continue with division by 2.

$2 \times 9 = 18$ Our list now contains 1, 2, 9, 18.

We check divisibility by each subsequent whole number until we get to a number that is already in the factor list.

$3 \times 6 = 18$ The list is now 1, 2, 3, 6, 9, 18.

Our number, 18, is not divisible by either 4 or 5. It is divisible by 6, but 6 is already in the list. The factor list is complete.

**NOTE**

This is a complete list of the factors. There are no other whole numbers that divide 18 exactly. Note that the factors of 18, except for 18 itself, are *smaller* than 18.

### Check Yourself 1

List all the factors of 24.

Listing factors leads us to an important classification of whole numbers. Any whole number larger than 1 is either a *prime* or a *composite* number.

A whole number greater than 1 always has itself and 1 as factors. Sometimes these are the *only* factors. For instance, 1 and 3 are the only factors of 3.

**Definition**

**Prime Number**

A **prime number** is any whole number that has exactly two factors, 1 and itself.

As examples, 2, 3, 5, and 7 are prime numbers. Their only factors are 1 and themselves.

To check whether a number is prime, one approach is simply to divide the smaller primes—2, 3, 5, 7, and so on—into the given number. If no factors other than 1 and the given number are found, the number is prime.

Here is the method known as the **sieve of Eratosthenes** for identifying prime numbers.

1. Write down a series of counting numbers, starting with the number 2. In this example, we stop at 50.
2. Start at the number 2. Delete every second number after 2.
3. Move to the number 3. Delete every third number after 3 (some numbers will be deleted twice).
4. Continue this process, deleting every fourth number after 4, every fifth number after 5, and so on.
5. When you have finished, the undeleted numbers are the prime numbers.

**NOTE**

How large can a prime number be? There is no largest prime number. To date, the largest *known* prime is $2^{13,466,917} - 1$. This is a number with 4,053,946 digits, if you are curious. Of course, a computer had to be used to verify that a number of this size is prime. By the time you read this, someone may very well have found an even larger prime number.

| | 2 | 3 | ~~4~~ | 5 | ~~6~~ | 7 | ~~8~~ | ~~9~~ | ~~10~~ |
|---|---|---|---|---|---|---|---|---|---|
| 11 | ~~12~~ | 13 | ~~14~~ | ~~15~~ | ~~16~~ | 17 | ~~18~~ | 19 | ~~20~~ |
| ~~21~~ | ~~22~~ | 23 | ~~24~~ | ~~25~~ | ~~26~~ | ~~27~~ | ~~28~~ | 29 | ~~30~~ |
| 31 | ~~32~~ | ~~33~~ | ~~34~~ | ~~35~~ | ~~36~~ | 37 | ~~38~~ | ~~39~~ | ~~40~~ |
| 41 | ~~42~~ | 43 | ~~44~~ | ~~45~~ | ~~46~~ | 47 | ~~48~~ | ~~49~~ | ~~50~~ |

The prime numbers less than 50 are 2, 3, 5, 7, 11, 13, 17, 19, 23, 29, 31, 37, 41, 43, 47.

## Example 2 Identifying Prime Numbers

< Objective 2 >

Which of the numbers 17, 29, and 33 are prime?

17 is a prime number. 1 and 17 are the only factors.

29 is a prime number. 1 and 29 are the only factors.

33 is *not* prime. 1, 3, 11, and 33 are all factors of 33.

**Note:** For two-digit numbers, if the number is *not* a prime, it has one or more of the numbers 2, 3, 5, or 7 as factors.

**NOTE**

The definition of a composite number tells us that a composite number *does* have factors other than 1 and itself.

### Check Yourself 2

**Which of the following numbers are prime numbers?**

**2, 6, 9, 11, 15, 19, 23, 35, 41**

We can now define a second class of whole numbers.

**Definition**

**Composite Number**

A **composite number** is any whole number greater than 1 that is not prime. Every composite number has more than two factors.

## Example 3 Identifying Composite Numbers

Which of the numbers 18, 23, 25, and 38 are composite?

18 is a composite number. 1, 2, 3, 6, 9, and 18 are all factors of 18.

23 is *not* a composite number. 1 and 23 are the only factors. This means that 23 is a *prime number.*

25 is a composite number. 1, 5, and 25 are factors.

38 is a composite number. 1, 2, 19, and 38 are factors.

**Check Yourself 3**

Which of the following numbers are composite numbers?

2, 6, 10, 13, 16, 17, 22, 27, 31, 35

By the definitions of prime and composite numbers:

**Definition**

**Zero and One** The whole numbers 0 and 1 are neither prime nor composite.

This is simply a matter of the way in which prime and composite numbers are defined in mathematics. The numbers 0 and 1 are the *only* two whole numbers that cannot be classified as one or the other.

**NOTE**

Divisibility by 2 indicates that a number is *even*.

For our work in this and the following sections, it is very useful to be able to tell whether a given number is divisible by 2, 3, or 5. The tests that follow will give you some tools to check divisibility without actually having to divide.

Tests for divisibility by other numbers are also available. However, we have limited this section to those tests involving 2, 3, 4, 5, 6, 9, and 10 because they are very easy to use and occur frequently in our work.

**Property**

**Divisibility by 2** A whole number is divisible by 2 if its last digit is 0, 2, 4, 6, or 8.

**Example 4** Determining If a Number Is Divisible by 2

< Objective 3 >

Which of the numbers 2,346, 13,254, 23,573, and 57,085 are divisible by 2?

2,346 is divisible by 2. The final digit is 6.

13,254 is divisible by 2. The final digit is 4.

23,573 is *not* divisible by 2. The final digit is not 0, 2, 4, 6, or 8.

57,085 is *not* divisible by 2.

**Check Yourself 4**

Which of the following are divisible by 2?

274 3,587 7,548 13,593

**Property**

**Divisibility by 3** A whole number is divisible by 3 if the sum of its digits is divisible by 3.

## Example 5 Determining If a Number Is Divisible by 3

Which of the numbers 345, 1,243, and 25,368 are divisible by 3?

345 is divisible by 3. The sum of the digits, 3 + 4 + 5, is 12, and 12 is divisible by 3.

1,243 is *not* divisible by 3. The sum of the digits, 1 + 2 + 4 + 3, is 10, and 10 is not divisible by 3.

25,368 is divisible by 3. The sum of the digits, 2 + 5 + 3 + 6 + 8, is 24, and 24 is divisible by 3. Note that 25,368 is also divisible by 2.

### Check Yourself 5

**(a)** Is 372 divisible by 2? By 3? **(b)** Is 5,493 divisible by 2? By 3?

**Property**

**Divisibility by 5**

A whole number is divisible by 5 if its last digit is 0 or 5.

## Example 6 Determining If a Number Is Divisible by 5

Determine which of the following are divisible by 5.

2,435 is divisible by 5. Its last digit is 5.

23,123 is *not* divisible by 5. Its last digit is 3.

123,240 is divisible by 5. Its last digit is 0. Do you see that 123,240 is also divisible by 2 and 3?

### Check Yourself 6

**(a)** Is 12,585 divisible by 5? By 2? By 3?

**(b)** Is 5,890 divisible by 5? By 2? By 3?

Combining some of the techniques we have developed in this section, we can come up with divisibility tests for composite numbers as well.

**Property**

**Divisibility by 4**

A whole number is divisible by 4 if its final two digits are divisible by 4.

## Example 7 Determining If a Number Is Divisible by 4

Determine if the numbers 1,464 and 2,434 are divisible by 4.

1,464 is divisible by 4. 64 is divisible by 4.

2,434 is *not* divisible by 4. 34 is not divisible by 4.

### Check Yourself 7

Determine if each number is divisible by 4.

(a) 6,456 (b) 242 (c) 22,100

**Property**

**Divisibility by 6** A whole number is divisible by 6 if it is an even number that is divisible by 3.

## Example 8 Determining If a Number Is Divisible by 6

Determine if the numbers 1,464 and 2,434 are divisible by 6.

1,464 is divisible by 6. It is an even number, and the sum of the digits, 15, indicates that it is divisible by 3.

2,434 is *not* divisible by 6. Although it is an even number, the sum of the digits, 13, indicates that it is not divisible by 3.

### Check Yourself 8

Determine if each number is divisible by 6.

(a) 6,456 (b) 242 (c) 22,100

**Property**

**Divisibility by 9** A whole number is divisible by 9 if the sum of its digits is divisible by 9.

## Example 9 Determining If a Number Is Divisible by 9

Determine if the numbers 1,494 and 2,634 are divisible by 9.

1,494 is divisible by 9. The sum of the digits, 18, indicates that it is divisible by 9.

2,634 is *not* divisible by 9. The sum of the digits, 15, indicates that it is not divisible by 9.

### Check Yourself 9

Determine if each number is divisible by 9.

(a) 3,456 (b) 243,000 (c) 22,200

**Property**

**Divisibility by 10** A whole number is divisible by 10 if it ends with a zero.

### Example 10 Determining If a Number Is Divisible by 10

Determine if the numbers 4,390,005 and 6,420 are divisible by 10.

4,390,005 is *not* divisible by 10. The number does not end with a zero.

6,420 is divisible by 10. The number does end with a zero.

### Check Yourself 10

Determine whether each number is divisible by 10.

(a) 2,000,020 (b) 2,000,002 (c) 3,571,110

### Check Yourself ANSWERS

**1.** 1, 2, 3, 4, 6, 8, 12, and 24 **2.** 2, 11, 19, 23, and 41 are prime numbers
**3.** 6, 10, 16, 22, 27, and 35 are composite numbers **4.** 274 and 7,548
**5.** **(a)** Yes in both cases; **(b)** only by 3 **6.** **(a)** By 5 and by 3; **(b)** by 5 and by 2
**7.** **(a)** Yes; **(b)** no; **(c)** yes **8.** **(a)** Yes; **(b)** no; **(c)** no
**9.** **(a)** Yes; **(b)** yes; **(c)** no **10.** **(a)** Yes; **(b)** no; **(c)** yes

### Reading Your Text

The following fill-in-the-blank exercises are designed to ensure that you understand some of the key vocabulary used in this section.

SECTION 2.1

**(a)** ____________ is a factor of every number.

**(b)** A ____________ number is any whole number that has exactly two factors, 1 and itself.

**(c)** A ____________ number is any whole number greater than 1 that is not prime.

**(d)** When a whole number is divisible by 2, we call it an ____________ number.

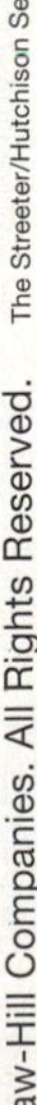

# 2.1 exercises

Basic Skills | Advanced Skills | Vocational-Technical Applications | Calculator/Computer | Above and Beyond

MathZone

Boost your grade at mathzone.com!
> Practice Problems
> NetTutor
> Self-Tests
> e-Professors
> Videos

Name ______

Section ______ Date ______

## Answers

1. ______ 2. ______
3. ______
4. ______
5. ______ 6. ______
7. ______
8. ______
9. ______
10. ______
11. ______ 12. ______
13. ______
14. ______
15. ______
16. ______
17. ______
18. ______
19. ______
20. ______

< Objective 1 >

*List the factors of each of the following numbers.*

**1.** 4 **2.** 6

**3.** 10 **4.** 12

**5.** 15 **6.** 21

**7.** 24 > Videos **8.** 32

**9.** 64 **10.** 66

**11.** 11 > Videos **12.** 37

**13.** 135 **14.** 236

**15.** 256 **16.** 512

< Objective 2 >

*Use the following list of numbers for exercises 17 and 18.*

0, 1, 15, 19, 23, 31, 49, 55, 59, 87, 91, 97, 103, 105

**17.** Which of the given numbers are prime?

**18.** Which of the given numbers are composite?

**19.** List all the prime numbers between 30 and 50.

**20.** List all the prime numbers between 55 and 75.

< Objective 3 >

*Use the following list of numbers for exercises 21 through 26.*

45, 72, 158, 260, 378, 569, 570, 585, 3,541, 4,530, 8,300

**21.** Which of the given numbers are divisible by 2?

**22.** Which of the given numbers are divisible by 3?

**23.** Which of the given numbers are divisible by 6?

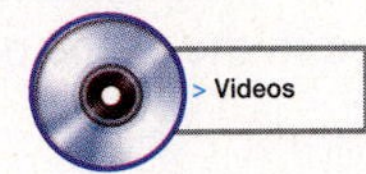


**24.** Which of the given numbers are divisible by 9?

**25.** Which of the given numbers are divisible by 4?

**26.** Which of the given numbers are divisible by 10?

**27.** **Number Problem** A school auditorium is to have 350 seats. The principal wants to arrange them in rows with the same number of seats in each row. Use divisibility tests to determine if it is possible to have rows of 10 seats each. Are 15 rows of seats possible?

**28.** **Social Science** Dr. Mento has a class of 80 students. For a group project, she wants to divide the students into groups of 6, 8, or 10. Is this possible? Explain your answer.

Basic Skills | Advanced Skills | Vocational-Technical Applications | Calculator/Computer | **Above and Beyond**

**29.** **Number Problem** Use the *sieve of Eratosthenes* to determine all the prime numbers less than 100.

| | 2 | 3 | 4 | 5 | 6 | 7 | 8 | 9 | 10 |
|---|---|---|---|---|---|---|---|---|---|
| 11 | 12 | 13 | 14 | 15 | 16 | 17 | 18 | 19 | 20 |
| 21 | 22 | 23 | 24 | 25 | 26 | 27 | 28 | 29 | 30 |
| 31 | 32 | 33 | 34 | 35 | 36 | 37 | 38 | 39 | 40 |
| 41 | 42 | 43 | 44 | 45 | 46 | 47 | 48 | 49 | 50 |
| 51 | 52 | 53 | 54 | 55 | 56 | 57 | 58 | 59 | 60 |
| 61 | 62 | 63 | 64 | 65 | 66 | 67 | 68 | 69 | 70 |
| 71 | 72 | 73 | 74 | 75 | 76 | 77 | 78 | 79 | 80 |
| 81 | 82 | 83 | 84 | 85 | 86 | 87 | 88 | 89 | 90 |
| 91 | 92 | 93 | 94 | 95 | 96 | 97 | 98 | 99 | 100 |

**30.** Why is the following not a valid divisibility test for 8?

"A number is divisible by 8 if it is divisible by 2 and 4."

Support your answer with an example. Give a valid divisibility test for 8.

**31.** Prime numbers that differ by 2 are called *twin primes*. Examples are 3 and 5, 5 and 7, and so on. Find one pair of twin primes between 85 and 105.

## Answers

21. ______

22. ______

23. ______

24. ______

25. ______

26. ______

27. ______

28. ______

29. ______

30. ______

31. ______

## Answers

32. ______

33. ______

34. ______

35. ______

**32.** The following questions refer to *twin primes* (see exercise 31).

**(a)** Search for, and make a list of, several pairs of twin primes in which the primes are greater than 3.

**(b)** What do you notice about each number that lies *between* a pair of twin primes?

**(c)** Write an explanation for your observation in part **(b)**.

**33.** Obtain (or imagine that you have) a quantity of square tiles. Six tiles can be arranged in the shape of a rectangle in two different ways:

| | | | | | |
|---|---|---|---|---|---|

| | | |
|---|---|---|
| | | |

**(a)** Record the dimensions of the rectangles shown.

**(b)** If you use 7 tiles, how many different rectangles can you form?

**(c)** If you use 10 tiles, how many different rectangles can you form?

**(d)** What kind of number (of tiles) permits *only one* arrangement into a rectangle? *More than one* arrangement?

**34.** The number 10 has 4 factors: 1, 2, 5, and 10. We can say that 10 has an even number of factors. Investigate several numbers to determine which numbers have an *even number* of factors and which numbers have an *odd number* of factors.

**35.** **Number Problem** Suppose that a school has 1,000 lockers and that they are all closed. A person passes through, opening every other locker, beginning with locker 2. Then another person passes through, changing every third locker (closing it if it is open, opening it if it is closed), starting with locker 3. Yet another person passes through, changing every fourth locker, beginning with locker 4. This process continues until 1,000 people pass through.

**(a)** At the end of this process, which locker numbers are closed?

**(b)** Write an explanation for your answer to part **(a)**. (*Hint:* It may help to attempt exercise 34 first.)

## Answers

**1.** 1, 2, 4 **3.** 1, 2, 5, 10 **5.** 1, 3, 5, 15 **7.** 1, 2, 3, 4, 6, 8, 12, 24
**9.** 1, 2, 4, 8, 16, 32, 64 **11.** 1, 11 **13.** 1, 3, 5, 9, 15, 27, 45, 135
**15.** 1, 2, 4, 8, 16, 32, 64, 128, 256 **17.** 19, 23, 31, 59, 97, 103
**19.** 31, 37, 41, 43, 47 **21.** 72, 158, 260, 378, 570, 4,530, 8,300
**23.** 72, 378, 570, 4,530 **25.** 72, 260, 8,300 **27.** Yes; no
**29.** Above and Beyond **31.** Above and Beyond **33.** Above and Beyond
**35.** Above and Beyond

# 2.2 Factoring Whole Numbers

< 2.2 Objectives >

1 > Find the factors of a whole number

2 > Find the prime factorization for any number

3 > Find the greatest common factor (GCF) of two numbers

4 > Find the GCF for a group of numbers

To **factor a number** means to write the number as a product of its whole-number factors.

## Example 1 Factoring a Composite Number

< Objective 1 >

Factor the number 10.

$10 = 2 \times 5$ The order in which you write the factors does not matter, so $10 = 5 \times 2$ would also be correct.

Of course, $10 = 10 \times 1$ is also a correct statement. However, in this section we are interested in factors other than 1 and the given number.

Factor the number 21.

$21 = 3 \times 7$

### Check Yourself 1

Factor 35.

In writing composite numbers as a product of factors, there are a number of different possible factorizations.

## Example 2 Factoring a Composite Number

**NOTE**

There have to be at least two different factorizations, because a composite number has factors other than 1 and itself.

Find three different factorizations of 72.

$72 = 8 \times 9$ (1)

$= 6 \times 12$ (2)

$= 3 \times 24$ (3)

### Check Yourself 2

Find three different factorizations of 42.

We now want to write composite numbers as a product of their **prime factors.** Look again at the first factored line of Example 2. The process of factoring can be continued until all the factors are prime numbers.

## Example 3 Factoring a Composite Number

< Objective 3 >

**NOTES**

This is often called a **factor tree.**

Finding the prime factorization of a number will be important in our later work in adding fractions.

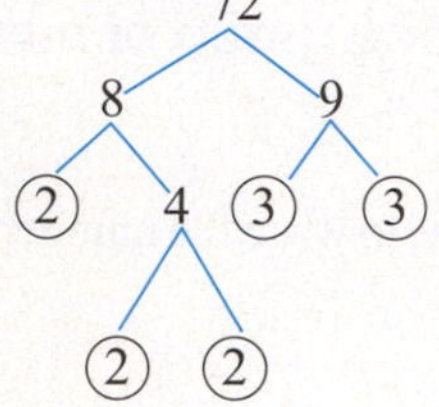


4 is still not prime, and so we continue by factoring 4.

72 is now written as a product of prime factors.

When we write 72 as $2 \times 2 \times 2 \times 3 \times 3$, no further factorization is possible. This is called the *prime factorization* of 72.

Now, what if we start with a different factored line from the same example, $72 = 6 \times 12$?

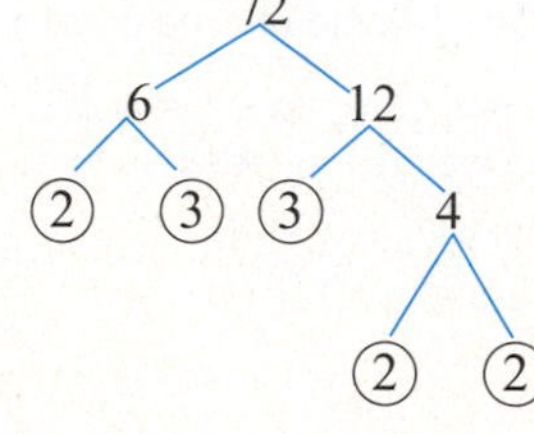


Continue to factor 6 and 12.

Continue again to factor 4. Other choices for the factors of 12 are possible. As we shall see, the end result will be the same.

No matter which pair of factors you start with, you will find the same prime factorization. In this case, there are three factors of 2 and two factors of 3. Because **multiplication is commutative,** the order in which we write the factors does not matter.

### Check Yourself 3

We could also begin

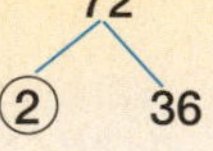


Continue the factorization.

**Property**

**The Fundamental Theorem of Arithmetic** There is exactly one prime factorization for any composite number.

The method of Example 3 always works. However, another method for factoring composite numbers exists. This method is particularly useful when numbers get large, in which case a factor tree becomes unwieldy.

Property

**Factoring by Division**

To find the prime factorization of a number, divide the number by a series of primes until the final quotient is a prime number.

The prime factorization is then the product of all the prime divisors and the final quotient, as we see in Example 4.

## Example 4 Finding the Prime Factorization

< Objective 2 >

**NOTE**

Do you see how the divisibility tests are used here? 60 is divisible by 2, 30 is divisible by 2, and 15 is divisible by 3.

To write 60 as a product of prime factors, divide 2 into 60 for a quotient of 30. Continue to divide by 2 again for the quotient of 15. Because 2 won't divide evenly into 15, we try 3. Because the quotient 5 is prime, we are done.

$2\overline{)60}$ = 30 → $2\overline{)30}$ = 15 → $3\overline{)15}$ = 5 Prime

Our factors are the prime divisors and the final quotient. We have

$60 = 2 \times 2 \times 3 \times 5$

### Check Yourself 4

Complete the process to find the prime factorization of 90.

$2\overline{)90}$ = 45 → $?\overline{)45}$ = ?

Remember to continue until the final quotient is prime.

Writing composite numbers in their completely factored form can be simplified if we use a format called **continued division.**

## Example 5 Finding Prime Factors Using Continued Division

**NOTE**

In each short division, we write the quotient *below* rather than above the dividend. This is just a convenience for the next division.

Use the continued-division method to divide 60 by a series of prime numbers.

Primes →
$2\overline{)60}$
$2\overline{)30}$
$3\overline{)15}$
5 Stop when the final quotient is prime.

To write the factorization of 60, we list each divisor used and the final prime quotient. In our example, we have

$60 = 2 \times 2 \times 3 \times 5$

**Check Yourself 5**

Find the prime factorization of 234.

**NOTE**

The factors of 20, other than 20 itself, are less than 20.

We know that a factor or a divisor of a whole number divides that number exactly. The factors or divisors of 20 are

1, 2, 4, 5, 10, 20

Each of these numbers divides 20 exactly; that is, with no remainder.

Our work in this section involves common factors or divisors. A **common factor** or **divisor** for two numbers is any factor that divides both numbers exactly.

## Example 6 Finding Common Factors

< Objective 3 >

Look at the numbers 20 and 30. Is there a common factor for the two numbers?

First, we list the factors. Then we circle the ones that appear in both lists.

Factors

20: (1), (2), 4, (5), (10), 20

30: (1), (2), 3, (5), 6, (10), 15, 30

We see that 1, 2, 5, and 10 are common factors of 20 and 30. Each of these numbers divides both 20 and 30 exactly.

Our later work with fractions will require that we find the greatest common factor of a group of numbers.

**Definition**

**Greatest Common Factor**

The **greatest common factor** (GCF) of a group of numbers is the *largest* number that divides each of the given numbers exactly.

## Example 6 (Continued) Finding the Greatest Common Factor

In the first part of Example 6, the common factors of the numbers 20 and 30 were listed as

1, 2, 5, 10 Common factors of 20 and 30

The greatest common factor of the two numbers is then 10, because 10 is the *largest* of the four common factors.

**Check Yourself 6**

List the factors of 30 and 36 and then find the greatest common factor.

The method of Example 6 also works in finding the greatest common factor of a group of more than two numbers.

## Example 7 Finding the Greatest Common Factor by Listing Factors

< Objective 4 >

**NOTE**

Looking at the three lists, we see that 1, 2, 3, and 6 are common factors.

Find the GCF of 24, 30, and 36. We list the factors of each of the three numbers.

24: (1), (2), (3), 4, (6), 8, 12, 24

30: (1), (2), (3), 5, (6), 10, 15, 30

36: (1), (2), (3), 4, (6), 9, 12, 18, 36

So 6 is the greatest common factor of 24, 30, and 36.

**Check Yourself 7**

Find the greatest common factor of 16, 24, and 32.

**NOTE**

If there are no common prime factors, the GCF is 1.

The process shown in Example 7 is very time-consuming when larger numbers are involved. A better approach to the problem of finding the GCF of a group of numbers uses the prime factorization of each number. We outline the process here.

**Step by Step**

**Finding the Greatest Common Factor**

**Step 1** Write the prime factorization for each of the numbers in the group.
**Step 2** Locate the prime factors that are *common* to all the numbers.
**Step 3** The greatest common factor (GCF) is the *product* of all the common prime factors.

## Example 8 Finding the Greatest Common Factor

Find the GCF of 20 and 30.

**Step 1** Write the prime factorization of 20 and 30.

$20 = 2 \times 2 \times 5$

$30 = 2 \times 3 \times 5$

**Step 2** Find the prime factors common to each number.

$20 = (2) \times 2 \times (5)$

$30 = (2) \times 3 \times (5)$

2 and 5 are the common prime factors.

**Step 3** Form the product of the common prime factors.

$2 \times 5 = 10$

So 10 is the greatest common factor.

### Check Yourself 8

**Find the GCF of 30 and 36.**

To find the greatest common factor of a group of more than two numbers, we use the same process.

## Example 9 Finding the Greatest Common Factor

Find the GCF of 24, 30, and 36.

$24 = \textcircled{2} \times 2 \times 2 \times \textcircled{3}$

$30 = \textcircled{2} \times \textcircled{3} \times 5$

$36 = \textcircled{2} \times 2 \times \textcircled{3} \times 3$

So 2 and 3 are the prime factors common to *all three numbers.*

And $2 \times 3 = 6$ is the GCF.

### Check Yourself 9

**Find the GCF of 15, 30, and 45.**

## Example 10 Finding the Greatest Common Factor

Find the greatest common factor of 15 and 28.

$15 = 3 \times 5$

$28 = 2 \times 2 \times 7$

There are no common prime factors listed. But remember that 1 is a factor of every whole number.

The greatest common factor of 15 and 28 is 1.

**NOTE**

If two numbers, such as 15 and 28, have no common factor other than 1, they are called **relatively prime**.

### Check Yourself 10

**Find the greatest common factor of 30 and 49.**

## Check Yourself ANSWERS

**1.** $5 \times 7$ **2.** $2 \times 21, 3 \times 14, 6 \times 7$ **3.** $2 \times 2 \times 2 \times 3 \times 3$

**4.** $2\overline{)90}$ = 45 → $3\overline{)45}$ = 15 → $3\overline{)15}$ = 5 **5.** $2 \times 3 \times 3 \times 13$

$90 = 2 \times 3 \times 3 \times 5$

**6.** 30: ⓵, ⓶, ⓷, 5, ⓺, 10, 15, 30

36: ⓵, ⓶, ⓷, 4, ⓺, 9, 12, 18, 36

6 is the greatest common factor.

**7.** 16: ⓵, ⓶, ⓸, ⓼, 16

24: ⓵, ⓶, 3, ⓸, 6, ⓼, 12, 24

32: ⓵, ⓶, ⓸, ⓼, 16, 32

The GCF is 8.

**8.** $30 = (2) \times (3) \times 5$

$36 = (2) \times 2 \times (3) \times 3$

The GCF is $2 \times 3 = 6$.

**9.** $15 = (3) \times (5)$

$30 = 2 \times (3) \times (5)$

$45 = (3) \times 3 \times (5)$

The GCF is 15.

**10.** GCF is 1; 30 and 49 are relatively prime

## Reading Your Text

The following fill-in-the-blank exercises are designed to ensure that you understand some of the key vocabulary used in this section.

### SECTION 2.2

**(a)** Because multiplication is ____________, the order in which we write factors does not matter.

**(b)** There is exactly one ____________ factorization for any whole number.

**(c)** A ____________ factor for two numbers is any factor that divides both numbers exactly.

**(d)** GCF is an abbreviation for ____________ common factor.

# 2.2 exercises

Basic Skills | Advanced Skills | Vocational-Technical Applications | Calculator/Computer | Above and Beyond

MathZone

Boost your grade at mathzone.com!
- > Practice Problems
- > NetTutor
- > Self-Tests
- > e-Professors
- > Videos

Name ____________

Section ________ Date ________

## Answers

1. ______ 2. ______
3. ______ 4. ______
5. ______ 6. ______
7. ______
8. ______
9. ______
10. ______
11. ______
12. ______
13. ______
14. ______
15. ______
16. ______
17. ______ 18. ______
19. ______ 20. ______
21. ______ 22. ______
23. ______ 24. ______

< Objectives 1 and 2 >

*Find the prime factorization of each number.*

**1.** 18 **2.** 22

**3.** 30 **4.** 35

**5.** 51 **6.** 42

**7.** 66 **8.** 100

**9.** 130 **10.** 88

**11.** 315 **12.** 400

**13.** 225 **14.** 132

**15.** 189 

**16.** 330

*In later mathematics courses, you often will want to find factors of a number with a given sum or difference. The following exercises use this technique.*

**17.** Find two factors of 48 with a sum of 14.

**18.** Find two factors of 48 with a sum of 26.

**19.** Find two factors of 48 with a difference of 8.

**20.** Find two factors of 48 with a difference of 2.

**21.** Find two factors of 24 with a sum of 10. 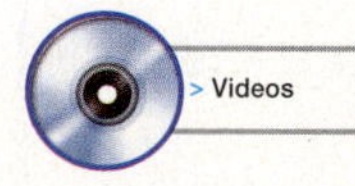


**22.** Find two factors of 15 with a difference of 2.

**23.** Find two factors of 30 with a difference of 1.

**24.** Find two factors of 28 with a sum of 11.

< Objective 3 >

*Find the greatest common factor for each of the following groups of numbers.*

**25.** 4 and 6 

**26.** 6 and 9

**27.** 10 and 15

**28.** 12 and 14

**29.** 21 and 24

**30.** 22 and 33

**31.** 20 and 21

**32.** 28 and 42

**33.** 18 and 24

**34.** 35 and 36

**35.** 18 and 54 

**36.** 12 and 48

Basic Skills | **Advanced Skills** | Vocational-Technical Applications | Calculator/Computer | Above and Beyond

< Objective 4 >

*Find the GCF for each of the following groups of numbers.*

**37.** 12, 36, and 60

**38.** 15, 45, and 90

**39.** 105, 140, and 175

**40.** 17, 19, and 31

**41.** 25, 75, and 150 

**42.** 36, 72, and 144

*For exercises 43 to 46 fill in each blank with either* **always, sometimes,** *or* **never.**

**43.** Factors of a composite number __________ include 1 and the number itself.

**44.** Factors of a prime number __________ include 1 and the number itself.

**45.** A number with a repeated factor is ____________ a prime number.

**46.** Factors of an even number are ________________ even numbers.

## Answers

25. ______
26. ______
27. ______
28. ______
29. ______
30. ______
31. ______
32. ______
33. ______
34. ______
35. ______
36. ______
37. ______
38. ______
39. ______
40. ______
41. ______
42. ______
43. ______
44. ______
45. ______
46. ______

Basic Skills | Advanced Skills | Vocational-Technical Applications | Calculator/Computer | Above and Beyond

## Answers

47. ______

48. ______

49. ______

50. ______

**47.** A natural number is said to be *perfect* if it is equal to the sum of its factors, except itself.

**(a)** Show that 28 is a perfect number.

**(b)** Identify another perfect number less than 28.

**48.** Find the smallest natural number that is divisible by all the following: 2, 3, 4, 6, 8, 9.

**49.** **Social Science** Tom and Dick both work the night shift at the steel mill. Tom has every sixth night off, and Dick has every eighth night off. If they both have August 1 off, when will they both be off together again?

**50.** **Science and Medicine** Mercury, Venus, and Earth revolve around the sun once every 3, 7, and 12 months, respectively. If the three planets are now in the same straight line, what is the smallest number of months that must pass before they line up again?

## Answers

**1.** $2 \times 3 \times 3$ **3.** $2 \times 3 \times 5$ **5.** $3 \times 17$ **7.** $2 \times 3 \times 11$
**9.** $2 \times 5 \times 13$ **11.** $3 \times 3 \times 5 \times 7$ **13.** $3 \times 3 \times 5 \times 5$
**15.** $3 \times 3 \times 3 \times 7$ **17.** 6, 8 **19.** 4, 12 **21.** 4, 6 **23.** 5, 6
**25.** 2 **27.** 5 **29.** 3 **31.** 1 **33.** 6 **35.** 18 **37.** 12
**39.** 35 **41.** 25 **43.** always **45.** never
**47.** **(a)** $1 + 2 + 4 + 7 + 14 = 28$; **(b)** 6 **49.** August 25

# 2.3 Fraction Basics

< **2.3 Objectives** >

1 > Identify the numerator and denominator of a fraction

2 > Use fractions to name parts of a whole

3 > Identify proper and improper fractions

4 > Write improper fractions as mixed numbers

5 > Write mixed numbers as improper fractions

Previous sections dealt with whole numbers and the operations that are performed on them. We are now ready to consider a new kind of number, a **fraction.**

**Definition**

**Fraction** Whenever a unit or a whole quantity is divided into parts, we call those parts **fractions** of the unit.

> **NOTE**
>
> Our word *fraction* comes from the Latin stem *fractio,* which means "breaking into pieces."

In Figure 1, the whole has been divided into five equal parts. We use the symbol $\frac{2}{5}$ to represent the shaded portion of the whole.

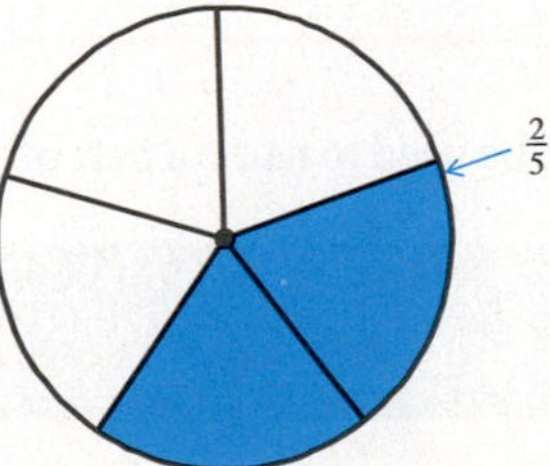


Figure 1

> **NOTE**
>
> *Common fraction* is technically the correct term. We just use *fraction* in these materials if there is no room for confusion.

The symbol $\frac{2}{5}$ is called a **common fraction,** or more simply a fraction. A fraction is written in the form $\frac{a}{b}$, in which $a$ and $b$ represent whole numbers and $b$ cannot be equal to 0.

We give the numbers $a$ and $b$ special names. The **denominator,** $b$, is the number on the bottom. This tells us into how many equal parts the unit or whole has been divided. The **numerator,** $a$, is the number on the top. This tells us how many parts of the unit are used.

In the fraction $\frac{2}{5}$, the *denominator* is 5; the unit or whole (the circle) has been divided into five equal parts. The *numerator* is 2. We have taken two parts of the unit.

$\frac{2}{5}$ ⟵ Numerator / ⟵ Denominator

## Example 1 Labeling Fraction Components

< Objectives 1, 2 >

The fraction $\frac{4}{7}$ names the shaded part of the rectangle in Figure 2.

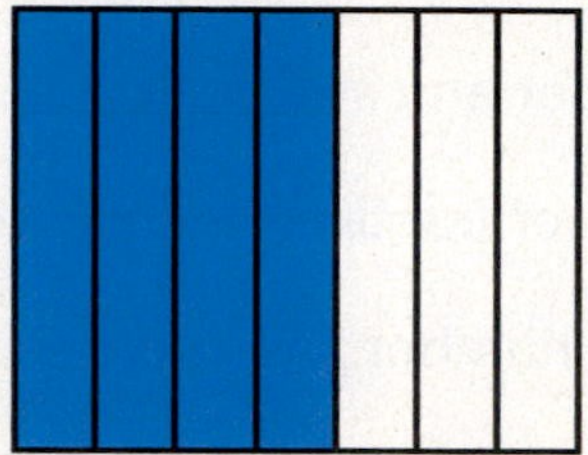

Figure 2

The unit or whole is divided into seven equal parts, so the denominator is 7. We have shaded four of those parts, and so we have a numerator of 4.

### Check Yourself 1

**What fraction names the shaded part of this diagram? Identify the numerator and denominator.**

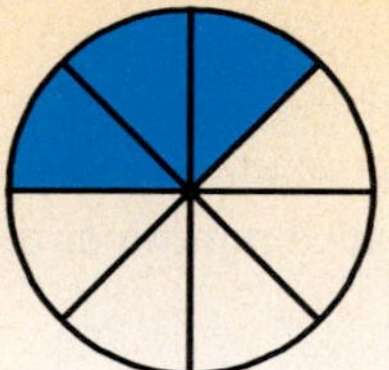

Fractions can also be used to name a part of a collection or a set of identical objects.

## Example 2 Naming a Fractional Part

The fraction $\frac{5}{6}$ names the shaded part of Figure 3. We have shaded five of the six identical objects.

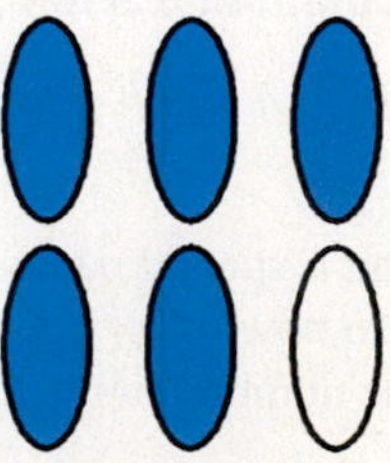

Figure 3

## Check Yourself 2

What fraction names the shaded part of this diagram?

## Example 3 Naming a Fractional Part

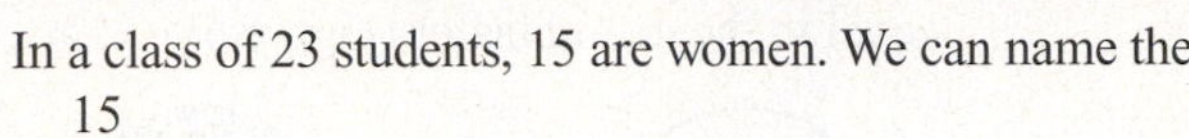

In a class of 23 students, 15 are women. We can name the part of the class that is women as $\frac{15}{23}$.

**NOTES**

The fraction $\frac{8}{23}$ names the part of the class that is not women.

$\frac{a}{b}$ names the *quotient* when $a$ is divided by $b$. Of course, $b$ cannot be 0.

## Check Yourself 3

**Seven replacement parts out of a shipment of 50 were faulty. What fraction names the portion of the shipment that was faulty?**

A fraction can also be thought of as indicating division. The symbol $\frac{a}{b}$ also means $a \div b$.

## Example 4 Interpreting Division as a Fraction

The fraction $\frac{2}{3}$ names the quotient when 2 is divided by 3. So $\frac{2}{3} = 2 \div 3$.

**Note:** $\frac{2}{3}$ can be read as "two-thirds" or as "2 divided by 3."

## Check Yourself 4

**Write $\frac{5}{9}$ using division.**

We can use the relative size of the numerator and denominator of a fraction to separate fractions into two different categories.

**Definition**

**Proper Fraction**

If the numerator is *less than* the denominator, the fraction names a number less than 1 and is called a **proper fraction.**

**Definition**

**Improper Fraction**

If the numerator is *greater than or equal to* the denominator, the fraction names a number greater than or equal to 1 and is called an **improper fraction.**

## Example 5 Categorizing Fractions

< Objective 3 >

**(a)** $\frac{2}{3}$ is a proper fraction because the numerator is less than the denominator (Figure 4).

**(b)** $\frac{4}{3}$ is an improper fraction because the numerator is larger than the denominator (Figure 5).

**(c)** Also, $\frac{6}{6}$ is an improper fraction because it names exactly 1 unit; the numerator is equal to the denominator (Figure 6).

**NOTE**

In Figure 5, the circle on the left is divided into 3 parts, so it represents $\frac{3}{3}$.

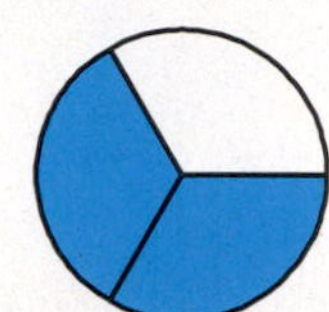

$\frac{2}{3}$ names less than 1 unit and $2 < 3$.

Numerator Denominator

Figure 4

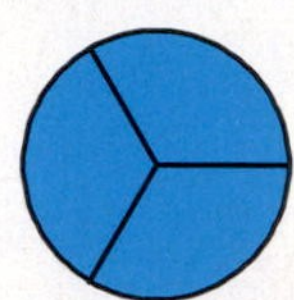

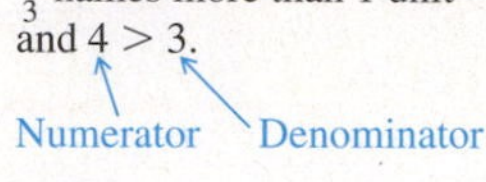

$\frac{4}{3}$ names more than 1 unit and $4 > 3$.

Numerator Denominator

Figure 5

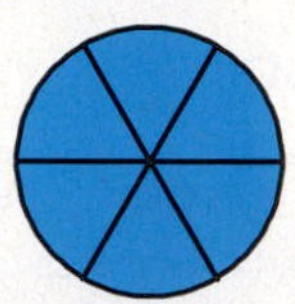

$\frac{6}{6} = 1$

Figure 6

### Check Yourself 5

List the proper fractions and the improper fractions in the following list:

$$\frac{5}{4}, \frac{10}{11}, \frac{3}{4}, \frac{8}{5}, \frac{6}{6}, \frac{13}{10}, \frac{7}{8}, \frac{15}{8}$$

Another way to write a fraction that is larger than 1 is as a **mixed number.**

**Definition**

**Mixed Number**

A **mixed number** is the sum of a whole number and a proper fraction.

## Example 6 Identifying a Mixed Number

**NOTE**

$2\frac{3}{4}$ means $2 + \frac{3}{4}$. In fact, we read the mixed number as "two *and* three-fourths." The plus sign is usually not written.

The number $2\frac{3}{4}$ is a mixed number. It represents the sum of the whole number 2 and the fraction $\frac{3}{4}$. Look at the following diagram, which represents $2\frac{3}{4}$.

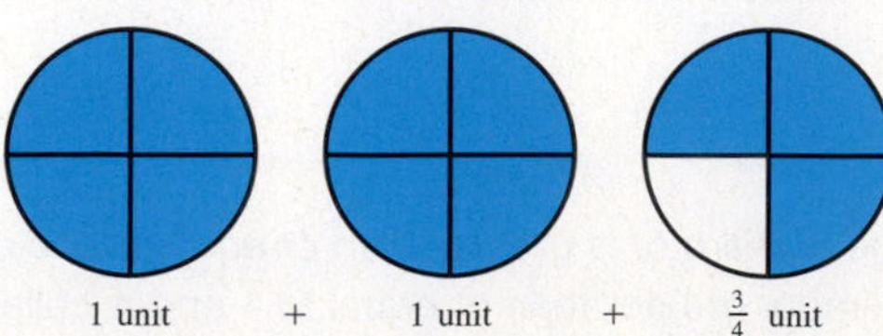

### Check Yourself 6

Give the mixed number that names the shaded portion of the given diagram.

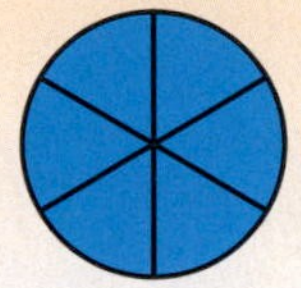 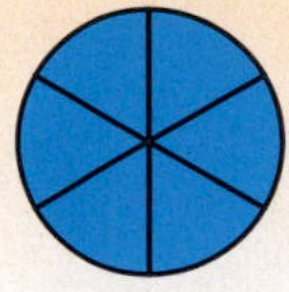 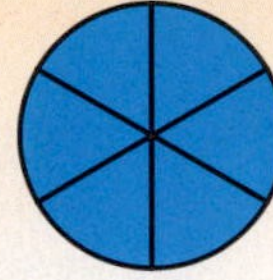 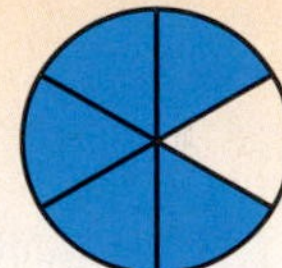

**NOTE**

In subsequent courses, you will find that improper fractions are preferred to mixed numbers.

For our later work it will be important to be able to change back and forth between improper fractions and mixed numbers. Because an improper fraction represents a number that is greater than or equal to 1, we have the following property:

**Property**

**Improper Fractions to Mixed Numbers**

An improper fraction can always be written as either a mixed number or a whole number.

To do this, remember that you can think of a fraction as indicating division. The numerator is divided by the denominator. This leads us to the following process:

**Step by Step**

**To Change an Improper Fraction to a Mixed Number**

**Step 1** Divide the numerator by the denominator.

**Step 2** If there is a remainder, write the remainder over the original denominator.

### Example 7 Converting a Fraction to a Mixed Number

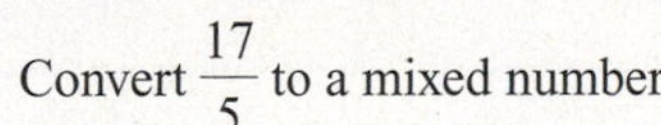

< Objective 4 >

Convert $\frac{17}{5}$ to a mixed number.

Divide 17 by 5.

$$5\overline{)17}\ \ 3 \qquad 15 \qquad 2$$

$$\frac{17}{5} = 3\frac{2}{5}$$

Remainder (2); Original denominator (5); Quotient (3)

**NOTES**

You can write the fraction $\frac{17}{5}$ as 17 ÷ 5. We divide the numerator by the denominator.

In step 1, the quotient gives the whole-number portion of the mixed number. Step 2 gives the fractional portion of the mixed number.

In diagram form:

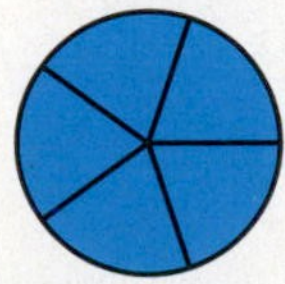 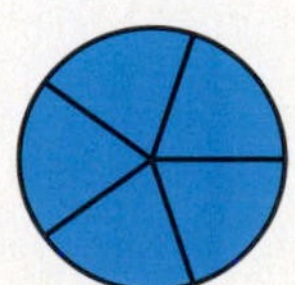 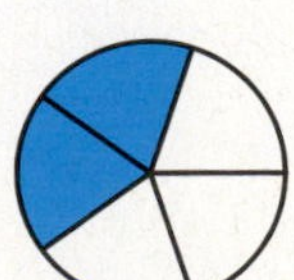

$$\frac{17}{5} = 3\frac{2}{5}$$

### Check Yourself 7

Convert $\frac{32}{5}$ to a mixed number.

## Example 8 Converting a Fraction to a Mixed Number

**NOTE**

If there is *no* remainder, the improper fraction is equal to some whole number, in this case 3.

Convert $\frac{21}{7}$ to a mixed or a whole number.

Divide 21 by 7.

$$7\overline{)21}\quad \text{quotient } 3,\quad \frac{21}{0} \qquad \frac{21}{7} = 3$$

### Check Yourself 8

Convert $\frac{48}{6}$ to a mixed or a whole number.

It is also easy to convert mixed numbers to improper fractions. Just use the following rule:

**Step by Step**

**To Change a Mixed Number to an Improper Fraction**

**Step 1** Multiply the denominator of the fraction by the whole-number portion of the mixed number.

**Step 2** Add the numerator of the fraction to that product.

**Step 3** Write that sum over the original denominator to form the improper fraction.

## Example 9 Converting Mixed Numbers to Improper Fractions

< Objective 5 >

**(a)** Convert $3\frac{2}{5}$ to an improper fraction.

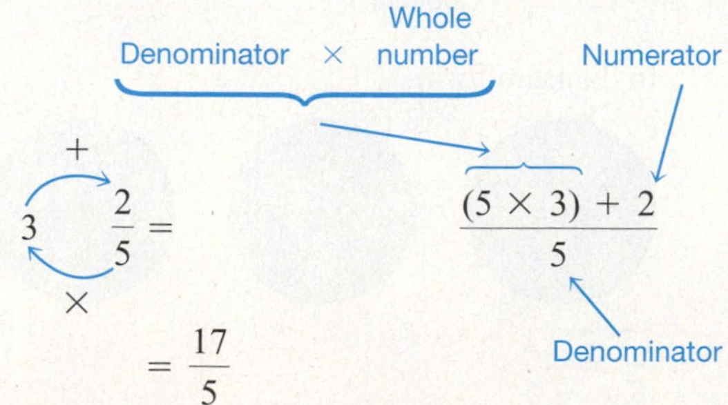


$$3\frac{2}{5} = \frac{(5 \times 3) + 2}{5}$$

$$= \frac{17}{5}$$

Multiply the denominator by the whole number ($5 \times 3 = 15$). Add the numerator. We now have 17.

Write 17 over the original denominator.

In diagram form:

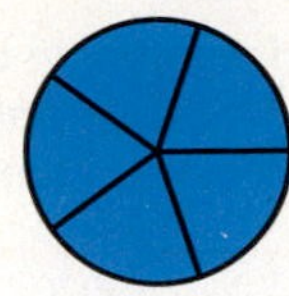 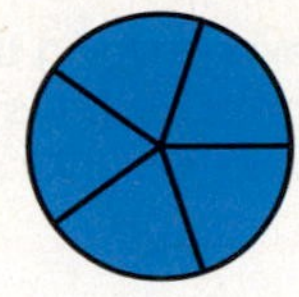 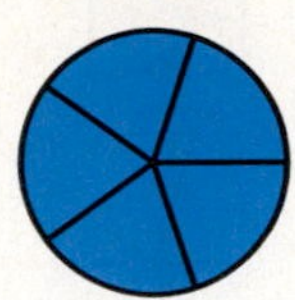 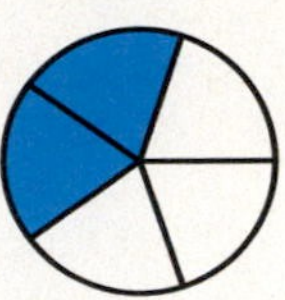

Each of the three units has 5 fifths, so the whole-number portion is $5 \times 3$, or 15, fifths. Then add the $\frac{2}{5}$ from the fractional portion for $\frac{17}{5}$.

**NOTE**

Multiply the denominator, 7, by the whole number, 4, and add the numerator, 5.

**(b)** Convert $4\frac{5}{7}$ to an improper fraction.

$$4\frac{5}{7} = \frac{(7 \times 4) + 5}{7} = \frac{33}{7}$$

**Check Yourself 9**

Convert $5\frac{3}{8}$ to an improper fraction.

**NOTE**

The definition below is the same as our earlier rule for dividing by 1.

One special kind of improper fraction should be mentioned at this point: a fraction with a denominator of 1.

**Definition**

**Fractions with a Denominator of 1**

Any fraction with a denominator of 1 is equal to the numerator alone. For example,

$$\frac{5}{1} = 5 \quad \text{and} \quad \frac{12}{1} = 12$$

You probably do many conversions between mixed and whole numbers without even thinking about the process that you follow, as Example 10 illustrates.

### Example 10 Converting Quarter-Dollars to Dollars

Maritza has 53 quarters in her bank. How many dollars does she have?

Because there are 4 quarters in each dollar, 53 quarters can be written as

$$\frac{53}{4}$$

Converting the amount to dollars is the same as rewriting it as a mixed number.

$$\frac{53}{4} = 13\frac{1}{4}$$

She has $13\frac{1}{4}$ dollars, which you would probably write as \$13.25. (*Note:* We will discuss decimal point usage later in this text.)

### Check Yourself 10

**Kevin is doing the inventory in the convenience store in which he works. He finds there are 11 half-gallons (gal) of milk. Write the amount of milk as a mixed number of gallons.**

### Check Yourself ANSWERS

**1.** $\frac{3}{8}$ ← Numerator, ← Denominator

**2.** $\frac{2}{7}$ **3.** $\frac{7}{50}$ **4.** $5 \div 9$

**5.** Proper fractions: $\frac{10}{11}, \frac{3}{4}, \frac{7}{8}$ Improper fractions: $\frac{5}{4}, \frac{8}{5}, \frac{6}{6}, \frac{13}{10}, \frac{15}{8}$ **6.** $3\frac{5}{6}$ **7.** $6\frac{2}{5}$ **8.** 8

**9.** $\frac{43}{8}$ **10.** $5\frac{1}{2}$ gal

### Reading Your Text

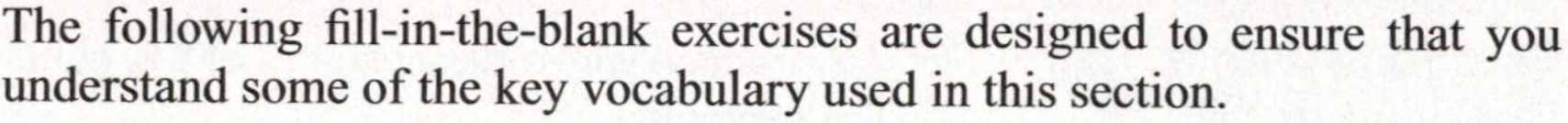

The following fill-in-the-blank exercises are designed to ensure that you understand some of the key vocabulary used in this section.

**SECTION 2.3**

**(a)** Given a fraction like $\frac{3}{4}$, we call 4 the ______________ of the fraction.

**(b)** If the numerator is less than the denominator, the fraction names a number less than 1 and is called a ______________ fraction.

**(c)** An improper fraction can always be written as either a ______________ number or a whole number.

**(d)** Any fraction with a denominator of 1 is equal to the ______________ alone.

Basic Skills | Advanced Skills | Vocational-Technical Applications | Calculator/Computer | Above and Beyond

# 2.3 exercises

< Objective 1 >

*Identify the numerator and denominator of each fraction.*

1. $\frac{6}{11}$

2. $\frac{5}{12}$

3. $\frac{3}{11}$

4. $\frac{9}{14}$

< Objective 2 >

*What fraction names the shaded part of each of the following figures?*

5. 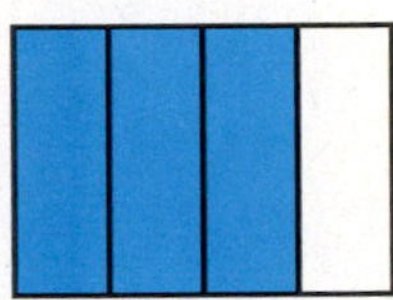

6. 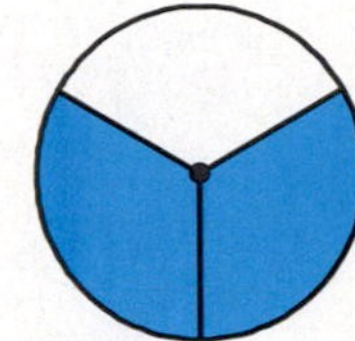

7. 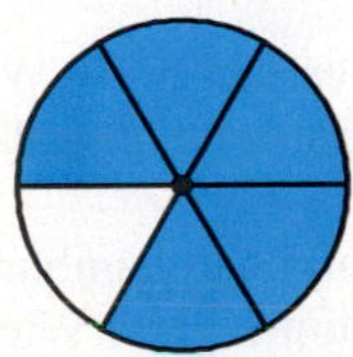

8. 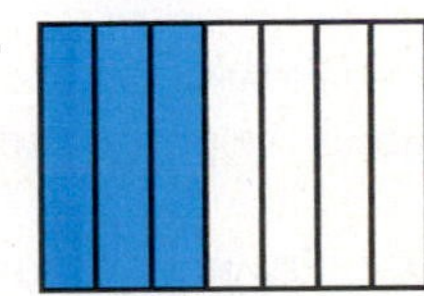

9. 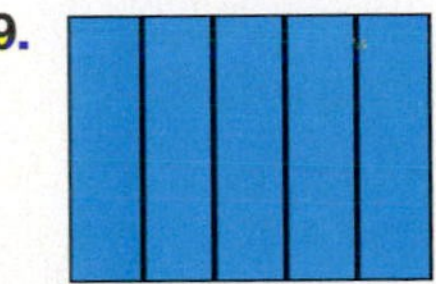

10. 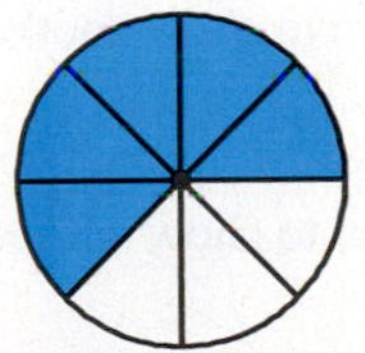

11. 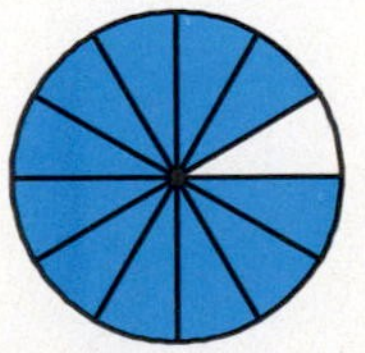

12. 

13. 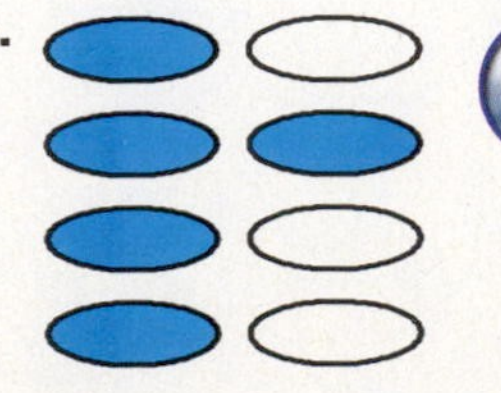

> Videos

14. 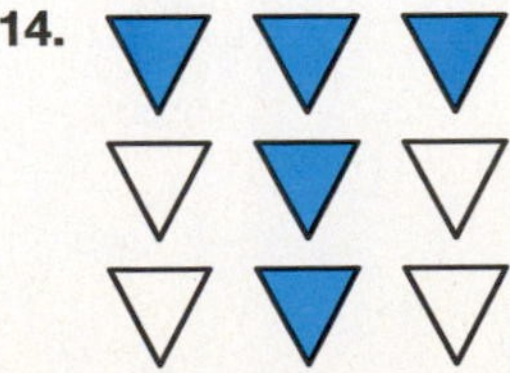



Name ____________

Section ________ Date ________

## Answers

1. ____________
2. ____________
3. ____________
4. ____________
5. ____________
6. ____________
7. ____________
8. ____________
9. ____________
10. ____________
11. ____________
12. ____________
13. ____________
14. ____________

Answers

15. ______

16. ______

17. ______

18. ______

19. ______

20. ______

21. ______

22. ______

23. ______

24. ______

25. ______

26. ______

27. ______

28. ______

*Solve the following applications.*

**15.** **Statistics** You missed 7 questions on a 20-question test. What fraction names the part you got correct? The part you got wrong?

> Videos

**16.** **Statistics** Of the 5 starters on a basketball team, 2 fouled out of a game. What fraction names the part of the starting team that fouled out?

**17.** **Business and Finance** A used-car dealer sold 11 of the 17 cars in stock. What fraction names the portion sold? What fraction names the portion *not* sold?

**18.** **Business and Finance** At lunch, 5 people out of a group of 9 had hamburgers. What fraction names the part of the group who had hamburgers? What fraction names the part who did *not* have hamburgers?

**19.** Use division to show another way of writing $\frac{2}{5}$.

**20.** Use division to show another way of writing $\frac{4}{5}$.

< Objective 3 >

*Identify each number as a proper fraction, an improper fraction, or a mixed number.*

**21.** $\frac{3}{5}$

**22.** $\frac{9}{5}$

**23.** $2\frac{3}{5}$

**24.** $\frac{7}{9}$

**25.** $\frac{6}{6}$

**26.** $1\frac{1}{5}$

**27.** $\frac{13}{17}$

**28.** $\frac{16}{15}$

*Give the mixed number that names the shaded portion of each diagram. Also write each as an improper fraction.*

**29.** 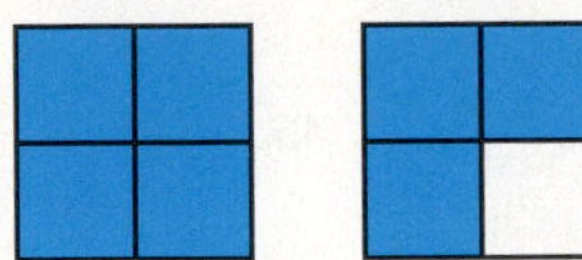

**30.** 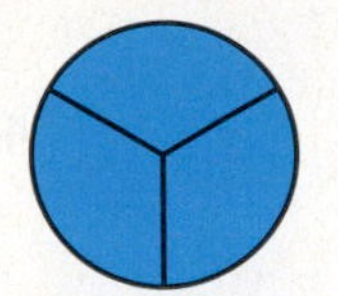 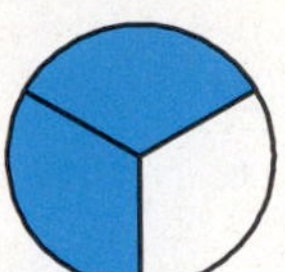

**31.** 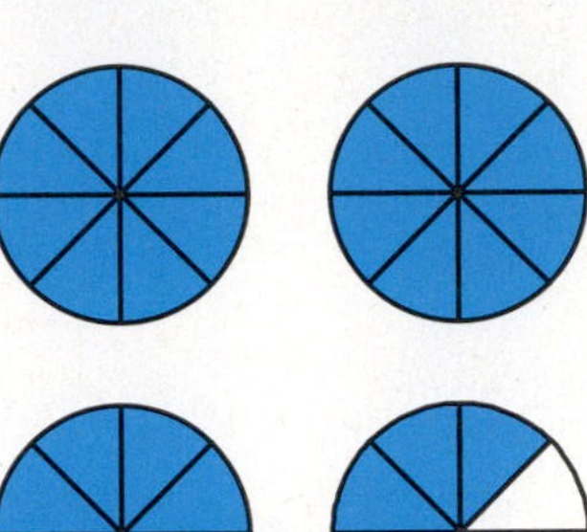

**32.**  

*Solve the following applications.*

**33.** **Business and Finance** Clayton has 64 quarters in his bank. How many dollars does he have?

**34.** **Business and Finance** Amy has 19 quarters in her purse. How many dollars does she have?

**35.** **Business and Finance** Manuel counted 35 half-gallons of orange juice in his store. Write the amount of orange juice as a mixed number of gallons.

**36.** **Business and Finance** Sarah has 19 half-gallons of turpentine in her paint store. Write the amount of turpentine as a mixed number of gallons.

< Objective 4 >

*Change to a mixed or whole number.*

**37.** $\frac{22}{5}$ 

**38.** $\frac{27}{8}$

**39.** $\frac{34}{5}$

## Answers

29. ______

30. ______

31. ______

32. ______

33. ______

34. ______

35. ______

36. ______

37. ______

38. ______

39. ______

**Answers**

40. ______
41. ______
42. ______
43. ______ 44. ______
45. ______ 46. ______
47. ______
48. ______
49. ______
50. ______
51. ______
52. ______
53. ______
54. ______
55. ______
56. ______
57. ______
58. ______
59. ______
60. ______

**40.** $\frac{25}{6}$ **41.** $\frac{73}{8}$ **42.** $\frac{151}{12}$

**43.** $\frac{24}{6}$ **44.** $\frac{160}{8}$ **45.** $\frac{9}{1}$

**46.** $\frac{8}{1}$

< **Objective 5** >

*Change to an improper fraction.*

**47.** $4\frac{2}{3}$ **48.** $2\frac{5}{6}$ **49.** 8

**50.** $4\frac{5}{8}$ **51.** $7\frac{6}{13}$ **52.** $7\frac{3}{10}$

**53.** $10\frac{2}{5}$ **54.** $13\frac{2}{5}$ **55.** $118\frac{3}{4}$

**56.** $250\frac{3}{4}$

Basic Skills | Advanced Skills | **Vocational-Technical Applications** | Calculator/Computer | Above and Beyond

**57. Allied Health** A dilution contains 3 parts blood serum out of a total of 10 parts. Write this number as a fraction.

**58. Electronics** Write all the fractional values described in the following paragraph as fractions.

Betsy ordered electronic components from her favorite supplier. She bought two dozen, one-quarter watt resistors, 10 light-emitting diodes (LEDs) that require ten-thirds of a volt (forward voltage) to illuminate, and one, three-eighths henry inductor.

**59. Manufacturing Technology** In the packaging division of Early Enterprises, there are 36 packaging machines. At any given time, five of the machines are shut down for scheduled maintenance and service. What is the fraction of machines that are operating at one time?

**60. Manufacturing Technology** A run of parts requires five machines. The line used to run the parts has nine machines. What is the fraction of machines in the line to the ones needed to run the parts? Express this as both an improper fraction and a mixed number.

61. **Information Technology** On a visit to a wiring closet, Joseph finds a rack of servers that has eight slots. He wants to know what fraction names the two already in the slots. Also, if he buys five more servers, what fraction names the total servers installed in the slots? If he has to remove two servers because of failure, what fraction names the total servers installed in the slots?

62. **Information Technology** Paul needs to make 3-foot (ft) cables for a wiring closet project from a 1,000-ft roll. What fraction names the cable of 3-ft pieces? If Paul makes 10 cables, what fraction names the total length of cables made from the 1,000-ft roll?

Basic Skills | Advanced Skills | Vocational-Technical Applications | Calculator/Computer | **Above and Beyond**

63. **Social Science** The U.S. Census information can be found in your library, or on the Web, at www.census.gov. Use the 2000 census to determine the following:

(a) Fraction of the population of the United States contained in your state

(b) Fraction of the population of the United States 65 years of age or older

(c) Fraction of the United States that is female

64. **Social Science** Suppose the national debt had to be paid by individuals.

(a) How would the amount each individual owed be determined?

(b) Would this be a proper or an improper fraction?

## Answers

61. ______

62. ______

63. ______

64. ______

## Answers

1. 6 is the numerator; 11 is the denominator
3. 3 is the numerator; 11 is the denominator   5. $\frac{3}{4}$   7. $\frac{5}{6}$
9. $\frac{5}{5}$   11. $\frac{11}{12}$   13. $\frac{5}{8}$   15. Correct: $\frac{13}{20}$; wrong: $\frac{7}{20}$
17. Sold: $\frac{11}{17}$; not sold: $\frac{6}{17}$   19. $2 \div 5$   21. Proper
23. Mixed number   25. Improper   27. Proper
29. $1\frac{3}{4}$ or $\frac{7}{4}$   31. $3\frac{5}{8}$ or $\frac{29}{8}$   33. \$16   35. $17\frac{1}{2}$ gal   37. $4\frac{2}{5}$
39. $6\frac{4}{5}$   41. $9\frac{1}{8}$   43. 4.   45. 9   47. $\frac{14}{3}$   49. $\frac{8}{1}$   51. $\frac{97}{13}$
53. $10\frac{2}{5} = \frac{(5 \times 10) + 2}{5} = \frac{52}{5}$   55. $\frac{475}{4}$   57. $\frac{3}{10}$   59. $\frac{31}{36}$
61. $\frac{2}{8}, \frac{7}{8}, \frac{5}{8}$   63. Above and Beyond

# 2.4 Simplifying Fractions

< 2.4 Objectives >

1 > Determine whether two fractions are equivalent

2 > Use the fundamental principle to simplify fractions

It is possible to represent the same portion of a whole by different fractions. Look at Figure 1, representing $\frac{3}{6}$ and $\frac{1}{2}$. The two fractions are simply different names for the same number. They are called **equivalent fractions** for this reason.

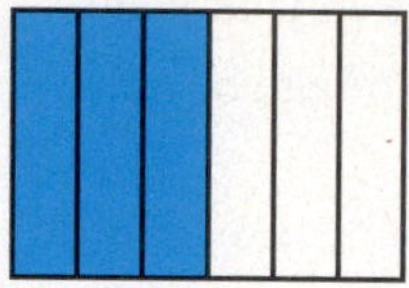
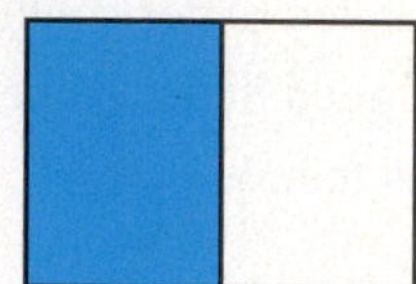

Figure 1

Any fraction has many equivalent fractions. For instance, $\frac{2}{3}$, $\frac{4}{6}$, and $\frac{6}{9}$ are all equivalent fractions because they name the same part of a unit. This is illustrated in Figure 2.

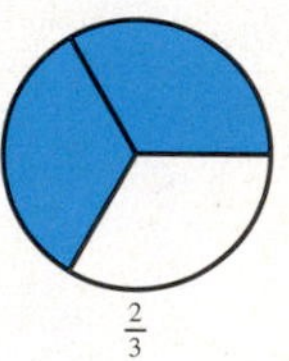
$\frac{2}{3}$

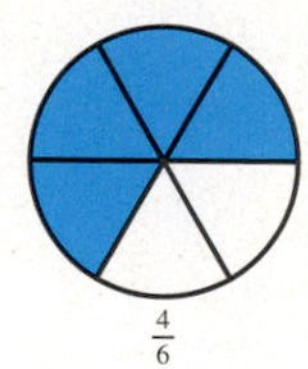
$\frac{4}{6}$

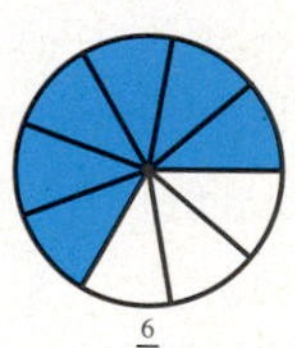
$\frac{6}{9}$

Figure 2

Many more fractions are equivalent to $\frac{2}{3}$. All these fractions can be used interchangeably. An easy way to find out if two fractions are equivalent is to use cross products.

$$\frac{a}{b} \quad \frac{c}{d}$$

We call $a \times d$ and $b \times c$ the cross products.

**Property**

**Testing for Equivalence** If the cross products for two fractions are equal, the two fractions are equivalent.

## Example 1 Identifying Equivalent Fractions Using Cross Products

< Objective 1 >

**(a)** Are $\frac{3}{24}$ and $\frac{4}{32}$ equivalent fractions?

The cross products are $3 \times 32$, or 96, and $24 \times 4$, or 96. Because the cross products are equal, the fractions are equivalent.

**(b)** Are $\frac{2}{5}$ and $\frac{3}{7}$ equivalent fractions?

The cross products are $2 \times 7$ and $5 \times 3$.

$2 \times 7 = 14$ and $5 \times 3 = 15$

Because $14 \neq 15$, the fractions are *not* equivalent.

### Check Yourself 1

**(a)** Are $\frac{3}{8}$ and $\frac{9}{24}$ equivalent fractions?

**(b)** Are $\frac{7}{8}$ and $\frac{8}{9}$ equivalent fractions?

In writing equivalent fractions, we use the following important principle.

**Property**

**The Fundamental Principle of Fractions**

For the fraction $\frac{a}{b}$ and any nonzero number $c$,

$$\frac{a}{b} = \frac{a \div c}{b \div c}$$

We are actually dividing by $\frac{c}{c}$, or 1, and dividing by 1 does not change the value of a number.

The Fundamental Principle of Fractions tells us that we can divide the numerator and denominator by the same nonzero number. The result is an equivalent fraction. For instance,

**NOTES**

Divide the numerator and denominator by 2, 3, and 4.

Divide the numerator and denominator by 5, 6, and 7.

$$\frac{2}{4} = \frac{2 \div 2}{4 \div 2} = \frac{1}{2} \qquad \frac{3}{6} = \frac{3 \div 3}{6 \div 3} = \frac{1}{2} \qquad \frac{4}{8} = \frac{4 \div 4}{8 \div 4} = \frac{1}{2}$$

$$\frac{5}{10} = \frac{5 \div 5}{10 \div 5} = \frac{1}{2} \qquad \frac{6}{12} = \frac{6 \div 6}{12 \div 6} = \frac{1}{2} \qquad \frac{7}{14} = \frac{7 \div 7}{14 \div 7} = \frac{1}{2}$$

**Simplifying a fraction** or *reducing a fraction to lower terms* means finding an equivalent fraction with a *smaller* numerator and denominator than those of the original fraction. Dividing the numerator and denominator by the same nonzero number does exactly that.

Consider Example 2.

## Example 2 Simplifying Fractions

< Objective 2 >

Simplify each fraction.

**NOTES**

We apply the fundamental principle to divide the numerator and denominator by 5.

We divide the numerator and denominator by 2.

**(a)** $\frac{5}{15} = \frac{5 \div 5}{15 \div 5} = \frac{1}{3}$

$\frac{5}{15}$ and $\frac{1}{3}$ are equivalent fractions.

Check this by finding the cross products.

**(b)** $\frac{4}{8} = \frac{4 \div 2}{8 \div 2} = \frac{2}{4}$

$\frac{4}{8}$ and $\frac{2}{4}$ are equivalent fractions.

### Check Yourself 2

Write two fractions that are equivalent to $\frac{30}{45}$.

**(a)** Divide the numerator and denominator by 5.
**(b)** Divide the numerator and denominator by 15.

We say that a fraction is in **simplest form,** or in **lowest terms,** if the numerator and denominator have no common factors other than 1. This means that the fraction has the smallest possible numerator and denominator.

**NOTE**

In this case, the numerator and denominator are *not* as small as possible. The numerator and denominator have a common factor of 2.

In Example 2, $\frac{1}{3}$ is in simplest form because the numerator and denominator have no common factors other than 1. The fraction is in lowest terms.

$\frac{2}{4}$ is *not* in simplest form.

Do you see that $\frac{2}{4}$ can also be written as $\frac{1}{2}$?

To write a fraction in simplest form or to *reduce a fraction to lowest terms,* divide the numerator and denominator by their greatest common factor (GCF).

## Example 3 Simplifying Fractions

Write $\frac{10}{15}$ in simplest form.

From our work earlier in this chapter, we know that the greatest common factor of 10 and 15 is 5. To write $\frac{10}{15}$ in simplest form, divide the numerator and denominator by 5.

$$\frac{10}{15} = \frac{10 \div 5}{15 \div 5} = \frac{2}{3}$$

The resulting fraction, $\frac{2}{3}$, is in lowest terms.

### Check Yourself 3

Write $\frac{12}{18}$ in simplest form by dividing the numerator and denominator by the GCF.

Many students prefer to simplify fractions by using the prime factorizations of the numerator and denominator. Example 4 uses this method.

## Example 4 Factoring to Simplify a Fraction

**(a)** Simplify $\frac{24}{42}$.

To simplify $\frac{24}{42}$, factor.

$$\frac{24}{42} = \frac{\overset{1}{\cancel{2}} \times 2 \times 2 \times \overset{1}{\cancel{3}}}{\underset{1}{\cancel{2}} \times \underset{1}{\cancel{3}} \times 7} = \frac{4}{7}$$

**NOTE**

From the prime factorization of 24 and 42, we divide by the common factors of 2 and 3.

**Note:** The numerator of the simplified fraction is the *product* of the prime factors remaining in the numerator after the division by 2 and 3.

**(b)** Simplify $\frac{120}{180}$.

To reduce $\frac{120}{180}$ to lowest terms, write the prime factorizations of the numerator and denominator. Then divide by any common factors.

$$\frac{120}{180} = \frac{\overset{1}{\cancel{2}} \times \overset{1}{\cancel{2}} \times 2 \times \overset{1}{\cancel{3}} \times \overset{1}{\cancel{5}}}{\underset{1}{\cancel{2}} \times \underset{1}{\cancel{2}} \times \underset{1}{\cancel{3}} \times 3 \times \underset{1}{\cancel{5}}} = \frac{2}{3}$$

### Check Yourself 4

Write each of the following fractions in simplest form.

(a) $\frac{60}{75}$

(b) $\frac{210}{252}$

There is another way to organize your work in simplifying fractions. It again uses the fundamental principle to divide the numerator and denominator by any common factors. We illustrate this with the fractions considered in Example 4.

## Example 5 Using Common Factors to Simplify Fractions

**(a)** Simplify $\frac{24}{42}$.

$$\frac{24}{42} = \frac{\overset{12}{\cancel{24}}}{\underset{21}{\cancel{42}}} = \frac{\overset{4}{\cancel{12}}}{\underset{7}{\cancel{21}}} = \frac{4}{7}$$

Divide by the common factor of 2. Divide by the common factor of 3.

The original numerator and denominator are divisible by 2, and so we divide by that factor to arrive at $\frac{12}{21}$. Our divisibility tests tell us that a common factor of 3 still exists. (Do you remember why?) Divide again for the result $\frac{4}{7}$, which is in lowest terms.

**Note:** If we had seen the GCF of 6 at first, we could have divided by 6 and arrived at the same result in one step.

**(b)** Simplify $\frac{120}{180}$.

$$\frac{120}{180} = \frac{\overset{\overset{2}{\cancel{20}}}{\cancel{120}}}{\underset{\underset{3}{\cancel{30}}}{\cancel{180}}} = \frac{2}{3}$$

Our first step is to divide by the common factor of 6. We then have $\frac{20}{30}$. There is still a common factor of 10, so we again divide.

Again, we could have divided by the GCF of 60 in one step if we had recognized it.

### Check Yourself 5

**Using the method of Example 5, write each of the fractions in simplest form.**

**(a)** $\frac{60}{75}$ **(b)** $\frac{84}{196}$

### Check Yourself ANSWERS

**1. (a)** Yes; **(b)** no **2. (a)** $\frac{6}{9}$; **(b)** $\frac{2}{3}$

**3.** 6 is the GCF of 12 and 18, so $\frac{12}{18} = \frac{12 \div 6}{18 \div 6} = \frac{2}{3}$

**4. (a)** $\frac{60}{75} = \frac{2 \times 2 \times \overset{1}{\cancel{3}} \times \overset{1}{\cancel{5}}}{\underset{1}{\cancel{3}} \times \underset{1}{\cancel{5}} \times 5} = \frac{4}{5}$; **(b)** $\frac{210}{252} = \frac{\overset{1}{\cancel{2}} \times \overset{1}{\cancel{3}} \times 5 \times \overset{1}{\cancel{7}}}{\underset{1}{\cancel{2}} \times 2 \times 3 \times \underset{1}{\cancel{3}} \times \underset{1}{\cancel{7}}} = \frac{5}{6}$

**5. (a)** Divide by the common factors of 3 and 5, $\frac{60}{75} = \frac{4}{5}$

**(b)** Divide by the common factors of 4 and 7, $\frac{84}{196} = \frac{3}{7}$

## Reading Your Text

The following fill-in-the-blank exercises are designed to ensure that you understand some of the key vocabulary used in this section.

### SECTION 2.4

**(a)** If the ______________ products of two fractions are equal, the two fractions are equivalent.

**(b)** Two fractions that are simply different names for the same fraction are called ______________ fractions.

**(c)** In writing equivalent fractions, we use the ______________ Principle of Fractions.

**(d)** We say that a fraction is in simplest form if the numerator and denominator have no ______________ factors other than 1.

# 2.4 exercises

Basic Skills | Advanced Skills | Vocational-Technical Applications | Calculator/Computer | Above and Beyond

MathZone

Boost your grade at mathzone.com!
- > Practice Problems
- > NetTutor
- > Self-Tests
- > e-Professors
- > Videos

Name ______________

Section ________ Date ________

## Answers

| | |
|---|---|
| 1. | 2. |
| 3. | 4. |
| 5. | 6. |
| 7. | 8. |
| 9. | 10. |
| 11. | 12. |
| 13. | 14. |
| 15. | 16. |
| 17. | 18. |
| 19. | 20. |
| 21. | 22. |
| 23. | 24. |

< Objective 1 >

*Are the pairs of fractions equivalent?*

**1.** $\frac{1}{3}, \frac{3}{5}$

**2.** $\frac{3}{5}, \frac{9}{15}$

**3.** $\frac{1}{7}, \frac{4}{28}$

**4.** $\frac{2}{3}, \frac{3}{5}$

**5.** $\frac{5}{6}, \frac{15}{18}$

**6.** $\frac{3}{4}, \frac{16}{20}$

**7.** $\frac{2}{21}, \frac{4}{25}$

**8.** $\frac{20}{24}, \frac{5}{6}$

**9.** $\frac{2}{7}, \frac{3}{11}$

**10.** $\frac{12}{15}, \frac{36}{45}$

**11.** $\frac{16}{24}, \frac{40}{60}$

**12.** $\frac{15}{20}, \frac{20}{25}$

< Objective 2 >

*Write each fraction in simplest form.*

**13.** $\frac{15}{30}$

**14.** $\frac{100}{200}$

**15.** $\frac{8}{12}$

**16.** $\frac{12}{15}$

**17.** $\frac{10}{14}$

**18.** $\frac{15}{50}$

**19.** $\frac{12}{18}$

**20.** $\frac{28}{35}$

**21.** $\frac{35}{40}$

**22.** $\frac{21}{24}$

**23.** $\frac{11}{44}$

**24.** $\frac{10}{25}$

25. $\frac{12}{36}$ 

26. $\frac{18}{48}$

27. $\frac{24}{27}$

28. $\frac{30}{50}$

29. $\frac{32}{40}$

30. $\frac{17}{51}$

31. **Statistics** On a test of 72 questions, Sam answered 54 correctly. On another test, Sam answered 66 correctly out of 88. Did Sam get the same portion of each test correct?

32. **Statistics** Jeff Bagwell of the Houston Astros has 104 hits in 325 times at bat. Matt Williams of the Arizona Diamondbacks has 88 hits in 275 times at bat. Do they have the same batting average?

*Solve the following applications.*

33. **Number Problem** A quarter is what fractional part of a dollar? Simplify your result.

> Videos

34. **Number Problem** A dime is what fractional part of a dollar? Simplify your result.

Basic Skills |  Advanced Skills | Vocational-Technical Applications | Calculator/Computer |  Above and Beyond

35. **Statistics** What fractional part of an hour is 15 minutes (min)? Simplify your result.

> Videos

36. **Statistics** What fractional part of a day is 6 hours (h)? Simplify your result.

**Answers**

25. ______
26. ______
27. ______
28. ______
29. ______
30. ______
31. ______
32. ______
33. ______
34. ______
35. ______
36. ______

Answers

37. ______

38. ______

39. ______

40. ______

41. ______

42. ______

43. ______

44. ______

45. ______

46. ______

47. ______

48. ______

49. ______

50. ______

51. ______

52. ______

53. ______

54. ______

55. ______

56. ______

**37. Science and Medicine** One meter is equal to 100 centimeters (cm). What fractional part of a meter is 70 cm? Simplify your result.

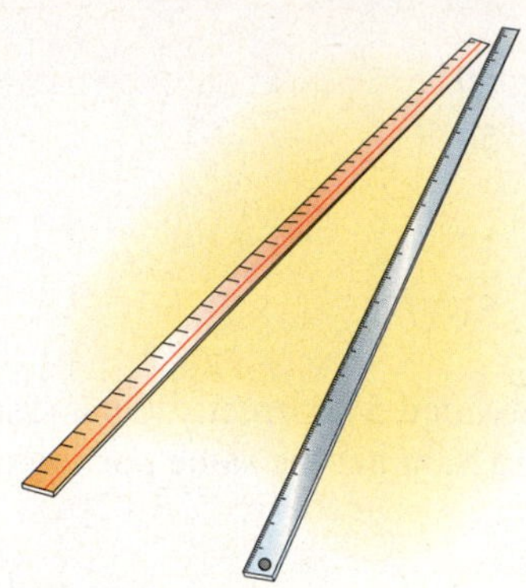

**38. Science and Medicine** One kilometer is equal to 1,000 meters (m). What fractional part of a kilometer is 300 m? Simplify your result.

**39. Technology** Susan did a tune-up on her automobile. She found that two of her eight spark plugs were fouled. What fraction represents the number of fouled plugs? Reduce to lowest terms.

**40. Statistics** Samantha answered 18 of 20 problems correctly on a test. What fractional part did she answer correctly? Reduce your answer to lowest terms.

*Write each fraction in simplest form.*

**41.** $\frac{75}{105}$

**42.** $\frac{62}{93}$

**43.** $\frac{48}{60}$

**44.** $\frac{48}{66}$

**45.** $\frac{105}{135}$

**46.** $\frac{54}{126}$

**47.** $\frac{66}{110}$

**48.** $\frac{280}{320}$

**49.** $\frac{16}{21}$

**50.** $\frac{21}{32}$

**51.** $\frac{31}{52}$

**52.** $\frac{42}{55}$

**53.** $\frac{96}{132}$

**54.** $\frac{33}{121}$

**55.** $\frac{85}{102}$

**56.** $\frac{133}{152}$

Basic Skills | Advanced Skills | **Vocational-Technical Applications** | Calculator/Computer | Above and Beyond

**57. Allied Health** Pepto-Bismol tablets contain 300 milligrams (mg) of medication; however, children 6 to 8 years of age should only take 200 mg at a time. What fractional part of a tablet should a 6-year-old child be given?

**58. Allied Health** The recommended adult dose of the laxative docusate is 500 milligrams (mg) per day, and the recommended dose for children 3 to 5 years old is 40 mg per day. The dose for a 4-year-old child is what fractional part of the adult dose?

**59. Manufacturing Technology** Express the width of the piece shown in the figure as a fraction of the length.

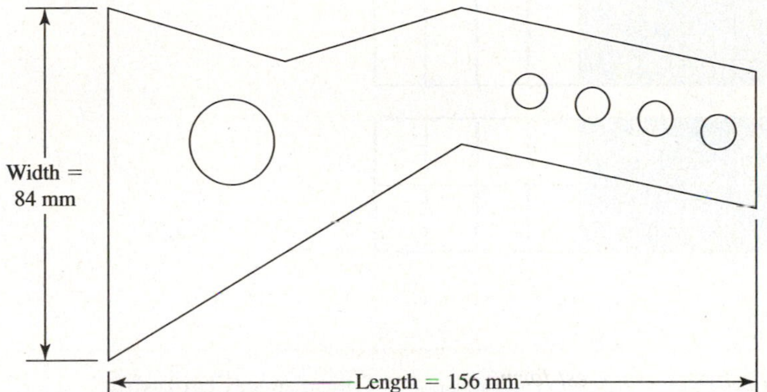


**60. Manufacturing Technology** In the packaging division of Early Enterprises, there are 36 packaging machines. At any given time, 4 of the machines are shut down for scheduled maintenance and service. What is the fraction of machines that are operating at one time?

**61. Information Technology** Jo, an executive vice president of information technology, has 10 people on staff, and she needs to hire two more people. What fraction names the new people of the total staff? Simplify your answer.

**62. Information Technology** Jason is responsible for the administration of the servers at his company. He measures that the average arrival rate of requests to the server is 50,000 requests per second, and he also finds out the servers can service 100,000 requests per second. The intensity of the traffic is measured by $x$, which is the quotient of the average arrival rate and the service rate. The intensity also shows how busy the servers are. What is the fraction that names this situation? Simplify your answer.

Basic Skills | Advanced Skills | Vocational-Technical Applications | Calculator/Computer | **Above and Beyond**

**63.** Can any of the following fractions be simplified?

**(a)** $\frac{824}{73}$ **(b)** $\frac{59}{11}$ **(c)** $\frac{135}{17}$

What characteristic do you notice about the denominator of each fraction? What rule would you make up based on your observations?

## Answers

57. ______

58. ______

59. ______

60. ______

61. ______

62. ______

63. ______

**Answers**

64. ______________________

65. ______________________

**64.** Consider the given figure.

**(a)** Give the fraction that represents the shaded region.

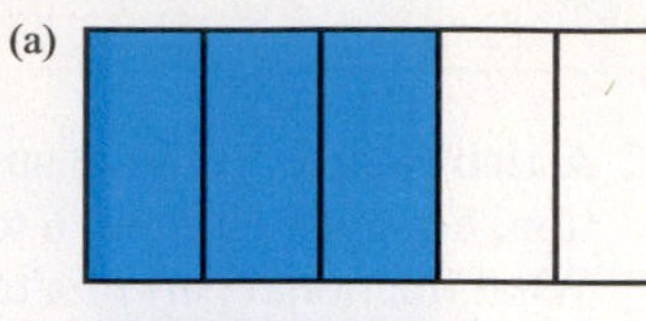

**(b)** Draw a horizontal line through the figure, as shown. Now give the fraction representing the shaded region.

**65.** Repeat exercise 65, using the following figures.

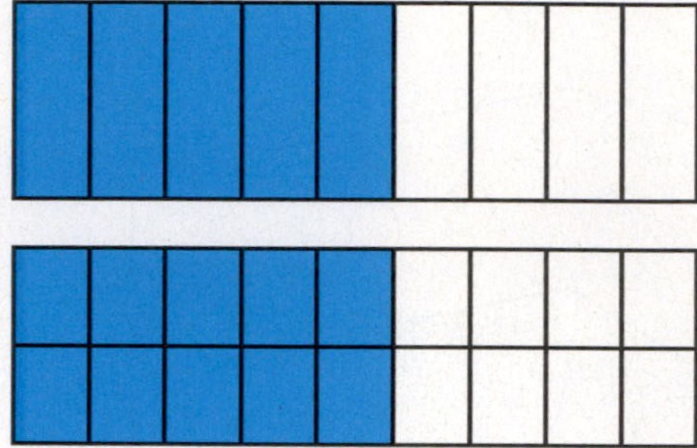

Basic Skills | Advanced Skills | Vocational-Technical Applications | **Calculator/Computer** | Above and Beyond

## Using Your Calculator to Simplify Fractions

If you have a calculator that supports fraction arithmetic, you should learn to use it to check your work. Here we look at two different types of these calculators.

### Scientific Calculator

Scientific calculators include the TI-34, the Casio fx-280 or fx-250, and the Sharp 506 VB or 520 VB.

Before doing the following example, find the button on your scientific calculator that is labeled **a b/c**. This is the button that will be used to enter fractions.

Simplify the fraction $\frac{24}{68}$.

There are four steps in simplifying fractions using a scientific calculator.

**(a)** Enter the numerator, 24.

**(b)** Press the **a b/c** key.

**(c)** Enter the denominator, 68.

**(d)** Press **=**.

The calculator will display the simplified fraction, $\frac{6}{17}$.

### Graphing Calculator

We can simplify the same fraction, $\frac{24}{68}$, using a graphing calculator, such as the TI-83 or TI-84.

(a) Enter the fraction as a division problem: 24 ÷ 68. The calculator will display $\frac{24}{68}$ as 24/68.

(b) Press the [MATH] key.

(c) Select [1: ▶ Frac].

(d) Press [Enter].

The calculator displays the simplified fraction, $\frac{6}{17}$.

The graphing calculator is particularly useful for simplifying fractions with large values in the numerator and denominator. Some scientific calculators cannot handle denominators larger than 999.

*Use your calculator to simplify the following fractions.*

**66.** $\frac{28}{40}$ **67.** $\frac{121}{132}$

**68.** $\frac{96}{144}$ **69.** $\frac{445}{623}$

**70.** $\frac{299}{391}$ **71.** $\frac{289}{459}$

Basic Skills | Advanced Skills | Vocational-Technical Applications | Calculator/Computer | **Above and Beyond**

**72.** A student is attempting to reduce the fraction $\frac{8}{12}$ to lowest terms. He produces the following argument:

$$\frac{8}{12} = \frac{4+4}{8+4} = \frac{4}{8} = \frac{1}{2}$$

What is the fallacy in this argument? What is the correct answer?

## Answers

66. ______

67. ______

68. ______

69. ______

70. ______

71. ______

72. ______

## Answers

**1.** 1 × 5 = 5; 3 × 3 = 9. The fractions are not equivalent.
**3.** Yes **5.** Yes **7.** No **9.** No
**11.** 16 × 60 = 960, and 24 × 40 = 960. The fractions are equivalent.
**13.** $\frac{1}{2}$ **15.** $\frac{2}{3}$ **17.** $\frac{5}{7}$ **19.** $\frac{2}{3}$ **21.** $\frac{7}{8}$ **23.** $\frac{1}{4}$ **25.** $\frac{1}{3}$
**27.** $\frac{8}{9}$ **29.** $\frac{4}{5}$ **31.** Yes **33.** $\frac{1}{4}$ **35.** $\frac{1}{4}$ **37.** $\frac{7}{10}$ **39.** $\frac{1}{4}$
**41.** $\frac{5}{7}$ **43.** $\frac{4}{5}$ **45.** $\frac{7}{9}$ **47.** $\frac{3}{5}$ **49.** $\frac{16}{21}$ **51.** $\frac{31}{52}$ **53.** $\frac{8}{11}$

**55.** $\frac{5}{6}$ **57.** $\frac{2}{3}$ **59.** $\frac{7}{13}$ **61.** $\frac{1}{6}$ **63.** Above and Beyond

**65.** (a) $\frac{5}{9}$; (b) $\frac{10}{18}$ **67.** $\frac{11}{12}$ **69.** $\frac{5}{7}$ **71.** $\frac{17}{27}$

# Activity 4 :: Recommended Daily Allowance

According to the Food and Drug Administration (FDA) in 2003, the following table represented the recommended daily allowance (RDA) for each dietary element, based on a 2,000-calorie diet.

| | RDA |
|---|---|
| Total fat | 65 grams |
| Saturated fat | 20 grams |
| Cholesterol | 300 milligrams |
| Sodium | 2,400 milligrams |
| Carbohydrates | 300 grams |
| Dietary fiber | 25 grams (minimum) |

A high-performance energy bar made by PowerBar has the following amounts of each food type.

| | RDA |
|---|---|
| Total fat | 3 grams |
| Saturated fat | 1 gram |
| Cholesterol | 0 milligrams |
| Sodium | 1,000 milligrams |
| Carbohydrates | 45 grams |
| Dietary fiber | 3 grams |

Use the tables to answer each of the following questions.

1. What fraction of the RDA for sodium is contained in the PowerBar?
2. What fraction of the RDA for carbohydrates is contained in the PowerBar?
3. What fraction of the RDA for dietary fiber is contained in the PowerBar?
4. What fraction of the RDA for total fat is contained in the PowerBar?
5. What fraction of the RDA for saturated fat is contained in the PowerBar?

# 2.5 Multiplying Fractions

< 2.5 Objectives >

1 > Multiply two fractions

2 > Multiply mixed numbers and fractions

3 > Simplify before multiplying fractions

4 > Estimate products by rounding

5 > Solve applications involving multiplication of fractions

Multiplication is the easiest of the four operations with fractions. We can illustrate multiplication by picturing fractions as parts of a whole or unit. Using this idea, we show the fractions $\frac{4}{5}$ and $\frac{2}{3}$ in Figure 1.

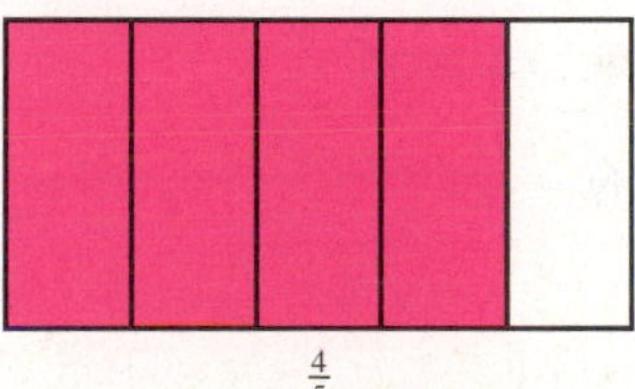


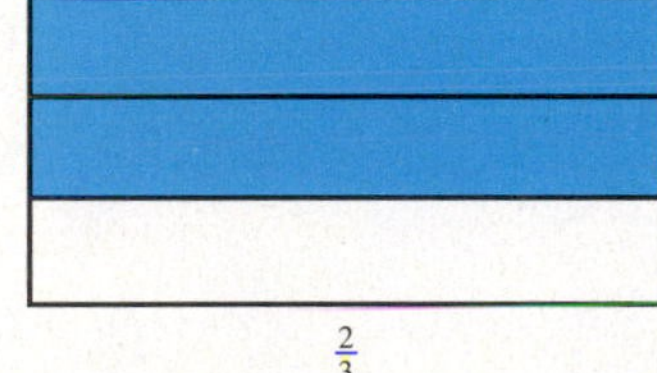


Figure 1

**NOTE**

A fraction followed by the word *of* means that we want to multiply by that fraction.

Suppose now that we wish to find $\frac{2}{3}$ of $\frac{4}{5}$. We can combine the diagrams as shown in Figure 2. The part of the whole representing the product $\frac{2}{3} \times \frac{4}{5}$ is the purple region in Figure 2. The unit has been divided into 15 parts, and 8 of those parts are purple, so $\frac{2}{3} \times \frac{4}{5}$ must be $\frac{8}{15}$.

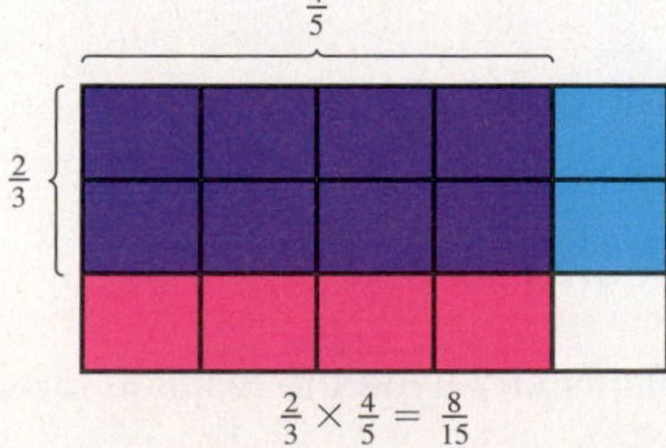


The purple parts represent $\frac{2}{3}$ of the red area, or $\frac{2}{3} \times \frac{4}{5}$.

Figure 2

The following rule is suggested by the diagrams.

Step by Step

**To Multiply Fractions**

**Step 1** Multiply the numerators to find the numerator of the product.
**Step 2** Multiply the denominators to find the denominator of the product.
**Step 3** Simplify the resulting fraction if possible.

We need only use the first two steps in Example 1.

## Example 1 Multiplying Two Fractions

< Objective 1 >

**NOTE**

We multiply fractions in this way *not* because it is easy, but because it works!

Multiply.

(a) $\frac{2}{3} \times \frac{4}{5} = \frac{2 \times 4}{3 \times 5} = \frac{8}{15}$

(b) $\frac{5}{8} \times \frac{7}{9} = \frac{5 \times 7}{8 \times 9} = \frac{35}{72}$

(c) $\left(\frac{2}{3}\right)^2 = \frac{2}{3} \times \frac{2}{3} = \frac{4}{9}$

### Check Yourself 1

Multiply.

(a) $\frac{7}{8} \times \frac{3}{10}$ (b) $\frac{5}{7} \times \frac{3}{4}$ (c) $\left(\frac{3}{4}\right)^2$

Step 3 indicates that the product of fractions should always be simplified to lowest terms. Consider the following.

## Example 2 Multiplying Two Fractions

Multiply and write the result in lowest terms.

$\frac{3}{4} \times \frac{2}{9} = \frac{3 \times 2}{4 \times 9} = \frac{6}{36} = \frac{1}{6}$

Noting that $\frac{6}{36}$ is not in simplest form, we divide numerator and denominator by 6 to write the product in lowest terms.

### Check Yourself 2

Multiply and write the result in lowest terms.

$\frac{5}{7} \times \frac{3}{10}$

To find the product of a fraction and a whole number, write the whole number as a fraction (the whole number divided by 1) and apply the multiplication rule as before. Example 3 illustrates this approach.

**Example 3** Multiplying a Whole Number and a Fraction

Do the indicated multiplication.

Remember that $5 = \frac{5}{1}$.

(a) $5 \times \frac{3}{4} = \frac{5}{1} \times \frac{3}{4} = \frac{5 \times 3}{1 \times 4}$

$= \frac{15}{4} = 3\frac{3}{4}$

**NOTES**

We have written the resulting improper fraction as a mixed number.

Write the product as a mixed number, and then reduce the fractional portion to simplest form.

(b) $\frac{5}{12} \times 6 = \frac{5}{12} \times \frac{6}{1}$

$= \frac{5 \times 6}{12 \times 1}$

$= \frac{30}{12} = 2\frac{6}{12}$

$= 2\frac{1}{2}$

**Check Yourself 3**

Multiply.

(a) $\frac{3}{16} \times 8$ (b) $4 \times \frac{5}{7}$

When mixed numbers are involved in multiplication, the problem requires an additional step. First change any mixed numbers to improper fractions. Then apply our multiplication rule for fractions.

**Example 4** Multiplying a Mixed Number and a Fraction

< Objective 2 >

$1\frac{1}{2} \times \frac{3}{4} = \frac{3}{2} \times \frac{3}{4}$ Change the mixed number to an improper fraction.

Here $1\frac{1}{2} = \frac{3}{2}$.

$= \frac{3 \times 3}{2 \times 4}$ Multiply as before.

$= \frac{9}{8} = 1\frac{1}{8}$ The product is usually written in mixed-number form.

### Check Yourself 4

Multiply.

$$\frac{5}{8} \times 3\frac{1}{2}$$

If two mixed numbers are involved, change both of the mixed numbers to improper fractions. Example 5 illustrates this.

### Example 5 Multiplying Two Mixed Numbers

Multiply.

$$3\frac{2}{3} \times 2\frac{1}{2} = \frac{11}{3} \times \frac{5}{2}$$

Change the mixed numbers to improper fractions.

$$= \frac{11 \times 5}{3 \times 2} = \frac{55}{6} = 9\frac{1}{6}$$

> CAUTION

**Be Careful!** Students sometimes think of

$$3\frac{2}{3} \times 2\frac{1}{2} \quad \text{as} \quad (3 \times 2) + \left(\frac{2}{3} \times \frac{1}{2}\right)$$

This is *not* the correct multiplication pattern. You must first change the mixed numbers to improper fractions.

### Check Yourself 5

Multiply.

$$2\frac{1}{3} \times 3\frac{1}{2}$$

When you multiply fractions, it is usually easier to simplify, that is, remove any common factors in the numerator and denominator, *before multiplying*. Remember that to simplify means to *divide* by the same common factor.

### Example 6 Simplifying Before Multiplying Two Fractions

< Objective 3 >

**NOTE**

Once again we are applying the fundamental principle to divide the numerator and denominator by 3.

Simplify and then multiply.

$$\frac{3}{5} \times \frac{4}{9} = \frac{\overset{1}{\cancel{3}} \times 4}{5 \times \underset{3}{\cancel{9}}}$$

To simplify, we divide the *numerator* and *denominator* by the common factor 3. Remember that $\overset{1}{\cancel{3}}$ means $3 \div 3 = 1$, and $\underset{3}{\cancel{9}}$ means $9 \div 3 = 3$.

$$= \frac{1 \times 4}{5 \times 3}$$

$$= \frac{4}{15}$$

Because we divide by any common factors before we multiply, the resulting product is *in simplest form.*

### Check Yourself 6

**Simplify and then multiply.**

$$\frac{7}{8} \times \frac{5}{21}$$

Our work in Example 6 leads to the following general rule about simplifying fractions in multiplication.

**Property**

**Simplifying Fractions Before Multiplying**

In multiplying two or more fractions, we can divide any factor of the numerator and any factor of the denominator by the same nonzero number to simplify the product.

When mixed numbers are involved, the process is similar. Consider Example 7.

### Example 7 Simplifying Before Multiplying Two Mixed Numbers

Multiply.

$$2\frac{2}{3} \times 2\frac{1}{4} = \frac{8}{3} \times \frac{9}{4}$$ First, convert the mixed numbers to improper fractions.

$$= \frac{\overset{2}{\cancel{8}} \times \overset{3}{\cancel{9}}}{\underset{1}{\cancel{3}} \times \underset{1}{\cancel{4}}}$$ To simplify, divide by the common factors of 3 and 4.

$$= \frac{2 \times 3}{1 \times 1}$$ Multiply as before.

$$= \frac{6}{1} = 6$$

### Check Yourself 7

**Simplify and then multiply.**

$$3\frac{1}{3} \times 2\frac{2}{5}$$

The ideas of our previous examples also allow us to find the product of more than two fractions.

## Example 8 Simplifying Before Multiplying Three Numbers

**RECALL**

We can divide *any* factor of the numerator and *any* factor of the denominator by the same nonzero number.

Simplify and then multiply.

$$\frac{2}{3} \times 1\frac{4}{5} \times \frac{5}{8} = \frac{2}{3} \times \frac{9}{5} \times \frac{5}{8}$$

Write any mixed or whole numbers as improper fractions.

$$= \frac{\overset{1}{\cancel{2}} \times \overset{3}{\cancel{9}} \times \overset{1}{\cancel{5}}}{\underset{1}{\cancel{3}} \times \underset{1}{\cancel{5}} \times \underset{4}{\cancel{8}}}$$

To simplify, divide by the common factors in the numerator and denominator.

$$= \frac{3}{4}$$

### Check Yourself 8

Simplify and then multiply.

$$\frac{5}{8} \times 4\frac{4}{5} \times \frac{1}{6}$$

We encountered estimation by rounding in our earlier work with whole numbers. Estimation can also be used to check the "reasonableness" of an answer when working with fractions or mixed numbers.

## Example 9 Estimating the Product of Two Mixed Numbers

< Objective 4 >

Estimate the product of

$$3\frac{1}{8} \times 5\frac{5}{6}$$

Round each mixed number to the nearest whole number.

$$3\frac{1}{8} \rightarrow 3$$

$$5\frac{5}{6} \rightarrow 6$$

Our estimate of the product is then

$$3 \times 6 = 18$$

**Note:** The actual product in this case is $18\frac{11}{48}$, which certainly seems reasonable in view of our estimate.

### Check Yourself 9

Estimate the product.

$$2\frac{7}{8} \times 8\frac{1}{3}$$

## Units ANALYSIS

When you divide two denominate numbers, the units are also divided. This yields a unit in fraction form.

EXAMPLES:

$$250 \text{ mi} \div 10 \text{ gal} = \frac{250 \text{ mi}}{10 \text{ gal}} = \frac{25 \text{ mi}}{1 \text{ gal}} = 25 \text{ mi/gal (read "miles per gallon")}$$

$$360 \text{ ft} \div 30 \text{ s} = \frac{360 \text{ ft}}{30 \text{ s}} = 12 \text{ ft/s ("feet per second")}$$

When we multiply denominate numbers that have these units in fraction form, they behave just as fractions do.

EXAMPLES:

$$25 \text{ mi/gal} \times 12 \text{ gal} = \frac{25 \text{ mi}}{1 \cancel{\text{gal}}} \times \frac{12 \cancel{\text{gal}}}{1} = 300 \text{ mi}$$

(If we look at the units, we see that the gallons essentially "cancel" when one is in the numerator and the other in the denominator.)

$$12 \text{ ft/s} \times 60 \text{ s/min} = \frac{12 \text{ ft}}{1 \cancel{\text{s}}} \times \frac{60 \cancel{\text{s}}}{1 \text{ min}} = \frac{720 \text{ ft}}{1 \text{ min}} = 720 \text{ ft/min}$$

(Again, the seconds cancel, leaving feet in the numerator and minutes in the denominator.)

Now we can look at some applications of fractions that involve multiplication. In solving these word problems, we use the same approach we used earlier with whole numbers. Let's review the four-step process introduced in Section 1.2.

Step by Step

**Solving Applications Involving the Multiplication of Fractions**

**Step 1** Read the problem carefully to determine the given information and what you are asked to find.

**Step 2** Decide upon the operation or operations to be used.

**Step 3** Write down the complete statement necessary to solve the problem and do the calculations.

**Step 4** Check to make sure that you have answered the question of the problem and that your answer seems reasonable.

We can work through some examples, using these steps.

### Example 10 An Application Involving Multiplication

< Objective 5 >

Lisa worked $10\frac{1}{4}$ hours per day $\left(\frac{\text{h}}{\text{day}}\right)$ for 5 days. How many hours did she work?

**Step 1** We are looking for the total hours Lisa worked.

**Step 2** We will multiply the hours per day by the days.

**Step 3** $10\frac{1}{4}\frac{\text{h}}{\text{day}} \times 5 \text{ days} = \frac{41}{4}\frac{\text{h}}{\text{day}} \times 5 \text{ days} = \frac{205}{4}\text{ h} = 51\frac{1}{4}\text{ h}$

**Step 4** Note the days cancel, leaving only the unit of hours. The units should always be compared to the desired units from step 1. The answer also seems reasonable. An answer such as 5 h or 500 h would not seem reasonable.

### Check Yourself 10

**Carlos gets 30 mi/gal in his Miata. How far should he be able to drive with an 11-gal tank of gas?**

In Example 11, we follow the four steps for solving applications, but do not label the steps. You should still think about these steps as we solve the problem.

### Example 11 An Application Involving the Multiplication of Mixed Numbers

A sheet of notepaper is $6\frac{3}{4}$ inches (in.) wide and $8\frac{2}{3}$ in. long. Find the area of the paper.

Multiply the given length by the width. This gives the desired area. First, we will estimate the area.

$9 \text{ in.} \times 7 \text{ in.} = 63 \text{ in.}^2$

Now, we find the exact area.

**RECALL**

The area of a rectangle is the product of its length and its width.

$$8\frac{2}{3}\text{ in.} \times 6\frac{3}{4}\text{ in.} = \frac{26}{3}\text{ in.} \times \frac{27}{4}\text{ in.}$$
$$= \frac{117}{2}\text{ in.}^2$$
$$= 58\frac{1}{2}\text{ in.}^2$$

The units (square inches) are units of area. Note that from our estimate the result is reasonable.

### Check Yourself 11

**A window is $4\frac{1}{2}$ feet (ft) high by $2\frac{1}{3}$ ft wide. What is its area?**

Example 12 reminds us that an abstract number multiplied by a denominate number yields the units of the denominate number.

## Example 12 — An Application Involving the Multiplication of a Mixed Number and a Fraction

A state park contains $38\frac{2}{3}$ acres. According to the plan for the park, $\frac{3}{4}$ of the park is to be left as a wildlife preserve. How many acres is this?

We want to find $\frac{3}{4}$ of $38\frac{2}{3}$ acres. We then multiply as shown:

**RECALL**

The word *of* indicates multiplication.

$$\frac{3}{4} \times 38\frac{2}{3} = \frac{3}{4} \times \frac{116}{3} = \frac{\overset{1}{\cancel{3}} \times \overset{29}{\cancel{116}}}{\underset{1}{\cancel{4}} \times \underset{1}{\cancel{3}}} \text{ acres} = 29 \text{ acres}$$

### Check Yourself 12

A backyard has $25\frac{3}{4}$ square yards ($yd^2$) of open space. If Patrick wants to build a vegetable garden covering $\frac{2}{3}$ of the open space, how many square yards is this?

We have mentioned the word *of* indicates multiplication. You should also note that it indicates that the fraction preceding it is an abstract number (it has no units attached). There are even occasions, as in Example 13, when we are looking at the product of two abstract numbers.

## Example 13 — An Application Involving the Multiplication of Fractions

A grocery store survey shows that $\frac{2}{3}$ of the customers buy meat. Of these, $\frac{3}{4}$ will buy at least one package of beef. What portion of the store's customers buy beef?

**Step 1** We know that $\frac{2}{3}$ of the customers buy meat and that $\frac{3}{4}$ of these customers buy beef.

**NOTE**

In this problem, *of* means to multiply.

**Step 2** We wish to know $\frac{3}{4}$ of $\frac{2}{3}$. The operation here is multiplication.

**Step 3** Multiplying, we have

$$\frac{3}{4} \times \frac{2}{3} = \frac{\overset{1}{\cancel{3}} \times \overset{1}{\cancel{2}}}{\underset{2}{\cancel{4}} \times \underset{1}{\cancel{3}}} = \frac{1}{2}$$

**Step 4** From step 3 we have the result: $\frac{1}{2}$ of the store's customers buy beef.

### Check Yourself 13

A supermarket survey shows that $\frac{2}{5}$ of the customers buy lunch meat. Of these, $\frac{3}{4}$ buy boiled ham. What portion of the store's customers buy boiled ham?

## Example 14 An Application Involving the Multiplication of Mixed Numbers

Shirley drives at an average speed of 52 miles per hour (mi/h) for $3\frac{1}{4}$ h. How far has she traveled at the end of $3\frac{1}{4}$ h?

**NOTE**

Distance is the product of speed and time.

$$\underset{\text{Speed}}{52\ \frac{\text{mi}}{\text{h}}} \times \underset{\text{Time}}{3\frac{1}{4}\ \text{h}} = \frac{52\ \text{mi}}{1\ \cancel{\text{h}}} \times \frac{13}{4}\ \cancel{\text{h}}$$

$$= \frac{\overset{13}{\cancel{52}} \times 13}{1 \times \underset{1}{\cancel{4}}}\ \text{mi}$$

$$= 169\ \text{mi}$$

### Check Yourself 14

(a) The scale on a map is 1 inch (in.) = 60 miles (mi). What is the distance in miles between two towns that are $3\frac{1}{2}$ in. apart on the map?

(b) Maria is ordering concrete for a new sidewalk that is to be $\frac{1}{9}$ yd thick, $22\frac{1}{2}$ yd long, and $1\frac{1}{3}$ yd wide. How much concrete should she order if she must order a whole number of cubic yards?

### Check Yourself ANSWERS

**1.** **(a)** $\frac{7}{8} \times \frac{3}{10} = \frac{7 \times 3}{8 \times 10} = \frac{21}{80}$; **(b)** $\frac{5}{7} \times \frac{3}{4} = \frac{5 \times 3}{7 \times 4} = \frac{15}{28}$; **(c)** $\frac{9}{16}$

**2.** $\frac{5}{7} \times \frac{3}{10} = \frac{5 \times 3}{7 \times 10} = \frac{15}{70} = \frac{3}{14}$ **3.** **(a)** $1\frac{1}{2}$; **(b)** $2\frac{6}{7}$

**4.** $\frac{5}{8} \times 3\frac{1}{2} = \frac{5}{8} \times \frac{7}{2} = \frac{35}{16} = 2\frac{3}{16}$ **5.** $8\frac{1}{6}$

**6.** $\frac{7}{8} \times \frac{5}{21} = \frac{\overset{1}{\cancel{7}} \times 5}{8 \times \underset{3}{\cancel{21}}} = \frac{5}{24}$

**7.** $3\frac{1}{3} \times 2\frac{2}{5} = \frac{10}{3} \times \frac{12}{5} = \frac{\overset{2}{\cancel{10}} \times \overset{4}{\cancel{12}}}{\underset{1}{\cancel{3}} \times \underset{1}{\cancel{5}}} = \frac{8}{1} = 8$ **8.** $\frac{1}{2}$ **9.** 24

**10.** 330 mi **11.** $10\frac{1}{2}$ ft$^2$ **12.** $17\frac{1}{6}$ yd$^2$ **13.** $\frac{3}{10}$

**14.** **(a)** 210 mi; **(b)** the answer, $3\frac{1}{3}$ yd$^3$, is rounded up to 4 yd$^3$.

## Reading Your Text

The following fill-in-the-blank exercises are designed to ensure that you understand some of the key vocabulary used in this section.

SECTION 2.5

**(a)** The product of fractions should always be expressed in ___________ terms.

**(b)** When multiplying fractions, it is usually easier to ___________ before multiplying.

**(c)** Estimation can be used to check the ______________ of an answer when working with fractions or mixed numbers.

**(d)** The final step in solving an application is to make sure that your answer seems ______________.

# 2.5 exercises

Basic Skills | Advanced Skills | Vocational-Technical Applications | Calculator/Computer | Above and Beyond

MathZone

Boost your grade at mathzone.com!
- Practice Problems
- NetTutor
- Self-Tests
- e-Professors
- Videos

Name ______________________

Section ________ Date ________

## Answers

1. ________ 2. ________

3. ________ 4. ________

5. ________ 6. ________

7. ________ 8. ________

9. ________ 10. ________

11. ________ 12. ________

13. ________ 14. ________

15. ________ 16. ________

17. ________ 18. ________

19. ________ 20. ________

21. ________ 22. ________

23. ________

< Objective 1 >

*Multiply. Be sure to write each answer in simplest form.*

1. $\frac{3}{4} \times \frac{5}{11}$

2. $\frac{2}{7} \times \frac{5}{9}$

3. $\frac{3}{4} \times \frac{7}{11}$ > Videos

4. $\frac{2}{5} \times \frac{3}{7}$

5. $\frac{3}{5} \cdot \frac{5}{7}$

6. $\frac{6}{11} \cdot \frac{8}{6}$

7. $\left(\frac{4}{9}\right)^2$

8. $\left(\frac{5}{6}\right)^2$

9. $\frac{3}{11} \cdot \frac{7}{9}$

10. $\frac{7}{9} \cdot \frac{3}{5}$

11. $\frac{3}{10} \times \frac{5}{9}$

12. $\frac{5}{21} \times \frac{14}{25}$

13. $\frac{7}{9} \cdot \frac{6}{5}$

14. $\frac{48}{63} \cdot \frac{81}{60}$

15. $\frac{24}{33} \cdot \frac{55}{40}$ > Videos

< Objectives 2, 3 >

16. $3\frac{1}{3} \times \frac{9}{11}$

17. $\left(\frac{2}{3}\right)\left(2\frac{2}{5}\right)$ > Videos

18. $\left(3\frac{1}{3}\right)\left(\frac{3}{7}\right)$

19. $\frac{2}{5} \times 3\frac{1}{4}$

20. $2\frac{1}{3} \times 2\frac{1}{6}$

21. $2\frac{1}{3} \cdot 2\frac{1}{2}$

22. $\frac{3}{7} \cdot 14$

23. $\left(\frac{5}{6}\right)^3$ 

**24.** $\left(\frac{4}{5}\right)^3$

**25.** $\frac{12}{25} \times \frac{11}{18}$

**26.** $\frac{10}{12} \times \frac{16}{25}$

**27.** $\frac{14}{15} \cdot \frac{10}{21}$

**28.** $\frac{21}{25} \cdot \frac{30}{7}$

**29.** $\left(\frac{18}{28}\right)\left(\frac{35}{22}\right)$

**30.** $\left(3\frac{2}{3}\right)\left(\frac{9}{10}\right)$

Basic Skills | **Advanced Skills** | Vocational-Technical Applications | Calculator/Computer | Above and Beyond

### < Objective 5 >

*Evaluate the following. Be sure to use the proper units.*

**31.** 36 mi/h × 4 h

**32.** 80 cal/g × 5 g

**33.** 55 joules/s × 11 s

**34.** 5 lb/ft × 3 ft

**35.** 88 ft/s × 1 mi/5,280 ft × 3,600 s/h

**36.** 24 h/day × 3,600 s/h × 365 days/yr

*Solve the following applications.*

**37. Business and Finance** Maria earns $11 per hour. Last week, she worked 9 hours/day for 6 days. What was her gross pay?

**38. Statistics** The gas tank in Luigi's Toyota Camry holds 17 gal when full. The car gets 21 mi/gal. How far can he travel on 3 full tanks?

**39. Crafts** A recipe calls for $\frac{2}{3}$ cup of sugar for each serving. How much sugar is needed for 6 servings?

chapter 2 > Make the Connection

**40. Crafts** Mom-Mom's French toast requires $\frac{3}{4}$ cup of batter for each serving. If 5 people are expected for breakfast, how much batter is needed?

## Answers

24. ______
25. ______
26. ______
27. ______
28. ______
29. ______
30. ______
31. ______
32. ______
33. ______
34. ______
35. ______
36. ______
37. ______
38. ______
39. ______
40. ______

## Answers

41. ______

42. ______

43. ______

44. ______

45. ______

46. ______

47. ______

48. ______

49. ______

50. ______

51. ______

52. ______

53. ______

54. ______

55. ______

56. ______

57. ______

58. ______

59. ______

60. ______

*Multiply and simplify.*

**41.** $\frac{4}{9} \times 3\frac{3}{5}$

**42.** $5\frac{1}{3} \times \frac{7}{8}$

**43.** $\frac{10}{27} \times 3\frac{3}{5}$

**44.** $1\frac{1}{3} \times 1\frac{1}{5}$

**45.** $2\frac{2}{5} \cdot 3\frac{3}{4}$

**46.** $2\frac{2}{7} \cdot 2\frac{1}{3}$

**47.** $4\frac{1}{5} \times \frac{10}{21} \times \frac{9}{20}$

**48.** $\frac{7}{8} \times 5\frac{1}{3} \times \frac{5}{14}$

**49.** $3\frac{1}{3} \cdot \frac{4}{5} \cdot 1\frac{1}{8}$

**50.** $4\frac{1}{2} \cdot 5\frac{5}{6} \cdot \frac{8}{15}$

**51.** Find $\frac{2}{3}$ of $\frac{3}{7}$.

**52.** What is $\frac{5}{6}$ of $\frac{9}{10}$?

< Objective 4 >

*Estimate the following products.*

**53.** $3\frac{1}{5} \times 4\frac{2}{3}$

**54.** $5\frac{1}{7} \times 2\frac{2}{13}$

**55.** $11\frac{3}{4} \times 5\frac{1}{4}$

**56.** $3\frac{4}{5} \times 5\frac{6}{7}$

**57.** $8\frac{2}{9} \cdot 7\frac{11}{12}$

**58.** $\frac{9}{10} \cdot 2\frac{2}{7}$

**59.** **Science and Medicine** A jet flew at an average speed of 540 mi/h on a $4\frac{2}{3}$-h flight. What was the distance flown?

**60.** **Construction** A piece of land that has $11\frac{2}{3}$ acres is being subdivided for home lots. It is estimated that $\frac{2}{7}$ of the area will be used for roads. What amount remains to be used for lots?

**61. GEOMETRY** Find the volume of a box that measures $2\frac{1}{4}$ in. by $3\frac{7}{8}$ in. by $4\frac{5}{6}$ in.

**62. CONSTRUCTION** Nico wishes to purchase mulch to cover his garden. The garden measures $7\frac{7}{8}$ ft by $10\frac{1}{8}$ ft. He wants the mulch to be $\frac{1}{3}$ ft deep. How much mulch should Nico order if he must order a whole number of cubic feet?

Basic Skills | Advanced Skills | **Vocational-Technical Applications** | Calculator/Computer | Above and Beyond

**63. MANUFACTURING TECHNOLOGY** Calculate the distance from the center of hole A to the center of hole B.

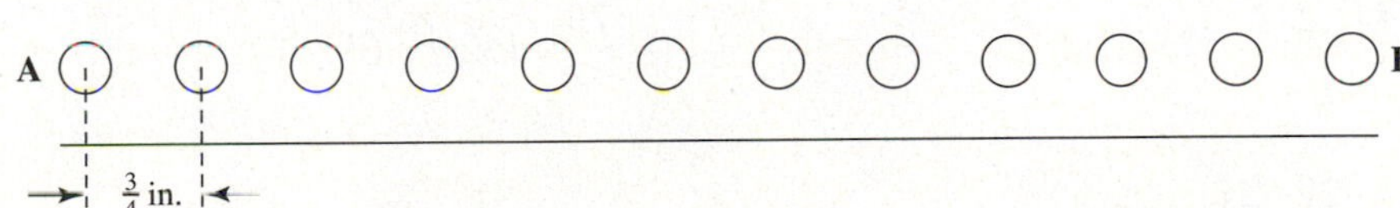


**64. MANUFACTURING TECHNOLOGY** A cut $3\frac{3}{8}$ in. long needs to be made in a piece of material. The cut rate is $\frac{3}{4}$ minute per in. How many minutes does it take to make the cut?

**65. MANUFACTURING TECHNOLOGY** A custom order requires $14\frac{3}{4}$ ounces of material that costs $7\frac{1}{2}$¢ per ounce. Find the total cost of the material.

**66. MANUFACTURING TECHNOLOGY** A customer order requires $8\frac{5}{8}$ ounces of material that costs $10\frac{1}{2}$¢ per ounce. Find the total cost of the material.

**Answers**

61. ______

62. ______

63. ______

64. ______

65. ______

66. ______

## Answers

67. ______

68. ______

69. ______

70. ______

71. ______

72. ______

Basic Skills | Advanced Skills | Vocational-Technical Applications | **Calculator/Computer** | Above and Beyond

## Using Your Calculator to Multiply Fractions

### Scientific Calculator

To multiply fractions on a scientific calculator, you enter the first fraction, using the [a b/c] key, then press the multiplication sign, next enter the second fraction, and then press the equal sign. It is always a good idea to separate the fractions by using parentheses. Note that we do that in the example below.

### Graphing Calculator

When using a graphing calculator, you must choose the fraction option [1:▶ Frac] from the [MATH] menu before pressing [Enter].

For the fraction problem $\frac{7}{15} \times \frac{5}{21}$, the keystroke sequence is

[(] 7 [÷] 15 [)] [×] [(] 5 [÷] 21 [)] [1:▶ Frac] [Enter]

The result is $\frac{1}{9}$.

*Find the following products, using your calculator.*

**67.** $\frac{7}{12} \times \frac{36}{63}$

**68.** $\frac{8}{27} \times \frac{45}{64}$

**69.** $\frac{12}{45} \times \frac{27}{72}$

**70.** $\frac{18}{132} \times \frac{36}{63}$

**71.** $\frac{27}{72} \cdot \frac{24}{45}$

**72.** $\frac{81}{136} \cdot \frac{84}{135}$

## Answers

**1.** $\frac{15}{44}$ **3.** $\frac{3}{4} \times \frac{7}{11} = \frac{21}{44}$ **5.** $\frac{3}{7}$ **7.** $\frac{16}{81}$ **9.** $\frac{7}{33}$
**11.** $\frac{3}{10} \times \frac{5}{9} = \frac{15}{90} = \frac{1}{6}$ **13.** $\frac{14}{15}$ **15.** 1 **17.** $1\frac{3}{5}$ **19.** $1\frac{3}{10}$
**21.** $5\frac{5}{6}$ **23.** $\frac{125}{216}$ **25.** $\frac{22}{75}$ **27.** $\frac{4}{9}$ **29.** $1\frac{1}{44}$ **31.** 144 mi
**33.** 605 joules **35.** 60 mi/h **37.** \$594 **39.** 4 cups
**41.** $1\frac{3}{5}$ **43.** $1\frac{1}{3}$ **45.** 9 **47.** $\frac{9}{10}$ **49.** 3 **51.** $\frac{2}{7}$ **53.** 15
**55.** 60 **57.** 64 **59.** 2,520 mi **61.** $42\frac{9}{64}$ in.$^3$ **63.** $8\frac{1}{4}$ in.
**65.** $110\frac{5}{8}$¢ or \$1.11 **67.** $\frac{1}{3}$ **69.** $\frac{1}{10}$ **71.** $\frac{1}{5}$

# Activity 5 ::
## Overriding a Presidential Veto

Read the following article and use the information you find there to answer the questions that follow.

A bill is sent to the President of the United States when it has passed both houses of Congress. A majority vote in both the House of Representatives (218 of 435 members) and the Senate (51 of 100 members) is needed for the bill to be passed on to the President. The majority vote is a majority of the members present, as long as more than one-half of all the members are present. More than one-half of the members makes up what is called a quorum.

Once the bill comes to him, the President may either sign the bill, making it a law, or he can veto the bill. His veto sends the bill back to Congress.

Congress can still make the bill a law by overriding the Presidential veto. To override the veto, $\frac{2}{3}$ of the members of each legislative body must vote to override it. Again, this is $\frac{2}{3}$ of a quorum of members.

**1.** Assume that 420 members of the House are available to vote. How many votes are necessary to make up a majority, which is anything over $\frac{1}{2}$?

**2.** If 90 members of the Senate are available to vote, how many votes would constitute a majority?

**3.** How many votes from a group of 420 members of the House would be necessary to overturn a veto?

**4.** How many votes from a group of 90 members of the Senate would be necessary to overturn a veto?

# 2.6 Dividing Fractions

< 2.6 Objectives >

1 > Find the reciprocal of a fraction

2 > Divide fractions

3 > Divide mixed numbers

4 > Solve applications involving division of fractions

We are now ready to look at the operation of division on fractions. First we need a new concept, the **reciprocal** of a fraction.

**Property**

**The Reciprocal of a Fraction**

We invert, or turn over, a fraction to write its **reciprocal.**

**Example 1** Finding the Reciprocal of a Fraction

< Objective 1 >

Find the reciprocal of **(a)** $\frac{3}{4}$, **(b)** 5, and **(c)** $1\frac{2}{3}$.

**NOTE**

In general, the reciprocal of the fraction $\frac{a}{b}$ is $\frac{b}{a}$. Neither $a$ nor $b$ can be 0.

**(a)** The reciprocal of $\frac{3}{4}$ is $\frac{4}{3}$.

Just invert, or turn over, the fraction.

**(b)** The reciprocal of 5, or $\frac{5}{1}$, is $\frac{1}{5}$.

Write 5 as $\frac{5}{1}$ and then turn over the fraction.

**(c)** The reciprocal of $1\frac{2}{3}$, or $\frac{5}{3}$, is $\frac{3}{5}$.

Write $1\frac{2}{3}$ as $\frac{5}{3}$ and then invert.

**Check Yourself 1**

Find the reciprocal of **(a)** $\frac{5}{8}$ and **(b)** $3\frac{1}{4}$.

An important property relating a number and its reciprocal follows.

Property

**Reciprocal Products** The product of any number and its reciprocal is 1. (Every number except zero has a reciprocal.)

We are now ready to use the reciprocal to find a rule for dividing fractions. Recall that we can represent the operation of division in several ways. We used the symbol ÷ earlier. Remember that a fraction also indicates division. For instance,

**NOTE**

$3 \div 5$ and $\frac{3}{5}$ both mean "3 divided by 5."

$$3 \div 5 = \frac{3}{5}$$

In this statement, 5 is called the *divisor*. It follows the division sign ÷ and is written *below* the fraction bar.

Using this information, we can write a statement involving fractions and division as a *complex fraction,* which has a fraction as both its numerator and its denominator, as Example 2 illustrates.

## Example 2 Writing a Quotient as a Complex Fraction

Write $\frac{2}{3} \div \frac{4}{5}$ as a complex fraction.

$$\frac{\frac{2}{3}}{\frac{4}{5}}$$

The numerator is $\frac{2}{3}$.

A *complex fraction* is written by placing the dividend in the numerator and the divisor in the denominator.

The denominator is $\frac{4}{5}$.

### Check Yourself 2

Write $\frac{2}{5} \div \frac{3}{4}$ as a complex fraction.

Let's continue with the same division problem you looked at in Check Yourself 2.

## Example 3 Rewriting a Division Problem

(I) $\frac{2}{5} \div \frac{3}{4} = \dfrac{\frac{2}{5}}{\frac{3}{4}}$

Write the original quotient as a complex fraction.

$$= \frac{\frac{2}{5} \times \frac{4}{3}}{\frac{3}{4} \times \frac{4}{3}}$$

Multiply the numerator and denominator by $\frac{4}{3}$, the reciprocal of the denominator. This does *not* change the value of the fraction.

$$= \frac{\frac{2}{5} \times \frac{4}{3}}{1}$$

The denominator becomes 1.

**(II)** $= \frac{2}{5} \times \frac{4}{3}$

Recall that a number divided by 1 is just that number.

**NOTE**

Do you see a rule suggested?

We see from **(I)** and **(II)** that

$$\frac{2}{5} \div \frac{3}{4} = \frac{2}{5} \times \frac{4}{3}$$

We would certainly like to be able to divide fractions easily without all the work of this example. Look carefully at the calculations. The following rule is suggested.

**Property**

**To Divide Fractions**

To divide one fraction by another, invert the divisor (the fraction after the division sign) and multiply.

**Check Yourself 3**

Write $\frac{3}{5} \div \frac{7}{8}$ as a multiplication problem.

Example 4 applies the rule for dividing fractions.

**Example 4** **Dividing Two Fractions**

< Objective 2 >

**RECALL**

The number inverted is the divisor. It *follows* the division sign.

Divide.

**(a)** $\frac{\frac{1}{3}}{\frac{4}{7}}$

$$\frac{\frac{1}{3}}{\frac{4}{7}} = \frac{1}{3} \div \frac{4}{7} = \frac{1}{3} \times \frac{7}{4}$$

We invert the divisor, $\frac{4}{7}$, and then multiply.

$$= \frac{1 \times 7}{3 \times 4} = \frac{7}{12}$$

Property

**Reciprocal Products** The product of any number and its reciprocal is 1. (Every number except zero has a reciprocal.)

**NOTE**

$3 \div 5$ and $\frac{3}{5}$ both mean "3 divided by 5."

We are now ready to use the reciprocal to find a rule for dividing fractions. Recall that we can represent the operation of division in several ways. We used the symbol ÷ earlier. Remember that a fraction also indicates division. For instance,

$$3 \div 5 = \frac{3}{5}$$

In this statement, 5 is called the *divisor*. It follows the division sign ÷ and is written *below* the fraction bar.

Using this information, we can write a statement involving fractions and division as a *complex fraction,* which has a fraction as both its numerator and its denominator, as Example 2 illustrates.

### Example 2 Writing a Quotient as a Complex Fraction

Write $\frac{2}{3} \div \frac{4}{5}$ as a complex fraction.

$$\frac{\frac{2}{3}}{\frac{4}{5}}$$

The numerator is $\frac{2}{3}$.

A *complex fraction* is written by placing the dividend in the numerator and the divisor in the denominator.

The denominator is $\frac{4}{5}$.

**Check Yourself 2**

**Write $\frac{2}{5} \div \frac{3}{4}$ as a complex fraction.**

Let's continue with the same division problem you looked at in Check Yourself 2.

### Example 3 Rewriting a Division Problem

**(I)** $\frac{2}{5} \div \frac{3}{4} = \frac{\frac{2}{5}}{\frac{3}{4}}$ Write the original quotient as a complex fraction.

$$= \frac{\frac{2}{5} \times \frac{4}{3}}{\frac{3}{4} \times \frac{4}{3}}$$ Multiply the numerator and denominator by $\frac{4}{3}$, the reciprocal of the denominator. This does *not* change the value of the fraction.

$$= \frac{\frac{2}{5} \times \frac{4}{3}}{1}$$ The denominator becomes 1.

**(II)** $$= \frac{2}{5} \times \frac{4}{3}$$ Recall that a number divided by 1 is just that number.

**NOTE**

Do you see a rule suggested?

We see from **(I)** and **(II)** that

$$\frac{2}{5} \div \frac{3}{4} = \frac{2}{5} \times \frac{4}{3}$$

We would certainly like to be able to divide fractions easily without all the work of this example. Look carefully at the calculations. The following rule is suggested.

**Property**

**To Divide Fractions**

To divide one fraction by another, invert the divisor (the fraction after the division sign) and multiply.

**Check Yourself 3**

Write $\frac{3}{5} \div \frac{7}{8}$ as a multiplication problem.

Example 4 applies the rule for dividing fractions.

## Example 4 Dividing Two Fractions

< Objective 2 >

**RECALL**

The number inverted is the divisor. It *follows* the division sign.

Divide.

**(a)** $\dfrac{\frac{1}{3}}{\frac{4}{7}}$

$$\frac{\frac{1}{3}}{\frac{4}{7}} = \frac{1}{3} \div \frac{4}{7} = \frac{1}{3} \times \frac{7}{4}$$ We invert the divisor, $\frac{4}{7}$, and then multiply.

$$= \frac{1 \times 7}{3 \times 4} = \frac{7}{12}$$

**(b)** $\dfrac{\frac{3}{4}}{6}$

$$\dfrac{\frac{3}{4}}{6} = \frac{3}{4} \div 6 = \frac{\overset{1}{\cancel{3}}}{4} \times \frac{1}{\underset{2}{\cancel{6}}}$$

$$= \frac{1}{8}$$

## Check Yourself 4

**Divide.**

**(a)** $\dfrac{\frac{2}{5}}{\frac{3}{4}}$ **(b)** $\dfrac{\frac{2}{7}}{4}$

Here is a similar example.

## Example 5 Dividing Two Fractions

Divide.

$$\frac{5}{8} \div \frac{3}{5} = \frac{5}{8} \times \frac{5}{3} = \frac{5 \times 5}{8 \times 3}$$

$$= \frac{25}{24} = 1\frac{1}{24}$$

## Check Yourself 5

**Divide.**

$$\frac{5}{6} \div \frac{3}{7}$$

Simplifying will also be useful in dividing fractions. Consider Example 6.

## Example 6 Dividing Two Fractions

> CAUTION

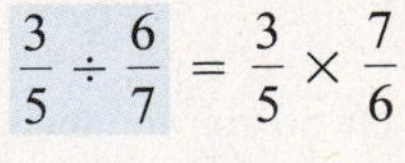

**NOTE**

**Be careful!** We must invert the divisor *before simplifying.*

Divide.

$$\frac{3}{5} \div \frac{6}{7} = \frac{3}{5} \times \frac{7}{6}$$

Invert the divisor *first!* Then you can divide by the common factor of 3.

$$= \frac{\overset{1}{\cancel{3}} \times 7}{5 \times \underset{2}{\cancel{6}}} = \frac{7}{10}$$

### Check Yourself 6

Divide.

$$\frac{4}{9} \div \frac{8}{15}$$

When mixed or whole numbers are involved, the process is similar. Simply change the mixed or whole numbers to improper fractions as the first step. Then proceed with the division rule. Example 7 illustrates this approach.

## Example 7 Dividing Two Mixed Numbers

< Objective 3 >

Divide.

$$2\frac{3}{8} \div 1\frac{3}{4} = \frac{19}{8} \div \frac{7}{4}$$ Write the mixed numbers as improper fractions.

$$\frac{19}{8} \times \frac{4}{7} = \frac{19 \times \overset{1}{\cancel{4}}}{\underset{2}{\cancel{8}} \times 7}$$ Invert the divisor and multiply as before.

$$= \frac{19}{14} = 1\frac{5}{14}$$

### Check Yourself 7

Divide.

$$3\frac{1}{5} \div 2\frac{2}{5}$$

Example 8 illustrates the division process when a whole number is involved.

## Example 8 Dividing a Mixed Number and a Whole Number

**NOTE**

Write the whole number 6 as $\frac{6}{1}$.

Divide and simplify.

$$1\frac{4}{5} \div 6 = \frac{9}{5} \div \frac{6}{1}$$

$$\frac{9}{5} \times \frac{1}{6} = \frac{\overset{3}{\cancel{9}} \times 1}{5 \times \underset{2}{\cancel{6}}}$$ Invert the divisor, and then divide by the common factor of 3.

$$= \frac{3}{10}$$

### Check Yourself 8

Divide.

$$8 \div 4\frac{4}{5}$$

## Units ANALYSIS

When dividing by denominate numbers that have fractional units, we multiply by the reciprocal of the number *and its units.*

EXAMPLES:

$$500 \text{ mi} \div \frac{25 \text{ mi}}{1 \text{ gal}} = \frac{500 \text{ mi}}{1} \times \frac{1 \text{ gal}}{25 \text{ mi}} = 20 \text{ gal}$$

$$\$24{,}000 \div \frac{\$400}{1 \text{ yr}} = \frac{24{,}000 \text{ dol}}{1} \div \frac{400 \text{ dol}}{1 \text{ yr}} = \frac{24{,}000 \text{ dol}}{1} \times \frac{1 \text{ yr}}{400 \text{ dol}} = 60 \text{ yr}$$

(As always, note that in each case, the arithmetic of the units produces the final units.)

As was the case with multiplication, dividing fractions can be used in the solution of a variety of applications. The steps of the problem-solving process remain the same.

## Example 9 An Application Involving the Division of Mixed Numbers

< Objective 4 >

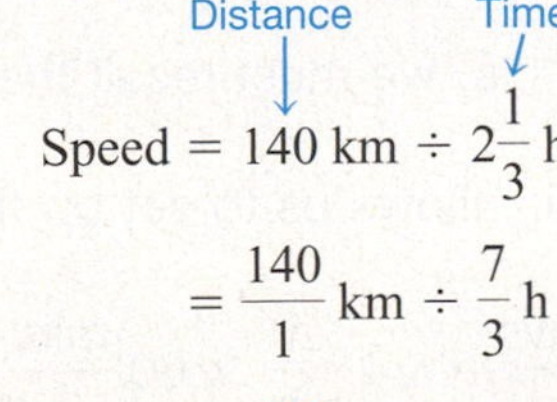


Jack traveled 140 kilometers (km) in $2\frac{1}{3}$ hours (h). What was his average speed?

Distance, Time

$$\text{Speed} = 140 \text{ km} \div 2\frac{1}{3} \text{ h}$$

$$= \frac{140}{1} \text{ km} \div \frac{7}{3} \text{ h}$$

We know the distance traveled and the time for that travel. To find the *average* speed, we must use division. Do you remember why?

$$\frac{140}{1} \times \frac{3}{7} = \frac{\overset{20}{\cancel{140}} \times 3}{1 \times \underset{1}{\cancel{7}}} \frac{\text{km}}{\text{h}}$$

$$= 60 \text{ km/h}$$

$\frac{\text{km}}{\text{h}}$ is read "kilometers per hour." This is a unit of speed.

### Check Yourself 9

A light plane flew **280** mi in $1\frac{3}{4}$ h. What was its average speed?

## Example 10 An Application Involving the Division of Mixed Numbers

**NOTE**

We must divide the length of the longer piece by the desired length of the shorter piece.

An electrician needs pieces of wire $2\frac{3}{5}$ in. long. If she has a $20\frac{4}{5}$-in. piece of wire, how many of the shorter pieces can she cut?

$$20\frac{4}{5}\text{ in.} \div 2\frac{3}{5}\,\frac{\text{in.}}{\text{pieces}} = \frac{104}{5}\text{ in.} \div \frac{13}{5}\,\frac{\text{in.}}{\text{pieces}}$$

$$\frac{104}{5}\text{ in.} \times \frac{5}{13}\,\frac{\text{pieces}}{\text{in.}} = \frac{\overset{8}{\cancel{104}} \times \overset{1}{\cancel{5}}}{\underset{1}{\cancel{5}} \times \underset{1}{\cancel{13}}}\text{ pieces}$$

$$= 8\text{ pieces}$$

### Check Yourself 10

A piece of plastic water pipe 63 in. long is to be cut into lengths of $3\frac{1}{2}$ in. How many of the shorter pieces can be cut?

### Units ANALYSIS

When you convert units, it is important, and helpful, to carry out all unit arithmetic.

EXAMPLES:

Convert $1\frac{1}{2}$ years into minutes.

To accomplish this, we must recall that there are $365\,\frac{\text{days}}{\text{yr}}$, $24\,\frac{\text{h}}{\text{day}}$, and $60\,\frac{\text{min}}{\text{h}}$. This allows us to set up the following expression.

$$1\frac{1}{2}\text{ yr} \times 365\,\frac{\text{days}}{\text{yr}} \times 24\,\frac{\text{h}}{\text{day}} \times 60\,\frac{\text{min}}{\text{h}}$$

Before we work with the numbers, we should check the units to make certain that our result will be in "min(utes)."

Because the years, days, and hours all cancel, we are left with only minutes, so we can go ahead with the computation.

$$\frac{3}{2}\text{ yr} \times 365\,\frac{\text{days}}{\text{yr}} \times 24\,\frac{\text{h}}{\text{day}} \times 60\,\frac{\text{min}}{\text{h}} = 788{,}400\text{ min}$$

There are 788,400 min in $1\frac{1}{2}$ yr.

Some applications require both multiplication and division. Example 11 is such an application.

**Example 11** An Application Involving the Division of Mixed Numbers

A parcel of land that is $2\frac{1}{2}$ mi long and $1\frac{1}{3}$ mi wide is to be divided into tracts that are each $\frac{1}{3}$ square mile (mi²). How many of these tracts will the parcel make?

The area of the parcel is its length times its width:

$$\begin{aligned}\text{Area} &= 2\frac{1}{2}\text{ mi} \times 1\frac{1}{3}\text{ mi}\\ &= \frac{5}{2}\text{ mi} \times \frac{4}{3}\text{ mi}\\ &= \frac{10}{3}\text{ mi}^2\end{aligned}$$

We need to divide the total area of the parcel into $\frac{1}{3}$-mi² tracts.

$$\frac{10}{3}\text{ mi}^2 \div \frac{1}{3}\text{ mi}^2 = \frac{10}{3}\text{ mi}^2 \times \frac{3}{1\text{ mi}^2} = 10$$

The land will provide 10 tracts, each with an area of $\frac{1}{3}$ mi².

**Check Yourself 11**

A parcel of land that is $3\frac{1}{3}$ mi long and $2\frac{1}{2}$ mi wide is to be divided into $\frac{1}{3}$-mi² tracts. How many of these tracts will the parcel make?

In our final example, we look at a case in which the divisor has fractional units.

**Example 12** An Application Involving Mixed Numbers

> **NOTE**
>
> We are using the gardener/contractor definition of a "yard" of mulch. It is actually 1 yd × 1 yd × 1 yd, or 1 yd³.

Jackson has $6\frac{1}{2}$ yards of mulch. His garden needs $\frac{2}{3}$ yards per row. How many rows can he cover with the mulch?

We have $6\frac{1}{2}$ yards and $\frac{2\text{ yards}}{3\text{ rows}}$. Even if you don't immediately see how to solve the problem, units analysis can help. The units of the answer will be "rows." To get there, we need to have the yards units cancel. That will happen if we divide $6\frac{1}{2}$ by $\frac{2}{3}$!

$$6\frac{1}{2}\text{ yards} \div \frac{2\text{ yards}}{3\text{ rows}} = \frac{13}{2}\text{ yards} \div \frac{2\text{ yards}}{3\text{ rows}}$$

$$= \frac{13}{2}\text{ yards} \times \frac{3\text{ rows}}{2\text{ yards}} = \frac{39}{4}\text{ rows} = 9\frac{3}{4}\text{ rows}$$

He can cover all of 9 rows and part $\left(\frac{3}{4}\right)$ of the tenth row.

### Check Yourself 12

**Tangela has $4,100 to invest in a certain stock. If the stock is selling at $\$25\frac{5}{8}$ per share, how many shares can she buy?**

### Check Yourself ANSWERS

**1.** **(a)** $\frac{8}{5}$; **(b)** $3\frac{1}{4}$ is $\frac{13}{4}$, so the reciprocal is $\frac{4}{13}$  **2.** $\dfrac{\frac{2}{5}}{\frac{3}{4}}$  **3.** $\frac{3}{5} \times \frac{8}{7}$

**4.** **(a)** $\frac{8}{15}$; **(b)** $\frac{1}{14}$  **5.** $1\frac{17}{18}$  **6.** $\frac{4}{9} \div \frac{8}{15} = \frac{4}{9} \times \frac{15}{8} = \frac{\overset{1}{\cancel{4}} \times \overset{5}{\cancel{15}}}{\underset{3}{\cancel{9}} \times \underset{2}{\cancel{8}}} = \frac{5}{6}$

**7.** $3\frac{1}{5} \div 2\frac{2}{5} = \frac{16}{5} \div \frac{12}{5} = \frac{16}{5} \times \frac{5}{12} = \frac{\overset{4}{\cancel{16}} \times \overset{1}{\cancel{5}}}{\underset{1}{\cancel{5}} \times \underset{3}{\cancel{12}}} = \frac{4}{3} = 1\frac{1}{3}$  **8.** $1\frac{2}{3}$

**9.** 160 mi/h  **10.** 18 pieces  **11.** 25 tracts  **12.** 160 shares

## Reading Your Text

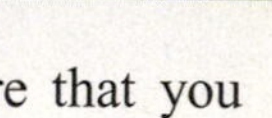

The following fill-in-the-blank exercises are designed to ensure that you understand some of the key vocabulary used in this section.

**SECTION 2.6**

**(a)** The product of any number and its ______________ is 1.

**(b)** An expression that has a fraction as both its numerator and denominator is called a ______________ fraction.

**(c)** To divide one fraction by another, invert the ______________ and multiply.

**(d)** When dividing by denominate numbers that have fractional units, we multiply by the reciprocal of the number and its ______________.

# 2.6 exercises

MathZone

Boost your grade at mathzone.com!
> Practice Problems
> NetTutor
> Self-Tests
> e-Professors
> Videos

Name ____________

Section ________ Date ________

< Objective 1 >

*Find the reciprocal of each of the following.*

1. $\frac{7}{8}$ 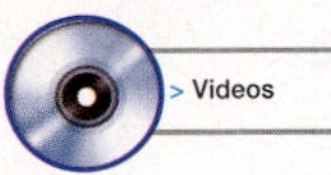


2. $\frac{9}{5}$

3. $\frac{1}{2}$

4. 6

5. $2\frac{1}{3}$ > Videos

6. $4\frac{3}{5}$

7. $9\frac{3}{4}$

8. $\frac{1}{8}$

< Objective 2 >

*Divide. Write each result in simplest form.*

9. $\frac{1}{5} \div \frac{3}{4}$ 

10. $\frac{2}{5} \div \frac{1}{3}$

11. $\dfrac{\frac{2}{5}}{\frac{3}{4}}$

12. $\dfrac{\frac{5}{8}}{\frac{3}{4}}$

13. $\frac{8}{9} \div \frac{4}{3}$

14. $\frac{5}{9} \div \frac{8}{11}$

15. $\frac{7}{10} \div \frac{5}{9}$

16. $\frac{8}{9} \div \frac{11}{15}$

17. $\frac{8}{15} \div \frac{2}{5}$

18. $\frac{5}{27} \div \frac{15}{54}$

19. $\dfrac{\frac{5}{27}}{\frac{25}{36}}$

20. $\dfrac{\frac{9}{28}}{\frac{27}{35}}$

21. $\frac{4}{5} \div 4$

22. $27 \div \frac{3}{7}$

23. $12 \div \frac{2}{3}$

24. $\frac{5}{8} \div 5$

25. $\dfrac{\frac{12}{17}}{\frac{6}{7}}$

26. $\dfrac{\frac{3}{4}}{\frac{9}{10}}$

## Answers

1. ______ 2. ______
3. ______ 4. ______
5. ______ 6. ______
7. ______ 8. ______
9. ______ 10. ______
11. ______ 12. ______
13. ______ 14. ______
15. ______ 16. ______
17. ______ 18. ______
19. ______ 20. ______
21. ______ 22. ______
23. ______ 24. ______
25. ______ 26. ______

## Answers

27. ______
28. ______
29. ______
30. ______
31. ______
32. ______
33. ______
34. ______
35. ______
36. ______
37. ______
38. ______
39. ______
40. ______
41. ______
42. ______
43. ______
44. ______

< Objective 3 >

**27.** $15 \div 3\frac{1}{3}$

**28.** $2\frac{4}{7} \div 12$

**29.** $1\frac{3}{5} \div \frac{4}{15}$

**30.** $\frac{9}{14} \div 2\frac{4}{7}$

**31.** $\frac{7}{12} \div 2\frac{1}{3}$

**32.** $1\frac{3}{8} \div \frac{5}{12}$

**33.** $5\frac{3}{5} \div \frac{7}{15}$

**34.** $\frac{7}{18} \div 5\frac{5}{6}$

Basic Skills | **Advanced Skills** | Vocational-Technical Applications | Calculator/Computer | Above and Beyond

**35.** $\left(\frac{2}{3}\right)^3 \div 4$

**36.** $\left(\frac{3}{5}\right)^3 \div 9$

**37.** $\frac{3}{5} \div \left(\frac{3}{10}\right)^2$

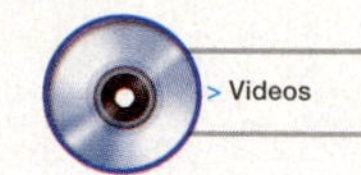


**38.** $\left(\frac{2}{7}\right)^2 \div \frac{3}{14}$

< Objective 4 >

*Divide the following. Be sure to attach the proper units.*

**39.** $900 \text{ mi} \div 15 \frac{\text{mi}}{\text{gal}}$

**40.** $1{,}500 \text{ joules} \div 75 \frac{\text{joules}}{\text{s}}$

**41.** $8{,}750 \text{ watts} \div 350 \frac{\text{watts}}{\text{s}}$

**42.** $\$75{,}744 \div \frac{\$3{,}156}{\text{month}}$

*Solve the following applications.*

**43. Construction** A $5\frac{1}{4}$ ft long wire is to be cut into 7 pieces of the same length. How long is each piece?

> Videos

**44. Crafts** A potter uses $\frac{2}{3}$ pound (lb) of clay in making a bowl. How many bowls can be made from 16 lb of clay?

**45.** **Statistics** Virginia made a trip of 95 mi in $1\frac{1}{4}$ h. What was her average speed?

**46.** **Business and Finance** A piece of land measures $3\frac{3}{4}$ acres and is for sale at \$60,000. What is the price per acre?

**47.** **Crafts** A roast weighs $3\frac{1}{4}$ lb. How many $\frac{1}{4}$-lb servings will the roast provide?

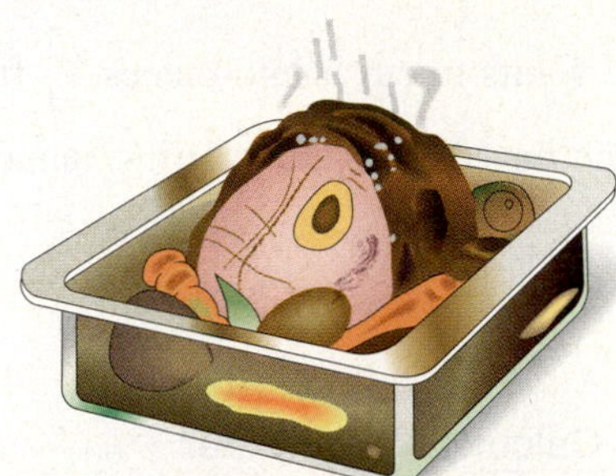

**48.** **Construction** A bookshelf is 55 in. long. If the books have an average thickness of $1\frac{1}{4}$ in., how many books can be put on the shelf?

Basic Skills | Advanced Skills | **Vocational-Technical Applications** | Calculator/Computer | Above and Beyond

**49.** **Manufacturing Technology** Calculate the distance from the center of hole A to the center of hole B.

A B

$8\frac{1}{4}$ in.

**50.** **Manufacturing Technology** A staircase is $95\frac{7}{8}$ in. tall and has 13 risers. What is the height of each riser? (*Hint:* Convert $95\frac{7}{8}$ into an improper fraction and divide by $\frac{13}{1}$.)

**51.** **Manufacturing Technology** A part that is $\frac{3}{4}$ in. wide is to be magnified for a detailed drawing. The scale is 1 in. = $\frac{3}{8}$ in. What is the width of the part in the drawing? (*Hint:* Divide the width by $\frac{3}{8}$.)

## Answers

45. ____________

46. ____________

47. ____________

48. ____________

49. ____________

50. ____________

51. ____________

 The Streeter/Hutchison Series in Mathematics Basic Mathematical Skills with Geometry

**Answers**

52. ______

53. ______

54. ______

55. ______

56. ______

57. ______

58. ______

59. ______

60. ______

61. ______

62. ______

**52. Manufacturing Technology** A typical unified threaded bolt has one thread every $\frac{1}{20}$ in. How many threads are in $\frac{1}{4}$ in.?

**53. Crafts** Manuel has $7\frac{1}{2}$ yd of cloth. He wants to cut it into strips $1\frac{3}{4}$ yd long. How many strips will he have? How much cloth remains, if any?

**54. Crafts** Evette has $41\frac{1}{2}$ ft of string. She wants to cut it into pieces $3\frac{3}{4}$ ft long. How many pieces of string will she have? How much string remains, if any?

Basic Skills | Advanced Skills | Vocational-Technical Applications | **Calculator/Computer** | Above and Beyond

## Using Your Calculator to Divide Fractions

Dividing fractions on your calculator is almost exactly the same as multiplying them. You simply press the [÷] key instead of the [×] key. Again, parentheses are important when using a calculator to work with fractions.

### Scientific Calculator

Dividing fractions on a scientific calculator requires only that you enter the problem followed by the equal sign. Recall that fractions are entered using the [**a b/c**] key.

### Graphing Calculator

When using a graphing calculator, you must choose the fraction option [**1:▶ Frac**] from the [**MATH**] menu before pressing [Enter].

*Find the following quotients, using your calculator.*

**55.** $\frac{1}{5} \div \frac{2}{15}$

**56.** $\frac{13}{17} \div \frac{39}{34}$

**57.** $\frac{5}{7} \div \frac{15}{28}$

**58.** $\frac{3}{7} \div \frac{9}{28}$

**59.** $\frac{15}{18} \div \frac{45}{27}$

**60.** $\frac{19}{63} \div \frac{38}{9}$

**61.** $\frac{25}{45} \div \frac{100}{135}$

**62.** $\frac{86}{24} \div \frac{258}{96}$

Basic Skills | Advanced Skills | Vocational-Technical Applications | Calculator/Computer | Above and Beyond

Answers

63. ______

64. ______

65. ______

66. ______

67. ______

68. ______

**63. Crafts** In squeezing oranges for fresh juice, 3 oranges yield about $\frac{1}{3}$ of a cup.

**(a)** How much juice could you expect to obtain from a bag containing 24 oranges?

**(b)** If you needed 8 cups of orange juice, how many bags of oranges should you buy?

**64. Number Problem** A farmer died and left 17 cows to be divided among three workers. The first worker was to receive $\frac{1}{2}$ of the cows, the second worker was to receive $\frac{1}{3}$ of the cows, and the third worker was to receive $\frac{1}{9}$ of the cows. The executor of the farmer's estate realized that 17 cows could not be divided into halves, thirds, or ninths and so added a neighbor's cow to the farmer's. With 18 cows, the executor gave 9 cows to the first worker, 6 cows to the second worker, and 2 cows to the third worker. This accounted for the 17 cows, so the executor returned the borrowed cow to the neighbor. Explain why this works.

**65.** In general division of fractions is not commutative.

For example, $\frac{3}{4} \div \frac{5}{6} \neq \frac{5}{6} \div \frac{3}{4}$.

There could be an exception. Can you think of a situation in which division of fractions would be commutative?

**66.** Josephine's boss tells her that her salary is to be divided by $\frac{1}{3}$. Should she quit?

**67.** Compare the English phrases "divide in half" and "divide by one-half." Do they say the same thing? Create examples to support your answer.

**68. (a)** Compute the following: $5 \div \frac{1}{10}$; $5 \div \frac{1}{100}$; $5 \div \frac{1}{1{,}000}$; $5 \div \frac{1}{10{,}000}$.

**(b)** As the divisor gets smaller (approaches 0), what happens to the quotient?

**(c)** What does this say about the answer to $5 \div 0$?

## Answers

**1.** $\frac{8}{7}$ **3.** 2 **5.** $\frac{3}{7}$ **7.** $\frac{4}{39}$ **9.** $\frac{4}{15}$ **11.** $\frac{8}{15}$ **13.** $\frac{2}{3}$
**15.** $1\frac{13}{50}$ **17.** $1\frac{1}{3}$ **19.** $\frac{4}{15}$ **21.** $\frac{1}{5}$ **23.** 18 **25.** $\frac{14}{17}$
**27.** $4\frac{1}{2}$ **29.** 6 **31.** $\frac{1}{4}$ **33.** 12 **35.** $\frac{2}{27}$ **37.** $6\frac{2}{3}$
**39.** 60 gal **41.** 25 s **43.** $\frac{3}{4}$ ft **45.** 76 $\frac{\text{mi}}{\text{h}}$ **47.** 13 servings
**49.** $2\frac{1}{4}$ in. **51.** 2 in. **53.** 4 strips; $\frac{1}{2}$ yd **55.** $\frac{3}{2}$
**57.** $\frac{4}{3}$ or $1\frac{1}{3}$ **59.** $\frac{1}{2}$ **61.** $\frac{3}{4}$ **63.** $2\frac{2}{3}$ cups; 3 bags
**65.** Above and Beyond **67.** Above and Beyond

# Activity 6 :: Adapting a Recipe

Tom and Susan like eating in ethnic restaurants, so they were thrilled when Marco's Cafe, an Indian restaurant, opened in their neighborhood. The first time they ate there, Susan had a bowl of Mulligatawny soup and she loved it. She decided that it would be a great soup to serve her friends, so she asked Marco for the recipe. Marco said that was no problem. He had already had so many requests for the recipe that he had made up a handout. A copy of it is reproduced here (try it if you are adventurous):

## Mulligatawny Soup

This recipe makes 10 gal; recommended serving size is a 12-ounce (12-oz) bowl. Sauté the following in a steam kettle until the onions are translucent:

| | |
|---|---|
| 10 lb | diced onion |
| 10 lb | diced celery |
| $\frac{1}{2}$ cup | garlic puree |
| 1 cup | madras curry |
| 2 cups | mild curry |

Add the following and bring to a boil:

| | |
|---|---|
| 4 cups | white wine |
| $\frac{1}{3}$ cup | sugar |
| 1 #10 can | diced tomato |
| 1 gal | fresh apple juice |
| $\frac{1}{3}$ cup | lemon juice |
| 2 gal | water |
| 1 #10 can | diced carrots |
| 16 oz | chicken stock |

Finish with:

Roux (1 lb butter and 1 lb flour) and 8 quarts cream (temper into hot liquid). Season to taste with salt, pepper, celery seed, basil, and garlic.

How many servings does this recipe make?

Visit local grocery stores to find out how much each item costs. Calculate the total cost for 10 gal of soup. What is the cost for each 12-oz serving? (This is called the *marginal cost*—it does not include the overhead for running the restaurant.)

What is roux? Call another restaurant to find out whether it would use the same definition.

Susan does want to make this soup for a dinner party she is having. Rewrite the recipe so that it will serve six 12-oz bowls. Use reasonable measures, such as teaspoons and cups. Answering the following questions will help. For some items you may have to experiment.

How many ounces in a #10 can?
How many cups in a gallon?
How many ounces in a pound?
How many teaspoons in a cup?
How many cups in a pound of diced onion?

| Definition/Procedure | Example | Reference |
|---|---|---|
| **Prime Numbers and Divisibility** | | Section 2.1 |
| **Prime Number** Any whole number that has exactly two factors, 1 and itself. | 7, 13, 29, and 73 are prime numbers. | *p.* 116 |
| **Composite Number** Any whole number greater than 1 that is not prime. | 8, 15, 42, and 65 are composite numbers. | *p.* 117 |
| **Zero and One** Zero and one are neither classified as prime nor composite numbers. | | *p.* 118 |
| ***Divisibility Tests*** | | |
| ***By 2*** A whole number is divisible by 2 if its last digit is 0, 2, 4, 6, or 8. | 932 is divisible by 2; 1,347 is not. | *p.* 118 |
| ***By 3*** A whole number is divisible by 3 if the sum of its digits is divisible by 3. | 546 is divisible by 3; 2,357 is not. | *p.* 118 |
| ***By 5*** A whole number is divisible by 5 if its last digit is 0 or 5. | 865 is divisible by 5; 23,456 is not. | *p.* 119 |
| **Factoring Whole Numbers** | | Section 2.2 |
| ***Prime Factorization*** | | |
| To find the prime factorization of a number, divide the number by a series of primes until the final quotient is a prime number. The prime factors include each prime divisor and the final quotient. | 2)630<br>3)315<br>3)105<br>5)35<br>7<br>So $630 = 2 \times 3 \times 3 \times 5 \times 7$. | *p.* 127 |
| **Greatest Common Factor (GCF)** The GCF is the *largest* number that is a factor of each of a group of numbers. | | *p.* 128 |

*Continued*

 The Streeter/Hutchison Series in Mathematics Basic Mathematical Skills with Geometry

| Definition/Procedure | Example | Reference |
|---|---|---|
| ***To Find the GCF*** | | |
| **Step 1** Write the prime factorization for each of the numbers in the group.<br>**Step 2** Locate the prime factors that are *common* to all the numbers.<br>**Step 3** The greatest common factor is the *product* of all of the common prime factors. If there are no common prime factors, the GCF is 1. | To find the GCF of 24, 30, and 36:<br>$24 = ② \times 2 \times 2 \times ③$<br>$30 = ② \times ③ \times 5$<br>$36 = ② \times 2 \times ③ \times 3$<br>The GCF is $2 \times 3 = 6$. | *p.* 129 |
| **Fraction Basics** | | **Section 2.3** |
| **Fraction** Fractions name a number of equal parts of a unit or whole. A fraction is written in the form $\frac{a}{b}$, in which $a$ and $b$ are whole numbers and $b$ cannot be zero. | | *p.* 135 |
| **Denominator** The number of equal parts into which the whole is divided.<br>**Numerator** The number of parts of the whole that are used. | $\frac{5}{8}$ ← Numerator (5), Denominator (8) | *p.* 135 |
| **Proper Fraction** A fraction whose numerator *is less than* its denominator. It names a number less than 1. | $\frac{2}{3}$ and $\frac{11}{15}$ are proper fractions. | *p.* 137 |
| **Improper Fraction** A fraction whose numerator *is greater than* or equal to its denominator. It names a number greater than or equal to 1. | $\frac{7}{5}$, $\frac{21}{20}$, and $\frac{8}{8}$ are improper fractions. | *p.* 137 |
| **Mixed Number** The sum of a whole number and a proper fraction. | $2\frac{1}{3}$ and $5\frac{7}{8}$ are mixed numbers.<br>Note that $2\frac{1}{3}$ means $2 + \frac{1}{3}$. | *p.* 138 |

| Definition/Procedure | Example | Reference |
|---|---|---|
| ***To Change an Improper Fraction into a Mixed Number*** | | |
| **Step 1** Divide the numerator by the denominator. The quotient is the whole-number portion of the mixed number.<br>**Step 2** If there is a remainder, write the remainder over the original denominator. This gives the fractional portion of the mixed number. | To change $\frac{22}{5}$ to a mixed number:<br>$5\overline{)22}$ gives 4 (Quotient); $20$; $2$ (Remainder)<br>$\frac{22}{5} = 4\frac{2}{5}$ | *p.* 139 |
| ***To Change a Mixed Number to an Improper Fraction*** | | |
| **Step 1** Multiply the denominator of the fraction by the whole-number portion of the mixed number.<br>**Step 2** Add the numerator of the fraction to that product.<br>**Step 3** Write that sum over the original denominator to form the improper fraction. | Denominator, Whole number, Numerator<br>$5\frac{3}{4} = \frac{(4 \times 5) + 3}{4} = \frac{23}{4}$<br>Denominator | *p.* 140 |
| **Simplifying Fractions** | | **Section 2.4** |
| **Equivalent Fractions** Two fractions that are equivalent (have equal value) are different names for the same number. | | *p.* 148 |
| **Cross Products**<br>$\frac{a}{b} \times \frac{c}{d}$   $a \times d$ and $b \times c$ are called the *cross products*.<br>If the cross products for two fractions are equal, the two fractions are equivalent. | $\frac{2}{3} = \frac{4}{6}$ because<br>$2 \times 6 = 3 \times 4$ | *p.* 148 |
| **The Fundamental Principle of Fractions**<br>For the fraction $\frac{a}{b}$, and any nonzero number $c$,<br>$\frac{a}{b} = \frac{a \div c}{b \div c}$<br>**In words:** We can divide the numerator and denominator of a fraction by the same nonzero number. The result is an equivalent fraction. | $\frac{8}{12} = \frac{8 \div 4}{12 \div 4} = \frac{2}{3}$<br>$\frac{8}{12}$ and $\frac{2}{3}$ are equivalent fractions. | *p.* 149 |

*Continued*

| Definition/Procedure | Example | Reference |
|---|---|---|
| **Simplest Form** A fraction is in simplest form, or in lowest terms, if the numerator and denominator have no common factors other than 1. This means that the fraction has the smallest possible numerator and denominator. | $\frac{2}{3}$ is in simplest form.<br>$\frac{12}{18}$ is *not* in simplest form.<br>The numerator and denominator have the common factor 6. | *p.* 150 |
| ***To Write a Fraction in Simplest Form*** | | |
| Divide the numerator and denominator by their greatest common factor. | $\frac{10}{15} = \frac{10 \div 5}{15 \div 5} = \frac{2}{3}$ | *p.* 150 |
| **Multiplying Fractions** | | **Section 2.5** |
| ***To Multiply Two Fractions*** | | |
| **Step 1** Multiply numerator by numerator. This gives the numerator of the product.<br>**Step 2** Multiply denominator by denominator. This gives the denominator of the product.<br>**Step 3** Simplify the resulting fraction if possible.<br>In multiplying fractions it is usually easiest to divide by any common factors in the numerator and denominator *before* multiplying. | $\frac{5}{8} \times \frac{3}{7} = \frac{5 \times 3}{8 \times 7} = \frac{15}{56}$<br>$\frac{5}{9} \times \frac{3}{10} = \frac{\overset{1}{\cancel{5}} \times \overset{1}{\cancel{3}}}{\underset{3}{\cancel{9}} \times \underset{2}{\cancel{10}}} = \frac{1}{6}$ | *p.* 162 |
| **Dividing Fractions** | | **Section 2.6** |
| ***To Divide Two Fractions*** | | |
| Invert the divisor and multiply. | $\frac{3}{7} \div \frac{4}{5} = \frac{3}{7} \times \frac{5}{4} = \frac{15}{28}$ | *p.* 180 |
| ***Multiplying or Dividing Mixed Numbers*** | | |
| Convert any mixed or whole numbers to improper fractions. Then multiply or divide the fractions as before. | $6\frac{2}{3} \times 3\frac{1}{5} = \frac{\overset{4}{\cancel{20}} \times 16}{3 \times \underset{1}{\cancel{5}}}$<br>$= \frac{64}{3} = 21\frac{1}{3}$ | *p.* 182 |

# summary exercises :: chapter 2

This summary exercise set is provided to give you practice with each of the objectives of this chapter. Each exercise is keyed to the appropriate chapter section. When you are finished, you can check your answers to the odd-numbered exercises against those presented in the back of the text. If you have difficulty with any of these questions, go back and reread the examples from that section. The answers to the even-numbered exercises appear in the *Instructor's Solutions Manual.* Your instructor will give you guidelines on how to best use these exercises in your instructional setting.

2.1 *List all the factors of the given numbers.*

**1.** 52

**2.** 41

*Use the group of numbers 2, 5, 7, 11, 14, 17, 21, 23, 27, 39, and 43.*

**3.** List the prime numbers; then list the composite numbers.

*Use the divisibility tests to determine which, if any, of the numbers 2, 3, and 5 are factors of the following numbers.*

**4.** 2,350

**5.** 33,451

2.2 *Find the prime factorization for the given numbers.*

**6.** 48

**7.** 420

**8.** 2,640

**9.** 2,250

*Find the greatest common factor (GCF).*

**10.** 15 and 20

**11.** 30 and 31

**12.** 24 and 40

**13.** 39 and 65

**14.** 49, 84, and 119

**15.** 77, 121, and 253

2.3 *Identify the numerator and denominator of each fraction.*

**16.** $\frac{5}{9}$

**17.** $\frac{17}{23}$

*Give the fractions that name the shaded portions of the following diagrams. Identify the numerator and the denominator.*

**18.**

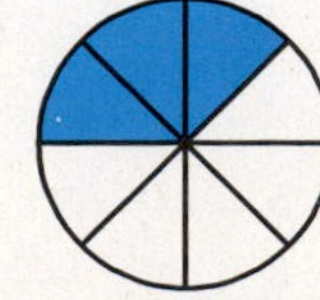

Fraction ________

Numerator ________

Denominator ________

**19.** 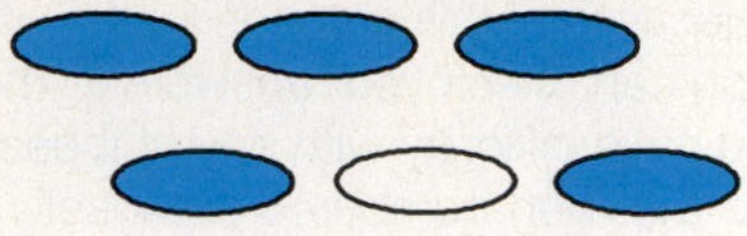

Fraction: ________

Numerator: ________

Denominator ________

**20.** From the following group of numbers:

$\frac{2}{3}, \frac{5}{4}, 2\frac{3}{7}, \frac{45}{8}, \frac{7}{7}, 3\frac{4}{5}, \frac{9}{1}, \frac{7}{10}, \frac{12}{5}, 5\frac{2}{9}$

List the proper fractions. ________

List the improper fractions. ________

List the mixed numbers. ________

*Convert to mixed or whole numbers.*

**21.** $\frac{41}{6}$ **22.** $\frac{32}{8}$ **23.** $\frac{23}{3}$ **24.** $\frac{47}{4}$

*Convert to improper fractions.*

**25.** $7\frac{5}{8}$ **26.** $4\frac{3}{10}$ **27.** $5\frac{2}{7}$ **28.** $12\frac{8}{13}$

**2.4** *Determine whether each of the following pairs of fractions is equivalent.*

**29.** $\frac{5}{8}, \frac{7}{12}$ **30.** $\frac{8}{15}, \frac{32}{60}$

*Write each fraction in simplest form.*

**31.** $\frac{24}{36}$ **32.** $\frac{45}{75}$ **33.** $\frac{140}{180}$ **34.** $\frac{16}{21}$

*Decide whether each is a true statement.*

**35.** $\frac{15}{25} = \frac{3}{5}$ **36.** $\frac{36}{40} = \frac{4}{5}$

**2.5** *Multiply.*

**37.** $\frac{7}{15} \times \frac{5}{21}$ **38.** $\frac{10}{27} \times \frac{9}{20}$ **39.** $4 \cdot \frac{3}{8}$ **40.** $3\frac{2}{5} \cdot \frac{5}{8}$

**41.** $5\frac{1}{3} \times 1\frac{4}{5}$

**42.** $1\frac{5}{12} \times 8$

**43.** $3\frac{1}{5} \times \frac{7}{8} \times 2\frac{6}{7}$

**2.1–2.5** *Solve the following applications.*

**44. Social Science** The scale on a map is 1 inch (in.) = 80 miles (mi). If two cities are $2\frac{3}{4}$ in. apart on the map, what is the actual distance between the cities?

**45. Construction** A kitchen measures $5\frac{1}{3}$ yards (yd) by $4\frac{1}{4}$ yd. If you purchase linoleum costing \$9 per square yard ($yd^2$), what will it cost to cover the floor?

**46. Construction** Your living room measures $6\frac{2}{3}$ yd by $4\frac{1}{2}$ yd. If you purchase carpeting at \$18 per square yard ($yd^2$), what will it cost to carpet the room?

**47. Business and Finance** Maria earns \$72 per day. If she works $\frac{5}{8}$ of a day, how much will she earn?

**48. Science and Medicine** David drove at an average speed of 65 mi/h for $2\frac{2}{5}$ h. How many miles did he travel?

**49. Social Science** The scale on a map is 1 in. = 120 mi. What actual distance, in miles, does $3\frac{2}{5}$ in. on the map represent?

**50. Social Science** At a college, $\frac{2}{5}$ of the students take a science course. Of the students taking science, $\frac{1}{4}$ take biology. What fraction of the students take biology?

**51. Social Science** A student survey found that $\frac{3}{4}$ of the students have jobs while going to school. Of those who have jobs, $\frac{5}{6}$ work more than 20 h/week. What fraction of those surveyed work more than 20 h/week?

**52. Construction** A living room has dimensions $5\frac{2}{3}$ yd by $4\frac{1}{2}$ yd. How much carpeting must be purchased to cover the room?

**2.6** *Divide.*

**53.** $\frac{5}{12} \div \frac{5}{8}$

**54.** $\dfrac{\frac{7}{15}}{\frac{14}{25}}$

**55.** $\dfrac{\frac{9}{20}}{\frac{12}{5}}$

**56.** $3\frac{3}{8} \div 2\frac{1}{4}$

**57.** $3\frac{3}{7} \div 8$

**58.** $6\frac{1}{7} \div \frac{3}{14}$

*Solve the following applications.*

**59. Construction** A piece of wire $3\frac{3}{4}$ ft long is to be cut into 5 pieces of the same length. How long will each piece be?

**60. Crafts** A blouse pattern requires $1\frac{3}{4}$ yd of fabric. How many blouses can be made from a piece of silk that is 28 yd long?

**61. Science and Medicine** If you drive 126 mi in $2\frac{1}{4}$ h, what is your average speed?

**62. Science and Medicine** If you drive 117 mi in $2\frac{1}{4}$ h, what is your average speed?

**63. Construction** An 18-acre piece of land is to be subdivided into home lots that are each $\frac{3}{8}$ acre. How many lots can be formed?

# self-test 2

Name ______

Section ______ Date ______

The purpose of this self-test is to help you check your progress so that you can find sections and concepts that you need to review before the next exam. Allow yourself about an hour to take this test. At the end of that hour, check your answers against those given in the back of the text. If you missed any, note the section reference that accompanies the answer. Go back to that section and reread the examples until you have mastered that particular concept.

**1.** Which of the numbers 5, 9, 13, 17, 22, 27, 31, and 45 are prime numbers? Which are composite numbers?

**2.** Use the divisibility tests to determine which, if any, of the numbers 2, 3, and 5 are factors of 54,204.

**3.** Find the prime factorization for 264.

*Find the greatest common factor (GCF) for the given numbers.*

**4.** 36 and 84

**5.** 16, 24, and 72

*What fraction names the shaded part of each diagram? Identify the numerator and denominator.*

**6.**

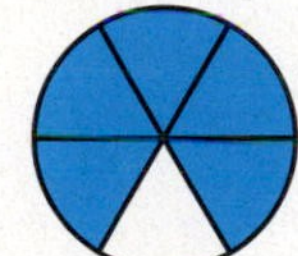

**7.**

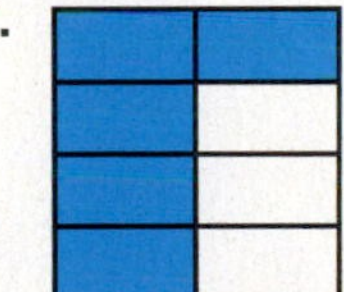

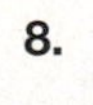

**9.** Identify the proper fractions, improper fractions, and mixed numbers in the following group.

$\frac{10}{11}, \frac{9}{5}, \frac{7}{7}, \frac{8}{1}, 2\frac{3}{5}, \frac{1}{8}$

Proper Improper Mixed number

## Answers

1. ______
2. ______
3. ______
4. ______
5. ______
6. ______
7. ______
8. ______
9. ______

Answers

10. ______

11. ______

12. ______

13. ______

14. ______

15. ______

16. ______

17. ______

18. ______

19. ______

20. ______

21. ______

22. ______

23. ______

24. ______

25. ______

26. ______

**10.** Give the mixed number that names the shaded part of the following diagram.

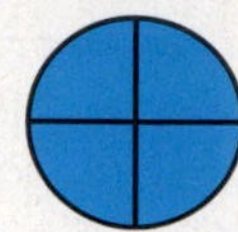
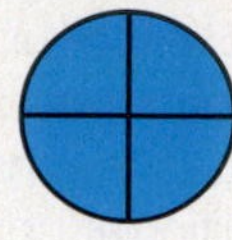
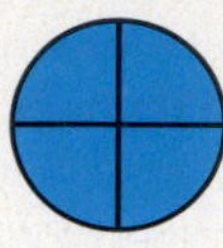
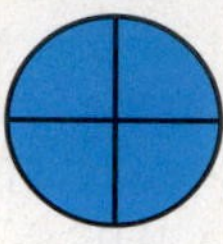
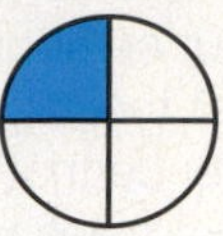

*Convert the fractions to mixed or whole numbers.*

**11.** $\frac{17}{4}$

**12.** $\frac{74}{8}$

**13.** $\frac{18}{6}$

**14.** $\frac{15}{1}$

*Convert the mixed numbers to improper fractions.*

**15.** $5\frac{2}{7}$

**16.** $4\frac{3}{8}$

**17.** $8\frac{2}{9}$

*Use the cross-product method to find out whether the pair of fractions is equivalent.*

**18.** $\frac{2}{7}, \frac{8}{28}$

**19.** $\frac{8}{20}, \frac{12}{30}$

**20.** $\frac{3}{20}, \frac{2}{15}$

*Write the fractions in simplest form.*

**21.** $\frac{21}{27}$

**22.** $\frac{36}{84}$

**23.** $\frac{8}{23}$

*Determine whether each statement is true.*

**24.** $\frac{28}{35} = \frac{4}{5}$

**25.** $\frac{42}{98} = \frac{6}{7}$

**26.** $\frac{105}{120} = \frac{7}{8}$

*Multiply.*

27. $\frac{2}{3} \times \frac{5}{7}$

28. $\frac{9}{10} \times \frac{5}{8}$

29. $2\frac{2}{3} \times 1\frac{2}{7}$

30. $3\frac{5}{6} \times 2\frac{2}{5}$

31. $\frac{16}{35} \times \frac{14}{24}$

32. $5\frac{1}{3} \times \frac{3}{4}$

*Solve each application.*

33. **Business and Finance** What is the cost of $2\frac{3}{4}$ pounds (lb) of apples if the price per pound is 48 cents?

34. **Construction** A room measures $5\frac{1}{3}$ yd by $3\frac{3}{4}$ yd. How many square yards ($yd^2$) of linoleum must be purchased to cover the floor?

35. **Social Science** The scale on a map is 1 inch (in.) = 80 miles (mi). If two towns are $2\frac{3}{8}$ in. apart on the map, what is the actual distance in miles between the towns?

*Divide.*

36. $\frac{6}{7} \div \frac{3}{4}$

37. $\frac{7}{12} \div \frac{14}{15}$

## Answers

27. ______

28. ______

29. ______

30. ______

31. ______

32. ______

33. ______

34. ______

35. ______

36. ______

37. ______

**Answers**

38. ______

39. ______

40. ______

41. ______

**38.** $1\frac{3}{4} \div 1\frac{3}{8}$

**39.** $5\frac{3}{5} \div 2\frac{1}{10}$

**40. Construction** A $31\frac{1}{3}$-acre piece of land is subdivided into home lots. Each home lot is to be $\frac{2}{3}$ acre. How many homes can be built?

**41. Crafts** A bookshelf is 66 in. long. If the thickness of each book on the shelf is $1\frac{3}{8}$ in., how many books can be placed on the shelf?

# cumulative review chapters 1-2

Name ______________________

Section ________ Date ________

The following exercises are presented to help you review concepts from earlier chapters. This is meant as review material and not as a comprehensive exam. The answers are presented in the back of the text. Beside each answer is a section reference for the concept. If you have difficulty with any of these exercises, be certain to at least read through the summary related to that section.

**1.** Give the place value of 7 in 3,738,500.

**2.** Give the word name for 302,525.

**3.** Write two million, four hundred thirty thousand as a numeral.

*Name the property of addition that is illustrated.*

**4.** $5 + 12 = 12 + 5$

**5.** $9 + 0 = 9$

**6.** $(7 + 3) + 8 = 7 + (3 + 8)$

*Perform the indicated operations.*

**7.**
$$\begin{array}{r} 593 \\ 275 \\ +\ 98 \\ \hline \end{array}$$

**8.** Find the sum of 58, 673, 5,325, and 17,295.

*Round the numbers to the indicated place value.*

**9.** 5,873 to the nearest hundred

**10.** 953,150 to the nearest ten thousand

*Estimate the sum by rounding each addend to the nearest hundred.*

**11.**
$$\begin{array}{r} 943 \\ 3,281 \\ 778 \\ 2,112 \\ +\ 570 \\ \hline \end{array}$$

*Complete the statements by using the symbol < or >.*

**12.** 49 ______ 47

**13.** 80 ______ 90

## Answers

1. ______________________

2. ______________________

3. ______________________

4. ______________________

5. ______________________

6. ______________________

7. ______________________

8. ______________________

9. ______________________

10. ______________________

11. ______________________

12. ______________________

13. ______________________

**Answers**

14. ______
15. ______
16. ______
17. ______
18. ______
19. ______
20. ______
21. ______
22. ______
23. ______
24. ______
25. ______
26. ______
27. ______
28. ______
29. ______

*Perform the indicated operations.*

**14.** $\begin{array}{r} 4{,}834 \\ -\ 973 \\ \hline \end{array}$

**15.** Find the difference of 25,000 and 7,535.

*Solve the applications.*

**16. STATISTICS** Attendance for five performances of a play was 172, 153, 205, 193, and 182. How many people in total attended those performances?

**17. BUSINESS AND FINANCE** Alan bought a Volkswagen with a list price of $18,975. He added stereo equipment for $439 and an air conditioner for $615. If he made a down payment of $2,450, what balance remained on the car?

*Name the property of addition and/or multiplication that is illustrated.*

**18.** $3 \times (4 \times 7) = (3 \times 4) \times 7$

**19.** $3 \times 4 = 4 \times 3$

**20.** $5 \times (2 + 4) = 5 \times 2 + 5 \times 4$

*Perform the indicated operations.*

**21.** $\begin{array}{r} 538 \\ \times\ 703 \\ \hline \end{array}$

**22.** $\begin{array}{r} 1{,}372 \\ \times\ 500 \\ \hline \end{array}$

*Solve the application.*

**23. CONSTRUCTION** A classroom is 8 yards (yd) wide by 9 yd long. If the room is to be recarpeted with material costing $14 per square yard, find the cost of the carpeting.

*Divide, using long division.*

**24.** $48\overline{)3{,}259}$

**25.** $458\overline{)47{,}350}$

*Evaluate the expressions.*

**26.** $3 + 5 \times 7$

**27.** $(3 + 5) \times 7$

**28.** $4 \times 3^2$

**29.** $2 + 8 \times 3 \div 4$

*Solve the applications.*

**30. Business and Finance** William bought a washer-dryer combination that, with interest charges, cost \$841. He paid \$145 down and agreed to pay the balance in 12 monthly payments. Find the amount of each payment.

**31. Number Problem** Which of the numbers 5, 9, 13, 17, 22, 27, 31, and 45 are prime numbers? Which are composite numbers?

**32.** Use the divisibility tests to determine which, if any, of the numbers 2, 3, and 5 are factors of 54,204.

**33.** Find the prime factorization for 264.

*Find the greatest common factor (GCF) for the given numbers.*

**34.** 36 and 96

**35.** 16, 40, and 72

*Identify the proper fractions, improper fractions, and mixed numbers from the following group.*

$\frac{7}{12}, \frac{10}{8}, 3\frac{1}{5}, \frac{9}{9}, \frac{7}{1}, \frac{3}{7}, 2\frac{2}{3}$

**36.** Proper: ___________ Improper: ___________

Mixed numbers: ___________

*Convert to mixed or whole numbers.*

**37.** $\frac{14}{5}$

**38.** $\frac{28}{7}$

*Convert to improper fractions.*

**39.** $4\frac{1}{3}$

**40.** $7\frac{7}{8}$

*Determine whether each pair of fractions is equivalent.*

**41.** $\frac{7}{21}, \frac{8}{24}$

**42.** $\frac{7}{12}, \frac{8}{15}$

## Answers

30. ___________

31. ___________

32. ___________

33. ___________

34. ___________

35. ___________

36. ___________

37. ___________

38. ___________

39. ___________

40. ___________

41. ___________

42. ___________

**Answers**

43. ____________

44. ____________

45. ____________

46. ____________

47. ____________

48. ____________

49. ____________

50. ____________

51. ____________

52. ____________

53. ____________

54. ____________

55. ____________

*Write the fraction in simplest form.*

**43.** $\frac{28}{42}$

**44.** $\frac{36}{96}$

*Multiply.*

**45.** $\frac{5}{9} \times \frac{8}{15}$

**46.** $\frac{20}{21} \cdot \frac{7}{25}$

**47.** $1\frac{1}{8} \cdot 4\frac{4}{5}$

**48.** $8 \times 2\frac{5}{6}$

**49.** $\frac{2}{3} \times 1\frac{4}{5} \times \frac{5}{8}$

*Divide.*

**50.** $\frac{5}{8} \div \frac{15}{32}$

**51.** $2\frac{5}{8} \div \frac{7}{12}$

**52.** $4\frac{1}{6} \div 5$

**53.** $2\frac{2}{7} \div 1\frac{11}{21}$

*Solve each application.*

**54. CONSTRUCTION** Your living room measures $6\frac{2}{3}$ yd by $4\frac{1}{2}$ yd. If you purchase carpeting at \$18 per square yard (yd$^2$), what will it cost to carpet the room?

**55. CONSTRUCTION** If a stack of $\frac{5}{8}$-in. plywood measures 55 in. high, how many sheets of plywood are in the stack?

CHAPTER

# Adding and Subtracting Fractions

## INTRODUCTION

Carpentry is one of the oldest and most important of all trades. It dates back to ancient times and the earliest use of primitive tools. It includes large-scale work such as architecture and smaller-scale work such as cabinetry and furniture making. Carpenters mostly work with wood, but they also use ceramic, metal, and plastic. Sometimes carpenters also do roofing, refinishing, remodeling, restoration, and flooring.

Carpenters need to have a vast knowledge of scale drawing and an understanding of blueprints. They have to be able to use a substantial amount of math for measuring and making scale drawings and drafts. They work with models that are hundreds of times smaller than the actual construction. Sometimes these are actual models, and sometimes they are drawings. Carpenters have to be extremely precise in their measuring, sometimes to tiny fractions of an inch. Errors in measurement can have dire consequences such as warping or cracking of materials later on.

Carpenters learn their trades from vocational schools or by serving as apprentices to a more experienced carpenter. Some carpenters are highly skilled and looked upon as artists, whereas others just do handy work.

## CHAPTER 3 OUTLINE

# 3 pretest



Name ______________________

Section ________ Date ________

This pretest provides a preview of the types of exercises you will encounter in each section of this chapter. The answers for these exercises can be found in the back of the text. If you are working on your own, or ahead of the class, this pretest can help you identify the sections in which you should focus more of your time.

## Answers

1. ______________________
2. ______________________
3. ______________________
4. ______________________
5. ______________________
6. ______________________
7. ______________________
8. ______________________
9. ______________________
10. ______________________
11. ______________________
12. ______________________

**3.1** **1.** **(a)** $\frac{3}{7} + \frac{2}{7}$ **(b)** $\frac{8}{9} - \frac{3}{9}$

**3.2** **2.** Find the least common denominator for fractions with the denominators 16 and 24.

**3.** Find the least common multiple of the following:

**(a)** 25 and 40 **(b)** 20, 24, and 30

**3.3** Add or subtract. Simplify when possible.

**4.** $\frac{3}{10} + \frac{4}{15}$ **5.** $\frac{4}{5} + \frac{7}{8} + \frac{9}{20}$ **6.** $\frac{7}{24} - \frac{2}{9}$

**3.4** **7.** $3\frac{5}{6} + 2\frac{7}{8}$ **8.** $2\frac{1}{9} - 1\frac{1}{12}$

**3.5** Evaluate and simplify.

**9.** $\left(\frac{3}{5}\right)^2 - \frac{1}{2} \cdot \frac{1}{5}$ **10.** $\frac{2}{3} \times \frac{3}{4} + \left(\frac{1}{6}\right)^2$

**3.6** **11.** **Construction** A house plan calls for $12\frac{3}{4}$ yd$^2$ of carpeting in the living room and $5\frac{1}{2}$ yd$^2$ in the hallway. How much carpeting will be needed?

**12.** **Business and Finance** A stock is listed at $50\frac{1}{4}$ points at the start of a week. By the end of the week it is at $54\frac{5}{8}$ points. How much did it gain during the week?

# 3.1 Adding and Subtracting Fractions with Like Denominators

< 3.1 Objectives >

1 > Add two like fractions

2 > Add a group of like fractions

3 > Subtract two like fractions

Recall from our work in Chapter 1 that adding can be thought of as combining groups of the *same kind* of objects. This is also true when you think about adding fractions.

Fractions can be added only if they name a number of the *same parts* of a whole. This means that we can add fractions only when they are **like fractions,** that is, when they have the *same* (*a common*) denominator. For instance, we can add two nickels and three nickels to get five nickels. We *cannot* directly add two nickels and three dimes!

As long as we are dealing with like fractions, addition is an easy matter. Just use the following rule.

**Step by Step**

**To Add Like Fractions**

**Step 1** Add the numerators.
**Step 2** Place the sum over the common denominator.
**Step 3** Simplify the resulting fraction when necessary.

Example 1 illustrates the use of this rule.

**Example 1** Adding Like Fractions

< Objective 1 >

Add.

$$\frac{1}{5} + \frac{3}{5}$$

**Step 1** Add the numerators.

$1 + 3 = 4$

**Step 2** Write that sum over the common denominator, 5. We are done at this point because the answer, $\frac{4}{5}$, is in the simplest possible form.

Step 1 Step 2

$$\frac{1}{5} + \frac{3}{5} = \frac{1 + 3}{5} = \frac{4}{5}$$




The Streeter/Hutchison Series in Mathematics Basic Mathematical Skills with Geometry

NOTE

Combining 1 of the 5 parts with 3 of the 5 parts gives a total of 4 of the 5 equal parts.

We illustrate this addition with a diagram.

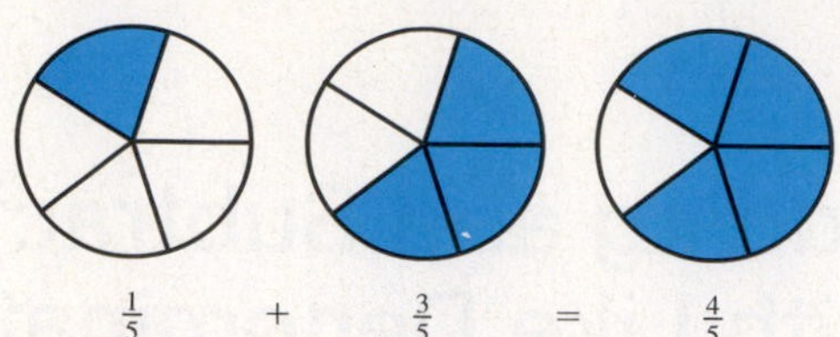

$\frac{1}{5} + \frac{3}{5} = \frac{4}{5}$

Check Yourself 1

Add.

$$\frac{2}{9} + \frac{5}{9}$$

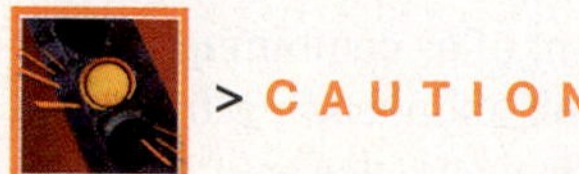

**Be Careful!** In adding fractions, *do not* follow the rule for multiplying fractions. To multiply $\frac{1}{5} \times \frac{3}{5}$, you would multiply both the numerators and the denominators:

$$\frac{1}{5} \times \frac{3}{5} = \frac{3}{25}$$

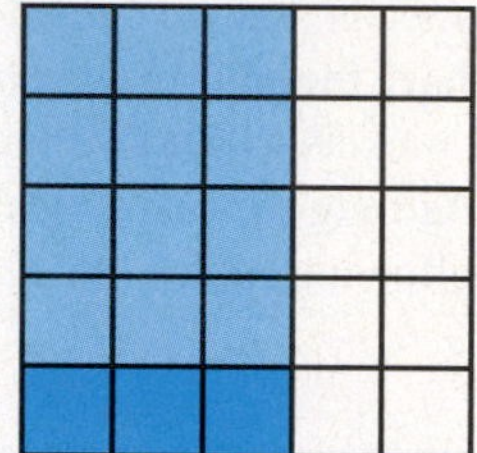

$\frac{1}{5}$ of $\frac{3}{5} = \frac{3}{25}$

However, when you add two fractions, the sum will have the same like denominator. So, you **do not** add the denominators.

$$\frac{1}{5} + \frac{3}{5} \neq \frac{4}{10}$$

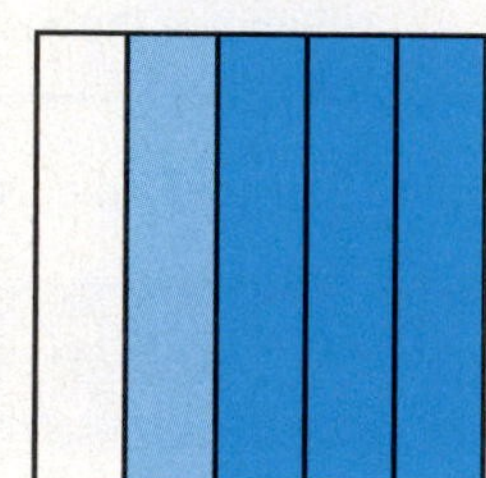

$\frac{1}{5} + \frac{3}{5} = \frac{4}{5}$

Step 3 of the addition rule for like fractions tells us to *simplify* the sum. Fractions should always be written in lowest terms. Consider Example 2.

## Example 2 Adding Like Fractions That Require Simplifying

Add and simplify.

Step 3

$$\frac{3}{12} + \frac{5}{12} = \frac{8}{12} = \frac{2}{3}$$

The sum $\frac{8}{12}$ is *not* in lowest terms.

Divide the numerator and denominator by 4 to simplify the result.

### Check Yourself 2

Add.

$$\frac{4}{15} + \frac{6}{15}$$

If the sum of two fractions is an improper fraction, we usually write that sum as a mixed number.

### Example 3 Adding Like Fractions That Result in Mixed Numbers

**NOTE**

Add as before. Then convert the sum to a mixed number.

Add.

$$\frac{5}{9} + \frac{8}{9} = \frac{13}{9} = 1\frac{4}{9}$$ Write the sum $\frac{13}{9}$ as a mixed number.

### Check Yourself 3

Add.

$$\frac{7}{12} + \frac{10}{12}$$

We can easily extend our addition rule to find the sum of more than two fractions as long as they all have the same denominator. This is shown in Example 4.

### Example 4 Adding a Group of Like Fractions

< Objective 2 >

Add.

$$\frac{2}{7} + \frac{3}{7} + \frac{6}{7} = \frac{11}{7}$$ Add the numerators: $2 + 3 + 6 = 11$.

$$= 1\frac{4}{7}$$

### Check Yourself 4

Add.

$$\frac{1}{8} + \frac{3}{8} + \frac{5}{8}$$

Many applications can be solved by adding fractions.

## Example 5 An Application Involving the Adding of Like Fractions

Noel walked $\frac{9}{10}$ miles (mi) to Jensen's house and then walked $\frac{7}{10}$ mi to school. How far did Noel walk?

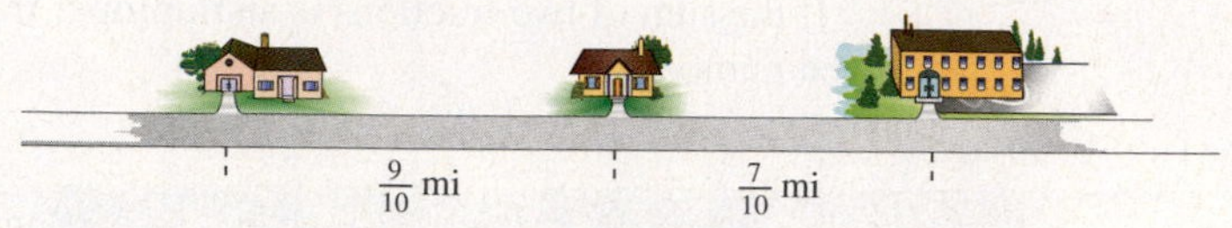


To find the total distance Noel walked, add the two distances.

$$\frac{9}{10} + \frac{7}{10} = \frac{16}{10} = 1\frac{6}{10} = 1\frac{3}{5}$$

Noel walked $1\frac{3}{5}$ mi.

### Check Yourself 5

Emir bought $\frac{7}{16}$ pounds (lb) of candy at one store and $\frac{11}{16}$ lb at another store. How much candy did Emir buy?

**NOTE**

Like fractions have the same denominator.

If a problem involves like fractions, then subtraction, like addition, is not difficult.

**Step by Step**

**To Subtract Like Fractions**

**Step 1** Subtract the numerators.
**Step 2** Place the difference over the common denominator.
**Step 3** Simplify the resulting fraction when necessary.

## Example 6 Subtracting Like Fractions

< Objective 3 >

Subtract.

**(a)** $\frac{4}{5} - \frac{2}{5} = \frac{4-2}{5} = \frac{2}{5}$

Step 1 (points to $\frac{4-2}{5}$); Step 2 (points to $\frac{2}{5}$)

Subtract the numerators: $4 - 2 = 2$. Write the difference over the common denominator, 5. Step 3 is not necessary because the difference is in simplest form.

> **NOTE**
>
> Subtracting 2 of the 5 parts from 4 of the 5 parts leaves 2 of the 5 parts.

Illustrating with a diagram:

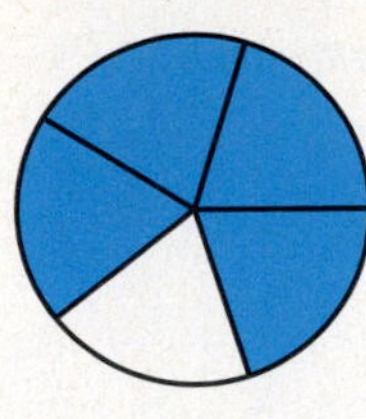
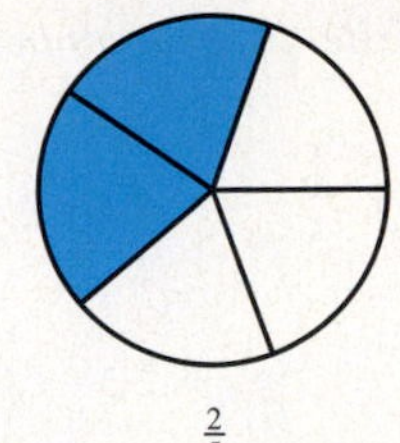
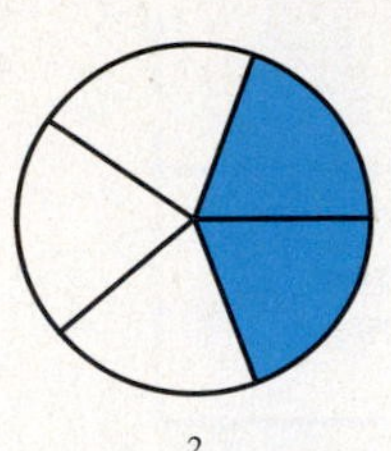

$\frac{4}{5} \quad - \quad \frac{2}{5} \quad = \quad \frac{2}{5}$

> **NOTE**
>
> Always write the result in lowest terms.

**(b)** $\frac{5}{8} - \frac{3}{8} = \frac{5-3}{8} = \frac{2}{8} = \frac{1}{4}$

## Check Yourself 6

Subtract.

$\frac{11}{12} - \frac{5}{12}$

## Check Yourself ANSWERS

**1.** $\frac{7}{9}$ **2.** $\frac{2}{3}$ **3.** $1\frac{5}{12}$ **4.** $1\frac{1}{8}$ **5.** $1\frac{1}{8}$ lb **6.** $\frac{1}{2}$

## Reading Your Text

The following fill-in-the-blank exercises are designed to ensure that you understand some of the key vocabulary used in this section.

### SECTION 3.1

**(a)** Fractions with the same (common) denominator are called ____________ fractions.

**(b)** When adding like fractions, add the ____________.

**(c)** After adding two fractions, ____________ the result when necessary.

**(d)** The result of subtraction is called the ____________.

# 3.1 exercises

Basic Skills | Advanced Skills | Vocational-Technical Applications | Calculator/Computer | Above and Beyond

MathZone

Boost your grade at mathzone.com!

> Practice Problems
> NetTutor
> Self-Tests
> e-Professors
> Videos

Name ____________

Section ________ Date ________

## Answers

1. ______ 2. ______
3. ______ 4. ______
5. ______ 6. ______
7. ______ 8. ______
9. ______ 10. ______
11. ______ 12. ______
13. ______ 14. ______
15. ______ 16. ______
17. ______ 18. ______
19. ______ 20. ______
21. ______ 22. ______
23. ______ 24. ______

### < Objective 1 >

*Add. Write all answers in lowest terms.*

**1.** $\frac{3}{5} + \frac{1}{5}$

**2.** $\frac{4}{7} + \frac{1}{7}$

**3.** $\frac{4}{11} + \frac{6}{11}$

**4.** $\frac{5}{16} + \frac{4}{16}$

**5.** $\frac{2}{10} + \frac{3}{10}$

> Videos

**6.** $\frac{5}{12} + \frac{1}{12}$

**7.** $\frac{3}{7} + \frac{4}{7}$

**8.** $\frac{13}{20} + \frac{17}{20}$

**9.** $\frac{29}{30} + \frac{11}{30}$

**10.** $\frac{4}{9} + \frac{5}{9}$

**11.** $\frac{13}{48} + \frac{23}{48}$

**12.** $\frac{17}{60} + \frac{31}{60}$

**13.** $\frac{3}{7} + \frac{6}{7}$

**14.** $\frac{3}{5} + \frac{4}{5}$

**15.** $\frac{7}{10} + \frac{9}{10}$

**16.** $\frac{5}{8} + \frac{7}{8}$

**17.** $\frac{11}{12} + \frac{10}{12}$

**18.** $\frac{13}{18} + \frac{11}{18}$

### < Objective 2 >

**19.** $\frac{1}{8} + \frac{1}{8} + \frac{3}{8}$

**20.** $\frac{1}{10} + \frac{3}{10} + \frac{3}{10}$

**21.** $\frac{1}{9} + \frac{4}{9} + \frac{5}{9}$

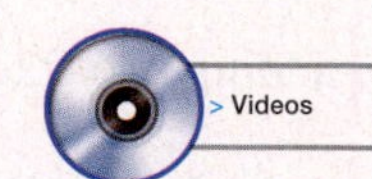


**22.** $\frac{7}{12} + \frac{11}{12} + \frac{1}{12}$

### < Objective 3 >

*Subtract. Write all answers in lowest terms.*

**23.** $\frac{3}{5} - \frac{1}{5}$

**24.** $\frac{5}{7} - \frac{2}{7}$

25. $\frac{7}{9} - \frac{4}{9}$

26. $\frac{7}{10} - \frac{3}{10}$

27. $\frac{13}{20} - \frac{3}{20}$

28. $\frac{19}{30} - \frac{17}{30}$

29. $\frac{19}{24} - \frac{5}{24}$

30. $\frac{25}{36} - \frac{13}{36}$

31. $\frac{11}{12} - \frac{7}{12}$

32. $\frac{9}{10} - \frac{6}{10}$

33. $\frac{8}{9} - \frac{3}{9}$

34. $\frac{5}{8} - \frac{1}{8}$

*Solve the following applications. Write each answer in lowest terms.*

35. **Number Problem** You work 7 hours (h) one day, 5 h the second day, and 6 h the third day. How long did you work, as a fraction of a 24-h day?

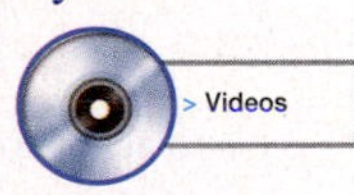


36. **Number Problem** One task took 7 minutes (min), a second task took 12 min, and a third task took 21 min. How long did the three tasks take, as a fraction of an hour?

37. **Geometry** What is the perimeter of a rectangle if the length is $\frac{7}{10}$ inches (in.) and the width is $\frac{2}{10}$ in.?

38. **Geometry** Find the perimeter of a rectangular picture if the width is $\frac{7}{9}$ yd and the length is $\frac{5}{9}$ yd.

## Answers

25. ______
26. ______
27. ______
28. ______
29. ______
30. ______
31. ______
32. ______
33. ______
34. ______
35. ______
36. ______
37. ______
38. ______

## Answers

39. ______

40. ______

41. ______

42. ______

43. ______

44. ______

45. ______

46. ______

**39.** **Statistics** Patrick spent $\frac{4}{9}$ of an hour (h) in the batting cages on Friday and $\frac{7}{9}$ of an hour on Saturday. He wants to spend 2 h total on the weekend. How much time should he spend on Sunday to accomplish this goal?

**40.** **Business and Finance** Maria, a road inspector, must inspect $\frac{17}{30}$ of a mile of road. If she has already inspected $\frac{11}{30}$ of a mile, how much more does she need to inspect?

Basic Skills | **Advanced Skills** | Vocational-Technical Applications | Calculator/Computer | Above and Beyond

**Geometry** *Find the perimeters of the following triangles.*

**41.**

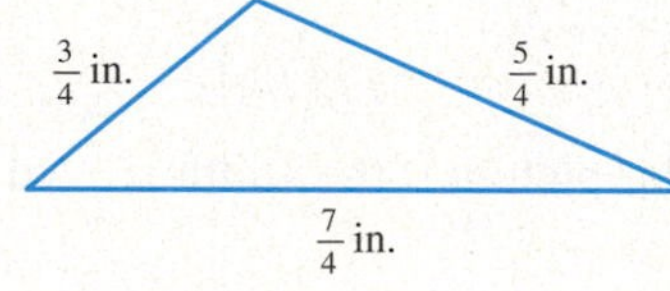


**42.**

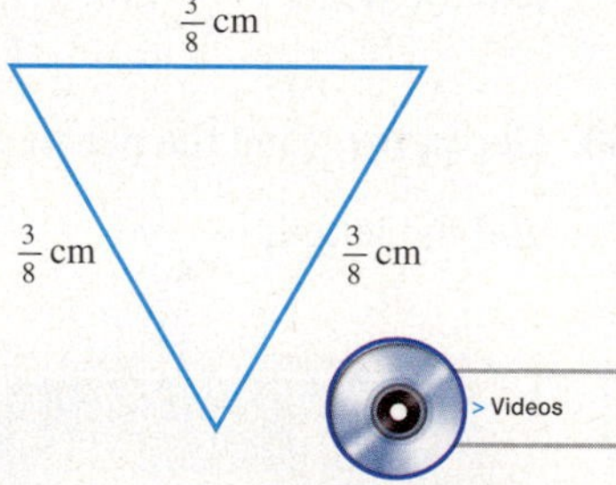


**43.**

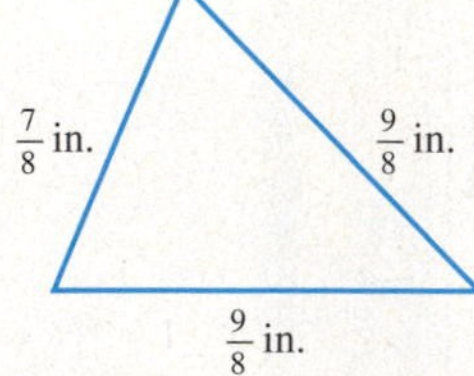


**44.**

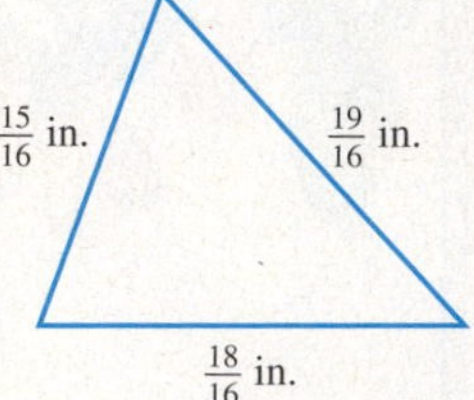


*Evaluate each of the following. Write all answers in lowest terms.*

**45.** $\frac{7}{12} - \frac{4}{12} + \frac{3}{12}$

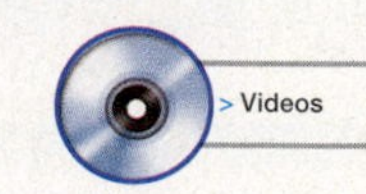


**46.** $\frac{8}{9} + \frac{3}{9} - \frac{5}{9}$

47. $\frac{6}{13} - \frac{3}{13} + \frac{11}{13}$

48. $\frac{9}{11} - \frac{3}{11} + \frac{7}{11}$

49. $\frac{18}{23} - \frac{13}{23} - \frac{3}{23}$

50. $\frac{17}{18} - \frac{11}{18} - \frac{5}{18}$

Basic Skills | Advanced Skills | **Vocational-Technical Applications** | Calculator/Computer | Above and Beyond

51. **ALLIED HEALTH** Carla is to be given $\frac{1}{8}$ gram (g) of medication in the morning, $\frac{1}{8}$ g at 3 P.M. in the afternoon, and $\frac{3}{8}$ g before she goes to bed. How many grams of medication will she receive in one day?

52. **ALLIED HEALTH** Prior to chemotherapy, a patient's malignant tumor weighed $\frac{7}{8}$ pound (lb). After the initial round of chemotherapy, the tumor's weight had been reduced by $\frac{3}{8}$ lb. How much did the tumor weigh after the chemotherapy treatment?

53. **MANUFACTURING TECHNOLOGY** Triplet Precision Machine has ordered $\frac{5}{8}$ ton of steel. The first truck arrived with $\frac{1}{8}$ ton of the steel. How much is yet to arrive?

54. **MANUFACTURING TECHNOLOGY** A sidewalk will require $\frac{15}{16}$ yard (yd) of concrete. The concrete mixer is capable of mixing $\frac{3}{16}$ yd at a time. How much concrete is still needed after one mixer load is used? chapter 3 > Make the Connection

**GEOMETRY** *Find the perimeters of the following polygons.*

55.

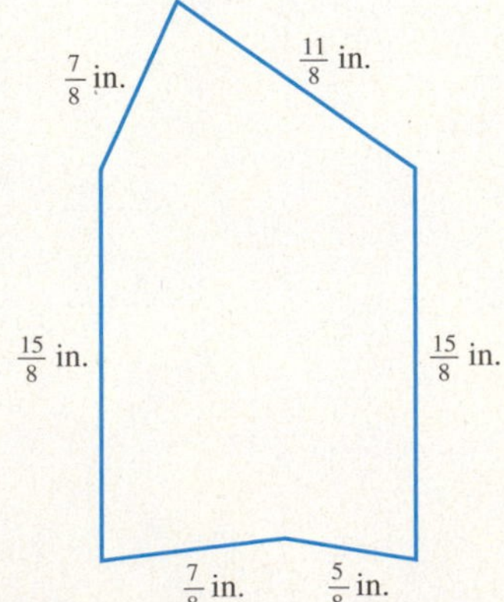


56.

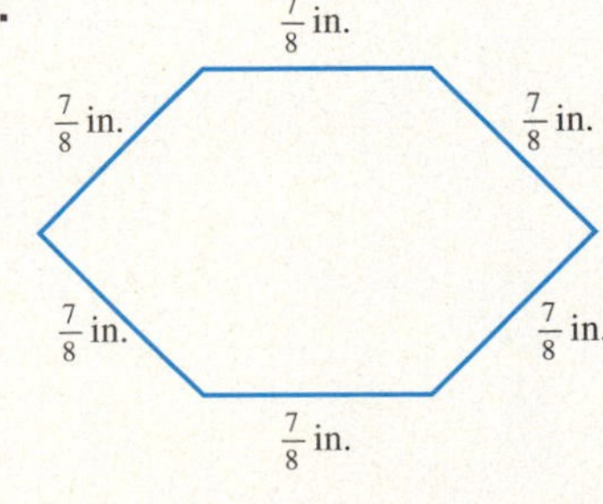


## Answers

47. ______
48. ______
49. ______
50. ______
51. ______
52. ______
53. ______
54. ______
55. ______
56. ______

## Answers

1. $\frac{4}{5}$ 3. $\frac{10}{11}$ 5. $\frac{1}{2}$ 7. 1 9. $\frac{4}{3} = 1\frac{1}{3}$ 11. $\frac{3}{4}$ 13. $1\frac{2}{7}$
15. $1\frac{3}{5}$ 17. $1\frac{3}{4}$ 19. $\frac{5}{8}$ 21. $1\frac{1}{9}$ 23. $\frac{2}{5}$ 25. $\frac{1}{3}$ 27. $\frac{1}{2}$
29. $\frac{7}{12}$ 31. $\frac{1}{3}$ 33. $\frac{5}{9}$ 35. $\frac{3}{4}$ day 37. $\frac{9}{5}$ in. $= 1\frac{4}{5}$ in.
39. $\frac{7}{9}$ h 41. $\frac{15}{4}$ in. $= 3\frac{3}{4}$ in. 43. $\frac{25}{8}$ in. $= 3\frac{1}{8}$ in. 45. $\frac{1}{2}$
47. $\frac{14}{13} = 1\frac{1}{13}$ 49. $\frac{2}{23}$ 51. $\frac{5}{8}$ g 53. $\frac{1}{2}$ ton
55. $\frac{15}{2}$ in. $= 7\frac{1}{2}$ in.

# 3.2 Common Multiples

< 3.2 Objectives >

1 > Find the least common multiple (LCM) of two numbers

2 > Find the LCM of a group of numbers

3 > Compare the size of two fractions

In this chapter, we discuss the process used for adding or subtracting fractions. One of the most important concepts we use when we add or subtract fractions is that of **multiples.**

**Definition**

**Multiples** The *multiples* of a number are the product of that number with the natural numbers 1, 2, 3, 4, 5, . . .

### Example 1 Listing Multiples

List the multiples of 3.

The multiples of 3 are

$3 \times 1, 3 \times 2, 3 \times 3, 3 \times 4, \ldots$

or

$3, 6, 9, 12, \ldots$ The three dots indicate that the list continues without stopping.

**NOTE**

Notice that the multiples, except for 3 itself, are *larger* than 3.

An easy way of listing the multiples of 3 is to think of *counting by threes.*

**Check Yourself 1**

List the first seven multiples of 4.

Sometimes we need to find common multiples of two or more numbers.

**Definition**

**Common Multiples** If a number is a multiple of each of a group of numbers, it is called a *common multiple* of the numbers; that is, it is a number that is evenly divisible by each of the numbers in the group.

## Example 2 Finding Common Multiples

**NOTE**

15, 30, 45, and 60 are multiples of *both* 3 and 5.

Find four common multiples of 3 and 5.

Some common multiples of 3 and 5 are

15, 30, 45, 60

**Check Yourself 2**

**List the first six multiples of 6. Then look at your list from Check Yourself 1 and list some common multiples of 4 and 6.**

For our later work, we will use the *least common multiple* of a group of numbers.

**Definition**

**Least Common Multiple**

The **least common multiple** (LCM) of a group of numbers is the *smallest* number that is divisible by each number in the group.

It is possible to simply list the multiples of each number and then find the LCM by inspection.

## Example 3 Finding the Least Common Multiple (LCM)

< Objective 1 >

**NOTE**

48 is also a common multiple of 6 and 8, but we are looking for the *smallest* such number.

Find the least common multiple of 6 and 8.

Multiples

6: 6, 12, 18, (24), 30, 36, 42, 48, . . .

8: 8, 16, (24), 32, 40, 48, . . .

We see that 24 is the smallest number common to both lists. So 24 is the LCM of 6 and 8.

**Check Yourself 3**

**Find the least common multiple of 20 and 30 by listing the multiples of each number.**

The technique used in Example 3 will work for any group of numbers. However, it becomes tedious for larger numbers. Let's outline a different approach.

**Step by Step**

**Finding the Least Common Multiple**

**Step 1** Write the prime factorization for each of the numbers in the group.

**Step 2** Find all the prime factors that appear in any one of the prime factorizations.

**Step 3** Form the product of those prime factors, using each factor the greatest number of times it occurs in any one factorization.

Note that if a number appears three times in the factorization of a number, it must be included at least three times in forming the least common multiple.

This method is illustrated in Example 4.

## Example 4 Finding the Least Common Multiple (LCM)

**NOTE**

Line up the *like* factors vertically.

To find the LCM of 10 and 18, factor:

$$\begin{aligned} 10 &= 2 \qquad\qquad\quad \times 5 \\ 18 &= \underline{2 \times 3 \times 3\quad\;\;} \\ &\;\;\; 2 \times 3 \times 3 \times 5 \end{aligned}$$

Bring down the factors.

The numbers 2 and 5 appear, at most, one time in any one factorization. And 3 appears two times in one factorization.

$2 \times 3 \times 3 \times 5 = 90$

So 90 is the LCM of 10 and 18.

### Check Yourself 4

**Use the method of Example 4 to find the LCM of 24 and 36.**

The procedure is the same for a group of more than two numbers.

## Example 5 Finding the Least Common Multiple (LCM)

< Objective 2 >

**NOTE**

The different factors that appear are 2, 3, and 5.

To find the LCM of 12, 18, and 20, we factor:

$$\begin{aligned} 12 &= 2 \times 2 \times 3 \\ 18 &= 2 \qquad\; \times 3 \times 3 \\ 20 &= \underline{2 \times 2 \qquad\qquad\quad \times 5} \\ &\;\;\; 2 \times 2 \times 3 \times 3 \times 5 \end{aligned}$$

The numbers 2 and 3 appear twice in one factorization, and 5 appears just once.

$2 \times 2 \times 3 \times 3 \times 5 = 180$

So 180 is the LCM of 12, 18, and 20.

### Check Yourself 5

**Find the LCM of 3, 4, and 6.**

The process of finding the least common multiple is very useful when we are adding, subtracting, or comparing unlike fractions (fractions with different denominators).

Suppose you are asked to compare the sizes of the fractions $\frac{3}{7}$ and $\frac{4}{7}$. Because each unit in the diagram is divided into seven parts, it is easy to see that $\frac{4}{7}$ is larger than $\frac{3}{7}$.

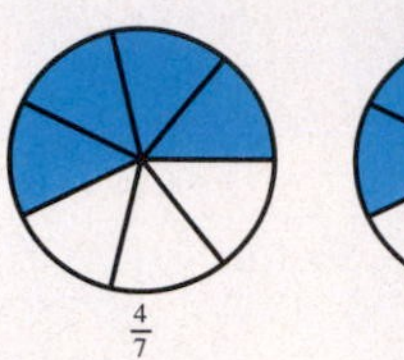

$\frac{4}{7}$

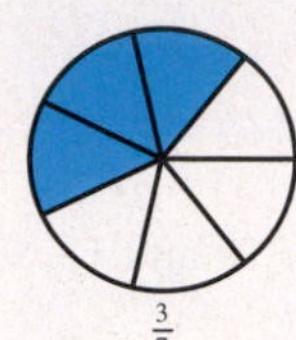

$\frac{3}{7}$

Four parts of seven are a greater portion than three parts. Now compare the size of the fractions $\frac{2}{5}$ and $\frac{3}{7}$.

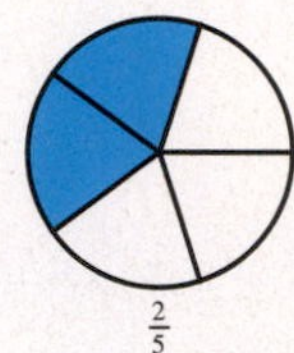

$\frac{2}{5}$

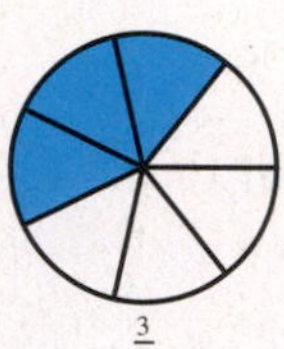

$\frac{3}{7}$

We *cannot* compare fifths with sevenths! The fractions $\frac{2}{5}$ and $\frac{3}{7}$ are *not* like fractions. Because they name different ways of dividing the whole, deciding which fraction is larger is not nearly so easy.

To compare the sizes of fractions, we change them to equivalent fractions having a *common denominator*. This common denominator must be a common multiple of the original denominators.

We will use the following property to form such fractions.

**Property**

**The Fundamental Principle of Fractions**

$$\frac{a}{b} = \frac{a \times c}{b \times c}$$

## Example 6 Comparing the Size of Fractions

< Objective 3 >

**RECALL**

$\frac{2}{5}$ and $\frac{14}{35}$ are equivalent **fractions.** They name the same part of a whole.

Compare the sizes of $\frac{2}{5}$ and $\frac{3}{7}$.

The original denominators are 5 and 7. Because 35 is a common multiple of 5 and 7, let's use 35 as our common denominator.

$$\frac{2}{5} = \frac{14}{35} \quad (\times 7 \text{ numerator and denominator})$$

Think, "What must we multiply 5 by to get 35?" The answer is 7. Multiply the numerator and denominator by that number.

$$\frac{3}{7} \overset{\times 5}{=} \frac{15}{35}$$

Multiply the numerator and denominator by 5.

**NOTE**

15 of 35 parts represents a greater portion of the whole than 14 parts.

Because $\frac{2}{5} = \frac{14}{35}$ and $\frac{3}{7} = \frac{15}{35}$, we see that $\frac{3}{7}$ is larger than $\frac{2}{5}$.

### Check Yourself 6

Which is larger, $\frac{5}{9}$ or $\frac{4}{7}$?

Now, consider an example that uses inequality notation.

## Example 7 Using an Inequality Symbol with Two Fractions

**RECALL**

The inequality symbol "points" to the smaller quantity.

Use the inequality symbol < or > to complete the following statement.

$$\frac{5}{8} \square \frac{3}{5}$$

Once again we must compare the sizes of the two fractions, and we do this by converting the fractions to equivalent fractions with a common denominator. Here we use 40 as that denominator.

**NOTE**

We use the Fundamental Principle of Fractions to convert these fractions.

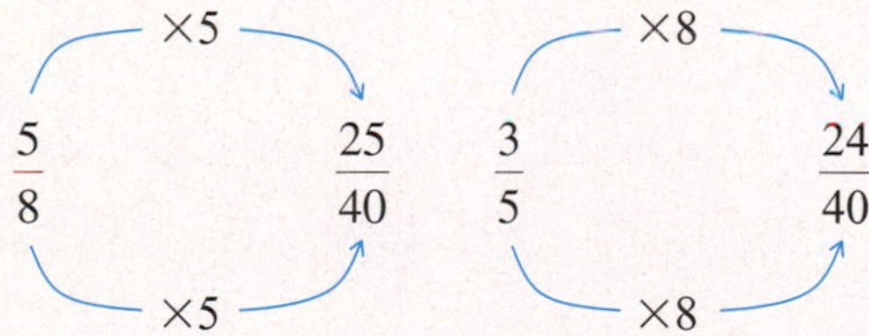

$$\frac{5}{8} \overset{\times 5}{=} \frac{25}{40} \qquad \frac{3}{5} \overset{\times 8}{=} \frac{24}{40}$$

Because $\frac{5}{8}\left(\text{or } \frac{25}{40}\right)$ is larger than $\frac{3}{5}\left(\text{or } \frac{24}{40}\right)$, we write

$$\frac{5}{8} > \frac{3}{5}$$

### Check Yourself 7

Use the symbol < or > to complete the statement.

$$\frac{5}{9} \square \frac{6}{11}$$

## Check Yourself ANSWERS

**1.** The first seven multiples of 4 are 4, 8, 12, 16, 20, 24, and 28.
**2.** 6, 12, 18, 24, 30, 36; some common multiples of 4 and 6 are 12 and 24.
**3.** The multiples of 20 are 20, 40, 60, 80, 100, 120, . . . ; the multiples of 30 are 30, 60, 90, 120, 150, . . . ; the least common multiple of 20 and 30 is 60, the smallest number common to both lists.
**4.** $2 \times 2 \times 2 \times 3 \times 3 = 72$ **5.** 12 **6.** $\frac{4}{7}$ is larger **7.** $\frac{5}{9} > \frac{6}{11}$

## Reading Your Text

The following fill-in-the-blank exercises are designed to ensure that you understand some of the key vocabulary used in this section.

### SECTION 3.2

**(a)** The _______________ of a number are the products of that number with the natural numbers.

**(b)** The LCM of a group of numbers is the _______________ number that is divisible by each number in that group.

**(c)** To compare the sizes of fractions, we change them to equivalent fractions having a common _______________.

**(d)** The statement $\frac{5}{8} > \frac{3}{5}$ is read "$\frac{5}{8}$ is _______________ than $\frac{3}{5}$."

Basic Skills | Advanced Skills | Vocational-Technical Applications | Calculator/Computer | Above and Beyond

# 3.2 exercises

MathZone

Boost your grade at mathzone.com!
> Practice Problems
> NetTutor
> Self-Tests
> e-Professors
> Videos

Name ____________

Section ______ Date ______

< Objective 1 >

*Find the least common multiple (LCM) for each of the following groups of numbers. Use whichever method you wish.*

**1.** 2 and 3

**2.** 3 and 5

**3.** 4 and 6

**4.** 6 and 9

**5.** 10 and 20

**6.** 12 and 36

**7.** 9 and 12

**8.** 20 and 30

**9.** 12 and 16

**10.** 10 and 15

**11.** 12 and 15

**12.** 12 and 21

**13.** 18 and 36

**14.** 25 and 50

**15.** 25 and 40

**16.** 10 and 14

< Objective 2 >

**17.** 3, 5, and 6

**18.** 2, 8, and 10

**19.** 18, 21, and 28

**20.** 8, 15, and 20

**21.** 20, 30, and 45

**22.** 12, 20, and 35

< Objective 3 >

*Arrange the given fractions from smallest to largest.*

**23.** $\frac{12}{17}, \frac{9}{10}$

**24.** $\frac{4}{9}, \frac{5}{11}$

**25.** $\frac{5}{8}, \frac{3}{5}$

**26.** $\frac{9}{10}, \frac{8}{9}$

## Answers

1. ______ 2. ______
3. ______ 4. ______
5. ______ 6. ______
7. ______ 8. ______
9. ______ 10. ______
11. ______ 12. ______
13. ______ 14. ______
15. ______ 16. ______
17. ______ 18. ______
19. ______ 20. ______
21. ______ 22. ______
23. ______
24. ______
25. ______
26. ______

 

Answers

27. ______

28. ______

29. ______

30. ______

31. ______

32. ______

33. ______

34. ______

35. ______

36. ______

37. ______

38. ______

39. ______

40. ______

41. ______

42. ______

**27.** $\frac{3}{8}, \frac{1}{3}, \frac{1}{4}$

**28.** $\frac{7}{12}, \frac{5}{18}, \frac{1}{3}$

**29.** $\frac{11}{12}, \frac{4}{5}, \frac{5}{6}$

**30.** $\frac{5}{8}, \frac{9}{16}, \frac{13}{32}$

*Solve the following applications.*

**31. Crafts** Three drill bits are marked $\frac{3}{8}, \frac{5}{16}$, and $\frac{11}{32}$. Which drill bit is largest?

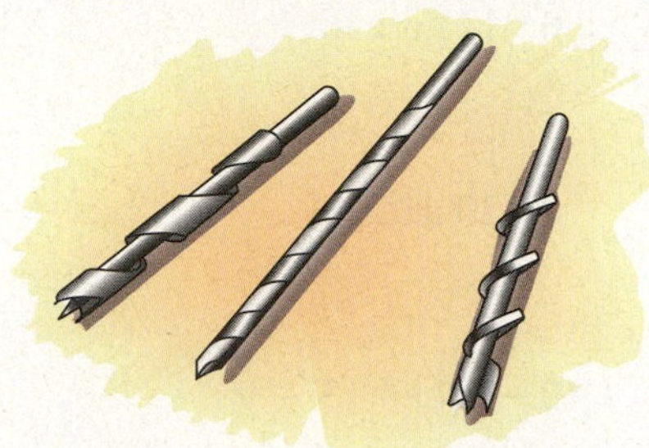

**32. Crafts** Bolts can be purchased with diameters of $\frac{3}{8}, \frac{1}{4}$, or $\frac{3}{16}$ inches (in.). Which is smallest?

chapter 3 > Make the Connection

**33. Construction** Plywood comes in thicknesses of $\frac{5}{8}, \frac{3}{4}, \frac{1}{2}$, and $\frac{3}{8}$ in. Which size is thickest?

chapter 3 > Make the Connection

**34. Construction** Dowels are sold with diameters of $\frac{1}{2}, \frac{9}{16}, \frac{5}{8}$, and $\frac{3}{8}$ in. Which size is smallest?

chapter 3 > Make the Connection

**35.** Elian is asked to create a fraction equivalent to $\frac{1}{4}$. His answer is $\frac{4}{7}$. What did he do wrong? What would be a correct answer?

**36.** A sign on a busy highway says Exit 5A is $\frac{3}{4}$ mile away and Exit 5B is $\frac{5}{8}$ mile away. Which exit is first?

*Complete each equivalent fraction.*

**37.** $\frac{4}{5} = \frac{\phantom{0}}{25}$

> Videos

**38.** $\frac{6}{13} = \frac{\phantom{0}}{26}$

**39.** $\frac{5}{6} = \frac{25}{\phantom{0}}$

**40.** $\frac{2}{5} = \frac{14}{\phantom{0}}$

**41.** $\frac{11}{37} = \frac{\phantom{0}}{111}$

**42.** $\frac{9}{31} = \frac{\phantom{0}}{248}$

Basic Skills | **Advanced Skills** | Vocational-Technical Applications | Calculator/Computer | Above and Beyond

*Complete the statements, using the symbol < or >.*

**43.** $\frac{5}{6} \square \frac{2}{5}$ 

**44.** $\frac{3}{4} \square \frac{10}{11}$

**45.** $\frac{4}{9} \square \frac{3}{7}$

**46.** $\frac{7}{10} \square \frac{11}{15}$

**47.** $\frac{7}{20} \square \frac{9}{25}$

**48.** $\frac{5}{12} \square \frac{7}{18}$

**49.** $\frac{5}{16} \square \frac{7}{20}$

**50.** $\frac{7}{12} \square \frac{9}{15}$

**51.** **Business and Finance** A company uses two types of boxes, 8 cm and 10 cm long. They are packed in larger cartons to be shipped. What is the shortest length of a container that will accommodate boxes of either size without any room left over? (Each container can contain only boxes of one size—no mixing allowed.)

**52.** There is an alternate approach to finding the least common multiple of two numbers. The LCM of two numbers can be found by dividing the product of the two numbers by the greatest common factor (GCF) of those two numbers. For example, the GCF of 24 and 36 is 12. If we use the given formula, we obtain

$$\text{LCM of 24 and 36} = \frac{240 \cdot 36}{12} = 72$$

**(a)** Use the given formula to find the LCM of 150 and 480.

**(b)** Verify the result by finding the LCM using the method of prime factorization.

Basic Skills | Advanced Skills | **Vocational-Technical Applications** | Calculator/Computer | Above and Beyond

**53.** **Electronics** Imagine a guitar string vibrating back and forth. Each back-and-forth vibration is called a **cycle.** The number of times that a string vibrates back and forth per second is called the **frequency.** Accordingly, frequency is measured in **cycles per second** [CPS or hertz (Hz)]. If you had a camera that could zoom in on a vibrating guitar string, you'd see that it doesn't simply vibrate back and forth but actually vibrates differently in different sections of the string. The loudest vibration is called the fundamental frequency; quieter vibrations are called harmonics, which are **multiples** of the fundamental frequency.

**I.** If a note's fundamental frequency is 440 Hz, 1,760 Hz is a harmonic.

**II.** If a note's fundamental frequency is 440 Hz, 771.76 Hz is a harmonic.

Which of the following conclusions is valid?

**(a)** I is true and II is true.
**(b)** I is true and II is false.
**(c)** I is false and II is true.
**(d)** I is false and II is false.
**(e)** None of the above.

## Answers

43. ______
44. ______
45. ______
46. ______
47. ______
48. ______
49. ______
50. ______
51. ______
52. ______
53. ______

 

Answers

54. ______________________

55. ______________________

**54. Electronics** Music producers know that, when mixing tracks containing instruments and vocals, it is usually necessary to add some reverb and/or delay (echo) from each track into the mix. For a good mix, every instrument or vocal track usually has different degrees of reverb and delay. Delay, measured in milliseconds (ms), is the time it takes to hear the "echo," which can be quiet or sometimes quite loud. It is usually desirable to have the delay times be "in beat." If we let BPM denote the number of beats per minute the drum machine is set at, then the delay time is a multiple of $\frac{60,000}{\text{BPM}}$.

**I.** If the drum machine is set at 120 BPM, then 1,200 ms will produce an in-beat delay.

**II.** If the drum machine is set at 120 BPM, then 1,000 ms will produce an in-beat delay.

Which of the following conclusions is valid?

**(a)** I is true and II is true.
**(b)** I is true and II is false.
**(c)** I is false and II is true.
**(d)** I is false and II is false.
**(e)** None of the above.

Basic Skills | Advanced Skills | Vocational-Technical Applications | Calculator/Computer | **Above and Beyond**

**55.** Complete the following crossword puzzle.

**Across**

**2.** The LCM of 11 and 13
**4.** The GCF of 120 and 300
**7.** The GCF of 13 and 52
**8.** The GCF of 360 and 540

**Down**

**1.** The LCM of 8, 14, and 21
**3.** The LCM of 16 and 12
**5.** The LCM of 2, 5, and 13
**6.** The GCF of 54 and 90

## Answers

**1.** 6 **3.** 12 **5.** 20 **7.** 36 **9.** 48
**11.** $12 = 2 \times 2 \times 3$; $15 = 3 \times 5$; the LCM is $2 \times 2 \times 3 \times 5 = 60$ **13.** 36
**15.** 200 **17.** 30 **19.** 252 **21.** 180 **23.** $\frac{12}{17}, \frac{9}{10}$ **25.** $\frac{3}{5}, \frac{5}{8}$
**27.** $\frac{1}{4}, \frac{1}{3}, \frac{3}{8}$ **29.** $\frac{4}{5}, \frac{5}{6}, \frac{11}{12}$ **31.** $\frac{3}{8}$ **33.** $\frac{3}{4}$ in.
**35.** He added 3 to both the numerator and denominator; $\frac{1 \times 3}{4 \times 3} = \frac{3}{12}$
**37.** 20 **39.** 30 **41.** 33 **43.** > **45.** > **47.** < **49.** <
**51.** Above and Beyond **53.** b

# 3.3 Adding and Subtracting Fractions with Unlike Denominators

< 3.3 Objectives >

1 > Add any two fractions

2 > Add any group of fractions

3 > Subtract any two fractions

In Section 3.1, you dealt with like fractions (fractions with a common denominator). What about a sum that deals with **unlike fractions,** such as $\frac{1}{3} + \frac{1}{4}$?

**NOTE**

Only *like* fractions can be added.

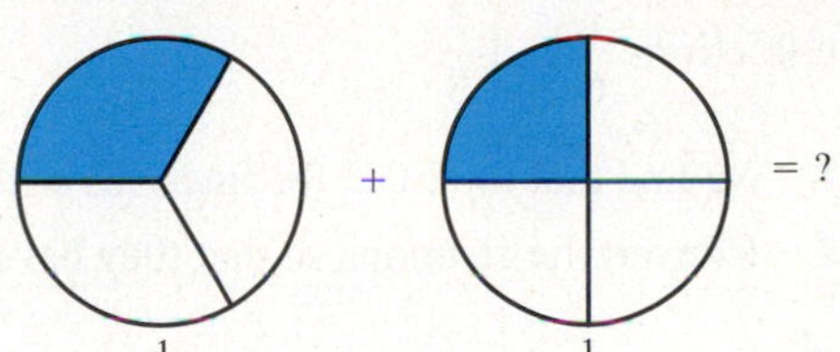


We cannot add unlike fractions because they have different denominators.

To add unlike fractions, write them as equivalent fractions with a common denominator. In this case, let's use 12 as the denominator.

**NOTE**

We can now add because we have like fractions.

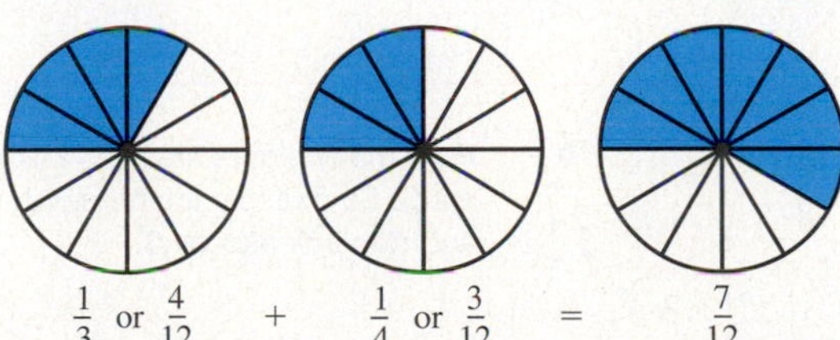


We have chosen 12 because it is a *multiple* of 3 and 4.
$\frac{1}{3}$ is equivalent to $\frac{4}{12}$.
$\frac{1}{4}$ is equivalent to $\frac{3}{12}$.

Any common multiple of the denominators will work in forming equivalent fractions. For instance, we can write $\frac{1}{3}$ as $\frac{8}{24}$ and $\frac{1}{4}$ as $\frac{6}{24}$. Our work is simplest, however, if we use the smallest possible number for the common denominator. This is called the **least common denominator (LCD).**

The LCD is the least common multiple of the denominators of the fractions. This is the *smallest* number that is a multiple of all the denominators. For example, the LCD for $\frac{1}{3}$ and $\frac{1}{4}$ is 12, *not* 24.

**Step by Step**

**To Find the Least Common Denominator**

**Step 1** Write the prime factorization for each of the denominators.

**Step 2** Find all the prime factors that appear in any one of the prime factorizations.

**Step 3** Form the product of those prime factors, using each factor the greatest number of times it occurs in any one factorization.

We are now ready to add unlike fractions. In this case, the fractions must be renamed as equivalent fractions that have the same denominator.

**Step by Step**

**To Add Unlike Fractions**

**Step 1** Find the LCD of the fractions.

**Step 2** Change each unlike fraction to an equivalent fraction with the LCD as a common denominator.

**Step 3** Add the resulting like fractions as before.

Example 1 shows this process.

## Example 1 Adding Unlike Fractions

< Objective 1 >

**NOTE**

See Section 3.2 to review how we arrived at 24.

Add the fractions $\frac{1}{6}$ and $\frac{3}{8}$.

**Step 1** We find that the LCD for fractions with denominators of 6 and 8 is 24.

**Step 2** Convert the fractions so that they have the denominator 24.

$$\frac{1}{6} = \frac{4}{24} \quad (\times 4)$$

How many sixes are in 24? There are 4. So multiply the numerator and denominator by 4.

$$\frac{3}{8} = \frac{9}{24} \quad (\times 3)$$

How many eights are in 24? There are 3. So multiply the numerator and denominator by 3.

**Step 3** We can now add the equivalent like fractions.

$$\frac{1}{6} + \frac{3}{8} = \frac{4}{24} + \frac{9}{24} = \frac{13}{24}$$

Add the numerators and place that sum over the common denominator.

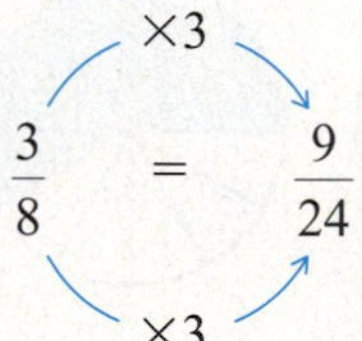

**Check Yourself 1**

Add.

$$\frac{3}{5} + \frac{1}{3}$$

Here is a similar example. Remember that the sum should always be written in simplest form.

## Example 2 Adding Unlike Fractions That Require Simplifying

Add the fractions $\frac{7}{10}$ and $\frac{2}{15}$.

**Step 1** The LCD for fractions with denominators of 10 and 15 is 30.

**Step 2** $\frac{7}{10} = \frac{21}{30}$

$\frac{2}{15} = \frac{4}{30}$

Do you see how the equivalent fractions are formed?

**Step 3** $\frac{7}{10} + \frac{2}{15} = \frac{21}{30} + \frac{4}{30}$

$= \frac{25}{30} = \frac{5}{6}$

Add the resulting like fractions. Be sure the sum is in simplest form.

### Check Yourself 2

**Add.**

$\frac{1}{6} + \frac{7}{12}$

We can use the same procedure to add more than two fractions. Example 3 illustrates this approach.

## Example 3 Adding a Group of Unlike Fractions

< Objective 2 >

Add $\frac{5}{6} + \frac{2}{9} + \frac{4}{15}$.

**Step 1** The LCD is 90.

**Step 2** $\frac{5}{6} = \frac{75}{90}$ Multiply the numerator and denominator by 15.

$\frac{2}{9} = \frac{20}{90}$ Multiply the numerator and denominator by 10.

$\frac{4}{15} = \frac{24}{90}$ Multiply the numerator and denominator by 6.

**Step 3** $\frac{75}{90} + \frac{20}{90} + \frac{24}{90} = \frac{119}{90}$ Now add.

$= 1\frac{29}{90}$

*Remember*, if the sum is an improper fraction, it should be changed to a mixed number.

### Check Yourself 3

Add.

$$\frac{2}{5} + \frac{3}{8} + \frac{7}{20}$$

Many of the measurements you deal with in everyday life involve fractions. Following are some typical situations.

### Example 4 An Application Involving the Addition of Unlike Fractions

Jack ran $\frac{1}{2}$ mi on Monday, $\frac{2}{3}$ mi on Wednesday, and $\frac{3}{4}$ mi on Friday. How far did he run during the week?

The three distances that Jack ran are the given information in the problem. We want to find a total distance, so we must add to find the solution.

$$\frac{1}{2} + \frac{2}{3} + \frac{3}{4} = \frac{6}{12} + \frac{8}{12} + \frac{9}{12}$$

$$= \frac{23}{12} = 1\frac{11}{12} \text{ mi}$$

Because we have no common denominator, we must convert to equivalent fractions before we can add.

Jack ran $1\frac{11}{12}$ mi during the week.

### Check Yourself 4

Susan is designing an office complex. She needs $\frac{2}{5}$ acre for buildings, $\frac{1}{3}$ acre for driveways and parking, and $\frac{1}{6}$ acre for walks and landscaping. How much land does she need?

## Example 5 An Application Involving the Addition of Unlike Fractions

Sam bought three packages of spices weighing $\frac{1}{4}$, $\frac{5}{8}$, and $\frac{1}{2}$ pound (lb). What was the total weight?

**NOTE**

The abbreviation for pounds is "lb" from the Latin *libra*, meaning "balance" or "scales."

We need to find the total weight, so we must add.

$$\frac{1}{4} + \frac{5}{8} + \frac{1}{2} = \frac{2}{8} + \frac{5}{8} + \frac{4}{8}$$

Write each fraction with the denominator 8.

$$= \frac{11}{8} = 1\frac{3}{8} \text{ lb}$$

The total weight was $1\frac{3}{8}$ lb.

### Check Yourself 5

**For three different recipes, Max needs $\frac{3}{8}$, $\frac{1}{2}$, and $\frac{3}{4}$ gallon (gal) of tomato sauce. How many gallons should he buy altogether?**

To subtract unlike fractions, which are fractions that do not have the same denominator, we use the following process.

**Step by Step**

**To Subtract Unlike Fractions**

**Step 1** Find the LCD of the fractions.

**Step 2** Change each unlike fraction to an equivalent fraction with the LCD as a common denominator.

**Step 3** Subtract the resulting like fractions as before.

## Example 6 Subtracting Unlike Fractions

< Objective 3 >

Subtract $\frac{5}{8} - \frac{1}{6}$.

**Step 1** The LCD is 24.

**Step 2** Convert the fractions so that they have the common denominator 24.

$$\frac{5}{8} = \frac{15}{24}$$

$$\frac{1}{6} = \frac{4}{24}$$

The first two steps are exactly the same as if we were adding.

**NOTE**

You can use your calculator to check your result.

**Step 3** Subtract the equivalent like fractions.

$$\frac{5}{8} - \frac{1}{6} = \frac{15}{24} - \frac{4}{24} = \frac{11}{24}$$

> CAUTION

**Be Careful!** You *cannot* subtract the numerators and subtract the denominators.

$\frac{5}{8} - \frac{1}{6}$ is *not* $\frac{4}{2}$

## Check Yourself 6

Subtract.

$$\frac{7}{10} - \frac{1}{4}$$

Following is an application that involves subtracting unlike fractions.

## Example 7 An Application Involving the Subtraction of Unlike Fractions

You have $\frac{7}{8}$ yards (yd) of a handwoven linen. A pattern for a placemat calls for $\frac{1}{2}$ yd. Will you have enough left for two napkins that will use $\frac{1}{3}$ yd?

First, find out how much fabric is left over after the placemat is made.

$$\frac{7}{8}\text{ yd} - \frac{1}{2}\text{ yd} = \frac{7}{8}\text{ yd} - \frac{4}{8}\text{ yd} = \frac{3}{8}\text{ yd}$$

**NOTE**

Remember that $\frac{3}{8}$ yd is left over and that $\frac{1}{3}$ yd is needed.

Now compare the sizes of $\frac{1}{3}$ and $\frac{3}{8}$.

$$\frac{3}{8}\text{ yd} = \frac{9}{24}\text{ yd} \qquad \text{and} \qquad \frac{1}{3}\text{ yd} = \frac{8}{24}\text{ yd}$$

Because $\frac{3}{8}$ yd is *more than* the $\frac{1}{3}$ yd that is needed, there is enough material for the placemat *and* two napkins.

### Check Yourself 7

**A concrete walk will require $\frac{3}{4}$ cubic yard (yd$^3$) of concrete. If you have mixed $\frac{8}{9}$ yd$^3$, will enough concrete remain to do a project that will use $\frac{1}{6}$ yd$^3$?**

Our next application involves measurements. Note that on a ruler or yardstick, the marks divide each inch into $\frac{1}{2}$-in., $\frac{1}{4}$-in., and $\frac{1}{8}$-in. sections, and on some rulers, $\frac{1}{16}$-in. sections. We will use denominators of 2, 4, 8, and 16 in our measurement applications.

### Example 8 — An Application Involving the Subtraction of Unlike Fractions

Alexei cut two $\frac{3}{16}$-in. slats from a piece of wood that is $\frac{3}{4}$ in. across. How much is left?

The two $\frac{3}{16}$-in. pieces total

$$2 \times \frac{3}{16} = \frac{6}{16} = \frac{3}{8}\text{ in.}$$

$$\frac{3}{4} = \frac{6}{8}$$

$$\frac{6}{8} - \frac{3}{8} = \frac{3}{8}$$

The remaining strip is $\frac{3}{8}$ in. wide.

## Check Yourself 8

Ricardo cut three strips from a 1-in. piece of metal. Each strip has a width of $\frac{3}{16}$ in. How much metal remains after the cuts?

## Check Yourself ANSWERS

**1.** $\frac{14}{15}$ **2.** $\frac{1}{6} + \frac{7}{12} = \frac{2}{12} + \frac{7}{12} = \frac{9}{12} = \frac{3}{4}$ **3.** $1\frac{1}{8}$ **4.** $\frac{9}{10}$ acre

**5.** $1\frac{5}{8}$ gal **6.** $\frac{9}{20}$

**7.** $\frac{5}{36}$ yd$^3$ remains. You do *not* have enough concrete for both projects.

**8.** $\frac{7}{16}$ in.

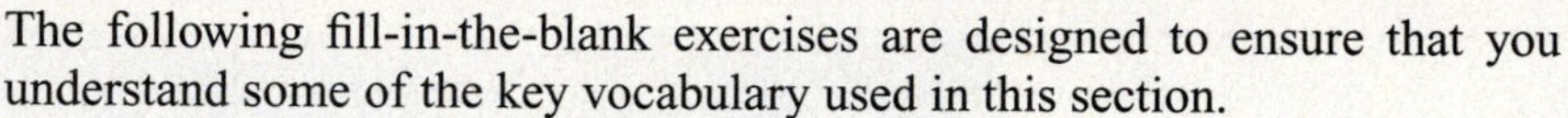

## Reading Your Text

The following fill-in-the-blank exercises are designed to ensure that you understand some of the key vocabulary used in this section.

**SECTION 3.3**

**(a)** The process for finding the LCD is nearly identical to the Step by Step for finding the ___________.

**(b)** To add ___________ fractions, we first find the LCD of the fractions.

**(c)** Two fractions with the same value but different denominators are called ___________ fractions.

**(d)** When adding fractions with a common denominator, we add the ___________ and put that sum over the common denominator.

# 3.3 exercises

MathZone

Boost your grade at mathzone.com!
- Practice Problems
- NetTutor
- Self-Tests
- e-Professors
- Videos

Name ____________

Section ________ Date ________

< Objective 1 >

*Find the least common denominator (LCD) for fractions with the given denominators.*

1. 3 and 4
2. 3 and 5
3. 4 and 8
4. 6 and 12
5. 9 and 27
6. 10 and 30
7. 8 and 12
8. 15 and 40
9. 14 and 21 > Videos
10. 15 and 20
11. 48 and 80
12. 60 and 84
13. 3, 4, and 5
14. 3, 4, and 6
15. 8, 10, and 15 > Videos
16. 6, 22, and 33
17. 5, 10, and 25
18. 8, 24, and 48

*Add.*

19. $\frac{2}{3} + \frac{1}{4}$
20. $\frac{3}{5} + \frac{1}{3}$
21. $\frac{1}{5} + \frac{3}{10}$
22. $\frac{1}{3} + \frac{1}{18}$
23. $\frac{3}{4} + \frac{1}{8}$
24. $\frac{4}{5} + \frac{1}{10}$
25. $\frac{1}{7} + \frac{3}{5}$ > Videos
26. $\frac{1}{6} + \frac{2}{15}$
27. $\frac{3}{7} + \frac{3}{14}$
28. $\frac{7}{20} + \frac{9}{40}$
29. $\frac{7}{15} + \frac{2}{35}$
30. $\frac{3}{10} + \frac{3}{8}$

## Answers

| | |
|---|---|
| 1. ______ | 2. ______ |
| 3. ______ | 4. ______ |
| 5. ______ | 6. ______ |
| 7. ______ | 8. ______ |
| 9. ______ | 10. ______ |
| 11. ______ | 12. ______ |
| 13. ______ | 14. ______ |
| 15. ______ | 16. ______ |
| 17. ______ | 18. ______ |
| 19. ______ | 20. ______ |
| 21. ______ | 22. ______ |
| 23. ______ | 24. ______ |
| 25. ______ | 26. ______ |
| 27. ______ | 28. ______ |
| 29. ______ | 30. ______ |

## Answers

31. ______

32. ______

33. ______

34. ______

35. ______

36. ______

37. ______

38. ______

39. ______

40. ______

**31.** $\frac{5}{8} + \frac{1}{12}$

**32.** $\frac{5}{12} + \frac{3}{10}$

Basic Skills | **Advanced Skills** | Vocational-Technical Applications | Calculator/Computer | Above and Beyond

*In exercises 33 to 36, complete each statement with either* **never, always,** *or* **sometimes.**

**33.** The sum of two like fractions is ______ the sum of the numerators over the common denominator.

**34.** The LCD for two unlike proper fractions is ______ the same as the GCF of their denominators.

**35.** The sum of two fractions can ______ be simplified.

**36.** The difference of two proper fractions is ______ less than either of the two fractions.

*Solve the following applications.*

**37. NUMBER PROBLEM** Paul bought $\frac{1}{2}$ pound (lb) of peanuts and $\frac{3}{8}$ lb of cashews. How many pounds of nuts did he buy?

**38. CONSTRUCTION** A countertop consists of a board $\frac{3}{4}$ inch (in.) thick and tile $\frac{3}{8}$ in. thick. What is the overall thickness? chapter 3 > Make the Connection

**39. BUSINESS AND FINANCE** Amy budgets $\frac{2}{5}$ of her income for housing and $\frac{1}{6}$ of her income for food. What fraction of her income is budgeted for these two purposes? What fraction of her income remains?

**40. SOCIAL SCIENCE** A person spends $\frac{3}{8}$ of a day at work and $\frac{1}{3}$ of a day sleeping. What fraction of a day do these two activities use? What fraction of the day remains?

< Objective 2 >

41. **Number Problem** Jose walked $\frac{3}{4}$ mile (mi) to the store, $\frac{1}{2}$ mi to a friend's house, and then $\frac{2}{3}$ mi home. How far did he walk?

42. **Geometry** Find the perimeter of, or the distance around, the accompanying figure.

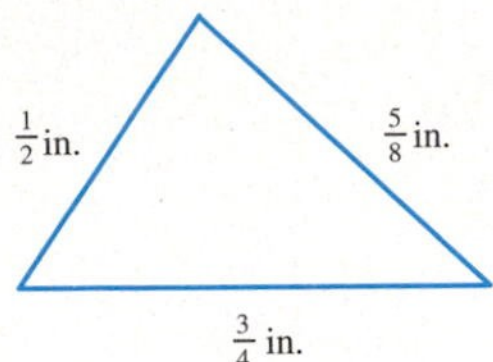


43. **Business and Finance** A budget guide states that you should spend $\frac{1}{4}$ of your salary for housing, $\frac{3}{16}$ for food, $\frac{1}{16}$ for clothing, and $\frac{1}{8}$ for transportation. What total portion of your salary do these four expenses account for?

44. **Business and Finance** Deductions from your paycheck are made roughly as follows: $\frac{1}{8}$ for federal tax, $\frac{1}{20}$ for state tax, $\frac{1}{20}$ for social security, and $\frac{1}{40}$ for a savings withholding plan. What portion of your pay is deducted?

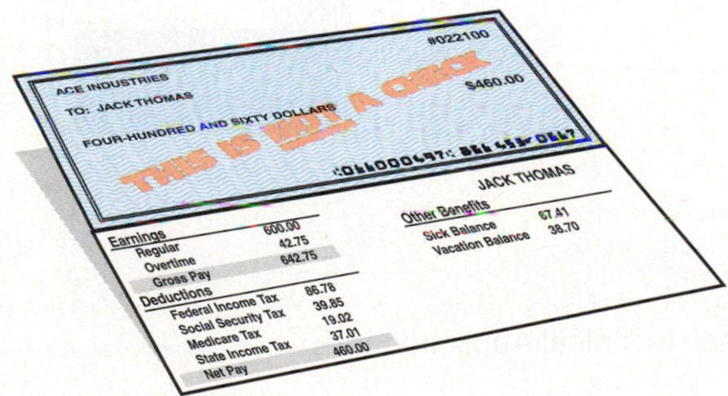


45. $\frac{1}{5} + \frac{7}{10} + \frac{4}{15}$

46. $\frac{2}{3} + \frac{1}{4} + \frac{3}{8}$

47. $\frac{1}{9} + \frac{7}{12} + \frac{5}{8}$

48. $\frac{1}{3} + \frac{5}{12} + \frac{4}{5}$

< Objective 3 >

*Subtract.*

49. $\frac{4}{5} - \frac{1}{3}$

50. $\frac{7}{9} - \frac{1}{6}$

## Answers

41. ____________

42. ____________

43. ____________

44. ____________

45. ____________

46. ____________

47. ____________

48. ____________

49. ____________

50. ____________

**Answers**

51. ______
52. ______
53. ______
54. ______
55. ______
56. ______
57. ______
58. ______
59. ______
60. ______
61. ______
62. ______
63. ______
64. ______

**51.** $\frac{11}{15} - \frac{3}{5}$

**52.** $\frac{5}{6} - \frac{2}{7}$

**53.** $\frac{3}{8} - \frac{1}{4}$

**54.** $\frac{9}{10} - \frac{4}{5}$

**55.** $\frac{5}{12} - \frac{3}{8}$

**56.** $\frac{13}{15} - \frac{11}{20}$

*Perform the following operations.*

**57.** $\frac{33}{40} - \frac{7}{24} + \frac{11}{30}$

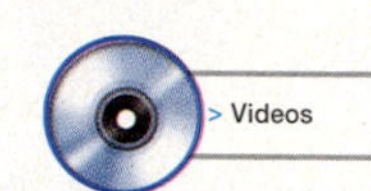


**58.** $\frac{13}{24} - \frac{5}{16} + \frac{3}{8}$

**59.** $\frac{15}{16} + \frac{5}{8} - \frac{1}{4}$

**60.** $\frac{9}{10} - \frac{1}{5} + \frac{1}{2}$

**GEOMETRY** *For exercises 61 and 62, find the missing dimension (?) in the given figure.*

**61.**

$\frac{7}{16}$ in. ? $\frac{3}{4}$ in.

**62.**

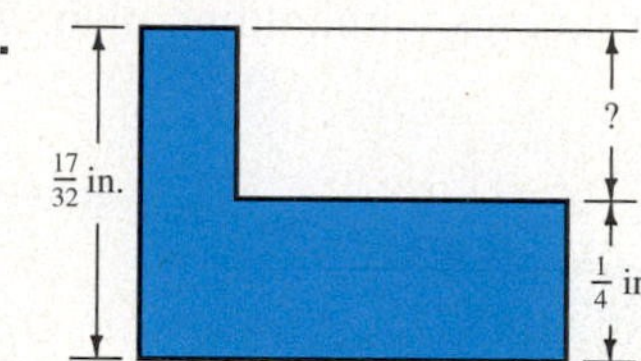


Basic Skills | Advanced Skills | **Vocational-Technical Applications** | Calculator/Computer | Above and Beyond

**63. ALLIED HEALTH** Prior to chemotherapy, a patient's malignant tumor weighed $\frac{5}{6}$ pound (lb). After a week of chemotherapy, the tumor's weight had been reduced by $\frac{1}{4}$ lb. How much did the tumor weigh at the end of the week?

**64. ALLIED HEALTH** Mrs. Lewis has three children who all need to take medication for the same illness. Charlie needs $\frac{3}{2}$ milliliters (mL), Sharon needs $\frac{2}{3}$ mL, and little Kevin needs $\frac{1}{4}$ mL of medication. How many milliliters of medication will Mrs. Lewis need to give each child a single dose of medicine?

**65. Manufacturing Technology** Find the total thickness of this part.

$\frac{3}{16}$ in.
$\frac{3}{16}$ in.
$\frac{1}{4}$ in.
$\frac{1}{8}$ in.

**66. Manufacturing Technology** Find the missing dimension ($x$).

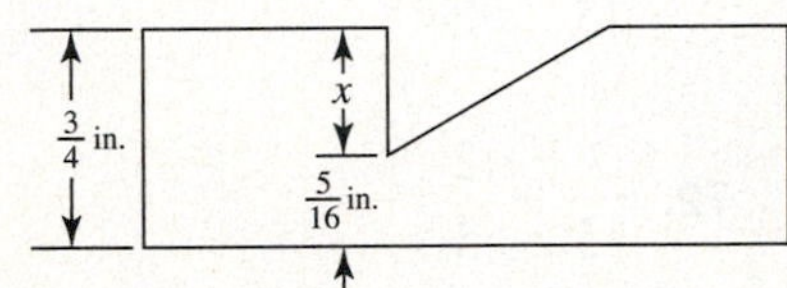

**67. Electronics**

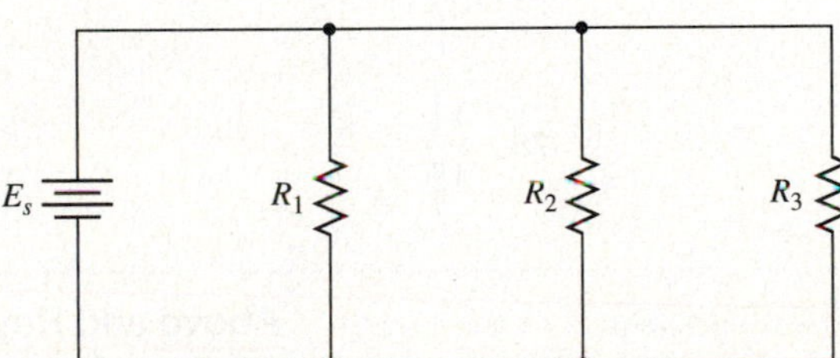


The circuit depicts three resistors ($R_1$, $R_2$, $R_3$) wired in parallel to a source, $E_s$. The circuit can be simplified by replacing the three separate resistors with a single equivalent resistor, $R_{eq}$, according to the following equation:

$$\frac{1}{R_{eq}} = \frac{1}{R_1} + \frac{1}{R_2} + \frac{1}{R_3}$$

If $R_1 = 10$ ohms ($\Omega$), $R_2 = 20\ \Omega$, and $R_3 = 40\ \Omega$, what is $\frac{1}{R_{eq}}$?

What is $R_{eq}$?

**68. Information Technology** Amin wants to figure out how long it takes to transmit 1,000 bits over two networks that have transmission rates of 50,000 and 10,000 bits per second, respectively. What is the total time after the transmission? Use the fraction of the number of bits to the transmission time to figure out your answer. Represent your answer as a simplified fraction.

## Answers

65. ______

66. ______

67. ______

68. ______

Answers

69. ________

70. ________

71. ________

72. ________

73. ________

74. ________

75. ________

76. ________

77. ________

78. ________

79. ________

80. ________

Basic Skills | Advanced Skills | Vocational-Technical Applications | **Calculator/Computer** | Above and Beyond

## Using Your Calculator to Add and Subtract Fractions

Adding or subtracting fractions on the calculator is very much like the multiplication and division you did in Chapter 2. The only thing that changes is the operation. Here's where the fraction calculator is a great tool for checking your work. No muss, no fuss, no searching for a common denominator. Just enter the fractions and get the right answer!

*Find the following sums or differences using your calculator.*

**69.** $\frac{1}{10} + \frac{7}{12}$

**70.** $\frac{7}{15} + \frac{17}{24}$

**71.** $\frac{8}{9} + \frac{6}{7}$

**72.** $\frac{7}{15} + \frac{2}{5}$

**73.** $\frac{11}{18} + \frac{5}{12}$

**74.** $\frac{5}{8} + \frac{4}{9}$

**75.** $\frac{15}{17} - \frac{9}{11}$

**76.** $\frac{31}{43} - \frac{18}{53}$

**77.** $\frac{4}{9} - \frac{2}{5}$

**78.** $\frac{11}{13} - \frac{2}{3}$

Basic Skills | Advanced Skills | Vocational-Technical Applications | Calculator/Computer | **Above and Beyond**

**79.** **Construction** A door is $4\frac{1}{4}$ ft wide. Two hooks are to be attached to the door so that they are $1\frac{1}{2}$ in. apart and the same distance from each edge. How far from the edge of the door should each hook be located? Give your answer in feet.

**80.** Complete the following:

$\frac{1}{2} + \frac{1}{4} =$ ________.

$\frac{1}{2} + \frac{1}{4} + \frac{1}{8} =$ ________.

$\frac{1}{2} + \frac{1}{4} + \frac{1}{8} + \frac{1}{16} =$ ________.

Based on these results, predict the answer to the following:

$\frac{1}{2} + \frac{1}{4} + \frac{1}{8} + \frac{1}{16} + \frac{1}{32} =$ ________.

Now, do the addition, and check your prediction.

## Answers

**1.** 12 **3.** 8 **5.** 27
**7.** $8 = 2 \times 2 \times 2$; $12 = 2 \times 2 \times 3$; the LCD is $2 \times 2 \times 2 \times 3 = 24$ **9.** 42
**11.** 240 **13.** 60 **15.** 120 **17.** 50 **19.** $\frac{11}{12}$ **21.** $\frac{1}{2}$
**23.** $\frac{7}{8}$ **25.** $\frac{26}{35}$ **27.** $\frac{9}{14}$ **29.** $\frac{11}{21}$ **31.** $\frac{17}{24}$ **33.** Always
**35.** Sometimes **37.** $\frac{7}{8}$ lb **39.** $\frac{17}{30}, \frac{13}{30}$ **41.** $1\frac{11}{12}$ mi **43.** $\frac{5}{8}$
**45.** $\frac{1}{5} + \frac{7}{10} + \frac{4}{15} = \frac{6}{30} + \frac{21}{30} + \frac{8}{30} = \frac{35}{30} = 1\frac{5}{30} = 1\frac{1}{6}$ **47.** $1\frac{23}{72}$
**49.** $\frac{4}{5} - \frac{1}{3} = \frac{12}{15} - \frac{5}{15} = \frac{7}{15}$ **51.** $\frac{2}{15}$ **53.** $\frac{1}{8}$ **55.** $\frac{1}{24}$
**57.** $\frac{33}{40} - \frac{7}{24} + \frac{11}{30} = \frac{99}{120} - \frac{35}{120} + \frac{44}{120} = \frac{108}{120} = \frac{9}{10}$ **59.** $1\frac{5}{16}$
**61.** $\frac{5}{16}$ in. **63.** $\frac{7}{12}$ lb **65.** $\frac{3}{4}$ in. **67.** $\frac{7}{40}; \frac{40}{7} = 5\frac{5}{7}$
**69.** $\frac{41}{60}$ **71.** $1\frac{47}{63}$ **73.** $\frac{37}{36} = 1\frac{1}{36}$ **75.** $\frac{12}{187}$ **77.** $\frac{2}{45}$
**79.** 2 ft $\frac{3}{4}$ in. $= 2\frac{1}{16}$ ft

# Activity 7 :: Kitchen Subflooring

Benjamin and Olivia are putting a new floor in their kitchen. To get the floor up to the desired height, they need to add $1\frac{1}{8}$ in. of subfloor. They can do this in one of two ways. They can put $\frac{1}{2}$-in. sheet on top of $\frac{5}{8}$-in. board (note that the total would be $\frac{9}{8}$ in. or $1\frac{1}{8}$ in.). They could also put $\frac{3}{8}$-in. board on top of $\frac{3}{4}$-in. sheet.

The following table gives the price for each sheet of plywood from a Home Depot store on 1/15/04.

| Thickness | Cost for a 4 ft × 8 ft Sheet |
|---|---|
| $\frac{1}{8}$ in. | \$9.15 |
| $\frac{1}{4}$ in. | 13.05 |
| $\frac{3}{8}$ in. | 14.99 |
| $\frac{1}{2}$ in. | 17.88 |
| $\frac{5}{8}$ in. | 19.13 |
| $\frac{3}{4}$ in. | 21.36 |
| $\frac{7}{8}$ in. | 25.23 |
| 1 in. | 28.49 |

**1.** What is the combined price for a $\frac{1}{2}$-in. sheet and a $\frac{5}{8}$-in. sheet?

**2.** What is the combined price for a $\frac{3}{8}$-in. sheet and a $\frac{3}{4}$-in. sheet?

**3.** What other combination of sheets of plywood, using two sheets, yields the needed $1\frac{1}{8}$-in. thickness?

**4.** Of the four combinations, which is most economical?

**5.** The kitchen is to be 12 ft × 12 ft. Find the total cost of the plywood you have suggested in question 4.

# 3.4 Adding and Subtracting Mixed Numbers

< 3.4 Objectives >

1 > Add any two mixed numbers

2 > Add any group of mixed numbers

3 > Subtract any two mixed numbers

4 > Solve an application that involves addition or subtraction of mixed numbers

Once you know how to add fractions, adding mixed numbers should be no problem if you keep in mind that addition involves combining groups of the *same kind* of objects. Because mixed numbers consist of two parts—a whole number and a fraction—we work with the whole numbers and the fractions separately.

**Step by Step**

**To Add Mixed Numbers**

**Step 1** Find the LCD.
**Step 2** Rewrite the fraction parts to match the LCD.
**Step 3** Add the fraction parts.
**Step 4** Add the whole number parts.
**Step 5** Simplify if necessary.

Example 1 illustrates the use of this rule.

**Example 1** Adding Mixed Numbers

< Objective 1 >

Add $3\frac{1}{5} + 4\frac{2}{5}$.

$$\begin{array}{r} 3\frac{1}{5} \\ +\ 4\frac{2}{5} \\ \hline 7\frac{3}{5} \end{array}$$

$$3\frac{1}{5} + 4\frac{2}{5} = 7\frac{3}{5}$$

### Check Yourself 1

Add $2\frac{3}{10} + 3\frac{4}{10}$.

When the fractional portions of the mixed numbers have different denominators, we must rename these fractions as equivalent fractions with the least common denominator to perform the addition in step 2. Consider Example 2.

## Example 2 Adding Mixed Numbers with Different Denominators

Add.

$$
\begin{array}{rcl}
3\frac{1}{6} & = & 3\frac{4}{24} \\
+\ 2\frac{3}{8} & = & 2\frac{9}{24} \\
\hline
 & & 5\frac{13}{24}
\end{array}
$$

The LCD of the fractions is 24. Rename them with that denominator.

### Check Yourself 2

Add $5\frac{1}{10} + 3\frac{1}{6}$.

You follow the same procedure if more than two mixed numbers are involved in the problem.

## Example 3 Adding Mixed Numbers with Different Denominators

< Objective 2 >

**NOTE**

The LCD of the three fractions is 40. Convert to equivalent fractions.

Add.

$$
\begin{array}{rcl}
2\frac{1}{5} & = & 2\frac{8}{40} \\
3\frac{3}{4} & = & 3\frac{30}{40} \\
+\ 4\frac{1}{8} & = & 4\frac{5}{40} \\
\hline
 & & 9\frac{43}{40}
\end{array}
$$

$$9\frac{43}{40} = 9 + \frac{43}{40} = 9 + 1 + \frac{3}{40} = 10\frac{3}{40}$$

### Check Yourself 3

Add $5\frac{1}{2} + 4\frac{2}{3} + 3\frac{3}{4}$.

We can use a similar technique for *subtracting* mixed numbers. The process is similar to the one stated earlier for adding mixed numbers.

**Step by Step**

**To Subtract Mixed Numbers**

**Step 1** Find the LCD.
**Step 2** Rewrite the fraction parts to match the LCD.
**Step 3** Subtract the fraction parts.
**Step 4** Subtract the whole number parts.
**Step 5** Simplify if necessary.

Example 4 illustrates the use of this process.

### Example 4 Subtracting Mixed Numbers with Like Denominators

< Objective 3 >

Subtract.

$$\begin{array}{r} 5\frac{7}{12} \\ -\ 3\frac{5}{12} \\ \hline 2\frac{2}{12} \end{array}$$

$$2\frac{2}{12} = 2\frac{1}{6}$$

**Check Yourself 4**

Subtract $8\frac{7}{8} - 5\frac{3}{8}$.

Again, we must rename the fractions if different denominators are involved. This approach is shown in Example 5.

### Example 5 Subtracting Mixed Numbers with Different Denominators

Subtract. $8\frac{7}{10} - 3\frac{3}{8}$.

$$8\frac{7}{10} = 8\frac{28}{40}$$ Write the fractions with denominator 40.

$$-\ 3\frac{3}{8} = 3\frac{15}{40}$$

$$5\frac{13}{40}$$ Subtract as before.

**Check Yourself 5**

Subtract $7\frac{11}{12} - 3\frac{5}{8}$.

To subtract a mixed number from a whole number, we must use a form of regrouping, or borrowing.

## Example 6 Subtracting Mixed Numbers

**NOTE**

$6 = 5 + 1$

$= 5 + \frac{4}{4}$

Subtract $6 - 2\frac{3}{4}$.

$$6 = 5\frac{4}{4}$$

$$\underline{-\,2\frac{3}{4}} = \underline{2\frac{3}{4}}$$

$$3\frac{1}{4}$$

There is no fraction from which we can subtract the $\frac{3}{4}$. We regroup so that we can borrow 1 from the 6 and write it as $\frac{4}{4}$.

**Check Yourself 6**

Subtract.

(a) $7 - 3\frac{2}{5}$ (b) $9 - 2\frac{5}{7}$

A similar technique is used whenever the fraction of the minuend is smaller than the fraction of the subtrahend.

## Example 7 Subtracting Mixed Numbers

Subtract.

$$5\frac{3}{8} = 5\frac{3}{8}$$

$$\underline{-\,3\frac{3}{4}} = \underline{3\frac{6}{8}}$$

Because you cannot subtract $\frac{6}{8}$ from $\frac{3}{8}$, borrow 1 from the 5 in the minuend.

$$5\frac{3}{8} = 5\frac{3}{8} = 4\frac{11}{8}$$

$$\underline{-\,3\frac{3}{4}} = \underline{3\frac{6}{8}} = \underline{3\frac{6}{8}}$$

$$1\frac{5}{8}$$

Because $5 = 4 + 1$, rewrite the 1 as $\frac{8}{8}$ and add it to $\frac{3}{8}$ to get $\frac{11}{8}$.

### Check Yourself 7

Subtract.

(a) $12\frac{5}{12} - 8\frac{2}{3}$ (b) $27\frac{7}{8} - 2\frac{9}{10}$

There is an alternative method that can be used when subtracting fractions. We can convert each mixed number into an improper fraction and then perform the subtraction. Although this technique requires an extra step, it eliminates the need to regroup.

### Example 8 Subtracting by Converting to Improper Fractions

Subtract.

$4\frac{5}{9} - 2\frac{11}{12}$ First, convert to improper fractions.

$\frac{41}{9} - \frac{35}{12}$ The LCD is 36. Find the equivalent fractions and subtract.

$\frac{164}{36} - \frac{105}{36} = \frac{59}{36}$ Rewrite as a mixed number.

$\frac{59}{36} = 1\frac{23}{36}$

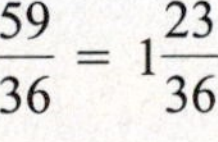

### Check Yourself 8

Subtract $5\frac{1}{4} - 2\frac{5}{18}$.

### Example 9 An Application of the Subtraction of Mixed Numbers

< Objective 4 >

Linda was $48\frac{1}{4}$ inches (in.) tall on her sixth birthday. By her seventh year she was $51\frac{5}{8}$ in. tall. How much did she grow during the year?

Because we want the difference in height, we must subtract $48\frac{1}{4}$ from $51\frac{5}{8}$.

$$\begin{array}{r} 51\frac{5}{8} = 51\frac{5}{8} \\ -\ 48\frac{1}{4} = 48\frac{2}{8} \\ \hline 3\frac{3}{8} \end{array}$$

Linda grew $3\frac{3}{8}$ in. during the year.

### Check Yourself 9

You use $4\frac{3}{4}$ yards (yd) of fabric from a 50-yd bolt. How much fabric remains on the bolt?

Often we will have to use more than one operation to find the solution to a problem. Consider Example 10.

### Example 10 An Application Involving Mixed Numbers

A rectangular poster is to have a total length of $12\frac{1}{4}$ in. We want a $1\frac{3}{8}$-in. border on the top and a 2-in. border on the bottom. What is the length of the printed part of the poster?

First, we draw a sketch of the poster:

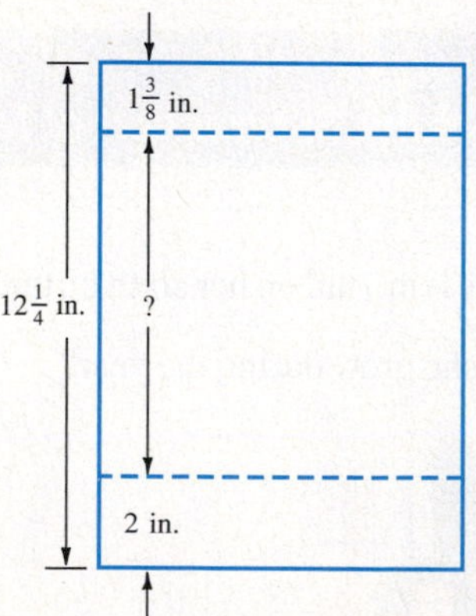


Now, we use this sketch to find the total width of the top and bottom borders.

$$\begin{array}{r} 1\frac{3}{8} \\ +\ 2 \\ \hline 3\frac{3}{8} \end{array}$$

Now *subtract* that sum (the top and bottom borders) from the total length of the poster.

$$\begin{array}{rcccc} & 12\frac{1}{4} & = 12\frac{2}{8} & = & 11\frac{10}{8} \\ - & 3\frac{3}{8} & = 3\frac{3}{8} & = & 3\frac{3}{8} \\ \hline & & & & 8\frac{7}{8} \end{array}$$

The length of the printed part is $8\frac{7}{8}$ in.

**Check Yourself 10**

You cut one shelf $3\frac{3}{4}$ feet (ft) long and one $4\frac{1}{2}$ ft long from a 12-ft piece of lumber. Can you cut another shelf 4 ft long?

**Check Yourself ANSWERS**

**1.** $5\frac{7}{10}$ **2.** $8\frac{4}{15}$ **3.** $13\frac{11}{12}$ **4.** $3\frac{1}{2}$
**5.** $7\frac{11}{12} - 3\frac{5}{8} = 7\frac{22}{24} - 3\frac{15}{24} = 4\frac{7}{24}$ **6.** **(a)** $3\frac{3}{5}$; **(b)** $6\frac{2}{7}$
**7.** **(a)** $3\frac{3}{4}$; **(b)** $24\frac{39}{40}$ **8.** $2\frac{35}{36}$ **9.** $45\frac{1}{4}$ yd **10.** No, only $3\frac{3}{4}$ ft remains.

## Reading Your Text

The following fill-in-the-blank exercises are designed to ensure that you understand some of the key vocabulary used in this section.

SECTION 3.4

**(a)** To add mixed numbers, we first find the ____________ of the fractions.

**(b)** To subtract a fraction from a whole number, we must use a form of ____________, or borrowing.

**(c)** When the fractional portions of mixed numbers have different denominators, we rename the fractions as ____________ fractions.

**(d)** We will always need to borrow when subtracting mixed numbers if the minuend is ____________ than the subtrahend.

# 3.4 exercises

Basic Skills | Advanced Skills | Vocational-Technical Applications | Calculator/Computer | Above and Beyond

MathZone

Boost your grade at mathzone.com!
- > Practice Problems
- > NetTutor
- > Self-Tests
- > e-Professors
- > Videos

Name ____________

Section ________ Date ________

## Answers

| 1. | 2. |
|---|---|
| 3. | 4. |
| 5. | 6. |
| 7. | 8. |
| 9. | 10. |
| 11. | 12. |
| 13. | 14. |
| 15. | 16. |
| 17. | 18. |
| 19. | 20. |
| 21. | 22. |
| 23. | 24. |
| 25. | 26. |
| 27. | 28. |

< Objective 1 >

*Perform the indicated operations.*

**1.** $2\frac{2}{9} + 3\frac{5}{9}$

**2.** $5\frac{2}{9} + 6\frac{4}{9}$

**3.** $2\frac{1}{9} + 5\frac{5}{9}$

**4.** $1\frac{1}{6} + 5\frac{5}{6}$

**5.** $6\frac{5}{9} + 4\frac{7}{9}$

**6.** $5\frac{8}{9} + 4\frac{4}{9}$

**7.** $1\frac{1}{3} + 2\frac{1}{5}$

**8.** $2\frac{1}{4} + 1\frac{1}{6}$

< Objective 2 >

**9.** $2\frac{1}{4} + 3\frac{5}{8} + 1\frac{1}{6}$

**10.** $3\frac{1}{5} + 2\frac{1}{2} + 5\frac{1}{4}$

**11.** $3\frac{3}{5} + 4\frac{1}{4} + 5\frac{3}{10}$

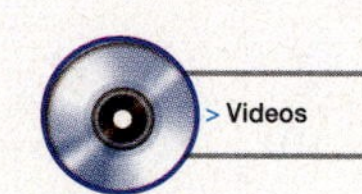


**12.** $4\frac{5}{6} + 3\frac{2}{3} + 7\frac{5}{9}$

< Objective 3 >

**13.** $7\frac{7}{8} - 3\frac{3}{8}$

**14.** $3\frac{5}{6} - 1\frac{1}{6}$

**15.** $3\frac{2}{5} - 1\frac{4}{5}$

**16.** $5\frac{3}{7} - 2\frac{5}{7}$

**17.** $3\frac{2}{3} - 2\frac{1}{4}$

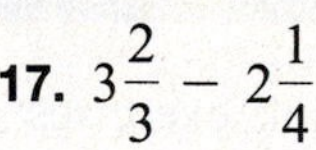

**18.** $5\frac{4}{5} - 2\frac{5}{6}$

**19.** $7\frac{5}{12} - 3\frac{11}{18}$

**20.** $9\frac{3}{7} - 2\frac{13}{21}$

**21.** $5 - 2\frac{1}{4}$

**22.** $4 - 1\frac{2}{3}$

**23.** $17 - 8\frac{3}{4}$

**24.** $23 - 11\frac{5}{8}$

**25.** $3\frac{3}{4} + 5\frac{1}{2} - 2\frac{3}{8}$

**26.** $1\frac{5}{6} + 3\frac{5}{12} - 2\frac{1}{4}$

**27.** $2\frac{3}{8} + 2\frac{1}{4} - 1\frac{5}{6}$

**28.** $1\frac{1}{15} + 3\frac{3}{10} - 2\frac{4}{5}$

Basic Skills | **Advanced Skills** | Vocational-Technical Applications | Calculator/Computer | Above and Beyond

*For exercises 29 to 32, label each statement as* **true** *or* **false.**

**29.** The LCM of 3, 6, and 12 is 24.

**30.** The GCF of 15, 21, and 300 is 3.

**31.** The LCD for $\frac{3}{4}$, $\frac{9}{10}$, and $\frac{3}{20}$ is 40.

**32.** The LCD for $\frac{5}{6}$, $\frac{13}{15}$, and $\frac{1}{21}$ is 3.

< Objective 4 >

*Solve the following applications.*

**33.** **Crafts** Senta is working on a project that uses three pieces of fabric with lengths of $\frac{3}{4}$, $1\frac{1}{4}$, and $\frac{5}{8}$ yd. She needs to allow for $\frac{1}{8}$ yd of waste. How much fabric should she buy?

**34.** **Construction** The framework of a wall is $3\frac{1}{2}$ in. thick. We apply $\frac{5}{8}$-in. wallboard and $\frac{1}{4}$-in. paneling to the inside. Siding that is $\frac{3}{4}$ in. thick is applied to the outside. What is the finished thickness of the wall?

chapter 3 > Make the Connection

**35.** **Business and Finance** A stock was listed at $34\frac{3}{8}$ points on Monday. By closing time Friday, it was at $28\frac{3}{4}$. How much did it drop during the week?

**36.** **Crafts** A roast weighed $4\frac{1}{4}$ lb before cooking and $3\frac{3}{8}$ lb after cooking. How many pounds were lost during the cooking?

**37.** **Crafts** A roll of paper contains $30\frac{1}{4}$ yd. If $16\frac{7}{8}$ yd is cut from the roll, how much paper remains?

**38.** **Geometry** Find the missing dimension in the given figure.

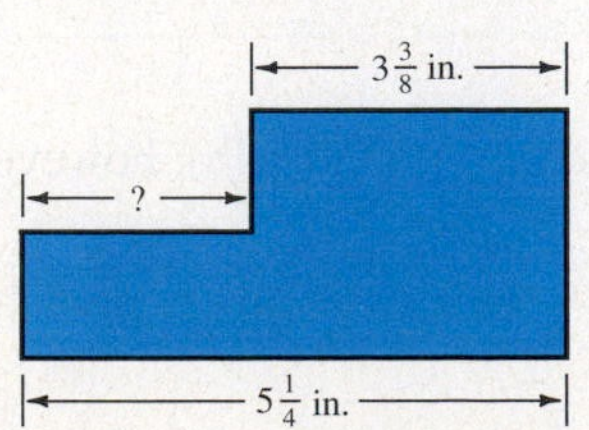


## Answers

29. ________

30. ________

31. ________

32. ________

33. ________

34. ________

35. ________

36. ________

37. ________

38. ________

Answers

39. ______

40. ______

41. ______

42. ______

43. ______

44. ______

45. ______

46. ______

47. ______

48. ______

49. ______

**39.** **Crafts** A $4\frac{1}{4}$-in. bolt is placed through a board that is $3\frac{1}{2}$ in. thick. How far does the bolt extend beyond the board? chapter 3 > Make the Connection

**40.** **Business and Finance** Ben can work 20 h per week on a part-time job. He works $5\frac{1}{2}$ h on Monday and $3\frac{3}{4}$ h on Tuesday. How many more hours can he work during the week?

**41.** **Geometry** Find the missing dimension in the given figure.

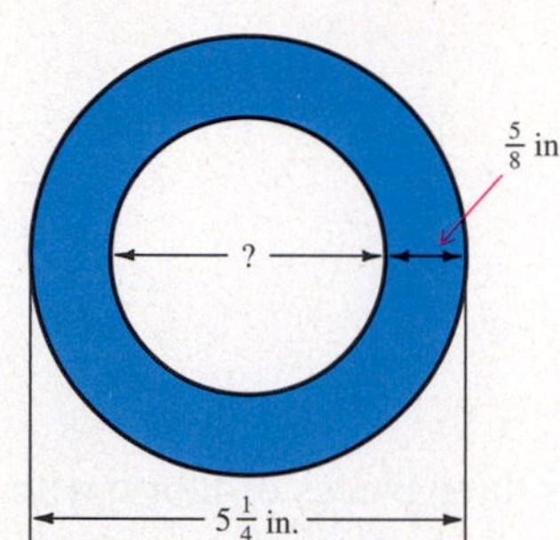


**42.** **Construction** The Whites used $20\frac{3}{4}$ square yards ($yd^2$) of carpet for their living room, $15\frac{1}{2}$ $yd^2$ for the dining room, and $6\frac{1}{4}$ $yd^2$ for a hallway. How much remains if a 50-$yd^2$ roll of carpet is used?

*Evaluate.*

**43.** $4\frac{1}{8} + \frac{3}{7} - 2\frac{23}{28}$

**44.** $5\frac{1}{3} + 1\frac{3}{7} - 5\frac{23}{42}$

**45.** $6\frac{1}{11} + \frac{2}{3} - 2\frac{1}{6}$

**46.** $3\frac{1}{5} + 1\frac{7}{8} - 5\frac{1}{20}$

**47.** $6\frac{1}{11} - \frac{2}{3} + 2\frac{1}{6}$

**48.** $3\frac{1}{5} - 1\frac{7}{8} + 5\frac{1}{20}$

Basic Skills | Advanced Skills | **Vocational-Technical Applications** | Calculator/Computer | Above and Beyond

**49.** **Allied Health** At 2 months old, Daniel weighed $12\frac{1}{2}$ pounds (lb); however, by the time he was 7 months old, he weighed $20\frac{1}{4}$ lb. How many pounds had Daniel gained?

50. **Allied Health** Carla suffers from growth hormone deficiency. The recommended dosage of Nutropin is $\frac{3}{50}$ milligrams per kilogram (mg/kg) of the patient's weight. How much Nutropin should the doctor prescribe if Carla weighs $20\frac{2}{3}$ kg?

51. **Manufacturing Technology** A factory floor is made up of several layers shown in the drawing.

What is the total thickness of the floor?

52. **Manufacturing Technology** Pieces that are $2\frac{1}{4}$, $5\frac{1}{16}$, $4\frac{3}{4}$, and $1\frac{7}{8}$ in. long need to be cut from round stock. How long of a piece of round stock is required? (Allow $\frac{3}{32}$ in. for each saw kerf.)

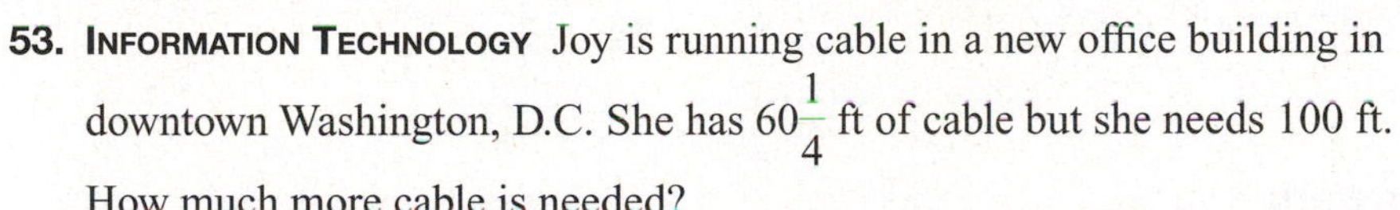

53. **Information Technology** Joy is running cable in a new office building in downtown Washington, D.C. She has $60\frac{1}{4}$ ft of cable but she needs 100 ft. How much more cable is needed?

54. **Information Technology** Abraham has a part-time technician job while he is going to college. He works the following hours in one week: $3\frac{1}{4}$, $5\frac{3}{4}$, $4\frac{1}{2}$, 4, and $2\frac{1}{4}$. For the week, how many hours did Abraham work?

Basic Skills | Advanced Skills | Vocational-Technical Applications | **Calculator/Computer** | Above and Beyond

## Using Your Calculator to Add and Subtract Mixed Numbers

We have already seen how to add, subtract, multiply, and divide fractions using our calculators. Now we will use our calculators to add and subtract mixed numbers.

### Scientific Calculator

To enter a mixed number on a scientific calculator, press the fraction key between both the whole number and the numerator and denominator. For example, to enter $3\frac{7}{12}$, press

3 [a b/c] 7 [a b/c] 12

**Answers**

50. ______

51. ______

52. ______

53. ______

54. ______

## Answers

55. ______

56. ______

57. ______

58. ______

59. ______

60. ______

61. ______

62. ______

### Graphing Calculator

As with multiplying and dividing fractions, when using a graphing calculator, you must choose the fraction option from the math menu before pressing [Enter].

For the problem $3\frac{7}{12} + 2\frac{11}{16}$, the keystroke sequence is

[(] 3 [+] 7 [÷] 12 [)] [+] [(] 2 [+] 11 [÷] 16 [)] [►Frac] [Enter]

Note that the parentheses are very important when doing subtraction. The display will read $\frac{301}{48}$. This is equivalent to $6\frac{13}{48}$.

*Add or subtract the following.*

**55.** $4\frac{7}{9} - 2\frac{11}{18}$

**56.** $7\frac{8}{11} - 4\frac{13}{22}$

**57.** $5\frac{11}{16} - 2\frac{5}{12}$

**58.** $18\frac{5}{24} - 11\frac{3}{40}$

**59.** $6\frac{2}{3} - 1\frac{5}{6}$

**60.** $131\frac{43}{45} - 99\frac{27}{60}$

**61.** $10\frac{2}{3} + 4\frac{1}{5} + 7\frac{2}{15}$

**62.** $7\frac{1}{5} + 3\frac{2}{3} + 1\frac{1}{5}$

## Answers

**1.** $5\frac{7}{9}$ **3.** $7\frac{2}{3}$ **5.** $11\frac{1}{3}$ **7.** $3\frac{8}{15}$ **9.** $7\frac{1}{24}$ **11.** $13\frac{3}{20}$ **13.** $4\frac{1}{2}$ **15.** $1\frac{3}{5}$ **17.** $1\frac{5}{12}$ **19.** $3\frac{29}{36}$ **21.** $2\frac{3}{4}$ **23.** $8\frac{1}{4}$ **25.** $6\frac{7}{8}$ **27.** $2\frac{19}{24}$ **29.** False **31.** False **33.** $2\frac{3}{4}$ yd **35.** $5\frac{5}{8}$ points **37.** $13\frac{3}{8}$ yd **39.** $\frac{3}{4}$ in. **41.** 4 in. **43.** $1\frac{41}{56}$ **45.** $4\frac{13}{22}$ **47.** $7\frac{13}{22}$ **49.** $7\frac{3}{4}$ lb **51.** $13\frac{3}{16}$ in. **53.** $39\frac{3}{4}$ ft **55.** $2\frac{1}{6}$ **57.** $3\frac{13}{48}$ **59.** $4\frac{5}{6}$ **61.** 22

# Activity 8 :: Sharing Costs

The Associated Student Government at CCC is sending six students to the national conference in Washington, D.C. Two of the students, Mikaila and Courtney, are in charge of making the room reservations. Looking at the hotels that are either hosting or adjacent to the conference site, they come up with the following information. Accommodations were not available for places on the table that are blank.

| Hotel | Price for a Single | Price for a Double | Price for a Triple | Suite (Sleeps 6) |
|---|---|---|---|---|
| Wyndham | $180 | $180 | $210 | |
| St. Gregory | 168 | 198 | 240 | $450 |
| Hyatt | 190 | 190 | 222 | |
| Marriott | 159 | 174 | | |

**1.** If they get three double rooms at the Marriott and each pays $\frac{1}{6}$ of the bill, what is the cost per person each night?

**2.** If they stay at the Wyndham in a triple room, they need only two rooms. If they each pay $\frac{1}{6}$ of that total bill, what is the cost per night?

**3.** If they get the suite at the St. Gregory, what is the per-person cost per night?

**4.** The Hyatt and the St. Gregory each offer a free breakfast for each person staying there. Is it now cheaper to stay at the Hyatt in a triple room than at the Wyndham? What information would you need to make the decision?

**5.** What about the St. Gregory suite with a free breakfast? How do you now make the decision?

# 3.5 Order of Operations with Fractions

< 3.5 Objectives >

1 > Evaluate an expression with grouping symbols

2 > Solve an application that involves evaluating an expression

In Chapter 1, we introduced the order of operations. We repeat them here with a small addition to step 1.

**Step by Step**

**Evaluating an Expression**

**Step 1** Do any operations within parentheses or other grouping symbols.
**Step 2** Apply any exponents.
**Step 3** Do all multiplication and division in order from left to right.
**Step 4** Do all addition and subtraction in order from left to right.

The following examples demonstrate the use of the order of operations in evaluating expressions involving fractions. Refer to the steps as we evaluate each expression.

**Example 1** Evaluating an Expression

Evaluate.

$\frac{14}{15} - \left(\frac{1}{2}\right)^2 \cdot \left(\frac{2}{3} + \frac{4}{5}\right)$ Perform operations in parentheses.

$\frac{14}{15} - \left(\frac{1}{2}\right)^2 \cdot \left(\frac{22}{15}\right)$ The next step is to apply the exponents.

$\frac{14}{15} - \left(\frac{1}{4}\right) \cdot \left(\frac{22}{15}\right)$ The next step is to perform the multiplication.

$\frac{14}{15} - \frac{11}{30}$ The final step is to perform addition and subtraction.

$\frac{17}{30}$

**Check Yourself 1**

Evaluate.

$$\left(\frac{2}{3}\right)^3 + \left(\frac{1}{3}\right)^2 \cdot \left(\frac{1}{2} + \frac{2}{3}\right)$$

Recall that, once parentheses are removed, multiplication and division are always performed left to right. Example 2 illustrates.

## Example 2 Evaluating an Expression

Evaluate.

$\left(\frac{1}{13}\right)^2 \cdot \left(\frac{1}{4} + \frac{1}{6}\right) \div \frac{5}{13}$ Add the expression inside the parentheses.

$\left(\frac{1}{13}\right)^2 \cdot \left(\frac{5}{12}\right) \div \frac{5}{13}$ Apply the exponent.

$\frac{1}{169} \cdot \frac{5}{12} \div \frac{5}{13}$ Invert and multiply.

$\frac{1}{\underset{13}{\cancel{169}}} \cdot \frac{\overset{1}{\cancel{5}}}{12} \cdot \frac{\overset{1}{\cancel{13}}}{\underset{1}{\cancel{5}}} = \frac{1}{156}$

### Check Yourself 2

Evaluate.

$\left(\frac{1}{5}\right)^2 \cdot \left(\frac{3}{10} + \frac{1}{2}\right) \div \frac{2}{5}$

When mixed numbers are involved in a complex expression, it is almost always best to convert them to improper fractions before continuing.

## Example 3 Evaluating an Expression with Mixed Numbers

Evaluate.

$3\frac{3}{4} + 5 \cdot \left(2\frac{1}{2}\right)$ Rewrite the mixed numbers as improper fractions.

$\frac{15}{4} + 5 \cdot \left(\frac{5}{2}\right)$ Multiplication precedes addition in the order of operations.

$\frac{15}{4} + \frac{25}{2}$ To add, use the common denominator of 4.

$\frac{15}{4} + \frac{50}{4} = \frac{65}{4} = 16\frac{1}{4}$

### Check Yourself 3

Evaluate.

$4\frac{1}{3} - \frac{1}{2} \cdot \left(5\frac{1}{8}\right)$

Example 4 illustrates an application of the material in this section.

## Example 4 Solving an Application

One formula for calculating children's dosages based on the recommended adult dosage and the child's age in years is Young's rule:

$$\text{Child's dose} = \left(\frac{\text{age}}{\text{age} + 12}\right) \times \text{adult dose}$$

According to this rule, the dose prescribed to a 3-year-old child if the recommended adult dose is 24 milligrams (mg) can be found by evaluating the expression

$$\left(\frac{3}{3 + 12}\right) \times 24 \text{ mg}$$

To evaluate this expression, first we do the operations inside the parentheses.

$$\left(\frac{3}{3 + 12}\right) \times 24 \text{ mg}$$

$$= \left(\frac{3}{15}\right) \times 24 \text{ mg}$$

$$= \frac{1}{5} \times \frac{24}{1} \text{ mg} = \frac{24}{5} \text{ mg} = 4\frac{4}{5} \text{ mg}$$

### Check Yourself 4

The approximate length of the belt pictured is given by

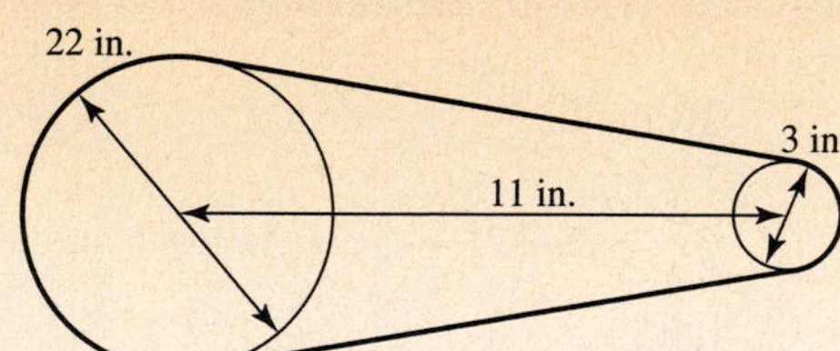


$$\frac{22}{7}\left(\frac{1}{2} \cdot 22 + \frac{1}{2} \cdot 3\right) + 2 \cdot 16$$

Find the length of the belt.

### Check Yourself ANSWERS

1. $\frac{23}{54}$ 2. $\frac{2}{25}$ 3. $1\frac{37}{48}$ 4. $71\frac{2}{7}$ in.

## Reading Your Text

The following fill-in-the-blank exercises are designed to ensure that you understand some of the key vocabulary used in this section.

### SECTION 3.5

**(a)** In evaluating an expression, first do operations inside parentheses or other ______________ symbols.

**(b)** The second step in evaluating an expression is to evaluate all ______________.

**(c)** When dividing by a fraction, ______________ and multiply by that fraction.

**(d)** When mixed numbers are involved in a complex expression, it is almost always best to convert them to __________ fractions before continuing.

# 3.5 exercises

MathZone

Boost your grade at mathzone.com!

- Practice Problems
- NetTutor
- Self-Tests
- e-Professors
- Videos

Name ____________

Section ________ Date ________

**Basic Skills** | Advanced Skills | Vocational-Technical Applications | Calculator/Computer | Above and Beyond

*Evaluate.*

**1.** $\frac{1}{3} - \left(\frac{1}{2} - \frac{1}{4}\right)$

**2.** $\frac{2}{3} - \left(\frac{3}{4} - \frac{1}{2}\right)$

**3.** $\frac{3}{4} - \left(\frac{1}{2}\right)\left(\frac{1}{3}\right)$

**4.** $\frac{5}{6} - \left(\frac{3}{4}\right)\left(\frac{2}{3}\right)$

**5.** $\left(\frac{1}{2}\right)^2 - \left(\frac{1}{4} - \frac{1}{5}\right)$

**6.** $\left(\frac{3}{4}\right)^2 - \left(\frac{1}{2} - \frac{1}{3}\right)$

**7.** $\frac{1}{2} + \frac{2}{3} \cdot \frac{9}{16}$

**8.** $\frac{2}{3} + \frac{1}{2} \div \frac{3}{4}$

**9.** $\left(\frac{1}{3}\right)\left(\frac{1}{2} + \frac{1}{4}\right)$

**10.** $\left(\frac{3}{4}\right)\left(\frac{1}{2} - \frac{1}{3}\right)$

Basic Skills | **Advanced Skills** | Vocational-Technical Applications | Calculator/Computer | Above and Beyond

**11.** $\left(\frac{1}{2}\right)^3 + \left(\frac{1}{3}\right) \cdot \left(\frac{1}{2} + \frac{1}{4}\right)$

**12.** $\left(\frac{1}{3}\right) + \left(\frac{1}{2}\right)^2 \cdot \left(\frac{1}{3} + \frac{1}{6}\right)$

**13.** $\left(\frac{3}{4}\right)^2 + \left(\frac{1}{2}\right)^3 \cdot \left(\frac{1}{2} + \frac{3}{4}\right)$

**14.** $\left(\frac{1}{10}\right) + \left(\frac{1}{2}\right)^3 \cdot \left(\frac{1}{5} + \frac{1}{15}\right)$

**15.** $\left(\frac{2}{3}\right)^2 \cdot \left(\frac{2}{15} + \frac{1}{2}\right) \div \frac{3}{5}$

**16.** $\left(\frac{1}{3}\right)^3 \cdot \left(\frac{1}{4} + \frac{1}{2}\right) \div \frac{2}{9}$

**17.** $2\frac{1}{5} - \frac{1}{2} \cdot \frac{1}{4}$

**18.** $4\frac{3}{8} - \frac{1}{2} \cdot \frac{1}{5}$

**19.** $11\frac{1}{10} - \frac{1}{2} \cdot \left(6\frac{2}{3}\right)$

**20.** $12\frac{2}{7} - \frac{1}{3} \cdot \left(3\frac{3}{7}\right)$

**21.** **Construction** A construction company has bids for paving roads of $1\frac{1}{2}$, $\frac{3}{4}$, and $3\frac{1}{3}$ miles (mi) for the month of July. With their present equipment,

## Answers

1. ______ 2. ______
3. ______ 4. ______
5. ______ 6. ______
7. ______ 8. ______
9. ______ 10. ______
11. ______ 12. ______
13. ______ 14. ______
15. ______ 16. ______
17. ______
18. ______
19. ______
20. ______
21. ______

they can pave 8 mi in 1 month. How much more work can they take on in July?

**22.** **Statistics** On an 8-h trip, Jack drives $2\frac{3}{4}$ h and Pat drives $2\frac{1}{2}$ h. How many hours are left to drive?

**23.** **Statistics** A runner has told herself that she will run 20 mi each week. She runs $5\frac{1}{2}$ mi on Sunday, $4\frac{1}{4}$ mi on Tuesday, $4\frac{3}{4}$ mi on Wednesday, and $2\frac{1}{8}$ mi on Friday. How far must she run on Saturday to meet her goal?

**24.** **Science and Medicine** If paper takes up $\frac{1}{2}$ of the space in a landfill and plastic takes up $\frac{1}{10}$ of the space, how much of the landfill is used for other materials?

**25.** **Science and Medicine** If paper takes up $\frac{1}{2}$ of the space in a landfill and organic waste takes up $\frac{1}{8}$ of the space, how much of the landfill is used for other materials?

## Answers

22. ____________________

23. ____________________

24. ____________________

25. ____________________

Answers

26. 

27. 

28. 

29. 

30. 

31. 

32. 

**26. Business and Finance** The interest rate on an auto loan in May was $12\frac{3}{8}\%$. By September the rate was up to $14\frac{1}{4}\%$. By how many percentage points did the interest rate increase over the period?

Basic Skills | Advanced Skills | **Vocational-Technical Applications** | Calculator/Computer | Above and Beyond

**27. Allied Health** Simone suffers from Gaucher's disease. The recommended dosage of Cerezyme is $2\frac{1}{2}$ units per kilogram (kg) of the patient's weight. How much Cerezyme should the doctor prescribe if Simone weighs $15\frac{1}{3}$ kg?

**28. Information Technology** Kendra is running cable in a new office building in downtown Kansas City. She has $110\frac{1}{4}$ ft of cable, but she needs 150 ft. How much more cable is needed?

**29. Manufacturing Technology** A cut $3\frac{3}{8}$ in. long needs to be made in a piece of material. The cut rate is $\frac{3}{4}$ in. per minute. How many minutes does it take to make the cut?

**30. Manufacturing Technology** Calculate the distance from the center of hole A to the center of hole B.

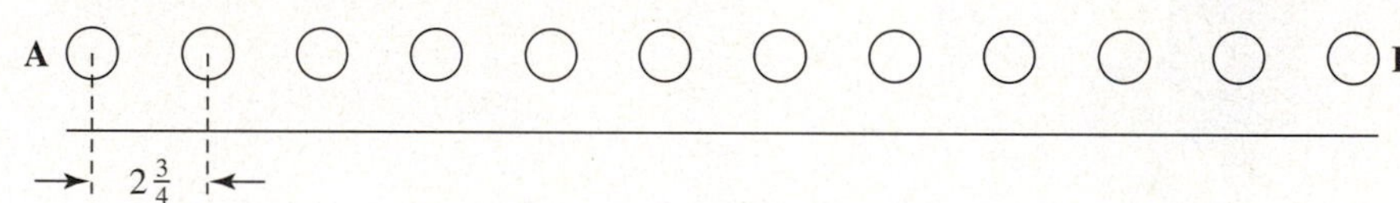


**31. Information Technology** On average, $18\frac{1}{4}$ printed circuit boards can be completely populated with components (with all parts soldered to the board) by Amara in 1 h. Burt can complete $34\frac{2}{3}$ printed circuit boards in 2 h. If both workers continue at their respective average paces, how many total printed circuit boards can be populated in 8 h?

**32. Information Technology** If Carlos joins Amara and Burt in soldering components on circuit boards, the three workers can average 410 complete boards in 8 h. Assuming the other two workers perform at their respective averages stated in exercise 31, how many boards can Carlos average in 8 h? How many boards does Carlos average in 1 h?

## Answers

1. $\frac{1}{12}$ 3. $\frac{7}{12}$ 5. $\frac{1}{5}$ 7. $\frac{7}{8}$ 9. $\frac{1}{4}$
11. $\frac{3}{8}$ 13. $\frac{23}{32}$ 15. $\frac{38}{81}$ 17. $2\frac{3}{40}$ 19. $7\frac{23}{30}$
21. $2\frac{5}{12}$ mi 23. $3\frac{3}{8}$ mi 25. $\frac{3}{8}$ 27. $38\frac{1}{3}$ units
29. $4\frac{1}{2}$ min 31. $284\frac{2}{3}$ or 284 completed boards

# 3.6 Estimation Applications

< 3.6 Objective >

1 > Use estimation to solve application problems

**Units ANALYSIS**

Every denominate number with a fractional unit, such as

$$25\ \frac{\text{mi}}{\text{gal}}$$

has a dual denominate number, which is the reciprocal number and the reciprocal units. In this case we have

$$\frac{1}{25}\ \frac{\text{gal}}{\text{mi}}$$

The $25\ \frac{\text{mi}}{\text{gal}}$ indicates that we can drive 25 mi on 1 gal. The $\frac{1}{25}\ \frac{\text{gal}}{\text{mi}}$ indicates that we use $\frac{1}{25}$ of a gallon for each mile.

EXAMPLES:

| Denominate Number | Dual |
|---|---|
| $55\ \frac{\text{mi}}{\text{h}}$ | $\frac{1}{55}\ \frac{\text{h}}{\text{mi}}$ |
| $\frac{2}{3}\ \frac{\text{page}}{\text{min}}$ | $\frac{3}{2}\ \frac{\text{min}}{\text{page}}$ |

Of all the skills you develop in the study of arithmetic, perhaps the most useful is that of estimation. Every day, you have occasion to estimate. Here are a few estimation exercises that you may have gone through this morning.

How much flour should I put in the pancakes?

How much cash am I likely to need today?

How long will it take me to walk to the bus stop?

How long is the ride to school?

How many miles can I get on just over $\frac{1}{4}$ tank of gas?

Although you may not think of these as math problems, they are. As your estimation skills improve, so will your ability to come up with good answers to everyday problems.

## Example 1 Estimating Measurements

< Objective 1 >

Based on the gauge, is the gas tank closer to $\frac{3}{4}$, $\frac{2}{3}$, $\frac{1}{3}$, or $\frac{1}{4}$ full?

At slightly more than $\frac{1}{2}$, the gauge indicates close to $\frac{2}{3}$ of a tank remains.

### Check Yourself 1

Is the reading of the thermometer closer to 72°, 74°, or 76°?

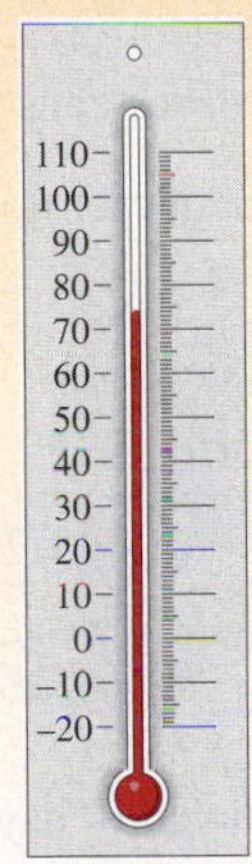


Cooking provides many opportunities for estimation. Example 2 illustrates one.

## Example 2 Estimating Measurement

A recipe calls for $\frac{1}{3}$ cup of sugar. The only measuring cups you have are 1 cup, $\frac{1}{2}$ cup, and $\frac{1}{4}$ cup. What should you do?

To solve this problem you must believe that recipes are only approximations anyway. Once you accept that idea, you can estimate $\frac{1}{3}$ cup. It is between $\frac{1}{4}$ cup and $\frac{1}{2}$ cup, so fill the $\frac{1}{4}$ cup, dump it into the $\frac{1}{2}$ cup, and add a little more sugar.

**Check Yourself 2**

**A recipe calls for $\frac{3}{4}$ cup of flour, but you have only a 1-cup, a $\frac{2}{3}$-cup, and a $\frac{1}{3}$-cup measure. How should you proceed?**

Almost every shopping trip presents many estimation opportunities. Whether you are estimating the impact of car payments on your monthly budget or estimating the number of bananas you can buy for \$3, you are practicing your arithmetic skills. Example 3 provides an opportunity to practice these skills.

## Example 3 Estimating Total Cost

Dog food is on special at three cans for a dollar. Your puppy eats about $\frac{1}{2}$ can per day. How much should you budget for dog food over the next semester (almost 5 months)?

First, estimate the number of days in the semester. At 30 days per month, we'll use

$$30\,\frac{\text{days}}{\text{month}} \times 5 \text{ months} = 150 \text{ days}$$

Next, estimate the amount of food to be consumed. We could try using $\frac{1}{2}$ can per day, which yields

$$\frac{1}{2}\,\frac{\text{can}}{\text{day}} \times 150 \text{ days} = 75 \text{ cans}$$

but remember that this is a puppy, so we will assume his food intake will increase over the next 5 months. To be safe, we will add 25 cans and estimate the total to be 100 cans.

Finally, we can estimate the total cost. If we buy all the food today, we can buy it at 3 cans per dollar. At $\frac{1}{3}$ dollar per can, we get

$$\frac{1}{3}\,\frac{\text{dollar}}{\text{can}} \times 100 \text{ cans} \approx \$33$$

The cost will be about \$33.

### Check Yourself 3

Cat food is also on special at seven cans for \$2 $\left(\frac{7 \text{ cans}}{2 \text{ dollars}}\right)$. If the cat eats about $\frac{1}{2}$ can per day, what is the approximate cost of the cat food for a 5-month semester?

### Check Yourself ANSWERS

1. It is closer to 74°.
2. Fill the $\frac{2}{3}$-cup measure, dump it into the 1-cup measure, and add a little flour.
3. The cost is a little over \$20.

## Reading Your Text

The following fill-in-the-blank exercises are designed to ensure that you understand some of the key vocabulary used in this section.

**SECTION 3.6**

**(a)** Every denominate number with a fractional unit has a ______________ denominate number.

**(b)** Of all the skills you develop in the study of arithmetic, perhaps the most useful is ______________.

**(c)** Measurements in a recipe are ______________.

**(d)** Almost every shopping trip represents many ______________ opportunities.

# 3.6 exercises

Basic Skills | Advanced Skills | Vocational-Technical Applications | Calculator/Computer | Above and Beyond

MathZone

Boost your grade at mathzone.com!
> Practice Problems
> NetTutor
> Self-Tests
> e-Professors
> Videos

Name ____________

Section ______ Date ______

## Answers

1. ____
2. ____
3. ____
4. ____
5. ____
6. ____
7. ____
8. ____
9. ____
10. ____
11. ____
12. ____
13. ____

< Objective 1 >

*Find the dual of each denominate number.*

1. $36 \frac{\text{ft}}{\text{s}}$

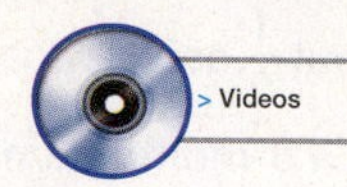


2. $42 \frac{\text{kilowatts}}{\text{h}}$

3. $5 \frac{\text{pages}}{\text{min}}$

4. $125 \frac{\text{joules}}{\text{s}}$

5. $18 \frac{\text{mi}}{\text{gal}}$

6. $84 \frac{\text{coulombs}}{\text{s}}$

7. $68 \frac{\text{lumens}}{\text{cm}^2}$

8. $55 \frac{\text{ergs}}{\text{min}}$

*Solve the following applications.*

9. **Crafts** Mark wants to make $300 by selling cupcakes at a bake sale, and his mother has offered to pay for the ingredients. He plans to sell them for $5 per dozen. A box of cake mix makes about 50 cupcakes. Approximately how many boxes should Mark buy?

10. **Statistics** Amy is traveling to a city 418 miles away at a speed of roughly 55 miles per hour. About how long should her trip take?

11. **Social Science** The map tells Manuel that it is 423 miles from Eastwick to West Goshen. He knows that, with mixed city and freeway driving, he can average about 50 miles per hour. He is currently in Eastwick and needs to be in West Goshen by noon. Make a rough estimate of the time that Manuel should leave Eastwick.

12. **Business and Finance** Sam works in a shipping department for a manufacturer. He is filling a customer's order for several small items. The items ordered weigh 21, 23, 18, and 7 ounces (oz). The company policy is to use a stronger box to ship products totaling more than 3 pounds (lb). Knowing 16 oz is 1 lb, estimate whether Sam should use the stronger box.

13. **Construction** Lauren is a contractor for a roofing job. She estimates that she will need about 4,800 roofing nails. According to a handbook, the roofing nails that are needed count out at about 189 nails per pound. Estimate how many pounds Lauren will need for the job.

**14. BUSINESS AND FINANCE** Julio works as a quality control expert in a beverage factory. The assembly line that he monitors produces about 20,000 bottles in a 24-hour period. Julio samples about 120 bottles an hour and rejects the line if he finds more than $\frac{1}{50}$ of the sample to be defective. About how many defective bottles should Julio allow before rejecting the entire line?

**15. CRAFTS** Based on the amount of liquid in the pitcher, is the pitcher closer to $\frac{1}{3}, \frac{2}{3}, \frac{1}{4}$, or $\frac{3}{4}$ full?

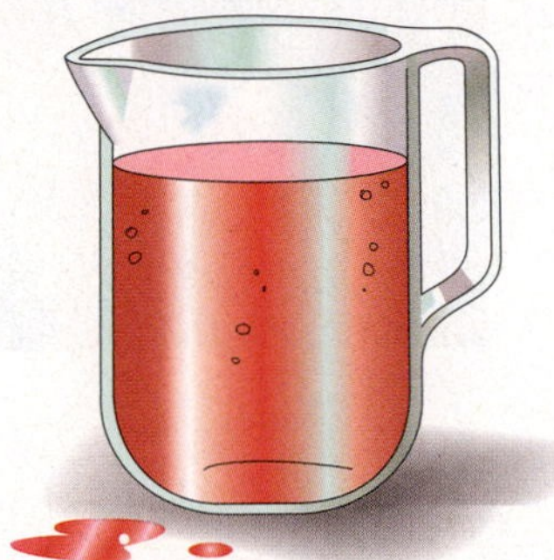

**16. BUSINESS AND FINANCE** Lunch meat sells for about \$3 per pound. You use about $\frac{1}{2}$ pound per day for sandwiches. How much should you budget for lunch meat over the next month (about 20 workdays)?

Basic Skills | **Advanced Skills** | Vocational-Technical Applications | Calculator/Computer | Above and Beyond

*Estimate the following sums or differences.*

**17.** $2\frac{1}{8} + 7\frac{9}{11} + 4\frac{6}{7} + 5\frac{1}{12}$

**18.** $4\frac{5}{6} + 3\frac{1}{5} + 8\frac{1}{7} + 11\frac{8}{9}$

**19.** $15\frac{4}{5} - 3\frac{9}{10} - 6\frac{1}{7}$

**20.** $18\frac{1}{5} + 11\frac{7}{8} - 14\frac{10}{11} - 10\frac{6}{7}$

*Estimate the following products or quotients.*

**21.** $2\frac{5}{6} \times 4\frac{1}{3}$

**22.** $3\frac{1}{7} \times 6\frac{8}{9}$

**23.** $9\frac{2}{9} \div 2\frac{6}{7}$

**24.** $17\frac{11}{12} \div 6\frac{1}{10}$

**25.** $15\frac{6}{7} \div 4\frac{1}{8} \times 1\frac{3}{4}$

## Answers

14. ____
15. ____
16. ____
17. ____
18. ____
19. ____
20. ____
21. ____
22. ____
23. ____
24. ____
25. ____

## Answers

**1.** $\frac{1 \text{ s}}{36 \text{ ft}}$ **3.** $\frac{1 \text{ min}}{5 \text{ page}}$ **5.** $\frac{1 \text{ gal}}{18 \text{ mi}}$ **7.** $\frac{1 \text{ cm}^2}{68 \text{ lumen}}$ **9.** 15 boxes **11.** 3 A.M. **13.** 24 lb **15.** $\frac{3}{4}$ **17.** 20 **19.** 6 **21.** 12 **23.** 3 **25.** 8

# Activity 9 :: Aerobic Exercise

According to the website www.Fitnesszone.com, **aerobic training zone** refers to the training intensity range that will produce improvement in your level of aerobic fitness without overtaxing your cardiorespiratory system. Your aerobic training zone is based on a percentage of your maximal heart rate. As a general rule, your maximal heart rate is measured directly or estimated by subtracting your age from 220. Depending upon how physically fit you are, the lower and upper limits of your aerobic training zone are then based on a fraction of the maximal heart rate, approximately $\frac{3}{5}$ to $\frac{9}{10}$, respectively.

The chart below uses this information to determine maximal heart rate (MHR) by subtracting the age from 220. Complete the table.

| Age | Maximal Heart Rate |
|---|---|
| 20 | 200 |
| 25 | |
| 30 | |
| 35 | |
| 40 | |
| 45 | |
| 50 | |
| 55 | |

Using the information provided in the first paragraph, complete the following table to find the lower and upper limits for the training zone for each age. Note that when we compute the upper limit of the training zone, we round up to the next integer. For example, we will compute the upper limit for age 25.

$$195 \times \frac{9}{10} = \frac{195}{1} \times \frac{9}{10} = \frac{39}{1} \times \frac{9}{2} = \frac{351}{2} = 175\frac{1}{2}$$

Rounding up, we get an upper limit of 176.

| Age | Maximal Heart Rate | Lower Limit of Zone $\left(\frac{3}{5}\text{ MHR}\right)$ | Upper Limit of Zone $\left(\frac{9}{10}\text{ MHR}\right)$ |
|---|---|---|---|
| 20 | 200 | 120 | 180 |
| 25 | 195 | 117 | 176 |
| 30 | 190 | | |
| 35 | 185 | | |
| 40 | 180 | | |
| 45 | 175 | | |
| 50 | 170 | | |
| 55 | 165 | | |

| Definition/Procedure | Example | Reference |
|---|---|---|
| **Adding and Subtracting Fractions with Like Denominators** | | Section 3.1 |
| *To Add Like Fractions* | | |
| **Step 1** Add the numerators.<br>**Step 2** Place the sum over the common denominator.<br>**Step 3** Simplify the resulting fraction if necessary. | $\frac{5}{18} + \frac{7}{18} = \frac{12}{18} = \frac{2}{3}$ | p. 213 |
| *To Subtract Like Fractions* | | |
| **Step 1** Subtract the numerators.<br>**Step 2** Place the difference over the common denominator.<br>**Step 3** Simplify the resulting fraction when necessary. | $\frac{17}{20} - \frac{7}{20} = \frac{10}{20} = \frac{1}{2}$ | p. 216 |
| **Common Multiples** | | Section 3.2 |
| **Least Common Multiple (LCM)** The LCM is the *smallest* number that is a multiple of each of a group of numbers. | | p. 224 |
| *To Find the LCM* | | |
| **Step 1** Write the prime factorization for each of the numbers in the group.<br>**Step 2** Find all the prime factors that appear in any one of the prime factorizations.<br>**Step 3** Form the product of those prime factors, using each factor the greatest number of times it occurs in any one factorization. | To find the LCM of 12, 15, and 18:<br>$12 = 2 \times 2 \times 3$<br>$15 = 3 \times 5$<br>$18 = 2 \times 3 \times 3$<br>$2 \times 2 \times 3 \times 3 \times 5$<br>The LCM is $2 \times 2 \times 3 \times 3 \times 5$, or 180. | p. 224 |
| **Adding and Subtracting Fractions with Unlike Denominators** | | Section 3.3 |
| *To Find the LCD of a Group of Fractions* | | |
| **Step 1** Write the prime factorization for each of the denominators.<br>**Step 2** Find all the prime factors that appear in any one of the prime factorizations.<br>**Step 3** Form the product of those prime factors, using each factor the greatest number of times it occurs in any one factorization. | To find the LCD of fractions with denominators 4, 6, and 15:<br>$4 = 2 \times 2$<br>$6 = 2 \times 3$<br>$15 = 3 \times 5$<br>$2 \times 2 \times 3 \times 5$<br>The LCD $= 2 \times 2 \times 3 \times 5$, or 60. | p. 233 |

*Continued*

| Definition/Procedure | Example | Reference |
|---|---|---|
| ***To Add Unlike Fractions*** | | |
| **Step 1** Find the LCD of the fractions.<br>**Step 2** Change each unlike fraction to an equivalent fraction with the LCD as a common denominator.<br>**Step 3** Add the resulting like fractions as before. | $\frac{3}{4} + \frac{7}{10} = \frac{15}{20} + \frac{14}{20}$<br>$= \frac{29}{20} = 1\frac{9}{20}$ | *p.* 234 |
| ***To Subtract Unlike Fractions*** | | |
| **Step 1** Find the LCD of the fractions.<br>**Step 2** Change each unlike fraction to an equivalent fraction with the LCD as a common denominator.<br>**Step 3** Subtract the resulting like fractions as before. | $\frac{8}{9} - \frac{5}{6} = \frac{16}{18} - \frac{15}{18} = \frac{1}{18}$ | *p.* 237 |
| **Adding and Subtracting Mixed Numbers** | | **Section 3.4** |
| ***To Add or Subtract Mixed Numbers*** | | |
| **Step 1** Find the LCD.<br>**Step 2** Rewrite the fraction parts to match the LCD.<br>**Step 3** Add or subtract the fractions.<br>**Step 4** Add or subtract the whole number part.<br>**Step 5** Simplify if necessary. | $2\frac{1}{4} + 3\frac{4}{5} = 2\frac{5}{20} + 3\frac{16}{20}$<br>$= 2 + 3 + \frac{5}{20} + \frac{16}{20} = 5 + \frac{21}{20} = 6\frac{1}{20}$ | *p.* 249 |
| **Order of Operations with Fractions** | | **Section 3.5** |
| ***Order of Operations*** | | |
| **Step 1** Do any operations within parentheses or other grouping symbols.<br>**Step 2** Evaluate all powers.<br>**Step 3** Do all multiplication and division in order from left to right.<br>**Step 4** Do all addition and subtraction in order from left to right. | $\frac{2}{3} + \left(\frac{1}{2}\right)^2\left(\frac{1}{3} + \frac{1}{2}\right) = \frac{2}{3} + \left(\frac{1}{2}\right)^2\left(\frac{5}{6}\right)$<br>$= \frac{2}{3} + \left(\frac{1}{4}\right)\left(\frac{5}{6}\right)$<br>$= \frac{2}{3} + \frac{5}{24}$<br>$= \frac{16}{24} + \frac{5}{24}$<br>$= \frac{21}{24} = \frac{7}{8}$ | *p.* 262 |
| **Estimation Applications** | | **Section 3.6** |
| Every denominate number with a fractional unit has a dual denominate number that is the reciprocal number and the reciprocal units. | Denominate Number — Dual<br>$50\,\frac{\text{mi}}{\text{h}}$ — $\frac{1}{50}\,\frac{\text{h}}{\text{mi}}$<br>$\frac{10 \text{ dollars}}{\text{ticket}}$ — $\frac{1}{10}\,\frac{\text{ticket}}{\text{dollars}}$ | *p.* 270 |

This summary exercise set is provided to give you practice with each of the objectives of this chapter. Each exercise is keyed to the appropriate chapter section. When you are finished, you can check your answers to the odd-numbered exercises against those presented in the back of the text. If you have difficulty with any of these questions, go back and reread the examples from that section. The answers to the even-numbered exercises appear in the *Instructor's Solutions Manual.* Your instructor will give you guidelines on how to best use these exercises in your instructional setting.

**3.1** *Add. Simplify when possible.*

**1.** $\frac{8}{15} + \frac{2}{15}$

**2.** $\frac{4}{7} + \frac{3}{7}$

**3.** $\frac{8}{13} + \frac{7}{13}$

**4.** $\frac{17}{18} + \frac{5}{18}$

**5.** $\frac{19}{24} + \frac{13}{24}$

**6.** $\frac{1}{9} + \frac{2}{9} + \frac{4}{9}$

**7.** $\frac{2}{9} + \frac{5}{9} + \frac{4}{9}$

**8.** $\frac{4}{15} + \frac{7}{15} + \frac{7}{15}$

**3.2** *Find the least common multiple (LCM) for each of the following groups of numbers.*

**9.** 4 and 12

**10.** 8 and 16

**11.** 18 and 24

**12.** 12 and 18

**13.** 15 and 20

**14.** 14 and 21

**15.** 9, 12, and 24

**16.** 14, 21, and 28

*Arrange the fractions in order from smallest to largest.*

**17.** $\frac{5}{8}, \frac{7}{12}$

**18.** $\frac{5}{6}, \frac{4}{5}, \frac{7}{10}$

*Complete the following statements, using the symbol $<$, $=$, or $>$.*

**19.** $\frac{5}{12} \square \frac{3}{8}$

**20.** $\frac{3}{7} \square \frac{9}{21}$

**21.** $\frac{9}{16} \square \frac{7}{12}$

**3.3** *Write as equivalent fractions with the LCD as a common denominator.*

**22.** $\frac{1}{6}, \frac{7}{8}$

**23.** $\frac{3}{10}, \frac{5}{8}, \frac{7}{12}$

*Find the least common denominator (LCD) for fractions with the given denominators.*

**24.** 6 and 24

**25.** 12 and 18

**26.** 20 and 24

**27.** 25 and 40

**28.** 4, 5, and 9

**29.** 3, 4, and 11

**30.** 2, 5, and 8

**31.** 3, 6, and 8

*Add.*

**32.** $\frac{3}{10} + \frac{7}{12}$

**33.** $\frac{3}{8} + \frac{5}{12}$

**34.** $\frac{5}{36} + \frac{7}{24}$

**35.** $\frac{2}{15} + \frac{9}{20}$

**36.** $\frac{9}{14} + \frac{10}{21}$

**37.** $\frac{7}{15} + \frac{13}{18}$

**38.** $\frac{12}{25} + \frac{19}{30}$

**39.** $\frac{1}{2} + \frac{1}{4} + \frac{1}{8}$

**40.** $\frac{1}{3} + \frac{1}{5} + \frac{1}{10}$

**41.** $\frac{3}{8} + \frac{5}{12} + \frac{7}{18}$

**42.** $\frac{5}{6} + \frac{8}{15} + \frac{9}{20}$

*Subtract.*

**43.** $\frac{8}{9} - \frac{3}{9}$

**44.** $\frac{9}{10} - \frac{6}{10}$

**45.** $\frac{5}{8} - \frac{1}{8}$

**46.** $\frac{11}{12} - \frac{7}{12}$

**47.** $\frac{7}{8} - \frac{2}{3}$

**48.** $\frac{5}{6} - \frac{3}{5}$

**49.** $\frac{11}{18} - \frac{2}{9}$

**50.** $\frac{5}{6} - \frac{1}{4}$

**51.** $\frac{5}{8} - \frac{1}{6}$

**52.** $\frac{13}{18} - \frac{5}{12}$

**53.** $\frac{8}{21} - \frac{1}{14}$

**54.** $\frac{13}{18} - \frac{7}{15}$

**55.** $\frac{11}{12} - \frac{1}{4} - \frac{1}{3}$

**56.** $\frac{13}{15} + \frac{2}{3} - \frac{3}{5}$

**3.4** *Perform the indicated operations.*

**57.** $4\frac{5}{8} - 4\frac{3}{8}$

**58.** $2\frac{11}{18} - 2\frac{2}{9}$

**59.** $3\frac{7}{10} - 3\frac{7}{12}$

**60.** $9\frac{11}{27} - 9\frac{5}{18}$

**61.** $5\frac{4}{9} + 2\frac{5}{12} - 7\frac{3}{8}$

**62.** $6\frac{5}{7} + 3\frac{4}{7}$

**63.** $5\frac{7}{10} + 3\frac{11}{12}$

**64.** $2\frac{1}{2} + 3\frac{5}{6} + 3\frac{3}{8}$

**65.** $7\frac{7}{9} - 3\frac{4}{9}$

**66.** $9\frac{1}{6} - 3\frac{1}{8}$

**67.** $6\frac{5}{12} - 3\frac{5}{8}$

**68.** $2\frac{1}{3} + 5\frac{1}{6} - 2\frac{4}{5}$

*Solve the following applications.*

**69.** **CRAFTS** A recipe calls for $\frac{1}{3}$ cup of milk. You have $\frac{3}{4}$ cup. How much milk will be left over?

**70.** **CONSTRUCTION** Bradley needs two shelves, one $32\frac{3}{8}$ in. long and the other $36\frac{11}{16}$ in. long. What is the total length of shelving that is needed?

**71.** **GEOMETRY** Find the perimeter of the triangle.

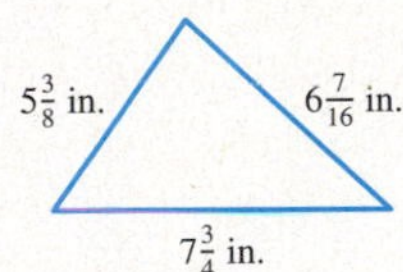


**72.** **STATISTICS** At the beginning of a year Miguel was $51\frac{3}{4}$ in. tall. In June, he measured $53\frac{1}{8}$ in. How much did he grow during that period?

**73.** **CONSTRUCTION** A bookshelf that is $42\frac{5}{16}$ in. long is cut from a board with a length of 8 ft. If $\frac{1}{8}$ in. is wasted in the cut, what length board remains?

**74.** **CRAFTS** Amelia buys an 8-yd roll of wallpaper on sale. After measuring, she finds that she needs the following amounts of the paper: $2\frac{1}{3}$, $1\frac{1}{2}$, and $3\frac{3}{4}$ yd. Does she have enough for the job? If so, how much will be left over?

**75.** **CRAFTS** Roberto used $1\frac{3}{4}$ gallon (gal) of paint in his living room, $1\frac{1}{3}$ gal in the dining room, and $\frac{1}{2}$ gal in a hallway. How much paint did he use?

**76.** **CONSTRUCTION** A sheet of plywood consists of two outer sections that are $\frac{3}{16}$ in. thick and a center section that is $\frac{3}{8}$ in. thick. How thick is the plywood overall?

**77.** **CRAFTS** A pattern calls for four pieces of fabric with lengths $\frac{1}{2}$, $\frac{3}{8}$, $\frac{1}{4}$, and $\frac{5}{8}$ yd. How much fabric must be purchased to use the pattern?

**3.5** *Evaluate.*

**78.** $\frac{3}{4} + \left(\frac{1}{2}\right)\left(\frac{1}{3}\right)$

**79.** $\frac{2}{3} - \left(\frac{3}{4}\right)^2\left(\frac{1}{2}\right)$

**80.** $\left(4\frac{3}{8}\right)\left(1\frac{1}{2}\right) \div \left(\frac{3}{4} + \frac{1}{2}\right)$

**81.** $\frac{1}{3} \cdot \left(\frac{3}{4} - \frac{1}{2}\right)^2$

**82.** $\left(\frac{2}{3}\right)^3 - \frac{1}{2} \cdot \frac{1}{9}$

**83.** $2\frac{1}{3} - \left[\frac{1}{2} \cdot \left(2 - \frac{1}{3}\right)\right]$

**3.6**

**84. Business and Finance** Jared had about \$2,800 in his checking account. He wrote five checks for equal amounts, and he knows his balance is about \$200. Estimate the amount of each check that he wrote.

**85. Business and Finance** An appliance store has the following items on sale:

19″ color television for \$399
Entertainment center for \$509.95
A mini sound system for \$369.95

Estimate the cost of one entertainment center, two color televisions, and three mini sound systems.

*Estimate the following.*

**86.** $3\frac{5}{6} + 4\frac{1}{7} - 2\frac{6}{7}$

**87.** $4\frac{1}{8} \times 5\frac{11}{12}$

**88.** $18\frac{1}{9} \div 1\frac{8}{9} \times 4\frac{5}{6}$

# self-test 3

Name ______

Section ______ Date ______

The purpose of this self-test is to help you check your progress so that you can find sections and concepts that you need to review before the next exam. Allow yourself about an hour to take this test. At the end of that hour, check your answers against those given in the back of the text. If you missed any, note the section reference that accompanies the answer. Go back to that section and reread the examples until you have mastered that particular concept.

*Add.*

**1.** $\frac{3}{10} + \frac{6}{10}$

**2.** $\frac{5}{12} + \frac{3}{12}$

**3.** Find the least common multiple of 18, 24, and 36.

*Find the least common denominator for fractions with the given denominators.*

**4.** 12 and 15

**5.** 3, 4, and 18

*Add.*

**6.** $\frac{2}{5} + \frac{4}{10}$

**7.** $\frac{1}{6} + \frac{3}{7}$

**8.** $\frac{3}{8} + \frac{5}{12}$

**9.** $\frac{11}{15} + \frac{9}{20}$

**10.** $\frac{1}{4} + \frac{5}{8} + \frac{7}{10}$

**11.** $\frac{5}{24} + \frac{3}{8}$

*Solve the following application.*

**12.** **Crafts** A recipe calls for $\frac{1}{2}$ cup of raisins, $\frac{1}{4}$ cup of walnuts, and $\frac{2}{3}$ cup of rolled oats. What is the total amount of these ingredients?

*Subtract.*

**13.** $\frac{7}{9} - \frac{4}{9}$

**14.** $\frac{7}{18} - \frac{5}{18}$

**15.** $\frac{11}{12} - \frac{3}{20}$

*Solve the following application.*

**16.** **Statistics** You have $\frac{5}{6}$ hour (h) to take a three-part test. You use $\frac{1}{3}$ h for the first section and $\frac{1}{4}$ h for the second. How much time do you have left to finish the last section of the test?

## Answers

1. ______
2. ______
3. ______
4. ______
5. ______
6. ______
7. ______
8. ______
9. ______
10. ______
11. ______
12. ______
13. ______
14. ______
15. ______
16. ______

## Answers

17. ____

18. ____

19. ____

20. ____

21. ____

22. ____

23. ____

24. ____

25. ____

26. ____

27. ____

28. ____

29. ____

30. ____

*Perform the indicated operations.*

**17.** $5\frac{3}{10} + 2\frac{4}{10}$ **18.** $7\frac{3}{8} + 2\frac{7}{8}$ **19.** $4\frac{1}{6} + 3\frac{3}{4}$

**20.** $6\frac{3}{8} + 5\frac{7}{10}$ **21.** $7\frac{3}{8} - 5\frac{5}{8}$ **22.** $3\frac{5}{6} - 2\frac{2}{9}$

**23.** $7\frac{1}{8} - 3\frac{1}{6}$ **24.** $7 - 5\frac{7}{15}$ **25.** $4\frac{2}{7} + 3\frac{3}{7} + 1\frac{3}{7}$

**26.** $3\frac{1}{2} + 4\frac{3}{4} + 5\frac{3}{10}$

**27. BUSINESS AND FINANCE** A worker has $2\frac{1}{6}$ h of overtime on Tuesday, $1\frac{3}{4}$ h on Wednesday, and $1\frac{5}{6}$ h on Friday. What is the total overtime for the week?

*Evaluate.*

**28.** $\frac{1}{4} + \left(\frac{1}{2}\right)^2 - \left(\frac{1}{3} + \frac{1}{12}\right)$ **29.** $\left(1\frac{1}{3} + 2\frac{1}{2}\right) \cdot 3\frac{1}{4}$

**30. STATISTICS** The average person drinks about $3\frac{1}{5}$ cups of coffee per day. If a person works 5 days a week for 50 weeks every year, estimate how many cups of coffee that person will drink in a working lifetime of $51\frac{3}{4}$ years.

# cumulative review chapters 1-3

Name ______________

Section ________ Date ________

The following exercises are presented to help you review concepts from earlier chapters. This is meant as review material and not as a comprehensive exam. The answers are presented in the back of the text. Beside each answer is a section reference for the concept. If you have difficulty with any of these exercises, be certain to at least read through the summary related to that section.

*Perform the indicated operations.*

**1.** $\begin{array}{r} 1{,}369 \\ +\ 5{,}804 \\ \hline \end{array}$

**2.** $\begin{array}{r} 489 \\ 562 \\ 613 \\ +\ 254 \\ \hline \end{array}$

**3.** $\begin{array}{r} 357 \\ 28 \\ +\ 2{,}346 \\ \hline \end{array}$

**4.** $\begin{array}{r} 13 \\ 2{,}543 \\ +\ 10{,}547 \\ \hline \end{array}$

**5.** $289 - 54$

**6.** $53{,}294 - 41{,}074$

**7.** $503 - 74$

**8.** $5{,}731 - 2{,}492$

**9.** $\begin{array}{r} 58 \\ \times\ 3 \\ \hline \end{array}$

**10.** Find the product of 273 and 7.

**11.** $\begin{array}{r} 89 \\ \times\ 56 \\ \hline \end{array}$

**12.** $\begin{array}{r} 538 \\ \times\ 103 \\ \hline \end{array}$

**13.** $281\overline{)6{,}935}$

**14.** $571\overline{)12{,}583}$

**15.** $293\overline{)61{,}382}$

*Evaluate each of the expressions.*

**16.** $12 \div 6 + 3$

**17.** $4 + 12 \div 4$

**18.** $3^3 \div 9$

**19.** $28 \div 7 \times 4$

**Answers**

1. ______________
2. ______________
3. ______________
4. ______________
5. ______________
6. ______________
7. ______________
8. ______________
9. ______________
10. ______________
11. ______________
12. ______________
13. ______________
14. ______________
15. ______________
16. ______________
17. ______________
18. ______________
19. ______________

**Answers**

20. __________

21. __________

22. __________

23. __________

24. __________

25. __________

26. __________

27. __________

28. __________

29. __________

30. __________

31. __________

32. __________

33. __________

34. __________

35. __________

**20.** $26 - 2 \times 3$

**21.** $36 \div (3^2 + 3)$

*Identify the proper fractions, improper fractions, and mixed numbers from the following group.*

$\frac{5}{7}, \frac{15}{9}, 4\frac{5}{6}, \frac{8}{8}, \frac{11}{1}, \frac{2}{5}, 3\frac{5}{6}$

**22.** Proper: ___ Improper: ______ Mixed numbers: ______

*Convert to mixed or whole numbers.*

**23.** $\frac{16}{9}$

**24.** $\frac{36}{5}$

*Convert to improper fractions.*

**25.** $5\frac{3}{4}$

**26.** $6\frac{1}{9}$

*Find out whether the pair of fractions is equivalent.*

**27.** $\frac{8}{32}, \frac{9}{36}$

**28.** $\frac{6}{11}, \frac{7}{9}$

*Multiply.*

**29.** $\frac{7}{15} \times \frac{5}{21}$

**30.** $\frac{10}{27} \times \frac{9}{20}$

**31.** $4 \times \frac{3}{8}$

**32.** $3\frac{2}{5} \times \frac{5}{8}$

**33.** $5\frac{1}{3} \times 1\frac{4}{5}$

**34.** $1\frac{5}{12} \times 8$

**35.** $3\frac{1}{5} \times \frac{7}{8} \times 2\frac{6}{7}$

Divide.

36. $\frac{5}{12} \div \frac{5}{8}$

37. $\frac{7}{15} \div \frac{14}{25}$

38. $\frac{9}{20} \div 2\frac{2}{5}$

Add.

39. $\frac{4}{15} + \frac{8}{15}$

40. $\frac{7}{25} + \frac{8}{15}$

41. $\frac{2}{5} + \frac{3}{4} + \frac{5}{8}$

Perform the indicated operations.

42. $\frac{17}{20} - \frac{7}{20}$

43. $\frac{5}{9} - \frac{5}{12}$

44. $\frac{5}{18} + \frac{4}{9} - \frac{1}{6}$

Perform the indicated operations.

45. $3\frac{5}{7} + 2\frac{4}{7}$

46. $4\frac{7}{8} + 3\frac{1}{6}$

47. $8\frac{1}{9} - 3\frac{5}{9}$

48. $7\frac{7}{8} - 3\frac{5}{6}$

49. $9 - 5\frac{3}{8}$

50. $3\frac{1}{6} + 3\frac{1}{4} - 2\frac{7}{8}$

**Answers**

36. ______
37. ______
38. ______
39. ______
40. ______
41. ______
42. ______
43. ______
44. ______
45. ______
46. ______
47. ______
48. ______
49. ______
50. ______

 

## Answers

51. ______

52. ______

53. ______

*Solve each application.*

**51. Business and Finance** In his part-time job, Manuel worked $3\frac{5}{6}$ hours (h) on Monday, $4\frac{3}{10}$ h on Wednesday, and $6\frac{1}{2}$ h on Friday. Find the number of hours that he worked during the week.

**52. Crafts** A $6\frac{1}{2}$-in. bolt is placed through a wall that is $5\frac{7}{8}$ in. thick. How far does the bolt extend beyond the wall?

**53. Statistics** On a 6-hour (h) trip, Carlos drove $1\frac{3}{4}$ h and then Maria drove for another $2\frac{1}{3}$ h. How many hours remained on the trip?

CHAPTER 4

# Decimals

## INTRODUCTION

When Barry is not at his job, he loves to be outside partaking in many outdoor activities. Some of these activities include bicycling, hiking, and tennis. Lately, Barry has been spending a lot of time on his bike. He bought a new road bike because of the long distances he was riding. Two years ago, Barry noticed a sign in his neighborhood advertising a 150-mile bike ride to raise money for muscular dystrophy. He thought it was a good cause and started training for the ride. Barry completed the ride and experienced a great feeling of accomplishment in doing so.

The MD ride inspired Barry to look for more rides that raised money for important causes. He has since ridden for cerebral palsy and cancer research. Barry likes the idea that he can raise money for a good cause while getting exercise doing something he enjoys.

Lance Armstrong is one of Barry's inspirational heroes. Armstrong is a cancer survivor who battled the disease and came back to win the Tour de France seven times. He founded the Lance Armstrong Foundation for Cancer Research and the yellow Livestrong wristbands.

You can learn more about the Tour de France by doing Activity 12 on page 354. For more information on the Tour de France or Lance Armstrong you can also go to www.letour.fr or lancewins.com.

## CHAPTER 4 OUTLINE

# 4 pretest



Name ____________________

Section __________ Date __________

This pretest provides a preview of the types of exercises you will encounter in each section of this chapter. The answers for these exercises can be found in the back of the text. If you are working on your own, or ahead of the class, this pretest can help you identify the sections in which you should focus more of your time.

**Answers**

1. ____________________
2. ____________________
3. ____________________
4. ____________________
5. ____________________
6. ____________________
7. ____________________
8. ____________________
9. ____________________
10. ____________________
11. ____________________
12. ____________________
13. ____________________
14. ____________________

4.1 **1.** Give the place value of 5 in the decimal 13.4658.

**2.** Write $2\frac{371}{1,000}$ in decimal form and in words.

4.2 **3.** **(a)** Add $56 + 5.16 + 1.8 + 0.33$ **(b)** Subtract $4.6 - 2.225$

**4.** **Business and Finance** You have \$20 in cash and make purchases of \$6.89 and \$10.75. How much money do you have left?

4.3 **5.** **(a)** Multiply $0.357 \times 2.41$ **(b)** Multiply $0.5362 \times 1,000$

**6.** Round 2.35878 to the nearest hundredth.

**7.** **Business and Finance** You fill up your gas tank with 9.2 gal of fuel at \$3.049 per gallon. What is the cost of the fill-up (to the nearest cent)?

4.3 **8.** Find the circumference of the given circle. Use 3.14 for $\pi$.

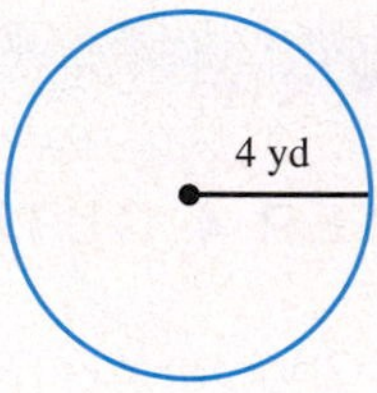


4.4 **9.** Divide $57\overline{)242.25}$.

**10.** Divide $1.6\overline{)3.896}$.

**11.** **Business and Finance** Manny worked 27.5 h in a week and earned \$209. What was his hourly rate of pay?

**12.** Divide $53.4 \div 1,000$.

4.5 **13.** Find the decimal equivalent of each of the following.

**(a)** $\frac{3}{8}$ **(b)** $\frac{7}{24}$ (to the nearest hundredth)

4.6 **14.** Write 0.625 as a common fraction in lowest terms.

4.1

# Place Value and Rounding

< 4.1 Objectives >

1 > Write a number in decimal form

2 > Identify place value in a decimal fraction

3 > Write a decimal in words

4 > Write a decimal as a fraction or mixed number

5 > Compare the size of several decimals

6 > Round a decimal to the nearest tenth

7 > Round a decimal to any specified decimal place

In Chapter 3, we looked at common fractions. Now we turn to a special kind of fraction called **decimal fractions.** A decimal fraction is a fraction whose denominator is a *power of* 10. Some examples of decimal fractions are $\frac{3}{10}$, $\frac{45}{100}$, and $\frac{123}{1,000}$.

Earlier we talked about the idea of place value. Recall that in our decimal place-value system, each place has *one-tenth* of the value of the place to its left.

## Example 1 Identifying Place Values

**RECALL**

The powers of 10 are 1, 10, 100, 1,000, and so on. You might want to review Section 1.7 before going on.

Label the place values for the number 538.

| 5 | 3 | 8 |
|---|---|---|
| ↑ | ↑ | ↑ |
| Hundreds | Tens | Ones |

The ones place value is $\frac{1}{10}$ of the tens place value; the tens place value is $\frac{1}{10}$ of the hundreds place value; and so on.

**Check Yourself 1**

Label the place values for the number 2,793.

**NOTE**

The decimal point separates the whole-number part and the fractional part of a decimal fraction.

We want to extend this idea *to the right* of the ones place. Write a period to the *right* of the ones place. This is called the **decimal point.** Each digit to the right of that decimal point will represent a fraction whose denominator is a power of 10. The first place to the right of the decimal point is the tenths place:

$$0.1 = \frac{1}{10}$$

## Example 2 Writing a Number in Decimal Form

< Objective 1 >

Write the mixed number $3\frac{2}{10}$ in decimal form.

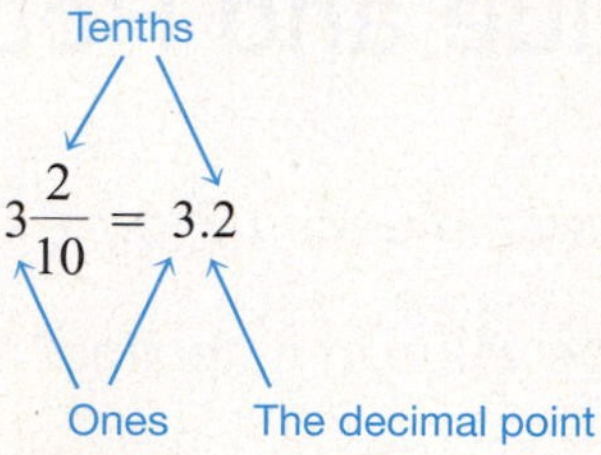


### Check Yourself 2

Write $5\frac{3}{10}$ in decimal form.

As you move farther to the *right,* each place value must be $\frac{1}{10}$ of the value before it. The second place value is hundredths $\left(0.01 = \frac{1}{100}\right)$. The next place is thousandths, the fourth position is the ten–thousandths place, and so on. Figure 1 illustrates the value of each position as we move to the right of the decimal point.

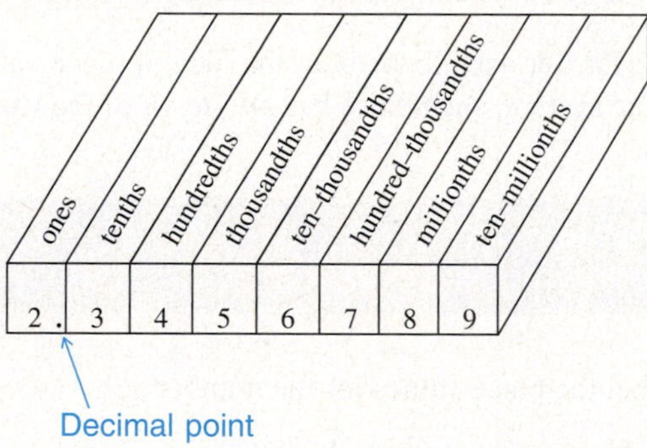


Figure 1

## Example 3 Identifying Place Values in a Decimal Fraction

< Objective 2 >

**NOTE**

For convenience we will shorten the term *decimal fraction* to *decimal* from this point on.

What are the place values of the 4 and 6 in the decimal 2.34567?

The place value of 4 is hundredths, and the place value of 6 is ten–thousandths.

### Check Yourself 3

What is the place value of 5 in the decimal of Example 3?

Understanding place values will allow you to read and write decimals. You can use the following steps.

Step by Step

**Reading or Writing Decimals in Words**

**Step 1** Read the digits *to the left* of the decimal point as a whole number.
**Step 2** Read the decimal point as the word *and*.
**Step 3** Read the digits *to the right* of the decimal point as a whole number followed by the place value of the rightmost digit.

## Example 4 Writing a Decimal Number in Words

< Objective 3 >

**NOTE**

If there are *no* nonzero digits to the left of the decimal point, start directly with step 3.

Write each decimal number in words.

5.03 is read "five and three hundredths."

Hundredths — The rightmost digit, 3, is in the hundredths position.

12.057 is read "twelve and fifty-seven thousandths."

Thousandths — The rightmost digit, 7, is in the thousandths position.

**NOTES**

An informal way of reading decimals is to simply read the digits in order and use the word *point* to indicate the decimal point. 2.58 can be read "two point five eight." 0.689 can be read "zero point six eight nine."

The number of digits to the right of the decimal point is called the number of **decimal places** in a decimal number. So, 0.35 has two decimal places.

0.5321 is read "five thousand three hundred twenty-one ten–thousandths."

When the decimal has no whole-number part, we have chosen to write a 0 to the left of the decimal point. This simply makes sure that you don't miss the decimal point. However, both 0.5321 and .5321 are correct.

### Check Yourself 4

**Write 2.58 in words.**

One quick way to write a decimal as a common fraction is to remember that the number of decimal places must be the same as the number of zeros in the denominator of the common fraction.

## Example 5 Writing a Decimal Number as a Mixed Number

< Objective 4 >

Write each decimal as a common fraction or mixed number.

$$0.35 = \frac{35}{100}$$

Two places — Two zeros

**NOTE**

The 0 to the right of the decimal point is a "placeholder" that is not needed in the common-fraction form.

The same method can be used with decimals that are greater than 1. Here the result will be a mixed number.

$$2.058 = 2\frac{58}{1{,}000}$$

Three places (under 058) — Three zeros (under 1,000)

## Check Yourself 5

**Write as common fractions or mixed numbers.**

**(a)** 0.528 **(b)** 5.08

**RECALL**

By the Fundamental Principle of Fractions, multiplying the numerator and denominator of a fraction by the same nonzero number does not change the value of the fraction.

It is often useful to compare the sizes of two decimal fractions. One approach to comparing decimals uses the following fact.

Adding zeros to the right *does not change* the value of a decimal. The number 0.53 is the same as 0.530. Look at the fractional form:

$$\frac{53}{100} = \frac{530}{1{,}000}$$

The fractions are equivalent because we multiplied both the numerator and denominator by 10.

This allows us to compare decimals as shown in Example 6.

## Example 6 Comparing the Sizes of Two Decimal Numbers

< Objective 5 >

Which is larger?

0.84 or 0.842

Write 0.84 as 0.840. Then we see that 0.842 (or 842 thousandths) is greater than 0.840 (or 840 thousandths), and we can write

$0.842 > 0.84$

## Check Yourself 6

**Complete the following statement, using the symbol $<$ or $>$.**

**0.588 ______ 0.59**

Whenever a decimal represents a measurement made by some instrument (a rule or a scale), the measurement is not exact. It is accurate only to a certain number of places and is called an **approximate number.** Usually, we want to make all decimals in a particular problem accurate to a specified decimal place or tolerance. This requires **rounding** the decimals. We can picture the process on a number line.

## Example 7 Rounding to the Nearest Tenth

< Objective 6 >

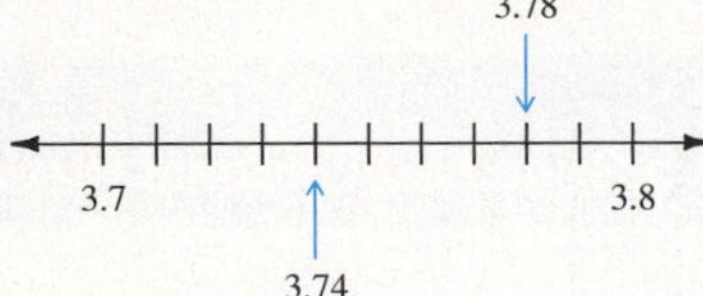


**NOTE**

3.74 is closer to 3.7 than it is to 3.8, while 3.78 is closer to 3.8.

3.74 is rounded down to the nearest tenth, 3.7, and 3.78 is rounded up to 3.8.

### Check Yourself 7

**Use the number line in Example 7 to round 3.77 to the nearest tenth.**

Rather than using the number line, the following rule can be applied.

**Step by Step**

### To Round a Decimal

**Step 1** Find the place to which the decimal is to be rounded.

**Step 2** If the next digit to the right is 5 or more, increase the digit in the place you are rounding to by 1. Discard remaining digits to the right.

**Step 3** If the next digit to the right is less than 5, just discard that digit and any remaining digits to the right.

## Example 8 Rounding to the Nearest Tenth

Round 34.58 to the nearest tenth.

34.58 Locate the digit you are rounding to. The 5 is in the tenths place.

Because the next digit to the right, 8, is 5 or more, increase the tenths digit by 1. Then discard the remaining digits.

34.58 is rounded to 34.6.

**NOTE**

Many students find it easiest to mark the digit they are rounded to with an arrow.

### Check Yourself 8

**Round 48.82 to the nearest tenth.**

## Example 9 Rounding to the Nearest Hundredth

< Objective 7 >

Round 5.673 to the nearest hundredth.

5.673 The 7 is in the hundredths place.

The next digit to the right, 3, is less than 5. Leave the hundredths digit as it is and discard the remaining digits to the right.

5.673 is rounded to 5.67.

### Check Yourself 9

Round 29.247 to the nearest hundredth.

## Example 10 Rounding to a Specified Decimal Place

**NOTE**

The fourth place to the *right* of the decimal point is the ten–thousandths place.

Round 3.14159 to four decimal places.

3.14159 The 5 is in the ten–thousandths place.

The next digit to the right, 9, is 5 or more, so increase the digit you are rounding to by 1. Discard the remaining digits to the right.

3.14159 is rounded to 3.1416.

### Check Yourself 10

Round 0.8235 to three decimal places.

### Check Yourself ANSWERS

1. 2 7 9 3 (2: Thousands; 7: Hundreds; 9: Tens; 3: Ones) 2. 5.3 3. Thousandths

4. Two and fifty-eight hundredths 5. (a) $\frac{528}{1,000}$; (b) $5\frac{8}{100}$

6. $0.588 < 0.59$ 7. 3.8 8. 48.8 9. 29.25 10. 0.824

## Reading Your Text

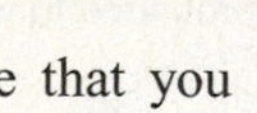

The following fill-in-the-blank exercises are designed to ensure that you understand some of the key vocabulary used in this section.

**SECTION 4.1**

(a) A ____________ fraction is a fraction whose denominator is a power of 10.

(b) The number of digits to the right of the decimal point is called the number of decimal ____________.

(c) Whenever a decimal represents a measure made by some instrument, the decimals are not ____________.

(d) The fourth place to the right of the decimal is called the ____________ place.

Basic Skills | Advanced Skills | Vocational-Technical Applications | Calculator/Computer | Above and Beyond

# 4.1 exercises

< Objective 2 >

*For the decimal 8.57932:*

**1.** What is the place value of 7?

**2.** What is the place value of 5?

**3.** What is the place value of 3? 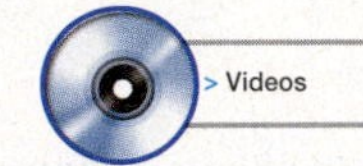


**4.** What is the place value of 2?

< Objective 1 >

*Write in decimal form.*

**5.** $\dfrac{23}{100}$

**6.** $\dfrac{371}{1,000}$

**7.** $\dfrac{209}{10,000}$ > Videos

**8.** $3\dfrac{5}{10}$

**9.** $23\dfrac{56}{1,000}$ > Videos

**10.** $7\dfrac{431}{10,000}$

< Objective 3 >

*Write in words.*

**11.** 0.23

**12.** 0.371

**13.** 0.071 > Videos

**14.** 0.0251

**15.** 12.07

**16.** 23.056

< Objective 1 >

*Write in decimal form.*

**17.** Fifty-one thousandths 

**18.** Two hundred fifty-three ten–thousandths

**19.** Seven and three tenths

**20.** Twelve and two hundred forty-five thousandths

MathZone

Boost your grade at mathzone.com!

> Practice Problems
> NetTutor
> Self-Tests
> e-Professors
> Videos

Name ______________

Section ________ Date ________

## Answers

1. ______
2. ______
3. ______
4. ______
5. ______ 6. ______
7. ______ 8. ______
9. ______ 10. ______
11. ______
12. ______
13. ______
14. ______
15. ______
16. ______
17. ______ 18. ______
19. ______ 20. ______

**Answers**

21. ____
22. ____
23. ____
24. ____
25. ____ 26. ____
27. ____ 28. ____
29. ____ 30. ____
31. ____ 32. ____
33. ____
34. ____
35. ____
36. ____
37. ____ 38. ____
39. ____ 40. ____
41. ____ 42. ____
43. ____ 44. ____

< **Objective 4** >

*Write each of the following as a common fraction or mixed number.*

**21.** 0.65

**22.** 0.00765

**23.** 5.231

**24.** 4.0171

< **Objective 5** >

*Complete each of the following statements, using the symbol <, =, or >.*

**25.** 0.69 ______ 0.689

**26.** 0.75 ______ 0.752

**27.** 1.23 ______ 1.230

**28.** 2.451 ______ 2.45

**29.** 10 ______ 9.9

**30.** 4.98 ______ 5

**31.** 1.459 ______ 1.46

**32.** 0.235 ______ 0.2350

*Arrange in order from smallest to largest.*

**33.** 4.0339, 4.034, $4\frac{3}{10}$, $\frac{432}{100}$, 4.33

**34.** $\frac{38}{1{,}000}$, 0.0382, 0.04, 0.37, $\frac{39}{100}$

**35.** 0.71, 0.072, $\frac{7}{10}$, 0.007, 0.0069, $\frac{7}{100}$, 0.0701, 0.0619, 0.0712

**36.** 2.05, $\frac{25}{10}$, 2.0513, 2.059, $\frac{251}{100}$, 2.0515, 2.052, 2.051

< **Objectives 6–7** >

*Round to the indicated place.*

**37.** 21.534 hundredths

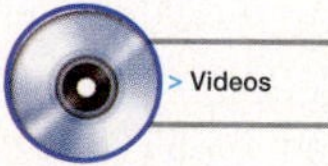


**38.** 5.842 tenths

**39.** 0.342 hundredths

**40.** 2.3576 thousandths

**41.** 2.71828 thousandths

**42.** 1.543 tenths

**43.** 0.0475 tenths

**44.** 0.85356 ten–thousandths

45. 4.85344 ten–thousandths

46. 52.8728 thousandths

47. 6.734 two decimal places

48. 12.5467 three decimal places

49. 6.58739 four decimal places

50. 503.824 two decimal places

*Round to the nearest cent or dollar, as indicated.*

51. $235.1457 cent

52. $1,847.9895 cent

53. $752.512 dollar

54. $5,642.4958 dollar

*Label exercises 55 to 58 as* **true** *or* **false.**

55. The only number between 8.6 and 8.8 is 8.7.

56. The smallest number that is greater than 98.6 would be 98.7.

57. The place value to the right of the decimal point is tenths.

58. The number 4.586 would always be rounded as 4.6.

59. Plot (draw a dot on the number line) the following: 3.2 and 3.7. Then estimate the location for 3.62.

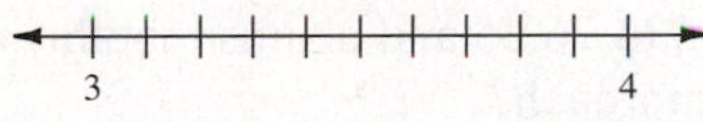


60. Plot the following on a number line: 12.51 and 12.58. Then estimate the location for 12.537.

12.5 12.6

61. Plot the following on a number line: 7.124 and 7.127. Then estimate the location of 7.1253.

62. Plot the following on a number line: 5.73 and 5.74. Then estimate the location for 5.782.

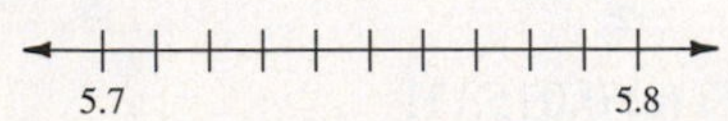


**Answers**

45. ______
46. ______
47. ______
48. ______
49. ______
50. ______
51. ______
52. ______
53. ______
54. ______
55. ______
56. ______
57. ______
58. ______
59. ______
60. ______
61. ______
62. ______

 

Basic Skills | Advanced Skills | **Vocational-Technical Applications** | Calculator/Computer | Above and Beyond

## Answers

63. ______

64. ______

65. ______

66. ______

67. ______

68. ______

69. ______

70. ______

**63. ALLIED HEALTH** A nurse calculates a child's dose of Reglan to be 1.53 milligrams (mg). Round this dose to the nearest tenth of a milligram.

**64. ALLIED HEALTH** A nurse calculates a young boy's dose of Dilantin to be 23.375 mg every 5 minutes. Round this dose to the nearest hundredth of a milligram.

**65. ELECTRONICS** Write in decimal form:

**(a)** Ten and thirty-five hundredths volts (V)

**(b)** Forty-seven hundred–thousandths of a farad (F)

**(c)** One hundred fifty-eight ten–thousandths of a henry (H)

**66. INFORMATION TECHNOLOGY** Josie needed to check connectivity of a PC on the network. She uses a tool called ping to see if the PC is configured properly. She receives three readings from ping: 2.1, 2.2, and 2.3 seconds. Convert the decimals to fractions and simplify if needed.

**67. MANUFACTURING** Put the following mill bits in order from smallest to largest:

0.308, 0.297, 0.31, 0.3, 0.311, 0.32

**68. MANUFACTURING** A drill size is listed as $\frac{372}{1{,}000}$. Express this as a decimal.

Basic Skills | Advanced Skills | Vocational-Technical Applications | Calculator/Computer | **Above and Beyond**

**69. (a)** What is the difference in the values of the following: 0.120, 0.1200, and 0.12000?

**(b)** Explain in your own words why placing zeros to the right of a decimal point does not change the value of the number.

**70.** Lula wants to round 76.24491 to the nearest hundredth. She first rounds 76.24491 to 76.245 and then rounds 76.245 to 76.25 and claims that this is the final answer. What is wrong with this approach?

## Answers

**1.** Hundredths **3.** Ten–thousandths **5.** 0.23 **7.** 0.0209
**9.** 23.056 **11.** Twenty-three hundredths **13.** Seventy-one thousandths
**15.** Twelve and seven hundredths **17.** 0.051 **19.** 7.3
**21.** $\frac{65}{100}\left(\text{or } \frac{13}{20}\right)$ **23.** $5\frac{231}{1{,}000}$ **25.** $0.69 > 0.689$ **27.** $1.23 = 1.230$
**29.** $10 > 9.9$ **31.** $1.459 < 1.46$ **33.** 4.0339, 4.034, $4\frac{3}{10}$, $\frac{432}{100}$, 4.33
**35.** 0.0069, 0.007, 0.0619, $\frac{7}{100}$, 0.0701, 0.0712, 0.072, $\frac{7}{10}$, 0.71 **37.** 21.53
**39.** 0.34 **41.** 2.718 **43.** 0.0 **45.** 4.8534 **47.** 6.73
**49.** 6.5874 **51.** $235.15 **53.** $753 **55.** False **57.** True
**59.** 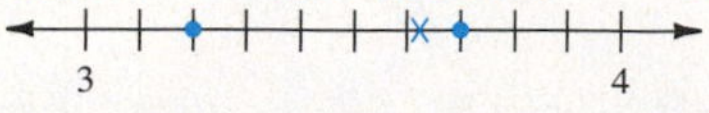

**61.** 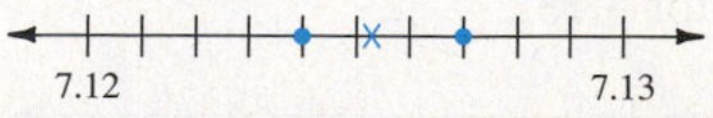

**63.** 1.5 mg **65. (a)** 10.35 V; **(b)** 0.00047 F; **(c)** 0.0158 H
**67.** 0.297, 0.3, 0.308, 0.31, 0.311, 0.32 **69.** Above and Beyond

# 4.2 Converting Between Fractions and Decimals

< 4.2 Objectives >

1 > Convert a common fraction to a decimal

2 > Convert a common fraction to a repeating decimal

3 > Convert a mixed number to a decimal

4 > Convert a decimal to a common fraction

Because a common fraction can be interpreted as division, you can divide the numerator of the common fraction by its denominator to convert a common fraction to a decimal. The result is called a **decimal equivalent.**

## Example 1 Converting a Fraction to a Decimal Equivalent

< Objective 1 >

**RECALL**

5 can be written as 5.0, 5.00, 5.000, and so on. In this case, we continue the division by adding zeros to the dividend until a zero remainder is reached.

Write $\frac{5}{8}$ as a decimal.

$$\begin{array}{r} 0.625 \\ 8\overline{)5.000} \\ \underline{4\,8\phantom{00}} \\ 20\phantom{0} \\ \underline{16\phantom{0}} \\ 40 \\ \underline{40} \\ 0 \end{array}$$

Because $\frac{5}{8}$ means $5 \div 8$, divide 8 into 5.

We see that $\frac{5}{8} = 0.625$; 0.625 is the decimal equivalent of $\frac{5}{8}$.

**Check Yourself 1**

Find the decimal equivalent of $\frac{7}{8}$.

Some fractions are used so often that we have listed their decimal equivalents here as a reference.

**NOTE**

The division used to find these decimal equivalents stops when a zero remainder is reached. The equivalents are called **terminating decimals.**

**Some Common Decimal Equivalents**

| | | | |
|---|---|---|---|
| $\frac{1}{2} = 0.5$ | $\frac{1}{4} = 0.25$ | $\frac{1}{5} = 0.2$ | $\frac{1}{8} = 0.125$ |
| | $\frac{3}{4} = 0.75$ | $\frac{2}{5} = 0.4$ | $\frac{3}{8} = 0.375$ |
| | | $\frac{3}{5} = 0.6$ | $\frac{5}{8} = 0.625$ |
| | | $\frac{4}{5} = 0.8$ | $\frac{7}{8} = 0.875$ |

If a decimal equivalent does not terminate, you can round the result to approximate the fraction to some specified number of decimal places. Consider Example 2.

## Example 2 Converting a Fraction to a Decimal Equivalent

Write $\frac{3}{7}$ as a decimal. Round the answer to the nearest thousandth.

$$\begin{array}{r} 0.4285 \\ 7\overline{)3.0000} \\ \underline{2\,8}\phantom{000} \\ 20\phantom{00} \\ \underline{14}\phantom{00} \\ 60\phantom{0} \\ \underline{56}\phantom{0} \\ 40 \\ \underline{35} \\ 5 \end{array}$$

In this example, we are choosing to round to three decimal places, so we must add enough zeros to carry the division to four decimal places.

So $\frac{3}{7} = 0.429$ (to the nearest thousandth).

### Check Yourself 2

Find the decimal equivalent of $\frac{5}{11}$ to the nearest thousandth.

If a decimal equivalent does *not* terminate, it will *repeat* a sequence of digits. These decimals are called **repeating decimals.**

## Example 3 Converting a Fraction to a Repeating Decimal

< Objective 2 >

**(a)** Write $\frac{1}{3}$ as a decimal.

$$\begin{array}{r} 0.333 \\ 3\overline{)1.000} \\ \underline{9} \\ 10 \\ \underline{9} \\ 10 \\ \underline{9} \end{array}$$

The digit 3 will just repeat indefinitely because each new remainder will be 1.

Adding more zeros and going on will simply lead to more 3s in the quotient.

We can say $\frac{1}{3} = 0.333\ldots$

The three dots mean "and so on" and tell us that 3 will repeat itself indefinitely.

**(b)** Write $\frac{5}{12}$ as a decimal.

$$\begin{array}{r} 0.4166\ldots \\ 12\overline{)5.0000} \\ \underline{4\,8} \\ 20 \\ \underline{12} \\ 80 \\ \underline{72} \\ 80 \\ \underline{72} \\ 8 \end{array}$$

In this example, the digit 6 will just repeat itself because the remainder, 8, will keep occurring if we add more zeros and continue the division.

so $\frac{5}{12} = 0.4166\ldots$

### Check Yourself 3

**Find the decimal equivalent of each fraction.**

**(a)** $\frac{2}{3}$ **(b)** $\frac{7}{12}$

Some important decimal equivalents (rounded to the nearest thousandth) are shown here as a reference.

$\frac{1}{3} = 0.333$ $\frac{1}{6} = 0.167$ $\frac{2}{3} = 0.667$ $\frac{5}{6} = 0.833$

Another way to write a repeating decimal is with a bar placed over the digit or digits that repeat. For example, we can write

$0.37373737\ldots$

as

$0.\overline{37}$

The bar placed over the digits indicates that 37 repeats indefinitely.

## Example 4 Converting a Fraction to a Repeating Decimal

Write $\frac{5}{11}$ as a decimal.

$$\begin{array}{r} 0.4545 \\ 11\overline{)5.0000} \\ \underline{4\,4}\phantom{000} \\ 60\phantom{00} \\ \underline{55}\phantom{00} \\ 50\phantom{0} \\ \underline{44}\phantom{0} \\ 60 \\ \underline{55} \\ 5 \end{array}$$

As soon as a remainder repeats itself, as 5 does here, the pattern of digits will repeat in the quotient.

$\frac{5}{11} = 0.\overline{45}$

$= 0.4545\ldots$

## Check Yourself 4

Use the bar notation to write the decimal equivalent of $\frac{5}{7}$. (Be patient. You'll have to divide for a while to find the repeating pattern.)

You can find the decimal equivalents for mixed numbers in a similar way. Find the decimal equivalent of the fractional part of the mixed number and then combine that with the whole-number part. Example 5 illustrates this approach.

## Example 5 Converting a Mixed Number to a Decimal Equivalent

< Objective 3 >

Find the decimal equivalent of $3\frac{5}{16}$.

$\frac{5}{16} = 0.3125$ First find the equivalent of $\frac{5}{16}$ by division.

$3\frac{5}{16} = 3.3125$ Add 3 to the result.

## Check Yourself 5

Find the decimal equivalent of $2\frac{5}{8}$.

We have learned something important in this section. To find the decimal equivalent of a fraction, we use long division. Because the remainder must be less than the divisor, the remainder must *either repeat or become zero.* Thus *every common fraction* has a *repeating* or a *terminating* decimal as its decimal equivalent.

Next, using what we learned about place values, you can easily write decimals as common fractions. The following rule is used.

**Step by Step**

**To Convert a Terminating Decimal Less Than 1 to a Common Fraction**

**Step 1** Write the digits of the decimal without the decimal point. This will be the numerator of the common fraction.

**Step 2** The denominator of the fraction is a 1 followed by as many zeros as there are places in the decimal.

## Example 6 Converting a Decimal to a Common Fraction

< Objective 4 >

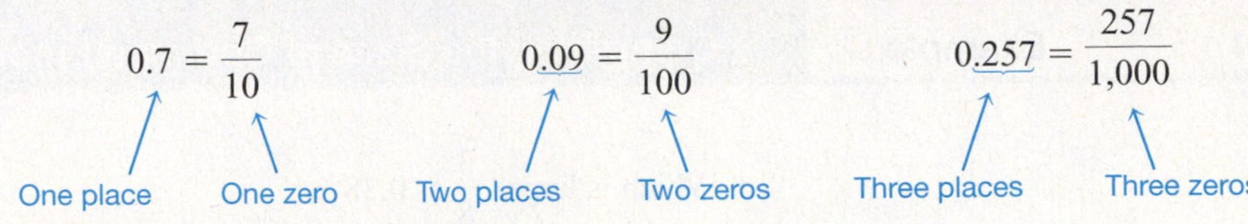

### Check Yourself 6

Write as common fractions.

(a) 0.3 (b) 0.311

When a decimal is converted to a common fraction, that resulting common fraction should be written in lowest terms.

## Example 7 Converting a Decimal to a Common Fraction

Convert 0.395 to a fraction and write the result in lowest terms.

$$0.395 = \frac{395}{1{,}000} = \frac{79}{200}$$

**NOTE**

Divide the numerator and denominator by 5.

### Check Yourself 7

Write 0.275 as a common fraction.

If the decimal has a whole-number portion, write the digits to the right of the decimal point as a proper fraction and then form a mixed number for your result.

## Example 8 Converting a Decimal to a Mixed Number

**NOTE**

Repeating decimals can also be written as common fractions, although the process is more complicated. We will limit ourselves to the conversion of terminating decimals in this textbook.

Write 12.277 as a mixed number.

$$0.277 = \frac{277}{1{,}000}$$

so

$$12.277 = 12\frac{277}{1{,}000}$$

### Check Yourself 8

Write 32.433 as a mixed number.

Comparing the sizes of common fractions and decimals requires finding the decimal equivalent of the common fraction and then comparing the resulting decimals.

## Example 9 Comparing the Sizes of Common Fractions and Decimals

Which is larger, $\frac{3}{8}$ or 0.38?

Write the decimal equivalent of $\frac{3}{8}$. That decimal is 0.375. Now comparing 0.375 and 0.38, we see that 0.38 is the larger of the numbers:

$0.38 > \frac{3}{8}$

### Check Yourself 9

Which is larger, $\frac{3}{4}$ or 0.8?

### Check Yourself ANSWERS

**1.** 0.875 **2.** $\frac{5}{11} = 0.455$ (to the nearest thousandth)
**3.** **(a)** 0.666 . . . ; **(b)** 0.583 . . . The digit 3 will continue indefinitely.
**4.** $\frac{5}{7} = 0.\overline{714285}$ **5.** 2.625 **6.** **(a)** $\frac{3}{10}$; **(b)** $\frac{311}{1,000}$ **7.** $\frac{11}{40}$
**8.** $32\frac{433}{1,000}$ **9.** $0.8 > \frac{3}{4}$

## Reading Your Text

The following fill-in-the-blank exercises are designed to ensure that you understand some of the key vocabulary used in this section.

**SECTION 4.2**

**(a)** You can ______________ the numerator of a fraction by its denominator to convert a common fraction to a decimal.

**(b)** We can write a repeating decimal with a ______________ placed over the digit or digits that repeat.

**(c)** Every common fraction has a repeating or a ______________ decimal as its equivalent.

**(d)** The denominator of a fraction that is equivalent to a given decimal is a 1 followed by as many zeros as there are ______________ in the decimal.

# 4.2 exercises

MathZone

Boost your grade at mathzone.com!
> Practice Problems
> NetTutor
> Self-Tests
> e-Professors
> Videos

Name ____________

Section ______ Date ______

## < Objective 1 >

*Find the decimal equivalents for each of the following fractions.*

1. $\frac{3}{4}$

2. $\frac{4}{5}$

3. $\frac{9}{20}$

4. $\frac{3}{10}$

5. $\frac{1}{5}$

6. $\frac{1}{8}$

7. $\frac{5}{16}$

8. $\frac{11}{20}$

9. $\frac{7}{10}$ > Videos

10. $\frac{7}{16}$

11. $\frac{27}{40}$

12. $\frac{17}{32}$

*Find the decimal equivalents rounded to the indicated place.*

13. $\frac{5}{6}$ thousandths > Videos

14. $\frac{7}{12}$ hundredths

15. $\frac{4}{15}$ thousandths

## < Objective 2 >

*Write the decimal equivalents, using the bar notation.*

16. $\frac{1}{18}$

17. $\frac{4}{9}$

18. $\frac{3}{11}$

## < Objective 3 >

*Find the decimal equivalents for each of the following mixed numbers.*

19. $5\frac{3}{5}$

20. $4\frac{7}{16}$

## < Objective 4 >

*Write each of the following as a common fraction or mixed number. Write your answer in lowest terms.*

21. 0.9

22. 0.3

23. 0.8

24. 0.6

## Answers

| | |
|---|---|
| 1. | 2. |
| 3. | 4. |
| 5. | 6. |
| 7. | 8. |
| 9. | 10. |
| 11. | 12. |
| 13. | 14. |
| 15. | 16. |
| 17. | 18. |
| 19. | 20. |
| 21. | 22. |
| 23. | 24. |

Answers

25. ______

26. ______

27. ______

28. ______

29. ______

30. ______

31. ______

32. ______

33. ______

34. ______

35. ______

36. ______

37. ______

38. ______

39. ______

40. ______

41. ______

42. ______

**25.** 0.37

**26.** 0.97

**27.** 0.587

**28.** 0.379

**29.** 0.48

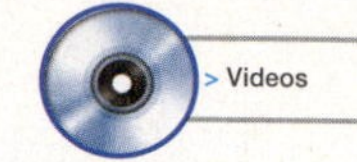


**30.** 0.75

**31.** 0.58

**32.** 0.65

**33.** 0.425

**34.** 0.116

**35.** 0.375

**36.** 0.225

**37.** 0.136

**38.** 0.575

**39.** 0.059

**40.** 0.067

**41.** **Statistics** In a weekend baseball tournament, Joel had 4 hits in 13 times at bat. That is, he hit safely $\frac{4}{13}$ of the time. Write the decimal equivalent for Joel's hitting, rounding to three decimal places. (That number is Joel's batting average.)

**42.** **Statistics** The following table gives the wins and losses of the teams in the National League East as of mid-September in a recent season. The winning percentage of each team is calculated by writing the number of wins over the total games played and converting this fraction to a decimal. Convert this fraction to a decimal for every team, rounding to three decimal places.

| Team | Wins | Losses |
|---|---|---|
| Atlanta | 92 | 56 |
| New York | 90 | 58 |
| Philadelphia | 70 | 77 |
| Montreal | 62 | 85 |
| Florida | 57 | 89 |

43. **Statistics** The following table gives the wins and losses of all the teams in the Western Division of the National Football Conference for a recent season. Determine the fraction of wins over total games played for every team, rounding to three decimal places for each of the teams.

| Team | Wins | Losses |
|---|---|---|
| San Francisco | 10 | 6 |
| St. Louis | 7 | 9 |
| Seattle | 7 | 9 |
| Arizona | 5 | 11 |

44. **Statistics** The following table gives the free throws attempted (FTA) and the free throws made (FTM) for the top five players in the NBA for a recent season. Calculate the free throw percentage for each player by writing the FTM over the FTA and converting this fraction to a decimal.

| Player | FTM | FTA |
|---|---|---|
| Allan Houston | 363 | 395 |
| Ray Allen | 316 | 345 |
| Steve Nash | 308 | 339 |
| Troy Hudson | 208 | 231 |
| Reggie Miller | 207 | 230 |

*Find the decimal equivalent for each fraction. Use the bar notation.*

45. $\frac{1}{11}$

46. $\frac{1}{111}$

47. $\frac{1}{1{,}111}$

48. From the pattern of exercises 45 to 47, can you guess the decimal representation for $\frac{1}{11{,}111}$?

*Insert > or < to form a true statement.*

49. $\frac{31}{34}$ □ 0.9118

50. $\frac{21}{37}$ □ 0.5664

51. $\frac{13}{17}$ □ 0.7657

Basic Skills | Advanced Skills | Vocational-Technical Applications | Calculator/Computer | **Above and Beyond**

52. Write the decimal equivalent of each fraction, using bar notation:

$\frac{1}{7}, \frac{2}{7}, \frac{3}{7}, \frac{4}{7}, \frac{5}{7}$

Describe any patterns that you see. Predict the decimal equivalent of $\frac{6}{7}$.

## Answers

43. ______

44. ______

45. ______

46. ______

47. ______

48. ______

49. ______

50. ______

51. ______

52. ______

## Answers

53. __________

54. __________

55. __________

56. __________

57. __________

58. __________

59. __________

60. __________

*Complete each of the following statements, using the symbol < or >.*

**53.** $\frac{7}{8} \square 0.87$

**54.** $\frac{5}{16} \square 0.313$

**55.** $\frac{9}{25} \square 0.4$

**56.** $\frac{11}{17} \square 0.638$

Basic Skills | Advanced Skills | **Vocational-Technical Applications** | Calculator/Computer | Above and Beyond

**57. Allied Health** The internal diameter, in millimeters (mm), of an endotracheal tube for a child is calculated using the formula $\frac{\text{Height}}{20}$, which is based on the child's height in centimeters (cm). Determine the size of endotracheal tube needed for a girl who is 110 cm tall. Since these tubes only come in 0.5-mm increments, the answer must be written as a decimal number rounded up to the nearest half or whole size.

**58. Allied Health** The stroke volume, which measures the average cardiac output per heartbeat (liters/beat), is based on a patient's cardiac output (CO), in liters per minute (L/min), and heart rate (HR), in beats per minute (beats/min). It is calculated using the fraction $\frac{\text{CO}}{\text{HR}}$. Determine the stroke volume for a patient whose cardiac output is 4 L/min and whose heart rate is 80 beats/min. Write your answer as a decimal number rounded to the nearest hundredth.

**59. Information Technology** The propagation delay for a satellite connection is 0.350 seconds (s). Convert to a fraction and simplify.

**60. Information Technology** From your computer, it takes 0.0021 s to transmit a ping packet to another computer. Convert to a fraction and simplify.

Basic Skills | Advanced Skills | Vocational-Technical Applications | **Calculator/Computer** | Above and Beyond

A calculator is very useful in converting common fractions to decimals. Just divide the numerator by the denominator, and the decimal equivalent will be in the display. Often, you will have to round the result in the display. For example, to find the decimal equivalent of $\frac{5}{24}$ to the nearest hundredth, enter

5 [÷] 24 [=]

The display may show 0.2083333, and rounding, we have $\frac{5}{24} = 0.21$.

When you are converting a mixed number to a decimal, addition must also be used. To change $7\frac{5}{8}$ to a decimal, for example, enter

7 [+] 5 [÷] 8 [=]

The result is 7.625.

*Use your calculator to find the decimal equivalents.*

**61.** $\frac{7}{8}$

**62.** $\frac{9}{16}$

**63.** $\frac{5}{32}$ to the thousandth

**64.** $\frac{11}{75}$ to the thousandth

**65.** $\frac{3}{11}$ using bar notation

**66.** $\frac{16}{33}$ using bar notation

**67.** $3\frac{7}{8}$

**68.** $8\frac{3}{16}$

## Answers

61. ______

62. ______

63. ______

64. ______

65. ______

66. ______

67. ______

68. ______

## Answers

**1.** 0.75 **3.** 0.45 **5.** 0.2 **7.** 0.3125 **9.** 0.7 **11.** 0.675
**13.** 0.833 **15.** 0.267 **17.** $0.\overline{4}$ **19.** 5.6 **21.** $\frac{9}{10}$ **23.** $\frac{4}{5}$
**25.** $\frac{37}{100}$ **27.** $\frac{587}{1,000}$ **29.** $\frac{12}{25}$ **31.** $\frac{29}{50}$ **33.** $\frac{17}{40}$ **35.** $\frac{3}{8}$
**37.** $\frac{17}{125}$ **39.** $\frac{59}{1,000}$ **41.** 0.308 **43.** 0.625, 0.438, 0.438, 0.313
**45.** $0.\overline{09}$ **47.** $0.\overline{0009}$ **49.** $<$ **51.** $<$ **53.** $>$ **55.** $<$
**57.** 5.5 mm **59.** $\frac{7}{20}$ **61.** 0.875 **63.** 0.156 **65.** $0.\overline{27}$
**67.** 3.875

# Activity 10 ::
## Terminate or Repeat?

Every fraction has a decimal equivalent that either terminates (for example, $\frac{1}{4} = 0.25$) or repeats (for example, $\frac{2}{9} = 0.\overline{2}$). Work with a group to discover which fractions have terminating decimals and which have repeating decimals. You may assume that the numerator of each fraction you consider is one and focus your attention on the denominator. As you complete the following table, you will find that the key to this question lies with the prime factorization of the denominator.

| Fraction | Decimal Form | Terminate? | Prime Factorization of the Denominator |
|---|---|---|---|
| $\frac{1}{2}$ | | | |
| $\frac{1}{3}$ | | | |
| $\frac{1}{4}$ | | | |
| $\frac{1}{5}$ | | | |
| $\frac{1}{6}$ | | | |
| $\frac{1}{7}$ | | | |
| $\frac{1}{8}$ | | | |
| $\frac{1}{9}$ | | | |
| $\frac{1}{10}$ | | | |
| $\frac{1}{11}$ | | | |
| $\frac{1}{12}$ | | | |

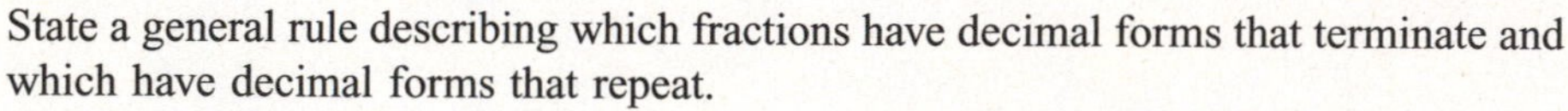

State a general rule describing which fractions have decimal forms that terminate and which have decimal forms that repeat.

Now test your rule on at least three new fractions. That is, be able to predict whether a fraction such as $\frac{1}{25}$ or $\frac{1}{30}$ has a terminating decimal or a repeating decimal. Then confirm your prediction.

# 4.3 Adding and Subtracting Decimals

< 4.3 Objectives >

1 > Add two or more decimals

2 > Use addition of decimals to solve application problems

3 > Subtract one decimal from another

4 > Use subtraction of decimals to solve application problems

Working with decimals rather than common fractions makes the basic operations much easier. First we will look at addition. One method for adding decimals is to write the decimals as common fractions, add, and then change the sum back to a decimal.

$$0.34 + 0.52 = \frac{34}{100} + \frac{52}{100} = \frac{86}{100} = 0.86$$

It is much more efficient to leave the numbers in decimal form and perform the addition in the same way as we did with whole numbers. You can use the following rule.

**Step by Step**

**To Add Decimals**

**Step 1** Write the numbers being added in column form *with their decimal points in a vertical line.*

**Step 2** Add just as you would with whole numbers.

**Step 3** Place the decimal point of the sum in line with the decimal points of the addends.

Example 1 illustrates the use of this rule.

**Example 1** **Adding Decimals**

< Objective 1 >

**NOTE**

Placing the decimal points in a vertical line ensures that we are adding digits of the same place value.

Add 0.13, 0.42, and 0.31.

$$\begin{array}{r} 0.13 \\ 0.42 \\ +\ 0.31 \\ \hline 0.86 \end{array}$$

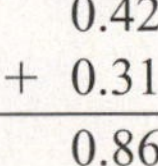

**Check Yourself 1**

Add 0.23, 0.15, and 0.41.

When adding decimals, you can use the *carrying process* just as you did when adding whole numbers. Consider the following.

## Example 2 Adding Decimals by Carrying

Add 0.35, 1.58, and 0.67.

```
  1 2 ←—— Carries
  0.35
  1.58
+ 0.67
------
  2.60
```

In the hundredths column:
5 + 8 + 7 = 20
Write 0 and carry 2 to the tenths column.
In the tenths column:
2 + 3 + 5 + 6 = 16
Write 6 and carry 1 to the ones column.

**Note:** The carrying process works with decimals, just as it did with whole numbers, because each place value is again $\frac{1}{10}$ the value of the place to its left.

### Check Yourself 2

Add 23.546, 0.489, 2.312, and 6.135.

When adding decimals, the numbers may not have the same number of decimal places. Just fill in as many zeros as needed so that all the numbers added have the same number of decimal places.

Recall that adding zeros to the right *does not change* the value of a decimal. The number 0.53 is the same as 0.530. We see this in Example 3.

## Example 3 Adding Decimals

**NOTE**

Be sure that the decimal points are in a vertical line.

Add 0.53, 4, 2.7, and 3.234.

```
  0.53
  4.
  2.7
+ 3.234
-------
```

Note that for a whole number, the decimal is understood to be to its right. So 4 = 4.

Now fill in the missing zeros and add as before.

```
  0.530
  4.000
  2.700
+ 3.234
-------
 10.464
```

Now all the numbers being added have *three* decimal places.

### Check Yourself 3

Add 6, 2.583, 4.7, and 2.54.

Many applied problems require working with decimals. For instance, filling up at a gas station means reading decimal amounts.

**Example 4** An Application of the Addition of Decimals

< Objective 2 >

NOTE

Because we want a total amount, addition is used for the solution.

On a trip the Chang family kept track of the gas purchases. If they bought 12.3, 14.2, 10.7, and 13.8 gallons (gal), how much gas did they use on the trip?

$$\begin{array}{r} 12.3 \\ 14.2 \\ 10.7 \\ +\ 13.8 \\ \hline 51.0 \text{ gal} \end{array}$$

**Check Yourself 4**

**The Higueras kept track of the gasoline they purchased on a recent trip. If they bought 12.4, 13.6, 9.7, 11.8, and 8.3 gal, how much gas did they buy on the trip?**

Every day you deal with amounts of money. Because our system of money is a decimal system, most problems involving money also involve operations with decimals.

**Example 5** An Application of the Addition of Decimals

Andre makes deposits of \$3.24, \$15.73, \$50, \$28.79, and \$124.38 during May. What is the total of his deposits for the month?

$$\begin{array}{r} \$\ \ 3.24 \\ 15.73 \\ 50.00 \\ 28.79 \\ +\ 124.38 \\ \hline \$222.14 \end{array}$$

Simply add the amounts of money deposited as decimals. Note that we write \$50 as \$50.00.

\$222.14 ← The total of deposits for May

**Check Yourself 5**

**Your textbooks for the fall term cost \$63.50, \$78.95, \$43.15, \$82, and \$85.85. What was the total cost of textbooks for the term?**

In Chapter 1, we defined *perimeter* as the distance around the outside of a straight-edged shape. Finding the perimeter often requires that we add decimal numbers.

### Example 6 An Application Involving the Addition of Decimals

Rachel is going to put a fence around the perimeter of her farm. Figure 1 is a picture of the land, measured in kilometers (km). How much fence does she need to buy?

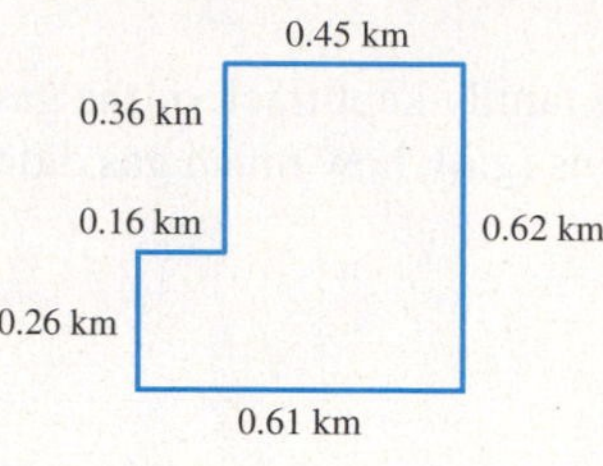


Figure 1

The perimeter is the sum of the lengths of the sides, so we add those lengths to find the total fencing needed.

$0.26 + 0.16 + 0.36 + 0.45 + 0.62 + 0.61 = 2.46$

Rachel needs 2.46 km of fence for the perimeter of her farm.

### Check Yourself 6

**Manuel intends to build a walkway around the perimeter of his garden (Figure 2). What will the total length of the walkway be?**

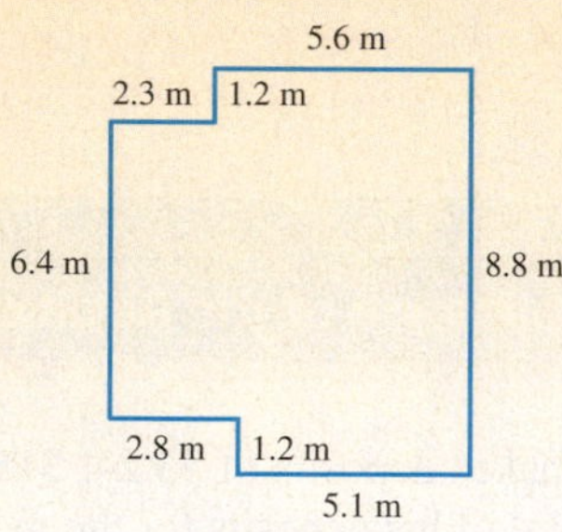


Figure 2

Much of what we have said about adding decimals is also true of subtraction. To subtract decimals, we use the following rule:

**Step by Step**

**To Subtract Decimals**

**Step 1** Write the numbers being subtracted in column form *with their decimal points in a vertical line.*

**Step 2** Subtract just as you would with whole numbers.

**Step 3** Place the decimal point of the difference in line with the decimal points of the numbers being subtracted.

Example 7 illustrates the use of this rule.

## Example 7 Subtracting a Decimal

< Objective 3 >

**RECALL**

The number that follows the word *from,* here 3.58, is written first. The number we are subtracting, here 1.23, is then written beneath 3.58.

Subtract 1.23 from 3.58.

$$\begin{array}{r} 3.58 \\ -\ 1.23 \\ \hline 2.35 \end{array}$$

Subtract in the hundredths, the tenths, and then the ones columns.

### Check Yourself 7

**Subtract 9.87 − 5.45.**

Because each place value is $\frac{1}{10}$ the value of the place to its left, borrowing, when you are subtracting decimals, works just as it did in subtracting whole numbers.

## Example 8 Decimal Subtraction That Involves Borrowing

Subtract 1.86 from 6.54.

$$\begin{array}{r} {}^{5}\,{}^{14}\,{}^{1} \\ \not{6}.\not{5}4 \\ -\ 1.86 \\ \hline 4.68 \end{array}$$

Here, borrow from the tenths and ones places to do the subtraction.

### Check Yourself 8

**Subtract 35.35 − 13.89.**

In subtracting decimals, as in adding, we can add zeros to the right of the decimal point so that both decimals have the same number of decimal places.

## Example 9 Subtracting a Decimal

**NOTES**

When you are subtracting, align the decimal points, and then add zeros to the right to align the digits.

9 has been rewritten as 9.000.

**(a)** Subtract 2.36 from 7.5.

$$\begin{array}{r} {}^{4}\,{}^{1} \\ 7.\not{5}0 \\ -\ 2.36 \\ \hline 5.14 \end{array}$$

We have added a 0 to 7.5. Next, borrow 1 tenth from the 5 tenths in the minuend.

**(b)** Subtract 3.657 from 9.

$$\begin{array}{r} {}^{8}\,{}^{9}{}^{9} \\ \not{9}.\not{0}\not{0}0 \\ -\ 3.657 \\ \hline 5.343 \end{array}$$

In this case, move left to the ones place to begin the borrowing process.

### Check Yourself 9

Subtract 5 − 2.345.

We can apply the subtraction methods of the previous examples in solving applications.

## Example 10 An Application of the Subtraction of a Decimal Number

< Objective 4 >

**NOTE**

We want to find the difference between the two measurements, so we subtract.

Jonathan was 98.3 centimeters (cm) tall on his sixth birthday. On his seventh birthday he was 104.2 cm. How much did he grow during the year?

$$\begin{array}{r} 104.2 \text{ cm} \\ -\ 98.3 \text{ cm} \\ \hline 5.9 \text{ cm} \end{array}$$

Jonathan grew 5.9 cm during the year.

### Check Yourself 10

A car's highway mileage before a tune-up was 28.8 miles per gallon (mi/gal). After the tune-up, it measured 30.1 mi/gal. What was the increase in mileage?

The same methods can be used in working with money.

## Example 11 An Application of the Subtraction of a Decimal Number

At the grocery store, Sally buys a roast that is marked \$12.37. She pays for her purchase with a \$20 bill. How much change does she receive?

$$\begin{array}{r} \$20.00 \\ -\ 12.37 \\ \hline \$\ 7.63 \end{array}$$

Add zeros to write \$20 as \$20.00. Then subtract as before.

Sally receives \$7.63 in change after her purchase.

**NOTE**

Sally's change is the *difference* between the price of the roast and the \$20 paid. We must use subtraction for the solution.

### Check Yourself 11

A stereo system that normally sells for \$549.50 is discounted (or marked down) to \$499.95 for a sale. What is the savings?

Balancing a checkbook requires addition and subtraction of decimal numbers.

## Example 12 An Application Involving the Addition and Subtraction of Decimals

For the following check register, find the running balance.

| Beginning balance | $234.15 |
|---|---|
| Check # 301 | 23.88 |
| Balance | ______ |
| Check # 302 | 38.98 |
| Balance | ______ |
| Check # 303 | 114.66 |
| Balance | ______ |
| Deposit | 175.75 |
| Balance | ______ |
| Check # 304 | 212.55 |
| Ending balance | ______ |

To keep a running balance, we add the deposits and subtract the checks.

| Beginning balance | $234.15 |
|---|---|
| Check # 301 | 23.88 |
| Balance | 210.27 |
| Check # 302 | 38.98 |
| Balance | 171.29 |
| Check # 303 | 114.66 |
| Balance | 56.63 |
| Deposit | 175.75 |
| Balance | 232.38 |
| Check # 304 | 212.55 |
| Ending balance | 19.83 |

### Check Yourself 12

For the following check register, add the deposits and subtract the checks to find the balance.

| | Beginning balance | $398.00 |
|---|---|---|
| | Check # 401 | 19.75 |
| (a) | Balance | ______ |
| | Check # 402 | 56.88 |
| (b) | Balance | ______ |
| | Check # 403 | 117.59 |
| (c) | Balance | ______ |
| | Deposit | 224.67 |
| (d) | Balance | ______ |
| | Check # 404 | 411.48 |
| (e) | Ending balance | ______ |

## Check Yourself ANSWERS

**1.** 0.79 **2.** 32.482 **3.**
$$\begin{array}{r} 6.000 \\ 2.583 \\ 4.700 \\ +\ 2.540 \\ \hline 15.823 \end{array}$$
**4.** 55.8 gal **5.** $353.45 **6.** 33.4 m

**7.** 4.42 **8.** 21.46 **9.** 2.655 **10.** 1.3 mi/gal **11.** $49.55

**12.** **(a)** $378.25; **(b)** $321.37; **(c)** $203.78; **(d)** $428.45; **(e)** $16.97

## Reading Your Text

The following fill-in-the-blank exercises are designed to ensure that you understand some of the key vocabulary used in this section.

### SECTION 4.3

**(a)** To add decimals, write the numbers being added in column form with their ______________ ______________ in a vertical line.

**(b)** Adding zeros to the right does not change the ______________ of a decimal.

**(c)** ______________ is the distance around the outside of a straight-edged shape.

**(d)** When subtracting one number from another, the number ____________ the word *from* is written first.

# 4.3 exercises

MathZone

Boost your grade at mathzone.com!
> Practice Problems
> NetTutor
> Self-Tests
> e-Professors
> Videos

Name ________ Section ________ Date ________

< Objective 1 >

*Add.*

1. $\begin{array}{r} 0.28 \\ +\ 0.79 \\ \hline \end{array}$

2. $\begin{array}{r} 2.59 \\ +\ 0.63 \\ \hline \end{array}$

3. $\begin{array}{r} 13.58 \\ 7.239 \\ +\ 1.5 \\ \hline \end{array}$

4. $\begin{array}{r} 8.625 \\ 2.45 \\ +\ 12.6 \\ \hline \end{array}$

5. $\begin{array}{r} 25.3582 \\ 6.5 \\ 1.898 \\ +\ 0.69 \\ \hline \end{array}$

6. $\begin{array}{r} 1.336 \\ 15.6857 \\ 7.9 \\ +\ 0.85 \\ \hline \end{array}$

7. $0.43 + 0.8 + 0.561$

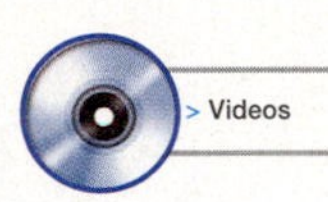


8. $1.25 + 0.7 + 0.259$

9. $42.731 + 1.058 + 103.24$

10. $27.4 + 213.321 + 39.38$

< Objective 3 >

*Subtract.*

11. $\begin{array}{r} 0.85 \\ -\ 0.59 \\ \hline \end{array}$

12. $\begin{array}{r} 5.68 \\ -\ 2.65 \\ \hline \end{array}$

13. $\begin{array}{r} 3.82 \\ -\ 1.565 \\ \hline \end{array}$

14. $\begin{array}{r} 8.59 \\ -\ 5.6 \\ \hline \end{array}$

15. $\begin{array}{r} 7.02 \\ -\ 4.7 \\ \hline \end{array}$

16. $\begin{array}{r} 45.6 \\ -\ 8.75 \\ \hline \end{array}$

17. $\begin{array}{r} 12 \\ -\ 5.35 \\ \hline \end{array}$

18. $\begin{array}{r} 15 \\ -\ 8.85 \\ \hline \end{array}$

19. Subtract 2.87 from 6.84.

20. Subtract 3.69 from 10.57.

21. Subtract 7.75 from 9.4.

22. Subtract 5.82 from 12.

23. Subtract 0.24 from 5.

24. Subtract 8.7 from 16.32.

## Answers

1. ________ 2. ________
3. ________ 4. ________
5. ________ 6. ________
7. ________ 8. ________
9. ________ 10. ________
11. ________ 12. ________
13. ________ 14. ________
15. ________ 16. ________
17. ________ 18. ________
19. ________ 20. ________
21. ________
22. ________
23. ________
24. ________

Basic Skills | **Advanced Skills** | Vocational-Technical Applications | Calculator/Computer | Above and Beyond

## Answers

25. ______

26. ______

27. ______

28. ______

29. ______

30. ______

< Objectives 2, 4 >

*Solve the following applications.*

**25. Business and Finance** On a 3-day trip, Dien bought 12.7, 15.9, and 13.8 gallons (gal) of gas. How many gallons of gas did he buy?

> Videos

**26. Science and Medicine** Felix ran 2.7 miles (mi) on Monday, 1.9 mi on Wednesday, and 3.6 mi on Friday. How far did he run during the week?

**27. Social Science** Rainfall was recorded in centimeters (cm) during the winter months, as indicated on the bar graph.

**(a)** How much rain fell during those months?

**(b)** How much more rain fell in December than in February?

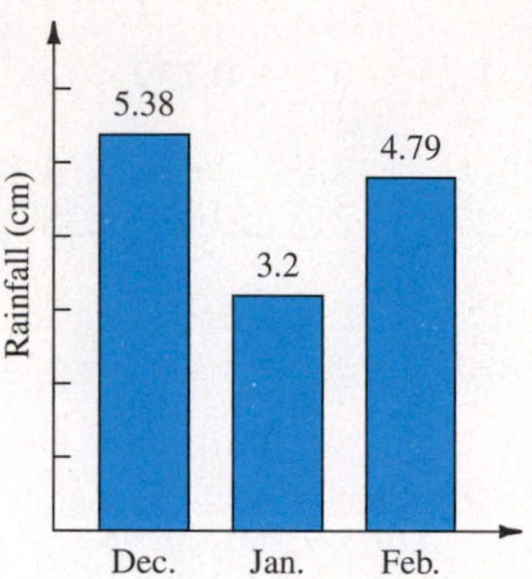


**28. Construction** A metal fitting has three sections, with lengths 2.5, 1.775, and 1.45 inches (in.). What is the total length of the fitting?

**29. Business and Finance** Nicole had the following expenses on a business trip: gas, \$45.69; food, \$123; lodging, \$95.60; and parking and tolls, \$8.65. What were her total expenses during the trip?

**30. Business and Finance** The deposit slip shown indicates the amounts that made up a deposit Peter Rabbit made. What was the total amount of his deposit?

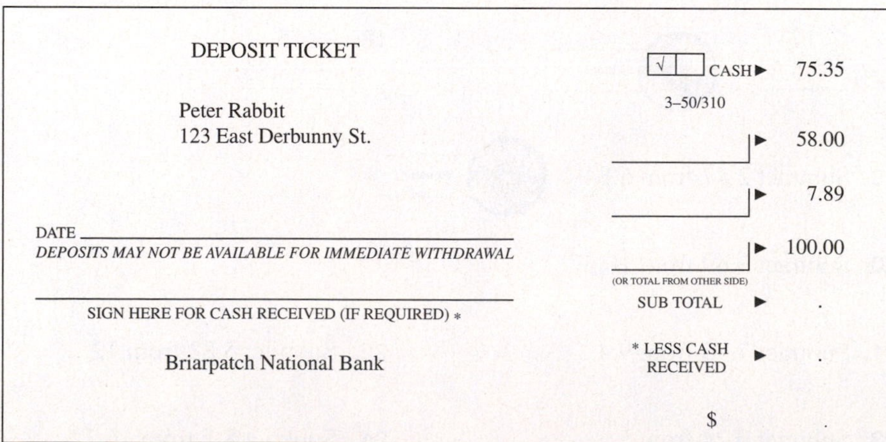
DEPOSIT TICKET

Peter Rabbit
123 East Derbunny St.

DATE ______
*DEPOSITS MAY NOT BE AVAILABLE FOR IMMEDIATE WITHDRAWAL*

______
SIGN HERE FOR CASH RECEIVED (IF REQUIRED) *

Briarpatch National Bank

| | |
|---|---|
| √ CASH ▶ | 75.35 |
| 3–50/310 | |
| ▶ | 58.00 |
| ▶ | 7.89 |
| ▶ | 100.00 |
| (OR TOTAL FROM OTHER SIDE) SUB TOTAL ▶ | . |
| * LESS CASH RECEIVED ▶ | . |
| $ | . |

31. **Construction** Lupe is putting a fence around her yard. Her yard is rectangular and measures 8.16 yards (yd) long and 12.68 yd wide. How much fence should Lupe purchase?

32. **Geometry** Find the perimeter of the given figure.

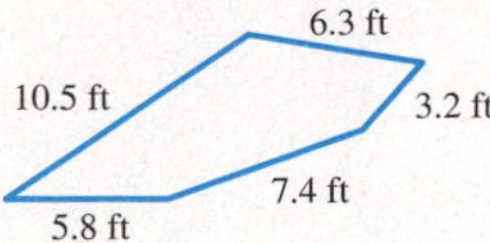


33. **Construction** The following figure gives the distance in miles (mi) of the boundary sections around a ranch. How much fencing is needed for the property?

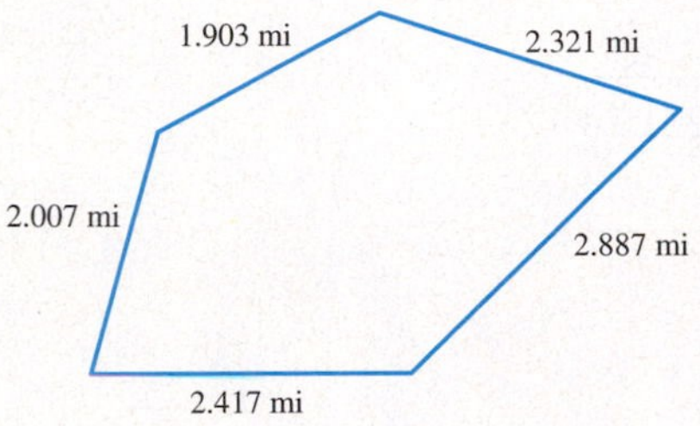


34. **Business and Finance** A television set selling for $399.50 is discounted (or marked down) to $365.75. What is the savings?

35. **Crafts** The outer radius of a piece of tubing is 2.8325 inches (in.). The inner radius is 2.775 in. What is the thickness of the wall of the tubing?

36. **Geometry** Given the following figure, find dimension $a$.

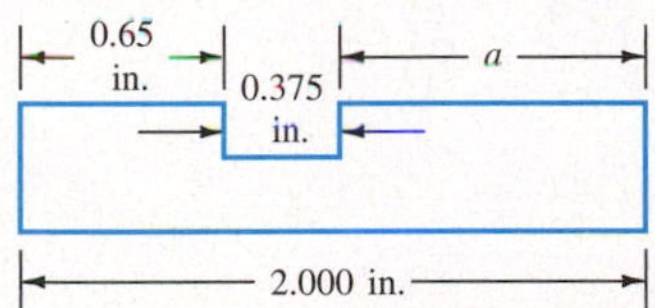


37. **Business and Finance** You make charges of $37.25, $8.78, and $53.45 on a credit card. If you make a payment of $73.50, how much do you still owe?

38. **Business and Finance** For the following check register, find the running balance.

| | |
|---|---|
| Beginning balance | $896.74 |
| Check # 501 | $425.69 |
| Balance | |
| Check # 502 | $ 56.34 |
| Balance | |
| Check # 503 | $ 41.89 |
| Balance | |
| Deposit | $123.91 |
| Balance | |
| Check # 504 | $356.98 |
| Ending balance | |

## Answers

31. ______

32. ______

33. ______

34. ______

35. ______

36. ______

37. ______

38. ______

Answers

39. ______

40. ______

41. ______

**39. Business and Finance** For the following check register, find the running balance.

| | |
|---|---|
| Beginning balance | $456.00 |
| Check # 601 | $199.29 |
| Balance | ______ |
| Service charge | $ 18.00 |
| Balance | ______ |
| Check # 602 | $ 85.78 |
| Balance | ______ |
| Deposit | $250.45 |
| Balance | ______ |
| Check # 603 | $201.24 |
| Ending balance | ______ |

**40. Business and Finance** For the following check register, find the running balance.

| | |
|---|---|
| Beginning balance | $589.21 |
| Check # 678 | $175.63 |
| Balance | ______ |
| Check # 679 | $ 56.92 |
| Balance | ______ |
| Deposit | $121.12 |
| Balance | ______ |
| Check # 680 | $345.99 |
| Ending balance | ______ |

**41. Business and Finance** For the following check register, find the running balance.

| | |
|---|---|
| Beginning balance | $1,345.23 |
| Check # 821 | $ 234.99 |
| Balance | ______ |
| Check # 822 | $ 555.77 |
| Balance | ______ |
| Deposit | $ 126.77 |
| Balance | ______ |
| Check # 823 | $ 53.89 |
| Ending balance | ______ |

Estimation can be a useful tool when working with decimal fractions. To estimate a sum, one approach is to round the addends to the nearest whole number and add for your estimate. For instance, to estimate the following sum:

| | Round | |
|---|---|---|
| 19.8 | ⟶ | 20 |
| 3.5 | | 4 |
| 24.2 | | 24 |
| + 10.4 | | + 10 |
| | | 58 |

Add to get the estimate.

*Use estimation to solve the following application.*

**42. Business and Finance** Alem's restaurant bill is pictured here. Estimate his total by rounding to the nearest dollar.

| Maggie's Kitchen | |
|---|---|
| calf's liver | $18.25 |
| caesar salad | $8.75 |
| 2 martinis | $6.80 |
| 2 chocolate cakes | $7.40 |
| coffee | $1.70 |
| | |
| | |
| Tip | |
| Total | |

Basic Skills | Advanced Skills | **Vocational-Technical Applications** | Calculator/Computer | Above and Beyond

**43. Allied Health** Respiratory therapists calculate the humidity deficit, in milligrams per liter (mg/L), for a patient by subtracting the actual humidity content of inspired air from 43.9 mg/L, which is the maximum humidity content at body temperature. Determine the humidity deficit if the inspired air has a humidity content of 32.7 mg/L.

**44. Allied Health** A patient is given three capsules of Tc99m sodium pertechnetate containing 79.4, 15.88, and 3.97 millicuries (mCi), respectively. What was the total activity, in millicuries, administered to the patient?

**45. Manufacturing** A pin is defined to have a clearance fit in a 0.618-in.-diameter hole. The clearance is called out to be 0.013 in. What is the diameter of the pin?

**46. Manufacturing** A dimension on a computer-aided design (CAD) drawing is given as 3.084 in. ± 0.125 in. What is the minimum and maximum length the feature may be?

**47. Electronics** Sandy purchased a lot of 12-volt (V) batteries from an online auction. There were 10 batteries in the lot. If the batteries were connected in series, the total open-voltage would be the sum of the battery voltages;

## Answers

42. ______

43. ______

44. ______

45. ______

46. ______

47. ______

## Answers

48. ______

49. ______

50. ______

51. ______

52. ______

therefore, she expected an open-voltage of 120 V. Unfortunately, her voltmeter (used to measure voltage) doesn't read above 100 V. So she measured each battery's voltage and recorded it in the following table. Calculate the actual open-voltage of the batteries if they are connected in series.

| Battery | Measured Voltage (in volts) |
|---|---|
| 1 | 12.20 |
| 2 | 13.84 |
| 3 | 11.42 |
| 4 | 13.00 |
| 5 | 12.45 |
| 6 | 12.82 |
| 7 | 11.93 |
| 8 | 11.01 |
| 9 | 12.77 |
| 10 | 12.03 |

Basic Skills | Advanced Skills | Vocational-Technical Applications | **Calculator/Computer** | Above and Beyond

Entering decimals in your calculator is similar to entering whole numbers. There is just one difference: The decimal point key [•] is used to place the decimal point as you enter the number.

Often both addition and subtraction are involved in a calculation. In this case, just enter the decimals and the operation signs, + or −, as they appear in the problem. To find 23.7 − 5.2 + 3.87 − 2.341, enter

23.7 [−] 5.2 [+] 3.87 [−] 2.341 [=]

The display should show 20.029.

*Solve the following exercises, using your calculator.*

**48.** 10,345.2 + 2,308.35 + 153.58

**49.** 8.7675 + 2.8 − 3.375 − 6

*Solve the following applications, using your calculator.*

**50. Business and Finance** Your checking account has a balance of $532.89. You write checks of $50, $27.54, and $134.75 and make a deposit of $50. What is your ending balance?

**51. Business and Finance** Your checking account has a balance of $278.45. You make deposits of $200 and $135.46. You write checks for $389.34, $249, and $53.21. What is your ending balance? Be careful with this exercise. A negative balance means that your account is overdrawn.

**52. Business and Finance** A small store makes a profit of $934.20 in the first week of a given month, $1,238.34 in the second week, and $853 in the third week. If the goal is a profit of $4,000 for the month, what profit must the store make during the remainder of the month?

Basic Skills | Advanced Skills | Vocational-Technical Applications | Calculator/Computer | **Above and Beyond**

53. **Business and Finance** Following are charges on a credit card account:

$8.97, $32.75, $15.95, $67.32, $215.78, $74.95, $83.90, and $257.28

(a) Estimate the total bill for the charges by rounding each number to the nearest dollar and adding the results.

(b) Estimate the total bill by adding the charges and then rounding to the nearest dollar.

(c) What are the advantages and disadvantages of the methods in (a) and (b)?

54. **Number Problem** Find the next number in the following sequence: 3.125, 3.375, 3.625, . . .

*Recall that a magic square is one in which the sum of every row, column, and diagonal is the same. Complete the following magic squares.*

55.

| | | |
|---|---|---|
| 2.4 | | 7.2 |
| 10.8 | | |
| 4.8 | | |

56.

| | | |
|---|---|---|
| 1.6 | | 1.2 |
| | 1 | |
| 0.8 | | |

57. (a) Determine the average amount of rainfall (to the nearest hundredth of an inch) in your town or city for each of the past 24 months.

(b) Determine the difference in rainfall amounts per month for each month from one year to the next.

58. Find the next two numbers in each of the following sequences:

(a) 0.75 0.62 0.5 0.39

(b) 1.0 1.5 0.9 3.5 0.8

## Answers

53. ______

54. ______

55. ______

56. ______

57. ______

58. ______

## Answers

1. 1.07 3. 22.319 5. 34.4462 7. 1.791 9. 147.029
11. 0.26 13. 2.255 15. 2.32 17. 6.65 19. 3.97
21. 1.65 23. 4.76 25. 42.4 gal 27. (a) 13.37 cm; (b) 0.59 cm
29. $272.94 31. 41.68 yd 33. 11.535 mi 35. 0.0575 in.
37. $25.98 39. $256.71; $238.71; $152.93; $403.38; $202.14
41. $1,110.24; $554.47; $681.24; $627.35 43. 11.2 mg/L
45. 0.605 in. 47. 123.47 V 49. 2.1925 51. $77.64 overdrawn
53. Above and Beyond 55.

| | | |
|---|---|---|
| 2.4 | 8.4 | 7.2 |
| 10.8 | 6 | 1.2 |
| 4.8 | 3.6 | 9.6 |

57. Above and Beyond

4.4

# Multiplying Decimals

< 4.4 Objectives >

1 > Multiply two or more decimals

2 > Use multiplication of decimals to solve application problems

3 > Multiply a decimal by a power of 10

4 > Use multiplication by a power of 10 to solve an application problem

To start our discussion of the multiplication of decimals, write the decimals in common-fraction form and then multiply.

## Example 1 Multiplying Two Decimals

< Objective 1 >

$$0.32 \times 0.2 = \frac{32}{100} \times \frac{2}{10} = \frac{64}{1{,}000} = 0.064$$

Here 0.32 has *two* decimal places, and 0.2 has *one* decimal place. The product 0.064 has *three* decimal places.

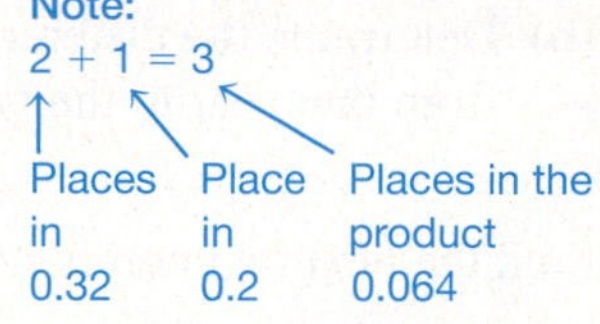

### Check Yourself 1

**Find the product and the number of decimal places.**

**0.14 × 0.054**

You do not need to write decimals as common fractions to multiply. Our work suggests the following rule.

Step by Step

**To Multiply Decimals**

Step 1 Multiply the decimals as though they were whole numbers.

Step 2 Add the number of decimal places in the numbers being multiplied.

Step 3 Place the decimal point in the product so that the number of decimal places in the product is the sum of the number of decimal places in the factors.

Example 2 illustrates this rule.

## Example 2 Multiplying Two Decimals

Multiply 0.23 by 0.7.

$$\begin{array}{r} 0.23 \\ \times\ 0.7 \\ \hline 0.161 \end{array}$$

0.23 ⟵ Two places
0.7 ⟵ One place
0.161 ⟵ Three places

### Check Yourself 2

Multiply 0.36 × 1.52.

You may have to affix zeros to the left in the product to place the decimal point. Consider Example 3.

## Example 3 Multiplying Two Decimals

Multiply.

$$\begin{array}{r} 0.136 \\ \times\ 0.28 \\ \hline 1088 \\ 272\phantom{0} \\ \hline 0.03808 \end{array}$$

0.136 ⟵ Three places
0.28 ⟵ Two places
0.03808 ⟵ Five places

$3 + 2 = 5$

Insert 0

Insert a 0 to mark off five decimal places.

### Check Yourself 3

Multiply 0.234 × 0.24.

Estimation is also helpful in multiplying decimals.

## Example 4 Estimating the Product of Two Decimals

Estimate the product 24.3 × 5.8.

$$\begin{array}{r} 24.3 \\ \times\ 5.8 \\ \hline \end{array} \xrightarrow{\text{Round}} \begin{array}{r} 24 \\ \times\ 6 \\ \hline 144 \end{array}$$

Multiply to get the estimate.

### Check Yourself 4

Estimate the product.

17.95 × 8.17

Many applications require decimal multiplication.

## Example 5 An Application Involving the Multiplication of Two Decimals

< Objective 2 >

A sheet of paper has dimensions 27.5 by 21.5 centimeters (cm). What is its area?

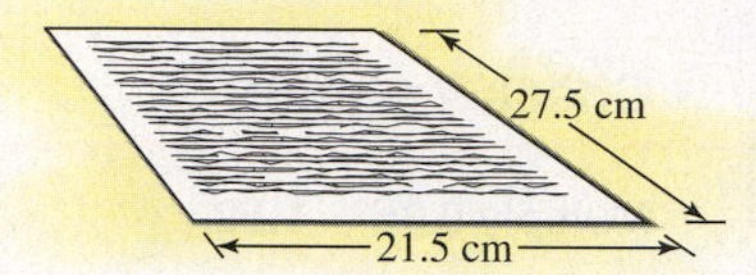


We multiply to find the required area.

**RECALL**

Area is length times width, so multiplication is the necessary operation.

$$\begin{array}{r} 27.5 \text{ cm} \\ \times\ 21.5 \text{ cm} \\ \hline 137\ 5\phantom{\text{ cm}} \\ 275\phantom{\text{ cm}} \\ 550\phantom{\text{ cm}} \\ \hline 591.25 \text{ cm}^2 \end{array}$$

The area of the paper is 591.25 cm$^2$.

### Check Yourself 5

**If 1 kilogram (kg) is 2.2 pounds (lb), how many pounds equal 5.3 kg?**

## Example 6 An Application Involving the Multiplication of Two Decimals

**NOTE**

Usually in problems dealing with money we round the result to the nearest cent (hundredth of a dollar).

Jack buys 8.7 gallons (gal) of propane at 98.9 cents per gallon. Find the cost of the propane.

We multiply the cost per gallon by the number of gallons. Then we round the result to the nearest cent. Note that the units of the answer will be cents.

$$\begin{array}{r} 98.9 \\ \times\ 8.7 \\ \hline 69\ 23 \\ 791\ 2\phantom{3} \\ \hline 860.43 \end{array}$$

The product 860.43 (cents) is rounded to 860 (cents), or \$8.60.

The cost of Jack's propane is \$8.60.

### Check Yourself 6

**One liter (L) is approximately 0.265 gal. On a trip to Europe, the Bernards purchased 88.4 L of gas for their rental car. How many gallons of gas did they purchase, to the nearest tenth of a gallon?**

Sometimes we will have to use more than one operation for a solution, as Example 7 shows.

**Example 7** **An Application Involving Two Operations**

Steve purchased a television set for \$299.50. He agreed to pay for the set by making payments of \$27.70 for 12 months. How much extra does he pay on the installment plan?

First we multiply to find the amount actually paid.

$$\begin{array}{r} \$\ 27.70 \\ \times\quad 12 \\ \hline 55\ 40 \\ 277\ 0\ \ \\ \hline \$332.40 \end{array} \longleftarrow \text{Amount paid}$$

Now subtract the listed price. The difference will give the extra amount Steve paid.

$$\begin{array}{r} \$332.40 \\ -\ 299.50 \\ \hline \$\ 32.90 \end{array} \longleftarrow \text{Extra amount}$$

Steve will pay an additional \$32.90 on the installment plan.

**Check Yourself 7**

Sandy's new car had a list price of \$10,985. She paid \$1,500 down and will pay \$305.35 per month for 36 months on the balance. How much extra will she pay with this loan arrangement?

There are enough applications involving multiplication by the powers of 10 to make it worthwhile to develop a special rule so you can do such operations quickly and easily. Look at the patterns in some of these special multiplications.

$$\begin{array}{r} 0.679 \\ \times\quad 10 \\ \hline 6.790, \text{ or } 6.79 \end{array} \qquad \begin{array}{r} 23.58 \\ \times\quad 10 \\ \hline 235.80, \text{ or } 235.8 \end{array}$$

Do you see that multiplying by 10 has moved the decimal point *one place to the right?* What happens when we multiply by 100?

**NOTE**

The digits remain the same. Only the *position* of the decimal point is changed.

$$\begin{array}{r} 0.892 \\ \times\quad 100 \\ \hline 89.200, \text{ or } 89.2 \end{array} \qquad \begin{array}{r} 5.74 \\ \times\ 100 \\ \hline 574.00, \text{ or } 574 \end{array}$$

Multiplying by 100 shifts the decimal point *two places to the right.* The pattern of these examples gives us the following rule:

**Property**

**To Multiply by a Power of 10**

Move the decimal point to the right the same number of places as there are zeros in the power of 10.

## Example 8 Multiplying by Powers of 10

< Objective 3 >

**NOTE**

Multiplying by 10, 100, or any other larger power of 10 makes the number *larger*. Move the decimal point to *the right*.

$2.356 \times 10 = 23.56$

One zero — The decimal point has moved one place to the right.

$34.788 \times 100 = 3{,}478.8$

Two zeros — The decimal point has moved two places to the right.

$3.67 \times 1{,}000 = 3{,}670.$

Three zeros — The decimal point has moved three places to the right. Note that we added a 0 to place the decimal point correctly.

**RECALL**

$10^5$ is just a 1 followed by five zeros.

$0.005672 \times 10^5 = 567.2$

Five zeros — The decimal point has moved five places to the right.

### Check Yourself 8

Multiply.

**(a)** 43.875 × 100 **(b)** 0.0083 × $10^3$

Example 9 is just one of many applications that require multiplying by a power of 10.

## Example 9 An Application Involving Multiplication by a Power of 10

< Objective 4 >

**NOTES**

There are 1,000 meters in 1 kilometer.

If the result is a whole number, there is no need to write the decimal point.

To convert from kilometers to meters, multiply by 1,000. Find the number of meters (m) in 2.45 kilometers (km).

2.45 km = 2,450. m

Just move the decimal point three places right to make the conversion. Note that we added a zero to place the decimal point correctly.

### Check Yourself 9

To convert from kilograms to grams, multiply by 1,000. Find the number of grams (g) in 5.23 kilograms (kg).

## Check Yourself ANSWERS

**1.** 0.00756, five decimal places **2.** 0.5472 **3.** 0.05616 **4.** 144 **5.** 11.66 lb **6.** 23.4 gal **7.** \$1,507.60 **8.** **(a)** 4,387.5; **(b)** 8.3 **9.** 5,230 g

## Reading Your Text

The following fill-in-the-blank exercises are designed to ensure that you understand some of the key vocabulary used in this section.

### SECTION 4.4

**(a)** When multiplying decimals, ____________ the number of decimal places in the numbers being multiplied.

**(b)** The decimal point in the ____________ is placed so that the number of places is the sum of the number of decimal places in the factors.

**(c)** It is sometimes necessary to affix ____________ to the left of the product of two decimals to accurately place the decimal point.

**(d)** Multiplying by 100 shifts the decimal point two places to the ____________.

# 4.4 exercises

Basic Skills | Advanced Skills | Vocational-Technical Applications | Calculator/Computer | Above and Beyond

MathZone

Boost your grade at mathzone.com!
- > Practice Problems
- > NetTutor
- > Self-Tests
- > e-Professors
- > Videos

Name ____________

Section ________ Date ________

## Answers

1. ________ 2. ________
3. ________ 4. ________
5. ________ 6. ________
7. ________ 8. ________
9. ________ 10. ________
11. ________ 12. ________
13. ________ 14. ________
15. ________ 16. ________
17. ________ 18. ________
19. ________ 20. ________
21. ________ 22. ________
23. ________ 24. ________
25. ________

< Objective 1 >

*Multiply.*

1. $2.3 \times 3.4$

2. $6.5 \times 4.3$

3. $8.4 \times 5.2$

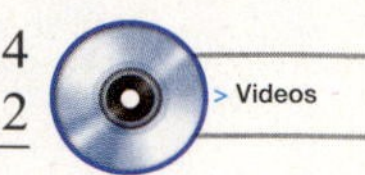


4. $9.2 \times 4.6$

5. $2.56 \times 72$

6. $56.7 \times 35$

7. $0.78 \times 2.3$

8. $9.5 \times 0.45$

9. $15.7 \times 2.35$

10. $28.3 \times 0.59$

11. $0.354 \times 0.8$

12. $0.624 \times 0.85$

13. $3.28 \times 5.07$

14. $0.582 \times 6.3$

15. $5.238 \times 0.48$

16. $0.372 \times 58$

17. $1.053 \times 0.552$

18. $2.375 \times 0.28$

19. $0.0056 \times 0.082$

20. $1.008 \times 0.046$

21. $0.8 \times 2.376$

22. $3.52 \times 58$

23. $0.3085 \times 4.5$

24. $0.028 \times 0.685$

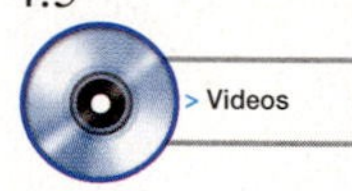


< Objective 2 >

*Solve the following applications.*

25. **Business and Finance** Kurt bought four shirts on sale as pictured. What was the total cost of the purchase?

Basic Skills | **Advanced Skills** | Vocational-Technical Applications | Calculator/Computer | Above and Beyond

**26. Business and Finance** Juan makes monthly payments of $123.65 on his car. What will he pay in 1 year?

**27. Science and Medicine** If 1 gallon (gal) of water weighs 8.34 pounds (lb), how much does 2.5 gal weigh?

**28. Business and Finance** Malik worked 37.4 hours (h) in 1 week. If his hourly rate of pay is $6.75, what was his pay for the week?

**29. Business and Finance** To find the amount of simple interest on a loan at $9\frac{1}{2}$ %, we have to multiply the amount of the loan by 0.095. Find the simple interest on a $1,500 loan for 1 year.

**30. Business and Finance** A beef roast weighing 5.8 lb costs $3.25 per pound. What is the cost of the roast?

**31. Business and Finance** Tom's state income tax is found by multiplying his income by 0.054. If Tom's income is $23,450, find the amount of his tax.

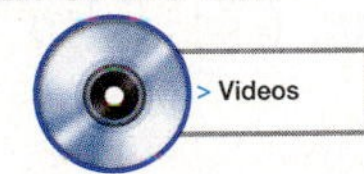


**32. Business and Finance** Claudia earns $6.40 per hour (h). For overtime (each hour over 40 h) she earns $9.60. If she works 48.5 h in a week, what pay should she receive?

**33. Geometry** A sheet of typing paper has dimensions as shown. What is its area?

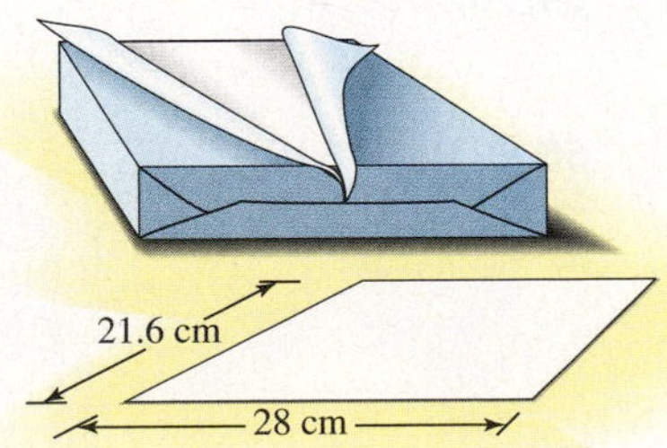


**34. Business and Finance** A rental car costs $24 per day plus 18 cents per mile (mi). If you rent a car for 5 days and drive 785 mi, what will the total car rental bill be?

## Answers

26. ______
27. ______
28. ______
29. ______
30. ______
31. ______
32. ______
33. ______
34. ______

## Answers

35. ______ 36. ______
37. ______ 38. ______
39. ______ 40. ______
41. ______ 42. ______
43. ______ 44. ______
45. ______
46. ______
47. ______
48. ______
49. ______
50. ______
51. ______
52. ______
53. ______
54. ______
55. ______
56. ______
57. ______
58. ______

*Answer exercises 35 to 40 with* **true** *or* **false.**

**35.** The decimal points must be aligned when finding the sum of two decimals.

**36.** The decimal points must be aligned when finding the difference of two decimals.

**37.** The decimal points must be aligned when finding the product of two decimals.

**38.** The decimal points must be aligned when finding the better looking of two decimals.

**39.** The number of decimal places in the product of two factors will be the product of the number of places in the two factors.

**40.** When multiplying a decimal by a power of 10, the decimal point is moved to the left the number of places as there are zeros in the power of 10.

< Objective 3 >

*Multiply.*

**41.** $5.89 \times 10$

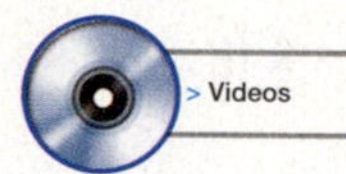


**42.** $0.895 \times 100$

**43.** $23.79 \times 100$

**44.** $2.41 \times 10$

**45.** $0.045 \times 10$

**46.** $5.8 \times 100$

**47.** $0.431 \times 100$

**48.** $0.025 \times 10$

**49.** $0.471 \times 100$

**50.** $0.95 \times 10{,}000$

**51.** $0.7125 \times 1{,}000$

**52.** $23.42 \times 1{,}000$

**53.** $4.25 \times 10^2$ > Videos

**54.** $0.36 \times 10^3$

**55.** $3.45 \times 10^4$

**56.** $0.058 \times 10^5$

< Objective 4 >

*Solve the following applications.*

**57. Business and Finance** A store purchases 100 items at a cost of $1.38 each. Find the total cost of the order.

**58. Science and Medicine** To convert from meters (m) to centimeters (cm), multiply by 100. How many centimeters are there in 5.3 m?

59. **Science and Medicine** How many grams (g) are there in 2.2 kilograms (kg)? Multiply by 1,000 to make the conversion.

60. **Business and Finance** An office purchases 1,000 pens at a cost of 17.8 cents each. What is the cost of the purchase in dollars?

Basic Skills | Advanced Skills | Vocational-Technical Applications | **Calculator/Computer** | Above and Beyond

The steps for finding the product of decimals on a calculator are similar to the ones we used for multiplying whole numbers. To multiply $2.8 \times 3.45 \times 3.725$, enter

2.8 [×] 3.45 [×] 3.725 [=]

The display should read 35.9835.

You can also easily find powers of decimals with your calculator by using a similar procedure. To find $(2.35)^3$, you can enter

2.35 [×] 2.35 [×] 2.35 [=]

The display should read 12.977875.

Some calculators have keys that will find powers more quickly. Look for keys marked [$x^2$] or [$y^x$]. Other calculators have a power key marked [∧]. To find $(2.35)^3$, enter 2.35 [∧] 3 [=] or 2.35 [$y^x$] 3 [=].

Again, the result is 12.977875.

*Use your calculator for the following exercises.*

61. $127.85 \times 0.055 \times 15.84$

62. $18.28 \times 143.45 \times 0.075$

63. $(3.95)^3$

64. $(0.521)^2$

65. **Geometry** Find the area of a rectangle with length 3.75 in. and width 2.35 in.

66. **Business and Finance** Mark works 38.4 h in a given week. If his hourly rate of pay is $7.85, what will he be paid for the week?

67. **Business and Finance** If fuel oil costs 87.5¢ per gallon, what will 150.4 gal cost?

68. **Business and Finance** To find the simple interest on a loan for 1 year at 12.5%, multiply the amount of the loan by 0.125. What simple interest will you pay on a loan of $1,458 at 12.5% for 1 year?

69. **Business and Finance** You are the office manager for Dr. Rogers. The increasing cost of making photocopies is a concern to Dr. Rogers. She wants

**Answers**

59. ______
60. ______
61. ______
62. ______
63. ______
64. ______
65. ______
66. ______
67. ______
68. ______
69. ______

## Answers

70. ____________

71. ____________

72. ____________

73. ____________

74. ____________

75. ____________

76. ____________

to examine alternatives to the current financing plan. The office currently leases a copy machine for $110 per month and pays $0.025 per copy. A 3-year payment plan is available that costs $125 per month and $0.015 per copy.

(a) If the office expects to run 100,000 copies per year, which is the better plan?

(b) How much money will the better plan save over the other plan?

**70.** **Business and Finance** In a bottling company, a machine can fill a 2-liter (L) bottle in 0.5 second (s) and move the next bottle into place in 0.1 s. How many 2-L bottles can be filled by the machine in 2 h?

What happens when your calculator wants to display an answer that is too big to fit in the display? Suppose you want to evaluate $10^{10}$. If you enter 10 [∧] 10 [=], your calculator will probably display 1 E 10 or perhaps $1^{10}$, both of which mean $1 \times 10^{10}$. Answers that are displayed in this way are said to be in **scientific notation.** This is a topic that you will study later. For now, we can use the calculator to see the relationship between numbers written in scientific notation and in decimal notation. For example, $3.485 \times 10^4$ is written in scientific notation. To write it in decimal notation, use your calculator to enter

3.485 [×] 10 [$y^x$] 4 [=] or 3.485 [×] 10 [∧] 4 [=]

The result should be 34,850. Note that the decimal point has moved four places (the power of 10) to the right.

*Write each of the following in decimal notation.*

**71.** $3.365 \times 10^3$

**72.** $4.128 \times 10^3$

**73.** $4.316 \times 10^5$

**74.** $8.163 \times 10^6$

**75.** $7.236 \times 10^8$

**76.** $5.234 \times 10^7$

## Answers

**1.** 7.82 **3.** 43.68 **5.** 184.32 **7.** 1.794 **9.** 36.895 **11.** 0.2832 **13.** 16.6296 **15.** 2.51424 **17.** 0.581256 **19.** 0.0004592 **21.** 1.9008 **23.** 1.38825 **25.** $39.92 **27.** 20.85 lb **29.** $142.50 **31.** $1,266.30 **33.** 604.8 $\text{cm}^2$ **35.** True **37.** False **39.** False **41.** 58.9 **43.** 2,379 **45.** 0.45 **47.** 43.1 **49.** 47.1 **51.** 712.5 **53.** 425 **55.** 34,500 **57.** $138 **59.** 2,200 g **61.** 111.38292 **63.** 61.629875 **65.** 8.8125 $\text{in.}^2$ **67.** $131.60 **69. (a)** 3-year plan; current plan: $11,460; 3-year lease: $9,000; **(b)** $2,460 **71.** 3,365 **73.** 431,600 **75.** 723,600,000

# Activity 11 :: Safe Dosages?

Chemotherapy drug dosages are generally calculated based on a patient's body surface area (BSA) in square meters ($m^2$). A patient's BSA is calculated using a nomogram and is based on the patient's height and weight. Print out the adult nomogram from the Science Museum of Minnesota's website www.smm.org/heart/lessons/nomogram_adult.htm. Draw a line connecting the patient's height with his or her weight. The point where this line crosses the middle column denotes the patient's BSA.

Dosages are then calculated using the formula

$$\text{Dose} = \text{recommended dose} \times \text{BSA}$$

For each of the following cases, determine if the prescribed dose falls within the recommended dose range for the given patient.

1. The doctor has prescribed Blenoxane to treat an adult male patient with testicular cancer. The patient is 70 inches (in.) tall and weighs 260 pounds (lb). According to the RxList website (www.rxlist.com), the recommended dose of Blenoxane in the treatment of testicular cancer should be between 10 to 20 units per square meter (units/$m^2$). The ordered dose is 50 units once per week.

2. The doctor has prescribed BiCNU to treat an adult patient with a brain tumor. The patient is 65 in. tall and weighs 150 lb. According to the RxList website, the recommended dose of BiCNU should be between 150 to 200 milligrams per square meter (mg/$m^2$). The ordered dose is 300 mg once every 6 weeks.

3. The doctor has prescribed Cisplatin to treat an adult patient with advanced bladder cancer. The patient is 73 in. tall and weighs 275 lb. According to the RxList website, the recommended dose of Cisplatin should be between 50 to 70 mg/$m^2$. The ordered dose is 150 mg once every 3 to 4 weeks.

4.5

# Dividing Decimals

< 4.5 Objectives >

1 > Divide a decimal by a whole number

2 > Use division of decimals to solve application problems

3 > Divide a decimal by a decimal

4 > Divide a decimal by a power of 10

Division of decimals is very similar to our earlier work dividing whole numbers. The only difference is in learning to place the decimal point in the quotient. Let's start with the case of dividing a decimal by a whole number. Here, placing the decimal point is easy. You can apply the following rule.

**Step by Step**

| To Divide a Decimal by a Whole Number | | |
|---|---|---|
| | Step 1 | Place the decimal point in the quotient *directly above* the decimal point of the dividend. |
| | Step 2 | Divide as you would with whole numbers. |

**Example 1** — **Dividing a Decimal by a Whole Number**

< Objective 1 >

**NOTE**

Do the division just as if you were dealing with whole numbers. Just remember to place the decimal point in the quotient *directly above* the one in the dividend.

Divide 29.21 by 23.

$$\begin{array}{r} 1.27 \\ 23\overline{)29.21} \\ \underline{23\phantom{.21}} \\ 6\ 2\phantom{1} \\ \underline{4\ 6\phantom{1}} \\ 1\ 61 \\ \underline{1\ 61} \\ 0 \end{array}$$

The quotient is 1.27.

**Check Yourself 1**

Divide 80.24 by 34.

Here is another example of dividing a decimal by a whole number.

## Example 2 Dividing a Decimal by a Whole Number

> **NOTE**
>
> Again place the decimal point of the quotient above that of the dividend.

Divide 122.2 by 52.

```
      2.3
52)122.2
   104
    18 2
    15 6
     2 6
```

We normally do not use a remainder when dealing with decimals. Add a 0 to the dividend and continue.

> **RECALL**
>
> Affixing a zero at the end of a decimal does not change the value of the dividend. It simply allows us to complete the division process in this case.

```
      2.35
52)122.20 ←—— Add a zero.
   104
    18 2
    15 6
     2 60
     2 60
        0
```

So $122.2 \div 52 = 2.35$. The quotient is 2.35.

### Check Yourself 2

**Divide 234.6 by 68.**

Often you will be asked to give a quotient to a certain place value. In this case, continue the division process to *one digit past* the indicated place value. Then round the result back to the desired accuracy.

When working with money, for instance, we normally give the quotient to the nearest hundredth of a dollar (the nearest cent). This means carrying the division out to the thousandths place and then rounding.

## Example 3 Dividing a Decimal by a Whole Number and Rounding the Result

> **NOTE**
>
> Find the quotient to *one place past* the desired place and then round the result.

Find the quotient of $25.75 \div 15$ to the nearest hundredth.

```
     1.716
15)25.750   ← Add a zero to carry the division
   15          to the thousandths place.
   10 7
   10 5
      25
      15
     100
      90
      10
```

So $25.75 \div 15 = 1.72$ (to the nearest hundredth).

### Check Yourself 3

**Find 99.26 ÷ 35 to the nearest hundredth.**

As we mentioned, problems similar to the one in Example 3 often occur when we are working with money. Example 4 is one of the many applications of this type of division.

## Example 4 An Application Involving the Division of a Decimal by a Whole Number

< Objective 2 >

**NOTE**

You might want to review the rules for rounding decimals in Section 4.1.

A carton of 144 items costs $56.10. What is the price per item to the nearest cent?

To find the price per item, divide the total price by 144.

```
        0.389
144 ) 56.100
      43 2
      12 90
      11 52
       1 380
       1 296
          84
```

Carry the division to the thousandths place and then round.

The cost per item is rounded to \$0.39, or 39¢.

### Check Yourself 4

**An office paid $26.55 for 72 pens. What was the cost per pen to the nearest cent?**

We now want to look at division *by* decimals. Here is an example using a fractional form.

## Example 5 Rewriting a Problem That Requires Dividing by a Decimal

< Objective 3 >

**NOTE**

It is always easy to rewrite a division problem so that you are dividing by a whole number. Dividing by a whole number makes it easy to place the decimal point in the quotient.

Divide.

$2.57 \div 3.4 = \dfrac{2.57}{3.4}$ Write the division as a fraction.

$= \dfrac{2.57 \times 10}{3.4 \times 10}$ We multiply the numerator and denominator by 10 so the divisor is a whole number. This *does not change* the value of the fraction.

$= \dfrac{25.7}{34}$ Multiplying by 10, shift the decimal point in the numerator and denominator *one place to the right*.

$= 25.7 \div 34$ Our division problem is rewritten so that the divisor is a whole number.

So

$2.57 \div 3.4 = 25.7 \div 34$ After we multiply the numerator and denominator by 10, we see that 2.57 ÷ 3.4 is the same as 25.7 ÷ 34.

**RECALL**

Multiplying by any whole-number power of 10 greater than 1 is just a matter of shifting the decimal point to the right.

### Check Yourself 5

Rewrite the division problem so that the divisor is a whole number.

3.42 ÷ 2.5

Do you see the rule suggested by example 5? We multiplied the numerator and the denominator (the dividend and the divisor) by 10. We made the divisor a whole number without altering the actual digits involved. All we did was shift the decimal point in the divisor and dividend the same number of places. This leads us to the following rule.

**Step by Step**

**To Divide by a Decimal**

**Step 1** Move the decimal point in the divisor to the *right,* making the divisor a whole number.

**Step 2** Move the decimal point in the dividend to the right *the same number of places.* Add zeros if necessary.

**Step 3** Place the decimal point in the quotient directly above the decimal point of the dividend.

**Step 4** Divide as you would with whole numbers.

Now look at an example of the use of our division rule.

### Example 6 Rounding the Result of Dividing by a Decimal

Divide 1.573 by 0.48 and give the quotient to the nearest tenth.

Write

$0.48\overline{)1.573}$

Shift the decimal points two places to the right to make the divisor a whole number.

**NOTE**

Once the division statement is rewritten, place the decimal point in the quotient above that in the dividend.

Now divide:

```
      3.27
48)157.30
   144
    13 3
     9 6
     3 70
     3 36
       34
```

Note that we add a zero to carry the division to the hundredths place. In this case, we want to find the quotient to the nearest tenth.

Round 3.27 to 3.3. So

1.573 ÷ 0.48 = 3.3 (to the nearest tenth)

### Check Yourself 6

Divide, rounding the quotient to the nearest tenth.

3.4 ÷ 1.24

Many applications involve decimal division.

## Example 7 Solving an Application Involving the Division of Decimals

Andrea worked 41.5 hours in a week and earned $239.87. What was her hourly rate of pay?

To find the hourly rate of pay we use division. We divide the number of hours worked into the total pay.

```
            5.78
41.5 )239.8 70
      207 5
       32 3 7
       29 0 5
        3 3 20
        3 3 20
             0
```

**NOTE**

We must add a zero to the dividend to complete the division process.

Andrea's hourly rate of pay was $5.78.

### Check Yourself 7

A developer wants to subdivide a 12.6-acre piece of land into 0.45-acre lots. How many lots are possible?

## Example 8 Solving an Application Involving the Division of Decimals

At the start of a trip the odometer read 34,563. At the end of the trip, it read 36,235. If 86.7 gallons (gal) of gas were used, find the number of miles per gallon (to the nearest tenth).

First, find the number of miles traveled by subtracting the initial reading from the final reading.

| | |
|---|---|
| 36,235 | Final reading |
| − 34,563 | Initial reading |
| 1,672 | Miles traveled |

Next, divide the miles traveled by the number of gallons used. This will give us the miles per gallon.

```
              1 9. 28
86.7 )1,672.0 00
       867
       805 0
       780 3
        24 7 0
        17 3 4
         7 3 60
         6 9 36
           4 24
```

Round 19.28 to 19.3 mi/gal.

### Check Yourself 8

**John starts his trip with an odometer reading of 15,436 and ends with a reading of 16,238. If he used 45.9 gallons (gal) of gas, find the number of miles per gallon (mi/gal) (to the nearest tenth).**

Recall that you can multiply decimals by powers of 10 by simply shifting the decimal point to the right. A similar approach works for division by powers of 10.

## Example 9 Dividing a Decimal by a Power of 10

< Objective 4 >

**(a)** Divide.

```
       3.53
  10)35.30
     30
      5 3
      5 0
        30
        30
         0
```

The dividend is 35.3. The quotient is 3.53. The decimal point has been shifted *one place to the left*. Note also that the divisor, 10, has *one* zero.

**(b)** Divide.

```
         3.785
  100)378.500
      300
       78 5
       70 0
        8 50
        8 00
          500
          500
            0
```

Here the dividend is 378.5, whereas the quotient is 3.785. The decimal point is now shifted *two places to the left*. In this case the divisor, 100, has *two* zeros.

### Check Yourself 9

**Perform each of the following divisions.**

**(a)** 52.6 ÷ 10 **(b)** 267.9 ÷ 100

Example 9 suggests the following rule.

**Property**

**To Divide a Decimal by a Power of 10**

Move the decimal point *to the left* the same number of places as there are zeros in the power of 10.

## Example 10 Dividing a Decimal by a Power of 10

Divide.

(a) 27.3 ÷ 10 = 2 7.3 Shift one place to the left.
= 2.73

(b) 57.53 ÷ 100 = 0 57.53 Shift two places to the left.
= 0.5753

(c) 39.75 ÷ 1,000 = 0 039.75 Shift three places to the left.
= 0.03975

(d) 85 ÷ 1,000 = 0 085. The decimal after the 85 is implied.
= 0.085

(e) 235.72 ÷ $10^4$ = 0 0235.72 Shift four places to the left.
= 0.023572

**NOTE**

We may have to add zeros to correctly place the decimal point.

**RECALL**

$10^4$ is a 1 followed by *four zeros.*

### Check Yourself 10

Divide.

(a) 3.84 ÷ 10

(b) 27.3 ÷ 1,000

Now, look at an application of our work in dividing by powers of 10.

## Example 11 Solving an Application Involving a Power of 10

To convert from millimeters (mm) to meters (m), we divide by 1,000. How many meters does 3,450 mm equal?

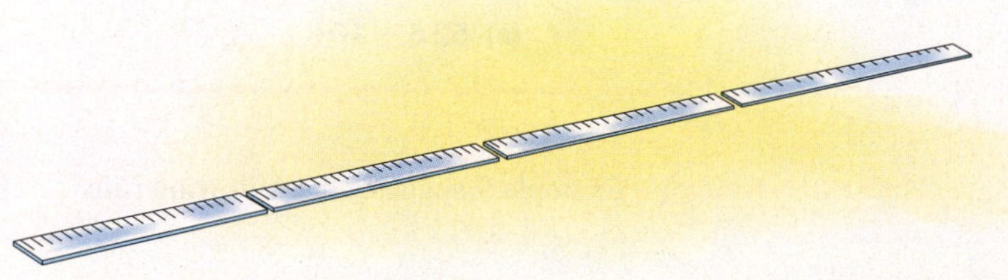

3,450 mm = 3 450. m Shift three places to the left to divide by 1,000.
= 3.450 m

## Check Yourself 11

A shipment of 1,000 notebooks cost a stationery store $658. What was the cost per notebook to the nearest cent?

Recall that the order of operations is always used to simplify a mathematical expression with several operations. You should recall the following order of operations.

**Property**

**The Order of Operations**

1. Perform any operations enclosed in **parentheses.**
2. Evaluate any **exponents.**
3. Do any **multiplication** and **division,** in order from left to right.
4. Do any **addition** and **subtraction,** in order from left to right.

## Example 12 Applying the Order of Operations

Simplify each expression.

**(a)** $4.6 + (0.5 \times 4.4)^2 - 3.93$ Parentheses

$= 4.6 + (2.2)^2 - 3.93$ Exponent

$= 4.6 + 4.84 - 3.93$ Add (left of the subtraction)

$= 9.44 - 3.93$ Subtract

$= 5.51$

**(b)** $16.5 - (2.8 + 0.2)^2 + 4.1 \times 2$ Parentheses

$= 16.5 - (3)^2 + 4.1 \times 2$ Exponent

$= 16.5 - 9 + 4.1 \times 2$ Multiply

$= 16.5 - 9 + 8.2$ Subtraction (left of the addition)

$= 7.5 + 8.2$ Add

$= 15.7$

## Check Yourself 12

Simplify each expression.

**(a)** $6.35 + (0.2 \times 8.5)^2 - 3.7$ **(b)** $2.5^2 - (3.57 - 2.14) + 3.2 \times 1.5$

## Check Yourself ANSWERS

**1.** 2.36 **2.** 3.45 **3.** 2.84 **4.** $0.37, or 37¢ **5.** $34.2 \div 25$
**6.** 2.7 **7.** 28 lots **8.** 17.5 mi/gal **9.** **(a)** 5.26; **(b)** 2.679
**10.** **(a)** 0.384; **(b)** 0.0273 **11.** 66¢ **12.** **(a)** 5.54; **(b)** 9.62

## Reading Your Text

The following fill-in-the-blank exercises are designed to ensure that you understand some of the key vocabulary used in this section.

### SECTION 4.5

**(a)** When dividing decimals, place the decimal point in the quotient directly ______________ the decimal point of the dividend.

**(b)** When asked to give a quotient to a certain place value, continue the division process to one digit ______________ the indicated place value.

**(c)** When dividing by a decimal, first move the decimal point in the divisor to the right, making the divisor a ______________ ______________.

**(d)** To divide a decimal by a power of 10, move the decimal point to the ______________ the same number of places as there are zeros in the power of 10.

Basic Skills | Advanced Skills | Vocational-Technical Applications | Calculator/Computer | Above and Beyond

# 4.5 exercises

MathZone

Boost your grade at mathzone.com!
> Practice Problems
> NetTutor
> Self-Tests
> e-Professors
> Videos

Name ____________

Section ________ Date ________

< Objectives 1–3 >

*Divide.*

**1.** $16.68 \div 6$ 

**2.** $43.92 \div 8$

**3.** $1.92 \div 4$

**4.** $5.52 \div 6$

**5.** $5.48 \div 8$

**6.** $2.76 \div 8$

**7.** $\frac{13.89}{6}$ 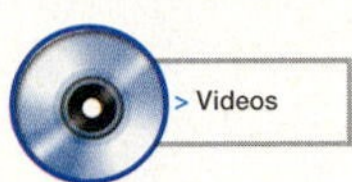

**8.** $\frac{21.92}{5}$

**9.** $\frac{185.6}{32}$

**10.** $\frac{165.6}{36}$

**11.** $79.9 \div 34$

**12.** $179.3 \div 55$

**13.** $52\overline{)13.78}$

**14.** $76\overline{)26.22}$

**15.** $0.6\overline{)11.07}$

**16.** $0.8\overline{)10.84}$

**17.** $\frac{7.22}{3.8}$

**18.** $\frac{13.34}{2.9}$

**19.** $\frac{11.622}{5.2}$

**20.** $\frac{3.616}{6.4}$

**21.** $0.27\overline{)1.8495}$ 

**22.** $0.038\overline{)0.8132}$

**23.** $0.046\overline{)1.587}$

**24.** $0.52\overline{)3.2318}$

**25.** $0.658 \div 2.8$

**26.** $0.882 \div 0.36$

< Objective 4 >

*Divide by moving the decimal point.*

**27.** $5.8 \div 10$

**28.** $5.1 \div 10$

**29.** $4.568 \div 100$

**30.** $3.817 \div 100$

## Answers

1. ______ 2. ______
3. ______ 4. ______
5. ______ 6. ______
7. ______ 8. ______
9. ______ 10. ______
11. ______ 12. ______
13. ______ 14. ______
15. ______ 16. ______
17. ______ 18. ______
19. ______ 20. ______
21. ______ 22. ______
23. ______ 24. ______
25. ______ 26. ______
27. ______ 28. ______
29. ______ 30. ______

## Answers

31. ______ 32. ______

33. ______ 34. ______

35. ______ 36. ______

37. ______ 38. ______

39. ______

40. ______

41. ______

42. ______

43. ______

44. ______

45. ______

46. ______

47. ______

48. ______

49. ______

50. ______

51. ______

52. ______

53. ______

54. ______

**31.** $24.39 \div 1{,}000$

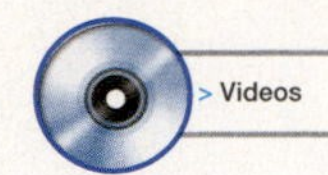


**32.** $8.41 \div 100$

**33.** $6.9 \div 1{,}000$

**34.** $7.2 \div 1{,}000$

**35.** $7.8 \div 10^2$

**36.** $3.6 \div 10^3$

**37.** $45.2 \div 10^5$

**38.** $57.3 \div 10^4$

*Divide and round the quotient to the indicated decimal place.*

**39.** $23.8 \div 9$ tenths

**40.** $5.27 \div 8$ hundredths

**41.** $38.48 \div 46$ hundredths

**42.** $3.36 \div 36$ thousandths

**43.** $125.4 \div 52$ tenths

**44.** $2.563 \div 54$ thousandths

**45.** $0.7\overline{)1.642}$ hundredths

**46.** $0.6\overline{)7.695}$ tenths

**47.** $4.5\overline{)8.415}$ tenths

**48.** $5.8\overline{)16}$ hundredths

**49.** $3.12\overline{)4.75}$ hundredths

**50.** $64.2\overline{)16.3}$ thousandths

< Objective 2 >

*Solve the following applications.*

**51.** **Business and Finance** Marv paid \$40.41 for three CDs on sale. What was the cost per CD?

**52.** **Business and Finance** Seven employees of an office donated a total of \$172.06 during a charity drive. What was the average donation?

**53.** **Business and Finance** A shipment of 72 paperback books cost a store \$190.25. What was the average cost per book to the nearest cent?

**54.** **Business and Finance** A restaurant bought 50 glasses at a cost of \$39.90. What was the cost per glass to the nearest cent?

55. **Business and Finance** The cost of a case of 48 items is \$28.20. What is the cost of an individual item to the nearest cent?

56. **Business and Finance** An office bought 18 handheld calculators for \$284. What was the cost per calculator to the nearest cent?

57. **Business and Finance** Al purchased a new refrigerator that cost \$736.12 with interest included. He paid \$100 as a down payment and agreed to pay the remainder in 18 monthly payments. What amount will he be paying per month?

58. **Business and Finance** The cost of a television set with interest is \$490.64. If you make a down payment of \$50 and agree to pay the balance in 12 monthly payments, what will be the amount of each monthly payment?

59. $79 - 28.2 + 13.7$

60. $15.9 - 4.2 \times 3.5$

61. $6.4 + 1.3^2$

62. $(6.4 + 1.3)^2$

*Simplify each expression.*

63. $5.2 - 3.1 \times 1.5 + (3.1 + 0.4)^2$

64. $150 + 4.1 \times 1.5 - (2.5 \times 1.6)^3 \times 2.4$

65. $17.9 \times 1.1 - (2.3 \times 1.1)^2 + (13.4 - 2.1 \times 4.6)$

66. $6.89^2 - 3.14 \times 2.5 + (4.1 - 3.2 \times 1.6)^2$

Basic Skills | Advanced Skills | **Vocational-Technical Applications** | Calculator/Computer | Above and Beyond

67. **Allied Health** Since people vary in body size, the cardiac index is used to normalize cardiac output measurements. The cardiac index, in liters per minute per square meters L/(min · m$^2$), is calculated by dividing a patient's cardiac output, in liters per minute (L/min), by his or her body surface area, in m$^2$. Calculate the cardiac index for a male patient whose cardiac output is 4.8 L/min if his body surface area is 2.03 m$^2$. Round your answer to the nearest hundredth.

## Answers

55. ______

56. ______

57. ______

58. ______

59. ______

60. ______

61. ______

62. ______

63. ______

64. ______

65. ______

66. ______

67. ______

## Answers

68. ______

69. ______

70. ______

71. ______

72. ______

73. ______

74. ______

75. ______

76. ______

77. ______

**68. Allied Health** The specific concentration of a radioactive drug, or radiopharmaceutical, is defined as the activity, in millicuries (mCi), divided by the volume, in milliliters (mL). Determine the specific concentration of a vial containing 7.3 mCi of I131 sodium iodide in 0.25 mL. Round your answer to the nearest tenth.

**69. Information Technology** A Web developer is responsible for designing a Web application for a customer. She uses a program called FTP to transmit pages from her local machine to a Web server. She needs to transmit 2.5 megabytes (Mbytes) of data. She notices it takes 10.2 seconds (s) to transmit the data. How fast is her connection to the Web server in Mbits/s? Round your answer to the nearest hundredth.

**70. Information Technology** After creating a presentation for a big customer, Joe sees that the size of the file is 1.6 Mbytes. Joe has a special application that allows him to save files across multiple disks. How many floppy disks will he need to store the file (a floppy disk can handle 1.4 Mbytes of data)?

Basic Skills | Advanced Skills | Vocational-Technical Applications | **Calculator/Computer** | Above and Beyond

Using your calculator to divide decimals is a good way to check your work and is also a reasonable way to solve applications. However, when using it for applications, we generally round off our answers to an appropriate place value.

*Using your calculator, divide and round to the indicated place.*

**71.** $2.546 \div 1.38$ hundredths

**72.** $45.8 \div 9.4$ tenths

**73.** $0.5782 \div 1.236$ thousandths

**74.** $1.25 \div 0.785$ hundredths

**75.** $1.34 \div 2.63$ two decimal places

**76.** $12.364 \div 4.361$ three decimal places

Basic Skills | Advanced Skills | Vocational-Technical Applications | Calculator/Computer | **Above and Beyond**

**77.** The blood alcohol content (BAC) of a person who has been drinking is determined by the Widmark formula. Find that formula using a search engine and use it to solve the following.

A 125-lb person is driving and is stopped by a policewoman on suspicion of driving under the influence (DUI). The driver claims that in the past 2 h he only consumed six 12-oz bottles of 3.9% beer. If he undergoes a breathalyzer test, what will his BAC be? Will this amount be under the legal limit for your state?

**78.** Four brands of soap are available in a local store.

| Brand | Ounces | Total Price | Unit Price |
|---|---|---|---|
| Squeaky Clean | 5.5 | $0.36 | |
| Smell Fresh | 7.5 | 0.41 | |
| Feel Nice | 4.5 | 0.31 | |
| Look Bright | 6.5 | 0.44 | |

Compute the unit price and decide which brand is the best buy.

**79.** Sophie is a quality control expert. She inspects boxes of number 2 pencils. Each pencil weighs 4.4 grams (g). The contents of a box of pencils weigh 66.6 g. If a box is labeled CONTENTS: 16 PENCILS, should Sophie approve the box as meeting specifications? Explain your answer.

**80.** Write a plan to determine the number of miles per gallon (mi/gal) your car (or your family car) gets. Use this plan to determine your car's actual mileage.

**81.** Express the width and length of a $1 bill in centimeters (cm). Then express this same length in millimeters (mm).

**82.** **Geometry** If the perimeter of a square is 19.2 cm, how long is each side?

$P = 19.2$ cm

*Solve the following applications.*

**83.** **Business and Finance** In 1 week, Tom earned $356.60 by working 36.25 hours (h). What was his hourly rate of pay to the nearest cent?

**84.** **Construction** An 80.5-acre piece of land is being subdivided into 0.35-acre lots. How many lots are possible in the subdivision?

## Answers

78. ______
79. ______
80. ______
81. ______
82. ______
83. ______
84. ______

## Answers

**1.** 2.78 **3.** 0.48 **5.** 0.685 **7.** 2.315 **9.** 5.8 **11.** 2.35 **13.** 0.265 **15.** 18.45 **17.** 1.9 **19.** 2.235 **21.** 6.85 **23.** 34.5 **25.** 0.235 **27.** 0.58 **29.** 0.04568 **31.** 0.02439 **33.** 0.0069 **35.** 0.078 **37.** 0.000452 **39.** 2.6 **41.** 0.84 **43.** 2.4 **45.** 2.35 **47.** 1.9 **49.** 1.52 **51.** $13.47 **53.** $2.64 **55.** $0.59, or 59¢ **57.** $35.34 **59.** 64.5 **61.** 8.09 **63.** 12.8 **65.** 17.0291 **67.** 2.36 L/(min · m$^2$) **69.** 0.25 Mbits/s **71.** 1.84 **73.** 0.468 **75.** 0.51 **77.** Above and Beyond **79.** Above and Beyond **81.** Above and Beyond **83.** $9.84

# Activity 12 :: The Tour de France

The Tour de France is perhaps the most grueling of all sporting events. It is a bicycle race that spans 22 days (including 2 rest days), and involves 20 stages of riding in a huge circuit around the country of France. This includes several stages that take the riders through two mountainous regions, the Alps and the Pyrenees. The following table presents the winners of each stage in the 2005 race, along with the winning time in hours and the length of the stage expressed in miles. For each stage, compute the winner's average speed by dividing the miles traveled by the winning time. Round your answers to the nearest tenth of a mile per hour.

| Stage | Winner | Length (mi) | Time (h) | Speed (mi/h) |
|---|---|---|---|---|
| Prologue | Zabriskie | 11.8 | 0.35 | |
| 1 | Boonen | 112.8 | 3.86 | |
| 2 | Boonen | 132 | 4.60 | |
| 3 | Discovery Channel | 41.9 | 1.18 | |
| 4 | McEwen | 113.7 | 3.77 | |
| 5 | Bernucci | 123.7 | 4.21 | |
| 6 | McEwen | 142 | 5.06 | |
| 7 | Weening | 143.8 | 5.07 | |
| 8 | Rasmussen | 106.3 | 4.14 | |
| 9 | Valverde | 110.9 | 4.84 | |
| 10 | Vinokourov | 107.5 | 4.79 | |
| 11 | Moncoutie | 116.2 | 4.34 | |
| 12 | McEwen | 107.8 | 3.72 | |
| 13 | Totsehnig | 137 | 5.73 | |
| 14 | Hincapie | 127.7 | 6.11 | |
| 15 | Pereiro | 112.2 | 4.64 | |
| 16 | Savoldelli | 148.8 | 5.69 | |
| 17 | Serrano | 117.4 | 4.63 | |
| 18 | Guerini | 95.4 | 3.55 | |
| 19 | Armstrong | 34.5 | 1.20 | |
| 20 | Vinokourov | 89.8 | 3.68 | |

Find the total number of miles traveled.

By examining the speeds of the stages, can you identify which stages occurred in the mountains?

The overall tour winner was Lance Armstrong with a total winning time of 86.25 hours. Compute his average speed for the entire race, again rounding to the nearest tenth of a mile per hour.

| Definition/Procedure | Example | Reference |
|---|---|---|
| **Place Value and Rounding** | | Section 4.1 |
| **Decimal Fraction** A fraction whose denominator is a power of 10. We call decimal fractions *decimals*. | $\frac{7}{10}$ and $\frac{47}{100}$ are decimal fractions. | p. 291 |
| **Decimal Place** Each position for a digit to the right of the decimal point. Each decimal place has a place value that is $\frac{1}{10}$ the value of the place to its left. | 2.3456 — 4: Tenths; 5: Hundredths; 6: Thousandths; 7: Ten–thousandths | p. 291 |
| ***Reading and Writing Decimals in Words*** | | |
| **Step 1** Read the digits *to the left* of the decimal point as a whole number.<br>**Step 2** Read the decimal point as the word *and*.<br>**Step 3** Read the digits *to the right* of the decimal point as a whole number followed by the place value of the rightmost digit. | Hundredths<br>8.15 is read "eight and fifteen hundredths." | p. 293 |
| ***Rounding Decimals*** | | |
| **Step 1** Find the place to which the decimal is to be rounded.<br>**Step 2** If the next digit to the right is 5 or more, increase the digit in the place you are rounding to by 1. Discard any remaining digits to the right.<br>**Step 3** If the next digit to the right is less than 5, just discard that digit and any remaining digits to the right. | To round 5.87 to the nearest tenth:<br>5.87 is rounded to 5.9<br>To round 12.3454 to the nearest thousandth:<br>12.3454 is rounded to 12.345. | p. 295 |
| **Converting Between Fractions and Decimals** | | Section 4.2 |
| ***To Convert a Common Fraction to a Decimal*** | | |
| **Step 1** Divide the numerator of the common fraction by its denominator.<br>**Step 2** The quotient is the decimal equivalent of the common fraction. | To convert $\frac{1}{2}$ to a decimal:<br>$2\overline{)1.0} = 0.5$; 1 0; 0 | p. 301 |
| ***To Convert a Terminating Decimal Less Than 1 to a Common Fraction*** | | |
| **Step 1** Write the digits of the decimal without the decimal point. This will be the numerator of the common fraction.<br>**Step 2** The denominator of the fraction is a 1 followed by as many zeros as there are places in the decimal. | To convert 0.275 to a common fraction:<br>$0.275 = \frac{275}{1{,}000} = \frac{11}{40}$ | p. 304 |

*Continued*

| Definition/Procedure | Example | Reference |
|---|---|---|
| **Adding and Subtracting Decimals** | | Section 4.3 |
| *To Add Decimals* | | |
| **Step 1** Write the numbers being added in column form with their decimal points in a vertical line.<br>**Step 2** Add just as you would with whole numbers.<br>**Step 3** Place the decimal point of the sum in line with the decimal points of the addends. | To add 2.7, 3.15, and 0.48:<br>$\begin{array}{r} 2.7 \\ 3.15 \\ +\ 0.48 \\ \hline 6.33 \end{array}$ | p. 313 |
| *To Subtract Decimals* | | |
| **Step 1** Write the numbers being subtracted in column form with their decimal points in a vertical line. You may have to place zeros to the right of the existing digits.<br>**Step 2** Subtract just as you would with whole numbers.<br>**Step 3** Place the decimal point of the difference in line with the decimal points of the numbers being subtracted. | To subtract 5.875 from 8.5:<br>$\begin{array}{r} 8.500 \\ -\ 5.875 \\ \hline 2.625 \end{array}$ | p. 316 |
| **Multiplying Decimals** | | Section 4.4 |
| *To Multiply Decimals* | | |
| **Step 1** Multiply the decimals as though they were whole numbers.<br>**Step 2** Add the number of decimal places in the factors.<br>**Step 3** Place the decimal point in the product so that the number of decimal places in the product is the sum of the number of decimal places in the factors. | To multiply $2.85 \times 0.045$:<br>$\begin{array}{r} 2.85 \\ \times\ 0.045 \\ \hline 1425 \\ 1140\phantom{0} \\ \hline 0.12825 \end{array}$<br>2.85 ⟵ Two places<br>0.045 ⟵ Three places<br>0.12825 ⟵ Five places | p. 328 |
| *Multiplying by Powers of 10* | | |
| Move the decimal point to the right the same number of places as there are zeros in the power of 10. | $2.37 \times 10 = 23.7$<br>$0.567 \times 1{,}000 = 567$ | p. 332 |
| **Dividing Decimals** | | Section 4.5 |
| *To Divide by a Decimal* | | |
| **Step 1** Move the decimal point in the divisor to the right, making the divisor a whole number.<br>**Step 2** Move the decimal point in the dividend to the right the same number of places. Add zeros if necessary.<br>**Step 3** Place the decimal point in the quotient directly above the decimal point of the dividend.<br>**Step 4** Divide as you would with whole numbers. | To divide 16.5 by 5.5, move the decimal points:<br>$\begin{array}{r} 3 \\ 5.5_\wedge \overline{)16.5_\wedge} \\ 16\ 5 \\ \hline 0 \end{array}$ | p. 343 |
| *To Divide by a Power of 10* | | |
| Move the decimal point to the left the same number of places as there are zeros in the power of 10. | $25.8 \div 10 = 2_\wedge5.8 = 2.58$ | p. 345 |

This summary exercise set is provided to give you practice with each of the objectives of this chapter. Each exercise is keyed to the appropriate chapter section. When you are finished, you can check your answers to the odd-numbered exercises against those presented in the back of the text. If you have difficulty with any of these questions, go back and reread the examples from that section. The answers to the even-numbered exercises appear in the *Instructor's Solutions Manual.* Your instructor will give you guidelines on how to best use these exercises in your instructional setting.

**4.1** *Find the indicated place values.*

**1.** 7 in 3.5742

**2.** 3 in 0.5273

*Write the fractions in decimal form.*

**3.** $\frac{37}{100}$

**4.** $\frac{307}{10,000}$

*Write the decimals in words.*

**5.** 0.071

**6.** 12.39

*Write the fractions in decimal form.*

**7.** Four and five tenths

**8.** Four hundred and thirty-seven thousandths

*Complete each statement, using the symbol* $<$, $=$, *or* $>$.

**9.** 0.79 ______ 0.785

**10.** 1.25 ______ 1.250

**11.** 12.8 ______ 13

**12.** 0.832 ______ 0.83

*Round to the indicated place.*

**13.** 5.837 hundredths

**14.** 9.5723 thousandths

**15.** 4.87625 three decimal places

*Write each of the following as a common fraction or a mixed number.*

**16.** 0.0067

**17.** 21.857

**4.2** *Find the decimal equivalents.*

**18.** $\frac{7}{16}$

**19.** $\frac{3}{7}$ (round to the thousandth)

**20.** $\frac{4}{15}$ (use bar notation)

**21.** $3\frac{3}{4}$

 

*Write as common fractions or mixed numbers. Simplify your answers.*

**22.** 0.21

**23.** 0.084

**24.** 5.28

4.3 *Add.*

**25.**
$$\begin{array}{r} 2.58 \\ +\ 0.89 \\ \hline \end{array}$$

**26.**
$$\begin{array}{r} 3.14 \\ 0.8 \\ 2.912 \\ +\ 12 \\ \hline \end{array}$$

**27.** 1.3, 25, 5.27, and 6.158

**28.** Add eight, forty-three thousandths, five and nineteen hundredths, and seven and three tenths.

*Subtract.*

**29.**
$$\begin{array}{r} 29.21 \\ -\ 5.89 \\ \hline \end{array}$$

**30.**
$$\begin{array}{r} 6.73 \\ -\ 2.485 \\ \hline \end{array}$$

**31.** 1.735 from 2.81

**32.** 12.38 from 19

*Solve the applications.*

**33. GEOMETRY** Find the perimeter (to the nearest hundredth of a centimeter) of a rectangle that has dimensions 5.37 cm by 8.64 cm.

**34. SCIENCE AND MEDICINE** Janice ran 4.8 miles (mi) on Sunday, 5.3 mi on Tuesday, 3.9 mi on Thursday, and 8.2 mi on Saturday. How far did she run during the week?

**35. GEOMETRY** Find dimension *a* in the following figure.

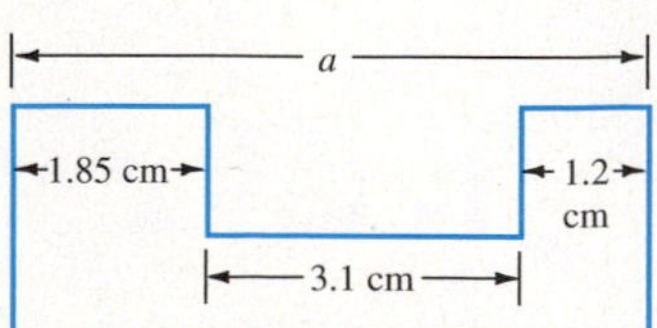


**36. BUSINESS AND FINANCE** A stereo system that normally sells for $499.50 is discounted (or marked down) to $437.75 for a sale. Find the savings.

**37. BUSINESS AND FINANCE** If you cash a $50 check and make purchases of $8.71, $12.53, and $9.83, how much money do you have left?

**4.4** *Multiply.*

**38.** $\begin{array}{r} 22.8 \\ \times\ 0.72 \\ \hline \end{array}$

**39.** $\begin{array}{r} 0.0045 \\ \times\ 0.058 \\ \hline \end{array}$

**40.** $1.24 \times 56$

**41.** $0.0025 \times 0.491$

**42.** $0.052 \times 1{,}000$

**43.** $0.045 \times 10^4$

*Solve the applications.*

**44.** **Business and Finance** Neal worked for 37.4 hours (h) during a week. If his hourly rate of pay was \$7.25, how much did he earn?

**45.** **Business and Finance** To find the simple interest on a loan at $11\frac{1}{2}\%$ for 1 year, we must multiply the amount of the loan by 0.115. Find the simple interest on a \$2,500 loan at $11\frac{1}{2}\%$ for 1 year.

**46.** **Business and Finance** A television set has an advertised price of \$499.50. You buy the set and agree to make payments of \$27.15 per month for 2 years. How much extra are you paying by buying with this installment plan?

**47.** **Business and Finance** A stereo dealer buys 100 portable radios for a promotion sale. If she pays \$57.42 per radio, what is her total cost?

**4.5** *Divide. Round answers to the nearest hundredth.*

**48.** $8\overline{)3.08}$

**49.** $58\overline{)269.7}$

**50.** $55\overline{)17.69}$

*Divide. Round answers to the nearest thousandth.*

**51.** $0.7\overline{)1.865}$

**52.** $3.042 \div 0.37$

**53.** $5.3\overline{)6.748}$

**54.** $0.2549 \div 2.87$

*Divide.*

**55.** $7.6 \div 10$

**56.** $80.7 \div 1{,}000$

**57.** $457 \div 10^4$

*Solve the applications.*

**58.** **Business and Finance** During a charity fund-raising drive 37 employees of a company donated a total of \$867.65. What was the donation per employee?

59. **Business and Finance** Faith always fills her gas tank as soon as the gauge hits the $\frac{1}{4}$ full mark. In six readings, Faith's gas mileage was 38.9, 35.3, 39.0, 41.2, 40.5, and 40.8 miles per gallon (mi/gal). What was the average mileage to the nearest tenth of a mile per gallon? (*Hint:* First find the sum of the mileages. Then divide the sum by 6, because there are 6 mileage readings.)

60. **Construction** A developer is planning to subdivide an 18.5-acre piece of land. She estimates that 5 acres will be used for roads and wants individual lots of 0.25 acre. How many lots are possible?

61. **Business and Finance** Paul drives 949 mi, using 31.8 gal of gas. What is his mileage for the trip (to the nearest tenth of a mile per gallon)?

62. **Business and Finance** A shipment of 1,000 videotapes cost a dealer $7,090. What was the cost per tape to the dealer?

# self-test 4

Name ______________________

Section __________ Date __________

The purpose of this self-test is to help you check your progress so that you can find sections and concepts that you need to review before the next in-class exam. Allow yourself about an hour to take this test. At the end of that hour, check your answers against those given in the back of the text. Section references accompany the answers. If you missed any, note the section reference that accompanies the answer. Go back to that section and reread the examples until you have mastered that particular concept.

**1.** Find the place value of 8 in 0.5248.

**2.** Write $\frac{49}{1,000}$ in decimal form.

**3.** Write 2.53 in words.

**4.** Write twelve and seventeen thousandths in decimal form.

*Complete the statement, using the symbol* $<$ *or* $>$.

**5.** 0.889 ________ 0.89

**6.** 0.531 ________ 0.53

*Add.*

**7.**
$$\begin{array}{r} 3.45 \\ 0.6 \\ +\ 12.59 \\ \hline \end{array}$$

**8.** 2.4, 35, 4.73, and 5.123.

**9.** Seven, seventy-nine hundredths, and five and thirteen thousandths.

**10.** **Business and Finance** On a business trip, Martin bought the following amounts of gasoline: 14.4, 12, 13.8, and 10 gallons (gal). How much gasoline did he purchase on the trip?

*Subtract.*

**11.**
$$\begin{array}{r} 18.32 \\ -\ 7.78 \\ \hline \end{array}$$

**12.**
$$\begin{array}{r} 40 \\ -\ 15.625 \\ \hline \end{array}$$

**13.** 1.742 from 5.63

**14.** **Business and Finance** You pay for purchases of \$13.99, \$18.75, \$9.20, and \$5 with a \$50 bill. How much cash will you have left?

*Multiply.*

**15.**
$$\begin{array}{r} 32.9 \\ \times\ 0.53 \\ \hline \end{array}$$

**16.**
$$\begin{array}{r} 0.049 \\ \times\ 0.57 \\ \hline \end{array}$$

**17.** $2.75 \times 0.53$

**18.** Find the area of a rectangle with length 3.5 inches (in.) and width 2.15 in.

*Multiply.*

**19.** $0.735 \times 1,000$

**20.** $1.257 \times 10^4$

## Answers

1. ______________________
2. ______________________
3. ______________________
4. ______________________
5. ______________________
6. ______________________
7. ______________________
8. ______________________
9. ______________________
10. ______________________
11. ______________________
12. ______________________
13. ______________________
14. ______________________
15. ______________________
16. ______________________
17. ______________________
18. ______________________
19. ______________________
20. ______________________

Answers

21. ______
22. ______
23. ______
24. ______
25. ______
26. ______
27. ______
28. ______
29. ______
30. ______
31. ______
32. ______
33. ______
34. ______
35. ______
36. ______
37. ______
38. ______
39. ______
40. ______
41. ______
42. ______

**21. Business and Finance** A college bookstore purchases 1,000 pens at a cost of 54.3 cents per pen. Find the total cost of the order in dollars.

*Round to the indicated place.*

**22.** 0.5977 thousandths

**23.** 23.5724 two decimal places

**24.** 36,139.0023 thousands

*Divide. When indicated, round to the given place value.*

**25.** $8\overline{)3.72}$

**26.** $27\overline{)63.45}$

**27.** $2.72 \div 53$ thousandths

**28.** $4.1\overline{)10.455}$

**29.** $0.6\overline{)1.431}$

**30.** $3.969 \div 0.54$

**31.** $0.263 \div 3.91$ three decimal places

**32. Construction** A 14-acre piece of land is being developed into home lots. If 2.8 acres of land will be used for roads and each home site is to be 0.35 acre, how many lots can be formed?

*Divide.*

**33.** $4.983 \div 1{,}000$

**34.** $523 \div 10^5$

**35. Construction** A street improvement project will cost $57,340, and that cost is to be divided among the 100 families in the area. What will be the cost to each individual family?

*Find the decimal equivalents of the common fractions. When indicated, round to the given place value.*

**36.** $\frac{7}{16}$

**37.** $\frac{3}{7}$ thousandths

**38.** $\frac{7}{11}$ use bar notation

*Write the decimals as common fractions or mixed numbers. Simplify your answer.*

**39.** 0.072

**40.** 4.44

**41.** Insert < or > to form a true statement.

0.168 ______ $\frac{3}{25}$

**42.** A baseball team has a winning percentage of 0.458. Write this as a fraction in simplest form.

# cumulative review chapters 1-4

Name ______________________

Section ________ Date ________

The following exercises are presented to help you review concepts from earlier chapters. This is meant as review material and not as a comprehensive exam. The answers are presented in the back of the text. Before each answer is a section reference for the concept. If you have difficulty with any of these exercises, be certain to at least read through the summary related to that section.

**1.** Write 286,543 in words.

**2.** What is the place value of 5 in the number 343,563?

*In exercises 3 to 8, perform the indicated operations.*

**3.**
$$\begin{array}{r} 2{,}340 \\ 685 \\ +\ 31{,}569 \\ \hline \end{array}$$

**4.**
$$\begin{array}{r} 75{,}363 \\ -\ 26{,}475 \\ \hline \end{array}$$

**5.** $83 \times 61$

**6.** $231 \times 305$

**7.** $21\overline{)357}$

**8.** $463\overline{)16{,}216}$

**9.** Evaluate the expression $18 \div 2 + 4 \times 2^3 - (18 - 6)$.

**10.** Round each number to the nearest hundred and find an estimated sum.

$294 + 725 + 2{,}321 + 689$

**11.** Find the perimeter and area of the given figure.

7 ft

5 ft

**12.** Write the fraction $\frac{15}{51}$ in simplest form.

**Answers**

1. ______________________

2. ______________________

3. ______________________

4. ______________________

5. ______________________

6. ______________________

7. ______________________

8. ______________________

9. ______________________

10. ______________________

11. ______________________

12. ______________________

 The Streeter/Hutchison Series in Mathematics Basic Mathematical Skills with Geometry

Answers

13. ______
14. ______
15. ______
16. ______
17. ______
18. ______
19. ______
20. ______
21. ______
22. ______
23. ______
24. ______
25. ______
26. ______
27. ______
28. ______

*Perform the indicated operations.*

**13.** $\frac{2}{3} \times \frac{9}{8}$ **14.** $1\frac{2}{3} \times 1\frac{5}{7}$ **15.** $\frac{3}{4} \div \frac{17}{12}$

**16.** $\frac{6}{7} - \frac{3}{7} + \frac{2}{7}$ **17.** $\frac{4}{5} - \frac{7}{10} + \frac{4}{30}$ **18.** $6\frac{3}{5} - 2\frac{7}{10}$

*Perform the indicated operations.*

**19.** $35.218 - 22.75$ **20.** $2.262 \div 0.58$

**21.** $523.8 \div 10^5$ **22.** $2.53 \times 0.45$

**23.** $1.53 \times 10^4$ **24.** Write 0.43 as a fraction.

**25.** Write the decimal equivalent of each of the following. Round to the given place value when indicated.

(a) $\frac{5}{8}$ (b) $\frac{9}{23}$ hundredths

**26.** Sam has had 15 hits in his last 35 at bats. That is, he has had a hit in $\frac{15}{35}$ of his times at bat. Write this as a decimal, rounding to the nearest thousandth.

*Evaluate the following:*

**27.** $18.4 - 3.16 \times 2.5 + 6.71$

**28.** $17.6 \div 2.3 \times 3.4 + 13.812$ (Round to the nearest thousandth.)

CHAPTER

5

# Ratios and Proportions



## INTRODUCTION

Sandra is a community college math instructor who loves teaching math and helping her students learn difficult concepts. When she is not teaching her classes, she spends a lot of time working out in the gym.

One day, during a particularly grueling yoga class, one of the participants asked Sandra to show him the right way to do the mountain pose. While Sandra was demonstrating this pose, she realized that she could become a personal trainer.

Sandra did some research and learned that she needed some coursework and that by passing an exam she could obtain a certificate through the American Fitness Professionals & Associates (AFPA). Some of the classes that personal trainers can teach include senior fitness, nutrition, group fitness, strength training, weight loss, and rehabilitative training.

Sandra learned some interesting facts about burning calories and weight loss such as the fact that a person has to burn off a lot more calories than he or she takes in just to lose one pound. In Activity 15 on page 414, we use ratios to determine the number of calories burned by various activities.

## CHAPTER 5 OUTLINE

# 5 pretest



Name ____________

Section ______ Date ______

This pretest provides a preview of the types of exercises you will encounter in each section of this chapter. The answers for these exercises can be found in the back of the text. If you are working on your own, or ahead of the class, this pretest can help you identify the sections in which you should focus more of your time.

**Answers**

1. ____________
2. ____________
3. ____________
4. ____________
5. ____________
6. ____________
7. ____________
8. ____________
9. ____________
10. ____________

**5.1**

**1.** Write the ratio of 7 to 10.

**2.** Write the ratio of 20 to 15 in lowest terms.

**5.2** Find each rate.

**3.** Find the rate equivalent to $\frac{138 \text{ mi}}{4 \text{ h}}$.

**4.** Augie eats 12 hamburgers in 48 minutes. How many minutes does it take Augie to eat one hamburger?

**5.3**

**5.** Is $\frac{4}{7} = \frac{12}{21}$ a true proportion?

**6.** Is $\frac{5}{9} = \frac{9}{16}$ a true proportion?

**5.4**

**7.** Solve for $x$: $\frac{x}{4} = \frac{5}{2}$

**8.** Solve for $n$: $\frac{\frac{1}{2}}{2} = \frac{3}{n}$

**9.** **Business and Finance** Cans of tomato juice are marked 2 for $1.05. At this price, what do 12 cans cost?

**10.** **Crafts** If 2 gal of paint covers 450 ft$^2$, how many gallons are needed to paint a room with 2,475 ft$^2$ of wall surface?

# 5.1 Ratios

< 5.1 Objectives >

1 > Write the ratio of two quantities

2 > Write the ratio of two quantities in simplest form

In Chapter 2, you learned two meanings for a fraction:

1. A fraction can name a certain number of parts of a whole. The fraction $\frac{3}{5}$ names 3 parts of a whole that has been divided into 5 equal parts.
2. A fraction can indicate division. The fraction $\frac{3}{5}$ is the quotient $3 \div 5$.

We now want to turn to a third meaning for a fraction—a fraction can be a **ratio.**

**Definition**

**Ratio** — A **ratio** is a comparison of two numbers or like quantities.

**NOTE**

In this text, we write ratios as simplified fractions.

The ratio $a$ to $b$ can also be written as $a : b$ and $\frac{a}{b}$. Ratios are always written in simplest form.

## Example 1 Writing a Ratio as a Fraction

< Objective 1 >

**NOTE**

Alternatively, the ratio of 3 to 5 can be written as 3:5.

Write the ratio 3 to 5 as a fraction.

To compare 3 to 5, we write the ratio of 3 to 5 as $\frac{3}{5}$.

So, $\frac{3}{5}$ also means "the ratio of 3 to 5."

**Check Yourself 1**

Write the ratio of 7 to 12 as a fraction.

**RECALL**

Numbers with units attached are called *denominate numbers*.

Ratios are often used to compare *like quantities* such as quarts to quarts, centimeters to centimeters, and apples to apples. In this case, we can simplify the fraction by "canceling" the units. In its simplest form, a ratio is always written without units.

## Example 2 Ratios of Denominate Numbers

A rectangle measures 7 cm wide and 19 cm long.

7 cm

19 cm

**(a)** Write the ratio of its width to its length, as a fraction.

$$\frac{7\text{ cm}}{19\text{ cm}} = \frac{7\text{ \sout{cm}}}{19\text{ \sout{cm}}} = \frac{7}{19}$$

We are comparing centimeters to centimeters, so we can simplify the fraction.

**(b)** Write the ratio of its length to its width, as a fraction.

We need to write the ratio in the order requested by the example, rather than in the order given in the preceding description.

$$\frac{19\text{ cm}}{7\text{ cm}} = \frac{19\text{ \sout{cm}}}{7\text{ \sout{cm}}} = \frac{19}{7}$$

**NOTE**

Ratios are *never* written as mixed numbers. Ratios are always written as improper fractions, in simplest terms, when necessary.

### Check Yourself 2

A basketball team wins 17 of its 29 games in a season.

**(a)** Write the ratio of wins to games played.
**(b)** Write the ratio of wins to losses.

Because a ratio is a fraction, we can simplify it, as in Example 3.

## Example 3 Simplifying a Ratio

< Objective 2 >

Write the ratio of 20 to 30 in simplest terms.

Begin by writing the fraction that represents the ratio: $\frac{20}{30}$. Now, simplify this fraction.

$$\frac{20}{30} = \frac{2}{3}$$

Simplify the fraction by dividing both the numerator and denominator by 10.

**RECALL**

This ratio may also be written as 2:3.

### Check Yourself 3

Write the ratio of 24 to 32 in simplest terms.

Because ratios are used to compare like quantities, a simplified ratio has no units. In Example 4, we need to simplify both the numbers and the units.

## Example 4 Simplifying the Ratio of Denominate Numbers

A common size for a movie screen is 32 ft by 18 ft. Write this ratio in simplest form.

**NOTE**

On DVDs, widescreen movies are often in 16:9 format.

$$\frac{32 \text{ ft}}{18 \text{ ft}} = \frac{32 \cancel{\text{ft}}}{18 \cancel{\text{ft}}} = \frac{32}{18} = \frac{16}{9}$$ The GCF of 32 and 18 is 2.

### Check Yourself 4

A common computer display mode is 640 pixels (picture elements) by 480 pixels. Write this as a ratio in simplest terms.

Often, the quantities in a ratio are given as fractions or decimals. In either of these cases, the ratio should be rewritten as an equivalent ratio comparing whole numbers.

## Example 5 Simplifying Ratios

**(a)** Loren sank a $22\frac{1}{2}$-ft putt and Carrie sank a 30-ft putt. Express the ratio of the two distances as a ratio of whole numbers.

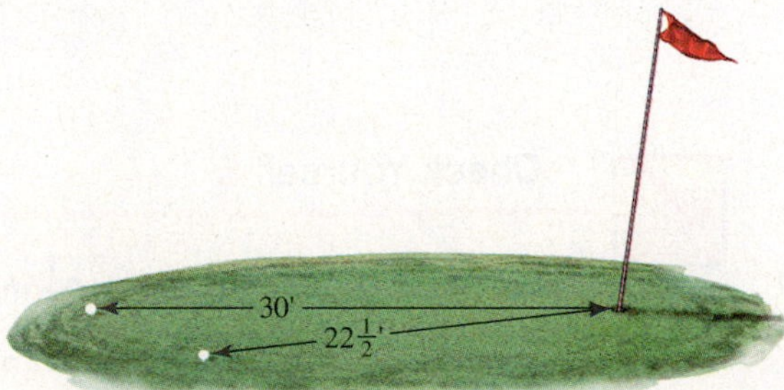


**RECALL**

In Section 2.3, you learned to rewrite $22\frac{1}{2}$ as $\frac{45}{2}$.

We begin by writing the ratio of the two distances: $\dfrac{22\frac{1}{2} \text{ ft}}{30 \text{ ft}}$. Then, we cancel and rewrite the mixed number as an improper fraction.

$$\frac{22\frac{1}{2} \text{ ft}}{30 \text{ ft}} = \frac{\frac{45}{2}}{30}$$

**RECALL**

In Section 2.6, you learned to simplify complex fractions.

In order to simplify this complex fraction, we rewrite it as a division problem.

$$\frac{\frac{45}{2}}{30} = \frac{45}{2} \div 30$$

$$= \frac{45}{2} \times \frac{1}{30} \quad \text{Invert, and multiply.}$$

$$= \frac{\overset{3}{\cancel{45}} \times 1}{2 \times \underset{2}{\cancel{30}}} \quad \text{The GCF of 45 and 30 is 15.}$$

$$= \frac{3}{4}$$

The ratio 3 to 4 is equivalent to the ratio $22\frac{1}{2}$ ft to 30 ft.

**(b)** The diameter of a 20-oz bottle is 2.8 in. The diameter of a 2-L bottle is 5.25 in. Express the ratio of the two diameters as a ratio of whole numbers.

$$\frac{2.8 \text{ in.}}{5.25 \text{ in.}} = \frac{2.8}{5.25}$$

**RECALL**

In Section 4.4, you learned to multiply a decimal by a power of 10 by moving the decimal point.

In order to simplify this fraction, we need to rewrite it as an equivalent fraction of whole numbers, that is, without the decimals.

If we multiply the numerator by 10, it would be a whole number. However, we need to multiply the denominator by 100 in order to make it a whole number. Because we want to write an equivalent fraction, we multiply it by $1 = \frac{100}{100}$.

$$\frac{2.8}{5.25} \cdot \frac{100}{100} = \frac{280}{525}$$

$$= \frac{8}{15} \quad \text{The GCF of 280 and 525 is 35.}$$

The ratio of the bottle diameters is 8 to 15.

**Check Yourself 5**

**(a)** One morning Rita jogged $3\frac{1}{2}$ mi, while Yi jogged $4\frac{1}{4}$ mi. Express the ratio of the two distances as a ratio of whole numbers.

**(b)** A standard newspaper column is 2.625 in. wide and 19.5 in. long. Express the ratio of the two measurements as a ratio of whole numbers.

**NOTE**

We will learn to convert measurements in more depth in Chapter 7.

Sometimes, we use a ratio to compare the same type of measurement using different units. In Example 6, both quantities are measures of time. In order to construct and simplify the ratio, we must express both quantities in the same units.

## Example 6 Rewriting Denominate Numbers to Find a Ratio

Joe took 2 hours (h) to complete his final exam. Jamie finished hers in 75 minutes (min). Write the ratio of the two times in simplest terms.

To find the ratio, both quantities must have the same units. Therefore, we rewrite 2 h as 120 min. This way, both quantities use minutes as their unit.

$$\frac{2\text{ h}}{75\text{ min}} = \frac{120\text{ min}}{75\text{ min}} \qquad \text{2 h = 120 min.}$$

$$= \frac{120}{75} = \frac{8}{5} \qquad \text{The GCF of 120 and 75 is 15.}$$

### Check Yourself 6

Find the ratio of whole numbers that is equivalent to the ratio of 15 ft to 9 yd.

### Check Yourself ANSWERS

1. $\frac{7}{12}$ 2. (a) $\frac{17}{29}$; (b) $\frac{17}{12}$ 3. $\frac{3}{4}$ 4. $\frac{4}{3}$ 5. (a) $\frac{14}{17}$; (b) $\frac{7}{52}$ 6. $\frac{5}{9}$

### Reading Your Text

The following fill-in-the-blank exercises are designed to ensure that you understand some of the key vocabulary used in this section.

SECTION 5.1

(a) A ______________ can indicate division.

(b) A ratio is a means of comparing two ______________ quantities.

(c) Because a ratio is a fraction, we can write it in ______________ terms.

(d) Ratios are never written as ______________ numbers.

# 5.1 exercises

Basic Skills | Advanced Skills | Vocational-Technical Applications | Calculator/Computer | Above and Beyond

Boost your grade at mathzone.com!

> Practice Problems
> NetTutor
> Self-Tests
> e-Professors
> Videos

Name ____________

Section ________ Date ________

**Answers**

1. ____
2. ____
3. ____
4. ____
5. ____
6. ____
7. ____
8. ____ 9. ____
10. ____ 11. ____
12. ____ 13. ____
14. ____ 15. ____
16. ____ 17. ____
18. ____ 19. ____
20. ____ 21. ____

< Objectives 1–2 >

*Write each of the following ratios in simplest form.*

**1.** The ratio of 9 to 13

**2.** The ratio of 5 to 4

**3.** The ratio of 9 to 4

**4.** The ratio of 5 to 12

**5.** The ratio of 10 to 15

**6.** The ratio of 16 to 12

**7.** The ratio of $3\frac{1}{2}$ to 14

**8.** The ratio of $5\frac{3}{5}$ to $2\frac{1}{10}$

**9.** The ratio of 10.5 to 2.7

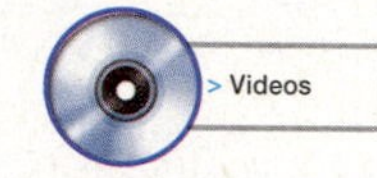


**10.** The ratio of 2.2 to 0.6

**11.** The ratio of 12 miles (mi) to 18 mi

**12.** The ratio of 100 centimeters (cm) to 90 cm

**13.** The ratio of 40 ft to 65 ft

**14.** The ratio of 12 oz to 18 oz

**15.** The ratio of \$48 to \$42

**16.** The ratio of 20 ft to 24 ft

*Solve the following applications.*

**17.** **Social Science** An algebra class has 7 men and 13 women. Write the ratio of men to women. Write the ratio of women to men.

**18.** **Statistics** A football team wins 9 of its 16 games with no ties. Write the ratio of wins to games played. Write the ratio of wins to losses.

**19.** **Social Science** In a school election 4,500 yes votes were cast, and 3,000 no votes were cast. Write the ratio of yes to no votes.

> Videos

**20.** **Business and Finance** One car has an $11\frac{1}{2}$-gal tank and another has a $17\frac{3}{4}$-gal tank. Express the ratio of the capacities as a ratio of whole numbers.

**21.** **Business and Finance** One refrigerator holds $2\frac{2}{3}$ cubic feet ($ft^3$) of food, and another holds $5\frac{3}{4}$ $ft^3$ of food. Express the ratio of the capacities as a ratio of whole numbers.

> Videos

22. **Science and Medicine** The price of an antibiotic in one drugstore is $12.50 although the price of the same antibiotic in another drugstore is $8.75. Write the ratio of the prices as a ratio of whole numbers.

23. **Social Science** A company employs 24 women and 18 men. Write the ratio of men to women employed by the company.

24. **Geometry** If a room is 30 ft long by 6 yd wide, write the ratio of the length to the width of the room.

*Determine whether each statement is* **true** *or* **false.**

25. We use ratios to compare like quantities.

26. We use ratios to compare different types of measurements.

*Fill in each blank with* **always, sometimes,** *or* **never.**

27. A ratio should __________ be written in simplest form.

28. A ratio is __________ used to compare numbers.

*Write each of the following ratios in simplest form.*

29. The ratio of 75 seconds (s) to 3 minutes (min)

30. The ratio of 7 oz to 3 lb

31. The ratio of 4 nickels to 5 dimes

32. The ratio of 8 in. to 3 ft

33. The ratio of 2 days to 10 h

34. The ratio of 4 ft to 4 yd

35. The ratio of 5 gallons (gal) to 12 quarts (qt)

36. The ratio of 7 dimes to 3 quarters

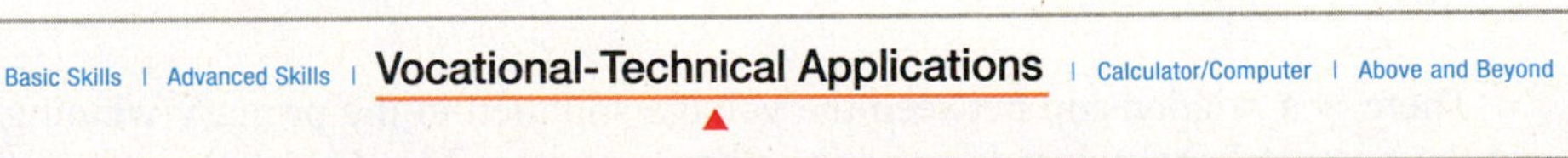


37. **Information Technology** Millicent can fix 5 cell phones per hour. Tyler can fix 4 cell phones per hour. Express the ratio of the number of cell phones that Millicent can fix to the number that Tyler can fix as a fraction.

38. **Allied Health** In preparing specimens for testing, it is often necessary to dilute the original solution. The dilution ratio is the ratio of the volume of original solution to the total volume. Each of these volumes is usually measured in milliliters (mL).

Determine the dilution ratio when 2.8 mL of blood serum is diluted with 47.2 mL of water.

**Answers**

22. __________
23. __________
24. __________
25. __________
26. __________
27. __________
28. __________
29. __________
30. __________
31. __________
32. __________
33. __________
34. __________
35. __________
36. __________
37. __________
38. __________

Answers

39. ______

40. ______

41. ______

42. ______

39. **Mechanical Engineering** A gear ratio is the ratio of the number of teeth on the driven gear to the number of teeth on the driving gear. In general, the driving gear is attached to the power source or motor. Write the gear ratio for the system shown.

40. **Manufacturing Operations Technology** Of the 384 parts manufactured during a shift, 26 were defective.

(a) Write the ratio of defective parts to total parts.

(b) Write the ratio of defective parts to good parts.

41. **Agricultural Technology** A soil test indicates that a field requires a fertilizer containing 400 lb of nitrogen and 500 lb of phosphorus. Write the ratio of nitrogen to phosphorus needed.

42. **Information Technology** Shakira connects to the Internet from home using a phone line. Her modem generally connects at a speed of 56,000 bits per second (bits/s). Her brother, Carl, connects with an older computer at 28,800 bits/s. Write the ratio of Shakira's connection speed to Carl's speed.

Basic Skills | Advanced Skills | Vocational-Technical Applications | Calculator/Computer | **Above and Beyond**

The accompanying image is a common symbol on a schematic (electrical diagram) for a transformer. A transformer uses electromagnetism to change voltage levels. Commonly, two coils of wire (or some conductor) are located in close proximity but kept from directly touching or conducting. Some sort of ferromagnetic core (such as iron) is typically used. When alternating current (AC) is applied to one conductor or coil, referred to as the primary winding, current is induced on the second or secondary winding.

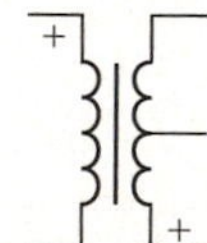


There is a relationship between the voltage supplied to the primary winding and the open-voltage induced in the secondary winding, based on the number of turns in each winding. This relationship is called the turns ratio ($a$):

$$a = \frac{N_p}{N_s}$$

in which $N_p$ represents the number of turns in the primary winding and $N_s$ represents the number of turns in the secondary winding.

Theoretically, the turns ratio is also equal to the voltage ratio:

$$a = \frac{V_p}{V_s}$$

in which $V_p$ represents the voltage supplied to the primary winding and $V_s$ represents the voltage induced in the secondary winding.

After setting the two ratios equal to one another and performing a multiplication manipulation to isolate $V_p$, this relationship can be expressed as

$$V_p = \frac{N_P}{N_s} V_s$$

**43.** Give three combinations of turns of the primary and secondary windings that achieve a turns ratio for a transformer of 3.2.

**44.** Using a turns ratio of 3.2 and a secondary voltage of 35 volts (V) AC, calculate the voltage supplied to the primary winding.

**45.** If the turns ratio is 4.5 and the primary voltage is 28 V AC, what is the induced open-voltage on the secondary winding?

**46.** If the turns ratio from exercise 45 is doubled, how will that affect the voltage on the secondary winding?

**47.** **(a)** Buy a 1.69-oz (medium-size) bag of M&M's. For each color, determine the ratio of M&M's that are that color to the total number of M&M's in the bag.

**(b)** Compare your ratios from part **(a)** to those of a classmate.

**(c)** Use the information from parts **(a)** and **(b)** to estimate the correct ratios for all the different color M&M's in a bag.

**(d)** Go to the M&M's manufacturer's website (Mars, Inc.) and see how your ratios compare to their claimed color distribution.

**48.** Sarah is a field service technician for ABC Networks, Inc. She has been asked to design a wireless home-network for a customer. This customer wants to have the fastest throughput for the wireless network in his home.

From your experience, you know that wireless networks come in different varieties: 802.11a, b, and g. The standard for most coffeehouses and restaurants is 802.11b.

If it takes 1 second to transmit a packet on 802.11b, how long does it take to transmit a packet on 802.11a and g?

Which technology will you recommend and why?

## Answers

**1.** $\frac{9}{13}$ **3.** $\frac{9}{4}$ **5.** $\frac{2}{3}$ **7.** $\frac{1}{4}$ **9.** $\frac{35}{9}$ **11.** $\frac{2}{3}$ **13.** $\frac{8}{13}$ **15.** $\frac{8}{7}$

**17.** $\frac{7}{13}; \frac{13}{7}$ **19.** $\frac{3}{2}$ **21.** $\frac{32}{69}$ **23.** $\frac{3}{4}$ **25.** True **27.** always

**29.** $\frac{5}{12}$ **31.** $\frac{2}{5}$ **33.** $\frac{24}{5}$ **35.** $\frac{5}{3}$ **37.** $\frac{5}{4}$ **39.** $\frac{5}{8}$ **41.** $\frac{4}{5}$

**43.** Answers will vary **45.** $6\frac{2}{9}$ V AC **47.** Above and Beyond

## Answers

43. ______

44. ______

45. ______

46. ______

47. ______

48. ______

# Activity 13 :: Working with Ratios Visually

To solve the following problems, we think that you and your group members will find one of these three approaches useful: (1) You may wish to use actual black and white markers; (2) you may wish to make sketches of such markers; or (3) you may wish to simply imagine the necessary markers. Each line in the following table is a new (and different) problem, and you are to fill in the missing parts in a given line. Be sure to express a ratio by using the smallest possible whole numbers.

| | Ratio of Black to White Markers | Number of Black Markers | Number of White Markers | Total Number of Markers |
|---|---|---|---|---|
| 1. | to | | 15 | 20 |
| 2. | to | 12 | | 30 |
| 3. | to | | 9 | 21 |
| 4. | to | 15 | | 33 |
| 5. | 2 to 5 | 6 | | |
| 6. | 5 to 3 | | 24 | |
| 7. | 4 to 1 | 28 | | |
| 8. | 3 to 7 | | 21 | |
| 9. | 1 to 3 | | | 36 |
| 10. | 3 to 5 | | | 40 |
| 11. | 7 to 2 | | | 360 |
| 12. | 4 to 7 | | | 550 |
| 13. | | | | |
| 14. | | | | |

For problems 13 and 14, create (and solve!) your own problems of the same sort. You might challenge another group with these.

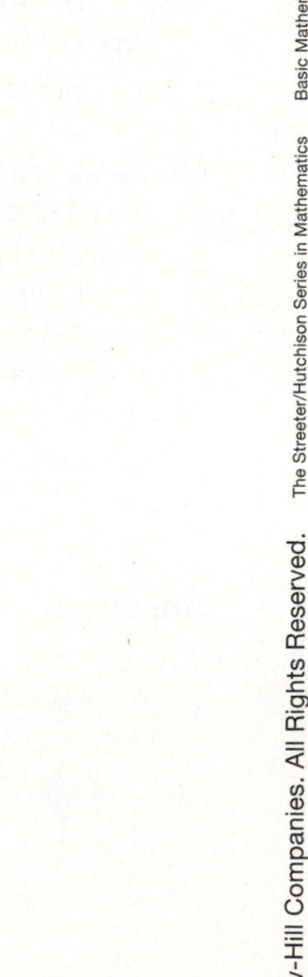

# 5.2 Rates and Unit Pricing

< 5.2 Objectives >

1 > Write a rate as a unit rate

2 > Interpret and compare unit rates

3 > Find a unit price

4 > Use unit prices to compare the cost of two items

In Section 5.1, we used ratios to compare two like quantities. For instance, the ratio of 9 seconds to 12 seconds is $\frac{3}{4}$.

$$\frac{9 \cancel{\text{sec}}}{12 \cancel{\text{sec}}} = \frac{\overset{3}{\cancel{9}}}{\underset{4}{\cancel{12}}} = \frac{3}{4}$$

**RECALL**

When simplified, a ratio has no units.

Because the units in the numerator and denominator are the same, we can "cancel" them and simplify the fraction.

We also learned that as long as the two quantities represented the same type of measurement, we could compare them using ratios. For example, if both quantities are measurements of length, then we can convert one of the measurements so that they are like quantities.

In Chapter 7, you will study measurement conversions in more depth. For now, we can do straightforward conversions. For example, we can use a ratio to compare 4 in. and 3 ft by converting 3 ft to 36 in.

$$\frac{4 \text{ in.}}{3 \text{ ft}} = \frac{4 \text{ in.}}{36 \text{ in.}} \qquad 3 \text{ ft} = 36 \text{ in.}$$

$$= \frac{4}{36}$$

$$= \frac{1}{9}$$

**RECALL**

*Denominate numbers* have units "attached."

Often, we want to compare denominate numbers with different types of units. For example, we might be interested in the gas mileage that a car gets. In such a case, we are comparing the miles driven (distance) and the gas used (volume). We make this comparison in part **(b)** of Example 1.

When we compare denominate numbers with different types of units, we get a *rate.*

**Definition**

| **Rate** | A rate is a comparison of two denominate numbers with different types of units. |
|---|---|

In general, rates are presented in simplified form as *unit rates.*

**Definition**

**Unit Rate**

A **unit rate** is a *rate* that is simplified so that the numerical value is given followed by the units, written as a fraction.

**NOTE**

We read the rate $30\frac{\text{mi}}{\text{gal}}$ as "thirty miles *per* gallon."

When reading a rate, the fraction bar is read, "per." In Example 1, we write each of the given rates as unit rates.

## Example 1 Finding a Unit Rate

< Objective 1 >

Simplify each rate.

**(a)** $\frac{12 \text{ feet}}{16 \text{ seconds}} = \frac{12}{16}\frac{\text{ft}}{\text{s}} = \frac{3}{4}\frac{\text{ft}}{\text{s}}$

We read this as "$\frac{3}{4}$ feet per second."

**(b)** $\frac{200 \text{ miles}}{10 \text{ gallons}} = \frac{200}{10}\frac{\text{mi}}{\text{gal}} = 20\frac{\text{mi}}{\text{gal}}$

**(c)** $\frac{10 \text{ gallons}}{200 \text{ miles}} = \frac{10}{200}\frac{\text{gal}}{\text{mi}} = \frac{1}{20}\frac{\text{gal}}{\text{mi}}$

### Check Yourself 1

Simplify each rate.

(a) $\frac{250 \text{ miles}}{10 \text{ hours}}$ (b) $\frac{\$60{,}000}{2 \text{ years}}$ (c) $\frac{2 \text{ years}}{\$60{,}000}$

Consider part **(a)** in Example 1. We begin with 12 ft (length) compared to 16 seconds (time). We simplify the rate so that we know the number of feet, $\frac{3}{4}$, per 1 second. In general, we simplify a rate so that we are comparing the quantity of the numerator's units per one of the denominator's units.

In Example 2, we consider parts **(b)** and **(c)** of Example 1.

## Example 2 Comparing Unit Rates

< Objective 2 >

**NOTE**

We could also write this rate as a decimal,

$\frac{1}{20}\frac{\text{gal}}{\text{mi}} = 0.05\frac{\text{gal}}{\text{mi}}$

**(a)** Write the rate $20\frac{\text{mi}}{\text{gal}}$ in words.

We write this rate as "twenty miles per gallon."

**(b)** Write the rate $\frac{1}{20}\frac{\text{gal}}{\text{mi}}$ in words.

We write this rate as "one-twentieth of a gallon per mile."

**(c)** Describe the difference between the rates in parts **(a)** and **(b).**

The rate $20\frac{\text{mi}}{\text{gal}}$ states that 20 miles can be traveled on a single gallon of fuel.

The rate $\frac{1}{20}\frac{\text{gal}}{\text{mi}}$ states that $\frac{1}{20}$ of a gallon of fuel is used when traveling 1 mile.

We can also interpret these as 20 miles are traveled for each gallon of fuel used and $\frac{1}{20}$ of a gallon of fuel is used for each mile traveled, respectively.

### Check Yourself 2

**(a)** Write the rate $30{,}000\frac{\text{dollars}}{\text{yr}}$ in words.

**(b)** Write the rate $\frac{1}{30{,}000}\frac{\text{yr}}{\text{dollar}}$ in words.

**(c)** Describe the difference between the rates in parts **(a)** and **(b)**.

Sometimes, we need to find the appropriate rate within a written statement, as in Example 3.

## Example 3 Finding a Unit Rate

**RECALL**

In Section 5.1 we stated that mixed numbers were inappropriate for ratios. When we write unit rates, mixed numbers and decimals are not just appropriate, they are preferred.

Randy Johnson had 320 strikeouts in 280 innings. What was his strikeout per inning rate?

$$\frac{320 \text{ strikeouts}}{280 \text{ innings}} = \frac{320}{280}\frac{\text{strikeouts}}{\text{inning}}$$

$$= 1\frac{1}{7}\frac{\text{strikeouts}}{\text{inning}}$$

### Check Yourself 3

**Chamique Holdsclaw scored 450 points in 15 games. What was her points per game rate?**

One purpose for computing a rate is for comparison. As we see in Example 4, it is often convenient to write a rate in decimal form.

### Example 4 Comparing Rates

Player A scores 50 points in 9 games and player B scores 260 points in 36 games. Which player scored at a higher rate?

$$\text{Player A's rate was } \frac{50 \text{ points}}{9 \text{ games}} = \frac{50}{9}\frac{\text{points}}{\text{game}}$$

$$\approx 5.56 \frac{\text{points}}{\text{game}} \qquad \frac{50}{9} = 50 \div 9 \approx 5.56$$

$$\text{Player B's rate was } \frac{260 \text{ points}}{36 \text{ games}} = \frac{260}{36}\frac{\text{points}}{\text{game}}$$

$$= \frac{65}{9}\frac{\text{points}}{\text{game}}$$

$$\approx 7.22 \frac{\text{points}}{\text{game}} \qquad 65 \div 9 \approx 7.22$$

Player B scored at a higher rate.

**NOTE**

A unit price is a price *per unit.* The unit used may be ounces, pints, pounds, or some other unit.

### Check Yourself 4

**Hassan scored 25 goals in 8 games and Lee scored 52 goals in 18 games. Which player scored at a higher rate?**

**Unit pricing** represents one of the most common uses of rates. Posted on nearly every item in a supermarket or grocery store is the price of the item as well as its *unit price.*

**Definition**

**Unit Price** — The **unit price** relates a price to some common unit.

### Example 5 Finding a Unit Price

< Objective 3 >

Find the unit price for each item.

**(a)** 8 ounces (oz) of cream cost \$1.53.

$$\frac{\$1.53}{8 \text{ oz}} = \frac{153 \text{ cents}}{8 \text{ oz}} = \frac{153}{8}\frac{\text{cents}}{\text{oz}}$$

$$\approx 19 \frac{\text{cents}}{\text{oz}}$$

**(b)** 20 pounds (lb) of potatoes cost $3.98.

$$\frac{\$3.98}{20 \text{ lb}} = \frac{398 \text{ cents}}{20 \text{ lb}}$$

$$= \frac{398}{20} \frac{\text{cents}}{\text{lb}}$$

$$\approx 20 \frac{\text{cents}}{\text{lb}}$$

**Check Yourself 5**

**Find the unit price for each item.**

**(a)** 12 soda cans cost $2.98.
**(b)** 25 pounds (lb) of dog food cost $9.99.

Similarly to ratios, rates are most often used for comparisons. For instance, unit pricing allows people to compare the cost of different size items.

In Example 6, we use unit prices to determine whether a glass of milk is less expensive when poured from a 128-oz container (gallon) or a 32-oz container (quart).

**Example 6** Using Unit Prices to Compare Cost

< Objective 4 >

A store sells a 1-gallon carton (128 oz) of organic whole milk for $4.89. They sell a 1-quart carton (32 oz) for $1.29. Which is the better buy?

We begin by determining the unit price of each item. To do this, we compute the cost per ounce for each carton of milk. We choose to use cents instead of dollars to make the decimal easier to work with.

*Gallon*

$$\frac{\$4.89}{128 \text{ oz}} = \frac{489 \text{ cents}}{128 \text{ oz}} = \frac{489}{128} \frac{\text{cents}}{\text{oz}} \approx 3.8203 \frac{\text{cents}}{\text{oz}}$$

*Quart*

$$\frac{\$1.29}{32 \text{ oz}} = \frac{129 \text{ cents}}{32 \text{ oz}} = \frac{129}{32} \frac{\text{cents}}{\text{oz}} \approx 4.0313 \frac{\text{cents}}{\text{oz}}$$

At these prices, the gallon of milk is the better buy.

**NOTE**

Usually, we round money to the nearest cent. When comparing unit prices, however, we may round to four decimal places (or more, if necessary).

## Check Yourself 6

**A store sells a 5-lb bag of Valencia oranges for $2.29. A 12-lb case sells for $5.69. Which is the better buy?**

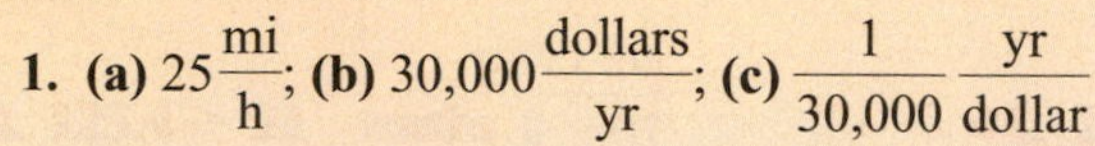

If we compare the two unit prices in Example 6, we see that both items round to 4¢ per ounce. However, milk sold by the gallon is about 0.211¢ cheaper per ounce than milk sold by the quart. We need to consider this small fraction of a cent because we are not buying 1 ounce of milk. Rather, we are buying cartons of milk, and for a whole quart of milk, the fraction adds up to nearly 7¢ (it adds up to 27¢ for a whole gallon).

## Check Yourself ANSWERS

1. **(a)** $25 \frac{\text{mi}}{\text{h}}$; **(b)** $30{,}000 \frac{\text{dollars}}{\text{yr}}$; **(c)** $\frac{1}{30{,}000} \frac{\text{yr}}{\text{dollar}}$
2. **(a)** Thirty-thousand dollars per year. **(b)** One thirty-thousandth of a year per dollar. **(c)** The rate in part **(a)** describes the amount of money for each year. The rate in part **(b)** describes the amount of time per dollar.
3. $30 \frac{\text{points}}{\text{game}}$
4. Hassan had a higher rate.
5. **(a)** $\approx 25 \frac{\text{cents}}{\text{can}}$; **(b)** $\approx 40 \frac{\text{cents}}{\text{pound}}$
6. The 5-lb bag is the better buy at 45.8¢ per pound compared to approximately 47.4¢ per pound for the case.

## Reading Your Text

The following fill-in-the-blank exercises are designed to ensure that you understand some of the key vocabulary used in this section.

### SECTION 5.2

**(a)** Ratios are used to compare ______________ quantities.

**(b)** When we compare measurements with different types of units, we get a ______________.

**(c)** ______________ numbers are preferable to improper fractions, when simplifying a rate.

**(d)** ______________ prices are used to compare the cost of items in different size packages.

Basic Skills | Advanced Skills | Vocational-Technical Applications | Calculator/Computer | Above and Beyond

# 5.2 exercises

MathZone

Boost your grade at mathzone.com!
> Practice Problems
> NetTutor
> Self-Tests
> e-Professors
> Videos

Name ______ Section ______ Date ______

## Answers

1. ______
2. ______
3. ______
4. ______
5. ______ 6. ______
7. ______
8. ______
9. ______
10. ______
11. ______
12. ______
13. ______
14. ______
15. ______ 16. ______
17. ______ 18. ______

< Objective 1 >

*Write each rate as a unit rate.*

1. $\frac{300 \text{ mi}}{4 \text{ h}}$

2. $\frac{95 \text{ cents}}{5 \text{ pencils}}$

3. $\frac{\$10{,}000}{5 \text{ yr}}$

4. $\frac{680 \text{ ft}}{17 \text{ s}}$

5. $\frac{7{,}200 \text{ revolutions}}{16 \text{ mi}}$

6. $\frac{57 \text{ oz}}{3 \text{ cans}}$

7. $\frac{\$2{,}000{,}000}{4 \text{ yr}}$

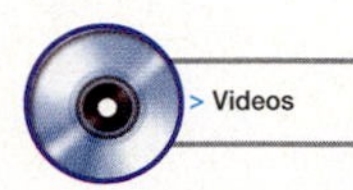


8. $\frac{150 \text{ cal}}{3 \text{ oz}}$

9. $\frac{240 \text{ lb of fertilizer}}{6 \text{ lawns}}$

10. $\frac{192 \text{ diapers}}{32 \text{ babies}}$

< Objective 2 >

*Write each rate in words.*

11. $120{,}000 \frac{\text{dollars}}{\text{yr}}$

12. $32 \frac{\text{ft}}{\text{s}}$

13. $\frac{1}{120{,}000} \frac{\text{yr}}{\text{dollar}}$

14. $\frac{1}{32} \frac{\text{s}}{\text{ft}}$

< Objective 3 >

*Find the unit price of each item.*

15. \$57.50 for 5 shirts

16. \$104.93 for 7 CDs

17. \$5.16 for a dozen oranges

18. \$10.44 for 18 bottles of water

## Answers

19. ______

20. ______

21. ______

22. ______

23. ______

24. ______

< Objective 4 >

*Find the best buy in each of the following exercises.*

**19.** Dishwashing liquid:

**(a)** 12 fl oz for $1.58
**(b)** 22 fl oz for $2.58

**20.** Canned corn:

**(a)** 10 oz for 42¢
**(b)** 17 oz for 78¢

**21.** Syrup:

**(a)** 12 oz for $1.98
**(b)** 24 oz for $3.18
**(c)** 36 oz for $4.38

**22.** Shampoo:

**(a)** 4 oz for $2.32
**(b)** 7 oz for $3.04
**(c)** 15 oz for $6.78

**23.** Salad oil (1 qt is 32 oz):

**(a)** 18 oz for $1.78
**(b)** 1 qt for $2.78
**(c)** 1 qt 16 oz for $4.38

**24.** Tomato juice (1 pt is 16 oz):

**(a)** 8 oz for 74¢
**(b)** 1 pt 10 oz for $2.38
**(c)** 1 qt 14 oz for $3.98

25. Peanut butter (1 lb is 16 oz):

**(a)** 12 oz for \$2.50
**(b)** 18 oz for \$3.44
**(c)** 1 lb 12 oz for \$5.08
**(d)** 2 lb 8 oz for \$7.52

26. Laundry detergent:

**(a)** 1 lb 2 oz for \$3.98
**(b)** 1 lb 12 oz for \$5.78
**(c)** 2 lb 8 oz for \$8.38
**(d)** 5 lb for \$15.98

*Solve each of the following applications.*

27. Trac uses 8 gallons of gasoline on a 256-mile drive. How many miles per gallon does his car get?

28. Seven pounds of fertilizer covers 1,400 square feet. How many square feet are covered by 1 pound of fertilizer?

29. A local college has 6,000 registered vehicles for 2,400 campus parking spaces. How many vehicles are there for each parking space? > Videos

30. A water pump can produce 280 gallons in 24 hours. How many gallons per hour is this?

31. The sum of \$5,992 was spent for 214 shares of stock. What was the cost per share?

32. A printer produces 4 pages of print in 6 seconds. How many pages are produced per second?

33. A 12-ounce can of tuna costs \$4.80. What is the cost of tuna per ounce?

34. The fabric for a dress costs \$76.45 for 9 yards. What is the cost per yard?

**Answers**

25. ____
26. ____
27. ____
28. ____
29. ____
30. ____
31. ____
32. ____
33. ____
34. ____

**Answers**

35. ______

36. ______

37. ______

38. ______

39. ______

40. ______

41. ______

42. ______

43. ______

44. ______

45. ______

46. ______

47. ______

48. ______

**35.** Gerry laid 634 bricks in 35 minutes and his friend Matt laid 515 bricks in 27 minutes. Who is the faster bricklayer?

*Solve each of the following applications.*

**36.** Mike drove 135 miles (mi) in 2.5 hours (h). Sam drove 91 mi in 1.75 h. Who drove faster?

**37.** Luis Gonzalez has 137 hits in 387 at bats. Larry Walker has 119 hits in 324 at bats. Who has the higher batting average?

**38.** Which is the better buy: 5 lb of sugar for \$4.75 or 20 lb of sugar for \$19.92?

*Determine whether each statement is* **true** *or* **false.**

**39.** We use rates to compare like quantities.

**40.** We use rates to compare different types of measurements.

*Fill in each blank with* **always, sometimes,** *or* **never.**

**41.** The units __________ cancel in a rate.

**42.** The units __________ cancel in a ratio.

Basic Skills | **Advanced Skills** | Vocational-Technical Applications | Calculator/Computer | Above and Beyond

*Find each rate.*

**43.** $\frac{69 \text{ ft}}{3 \text{ s}}$

**44.** $\frac{3 \text{ s}}{69 \text{ ft}}$

**45.** $\frac{5 \text{ yr}}{\$10{,}000}$

**46.** $\frac{480 \text{ mi}}{15 \text{ gal}}$

**47.** $\frac{15 \text{ gal}}{480 \text{ mi}}$

**48.** $\frac{657{,}200 \text{ library books}}{5{,}200 \text{ students}}$

Basic Skills | Advanced Skills | **Vocational-Technical Applications** | Calculator/Computer | Above and Beyond

**49. MECHANICAL ENGINEERING** The pitch of a gear is given by the quotient of the number of teeth on the gear and the diameter of the gear (distance from end to end, through the center). Calculate the pitch of the gear shown.

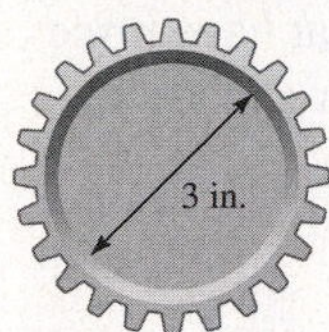


**50. ALLIED HEALTH** A patient's tidal volume, in milliliters (mL) per breath, is the quotient of his or her minute volume (mL/min) and his or her respiratory rate (breaths/min).

Report the tidal volume for an adult, female patient whose minute volume is 7,500 mL/min if her respiratory rate is 12 breaths/min.

chapter 5 > Make the Connection

**51. BUSINESS AND FINANCE** Determine the unit price of a 1,000-ft cable that costs $99.99.

**52. ELECTRICAL ENGINEERING** A 20-volt (V) DC pulse is sent down a 4,000-meter (m) length of conductor (see the figure). Because of resistance, when the pulse reaches the other end, the voltmeter measures the voltage as 4 V. What is the rate of voltage drop per meter of conductor?

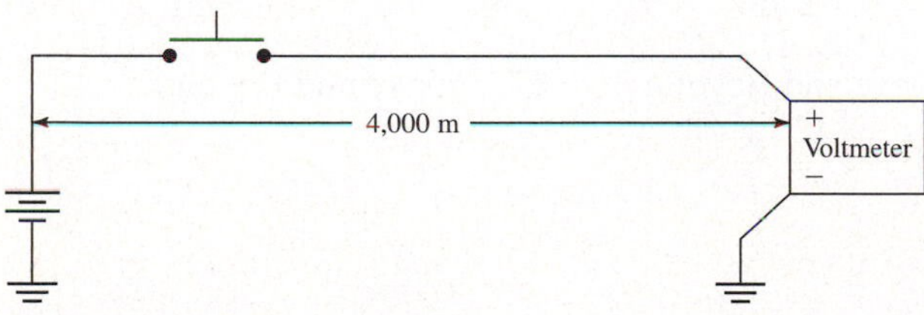


**53. BUSINESS AND FINANCE** A 200-bushel load of soybeans sells for $1,780. What is the price per bushel?

**54. MECHANICAL ENGINEERING** Stress is calculated as the rate of force applied compared to the cross-sectional area of a post. What is the stress on a post that supports 13,475 lb and has a cross-sectional area of 12.25 square inches?

Basic Skills | Advanced Skills | Vocational-Technical Applications | Calculator/Computer | **Above and Beyond**

**55.** Describe the difference between the rates $\frac{\$120{,}000}{\text{yr}}$ and $\frac{1}{120{,}000}\,\frac{\text{yr}}{\text{dollar}}$.

**56.** Describe the difference between the rates $32\,\frac{\text{ft}}{\text{s}}$ and $\frac{1}{32}\,\frac{\text{s}}{\text{ft}}$.

**Answers**

49. ______

50. ______

51. ______

52. ______

53. ______

54. ______

55. ______

56. ______

57. In your own words, explain the difference between a ratio and a rate.

58. Find several real-world examples of ratios and of rates.

59. Explain why unit pricing is useful.

60. Go to a supermarket or grocery store. Choose five items that have price and unit price listed. Check to see if the unit price given for each item is accurate.

## Answers

57. ______

58. ______

59. ______

60. ______

## Answers

**1.** $75\frac{\text{mi}}{\text{h}}$ **3.** $2{,}000\frac{\text{dollars}}{\text{yr}}$ **5.** $450\frac{\text{rev}}{\text{mi}}$ **7.** $500{,}000\frac{\text{dollars}}{\text{yr}}$

**9.** $40\frac{\text{lb}}{\text{lawn}}$ **11.** One hundred twenty thousand dollars per year

**13.** One one-hundred twenty thousandth of a year per dollar **15.** $\frac{\$11.50}{\text{shirt}}$

**17.** $\frac{\$0.43}{\text{orange}}$ **19.** (b) **21.** (c) **23.** (b) **25.** (c) **27.** $32\frac{\text{mi}}{\text{gal}}$

**29.** $2.5\frac{\text{vehicles}}{\text{space}}$ **31.** $28\frac{\text{dollars}}{\text{share}}$ **33.** $40\frac{\text{cents}}{\text{oz}}$ **35.** Matt

**37.** Larry Walker **39.** False **41.** never **43.** $23\frac{\text{ft}}{\text{s}}$

**45.** $0.0005\frac{\text{yr}}{\text{dollar}}$ **47.** $\frac{1}{32}\frac{\text{gal}}{\text{mi}}$ **49.** $4\frac{\text{teeth}}{\text{in.}}$ **51.** $\approx\frac{\$0.10}{\text{ft}}$

**53.** $\frac{\$8.90}{\text{bushel}}$ **55.** Above and Beyond **57.** Above and Beyond

**59.** Above and Beyond

# Activity 14 :: Baseball Statistics

There are many statistics in the sport of baseball that are expressed in decimal form. Two of these are batting average and earned run average. Both are actually examples of rates.

A batting average is a rate for which the units are "hits per at bat." To compute the batting average for a hitter, divide the number of hits (H) by the number of times at bat (AB). The result will be a decimal less than 1 (unless the batter always gets a hit!), and it is always expressed to the nearest thousandth. For example, if a hitter has 2 hits in 7 at bats, we divide 2 by 7, getting 0.285714. . . . The batting average is then rounded to 0.286.

Compute the batting average for each of the following major league players.

| | Player | Hits | At Bats | Average |
|---|---|---|---|---|
| 1 | Pujols | 139 | 373 | |
| 2 | Helton | 134 | 384 | |
| 3 | Guillen | 99 | 290 | |
| 4 | Suzuki | 142 | 419 | |
| 5 | Mora | 99 | 293 | |

The earned run average (ERA) for a pitcher is also a rate; its units are "earned runs per 9 innings." It represents the number of earned runs the pitcher gives up in 9 innings. To compute the ERA for a pitcher, multiply the number of earned runs allowed by the pitcher by 9, and then divide by the number of innings pitched. The result is always rounded to the nearest hundredth.

Compute the earned run average for each of the following major league players.

| | Player | Earned Runs | Innings | ERA |
|---|---|---|---|---|
| 6 | Loaiza | 33 | $137\frac{1}{3}$ | |
| 7 | Martinez | 27 | 110 | |
| 8 | Brown | 31 | $123\frac{1}{3}$ | |

*Challenge:* Suppose a hitter has 54 hits in 200 times at bat. How many hits in a row must the hitter get in order to raise his average to at least 0.300?

# 5.3 Proportions

< 5.3 Objectives >

1 > Write a proportion

2 > Determine whether two fractions are proportional

3 > Determine whether two rates are proportional

**Definition**

**Proportion** — A statement that two fractions or rates are equal is called a **proportion**.

**NOTES**

This is the same as saying the fractions are equivalent. They name the same number.

We call a letter representing an unknown value a *variable*. Here *a*, *b*, *c*, and *d* are variables. We could have chosen any other letter.

Because the ratio of 1 to 3 is equal to the ratio of 2 to 6, we can write the proportion

$$\frac{1}{3} = \frac{2}{6}$$

The proportion $\frac{a}{b} = \frac{c}{d}$ is read "*a* is to *b* as *c* is to *d*." We read the proportion $\frac{1}{3} = \frac{2}{6}$ as "one is to three as two is to six."

**Example 1 — Writing a Proportion**

< Objective 1 >

Write the proportion 3 is to 7 as 9 is to 21.

$$\frac{3}{7} = \frac{9}{21}$$

**Check Yourself 1**

Write the proportion 4 is to 12 as 6 is to 18.

When you write a proportion for two rates, placement of similar units is important.

## Example 2 Writing a Proportion with Two Rates

Write a proportion that is equivalent to the statement: If it takes 3 hours to mow 4 acres of grass, it will take 6 hours to mow 8 acres.

$$\frac{3 \text{ hours}}{4 \text{ acres}} = \frac{6 \text{ hours}}{8 \text{ acres}}$$

Note that, in both fractions, the hours units are in the numerator and the acres units are in the denominator.

### Check Yourself 2

**Write a proportion that is equivalent to the statement: If it takes 5 rolls of wallpaper to cover 400 square feet, it will take 7 rolls to cover 560 square feet.**

If two fractions form a true proportion, we say that they are **proportional.**

**Property**

### The Proportion Rule

If $\frac{a}{b} = \frac{c}{d}$, then $a \cdot d = b \cdot c$.

We say that the fractions $\frac{a}{b}$ and $\frac{c}{d}$ are proportional.

## Example 3 Determining Whether Two Fractions Are Proportional

< Objective 2 >

**NOTE**

Use the centered dot (·) for multiplication rather than the cross (×), so that the cross won't be confused with the letter *x*.

Determine whether each pair of fractions is proportional.

**(a)** $\frac{5}{6} \stackrel{?}{=} \frac{10}{12}$

Multiply:

$5 \cdot 12 = 60$
$6 \cdot 10 = 60$ Equal

Because $a \cdot d = b \cdot c$, $\frac{5}{6}$ and $\frac{10}{12}$ are proportional.

**(b)** $\frac{3}{7} \stackrel{?}{=} \frac{4}{9}$

Multiply:

$3 \cdot 9 = 27$
$7 \cdot 4 = 28$ Not equal

The products are not equal, so $\frac{3}{7}$ and $\frac{4}{9}$ are not proportional.

### Check Yourself 3

**Determine whether each pair of fractions is proportional.**

**(a)** $\dfrac{5}{8} \stackrel{?}{=} \dfrac{20}{32}$ **(b)** $\dfrac{7}{9} \stackrel{?}{=} \dfrac{3}{4}$

The proportion rule can also be used when fractions or decimals are involved.

## Example 4 Verifying a Proportion

Determine whether each pair of fractions is proportional.

**(a)** $\dfrac{3}{\frac{1}{2}} \stackrel{?}{=} \dfrac{30}{5}$

$$3 \cdot 5 = 15$$

$$\frac{1}{2} \cdot 30 = 15$$

Because the products are equal, the fractions are proportional.

**(b)** $\dfrac{0.4}{20} = \dfrac{3}{100}$

$$0.4 \cdot 100 = 40$$

$$20 \cdot 3 = 60$$

Because the products are *not* equal, the fractions are not proportional.

### Check Yourself 4

**Determine whether each pair of fractions is proportional.**

**(a)** $\dfrac{0.5}{8} \stackrel{?}{=} \dfrac{3}{48}$ **(b)** $\dfrac{\frac{1}{4}}{6} \stackrel{?}{=} \dfrac{3}{80}$

The proportion rule can also be used to verify that rates are proportional.

## Example 5 Determining Whether Two Rates are Proportional

< Objective 3 >

**NOTE**

Colones are the monetary unit of Costa Rica.

Is the rate $\dfrac{5 \text{ U.S. dollars}}{15{,}000 \text{ colones}}$ equivalent to the rate $\dfrac{27 \text{ U.S. dollars}}{81{,}000 \text{ colones}}$?

| US Dollars | Colones |
| --- | --- |
| $1.00 | 3000 |
| $0.00033 | 1.0 |

We want to know if the following is true.

$$\frac{5}{15{,}000} \stackrel{?}{=} \frac{27}{81{,}000}$$

$$5 \cdot 81{,}000 = 405{,}000$$
$$27 \cdot 15{,}000 = 405{,}000$$

The rates are equivalent.

**Check Yourself 5**

Is the rate $\frac{50 \text{ pages}}{45 \text{ minutes}}$ equivalent to the rate $\frac{30 \text{ pages}}{25 \text{ minutes}}$?

In Section 5.4, we will use proportions to solve many applications. For instance, if a 12-ft piece of steel stock weighs 27.6 lb, how much would a 25-ft piece weigh? Here, we check the accuracy of such a proportion.

## Example 6 An Application of Proportions

A 12-ft piece of steel stock weighs 27.6 lb. If 57.5 lb is the weight of a 25-ft piece, do the two pieces have the same density?

We check that the two rates are proportional.

> **RECALL**
>
> Be sure to align the units, regardless of the order in which the information appears in the problem.

$$\frac{12 \text{ ft}}{27.6 \text{ lb}} \stackrel{?}{=} \frac{25 \text{ ft}}{57.5 \text{ lb}}$$

$$12 \cdot 57.5 = 690$$
$$27.6 \cdot 25 = 690$$

Because the two products are equal, we have a true proportion, so the two pieces have the same density.

**Check Yourself 6**

One supplier sells a 200-lb lot of steel for $522.36. A second supplier charges $789.09 for a 300-lb lot of steel. Determine whether the suppliers are offering steel for the same price per pound.

**Check Yourself ANSWERS**

**1.** $\frac{4}{12} = \frac{6}{18}$ **2.** $\frac{5 \text{ rolls}}{400 \text{ square feet}} = \frac{7 \text{ rolls}}{560 \text{ square feet}}$ **3.** **(a)** Yes; **(b)** no
**4.** **(a)** Yes; **(b)** no **5.** No **6.** No

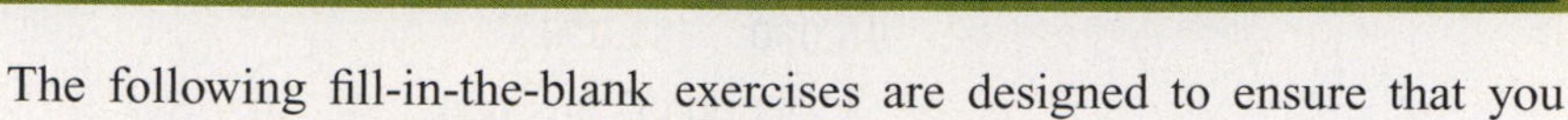

## Reading Your Text

The following fill-in-the-blank exercises are designed to ensure that you understand some of the key vocabulary used in this section.

### SECTION 5.3

**(a)** A statement that two rates are ______________ is called a proportion.

**(b)** A letter used to represent an unknown value is called a ______________.

**(c)** If two fractions form a true ______________, we say that they are proportional.

**(d)** When writing a proportion for two ______________, corresponding units must be similarly placed.

Basic Skills | Advanced Skills | Vocational-Technical Applications | Calculator/Computer | Above and Beyond

# 5.3 exercises

MathZone

Boost your grade at mathzone.com!
> Practice Problems
> NetTutor
> Self-Tests
> e-Professors
> Videos

Name ______________

Section ________ Date ________

< Objective 1 >

*Write each of the following as a proportion.*

**1.** 4 is to 9 as 8 is to 18.

**2.** 6 is to 11 as 18 is to 33.

**3.** 2 is to 9 as 8 is to 36.

**4.** 10 is to 15 as 20 is to 30.

**5.** 3 is to 5 as 15 is to 25.

**6.** 8 is to 11 as 16 is to 22.

**7.** 9 is to 13 as 27 is to 39.

**8.** 15 is to 21 as 60 is to 84.

< Objective 2 >

*Determine whether each pair of fractions is proportional.*

**9.** $\frac{3}{4} \stackrel{?}{=} \frac{9}{12}$

**10.** $\frac{6}{7} \stackrel{?}{=} \frac{18}{21}$

**11.** $\frac{3}{4} \stackrel{?}{=} \frac{15}{20}$

**12.** $\frac{3}{5} \stackrel{?}{=} \frac{6}{10}$

**13.** $\frac{11}{15} \stackrel{?}{=} \frac{9}{13}$

**14.** $\frac{9}{10} \stackrel{?}{=} \frac{2}{7}$

**15.** $\frac{8}{3} \stackrel{?}{=} \frac{24}{9}$

**16.** $\frac{5}{8} \stackrel{?}{=} \frac{15}{24}$

**17.** $\frac{6}{17} \stackrel{?}{=} \frac{9}{11}$

**18.** $\frac{5}{12} \stackrel{?}{=} \frac{8}{20}$

**19.** $\frac{7}{16} \stackrel{?}{=} \frac{21}{48}$

**20.** $\frac{2}{5} \stackrel{?}{=} \frac{7}{9}$

**21.** $\frac{10}{3} \stackrel{?}{=} \frac{150}{50}$

**22.** $\frac{5}{8} \stackrel{?}{=} \frac{75}{120}$

**23.** $\frac{3}{7} \stackrel{?}{=} \frac{18}{42}$

**24.** $\frac{12}{7} \stackrel{?}{=} \frac{96}{50}$

**25.** $\frac{7}{15} \stackrel{?}{=} \frac{84}{180}$

**26.** $\frac{76}{24} \stackrel{?}{=} \frac{19}{6}$

## Answers

1. ________ 2. ________
3. ________ 4. ________
5. ________ 6. ________
7. ________ 8. ________
9. ________ 10. ________
11. ________ 12. ________
13. ________ 14. ________
15. ________ 16. ________
17. ________ 18. ________
19. ________ 20. ________
21. ________ 22. ________
23. ________ 24. ________
25. ________ 26. ________

## Answers

27. ____

28. ____

29. ____

30. ____

31. ____

32. ____

33. ____

34. ____

35. ____

36. ____

37. ____

38. ____

< Objective 3 >

*Determine if the given rates are equivalent.*

27. $\frac{\text{7 cups of flour}}{\text{4 loaves of bread}} \stackrel{?}{=} \frac{\text{4 cups of flour}}{\text{3 loaves of bread}}$

28. $\frac{\text{6 U.S. dollars}}{\text{50 Krone}} \stackrel{?}{=} \frac{\text{15 U.S. dollars}}{\text{125 Krone}}$

29. $\frac{\text{22 miles}}{\text{15 gallons}} \stackrel{?}{=} \frac{\text{55 miles}}{\text{35 gallons}}$

30. $\frac{\text{46 pages}}{\text{30 minutes}} \stackrel{?}{=} \frac{\text{18 pages}}{\text{8 minutes}}$

31. $\frac{\text{9 inches}}{\text{57 miles}} \stackrel{?}{=} \frac{\text{6 inches}}{\text{38 miles}}$

32. $\frac{\text{12 yen}}{\text{5 pesos}} \stackrel{?}{=} \frac{\text{108 yen}}{\text{45 pesos}}$

*Write the proportion that is equivalent to the given statement.*

33. If 15 pounds (lb) of string beans cost \$4, then 45 lb will cost \$12.

34. If Maria hit 8 home runs in 15 softball games, then she should hit 24 home runs in 45 games.

35. If 3 credits at Bucks County Community College cost \$216, then 12 credits cost \$864.

36. If 16 pounds (lb) of fertilizer cover 1,520 square feet ($ft^2$), then 21 lb should cover 1,995 $ft^2$.

37. If Audrey travels 180 miles (mi) on interstate I-95 in 3 hours (h), then he should travel 300 mi in 5 h.

38. If 2 vans can transport 18 people, then 5 vans can transport 45 people.

*Determine whether each statement is* **true** *or* **false.**

**39.** Two ratios must be equal in order for the ratios to be proportional.

**40.** If $\frac{a}{b} = \frac{c}{d}$, then $a \cdot c = b \cdot d$.

*Fill in each blank with* **always, sometimes,** *or* **never.**

**41.** Proportions are ______________ used to compare two rates.

**42.** When writing a proportion for two rates, the placement of units is ______________ important.

Basic Skills | **Advanced Skills** | Vocational-Technical Applications | Calculator/Computer | Above and Beyond

*Determine whether each pair of fractions or rates is proportional.*

**43.** $\frac{60}{36} \stackrel{?}{=} \frac{25}{15}$

**44.** $\frac{\frac{1}{2}}{4} \stackrel{?}{=} \frac{5}{40}$

**45.** $\frac{3}{\frac{1}{5}} \stackrel{?}{=} \frac{30}{6}$

**46.** $\frac{\frac{2}{3}}{6} \stackrel{?}{=} \frac{1}{12}$

**47.** $\frac{\frac{3}{4}}{12} \stackrel{?}{=} \frac{1}{16}$

**48.** $\frac{0.3}{4} \stackrel{?}{=} \frac{1}{20}$

**49.** $\frac{3}{60} \stackrel{?}{=} \frac{0.3}{6}$

**50.** $\frac{0.6}{0.12} \stackrel{?}{=} \frac{2}{0.4}$

**51.** $\frac{0.6}{15} \stackrel{?}{=} \frac{2}{75}$

**52.** $\frac{\text{12 gallons of paint}}{8{,}329\text{ ft}^2} \stackrel{?}{=} \frac{\text{9 gallons of paint}}{1{,}240\text{ ft}^2}$

**53.** $\frac{\text{12 inches of snow}}{\text{1.4 inches of rain}} \stackrel{?}{=} \frac{\text{36 inches of snow}}{\text{7 inches of rain}}$

**54.** $\frac{\text{9 people}}{\text{2 cars}} \stackrel{?}{=} \frac{\text{11 people}}{\text{3 cars}}$

## Answers

39. ______________
40. ______________
41. ______________
42. ______________
43. ______________
44. ______________
45. ______________
46. ______________
47. ______________
48. ______________
49. ______________
50. ______________
51. ______________
52. ______________
53. ______________
54. ______________

Answers

55. ______

56. ______

57. ______

58. ______

Basic Skills | Advanced Skills | **Vocational-Technical Applications** | Calculator/Computer | Above and Beyond

55. **Allied Health** Quinidine is an antiarrhythmic heart medication. It is available for injection as an 80 milligrams per milliliter (mg/mL) solution. A patient receives a prescription for 300 mg of quinidine dissolved in 3.75 mL of solution. Are these rates proportional?

56. **Information Technology** A computer transmits 5 Web pages in 2 seconds (s) to a Web server. A second computer transmits 20 pages in 10 s to the server. Are these two computers transmitting at the same speed?

57. **Mechanical Engineering** A gear has a pitch diameter of 5 in. and 20 teeth. A second gear has a pitch diameter of 18 in. and 68 teeth. In order to mesh, the teeth to diameter ratios must be proportional. Will these two gears mesh?

58. **Agricultural Technology** A 13-acre field requires 7,020 lb of fertilizer. Will 11,340 lb of fertilizer cover a 21-acre field?

## Answers

1. $\frac{4}{9} = \frac{8}{18}$ 3. $\frac{2}{9} = \frac{8}{36}$ 5. $\frac{3}{5} = \frac{15}{25}$ 7. $\frac{9}{13} = \frac{27}{39}$ 9. Yes
11. Yes 13. No 15. Yes 17. No 19. Yes 21. No
23. Yes 25. Yes 27. No 29. No 31. Yes
33. $\frac{15 \text{ lb}}{\$4} = \frac{45 \text{ lb}}{\$12}$ 35. $\frac{3 \text{ credits}}{\$216} = \frac{12 \text{ credits}}{\$864}$ 37. $\frac{180 \text{ mi}}{3 \text{ h}} = \frac{300 \text{ mi}}{5 \text{ h}}$
39. True 41. sometimes 43. Yes 45. No 47. Yes 49. Yes
51. No 53. No 55. Yes 57. No

# 5.4 Solving Proportions

< 5.4 Objectives >

1 > Solve a proportion for an unknown value

2 > Solve an application involving a proportion

A proportion consists of four values. If three of the four values of a proportion are known, you can always find the missing or unknown value.

In the proportion $\frac{a}{3} = \frac{10}{15}$, the first value is unknown. We have chosen to represent the unknown value with the letter $a$. Using the proportion rule, we can proceed as follows.

**NOTE**

$\frac{?}{3} = \frac{10}{15}$ is a proportion in which the first value is unknown. Our work in this section involves learning how to find that unknown value.

$$\frac{a}{3} = \frac{10}{15}$$

$15 \cdot a = 3 \cdot 10$ or $15 \cdot a = 30$

The equal sign tells us that $15 \cdot a$ and 30 are just different names for the same number. This type of statement is called an **equation.**

**Definition**

| Equation | An **equation** is a statement that two expressions are equal. |
|---|---|

One important property of an equation is that we can divide both sides by the same nonzero number. Here we divide by 15.

**NOTE**

We always divide by the number multiplying the variable. This is called the *coefficient* of the variable.

$$15 \cdot a = 30$$

$$\frac{15 \cdot a}{15} = \frac{30}{15}$$

$$\frac{\overset{1}{\cancel{15}} \cdot a}{\underset{1}{\cancel{15}}} = \frac{\overset{2}{\cancel{30}}}{\underset{1}{\cancel{15}}}$$

Divide by the coefficient of the variable. Do you see why we divided by 15? It leaves our unknown $a$ by itself in the left term.

$$a = 2$$

You should always check your result. It is easy in this case. We found a value of 2 for $a$. Replace the unknown $a$ with that value. Then verify that the fractions are proportional. We started with $\frac{a}{3} = \frac{10}{15}$ and found a value of 2 for $a$. So we write

**NOTE**

Replace $a$ with 2 and multiply.

$$\frac{2}{3} \stackrel{?}{=} \frac{10}{15}$$

$$2 \cdot 15 \stackrel{?}{=} 3 \cdot 10$$

$$30 = 30$$

The value of 2 for $a$ is correct.

The procedure for solving a proportion is summarized as follows.

**Step by Step**

**To Solve a Proportion**

**Step 1** Use the proportion rule to write the equivalent equation $a \cdot d = b \cdot c$.
**Step 2** Divide both terms of the equation by the coefficient of the variable.
**Step 3** Use the value found to replace the unknown in the original proportion. Check that the ratios or the rates are proportional.

## Example 1 Solving Proportions for Unknown Values

< Objective 1 >

**NOTE**

You are really using algebra to solve these proportions. In algebra, we write the product $6 \cdot x$ as $6x$, omitting the dot. Multiplication of the number and the variable is understood.

**NOTE**

This gives us the unknown value. Now check the result.

Find the unknown value.

**(a)** $\frac{8}{x} = \frac{6}{9}$

**Step 1** Using the proportion rule, we have the following:

$$6 \cdot x = 8 \cdot 9$$

or $$6x = 72$$

**Step 2** Locate the coefficient of the variable, 6, and divide both sides of the equation by that coefficient.

$$\frac{\overset{1}{\cancel{6}}x}{\underset{1}{\cancel{6}}} = \frac{\overset{12}{\cancel{72}}}{\underset{1}{\cancel{6}}}$$

$$x = 12$$

**Step 3** To check, replace $x$ with 12 in the original proportion.

$$\frac{8}{12} \stackrel{?}{=} \frac{6}{9}$$

Multiply:

$$12 \cdot 6 \stackrel{?}{=} 8 \cdot 9$$

$$72 = 72$$ The value of 12 checks for $x$.

**(b)** $\frac{3}{4} = \frac{c}{25}$

**Step 1** Use the proportion rule.

$$4 \cdot c = 3 \cdot 25$$

or $$4c = 75$$

**Step 2** Locate the coefficient of the variable, 4, and divide both sides of the equation by that coefficient.

$$\frac{\overset{1}{\cancel{4}}c}{\underset{1}{\cancel{4}}} = \frac{75}{4}$$

$$c = \frac{75}{4}$$

**Step 3** To check, replace $c$ with $\frac{75}{4}$ in the original proportion.

**RECALL**

In the proportion equation $\frac{a}{b} = \frac{c}{d}$, we check to see if $ad = bc$.

$$\frac{3}{4} \stackrel{?}{=} \frac{\left(\frac{75}{4}\right)}{25}$$

Multiply:

$$3 \cdot 25 = 75$$

$$4 \cdot \frac{75}{4} = 75$$ The products are the same, so the value of $\frac{75}{4}$ checks for $c$.

## Check Yourself 1

Solve the proportions for $n$. Check your result.

**(a)** $\frac{4}{5} = \frac{n}{25}$ **(b)** $\frac{5}{9} = \frac{12}{n}$

In solving for a missing term in a proportion, we may find an equation involving fractions or decimals. Example 2 involves finding the unknown value in such cases.

## Example 2 Solving Proportions for Unknown Values

**(a)** Solve the proportion for $x$.

$$\frac{\frac{1}{4}}{3} = \frac{4}{x}$$

$$\frac{1}{4}x = 12$$

$$\frac{\frac{1}{4}x}{\frac{1}{4}} = \frac{12}{\frac{1}{4}}$$ We divide by the coefficient of $x$. In this case it is $\frac{1}{4}$.

**RECALL**

The coefficient is the numerical factor, in this case $\frac{1}{4}$.

$$x = \frac{12}{\frac{1}{4}}$$ Remember: $\frac{12}{\frac{1}{4}}$ is $12 \div \frac{1}{4}$.

$$x = 48$$ Invert the divisor and multiply.

**NOTE**

$x = 12 \cdot \frac{4}{1} = 48$

To check, replace $x$ with 48 in the original proportion.

$$\frac{\frac{1}{4}}{3} \stackrel{?}{=} \frac{4}{48}$$

$$\frac{1}{4} \cdot 48 \stackrel{?}{=} 3 \cdot 4$$

$$12 = 12$$

**NOTE**

Here we must divide 6 by 0.5 to find the unknown value. The steps of that division are shown here for review.

$$\begin{array}{r} 12. \\ 0.5\overline{)6.0} \\ 5\phantom{.} \\ \hline 10 \\ 10 \\ \hline 0 \end{array}$$

**(b)** Solve the proportion for $d$.

$$\frac{0.5}{2} = \frac{3}{d}$$

$$0.5d = 6$$

$$\frac{0.5d}{0.5} = \frac{6}{0.5}$$ Divide by the coefficient, 0.5.

$$d = 12$$

We leave it to you to confirm that $0.5 \cdot 12 = 2 \cdot 3$.

## Check Yourself 2

**(a)** Solve for $d$.

$$\frac{\frac{1}{2}}{5} = \frac{3}{d}$$

**(b)** Solve for $x$.

$$\frac{0.4}{x} = \frac{2}{30}$$

Now that we have learned how to find an unknown value in a proportion, we can solve many applications.

**Step by Step**

**Solving Applications of Proportions**

**Step 1** Read the problem carefully to determine the given information.

**Step 2** Write the proportion necessary to solve the problem. Use a letter to represent the unknown quantity. Be sure to include the units in writing the proportion.

**Step 3** Solve, answer the question of the original problem, and check the proportion.

## Example 3 Solve an Application Involving a Proportion

< Objective 2 >

**(a)** In a shipment of 400 parts, 14 are found to be defective. How many defective parts should be expected in a shipment of 1,000?

Assume that the ratio of defective parts to the total number remains the same.

$$\frac{14 \text{ defective}}{400 \text{ total}} = \frac{x \text{ defective}}{1{,}000 \text{ total}}$$ We decided to let $x$ be the unknown number of defective parts.

Multiply:

$$400x = 14{,}000$$

Divide by the coefficient, 400.

$$x = 35$$

So 35 defective parts should be expected in the shipment.

Checking the original proportion, we get

$$14 \cdot 1{,}000 \stackrel{?}{=} 400 \cdot 35$$

$$14{,}000 = 14{,}000$$

**(b)** Jill works 4.2 h and receives \$21. How much will she get if she works 10 h?

The ratio of hours worked to the amount of pay remains the same.

$$\frac{4.2 \text{ h}}{\$21} = \frac{10 \text{ h}}{\$a} \qquad \text{Let } a \text{ be the unknown amount of pay.}$$

$$4.2a = 210$$

$$\frac{4.2a}{4.2} = \frac{210}{4.2} \qquad \text{Divide both sides by 4.2.}$$

$$a = \$50$$

### Check Yourself 3

**(a)** An investment of \$3,000 earned \$330 for 1 year. How much will an investment of \$10,000 earn at the same rate for 1 year?

**(b)** A piece of cable 8.5 centimeters (cm) long weighs 68 grams (g). What does a 10-cm length of the same cable weigh?

## Example 4 Using Proportions to Solve a Map-Scale Application

The scale on a map is given as $\frac{1}{4}$ inch (in.) = 3 miles (mi). The distance between two towns is 4 in. on the map. How far apart are the towns in miles?

For this solution we use the fact that the ratio of inches (on the map) to miles remains the same. We also use another important property of an equation: We can multiply both sides of the equation by the same non-zero number.

$$\frac{\frac{1}{4} \text{ in.}}{3 \text{ mi}} = \frac{4 \text{ in.}}{x \text{ mi}}$$

$$\frac{1}{4} \cdot x = 3 \cdot 4$$

$$4\left(\frac{1}{4}\right) \cdot x = 4 \cdot 3 \cdot 4$$

$$1 \cdot x = 4 \cdot 3 \cdot 4$$

$$x = 48 \text{ (mi)}$$

**NOTE**

We could divide both sides by $\frac{1}{4}$:

$$\frac{\frac{1}{4} \cdot x}{\frac{1}{4}} = \frac{3 \cdot 4}{\frac{1}{4}}$$

$$x = \frac{3 \cdot 4}{\frac{1}{4}}$$

$$x = \frac{12}{\frac{1}{4}}$$

then invert and multiply.

$$x = \frac{12}{1} \cdot \frac{4}{1}$$

$$= 48$$

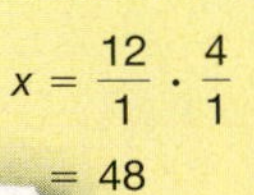

### Check Yourself 4

Jack drives 125 mi in $2\frac{1}{2}$ hours (h). At the same rate, how far will he be able to travel in 4 h? (Hint: Write $2\frac{1}{2}$ as an improper fraction.)

In Example 5 we must convert the units stated in the problem.

## Example 5 Using Proportions to Solve an Application

A machine can produce 15 tin cans in 2 minutes (min). At this rate how many cans can it make in an 8-h period?

In writing a proportion for this problem, we must write the times involved in terms of the same units.

**NOTE**

Always check that your units are properly placed.

$$\frac{15 \text{ cans}}{2 \text{ min}} = \frac{x \text{ cans}}{480 \text{ min}}$$

Because 1 h is 60 min, convert 8 h to 480 min.

$$15 \cdot 480 = 2x$$

or

$$7{,}200 = 2x$$

$$x = 3{,}600 \text{ cans}$$

### Check Yourself 5

Instructions on a can of film developer call for 2 ounces (oz) of concentrate to 1 quart (qt) of water. How much of the concentrate is needed to mix with 1 gallon (gal) of water? (4 qt = 1 gal.)

An important use of proportions is in solving problems involving *similar* geometric figures. These are figures that have the same shape and whose corresponding sides are proportional. For instance, in the similar triangles shown here,

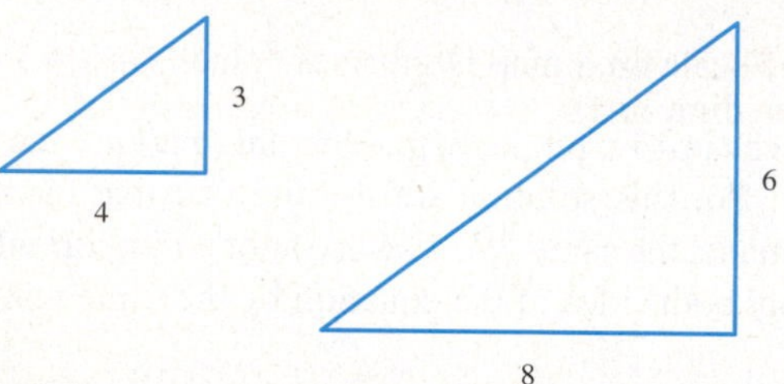


a proportion involving corresponding sides is

$$\frac{3}{4} = \frac{6}{8}$$

## Example 6 Solving an Application Using Similar Triangles

If a 6-foot-tall man casts a shadow that is 10 feet (ft) long, how tall is a tree that casts a shadow that is 140 ft long?

Look at a picture of the two triangles involved.

**NOTE**

Connect the top of the tree to the end of the shadow to create a triangle. Connecting the top of the man to the end of his shadow creates a similar triangle.

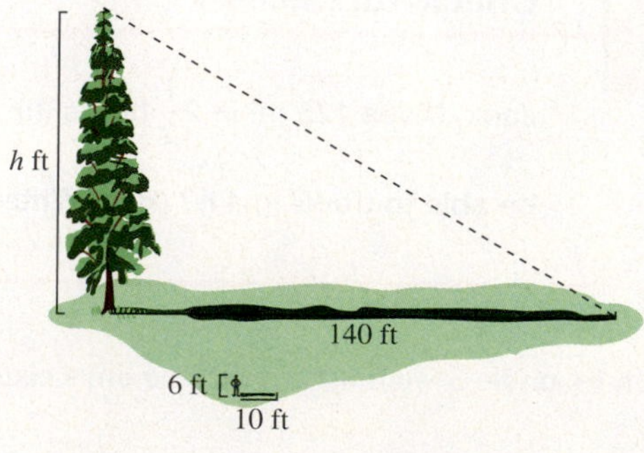

From the similar triangles, we have the proportion

$$\frac{6}{10} = \frac{h}{140}$$

Using the proportion rule, we have $6 \cdot 140 = 10 \cdot h$, so

$$10 \cdot h = 840$$

$$\frac{10 \cdot h}{10} = \frac{840}{10}$$

$$h = 84$$

The tree is 84 ft tall.

**Check Yourself 6**

If a woman who is $5\frac{1}{2}$ ft tall casts a shadow that is 3 ft long, how tall is a building that casts a shadow that is 90 ft long?

Proportions are used in solving a variety of problems such as the allied health application in Example 7.

### Example 7 An Application of Proportions

> CAUTION

Solving for $x$ does not give us an answer directly. $x$ represents the total volume, which includes both water and serum. We still need to subtract 8.5 from $x$ to get a final answer.

In preparing specimens for testing, it is often necessary to dilute the original solution. The dilution ratio is the ratio of the volume, in milliliters (mL), of original solution to the total volume, also in mL, of the diluted solution. How much water is required to make a $\frac{1}{20}$ dilution from 8.5 mL of serum?

We set up a proportion equation. The original solution is the 8.5 mL of serum, so that should be in the numerator. We name the denominator $x$.

$$\frac{1}{20} = \frac{8.5}{x}$$

$$1 \cdot x = 20 \cdot 8.5$$

$$x = 170$$

Therefore, the total volume should be 170 mL. Because 8.5 mL of the total solution is serum, we need to add $170 - 8.5 = 161.5$ mL of water.

**Check Yourself 7**

How much water is required to make a $\frac{3}{50}$ dilution from 11.25 mL of serum?

**Check Yourself ANSWERS**

**1.** **(a)** $n = 20$; **(b)** $n = \frac{108}{5}$ **2.** **(a)** $d = 30$; **(b)** $x = 6$ **3.** **(a)** \$1,100; **(b)** 80 g **4.** 200 mi **5.** 8 oz **6.** 165 ft **7.** 176.25 mL

## Reading Your Text

The following fill-in-the-blank exercises are designed to ensure that you understand some of the key vocabulary used in this section.

### SECTION 5.4

**(a)** An ______________ is a statement that two expressions are equal.

**(b)** When a number and a variable are multiplied, the number is called a ______________.

**(c)** The first step to solving application problems is to ______________ the problem carefully.

**(d)** Two triangles are similar if corresponding sides are ______________.

# 5.4 exercises

MathZone

Boost your grade at mathzone.com!

> Practice Problems
> NetTutor
> Self-Tests
> e-Professors
> Videos

Name ____________

Section ________ Date ________

< Objective 1 >

*Solve for the unknown in each of the following proportions.*

1. $\frac{x}{3} = \frac{6}{9}$ 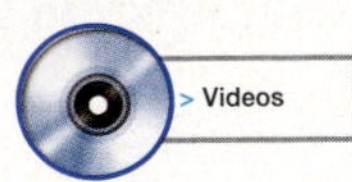

> Videos

2. $\frac{x}{6} = \frac{3}{9}$

3. $\frac{10}{n} = \frac{15}{6}$

4. $\frac{4}{3} = \frac{8}{n}$

5. $\frac{4}{7} = \frac{y}{14}$

6. $\frac{5}{8} = \frac{a}{16}$

7. $\frac{5}{7} = \frac{x}{35}$

8. $\frac{4}{15} = \frac{8}{n}$

9. $\frac{11}{a} = \frac{2}{44}$

10. $\frac{35}{40} = \frac{7}{n}$

11. $\frac{x}{8} = \frac{15}{24}$

12. $\frac{7}{12} = \frac{m}{24}$

13. $\frac{18}{12} = \frac{12}{p}$ > Videos

14. $\frac{20}{15} = \frac{100}{a}$

15. $\frac{5}{35} = \frac{a}{28}$

16. $\frac{20}{24} = \frac{p}{18}$

17. $\frac{12}{100} = \frac{3}{x}$

18. $\frac{b}{7} = \frac{21}{49}$

19. $\frac{p}{24} = \frac{25}{120}$

20. $\frac{5}{x} = \frac{20}{88}$

< Objective 2 >

*Solve the following applications.*

21. **Business and Finance** If 12 books are purchased for \$80, how much will you pay for 18 books at the same rate? > Videos

22. **Construction** If an 8-foot (ft) two-by-four costs \$1.92, what should a 12-ft two-by-four cost?

23. **Business and Finance** A box of 18 tea bags is marked \$2.70. At that price, what should a box of 48 tea bags cost?

24. **Business and Finance** A worker can complete the assembly of 15 MP3 players in 6 hours (h). At this rate, how many can the worker complete in a 40-h workweek?

## Answers

1. ______ 2. ______
3. ______ 4. ______
5. ______ 6. ______
7. ______ 8. ______
9. ______ 10. ______
11. ______ 12. ______
13. ______ 14. ______
15. ______ 16. ______
17. ______ 18. ______
19. ______ 20. ______
21. ______
22. ______
23. ______
24. ______

## Answers

25. ______

26. ______

27. ______

28. ______

29. ______

30. ______

**25.** **Social Science** The ratio of yes to no votes in an election was 3 to 2. How many no votes were cast if there were 2,880 yes votes?

**26.** **Social Science** The ratio of men to women at a college is 7 to 5. How many women students are there if there are 3,500 men?

**27.** **Crafts** A photograph 5 inches (in.) wide by 6 in. high is to be enlarged so that the new width is 15 in. What will the height of the enlargement be?

**28.** **Business and Finance** Christy can travel 110 miles (mi) in her new car on 5 gallons (gal) of gas. How far can she travel on a full (12 gal) tank?

**29.** **Business and Finance** The Changs purchased a \$120,000 home, and the property taxes were \$2,100. If they make improvements and the house is now valued at \$150,000, what will the new property tax be?

**30.** **Business and Finance** A car travels 165 mi in 3 h. How far will it travel in 8 h if it continues at the same speed?

*Using the given map, find the distances between the cities named in exercises 31 to 34. Measure distances to the nearest sixteenth of an inch.*

© MAGELLAN Geographix[SM] Santa Barbara, CA (800) 929-4MAP

31. Find the distance from Harrisburg to Philadelphia.

32. Find the distance from Punxsutawney (home of the groundhog) to State College (home of the Nittany Lions).

33. Find the distance from Gettysburg to Meadville.

34. Find the distance from Scranton to Waynesburg.

35. **Business and Finance** An inspection reveals 30 defective parts in a shipment of 500. How many defective parts should be expected in a shipment of 1,200?

36. **Business and Finance** You invest $4,000 in a stock that pays a $180 dividend in 1 year. At the same rate, how much will you need to invest to earn $270?

37. **Construction** A 6-ft fence post casts a 9-ft shadow. How tall is a nearby pole that casts a 15-ft shadow?

38. **Construction** A 9-ft light pole casts a 15-ft shadow. Find the height of a nearby tree that is casting a 40-ft shadow.

39. **Construction** On the blueprint of the Wilsons' new home, the scale is 5 in. equals 7 ft. What will the actual length of a bedroom be if it measures 10 in. long on the blueprint?

> Videos

40. **Social Science** The scale on a map is $\frac{1}{2}$ in. = 50 mi. If the distance between two towns on the map is 6 in., how far apart are they in miles?

41. **Science and Medicine** A metal bar expands $\frac{1}{4}$ in. for each 12°F rise in temperature. How much will it expand if the temperature rises 48°F?

42. **Business and Finance** Your car burns $2\frac{1}{2}$ quarts (qt) of oil on a trip of 5,000 mi. How many quarts should you expect to use when driving 7,200 mi?

## Answers

31. ______

32. ______

33. ______

34. ______

35. ______

36. ______

37. ______

38. ______

39. ______

40. ______

41. ______

42. ______

## Answers

43. ______________

44. ______________

45. ______________

46. ______________

47. ______________

48. ______________

49. ______________

50. ______________

51. ______________

52. ______________

53. ______________

54. ______________

**43. Social Science** Approximately 7 out of every 10 people in the U.S. workforce drive to work alone. During morning rush hour there are 115,000 cars on the streets of a medium-sized city. How many of these cars have one person in them?

**44. Social Science** Approximately 15 out of every 100 people in the U.S. workforce carpool to work. There are an estimated 320,000 people in the workforce of a given city. How many of these people are in car pools?

*Use a proportion to find the unknown side, labeled x, in each of the following pairs of similar figures.*

**45.**

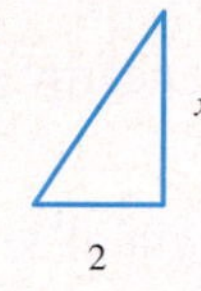


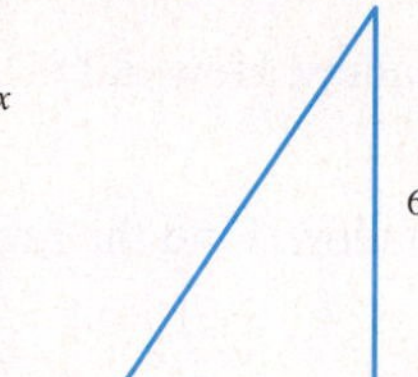


Videos

**46.**

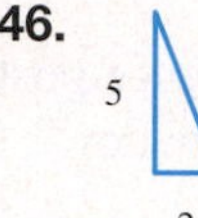


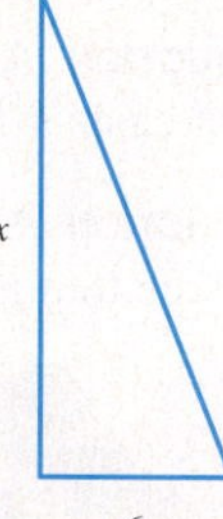


**47.**

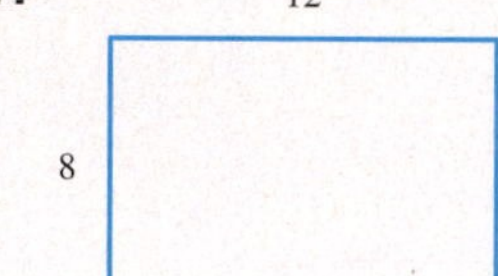


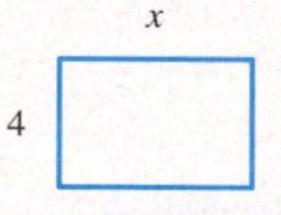


**48.**

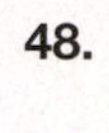


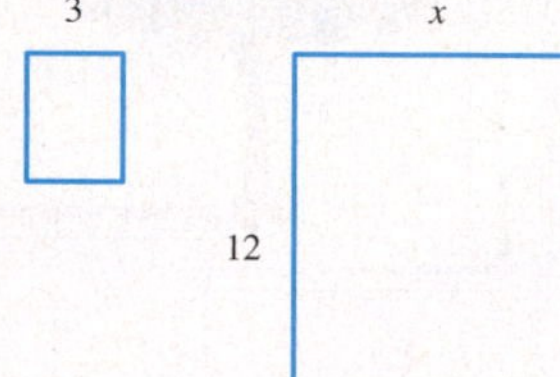


*Determine whether each statement is* **true** *or* **false.**

**49.** Given the product of a number and a variable, the variable factor is called a coefficient.

**50.** Given the product of a number and a variable, the numerical factor is called a coefficient.

*Fill in each blank with* **always, sometimes,** *or* **never.**

**51.** Equations are ____________ true.

**52.** Two similar triangles _________ have corresponding sides that are proportional.

Basic Skills | **Advanced Skills** | Vocational-Technical Applications | Calculator/Computer | Above and Beyond

*Solve for the unknown in each of the following proportions.*

**53.** $\dfrac{\frac{1}{2}}{2} = \dfrac{3}{a}$

**54.** $\dfrac{x}{5} = \dfrac{2}{\frac{1}{3}}$

55. $\dfrac{\frac{1}{4}}{12} = \dfrac{m}{40}$

56. $\dfrac{12}{\frac{1}{3}} = \dfrac{80}{y}$

57. $\dfrac{1}{6} = \dfrac{\frac{2}{x}}{18}$

58. $\dfrac{3}{4} = \dfrac{4}{\frac{x}{10}}$

59. $\dfrac{\frac{2}{5}}{8} = \dfrac{1.2}{n}$

60. $\dfrac{4}{a} = \dfrac{\frac{1}{4}}{0.8}$

61. $\dfrac{0.2}{2} = \dfrac{1.2}{a}$

62. $\dfrac{0.5}{x} = \dfrac{1.25}{5}$

63. $\dfrac{x}{3.3} = \dfrac{1.1}{6.6}$

64. $\dfrac{2.4}{5.7} = \dfrac{m}{1.1}$

65. $\dfrac{3}{2} = \dfrac{1}{x}$

66. $\dfrac{4}{2} = \dfrac{2}{x}$

67. $\dfrac{2}{t} = \dfrac{12}{5}$

68. $\dfrac{4}{x} = \dfrac{14}{15}$

69. $\dfrac{n}{5} = \dfrac{1}{20}$

70. $\dfrac{m}{2} = \dfrac{5}{24}$

71. $\dfrac{3}{4} = \dfrac{x}{6}$

72. $\dfrac{1}{14} = \dfrac{c}{7}$

Basic Skills | Advanced Skills | **Vocational-Technical Applications** | Calculator/Computer | Above and Beyond

73. **Information Technology** A computer transmits 5 Web pages in 2 seconds to a Web server. How many pages can the computer transmit in 1 min?

74. **Automotive Technology** A tire shows $\frac{1}{8}$ in. of tread wear after 32,000 mi. Assuming that the rate of wear remains constant, how much tread wear would you expect the tire to show after 48,000 mi?

75. **Manufacturing Technology** Cutting 7 holes removes 0.322 pounds of material from a frame. How much weight would 43 holes remove?

## Answers

55. ______
56. ______
57. ______
58. ______
59. ______
60. ______
61. ______
62. ______
63. ______
64. ______
65. ______
66. ______
67. ______
68. ______
69. ______
70. ______
71. ______
72. ______
73. ______
74. ______
75. ______

## Answers

76. ______

77. ______

78. ______

79. ______

80. ______

81. ______

82. ______

76. **Electrical Engineering** The voltage output $V_{out}$ of a transformer is given by the proportion

$$\frac{N_{out}}{N_{in}} = \frac{V_{out}}{V_{in}}$$

in which $N$ gives the number of turns in the coil (see the figure).

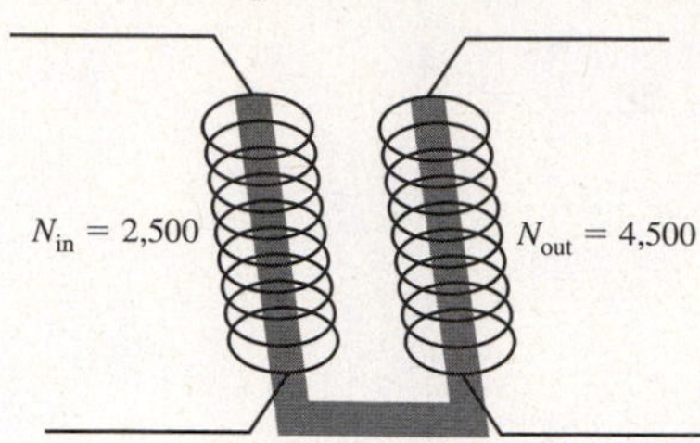


In the system shown, what voltage input is required in order for the output voltage to reach 630 volts?

Basic Skills | Advanced Skills | Vocational-Technical Applications | **Calculator/Computer** | Above and Beyond

When dealing with practical applications, the numbers involved in a proportion may be large or contain inconvenient decimals. A calculator is likely to be the tool of choice for solving such proportions. Typically, we set up the solution without doing any calculating, and then we put the calculator to work, usually rounding the result to an appropriate place value. For example, suppose you drive 278 miles (mi) on 13.6 gallons (gal) of gas. If the gas tank holds 21 gal, and you want to know how far you can travel on a full tank of gas, you write

$$\frac{278 \text{ mi}}{13.6 \text{ gal}} = \frac{x \text{ mi}}{21 \text{ gal}}$$

Solving for $x$, you obtain

$$x = \frac{(278)(21)}{13.6}$$

With your calculator, you enter

278 [×] 21 [÷] 13.6 [=]

The display shows 429.26471. Rounding to the nearest mile, you can travel 429 mi.

*Use your calculator to solve the following.*

77. $\frac{630}{1,365} = \frac{15}{a}$

78. $\frac{770}{1,988} = \frac{n}{71}$

79. $\frac{x}{4.7} = \frac{11.8}{16.9}$ (to nearest tenth)

80. $\frac{13.9}{8.4} = \frac{n}{9.2}$ (to nearest hundredth)

81. $\frac{2.7}{3.8} = \frac{5.9}{n}$ (to nearest tenth)

82. $\frac{12.2}{0.042} = \frac{x}{0.08}$ (to nearest hundredth)

*Solve the following applications.*

**83.** **Business and Finance** Bill earns \$248.40 for working 34.5 hours (h). How much will he receive if he works at the same pay rate for 31.75 h?

**84.** **Construction** Construction-grade lumber costs \$384.50 per 1,000 board-feet. What will be the cost of 686 board-feet? Round your answer to the nearest cent.

**85.** **Science and Medicine** A speed of 88 feet per second (ft/s) is equal to a speed of 60 miles per hour (mi/h). If the speed of sound is 750 mi/h, what is the speed of sound in feet per second?

**86.** **Business and Finance** A shipment of 75 parts is inspected, and 6 are found to be faulty. At the same rate, how many defective parts should be found in a shipment of 139? Round your result to the nearest whole number.

Basic Skills | Advanced Skills | Vocational-Technical Applications | Calculator/Computer | **Above and Beyond**

**87.** A recipe for 12 servings lists the following ingredients:

| | | |
|---|---|---|
| 12 cups ziti | 7 cups spaghetti sauce | 4 cups ricotta cheese |
| $\frac{1}{2}$ cup parsley | 1 teaspoon garlic powder | $\frac{1}{2}$ teaspoon pepper |
| 4 cups mozzarella cheese | 2 tablespoons parmesan cheese | |

Determine the amount of ingredients necessary to serve 5 people.

## Answers

**1.** $x = 2$ **3.** $n = 4$ **5.** $y = 8$ **7.** $x = 25$ **9.** $a = 242$
**11.** $x = 5$ **13.** $p = 8$ **15.** $a = 4$ **17.** $x = 25$ **19.** $p = 5$
**21.** \$120 **23.** \$7.20 **25.** 1,920 no votes **27.** 18 in.
**29.** \$2,625 **31.** 110 mi **33.** 215 mi **35.** 72 defective parts
**37.** 10 ft **39.** 14 ft **41.** 1 in. **43.** 80,500 cars with one person
**45.** 3 **47.** 6 **49.** False **51.** sometimes **53.** $a = 12$
**55.** $m = \frac{5}{6}$ **57.** $x = \frac{2}{3}$ **59.** $n = 24$ **61.** $a = 12$ **63.** $x = 0.55$
**65.** $x = \frac{2}{3}$ **67.** $t = \frac{5}{6}$ **69.** $n = \frac{1}{4}$ **71.** $x = \frac{9}{2}$
**73.** 150 Web pages **75.** 1.978 lb **77.** $a = 32.5$ **79.** $x \approx 3.3$
**81.** $n \approx 8.3$ **83.** \$228.60 **85.** 1,100 ft/s **87.** Above and Beyond

Answers

83. ____________

84. ____________

85. ____________

86. ____________

87. ____________

# Activity 15 :: Burning Calories

Many people are interested in losing weight through exercise. An important fact to consider is that a person needs to burn off 3,500 calories more than he or she takes in to lose 1 pound, according to the American Dietetic Association.

The following table shows the number of calories burned per hour (cal/h) for a variety of activities, where the figures are based on a 150-pound person.

| Activity | Cal/h | Activity | Cal/h |
|---|---|---|---|
| Bicycling 6 mi/h | 240 | Running 10 mi/h | 1,280 |
| Bicycling 12 mi/h | 410 | Swimming 25 yd/min | 275 |
| Cross-country skiing | 700 | Swimming 50 yd/min | 500 |
| Jogging $5\frac{1}{2}$ mi/h | 740 | Tennis (singles) | 400 |
| Jogging 7 mi/h | 920 | Walking 2 mi/h | 240 |
| Jumping rope | 750 | Walking 3 mi/h | 320 |
| Running in place | 650 | Walking $4\frac{1}{2}$ mi/h | 440 |

Work with your group members to solve the following problems. You may find that setting up proportions is helpful.

For problems 1 through 4, assume a 150-pound person.

**1.** If a person jogs at a rate of $5\frac{1}{2}$ mi/h for $3\frac{1}{2}$ h in a week, how many calories will the person burn?

**2.** If a person runs in place for 15 minutes, how many calories will the person burn?

**3.** If a person cross-country skis for 35 minutes, how many calories will the person burn?

**4.** How many hours would a person have to jump rope in order to lose 1 pound? (Assume calorie consumption is just enough to maintain weight, with no activity.)

Heavier people burn more calories (for the same activity), and lighter people burn fewer. In fact, you can calculate similar figures for burning calories by setting up the appropriate proportion.

**5.** At what rate would a 120-pound person burn calories while bicycling at 12 mi/h?

**6.** At what rate would a 180-pound person burn calories while bicycling at 12 mi/h?

**7.** How many hours of jogging at $5\frac{1}{2}$ mi/h would be needed for a 200-pound person to lose 5 pounds? (Again, assume calorie consumption is just enough to maintain weight, with no activity.)

| Definition/Procedure | Example | Reference |
|---|---|---|
| **Ratios** | | **Section 5.1** |
| **Ratio** A means of comparing two numbers or like quantities. A ratio can be written as a fraction. | $\frac{4}{7}$ can be thought of as "the ratio of 4 to 7." | p. 367 |
| **Rates and Unit Pricing** | | **Section 5.2** |
| **Rate** A fraction involving two unlike denominate numbers.<br>**Unit Rate** A rate that has been simplified and read so that the denominator is one unit. | $\frac{50 \text{ home runs}}{150 \text{ games}} = \frac{1}{3} \frac{\text{home run}}{\text{game}}$ | pp. 377–378 |
| **Unit Price** The cost per unit. | $\frac{\$2}{5 \text{ rolls}} = \$0.40 \text{ per roll}$<br>$= 40 \frac{\text{cents}}{\text{roll}}$ | p. 380 |
| **Proportions** | | **Section 5.3** |
| **Proportion** A statement that two fractions or ratios are equal. | $\frac{3}{5} = \frac{6}{10}$ is a proportion read "three is to five as six is to ten." | p. 390 |
| **The Proportion Rule** If $\frac{a}{b} = \frac{c}{d}$, then $a \cdot d = b \cdot c$. | If $\frac{3}{5} = \frac{6}{10}$, then $3 \cdot 10 = 5 \cdot 6$ | p. 391 |
| **Solving Proportions** | | **Section 5.4** |
| ***To Solve a Proportion***<br>**Step 1** Use the proportion rule to write the equivalent equation $a \cdot d = b \cdot c$.<br>**Step 2** Divide both terms of the equation by the coefficient of the variable.<br>**Step 3** Use the value found to replace the unknown in the original proportion. Check that the ratios or rates are proportional. | To solve: $\frac{x}{5} = \frac{16}{20}$<br>$20x = 5 \cdot 16$<br>$20x = 80$<br>$\frac{\overset{1}{\cancel{20}}x}{\underset{1}{\cancel{20}}} = \frac{80}{20}$<br>$x = 4$<br>Check:<br>$\frac{4}{5} \overset{?}{=} \frac{16}{20}$<br>$4 \times 20 \overset{?}{=} 5 \times 16$<br>$80 = 80$ | p. 400 |

*Continued*

| Definition/Procedure | Example | Reference |
|---|---|---|
| ***To Solve a Problem by Using Proportions*** | | |
| **Step 1** Read the problem carefully to determine the given information.<br>**Step 2** Write the proportion necessary to solve the problem, using a letter to represent the unknown quantity. Be sure to include the units in writing the proportion.<br>**Step 3** Solve, answer the question of the original problem, and check the proportion as before. | A machine can produce 250 units in 5 min. At this rate, how many can it produce in a 12 h period?<br>$\frac{250 \text{ units}}{5 \text{ min}} = \frac{x \text{ units}}{12 \text{ h}}$<br>$\frac{250 \text{ units}}{5 \text{ min}} = \frac{x \text{ units}}{720 \text{ min}}$<br>$250 \cdot 720 = 5x$<br>or<br>$5x = 180{,}000$<br>$x = 36{,}000$<br>The machine can produce 36,000 units in 12 h. | *p.* 402 |


The Streeter/Hutchison Series in Mathematics Basic Mathematical Skills with Geometry

# summary exercises :: chapter 5

This summary exercise set is provided to give you practice with each of the objectives of this chapter. Each exercise is keyed to the appropriate chapter section. When you are finished, you can check your answers to the odd-numbered exercises against those presented in the back of the text. If you have difficulty with any of these questions, go back and reread the examples from that section. The answers to the even-numbered exercises appear in the *Instructor's Solutions Manual.* Your instructor will give you guidelines on how to best use these exercises in your instructional setting.

**5.1** *Write each ratio in simplest form.*

**1.** The ratio of 4 to 17

**2.** The ratio of 28 to 42

**3.** For a football team that has won 10 of its 16 games, the ratio of wins to games played

**4.** For a rectangle of length 30 inches and width 18 inches, the ratio of its length to its width

**5.** The ratio of $2\frac{1}{3}$ to $5\frac{1}{4}$

**6.** The ratio of 7.5 to 3.25

**7.** The ratio of 7 in. to 3 ft

**8.** The ratio of 72 h to 4 days

**5.2** *Express each rate in simplest form.*

**9.** $\frac{\text{600 miles}}{\text{6 hours}}$

**10.** $\frac{\text{270 miles}}{\text{9 gallons}}$

**11.** $\frac{\text{350 calories}}{\text{7 ounces}}$

**12.** $\frac{\text{36,000 dollars}}{\text{9 years}}$

**13.** $\frac{\text{5,000 feet}}{\text{25 seconds}}$

**14.** $\frac{\text{10,500 revolutions}}{\text{3 minutes}}$

**15.** A baseball team has had 117 hits in 18 games. Find the team's hits per game rate.

**16.** A basketball team has scored 216 points in 8 quarters. Find the team's points per quarter rate.

**17.** Taniko scored 246 points in 20 games. Marisa scored 216 points in 16 games. Which player has the highest points per game rate?

18. One shop will charge $306 for a job that takes $4\frac{1}{2}$ hours. A second shop can do the same job in 4 hours and will charge $290. Which shop has the higher cost per hour rate?

5.2 *Find the unit price for each item.*

19. A 32-oz bottle of dishwashing liquid costs $2.88.

20. A 35-oz box of breakfast cereal costs $5.60.

21. A 24-oz loaf of bread costs $2.28.

22. Five large jars of fruit cost $67.30.

23. Three CDs cost $44.85.

24. Six tickets cost $267.60.

5.3 *Write each proportion.*

25. 4 is to 9 as 20 is to 45.

26. 7 is to 5 as 56 is to 40.

27. If Jorge can travel 110 miles (mi) in 2 hours (h), he can travel 385 mi in 7 h.

28. If it takes 4 gallons (gal) of paint to cover 1,000 square feet ($ft^2$), it takes 10 gal of paint to cover 2,500 $ft^2$.

*Determine whether the given fractions are proportional.*

29. $\frac{4}{13} \stackrel{?}{=} \frac{7}{22}$

30. $\frac{8}{11} \stackrel{?}{=} \frac{24}{33}$

31. $\frac{9}{24} \stackrel{?}{=} \frac{12}{32}$

32. $\frac{7}{18} \stackrel{?}{=} \frac{35}{80}$

**33.** $\dfrac{5}{\frac{1}{6}} \stackrel{?}{=} \dfrac{120}{4}$

**34.** $\dfrac{0.8}{4} \stackrel{?}{=} \dfrac{12}{50}$

**35.** Is $\dfrac{35 \text{ Euros}}{40 \text{ dollars}}$ equivalent to $\dfrac{75.25 \text{ Euros}}{86 \text{ dollars}}$?

**36.** Is $\dfrac{188 \text{ words}}{8 \text{ minutes}}$ equivalent to $\dfrac{121 \text{ words}}{5 \text{ minutes}}$?

**5.4** *Solve for the unknown in each proportion.*

**37.** $\dfrac{16}{24} = \dfrac{m}{3}$

**38.** $\dfrac{6}{a} = \dfrac{27}{18}$

**39.** $\dfrac{14}{35} = \dfrac{t}{10}$

**40.** $\dfrac{55}{88} = \dfrac{10}{p}$

**41.** $\dfrac{\frac{1}{2}}{18} = \dfrac{5}{w}$

**42.** $\dfrac{\frac{3}{2}}{9} = \dfrac{5}{a}$

**43.** $\dfrac{5}{x} = \dfrac{0.6}{12}$

**44.** $\dfrac{s}{2.5} = \dfrac{1.5}{7.5}$

*Solve each application.*

**45. Business and Finance** If 4 tickets to a civic theater performance cost $90, what is the price for 6 tickets?

**46. Social Science** The ratio of first-year to second-year students at a school is 8 to 7. If there are 224 second-year students, how many first-year students are there?

**47. Crafts** A photograph that is 5 inches (in.) wide by 7 in. tall is to be enlarged so that the new height will be 21 in. What will be the width of the enlargement?

**48. Business and Finance** Marcia assembles disk drives for a computer manufacturer. If she can assemble 11 drives in 2 hours (h), how many can she assemble in a workweek (40 h)?

49. **Business and Finance** A firm finds 14 defective parts in a shipment of 400. How many defective parts can be expected in a shipment of 800 parts?

50. **Social Science** The scale on a map is $\frac{1}{4}$ in. = 10 miles (mi). How many miles apart are two towns that are 3 in. apart on the map?

51. **Business and Finance** A piece of tubing that is 16.5 centimeters (cm) long weighs 55 grams (g). What is the weight of a piece of the same tubing that is 42 cm long?

52. **Crafts** If 1 quart (qt) of paint covers 120 square feet ($ft^2$), how many square feet does 2 gallons (gal) cover? (1 gal = 4 qt.)

# self-test 5

Name ______________________

Section __________ Date __________

The purpose of this chapter test is to help you check your progress so that you can find sections and concepts that you need to review before the next exam. Allow yourself about an hour to take this test. At the end of that hour, check your answers against those given in the back of this text. If you missed any, note the section reference that accompanies the answer. Go back to that section and reread the examples until you have mastered that particular concept.

*Write each ratio in simplest form.*

**1.** The ratio of 7 to 19

**2.** The ratio of 75 to 45

**3.** The ratio of 8 ft to 4 yd

**4.** The ratio of 6 h to 3 days

**5.** A basketball team wins 26 of its 33 games during a season. What is the ratio of wins to games played? What is the ratio of wins to losses?

*Express each rate in simplest form.*

**6.** $\dfrac{840 \text{ miles}}{175 \text{ gallons}}$

**7.** $\dfrac{132 \text{ dollars}}{16 \text{ hours}}$

**8.** The unit price, if 11 gallons of milk cost $28.16.

*Determine whether the given fractions are proportional.*

**9.** $\dfrac{3}{9} \stackrel{?}{=} \dfrac{27}{81}$

**10.** $\dfrac{6}{7} \stackrel{?}{=} \dfrac{9}{11}$

**11.** $\dfrac{9}{10} \stackrel{?}{=} \dfrac{27}{30}$

**12.** $\dfrac{\frac{1}{2}}{5} \stackrel{?}{=} \dfrac{2}{18}$

*Solve for the unknown in each proportion.*

**13.** $\dfrac{45}{75} = \dfrac{12}{x}$

**14.** $\dfrac{a}{24} = \dfrac{45}{65}$

**15.** $\dfrac{\frac{1}{2}}{p} = \dfrac{5}{30}$

**16.** $\dfrac{3}{m} = \dfrac{0.9}{4.8}$

## Answers

1. ______________________
2. ______________________
3. ______________________
4. ______________________
5. ______________________
6. ______________________
7. ______________________
8. ______________________
9. ______________________
10. ______________________
11. ______________________
12. ______________________
13. ______________________
14. ______________________
15. ______________________
16. ______________________

## Answers

17. ____________________

18. ____________________

19. ____________________

20. ____________________

*Solve each application, using a proportion.*

**17.** **Business and Finance** If ballpoint pens are marked 5 for 95¢, how much does a dozen cost?

**18.** **Business and Finance** Your new compact car travels 324 miles on 9 gallons of gas. If the tank holds 16 usable gallons, how far can you drive on a tankful of gas?

**19.** **Business and Finance** An assembly line can install 5 car mufflers in 4 min. At this rate, how many mufflers can be installed in an 8-h shift?

**20.** **Crafts** Instructions on a package of concentrated plant food call for 2 teaspoons (tsp) to 1 qt of water. We wish to use 3 gal of water. How much of the plant food concentrate should be added to the 3 gal of water?

# cumulative review chapters 1-5

The following exercises are presented to help you review concepts from earlier chapters. This is meant as review material and not as a comprehensive exam. The answers are presented in the back of the text. Beside each answer is a section reference for the concept. If you have difficulty with any of these exercises, be certain to at least read through the summary related to that section.

Name ____________

Section ________ Date ________

**1.** Write 45,789 in words.

**2.** What is the place value of 2 in the number 621,487?

*Perform the indicated operations.*

**3.** $\begin{array}{r} 2{,}790 \\ 831 \\ +\ 22{,}683 \\ \hline \end{array}$

**4.** $\begin{array}{r} 84{,}793 \\ -\ 36{,}987 \\ \hline \end{array}$

**5.** $76 \times 58$

**6.** $72\overline{)5{,}683}$

**7.** Luis owes \$815 on a credit card after a trip. He makes payments of \$125, \$80, and \$90. Interest amounting to \$48 is charged. How much does he still owe on the account?

**8.** Evaluate the expression: $48 \div 8 \times 2 - 3^2$.

**9.** Find the perimeter and area of the given figure.

6 ft

2 ft

**10.** A room that measures 6 yd by 8 yd is to be carpeted. The carpet costs \$23 per square yard. What will be the cost of the carpet?

**11.** Write the prime factorization of 924.

**12.** Find the greatest common factor (GCF) of 42 and 56.

**13.** Write the fraction $\frac{42}{168}$ in simplest form.

*Perform the indicated operations.*

**14.** $\frac{3}{4} \times \frac{24}{15}$

**15.** $2\frac{2}{3} \times 3\frac{3}{4}$

**16.** $\frac{6}{7} \div \frac{4}{21}$

## Answers

1. ____________
2. ____________
3. ____________
4. ____________
5. ____________
6. ____________
7. ____________
8. ____________
9. ____________
10. ____________
11. ____________
12. ____________
13. ____________
14. ____________
15. ____________
16. ____________

## Answers

17. ______ 18. ______
19. ______ 20. ______
21. ______ 22. ______
23. ______ 24. ______
25. ______
26. ______
27. ______
28. ______
29. ______
30. ______
31. ______
32. ______
33. ______
34. ______
35. ______
36. ______
37. ______
38. ______
39. ______
40. ______

**17.** $5\frac{1}{2} \div 3\frac{1}{4}$ **18.** $\frac{9}{11} - \frac{3}{4} + \frac{1}{2}$ **19.** $4\frac{5}{6} + 2\frac{3}{4}$

**20.** $8\frac{2}{7} - 3\frac{11}{14}$

**21.** Find the least common multiple (LCM) of 36 and 60.

**22.** Maria drove at an average speed of 55 mi/h for $3\frac{1}{5}$ h. How many miles did she travel?

**23.** Stefan drove 132 miles in $2\frac{3}{4}$ h. What was his average speed?

*Perform the indicated operations.*

**24.** $36.169 - 28.341$ **25.** $3.1488 \div 2.56$ **26.** $4.89 \times 1.35$

**27.** Write 0.36 as a fraction and simplify.

**28.** Write the decimal equivalent of $\frac{7}{22}$ (to the nearest hundredth).

**29.** Find the perimeter of a rectangle that has dimensions 4.23 m by 2.8 m.

**30.** Find the area of a rectangle with dimensions 8 cm by 6.28 cm.

*Write each ratio in simplest form.*

**31.** 12 to 26 **32.** 60 to 18

**33.** 6 dimes to 3 quarters

*Determine whether the given fractions are proportional.*

**34.** $\frac{5}{6} \stackrel{?}{=} \frac{20}{24}$ **35.** $\frac{3}{7} \stackrel{?}{=} \frac{9}{22}$

*Solve for the unknown.*

**36.** $\frac{x}{3} = \frac{8}{12}$ **37.** $\frac{5}{x} = \frac{4}{12}$

**38.** Give the unit price for an item that weighs 20 oz and costs \$4.88.

**39.** On a map, 3 cm represents 250 km. How far apart are two cities if the distance between them on the map is 7.2 cm?

**40.** A company finds 15 defective items in a shipment of 600 items. How many defective items can be expected in a shipment of 2,000 items?

# 6

# Percents

## INTRODUCTION

When she retired, Roberta decided that she wanted to study the stock market and investments. She learned to watch small companies and to purchase stocks whose value is greater than listed.

Roberta enrolled in continuing education courses covering business and finance at her local community college. She learned how to track data and where to find financial advice articles on the Internet.

Now, after purchasing stock in a company, Roberta continues to monitor the company through quarterly financial statements and tries to gauge the best time to sell her stocks. Her goal is to sell her stocks when she thinks their value has peaked.

A less risky way to invest is to put money in a savings account or some other vehicle with a fixed interest rate. In these accounts, money grows continually as the interest is added at regular intervals.

In Activity 17 on page 460, you will have the opportunity to conduct a more in-depth study of interest calculations and the growth of investments.

## CHAPTER 6 OUTLINE

# 6 pretest



Name ______________

Section ________ Date ________

This pretest provides a preview of the types of exercises you will encounter in each section of this chapter. The answers for these exercises can be found in the back of the text. If you are working on your own, or ahead of the class, this pretest can help you identify the sections in which you should focus more of your time.

## Answers

1. ______________
2. ______________
3. ______________
4. ______________
5. ______________
6. ______________
7. ______________
8. ______________
9. ______________
10. ______________

**6.1** **1.** Write 7% as a fraction.

**2.** Write 23% as a decimal.

**6.2** **3.** Write 0.035 as a percent.

**4.** Write $\frac{4}{5}$ as a percent.

**6.3** Identify the rate, base, and amount in each problem.

**5.** 76.8 is 32% of 240.

**6.** In a shipment of 580 parts, 18 were found to be defective. What percent of the parts were defective?

**6.4** **7.** What is 25% of 252?

**8.** What percent of 500 is 45?

**9.** **Business and Finance** How much simple interest will you pay on a $4,000 loan for 1 year if the interest rate is 14%?

**10.** **Business and Finance** A salary increase of 5% amounts to a $60 monthly raise. What was the monthly salary before the increase?

# 6.1 Writing Percents as Fractions and Decimals

< 6.1 Objectives >

1 > Use percent notation

2 > Write a percent as a fraction or mixed number

3 > Write a percent as a decimal

When we considered parts of a whole in earlier chapters, we used fractions and decimals. *Percents* are another useful way of naming parts of a whole. We can think of percents as ratios whose denominators are 100. In fact, the word **percent** means "for each hundred." Consider the following figure:

The symbol for percent, %, represents multiplication by the number $\frac{1}{100}$. In the figure, 25 of 100 squares are shaded. As a fraction, we write this as $\frac{25}{100}$.

RECALL

Multiplying by $\frac{1}{100}$ is the same as dividing by 100.

$$\frac{25}{100} = 25\left(\frac{1}{100}\right) = 25\%$$

25 percent of the squares are shaded.

Example 1 Using Percent Notation

< Objective 1 >

**(a)** Four out of five geography students passed their midterm exams. Write this statement, using percent notation.

The ratio of passing students to all students is $\frac{4}{5}$, which we need to write as an equivalent fraction with a denominator of 100.

RECALL

You learned to solve proportions in Section 5.4.

$$\frac{4}{5} = \frac{x}{100}$$

Using the proportion rule, we have

$$5 \cdot x = 4 \cdot 100$$

or $\quad 5x = 400$

The coefficient of the variable is 5, so we divide both sides of the equation by 5.

$$\frac{5x}{5} = \frac{400}{5}$$

$$x = 80 \qquad 400 \div 5 = 80.$$

Therefore, we write

$$\frac{4}{5} = \frac{80}{100} = 80\left(\frac{1}{100}\right) = 80\%$$

and we say that 80% of the geography students passed their midterm exams.

**NOTE**

The ratio of compact cars to all cars is $\frac{35}{50}$.

**(b)** Of 50 automobiles sold by a dealer in 1 month, 35 were compact cars. Write this statement, using percent notation.

$$\frac{35}{50} = \frac{70}{100} = 70\left(\frac{1}{100}\right) = 70\%$$

We can say that 70% of the cars sold were compact cars.

**Check Yourself 1**

**Rewrite the following statement using percent notation: 4 of the 50 parts in a shipment were defective.**

Because there are different ways of naming the parts of a whole, you need to know how to change from one of these ways to another. First, we look at writing a percent as a fraction. Because a percent is a fraction or a ratio with denominator of 100, we can use the following rule.

**Property**

**Writing a Percent as a Fraction**

To change a percent to a common fraction, replace the percent symbol with $\frac{1}{100}$ and multiply.

The use of this rule is shown in Example 2.

**Example 2** **Writing a Percent as a Fraction**

< Objective 2 >

**NOTE**

You should write $\frac{25}{100}$ in simplest form.

Write each percent as a fraction.

**(a)** $7\% = 7\left(\frac{1}{100}\right) = \frac{7}{100}$

**(b)** $25\% = 25\left(\frac{1}{100}\right) = \frac{25}{100} = \frac{1}{4}$

**Check Yourself 2**

Write 12% as a fraction.

If a percent is *greater than 100,* the resulting fraction will be *greater than 1,* as shown in Example 3.

## Example 3 Writing a Percent as a Mixed Number

Write 150% as a mixed number.

$$150\% = 150\left(\frac{1}{100}\right) = \frac{150}{100} = 1\frac{50}{100} = 1\frac{1}{2}$$

### Check Yourself 3

Write 125% as a mixed number.

In Examples 2 and 3, we wrote percents as fractions by replacing the percent sign with $\frac{1}{100}$ and multiplying. How do we convert percents when we are working with decimals? Just move the decimal point two places to the left. This gives us a second rule for converting percents.

**Property**

**Writing a Percent as a Decimal**

To write a percent as a decimal, replace the percent symbol with $\frac{1}{100}$. As a result of multiplying by $\frac{1}{100}$, the decimal point moves two places to the left.

## Example 4 Writing a Percent as a Decimal

< Objective 3 >

Change each percent to its decimal equivalent.

(a) $25\% = 25\left(\frac{1}{100}\right) = 0.25$ The decimal point in 25% is understood to be after the 5.

(b) $8\% = 8\left(\frac{1}{100}\right) = 0.08$ We must add a zero to move the decimal point.

(c) $130\% = 130\left(\frac{1}{100}\right) = 1.30$

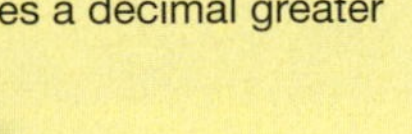

**NOTE**

A percent greater than 100 gives a decimal greater than 1.

### Check Yourself 4

Write as decimals.

(a) 5% (b) 32% (c) 115%

Example 5 involves fractions of a percent, in this case, decimal fractions.

### Example 5 Writing a Percent as a Decimal

Write as decimals.

(a) $4.5\% = 4.5\left(\frac{1}{100}\right) = 0.045$

(b) $0.5\% = 0.005$

**Check Yourself 5**

Write as decimals.

(a) 8.5%  (b) 0.3%

There are many situations in which common fractions are involved in a percent. Example 6 illustrates this situation.

### Example 6 Writing a Percent as a Decimal

**NOTE**

Write the common fractions as decimals. Then remove the percent symbol by our earlier rule.

Write as decimals.

(a) $9\frac{1}{2}\% = 9.5\% = 0.095$

(b) $\frac{3}{4}\% = 0.75\% = 0.0075$

**Check Yourself 6**

Write as decimals.

(a) $7\frac{1}{2}\%$  (b) $\frac{1}{2}\%$

Writing a percent in fraction or decimal form is required in many applications. One such application is presented in Example 7.

### Example 7 A Technology Application

A motor has an 86% efficiency rating. Express its efficiency rating as a fraction and as a decimal.

To express 86% as a fraction, we replace the % symbol with $\frac{1}{100}$ and simplify.

$$86\% = \frac{86}{100} = \frac{43}{50}$$

To express 86% as a decimal, we remove the % symbol and move the decimal point two places to the left.

86% = 0.86

### Check Yourself 7

**An inspection of Carina's hard drive reveals that it is 60% full. Write the amount of hard drive capacity that is full as a fraction and as a decimal.**

### Check Yourself ANSWERS

**1.** 8% were defective **2.** $12\% = \frac{12}{100} = \frac{3}{25}$ **3.** $1\frac{1}{4}$

**4.** **(a)** 0.05; **(b)** 0.32; **(c)** 1.15 **5.** **(a)** 0.085; **(b)** 0.003

**6.** **(a)** 0.075; **(b)** 0.005 **7.** $\frac{3}{5}$; 0.6

## Reading Your Text

The following fill-in-the-blank exercises are designed to ensure that you understand some of the key vocabulary used in this section.

### SECTION 6.1

**(a)** The word *percent* means, "for each ______________."

**(b)** To rewrite a percent as a ______________, divide the number before the percent symbol by 100 and simplify.

**(c)** To write a percent in decimal form, remove the % symbol and move the decimal two places to the ______________.

**(d)** A percent that is larger than 100% is equivalent to a number ______________ than 1.

# 6.1 exercises

Basic Skills | Advanced Skills | Vocational-Technical Applications | Calculator/Computer | Above and Beyond

**MathZone**

Boost your grade at mathzone.com!

- > Practice Problems
- > NetTutor
- > Self-Tests
- > e-Professors
- > Videos

Name ______________________

Section __________ Date __________

## Answers

1. ______________________
2. ______________________
3. ______________________
4. ______________________
5. ______________________
6. ______________________
7. ______________________
8. ______________________
9. ______________________
10. ______________________
11. ______________________
12. ______________________
13. ______________________
14. ______________________
15. ______________________
16. ______________________

< Objective 1 >

*Use percents to name the shaded portion of each drawing.*

1. 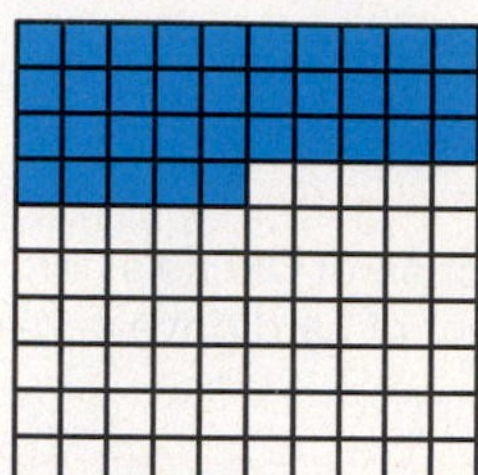

2. 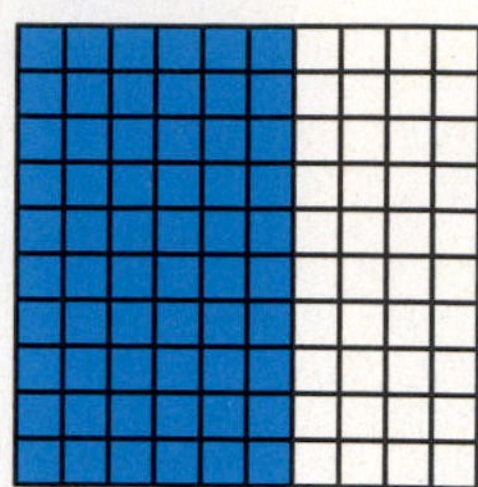

3. 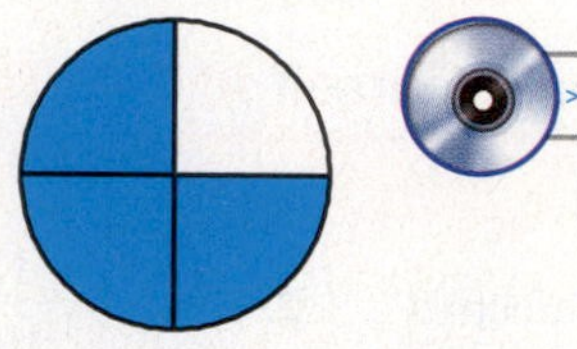


4. 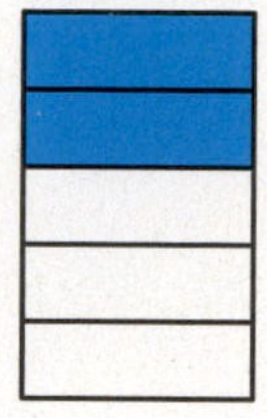

*Rewrite each statement, using percent notation.*

5. Out of every 100 eligible people, 53 voted in a recent election. 

6. You receive $5 in interest for every $100 saved for 1 year.

7. Out of every 100 entering students, 74 register for English composition.

8. Of 100 people surveyed, 29 watched a particular sports event on television.

9. Out of 10 voters in a state, 3 are registered as independents.

10. A dealer sold 9 of the 20 cars available during a 1-day sale.

11. Of 50 houses in a development, 27 are sold.

12. Of the 25 employees of a company, 9 are part-time.

13. Out of 50 people surveyed, 23 prefer decaffeinated coffee.

14. 17 out of 20 college students work at part-time jobs.

15. Of the 20 students in an algebra class, 5 receive a grade of A. 

16. Of the 50 families in a neighborhood, 31 have children in public schools.

< Objective 2 >

*Write as fractions.*

**17.** 6% > Videos

**18.** 17%

**19.** 75%

**20.** 20%

**21.** 65%

**22.** 48%

**23.** 50%

**24.** 52%

**25.** 46%

**26.** 35%

**27.** 66%

**28.** 4%

< Objective 3 >

*Write as decimals.*

**29.** 20%

**30.** 70%

**31.** 35%

**32.** 75%

**33.** 39%

**34.** 27%

**35.** 5% > Videos

**36.** 7%

**37.** 135%

**38.** 250%

**39.** 240%

**40.** 160%

**41.** **Social Science** Automobiles account for 85% of the travel between cities in the United States. What fraction does this percent represent?

**42.** **Social Science** Automobiles and small trucks account for 84% of the travel to and from work in the United States. What fraction does this percent represent?

**43.** Convert the discount shown to a decimal and to a simplified fraction.

Answers

17. ______
18. ______
19. ______
20. ______
21. ______
22. ______
23. ______
24. ______
25. ______
26. ______
27. ______
28. ______
29. ______ 30. ______
31. ______ 32. ______
33. ______ 34. ______
35. ______ 36. ______
37. ______ 38. ______
39. ______ 40. ______
41. ______ 42. ______
43. ______

Answers

44. ______

45. (a) ______

(b) ______

(c) ______

(d) ______

(e) ______

(f) ______

**44.** Convert the discount shown to a decimal and to a simplified fraction.

**45.** The minimum daily values (MDVs) for certain foods are given. They are based on a 2,000 calorie per day diet. Find decimal and fractional notation for the percent notation in each sentence.

**(a)** 1 ounce of Tostitos provides 9% of the MDV of fat.

**(b)** $\frac{1}{2}$ cup of B & M baked beans contains 15% of the MDV of sodium.

**(c)** $\frac{1}{2}$ cup of Campbells' New England clam chowder provides 6% of the MDV of iron.

**(d)** 2 ounces of Star Kist tuna provide 27% of the MDV of protein.

**(e)** Four 4-in. Aunt Jemima pancakes provide 33% of the MDV of sodium.

**(f)** 36 grams of Pop-Secret butter popcorn provide 2% of the MDV of sodium.

Basic Skills | **Advanced Skills** | Vocational-Technical Applications | Calculator/Computer | Above and Beyond

## Answers

46. ______

47. ______

48. ______

49. ______

50. ______

**46.** Complete the chart for the percentages given in the bar graph.

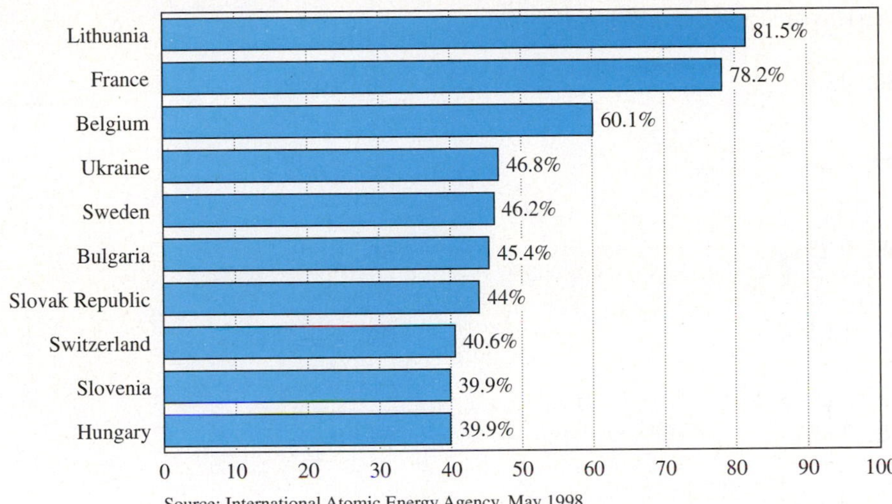

Source: International Atomic Energy Agency, May 1998

| Country | Decimal Equivalent | Fraction Equivalent |
|---|---|---|
| Lithuania | | |
| France | | |
| Belgium | | |
| Ukraine | | |
| Sweden | | |
| Bulgaria | | |
| Slovak Republic | | |
| Switzerland | | |
| Slovenia | | |
| Hungary | | |

*Rewrite each percent as a mixed number.*

**47.** 150%

**48.** 140%

**49.** 225%

**50.** 450%

Answers

51. ______ 52. ______
53. ______ 54. ______
55. ______ 56. ______
57. ______ 58. ______
59. ______
60. ______
61. ______
62. ______
63. ______
64. ______
65. ______
66. ______
67. ______
68. ______
69. ______
70. ______
71. ______
72. ______
73. ______
74. ______

**51.** $166\frac{2}{3}\%$

**52.** $233\frac{1}{3}\%$

**53.** $212\frac{1}{2}\%$

**54.** $116\frac{2}{3}\%$

*Rewrite each percent in decimal form.*

**55.** 23.6%

**56.** 10.5%

**57.** 6.4%

**58.** 3.5%

**59.** 0.2%

**60.** 0.5%

**61.** 1.05%

**62.** 0.023%

**63.** $7\frac{1}{2}\%$

**64.** $8\frac{1}{4}\%$

**65.** $87\frac{1}{2}\%$

**66.** $16\frac{2}{5}\%$

**67.** $128\frac{3}{4}\%$

**68.** $220\frac{3}{20}\%$

**69.** $\frac{1}{2}\%$

**70.** $\frac{3}{4}\%$

Basic Skills | Advanced Skills | **Vocational-Technical Applications** | Calculator/Computer | Above and Beyond

**71. Information Technology** An information technology project manager has a yearly budget of $49,000. She spent 10.2% of her budget in March. What fraction of her annual budget did she spend?

**72. Manufacturing Technology** A packaging company advertises that 87.5% of the machines that it produced in the last 20 years are still in operation. Express the proportion of machines still in service as a fraction.

**73. Manufacturing Technology** On an assembly line, 12% of all products are produced in the color red. Express this as a decimal and as a simplified fraction.

**74. Construction Technology** A board that starts with a width of 5.5 in. shrinks to 95% of its original width as it dries. What fraction is the dry width of the original width?

75. **Mechanical Engineering** As a piece of metal cools, it shrinks 3.125%. What fraction of its original size is lost due to shrinkage?

76. **Agricultural Technology** 12.5% of a growing season's 52 in. of rain fell in August. Write the percent of rain that fell in August as a decimal and as a simplified fraction.

Basic Skills | Advanced Skills | Vocational-Technical Applications | Calculator/Computer | **Above and Beyond**

77. Match each percent in column A with the equivalent fraction in column B.

| Column A | Column B |
|---|---|
| (a) $37\frac{1}{2}\%$ | (1) $\frac{3}{5}$ |
| (b) 5% | (2) $\frac{5}{8}$ |
| (c) $33\frac{1}{3}\%$ | (3) $\frac{1}{20}$ |
| (d) $83\frac{1}{3}\%$ | (4) $\frac{3}{8}$ |
| (e) 60% | (5) $\frac{5}{6}$ |
| (f) $62\frac{1}{2}\%$ | (6) $\frac{1}{3}$ |

78. Explain the difference between $\frac{1}{4}$ of a quantity and $\frac{1}{4}\%$ of a quantity.

## Answers

**1.** 35% **3.** 75% **5.** 53% **7.** 74% **9.** 30% **11.** 54% **13.** 46% **15.** 25% **17.** $\frac{3}{50}$ **19.** $\frac{3}{4}$ **21.** $\frac{13}{20}$ **23.** $\frac{1}{2}$ **25.** $\frac{23}{50}$ **27.** $\frac{33}{50}$ **29.** 0.2 **31.** 0.35 **33.** 0.39 **35.** 0.05 **37.** 1.35 **39.** 2.4 **41.** $\frac{17}{20}$ **43.** 0.15; $\frac{3}{20}$ **45.** **(a)** 0.09; $\frac{9}{100}$; **(b)** 0.15; $\frac{3}{20}$; **(c)** 0.06; $\frac{3}{50}$; **(d)** 0.27; $\frac{27}{100}$; **(e)** 0.33; $\frac{33}{100}$; **(f)** 0.02; $\frac{1}{50}$ **47.** $1\frac{1}{2}$ **49.** $2\frac{1}{4}$ **51.** $1\frac{2}{3}$ **53.** $2\frac{1}{8}$ **55.** 0.236 **57.** 0.064 **59.** 0.002 **61.** 0.0105 **63.** 0.075 **65.** 0.875 **67.** 1.2875 **69.** 0.005 **71.** $\frac{51}{500}$ **73.** 0.12; $\frac{3}{25}$ **75.** $\frac{1}{32}$ **77.** **(a)**-(4); **(b)**-(3); **(c)**-(6); **(d)**-(5); **(e)**-(1); **(f)**-(2)

## Answers

75. ______

76. ______

77. ______

78. ______

# 6.2 Writing Decimals and Fractions as Percents

< 6.2 Objectives >

1 > Write a decimal as a percent

2 > Write a fraction or mixed number as a percent

Writing a decimal as a percent is the opposite of writing a percent in decimal form. We simply reverse the process given in Section 6.1.

**Property**

**Writing a Decimal as a Percent**

To write a decimal as a percent, move the decimal point *two* places to the *right* and attach the percent symbol.

**Example 1** Writing a Decimal as a Percent

< Objective 1 >

**NOTES**

$0.18 = \frac{18}{100} = 18\left(\frac{1}{100}\right) = 18\%$

$0.03 = \frac{3}{100} = 3\left(\frac{1}{100}\right) = 3\%$

**(a)** Write 0.18 as a percent.

$0.18 = 18\%$

**(b)** Write 0.03 as a percent.

$0.03 = 3\%$

**Check Yourself 1**

Write each decimal in percent form.

**(a)** 0.27 **(b)** 0.05

There are many instances in which we use percents greater than 100%. For example, if a retailer marks goods up 25%, then the retail price is 125% of the wholesale price.

**Example 2** Writing a Decimal as a Percent

Write 1.25 as a percent.

$1.25 = 125\%$

NOTE

$$1.25 = \frac{125}{100} = 125\left(\frac{1}{100}\right) = 125\%$$

**Check Yourself 2**

Write 1.3 as a percent.

If a percent includes numbers to the right of the decimal point after the decimal is moved two places to the right, the fractional portion can be written as a decimal or as a fraction.

## Example 3 Writing a Decimal as a Percent

< Objective 1 >

NOTES

$$0.045 = \frac{45}{1{,}000} = \frac{45}{10}\left(\frac{1}{100}\right) = 4.5\%$$

$$0.003 = \frac{3}{1{,}000} = \frac{3}{10}\left(\frac{1}{100}\right) = 0.3\%$$

Write as a percent.

**(a)** $0.045 = 4.5\%$ or $4\frac{1}{2}\%$

**(b)** $0.003 = 0.3\%$ or $\frac{3}{10}\%$

**Check Yourself 3**

Write 0.075 as a percent.

The following rule allows us to change fractions to percents.

**Property**

**Writing a Fraction as a Percent**

To write a fraction as a percent, write the decimal equivalent of the fraction. Then use the previous rule to change the decimal to a percent.

## Example 4 Writing a Fraction as a Percent

< Objective 2 >

RECALL

We learned to write fractions as decimals in Section 4.2.

Write $\frac{3}{5}$ as a percent.

First write the decimal equivalent.

$\frac{3}{5} = 0.6$ To find the decimal equivalent, just divide the denominator into the numerator.

Now write the percent.

$\frac{3}{5} = 0.60 = 60\%$ Affix zeros to the right of the decimal if necessary.

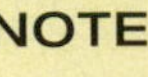

NOTE

Move the decimal point two places to the right and attach the percent symbol.

**Check Yourself 4**

Write $\frac{3}{4}$ as a percent.

Again, you will find both decimals and fractions used in writing percents. Consider Example 5.

## Example 5 Writing a Fraction as a Percent

Write $\frac{1}{8}$ as a percent.

$$\frac{1}{8} = 0.125 = 12.5\% \text{ or } 12\frac{1}{2}\%$$

### Check Yourself 5

Write $\frac{3}{8}$ as a percent.

### Units ANALYSIS

We rarely express the units when computing percentages. When we say "percent" we are essentially saying, "numerator units per 100 denominator units."

EXAMPLES:

Of 800 students, 200 were boys. What percent of the students were boys?

$$\frac{200 \text{ boys}}{800 \text{ students}} = 0.25 = 25\%$$

But what happened to our units? At the decimal, the units boys/student (0.25 boy per student) wouldn't make much sense, but we can read the % as

25 "boys per 100 students"

and have a reasonable unit phrase.

Of 500 computers sold, 180 were equipped with a scanner. What percent were equipped with a scanner?

$$\frac{180}{500} = 0.36 = 36\%$$

36 were equipped with a scanner for each (per) 100 computers sold.

To write a mixed number as a percent, we use exactly the same steps.

## Example 6 Writing a Mixed Number as a Percent

**NOTE**

The resulting percent must be greater than 100% because the original mixed number was greater than 1.

Write $1\frac{1}{4}$ as a percent.

$$1\frac{1}{4} = 1.25 = 125\%$$

### Check Yourself 6

Write $1\frac{2}{5}$ as a percent.

Some fractions have repeating-decimal equivalents. In writing these as percents, we either round to some indicated place or write the remainder as a fraction.

## Example 7 Writing a Fraction as a Percent

**NOTES**

$$\begin{array}{r} .33 \\ 3\overline{)1.00} \\ \underline{9} \\ 10 \\ \underline{9} \\ 1 \end{array} = 0.33\frac{1}{3}$$

In this case, we round the decimal equivalent. Then we write the percent.

**(a)** Write $\frac{1}{3}$ as a percent.

$$\frac{1}{3} = 0.33\frac{1}{3} = 33\frac{1}{3}\%$$

**(b)** Write $\frac{5}{7}$ as a percent.

$$\frac{5}{7} = 0.714 \qquad \text{(to the nearest thousandth)}$$

$$= 71.4\% \qquad \text{(to the nearest tenth of a percent)}$$

### Check Yourself 7

**(a)** Write $\frac{2}{3}$ as a percent.

**(b)** Write $\frac{2}{9}$ to the nearest tenth of a percent.

Some students prefer to use the proportion method when writing a fraction as a percent. The proportion method gives an easier way of writing a fraction as an *exact value* percent instead of as an approximation. We demonstrate this method in Example 8.

## Example 8 Using the Proportion Method

**RECALL**

You learned to solve proportions in Section 5.4.

**(a)** Write $\frac{7}{12}$ as a percent.

To use the proportion method, we write $\frac{7}{12}$ as an equivalent fraction whose denominator is 100.

$$\frac{7}{12} = \frac{x}{100}$$

Using the proportion rule, we have

$$12 \cdot x = 7 \cdot 100$$

$$12x = 700$$

Divide both sides by the coefficient of the variable, 12.

$$\frac{12x}{12} = \frac{700}{12} \qquad 700 \div 12 = 58\frac{4}{12} = 58\frac{1}{3}$$

$$x = 58\frac{1}{3}$$

Therefore,

$$\frac{7}{12} = \frac{58\frac{1}{3}}{100} = 58\frac{1}{3}\left(\frac{1}{100}\right) = 58\frac{1}{3}\%$$

**(b)** Write $\frac{1}{7}$ as a percent.

Again, we look for an equivalent fraction with a denominator of 100.

$$\frac{1}{7} = \frac{x}{100}$$

$$7x = 100$$

$$x = \frac{100}{7} = 14\frac{2}{7}$$

Therefore, $\frac{1}{7} = 14\frac{2}{7}\%$.

### Check Yourself 8

**Use the proportion method to write each fraction as a percent.**

**(a)** $\frac{5}{6}$ **(b)** $\frac{10}{9}$ **(c)** $\frac{3}{7}$

Applications often require that we write fractions and decimals in percent form, as in Example 9.

## Example 9 An Application of Percents

The hard drive in Emma's computer has a 74.5-gigabyte (GB) capacity. Currently, she is using 10.8 GB. What percent of her hard drive's capacity is Emma using?

We begin by constructing a fraction based on the given information.

$$\frac{\text{Used}}{\text{Capacity}} = \frac{10.8 \text{ GB}}{74.5 \text{ GB}}$$

$$= \frac{10.8 \not{\text{GB}}}{74.5 \not{\text{GB}}} \qquad \text{As with proportions, we can simply "cancel" the units.}$$

$$= \frac{108}{745} \qquad \text{Multiply the top and bottom by 10 to remove the decimal.}$$

$$\approx 0.145 \qquad \text{Divide and round the result to the nearest thousandth.}$$

$$= 14.5\%$$

Emma is using 14.5% of her hard drive's capacity.

## Check Yourself 9

**Three cylinders are not firing properly in an 8-cylinder engine. What percent of the cylinders are firing properly?**

Certain percents appear frequently enough that you should memorize their fraction and decimal equivalents.

| | | |
|---|---|---|
| $100\% = 1$ | $10\% = 0.1 = \frac{1}{10}$ | $12\frac{1}{2}\% = 0.125 = \frac{1}{8}$ |
| $200\% = 2$ | $20\% = 0.2 = \frac{1}{5}$ | $37\frac{1}{2}\% = 0.375 = \frac{3}{8}$ |
| | $30\% = 0.3 = \frac{3}{10}$ | $62\frac{1}{2}\% = 0.625 = \frac{5}{8}$ |
| $25\% = 0.25 = \frac{1}{4}$ | $40\% = 0.4 = \frac{2}{5}$ | $87\frac{1}{2}\% = 0.975 = \frac{7}{8}$ |
| $50\% = 0.5 = \frac{1}{2}$ | $60\% = 0.6 = \frac{3}{5}$ | |
| $75\% = 0.75 = \frac{3}{4}$ | $70\% = 0.7 = \frac{7}{10}$ | $33\frac{1}{3}\% = 0.\overline{3} = \frac{1}{3}$ |
| | $80\% = 0.8 = \frac{4}{5}$ | $66\frac{2}{3}\% = 0.\overline{6} = \frac{2}{3}$ |
| $5\% = 0.05 = \frac{1}{20}$ | $90\% = 0.9 = \frac{9}{10}$ | |

## Check Yourself ANSWERS

**1.** **(a)** 27%; **(b)** 5%  **2.** 130%  **3.** 7.5% or $7\frac{1}{2}\%$  **4.** 75%

**5.** 37.5% or $37\frac{1}{2}\%$  **6.** 140%  **7.** **(a)** $66\frac{2}{3}\%$; **(b)** 22.2%

**8.** **(a)** $83\frac{1}{3}\%$; **(b)** $111\frac{1}{9}\%$; **(c)** $42\frac{6}{7}\%$  **9.** 62.5%

## Reading Your Text

The following fill-in-the-blank exercises are designed to ensure that you understand some of the key vocabulary used in this section.

### SECTION 6.2

**(a)** To write a decimal in percent form, move the decimal two places to the ______________ and add the % symbol.

**(b)** To write a fraction as a percent, first convert the fraction to a ________.

**(c)** When writing a repeating decimal as a ______________, we round to some indicated place or write the remainder as a fraction.

**(d)** When writing a decimal as a percent, we may need to add ____________ to the right as placeholders.

# 6.2 exercises

Basic Skills | Advanced Skills | Vocational-Technical Applications | Calculator/Computer | Above and Beyond

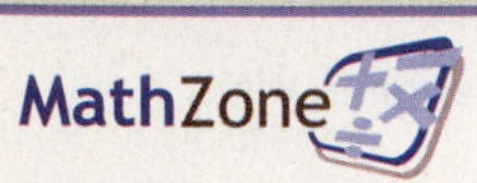

Boost your grade at mathzone.com!

- > Practice Problems
- > NetTutor
- > Self-Tests
- > e-Professors
- > Videos

Name ____________

Section ________ Date ________

**Answers**

1. ______ 2. ______
3. ______ 4. ______
5. ______ 6. ______
7. ______ 8. ______
9. ______ 10. ______
11. ______ 12. ______
13. ______ 14. ______
15. ______ 16. ______
17. ______
18. ______
19. ______
20. ______
21. ______ 22. ______
23. ______ 24. ______
25. ______ 26. ______

< Objective 1 >

*Write each decimal as a percent.*

**1.** 0.08 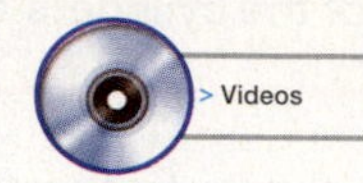 **2.** 0.09

**3.** 0.05 **4.** 0.13

**5.** 0.18 **6.** 0.63

**7.** 0.86 **8.** 0.45

**9.** 0.4 **10.** 0.3

**11.** 0.7 **12.** 0.6

**13.** 1.10 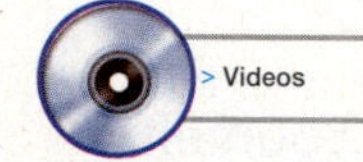 **14.** 2.50

**15.** 4.40 **16.** 5

**17.** 0.065  **18.** 0.095

**19.** 0.025 **20.** 0.085

< Objective 2 >

*Write each fraction or mixed number as a percent.*

**21.** $\frac{1}{4}$ **22.** $\frac{4}{5}$

**23.** $\frac{2}{5}$ **24.** $\frac{1}{2}$

**25.** $\frac{1}{5}$  **26.** $\frac{3}{4}$

27. $\frac{5}{8}$

28. $\frac{7}{8}$

29. $\frac{5}{16}$

30. $1\frac{1}{5}$

31. $3\frac{1}{2}$ > Videos

32. $\frac{2}{3}$

*In exercises 33 to 40, partially shaded decimal representations are given. Express the partially shaded region as a decimal, a fraction, and a percent.*

33.
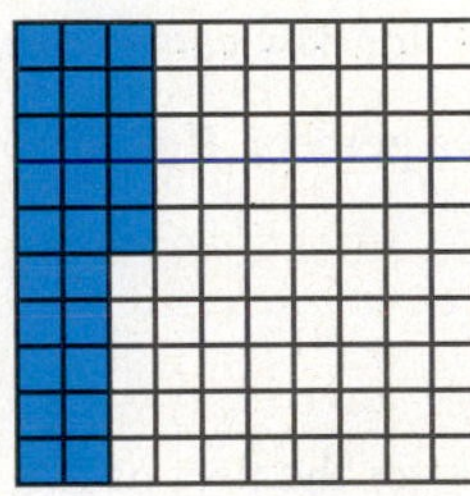

34.

35.

36.
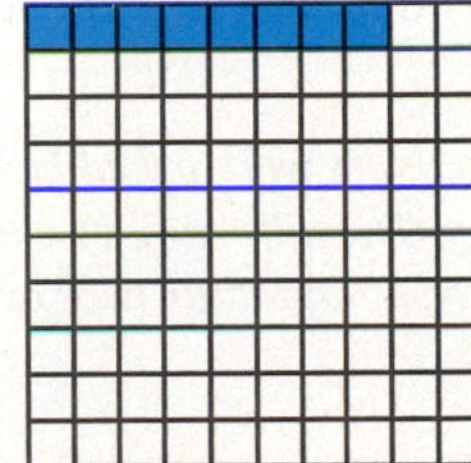

37.
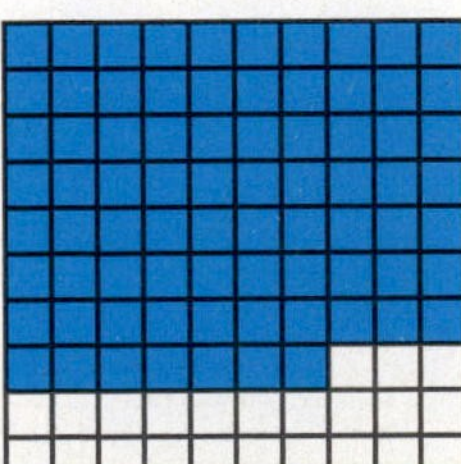

38.
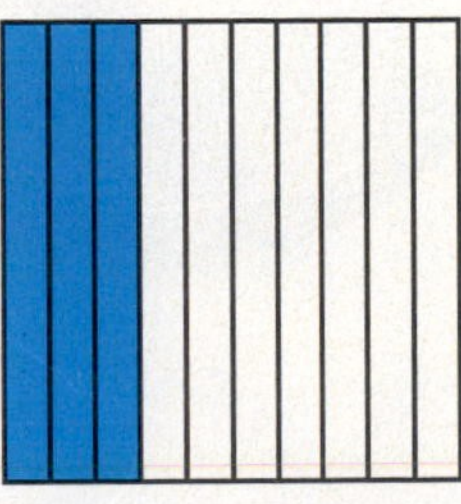

39.
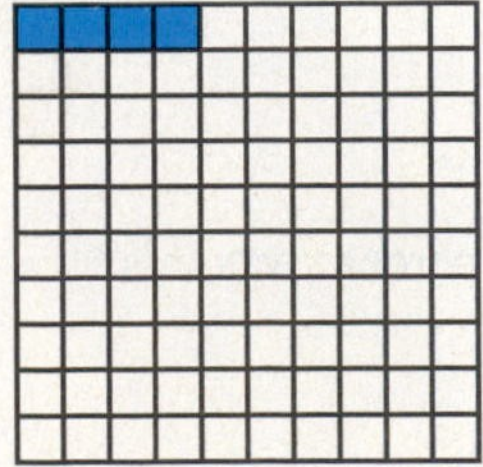

40.
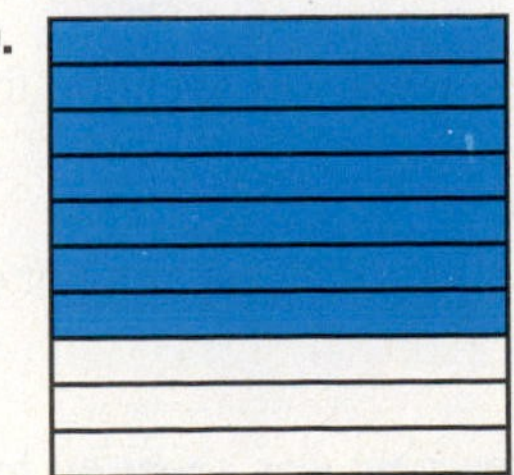

**Answers**

27. ____________

28. ____________

29. ____________

30. ____________

31. ____________

32. ____________

33. ____________

34. ____________

35. ____________

36. ____________

37. ____________

38. ____________

39. ____________

40. ____________

## Answers

41. ______

42. ______

43. ______

44. ______

45. ______

46. ______

*Between 1980 and 2001, the average fuel efficiency of new U.S. cars increased from 24.3 to 28.6 miles per gallon (mi/gal). During this same time, the average fuel efficiency for the entire fleet of U.S. cars rose from 16.0 to 22.1 mi/gal. For exercises 41 and 42, solve the applications involving fuel efficiency.*

**41. Business and Finance** The increase in fuel efficiency for new cars is given by the fraction $\frac{28.6 - 24.3}{24.3}$. Change this fraction to a percent. Round your answer to the nearest tenth.

**42. Business and Finance** The increase in fuel efficiency for the fleet of U.S. cars is given by the fraction $\frac{22.1 - 16.0}{16.0}$. Change this fraction to a percent. Round your answer to the nearest tenth.

*Business travelers were asked how much they spent on different items during a business trip. The following circle shows the results for every $1,000 spent. Use this information to answer exercises 43 to 46.*

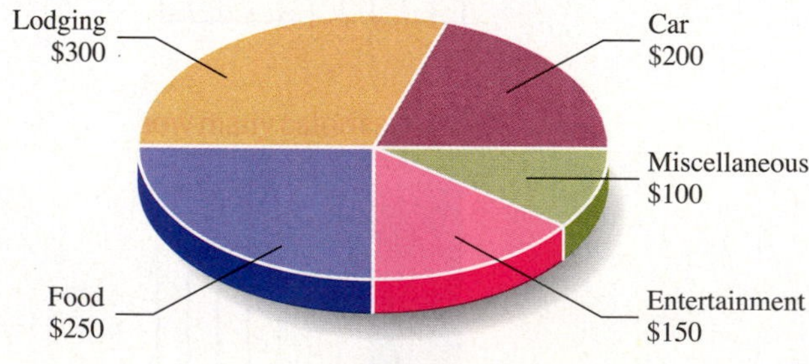


**43.** What percent was spent on car expenses?

**44.** What percent was spent on food?

**45.** Where was the least amount of money spent? What percent was this?

**46.** What percent was spent on food and lodging together?

*Determine whether each statement is* **true** *or* **false.**

**47.** To write a decimal as a percent, move the decimal two places to the right and add the % symbol.

**48.** To write a decimal as a percent, move the decimal two places to the left and add the % symbol.

*Fill in each blank with* **always, sometimes,** *or* **never.**

**49.** A decimal greater than 1 is ________ equivalent to a percent greater than 100%.

**50.** A percent less than 100% can ________ be written as a mixed number.

Basic Skills | **Advanced Skills** | Vocational-Technical Applications | Calculator/Computer | Above and Beyond

*Write each decimal as a percent.*

**51.** 0.002

**52.** 0.008

**53.** 0.004

**54.** 0.001

*Write each fraction or mixed number as a percent.*

**55.** $\frac{1}{6}$

**56.** $\frac{3}{16}$

**57.** $\frac{7}{9}$ (to nearest tenth of a percent)

**58.** $\frac{5}{11}$ (to nearest tenth of a percent)

**59.** $\frac{7}{9}$ (exact value)

**60.** $\frac{5}{11}$ (exact value)

**61.** $5\frac{1}{4}$

**62.** $1\frac{3}{4}$

**63.** $4\frac{1}{3}$ (exact value)

**64.** $2\frac{11}{12}$ (exact value)

**Answers**

47. ________

48. ________

49. ________

50. ________

51. ________

52. ________

53. ________

54. ________

55. ________

56. ________

57. ________

58. ________

59. ________

60. ________

61. ________

62. ________

63. ________

64. ________

## Answers

65. ______

66. ______

67. ______

68. ______

69. ______

70. ______

71. ______

Basic Skills | Advanced Skills | **Vocational-Technical Applications** | Calculator/Computer | Above and Beyond

65. **Automotive Technology** When an automobile engine is cold, its fuel efficiency is only $\frac{13}{16}$ of its maximum efficiency. Express this as a percent.

66. **Welding Technology** It was found that 165 of a total of 172 welds exceeded the required tensile strength. What percent of the welds exceeded the required tensile strength? Round your answer to the nearest whole percent.

67. **Electrical Engineering** The efficiency of an electric motor is defined as the output power divided by the input power. It is usually written as a percent. For a particular motor, the output power is measured to be 400 watts (W) given an input power level of 435 W. What is the efficiency of this motor? Round your answer to the nearest whole percent.

68. **Manufacturing Technology** A manufacturer determines that 2 out of every 27 products will be returned due to defects. What percent of the products will be returned? Round to the nearest tenth percent.

69. **Agricultural Technology** There were 52 in. of rainfall in one growing season. If 6.5 in. fell in August, what percent of the season's rainfall fell in August?

70. Complete the following table of equivalents. Round decimals to the nearest ten-thousandth. Round percents to the nearest hundredth of a percent.

| Fraction | Decimal | Percent |
|---|---|---|
| $\frac{7}{12}$ | | |
| | 0.08 | |
| | | 35% |
| | 0.265 | |
| | | $4\frac{3}{8}\%$ |
| $\frac{11}{18}$ | | |

Basic Skills | Advanced Skills | Vocational-Technical Applications | Calculator/Computer | **Above and Beyond**

71. When writing a fraction as a decimal, explain when you should convert the fraction to a decimal and then write it as a percent, rather than using the proportion method.

72. When writing a fraction as a decimal, explain when you should use the proportion method rather than converting the fraction to a decimal and then writing it as a percent.

## Answers

72. ______

## Answers

**1.** 8% **3.** 5% **5.** 18% **7.** 86% **9.** 40% **11.** 70%
**13.** 110% **15.** 440% **17.** 6.5% **19.** 2.5% **21.** 25%
**23.** 40% **25.** 20% **27.** 62.5% **29.** 31.25% **31.** 350%
**33.** 0.25; $\frac{1}{4}$; 25% **35.** 0.47; $\frac{47}{100}$; 47% **37.** 0.77; $\frac{77}{100}$; 77%
**39.** 0.04; $\frac{1}{25}$; 4% **41.** 17.7% **43.** 20% **45.** Miscellaneous; 10%
**47.** True **49.** always **51.** 0.2% **53.** 0.4% **55.** $16\frac{2}{3}$%
**57.** 77.8% **59.** $77\frac{7}{9}$% **61.** 525% **63.** $433\frac{1}{3}$% **65.** 81.25%
**67.** 92% **69.** 12.5% **71.** Above and Beyond

# Activity 16 :: M&M's

According to M&M's/MARS, M&M's are produced and packaged with approximately the following percentages.

| | |
|---|---|
| Brown: | 30% |
| Yellow: | 20% |
| Red: | 20% |
| Orange: | 10% |
| Green: | 10% |
| Blue: | 10% |

A typical pack was opened, yielding the following counts:

| Brown | Yellow | Red | Orange | Green | Blue |
|---|---|---|---|---|---|
| 14 | 8 | 15 | 6 | 6 | 10 |

Calculate the percent for each color M&M in this pack. Round to the nearest percent.

| Brown | Yellow | Red | Orange | Green | Blue |
|---|---|---|---|---|---|
| | | | | | |

Do any of these seem to differ markedly from the percents named by the M&M's/MARS company? If so, give reasons why this may have occurred.

Obtain your own package of M&M's and determine the percents for each color. Comment on how closely your percents agree with the percents named by the company.

# 6.3 Identifying the Parts of a Percent Problem

< 6.3 Objectives >

1 > Identify the rate in a percent problem

2 > Identify the base in a percent problem

3 > Identify the amount in a percent problem

There are many practical applications of our work with percents. All of these problems have three basic parts that need to be identified. Here are some definitions that will help with that process.

**Definition**

**Base, Amount, and Rate**

The **base** is the whole in a problem. It is the standard used for comparison.

The **amount** is the part of the whole being compared to the base.

The **rate** is the ratio of the amount to the base. It is written as a percent.

The following examples provide some practice in determining the parts of a percent problem.

### Example 1 Identifying Rates

< Objective 1 >

**NOTE**

The *rate R* is the easiest of the terms to identify. The rate is written with the percent symbol (%) or the word *percent*.

Identify each rate.

**(a)** What is 15% of 200?

*R*

Here 15% is the rate because it has the percent symbol attached.

**(b)** 25% of what number is 50?

*R*

25% is the rate.

**(c)** 20 is what percent of 40?

*R*

Here the rate is unknown.

**Check Yourself 1**

Identify the rate.

**(a)** 15% of what number is 75?
**(b)** What is 8.5% of 200?
**(c)** 200 is what percent of 500?

The *base* (*B*) is the whole, or 100%, in the problem. The base often follows the word *of,* as shown in Example 2.

## Example 2 Identifying Bases

< Objective 2 >

Identify each base.

**(a)** What is 15% of 200?

*B* — 200 is the base. It follows the word *of.*

**(b)** 25% of what number is 50?

*B* — Here the base is the unknown.

**(c)** 20 is what percent of 40?

*B* — 40 is the base.

### Check Yourself 2

**Identify the base.**

**(a)** 70 is what percent of 350? **(b)** What is 25% of 300?
**(c)** 14% of what number is 280?

The *amount* (*A*) is the part of the problem remaining once the rate and the base have been identified.

In many applications, the amount is found with the word *is.*

## Example 3 Identifying Amounts

< Objective 3 >

Identify the amount.

**(a)** What is 15% of 200?

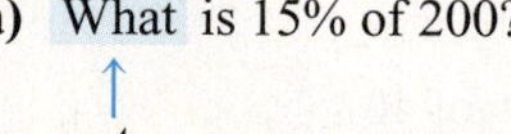

*A* — Here the amount is the unknown part of the problem. Note that the word *is* follows.

**(b)** 25% of what number is 50?

*A* — Here the amount, 50, follows the word *is.*

**(c)** 20 is what percent of 40?

*A* — Again the amount, here 20, can be found with the word *is.*

### Check Yourself 3

Identify the amount.

**(a)** 30 is what percent of 600? **(b)** What is 12% of 5,000?
**(c)** 24% of what number is 96?

In Example 4, we identify all three parts in a percent problem.

## Example 4 Identifying the Rate, Base, and Amount

Determine the rate, base, and amount in this problem:

12% of 800 is what number?

Finding the *rate* is not difficult. Just look for the percent symbol or the word *percent*. In this exercise, 12% is the rate.

The *base* is the whole. Here it follows the word *of.* 800 is the whole or the base.

The *amount* remains when the rate and the base have been found. Here the amount is the unknown. It follows the word *is*. "What number" asks for the unknown amount.

### Check Yourself 4

Find the rate, base, and amount in the following statements or questions.

**(a)** 75 is 25% of 300. **(b)** 20% of what number is 50?

We use percents to solve a variety of applied problems. In all these situations, you have to identify the three parts of the problem. Example 5 is intended to help you build that skill.

## Example 5 Identifying the Rate, Base, and Amount

**(a)** Determine the rate, base, and amount in the following application.

In an algebra class of 35 students, 7 received a grade of A. What percent of the class received an A?

The *base* is the whole in the problem, or the number of students in the class. 35 is the base.

The *amount* is the portion of the base, here the number of students that receive the A grade. 7 is the amount.

The *rate* is the unknown in this example. "What percent" asks for the unknown rate.

**(b)** Determine the rate, base, and amount in the following application:

Doyle borrows $2,000 for 1 year. If the interest rate is 12%, how much interest will he pay?

The *base* is again the whole, the size of the loan in this example. $2,000 is the base.

The *rate* is, of course, the interest rate. 12% is the rate.

The *amount* is the quantity left once the base and rate have been identified. Here the amount is the amount of interest that Doyle must pay. The amount is the unknown in this example.

## Check Yourself 5

**(a)** Determine the rate, base, and amount in the following application: In a shipment of 150 parts, 9 of the parts were defective. What percent were defective?
**(b)** Determine the rate, base, and amount in the following application: Robert earned $120 interest from a savings account paying 8% interest. What amount did he have invested?

**NOTE**

The *wholesale* price of an item is the amount the seller pays for the item. The *retail* price is the price that the seller charges for the item. The difference is the seller's *gross profit* on the item.

Earlier, we defined the *base* in a percent problem as the whole (100%) and the *amount* as the part of the whole being compared to the base. One of the most common applications of percents involves price changes. Whether a retail price is discounted or a wholesale price is marked up, percents are used to determine the new price.

In these applications, the *base* always represents the original price. This might be the price before the markdown, or it might be the wholesale price prior to a markup.

The *amount* in these applications is the quantity being compared to the base. This may be the new price (after markup or markdown), or it may be the difference between the new price and the old price. Which it is depends on the wording used in the application.

## Example 6 Percent Problems Involving Price Changes

**NOTE**

Many businesses use a 25% markup to determine the price to charge for an item. Often, a business then adds or subtracts a penny so that the price of an item ends with a nine or a five, such as $1.49, $99, and $499.95.

**(a)** To determine the retail price of an item, a grocer marks up the wholesale price by 25%. The markup on a bunch of organic Italian parsley that costs the grocer $1.20 is 30¢. Identify the rate, base, and amount.

The original or wholesale price is $1.20. This represents the base.

The markup is 30¢ ($0.30), which represents the amount.

The rate is 25%.

**(b)** In part **(a)** the grocer sells the parsley for $1.50 per bunch. Identify the rate, base, and amount using the selling price of $1.50 rather than the 30¢ markup.

The base is still the wholesale price of $1.20.

The amount is now the selling price not the markup. Therefore, the amount is $1.50.

The rate in this problem is used to determine the selling price not just the gross profit earned on the selling price. So the rate includes both the 25% markup (30¢) and 100% of the original price ($1.20).

Therefore, the rate is 125%.

## Check Yourself 6

Identify the base, amount, and rate in each of the following applications.

**(a)** A mason tender is paid $22.50 per hour at a construction site. The contractor marks up labor costs 10% when billing the client. In the case of a mason tender, the markup amounts to $2.25.
**(b)** In the application given in part **(a)**, the client is billed $24.75 per hour for labor provided by a mason tender.

## Check Yourself ANSWERS

**1.** **(a)** 15%; **(b)** 8.5%; **(c)** "what percent" (the unknown) **2.** **(a)** 350; **(b)** 300; **(c)** "what number" (the unknown) **3.** **(a)** 30; **(b)** "What" (the unknown); **(c)** 96 **4.** **(a)** $R$ = 25%, $B$ = 300, $A$ = 75; **(b)** $R$ = 20%, $B$ = "What number," $A$ = 50
**5.** **(a)** $R$ = "What percent" (the unknown), $B$ = 150, $A$ = 9; **(b)** $R$ = 8%, $B$ = "What amount," $A$ = \$120
**6.** **(a)** $R$ = 10%, $B$ = \$22.50, $A$ = \$2.25; **(b)** $R$ = 110%, $B$ = \$22.50, $A$ = \$24.75

## Reading Your Text

The following fill-in-the-blank exercises are designed to ensure that you understand some of the key vocabulary used in this section.

### SECTION 6.3

**(a)** In percent problems, the ______________ is the standard used for comparison.

**(b)** In percent problems, the ______________ is the ratio of the amount to the base.

**(c)** In percent problems, the ______________ is often written as a percent.

**(d)** The amount a seller pays for an item is referred to as the ______________ price of the item.

# 6.3 exercises

Basic Skills | Advanced Skills | Vocational-Technical Applications | Calculator/Computer | Above and Beyond

MathZone

Boost your grade at mathzone.com!

> Practice Problems
> NetTutor
> Self-Tests
> e-Professors
> Videos

Name ____________

Section ________ Date ________

## Answers

1. ____________
2. ____________
3. ____________
4. ____________
5. ____________
6. ____________
7. ____________
8. ____________
9. ____________
10. ____________
11. ____________
12. ____________
13. ____________
14. ____________
15. ____________

< Objectives 1–3 >

*Identify the rate, base, and amount in each statement or question.* Do not solve *the exercise at this point.*

**1.** 23% of 400 is 92.

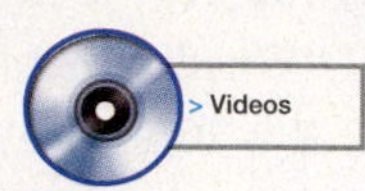

**2.** 150 is 20% of 750.

**3.** 40% of 600 is 240.

**4.** 200 is 40% of 500.

**5.** What is 7% of 325?

**6.** 80 is what percent of 400?

**7.** 16% of what number is 56?

**8.** What percent of 150 is 30?

**9.** 480 is 60% of what number?

**10.** What is 60% of 250?

**11.** What percent of 120 is 40?

**12.** 150 is 75% of what number?

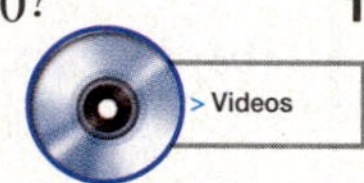

*Identify the rate, base, and amount in the following applications.* Do not solve *the applications at this point.*

**13. Business and Finance** Jan has a 5% commission rate on all her sales. If she sells $40,000 worth of merchandise in 1 month, what commission will she earn?

**14. Business and Finance** 22% of Shirley's monthly salary is deducted for withholding. If those deductions total $209, what is her salary?

**15. Science and Medicine** In a chemistry class of 30 students, 5 received a grade of A. What percent of the students received A's?

**16.** **Business and Finance** A can of mixed nuts contains 80% peanuts. If the can holds 16 ounces, how many ounces of peanuts does it contain?

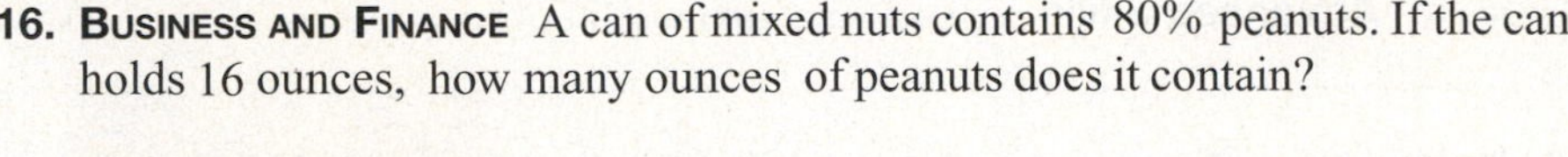

**17.** **Business and Finance** The sales tax rate in a state is 5.5%. If you pay a tax of $3.30 on an item that you purchase, what is its selling price?

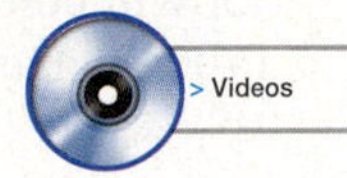


**18.** **Business and Finance** In a shipment of 750 parts, 75 were found to be defective. What percent of the parts were faulty?

**19.** **Social Science** A college had 9,000 students at the start of a school year. If there is an enrollment increase of 6% by the beginning of the next year, how many additional students are there?

**20.** **Business and Finance** Paul invested $5,000 in a time deposit. What interest will he earn for 1 year if the interest rate is 3.5%?

chapter 6 > Make the Connection

*Determine whether each statement is* **true** *or* **false.**

**21.** The amount that the seller charges for an item is referred to as the gross profit.

**22.** The base, in a percent problem, often follows the word *of.*

*Fill in each blank with* **always, sometimes,** *or* **never.**

**23.** The rate, in a percent problem, is __________ greater than 100%.

**24.** The new price, after a markup, is __________ the amount in a percent problem.

## Answers

16. ____________________

17. ____________________

18. ____________________

19. ____________________

20. ____________________

21. ____________________

22. ____________________

23. ____________________

24. ____________________

## Answers

25. ____________

26. ____________

27. ____________

28. ____________

29. ____________

30. ____________

31. ____________

32. ____________

33. ____________

34. ____________

Basic Skills | **Advanced Skills** | Vocational-Technical Applications | Calculator/Computer | Above and Beyond

**25.** **Business and Finance** A technician bills a client for parts and labor. When billing a client for parts, the technician marks up the price by 10%. If the technician pays $80 for a part, the client is billed $88. If $88 is the amount and $80 is the base, what is the rate?

**26.** **Business and Finance** A state charges 4.25% sales tax on goods sold. A customer pays $8.33 for a paperback book that lists for $7.99. Using $7.99 as the base and $8.33 as the amount, report the rate used.

**27.** **Business and Finance** An employee has 26% of his weekly salary deducted for withholding. If the base is his gross pay and the amount is his net pay (after deductions), report the rate.

**28.** **Manufacturing Technology** In a shipment of 450 parts, 8% of them (or 36) were found to be defective. What rate describes the percent of parts that were *not* defective?

Basic Skills | Advanced Skills | **Vocational-Technical Applications** | Calculator/Computer | Above and Beyond

*Identify the rate, base, and amount in each statement or question.* Do not solve *the problems at this point.*

**29.** **Allied Health** How much 25% alcohol solution can be prepared using 225 milliliters (mL) of ethyl alcohol?

**30.** **Information Technology** An Ethernet network transmits a maximum packet size of 1,500 bytes. 1.7% of each packet is "overhead." How much information (in bytes) in a maximum-size packet is overhead?

**31.** **Manufacturing Technology** The price of steel from a supplier increased 20% to $950 per ton. What was the old price for a ton of steel?

**32.** **Agricultural Technology** Milk that is labeled "3.5%" is made up of 3.5% butterfat. How many grams of butterfat are in 1 liter (938 g) of 3.5% milk?

**33.** **Environmental Technology** In some communities, "green" laws require that 40% of a lot remains green (covered in grass or other vegetation). How much green space is required in a 12,680-square-foot lot?

**34.** **Automotive Technology** Atkins Auto Repair marks up parts 25% above cost. How much would Atkins Auto Repair charge for a starter it bought for $62?

Basic Skills | Advanced Skills | Vocational-Technical Applications | Calculator/Computer | **Above and Beyond**

**35. Social Science** Using the latest census figures for your state, determine the percent of the following population groups: African-American, Hispanic, and Asian. These figures can be found on the World Wide Web at **www.census.gov/.**

## Answers

35. ______________

## Answers

1. 23% of 400 is 92
   ↑ 23% = R, ↑ 400 = B, ↑ 92 = A

3. 40% of 600 is 240
   ↑ 40% = R, ↑ 600 = B, ↑ 240 = A

5. What is 7% of 325
   ↑ What = A, ↑ 7% = R, ↑ 325 = B

7. 16% of what number is 56
   ↑ 16% = R, ↑ what number = B, ↑ 56 = A

9. 480 is 60% of what number
   ↑ 480 = A, ↑ 60% = R, ↑ what number = B

11. What percent of 120 is 40

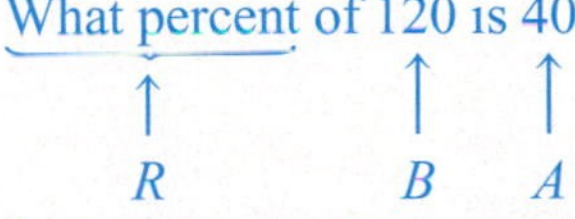

   ↑ What percent = R, ↑ 120 = B, ↑ 40 = A

13. \$40,000 is the base. 5% is the rate. Her commission, the unknown, is the amount.
15. 30 is the base. 5 is the amount. The unknown percent is the rate.
17. 5.5% is the rate. The tax, \$3.30, is the amount. The unknown selling price is the base.
19. The base is 9,000. The rate is 6%. The unknown number of additional students is the amount.
21. False
23. sometimes
25. 110%
27. 74%
29. $R = 25\%$, $B =$ unknown, $A = 225$ mL
31. $R = 120\%$, $B =$ unknown, $A = \$950$
33. $R = 40\%$, $B = 12{,}680\ \text{ft}^2$, $A =$ unknown
35. Above and Beyond

# Activity 17 ::
# A Matter of Interest

When you put your money into savings for investment purposes, you will often leave it for some time, allowing it to earn interest and grow. Typically, the interest your investment earns is computed periodically as a percent of your investment. The investment is referred to as the **principal,** and we use the following **interest** formula:

Interest = principal $\times$ rate $\times$ time

or $I = P \times R \times T$. If we use 1 year for time, the formula becomes $I = P \times R$. Note the following (for time = 1 year):

Interest = principal $\times$ rate

and

Amount = base $\times$ rate

which is the *percent relationship* seen in Section 6.4.

Suppose that your original investment is \$1,000 and that your money earns interest at a rate of 4% (per year). Let us further suppose that the interest is computed at the end of each year and then added onto the principal. The principal then grows, and the amount of interest in the following year will be greater. To see how this works, complete the following table.

| Year | Principal | Interest | Investment at End of Year |
|---|---|---|---|
| 1 | \$1,000 | \$40 | \$1,040 |
| 2 | 1,040 | | |
| 3 | | | |
| 4 | | | |
| 5 | | | |
| 6 | | | |
| 7 | | | |

After 7 years, how much has your investment grown?

For how many years must the money sit to earn total interest that is 50% of your original investment? (*Hint:* Continue the calculations in the table.)

# 6.4 Solving Percent Problems

< 6.4 Objectives >

1 > Solve a percent problem

2 > Solve an application of percents

3 > Solve applications that involve percent increase and decrease

4 > Solve percent applications involving interest

The concept of percent is perhaps the most frequently encountered arithmetic idea considered in this text.

In Section 6.3, you learned that statements and problems about percents consist of three basic parts: the rate, base, and amount. In fact, every problem involving percents consists of these three parts.

In nearly every percent problem, one of these three parts is missing. Solving a percent problem is a matter of identifying and computing the missing part. To do this, we use the **percent relationship.**

**NOTE**

This means that $r$ is the number before the percent symbol.

In this section, we will use the notation that $r = 100R$ for a rate $R$. In which case, we have $R = \frac{r}{100}$. So, for example, if the rate is $R = 54\% = \frac{54}{100}$, then we have $r = 54$.

**Property**

**The Percent Relationship**

In a percent problem, the quotient of the *amount* and the *base* is equal to the *rate*. In symbols, we have

$$\frac{A}{B} = R$$

With this notation, we can write this as a proportion problem

$$\frac{A}{B} = \frac{r}{100}$$

**NOTE**

When solving this proportion, we have $100A = r \cdot B$.

In this section, we illustrate how to solve percent problems. We first explain how to solve a problem of each type with an example. Then, we provide an application providing you with an example of how such a problem might occur in real life.

In Example 1, the amount is the unknown. In this case, we identify the base and rate and use the percent relationship, written as a proportion.

Instructor's note: In this section, we use proportions to solve percent problems. After introducing students to algebra, we use an equations approach in Section 11.5.

## Example 1 Finding an Unknown Amount

< Objective 1 >

What is 18% of 300?

In this statement, we know the rate, 18%, and the base, 300. The amount is unknown. Use the percent relationship to write the problem as a proportion.

$$\frac{A}{300} = \frac{18}{100}$$

Applying the proportion rule that we learned in Section 5.3 gives

$$100 \cdot A = 18 \cdot 300$$

or $$100A = 5{,}400$$

**RECALL**

If the rate is 18%, then $r = 18$ in the percent proportion $\frac{A}{B} = \frac{r}{100}$.

We identify the coefficient of the variable as 100 and divide by this number.

$$\frac{100A}{100} = \frac{5{,}400}{100}$$

$$A = 54$$

So, 54 is 18% of 300.

### Check Yourself 1

**Find 65% of 200.**

In Examples 2 and 3, we solve percent applications in which the amount is the unknown quantity.

If the rate is *less than* 100%, then the amount will be *less than* the base.

20 is 40% of 50 and $20 < 50$

If the rate is *greater than* 100%, then the amount will be *greater than* the base.

75 is 150% of 50 and $75 > 50$

## Example 2 Solving a Percent Application with an Unknown Amount

< Objective 2 >

A student needs to answer at least 70% of the questions correctly on a 50-question exam in order to pass. How many questions must the student get right?

In this case, the rate is 70%, so $r = 70$, and the base, or total, is 50 questions, so $B = 50$.

We write the percent relationship as a proportion and solve.

$$\frac{A}{50} = \frac{70}{100}$$

$$100A = 70 \cdot 50$$ If $\frac{a}{b} = \frac{c}{d}$, then $ad = bc$.

$$100A = 3{,}500$$ The coefficient of the variable is 100.

$$\frac{100A}{100} = \frac{3{,}500}{100}$$ Divide both sides by the coefficient of the variable.

$$A = 35$$ $3{,}500 \div 100 = 35$.

The student must answer 35 questions correctly in order to pass the exam.

## Check Yourself 2

**Generally, 72% of the students in a chemistry course pass the class. If there are 150 students in the class, how many are expected to pass?**

**NOTE**

Percent increase and percent decrease **always** describe the change starting from the base.

A common type of application involves *percent increase* or *percent decrease*. In this type of problem, the amount is described in terms of how much the base must be increased (or decreased) as a percent of the base.

A discount is usually presented in terms of a percent of the base rather than a fixed dollar amount because a discount needs to be reasonable given the cost of an item.

For instance, a \$20 discount on a \$100 item is a 20% discount and is a significant amount. Whereas, a \$20 discount on a \$20,000 car is not worth advertising as it is not a significant change in the price, amounting to a 0.1% difference. On the other hand, a 20% discount on a \$20,000 car amounts to a discount of \$4,000 which is much more noticeable.

In Example 3, we look at one such problem which involves an unknown amount.

## Example 3 Percent Increase with an Unknown Amount

< Objective 3 >

A state adds a 7.25% sales tax to the price of most goods. If a 30-GB iPod is listed for \$299, how much will it cost after the sales tax has been added?

**NOTE**

We could use $R = 7.25\%$, but then, after computing the amount, we would need to add it to the original price to get the actual selling price.

This problem is similar to the application in Example 1, in that we are missing the amount. In this problem, we are told that we need to increase the base by 7.25% in order to find the amount.

Because we need to add the sales tax to the original price, the base, we have a situation similar to one discussed near the end of Section 6.3. In this case, we use 107.25% as the rate.

As before, we use the percent relationship to solve the problem.

$$\frac{A}{299} = \frac{107.25}{100}$$

$$100A = (107.25) \cdot (299)$$

$$100A = 32{,}067.75$$

$$\frac{100A}{100} = \frac{32{,}067.75}{100}$$

$$A = 320.6775$$

**NOTE**

\$320.68 is a 7.25% increase of \$299.

Because our answer refers to money, we round to two decimal places. The iPod sells for \$320.68, after the sales tax has been included.

### Check Yourself 3

**In order to make room for the new fall line of merchandise, a proprietor offers to discount all existing stock by 15%. How much will the store charge for a Fendi handbag that normally sells for \$229?**

The second type of percent problem that we look at requires us to find an unknown rate. We solve these problems in the same way as before—we write the problem as a proportion and solve it.

## Example 4 Finding an Unknown Rate

30 is what percent of 150?

We know that 30 is the amount and 150 is the base. The rate is the unknown.

$$\frac{30}{150} = \frac{r}{100} \qquad \frac{A}{B} = \frac{r}{100}.$$

$$150 \cdot r = 30 \cdot 100 \qquad \text{Rewrite the equation using the proportion rule.}$$

$$150r = 3{,}000 \qquad \text{The coefficient of the variable is 150.}$$

$$\frac{150r}{150} = \frac{3{,}000}{150} \qquad \text{Divide both sides by the coefficient of the variable.}$$

$$r = 20 \qquad 3{,}000 \div 150 = 20.$$

Therefore, 30 is 20% of 150.

**RECALL**

If $r = 20$, then the rate is given by

$$R = \frac{r}{100} = \frac{20}{100} = 20\%$$

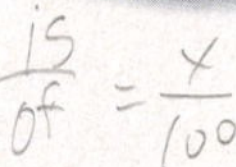

### Check Yourself 4

**75 is what percent of 300?**

We now work through a couple of applications involving an unknown rate.

If the amount is *less than* the base, then the rate will be *less than* 100%.

If the amount is *greater than* the base, then the rate will be *greater than* 100%.

## Example 5 Solving a Percent Application with an Unknown Rate

Simon waits tables at La Catalana, an upscale restaurant. A family left a \$45 tip on a \$250 meal. What percent of the bill did the family leave as a tip?

We are asked to find the rate given a base $B = \$250$ and an amount $A = \$45$. To find the rate, we use the percent relationship and the proportion rule.

$$\frac{45}{250} = \frac{r}{100}$$

$$250r = 4{,}500 \qquad 45 \cdot 100 = 4{,}500.$$

$$\frac{250r}{250} = \frac{4{,}500}{250}$$

$$r = 18 \qquad 4{,}500 \div 250 = 18.$$

Therefore, the family left a tip that was 18% of the bill.

**NOTE**

Remember to write your answer as a percent.

## Check Yourself 5

**Last year, Xian reported an income of $27,500 on her tax return. Of that, she paid $6,600 in taxes. What percent of her income did she pay in taxes?**

We often encounter percent increase and decrease problems involving an unknown rate. In Example 6, we look for the rate that describes a discount, or percent decrease.

## Example 6 Percent Decrease with an Unknown Rate

A kitchen store offers an All-Clad saucepan on sale for $92. The saucepan normally sells for $115. By what percent is the saucepan discounted?

> CAUTION

The amount is not $92. The amount is the size of the discount, or decrease.

We begin by identifying the base as the original price, $115, and the rate as the unknown. Because we are looking for the percent that the original price has been decreased, the amount is the difference between the original price and the sale price.

$A = \$115 - \$92 = \$23$

Now we are ready to apply the percent relationship to solve this problem.

$$\frac{23}{115} = \frac{r}{100}$$

$$115r = 2{,}300 \qquad 23 \cdot 100 = 2{,}300$$

$$r = \frac{2{,}300}{115} \qquad \text{Divide both sides by the coefficient of the variable, 115.}$$

$$= 20 \qquad 2{,}300 \div 115 = 20.$$

Therefore, the sale price represents a 20% decrease of the original price.

**NOTE**

The *markup* is the amount a store adds to the price of an item to cover expenses and profit. Retailers usually mark an item up by a percentage of its wholesale cost.

## Check Yourself 6

**An electronics store sells a certain Kicker amplifier for a car stereo system for $249.95. If the store pays $199.95 for the amplifier, what is the markup, or percent increase (to the nearest whole percent)?**

The final type of percent problem is one with a missing base.

## Example 7 Finding an Unknown Base

**RECALL**

$$\frac{A}{B} = \frac{r}{100}$$

28 is 40% of what number?

The amount is 28 and the rate is 40%. We use these to set up the percent proportion with a missing base.

$$\frac{28}{B} = \frac{40}{100}$$

As before, we use the proportion rule to solve this.

$$40B = 2{,}800 \qquad 28 \cdot 100 = 2{,}800.$$

$$\frac{40B}{40} = \frac{2{,}800}{40} \qquad \text{40 is the coefficient of the variable.}$$

$$B = 70 \qquad 2{,}800 \div 40 = 70.$$

Therefore, 28 is 40% of 70.

### Check Yourself 7

**70 is 35% of what number?**

We can now look at some applications involving an unknown base.

## Example 8 Solving a Percent Application with an Unknown Base

A computer ran 60% of a scan in 120 seconds (s). How long should it take to complete an entire scan?

In this case, the rate is 60% and the amount is 120 s. We employ the percent relationship, as in Example 7.

$$\frac{120}{B} = \frac{60}{100}$$

Next, we solve the percent proportion, as before.

$$60B = 12{,}000$$

$$B = \frac{12{,}000}{60} \qquad \text{Divide by the coefficient of the variable, 60.}$$

$$= 200$$

The entire scan should take 200 s.

### Check Yourself 8

**An indexing program takes 4 minutes to check 30% of the files on a laptop computer. How long should it take to index all the files (to the nearest second)?**

In some cases, the base amount is the unknown quantity in a percent increase or decrease problem.

## Example 9 Percent Increase with an Unknown Base

A grocery store adds a 30% markup to the wholesale price of an item to determine the retail price. If the store sells a half-gallon container of orange juice for \$2.99, what is the wholesale price of the juice?

The selling price is the amount, \$2.99.

To determine the rate, we use the fact that the selling price represents a 30% increase over the wholesale, or base, price. This means that it is 30% more than the full, 100%, wholesale price. So, the rate is 130%.

**RECALL**

$130\% = \frac{130}{100}$

$$\frac{2.99}{B} = \frac{130}{100} \qquad \frac{A}{B} = \frac{r}{100}.$$

$$130B = 299 \qquad 2.99 \cdot 100 = 299.$$

$$B = \frac{299}{130} \qquad \text{130 is the coefficient of the variable.}$$

$$= 2.3 \qquad 299 \div 130 = 2.3.$$

The wholesale price of a half-gallon container of juice is \$2.30.

### Check Yourself 9

A toy store is selling the Fisher-Price Rollin' Rumblin' Dump Truck at a 10% discount for \$16.19. How much does the toy normally sell for?

**NOTE**

The money borrowed or invested is called the **principal.**

Another common application of percent is interest. When you take out a home loan, you pay interest; when you invest money in a savings account, you earn interest. Interest is a percent of the whole, and the percent is called the **interest rate.**

When we work with interest on a certain amount of money (the principal) for a specific time period, the interest is called **simple interest.** For such applications, we use the *simple-interest formula.*

**Property**

**The Simple-Interest Formula**

$$\frac{\text{Interest}}{\text{Principal}} = \text{rate} \times \text{time}$$

or

$$\frac{I}{P} = R \cdot T$$

**NOTES**

Credit cards usually express time in days.

In fact, as long as we are working with a time of one unit, we can simplify the formula like this.

To use the formula, there must be *time agreement.* If the interest rate is 9% per year, then time must be expressed in years. If the rate is $1\frac{1}{2}\%$ per month, then time must be expressed in months.

If we are working with an annual rate and the time is 1 year, then the formula becomes

$$\frac{\text{Interest}}{\text{Principal}} = \text{rate}$$

**RECALL**

$R = \frac{r}{100}$

Using the notation from the beginning of this section, along with $I$ for interest and $P$ for principal, we have

$$\frac{I}{P} = \frac{r}{100}$$

This looks just like the percent proportion that we have been using throughout this section. In this case, the *interest* serves the same role as the "amount," and the *principal* takes the part of the "base."

## Example 10 Solving a Simple-Interest Application

< Objective 4 >

Find the interest you must pay if you borrow \$2,000 for 1 year at an interest rate of $9\frac{1}{2}\%$.

The time period is one unit (year, in this case), so we can use the simplified interest formula with $P = 2{,}000$ and $r = 9\frac{1}{2}$.

$$\frac{I}{2{,}000} = \frac{9\frac{1}{2}}{100}$$

As before, we use the proportion rule to rewrite the equation.

$$100 \cdot I = 2{,}000 \cdot 9\frac{1}{2}$$

$$100I = 19{,}000 \qquad 2{,}000 \cdot 9\frac{1}{2} = 19{,}000.$$

The variable here is $I$ and the number in front, 100, is its coefficient, so we divide both sides by 100.

$$\frac{\cancel{100}I}{\cancel{100}} = \frac{19{,}000}{100}$$

$$I = 190 \qquad 19{,}000 \div 100 = 190.$$

The interest (amount) is \$190.

**Check Yourself 10**

You invest \$5,000 for 1 year at an annual rate of $8\frac{1}{2}\%$. How much interest will you earn?

As with other percent problems, you can find the principal or the rate using the percent proportion relationship.

## Example 11 Solving a Simple-Interest Application

Ms. Hobson agrees to pay 11% interest on a loan for her new car. She is charged \$2,200 interest on the loan for 1 year. How much did she borrow?

The rate is 11% and the interest (or amount) is \$2,200. We need to find the principal (or base), which is the size of the loan.

As before, we set up the percent relationship and use the proportion rule to solve it.

$$\frac{2{,}200}{P} = \frac{11}{100} \qquad 11\% = \frac{11}{100}.$$

$$11P = 220{,}000 \qquad 2{,}200 \cdot 100 = 220{,}000.$$

$$\frac{11P}{11} = \frac{220{,}000}{11} \qquad \text{The coefficient of the variable is 11.}$$

$$P = 20{,}000 \qquad 220{,}000 \div 11 = 20{,}000.$$

Ms. Hobson borrowed \$20,000 to purchase her car.

### Check Yourself 11

**Sue pays \$210 in interest for a 1-year loan at 10.5% interest. What is the size of her loan?**

The true power of interest to earn money over time comes from the idea of **compound interest.** This means that once you earn (or are charged) interest, you start earning (or paying) interest on the interest. This is what is meant by compounding.

Compound interest is an exceptionally powerful idea. For many people, their retirement plans are not based on the amount of money invested, but rather on the fact that their retirement account grows with time.

The earnings on someone's retirement investments *compound* over decades and grow into much more than the original investments. This is why financial advisors always suggest that you should start saving for retirement as early as possible. The more time an investment has to grow, the larger it gets.

We conclude this section by looking at an example involving the compounding of interest.

## Example 12 Solving a Compound-Interest Application

Suppose you invest $1,000 at 5% (compounded annually) in a savings account for 2 years. How much will you have in the account at the end of the 2-year period?

After the first year, you earn 5% interest on your $1,000 principal.

$$\frac{I}{1{,}000} = \frac{5}{100}$$
$$100I = 5{,}000$$
$$I = \frac{5{,}000}{100}$$
$$= 50$$

So, you earn $50 after 1 year. The compounding comes into play in the second year when your account begins with $1,000 + $50 = $1,050 in it.

$$\underset{\text{Start}}{\$1{,}000} \xrightarrow{\text{At } 5\%} \underset{\text{Year 1}}{\$1{,}050}$$

In the second year, you earn interest on your original principal, plus you earn interest on the $50 that you earned in the first year. This process is called *compound interest.*

In the second year, you earn 5% interest on your account balance.

$$\frac{I}{1{,}050} = \frac{5}{100}$$
$$100I = 5{,}250 \qquad 5 \cdot 1{,}050 = 5{,}250.$$
$$I = \frac{5{,}250}{100}$$
$$= 52.5 \qquad 5{,}250 \div 100 = 52.5.$$

So, you earn $52.50 in the second year of your investment. Therefore, your account balance will be $1,050 + $52.50 = 1,102.50.

$$\underset{\text{Start}}{\$1{,}000} \xrightarrow{\text{At } 5\%} \underset{\text{Year 1}}{\$1{,}050} \xrightarrow{\text{At } 5\%} \underset{\text{Year 2}}{\$1{,}102.50}$$

### Check Yourself 12

**If you invest $6,000 at 4% (compounded annually) for 2 years, how much will you have after 2 years?**

### Check Yourself ANSWERS

**1.** 130 **2.** 108 students **3.** $194.65 **4.** 25% **5.** 24% **6.** 25% **7.** 200 **8.** 800 s **9.** $17.99 **10.** $425 **11.** $2,000 **12.** $6,489.60

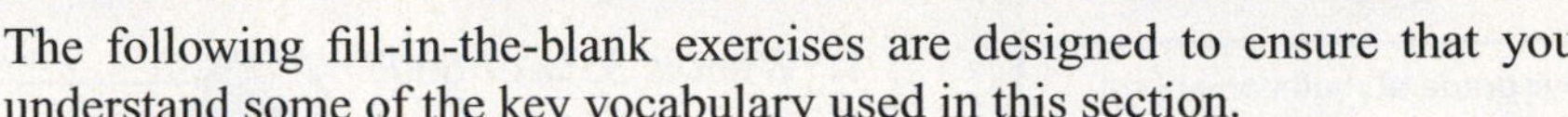

## Reading Your Text

The following fill-in-the-blank exercises are designed to ensure that you understand some of the key vocabulary used in this section.

### SECTION 6.4

**(a)** In the percent relationship, the ______________ is given by the product of the rate and the base.

**(b)** To use the percent relationship, the ______________ must be written as either a fraction or a decimal.

**(c)** If the amount is ______________ than the base, the rate will be less than 100%.

**(d)** In an investment, the money invested is called the ______________.

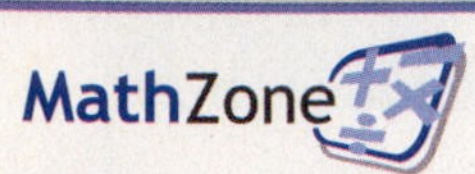

# 6.4 exercises

Basic Skills | Advanced Skills | Vocational-Technical Applications | Calculator/Computer | Above and Beyond

MathZone

Boost your grade at mathzone.com!

- > Practice Problems
- > NetTutor
- > Self-Tests
- > e-Professors
- > Videos

Name ____________

Section ______ Date ______

## Answers

| | |
|---|---|
| 1. | 2. |
| 3. | 4. |
| 5. | 6. |
| 7. | 8. |
| 9. | 10. |
| 11. | 12. |
| 13. | 14. |
| 15. | 16. |
| 17. | 18. |
| 19. | 20. |
| 21. | 22. |
| 23. | 24. |
| 25. | 26. |
| 27. | 28. |
| 29. | 30. |
| 31. | 32. |
| 33. | 34. |

< Objective 1 >

*Solve each of the following percent problems.*

**1.** What is 35% of 600?

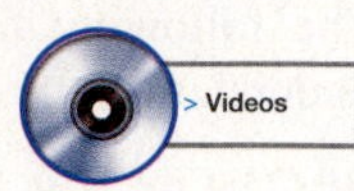


**2.** 20% of 400 is what number?

**3.** 45% of 200 is what number?

**4.** What is 40% of 1,200?

**5.** Find 40% of 2,500.

**6.** What is 75% of 120?

**7.** What percent of 50 is 4?

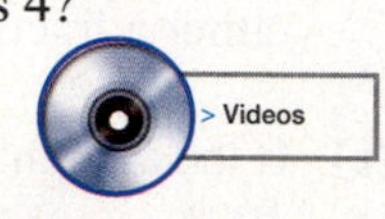


**8.** 51 is what percent of 850?

**9.** What percent of 500 is 45?

**10.** 14 is what percent of 200?

**11.** What percent of 200 is 340?

**12.** 392 is what percent of 2,800?

**13.** 46 is 8% of what number?

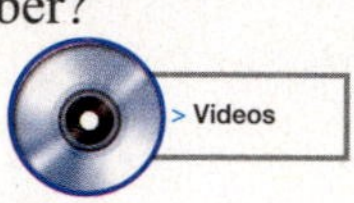


**14.** 7% of what number is 42?

**15.** Find the base if 11% of the base is 55.

**16.** 16% of what number is 192?

**17.** 58.5 is 13% of what number?

**18.** 21% of what number is 73.5?

**19.** Find 110% of 800.

**20.** What is 115% of 600?

**21.** What is 108% of 4,000?

**22.** Find 160% of 2,000.

**23.** 210 is what percent of 120?

**24.** What percent of 40 is 52?

**25.** 360 is what percent of 90?

**26.** What percent of 15,000 is 18,000?

**27.** 625 is 125% of what number?

**28.** 140% of what number is 350?

**29.** Find the base if 110% of the base is 935.

**30.** 130% of what number is 1,170?

< Objectives 2–4 >

*Solve each of the following applications.*

**31.** **Business and Finance** What interest will you pay on a $3,400 loan for 1 year if the interest rate is 12%?

**32.** **Science and Medicine** A chemist has 300 milliliters (mL) of solution that is 18% acid. How many milliliters of acid are in the solution?

**33.** **Business and Finance** If a salesman is paid a $140 commission on the sale of a $2,800 sailboat, what is his commission rate?

**34.** **Business and Finance** Ms. Jordan has been given a loan of $2,500 for 1 year. If the interest charged is $275, what is the interest rate on the loan?

35. **Statistics** On a test, Alice had 80% of the problems right. If she had 20 problems correct, how many questions were on the test?

> Videos

36. **Business and Finance** A state sales tax rate is 3.5%. If the tax on a purchase was \$7, what was the price of the purchase?

37. **Business and Finance** A state sales tax is levied at a rate of 6.4%. How much tax would one pay on a purchase of \$260?

38. **Business and Finance** Betty must make a $9\frac{1}{2}\%$ down payment on the purchase of a \$6,000 motorcycle. What is her down payment?

39. **Social Science** A study has shown that 102 of the 1,200 people in the work-force of a small town are unemployed. What is the town's unemployment rate?

> Videos

40. **Social Science** A survey of 400 people found that 66 were left-handed. What percent of those surveyed were left-handed?

41. **Statistics** In a recent survey, 65% of those responding were in favor of a freeway improvement project. If 780 people were in favor of the project, how many people responded to the survey?

42. **Social Science** A college finds that 42% of the students taking a foreign language are enrolled in Spanish. If 1,512 students are taking Spanish, how many foreign language students are there?

43. **Business and Finance** An appliance dealer marks up refrigerators 22% (based on cost). If the wholesale cost of one model is \$1,200, what should its selling price be?

44. **Social Science** A school had 900 students at the start of a school year. If there is an enrollment increase of 7% by the beginning of the next year, what is the new enrollment?

45. **Business and Finance** The price of a new van has increased \$2,030, which amounts to a 14% increase. What was the price of the van before the increase?

## Answers

35. ______
36. ______
37. ______
38. ______
39. ______
40. ______
41. ______
42. ______
43. ______
44. ______
45. ______

**Answers**

46. ______

47. ______

48. ______

49. ______

50. ______

51. ______

52. ______

53. ______

54. ______

46. **Business and Finance** A television set is marked down $75, to be placed on sale. If this is a 12.5% decrease from the original price, what was the selling price before the sale?

47. **Business and Finance** Carlotta received a monthly raise of $162.50. If this represented a 6.5% increase, what was her monthly salary before the raise?

48. **Business and Finance** Mr. Hernandez buys stock for $15,000. At the end of 6 months, the stock's value has decreased 7.5%. What is the stock worth at the end of the period?

chapter 6 > Make the Connection

49. **Social Science** The population of a town increases 14% in 2 years. If the population was 6,000 originally, what is the population after the increase?

50. **Business and Finance** A store marks up merchandise 25% to allow for profit. If an item costs the store $11, what will its selling price be?

51. **Business and Finance** A virus scanning program is checking every file for viruses. It has completed checking 40% of the files in 300 seconds. How long should it take to check all the files?

52. **Business and Finance** The price of a pair of shorts advertised for $48.75 is 25% off the original price. What was the original price?

53. **Social Science** In 1990, there were an estimated 145.0 million passenger cars registered in the United States. The total number of vehicles registered in the United States for 1990 was estimated at 194.5 million. What percent of the vehicles registered were passenger cars? Round to the nearest tenth percent.

54. **Social Science** Gasoline accounts for 85% of the motor fuel consumed in the United States every day. If 8,882 thousand barrels (bbl) of motor fuel is consumed each day, how much gasoline is consumed each day in the United States? Round to the nearest thousand barrels.

55. **Social Science** In 1999, transportation accounted for 67% of U.S. petroleum consumption. If 13.3 million bbl of petroleum were used each day for transportation in the United States, what was the approximate total daily petroleum consumption by all sources in the United States?

56. **Science and Medicine** Each year, 540 million metric tons (t) of carbon dioxide is added to the atmosphere by the United States. Burning gasoline and other transportation fuels is responsible for 35% of the carbon dioxide emissions in the United States. How much carbon dioxide is emitted each year by the burning of transportation fuels in the United States?

57. **Social Science** The progress of the local Lions club fund drive is shown here. What percent of the goal has been achieved so far?

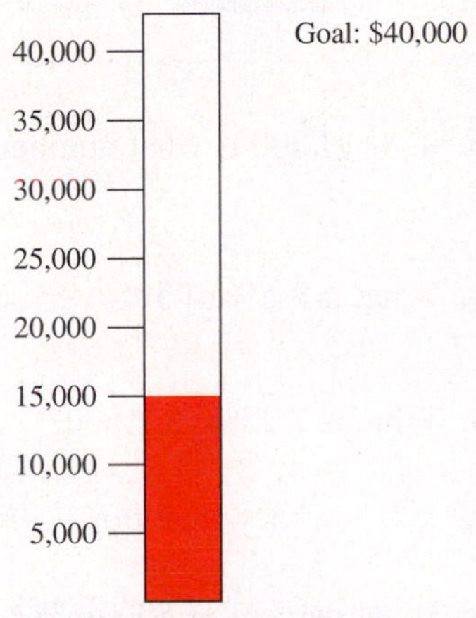


58. **Business and Finance** If the total bill at a restaurant, including a 15% tip, is $65.32, what was the cost of the meal alone?

*The following chart shows U.S. trade with Mexico from 1997 to 2002. Use this information for exercises 59 to 62.*

**U.S. Trade with Mexico, 1997–2002**

| | (millions of dollars) MEXICO | | |
|---|---|---|---|
| Year | Exports | Imports | Trade Balance[1] |
| 1997 | 71,389 | 85,938 | −14,549 |
| 1998 | 78,773 | 94,629 | −15,856 |
| 1999 | 86,909 | 109,721 | −22,812 |
| 2000 | 111,349 | 135,926 | −24,577 |
| 2001 | 101,297 | 131,338 | −30,041 |
| 2002 | 97,470 | 134,616 | −37,146 |

(1) Totals may not add due to rounding.
*Source:* U.S. Census Bureau.

## Answers

55. ______

56. ______

57. ______

58. ______

## Answers

59. ______
60. ______
61. ______
62. ______
63. ______
64. ______
65. ______
66. ______
67. ______
68. ______
69. ______
70. ______
71. ______
72. ______
73. ______ 74. ______
75. ______ 76. ______
77. ______ 78. ______
79. ______ 80. ______
81. ______ 82. ______
83. ______ 84. ______
85. ______ 86. ______

**59.** What is the rate of increase (to the nearest whole percent) of exports from 1997 to 2002?

**60.** What is the rate of increase (to the nearest whole percent) of imports from 1997 to 2002?

**61.** By what percent did imports exceed exports in 1997?

**62.** By what percent did imports exceed exports in 2002?

Basic Skills | **Advanced Skills** | Vocational-Technical Applications | Calculator/Computer | Above and Beyond

**63.** Find 8.5% of 300.

**64.** $8\frac{1}{4}$% of 800 is what number?

**65.** Find $11\frac{3}{4}$% of 6,000.

**66.** What is 3.5% of 500?

**67.** What is 5.25% of 3,000?

**68.** What is 7.25% of 7,600?

**69.** 60 is what percent of 800?

**70.** 500 is what percent of 1,500?

**71.** What percent of 180 is 120?

**72.** What percent of 800 is 78?

**73.** What percent of 1,200 is 750?

**74.** 68 is what percent of 800?

**75.** 10.5% of what number is 420?

**76.** Find the base if $11\frac{1}{2}$% of the base is 46.

**77.** 58.5 is 13% of what number?

**78.** 6.5% of what number is 325?

**79.** 195 is 7.5% of what number?

**80.** 21% of what number is 73.5?

*In exercises 81 to 84, assume the interest is compounded annually (at the end of each year) and find the amount in an account with the given interest rate and principal.*

**81.** \$4,000, 6%, 2 years

**82.** \$3,000, 7%, 2 years

**83.** \$4,000, 5%, 3 years

**84.** \$5,000, 6%, 3 years

*In exercises 85 to 88, use the following number line.*

**85.** Length $AC$ is what percent of length $AB$?

**86.** Length $AD$ is what percent of $AB$?

87. Length $AE$ is what percent of $AB$?

88. Length $AE$ is what percent of $AD$?

Basic Skills | Advanced Skills | **Vocational-Technical Applications** | Calculator/Computer | Above and Beyond

*Solve each application.*

89. **Allied Health** How much 25% alcohol solution can be prepared using 225 milliliters (mL) of ethyl alcohol?

90. **Information Technology** An Ethernet network transmits a maximum packet size of 1,500 bytes. 1.7% of each packet is "overhead." How much information (in bytes) in a maximum-size packet is overhead?

91. **Manufacturing Technology** The price of steel from a supplier increased 20% to \$950 per ton. What was the old price for a ton of steel?

92. **Agricultural Technology** Milk that is labeled "3.5%" is made up of 3.5% butterfat. How many grams of butterfat are in 1 liter (938 g) of 3.5% milk?

93. **Environmental Technology** In some communities, "green" laws require that 40% of a lot remains green (covered in grass or other vegetation). How much green space is required in a 12,680-square-foot lot?

94. **Automotive Technology** Atkins Auto Repair marks up parts 25% above cost. How much would Atkins Auto Repair charge for a starter it bought for \$62?

Basic Skills | Advanced Skills | Vocational-Technical Applications | **Calculator/Computer** | Above and Beyond

In many everyday applications of percent, the computations required can become quite messy. The calculator can be a great help. Whether we use the proportion approach or the equation approach in solving such an application, we typically set up the problem and isolate the desired variable before doing any calculations.

In some percent increase or percent decrease applications, we can set up the problem so it can be done in one step. (Previously, we did these as two-step problems.) For example, suppose that a store marks up an item 22.5%. If the original cost to the store was \$36.40, we want to know what the selling price will be. Since the selling price is the cost to the store plus the markup, the selling price will be 122.5% of the store's cost (100% + 22.5%). We can restate the problem as "What is 122.5% of \$36.40?" The base is \$36.40 and the rate is 122.5%, and we want the amount.

$$\frac{A}{36.40} = \frac{122.5}{100} \quad \text{so} \quad A = \frac{36.40 \times 122.5}{100}$$

We enter

36.40 [×] 122.5 [÷] 100 [=]

The selling price should be \$44.59.

## Answers

87. ______

88. ______

89. ______

90. ______

91. ______

92. ______

93. ______

94. ______

Answers

95. ______

96. ______

97. ______

98. ______

99. ______

100. ______

101. ______

Suppose now that a certain card collection *decreases* 8.2% in value from \$750. Note that 100% − 8.2% = 91.8%. To find the new value, we can restate the problem as: "What is 91.8% of \$750?" We set up the problem accordingly:

$$\frac{A}{750} = \frac{91.8}{100} \quad \text{so} \quad A = \frac{750 \times 91.8}{100}$$

Entering 750 [×] 91.8 [÷] 100 [=], we get \$688.50.

*Use your calculator to solve the following applications.*

95. **Social Science** The population of a town increases 4.2% in 1 year. If the original population was 19,500, what is the population after the increase?

96. **Business and Finance** A store marks up items 42.5% to allow for profit. If an item costs a store \$24.40, what will its selling price be?

97. **Business and Finance** A jacket that originally sold for \$98.50 is marked down by 12.5% for a sale. Find its sale price (to the nearest cent).

98. **Business and Finance** Jerry earned \$18,500 one year and then received a 10.5% raise. What is his new yearly salary?

99. **Business and Finance** Carolyn's salary is \$1,740 per month. If deductions average 24.6%, what is her take-home pay?

100. **Business and Finance** Yi Chen made a \$6,400 investment at the beginning of a year. By the end of the year, the value of the investment had decreased by 8.2%. What was its value at the end of the year?

chapter 6 > Make the Connection

Basic Skills | Advanced Skills | Vocational-Technical Applications | Calculator/Computer | **Above and Beyond**

101. **Business and Finance** The two ads pictured appeared last week and this week in the local paper. Is this week's ad accurate? Explain.

**102.** **Business and Finance** At True Grip hardware, you pay \$10 in tax for a barbecue grill, which is 6% of the purchase price. At Loose Fit hardware, you pay \$10 in tax for the same grill, but it is 8% of the purchase price. At which store do you get the better buy? Why?

**103.** It is customary when eating in a restaurant to leave a 15% tip.

**(a)** Outline a method to do a quick approximation for the amount of tip to leave.

**(b)** Use this method to figure a 15% tip on a bill of \$47.76.

**104.** The dean of enrollment management at a college states, "Last year was not a good year. Our enrollments were down 25%. But this year we increased our enrollment by 30% over last year. I think we have turned the corner." Evaluate the dean's analysis.

## Answers

102. ____

103. ____

104. ____

## Answers

**1.** 210 **3.** 90 **5.** 1,000 **7.** 8% **9.** 9% **11.** 170% **13.** 575 **15.** 500 **17.** 450 **19.** 880 **21.** 4,320 **23.** 175% **25.** 400% **27.** 500 **29.** 850 **31.** \$408 **33.** 5% **35.** 25 questions **37.** \$16.64 **39.** 8.5% **41.** 1,200 people **43.** \$1,464 **45.** \$14,500 **47.** \$2,500 **49.** 6,840 people **51.** 750 s **53.** 74.6% **55.** 19.8 million bbl **57.** 37.5% **59.** 37% **61.** 20% **63.** 25.5 **65.** 705 **67.** 157.5 **69.** 7.5% **71.** $66\frac{2}{3}\%$ **73.** 62.5% **75.** 4,000 **77.** 450 **79.** 2,600 **81.** \$4,494.40 **83.** \$4,630.50 **85.** 25% **87.** 37.5% **89.** 900 mL **91.** \$791.67 **93.** 5,072 $ft^2$ **95.** 20,319 people **97.** \$86.19 **99.** \$1,311.96 **101.** Above and Beyond **103.** Above and Beyond

# Activity 18 ::
## Population Changes Revisited

The following table gives the population for the United States and each of the six largest states from both the 1990 census and the 2000 census. Use this table to answer the questions that follow. Round all computations of percents to the nearest tenth of a percent.

| | 1990 Population | 2000 Population |
|---|---|---|
| United States | 248,709,873 | 281,421,906 |
| California | 29,760,021 | 33,871,648 |
| Texas | 16,986,510 | 20,851,820 |
| New York | 17,990,455 | 18,976,457 |
| Florida | 12,937,926 | 15,982,378 |
| Illinois | 11,430,602 | 12,419,293 |
| Pennsylvania | 11,881,643 | 12,281,054 |

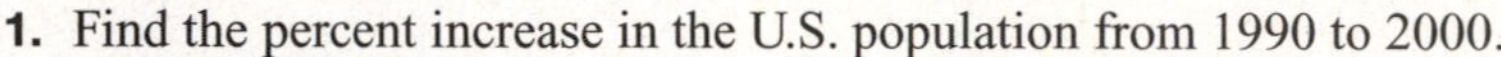

1. Find the percent increase in the U.S. population from 1990 to 2000.
2. By examining the table (no actual calculations yet!), predict which state had the greatest percent increase.
3. Predict which state had the smallest percent increase.
4. Now find the percent increase in population during this period for each state.

   California: Texas:

   New York: Florida:

   Illinois: Pennsylvania:
5. Which state had the greatest percent increase during this period? Which had the smallest?
6. The population of the six largest states combined represented what percent of the U.S. population in 1990?
7. The population of the six largest states combined represented what percent of the U.S. population in 2000?
8. Determine the percent increase in population from 1990 to 2000 for the combined six largest states.

| Definition/Procedure | Example | Reference |
|---|---|---|
| **Writing Percents as Fractions and Decimals** | | **Section 6.1** |
| **Percent** Another way of naming parts of a whole. *Percent* means per hundred. | Fractions and decimals are other ways of naming parts of a whole. $21\% = 21\left(\frac{1}{100}\right) = \frac{21}{100} = 0.21$ | p. 427 |
| *To write a percent as a fraction,* replace the percent symbol with $\frac{1}{100}$ and then multiply and simplify. | $40\% = 40\left(\frac{1}{100}\right) = \frac{40}{100} = \frac{2}{5}$ | p. 428 |
| *To write a percent as a decimal,* remove the percent symbol and move the decimal point two places to the left. | $37\% = 0.37$ | p. 429 |
| **Writing Decimals and Fractions as Percents** | | **Section 6.2** |
| *To write a decimal as a percent,* move the decimal point two places to the right and attach the percent symbol. | $0.581 = 58.1\%$ | p. 438 |
| There are two methods for writing a fraction as a percent.<br>*Method 1 The Proportion Method* Use proportions to write an equivalent fraction with a denominator of 100. Then, write the fraction as a percent. | $\frac{3}{5} = \frac{x}{100} \Rightarrow 3 \cdot 100 = 5x$<br>$\frac{300}{5} = \frac{5x}{5} \Rightarrow 60 = x$<br>Therefore, $\frac{3}{5} = 60\%$. | p. 441 |
| *Method 2 The Decimal Method* Write the decimal equivalent of the fraction, and then write that decimal as a percent. | $\frac{3}{5} = 0.6 = 60\%$ | p. 442 |
| **Identifying the Parts of a Percent Problem** | | **Section 6.3** |
| Every percent problem has the following three parts:<br>**1.** The *base, B.* This is the whole amount or starting amount in the problem. It is the standard used for comparison.<br>**2.** The *amount, A.* This is the part of the whole being compared to the base.<br>**3.** The *rate, R.* This is the ratio of the amount to the base. The rate is generally written as a percent. | 45 is 30% of 150.<br>↑ ↑ ↑<br>$A$ $R$ $B$ | p. 451 |

*Continued*

| Definition/Procedure | Example | Reference |
|---|---|---|
| **Solving Percent Problems**<br>Base, amount, and rate are related by the percent relationship,<br>$\frac{\text{Amount}}{\text{Base}} = \text{rate}$<br>If $R = \frac{r}{100}$, then we have the proportion<br>$\frac{A}{B} = \frac{r}{100}$ | $\frac{45}{150} = 30\%$<br>$\frac{45}{150} = \frac{30}{100}$ | **Section 6.4**<br>*p.* 461 |
| Use the percent relationship and the proportion rule to solve percent problems.<br>**Step 1** Substitute the two known values into the proportion.<br>**Step 2** Solve the proportion for the unknown value. | What is 24% of 300?<br>$\frac{A}{300} = \frac{24}{100}$<br>$100A = 7{,}200$<br>$A = 72$<br>72 is 24% of 300. | *p.* 462 |

This summary exercise set is provided to give you practice with each of the objectives of this chapter. Each exercise is keyed to the appropriate chapter section. When you are finished, you can check your answers to the odd-numbered exercises against those presented in the back of the text. If you have difficulty with any of these questions, go back and reread the examples from that section. The answers to the even-numbered exercises appear in the *Instructor's Solutions Manual.* Your instructor will give you guidelines on how to best use these exercises in your instructional setting.

### 6.1

**1.** Use a percent to name the shaded portion of the diagram.

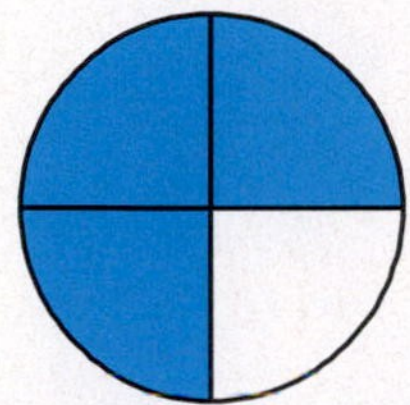

*Write the percent as a common fraction or a mixed number.*

**2.** 2%

**3.** 20%

**4.** 37.5%

**5.** 150%

**6.** $233\frac{1}{3}\%$

**7.** 300%

*Write the percents as decimals.*

**8.** 75%

**9.** 4%

**10.** 6.25%

**11.** 13.5%

**12.** 0.6%

**13.** 225%

### 6.2 *Write as percents.*

**14.** 0.06

**15.** 0.375

**16.** 2.4

**17.** 7

**18.** 0.035

**19.** 0.005

**20.** $\frac{43}{100}$

**21.** $\frac{7}{10}$

**22.** $\frac{2}{5}$

**23.** $1\frac{1}{4}$

**24.** $2\frac{2}{3}$

**25.** $\frac{3}{11}$ (to nearest tenth of a percent)

**6.4** *Find the unknown.*

**26.** 80 is 4% of what number?

**27.** 70 is what percent of 50?

**28.** 11% of 3,000 is what number?

**29.** 24 is what percent of 192?

**30.** Find the base if 12.5% of the base is 625.

**31.** 90 is 120% of what number?

**32.** What is 9.5% of 700?

**33.** Find 150% of 50.

**34.** Find the base if 130% of the base is 780.

**35.** 350 is what percent of 200?

**36.** 28.8 is what percent of 960?

*Solve each application.*

**37.** **Business and Finance** Joan works on a 4% commission basis. She sold $45,000 in merchandise during 1 month. What was the amount of her commission?

38. **Business and Finance** David buys a dishwasher that is marked down $77 from its original price of $350. What is the discount rate?

39. **Science and Medicine** A chemist prepares a 400-milliliter (mL) acid-water solution. If the solution contains 30 mL of acid, what percent of the solution is acid?

40. **Business and Finance** The price of a new compact car has increased $819 over that of the previous year. If this amounts to a 4.5% increase, what was the price of the car before the increase?

41. **Business and Finance** A store advertises, "Buy the red-tagged items at 25% off their listed price." If you buy a coat marked $136, what will you pay for the coat during the sale?

42. **Business and Finance** Tom has 6% of his salary deducted for a retirement plan. If that deduction is $168, what is his monthly salary?

43. **Social Science** A college finds that 35% of its science students take biology. If there are 252 biology students, how many science students are there altogether?

44. **Business and Finance** A company finds that its advertising costs increased from $72,000 to $76,680 in 1 year. What was the rate of increase?

45. **Business and Finance** A savings bank offers 5.25% interest on 1-year time deposits. If you place $3,000 in an account, how much will you have at the end of the year?

chapter 6 > Make the Connection

46. **Business and Finance** Maria's company offers her a 4% pay raise. This will amount to a $126 per month increase in her salary. What is her monthly salary before and after the raise?

47. **Business and Finance** A virus scanning program is checking every file for viruses. It has completed 30% of the files in 150 seconds. How long should it take to check all the files?

48. **Business and Finance** If the total bill at a restaurant for 10 people is $572.89, including an 18% tip, what was the cost of the food itself?

49. **Business and Finance** A pair of running shoes is advertised as selling at 30% off the original price, for $80.15. What was the original price?

50. **Business and Finance** A state sales tax rate is 7.5%. If the tax on a purchase is $9.75, what is the price of the purchase?

# self-test 6

Name ______________

Section ________ Date ________

The purpose of this chapter test is to help you check your progress so that you can find sections and concepts that you need to review before the next exam. Allow yourself about an hour to take this test. At the end of that hour, check your answers against those given in the back of this text. If you missed any, note the section reference that accompanies the answer. Go back to that section and reread the examples until you have mastered that particular concept.

**1.** Use a percent to name the shaded portion of the following diagram.

*Write as fractions.*

**2.** 7%

**3.** 72%

*Write as decimals.*

**4.** 42%

**5.** 6%

**6.** 160%

*Write as percents.*

**7.** 0.03

**8.** 0.042

**9.** $\frac{2}{5}$

**10.** $\frac{5}{8}$

*Identify the rate, base, and amount.* Do not solve *at this point.*

**11.** 50 is 25% of 200.

**12.** What is 8% of 500?

**13.** **Business and Finance** A state sales tax rate is 6%. If the tax on a purchase is $30, what is the amount of the purchase?

*Solve the percent problems.*

**14.** What is 4.5% of 250?

**15.** $33\frac{1}{3}$% of 1,500 is what number?

**16.** Find 125% of 600.

**17.** What percent of 300 is 60?

## Answers

1. ______________
2. ______________
3. ______________
4. ______________
5. ______________
6. ______________
7. ______________
8. ______________
9. ______________
10. ______________
11. ______________
12. ______________
13. ______________
14. ______________
15. ______________
16. ______________
17. ______________

Answers

18. ______

19. ______

20. ______

21. ______

22. ______

23. ______

24. ______

25. ______

26. ______

27. ______

28. ______

29. ______

30. ______

**18.** 4.5 is what percent of 60?

**19.** 875 is what percent of 500?

**20.** 96 is 12% of what number?

**21.** 8.5% of what number is 25.5?

*Solve each application.*

**22.** **Business and Finance** A state taxes sales at 6.2%. What tax will you pay on an item that costs $80?

**23.** **Statistics** You receive a grade of 75% on a test of 80 questions. How many questions did you have correct?

**24.** **Business and Finance** An item that costs a store $54 is marked up 30% (based on cost). Find its selling price.

**25.** **Business and Finance** Mrs. Sanford pays $300 in interest on a $2,500 loan for 1 year. What is the interest rate for the loan?

**26.** **Business and Finance** Jovita's monthly salary is $2,200. If the deductions for taxes from her monthly paycheck are $528, what percent of her salary goes for these deductions?

**27.** **Business and Finance** A car is marked down $1,552 from its original selling price of $19,400. What is the discount rate?

**28.** **Business and Finance** Sarah earns $540 in commissions in 1 month. If her commission rate is 3%, what were her total sales?

**29.** **Social Science** A community college has 480 more students in fall 2004 than in fall 2003. If this is a 7.5% increase, what was the fall 2003 enrollment?

**30.** **Business and Finance** Shawn arranges financing for his new car. The interest rate for the financing plan is 12%, and he will pay $2,220 interest for 1 year. How much money did he borrow to finance the car?

# cumulative review chapters 1-6

Name ____________

Section ________ Date ________

The following exercises are presented to help you review concepts from earlier chapters. This is meant as review material and not as a comprehensive exam. The answers are presented in the back of the text. Beside each answer is a section reference for the concept. If you have difficulty with any of these exercises, be certain to at least read through the summary related to that section.

**1.** What is the place value of 4 in the number 234,768?

*Perform the indicated operations.*

**2.** $56 \times 203$

**3.** $3{,}026 \div 34$

*Evaluate the following expressions.*

**4.** $8 - 5 + 2$

**5.** $15 - 3 \times 2$

**6.** $6 + 4 \times 3^2$

**7.** List the prime numbers between 50 and 70.

**8.** Write the prime factorization of 260.

**9.** Find the greatest common factor (GCF) of 84 and 140.

**10.** Find the least common multiple (LCM) of 18, 20, and 30.

*Perform the indicated operations.*

**11.** $3\frac{2}{5} \times 2\frac{1}{2}$

**12.** $5\frac{1}{3} \div 4$

**13.** $4\frac{3}{4} + 3\frac{5}{6}$

**14.** $7\frac{1}{6} - 2\frac{3}{8}$

**15.** A kitchen measures $5\frac{1}{2}$ yd by $3\frac{1}{4}$ yd. If vinyl flooring costs \$16 per yd$^2$, what will it cost to cover the floor?

**16.** If you drive 180 mi in $3\frac{1}{3}$ h, what is your average speed?

**17.** A bookshelf that is $54\frac{5}{8}$ in. long is cut from a board that is 8 ft long. If $\frac{1}{8}$ in. is wasted in the cut, what length board remains?

*Find the indicated place values.*

**18.** 8 in 4.2835

**19.** 4 in 6.09743

## Answers

1. ____________
2. ____________
3. ____________
4. ____________
5. ____________
6. ____________
7. ____________
8. ____________
9. ____________
10. ____________
11. ____________
12. ____________
13. ____________
14. ____________
15. ____________
16. ____________
17. ____________
18. ____________
19. ____________

**Answers**

20. ____
21. ____
22. ____
23. ____
24. ____
25. ____
26. ____
27. ____
28. ____
29. ____
30. ____
31. ____
32. ____
33. ____
34. ____
35. ____
36. ____
37. ____
38. ____
39. ____
40. ____

*Complete each statement, using the symbol <, =, or >.*

**20.** 6.28 ________ 6.3

**21.** 3.75 ________ 3.750

*Write as a common fraction or a mixed number. Simplify.*

**22.** 0.36

**23.** 5.125

*Perform the indicated operations.*

**24.** $2.8 \times 4.03$

**25.** $54.528 \div 3.2$

**26.** A television set has an advertised price of $599.95. You buy the set and agree to make payments of $29.50 per month for 2 years. How much extra are you paying on this installment plan?

**27.** Find the circumference of a circle ($C = 2\pi r$) with radius 12 ft. Use 3.14 for $\pi$.

**28.** Find the area of a circle ($A = \pi r^2$) with radius 12 ft. Use 3.14 for $\pi$.

*Write each ratio in simplest form.*

**29.** $8\frac{1}{2}$ to $12\frac{3}{4}$

**30.** 34 feet to 8 yards

*Solve for the unknown.*

**31.** $\frac{3}{7} = \frac{8}{x}$

**32.** $\frac{1.9}{y} = \frac{5.7}{1.2}$

**33.** On a map the scale is $\frac{1}{4}$ in. = 25 mi. How many miles apart are two towns that are $3\frac{1}{2}$ in. apart on the map?

**34.** Diane worked 23.5 hours on a part-time job and was paid $131.60. She is asked to work 25 hours the next week at the same pay rate. What salary will she receive?

**35.** Write 34% as a decimal and as a fraction.

**36.** Write $\frac{11}{20}$ as a decimal and as a percent.

**37.** Find 18% of 250.

**38.** 11% of what number is 55?

**39.** A company reduced the number of employees by 8% this year. There are now 115 employees. How many were there last year?

**40.** The sales tax on an item priced at $72 is $6.12. What percent is the tax rate?

CHAPTER

7

# Measurement

## INTRODUCTION

Rachael is a college sophomore studying anthropology. Rachael works after school and during the summers, in an optometry office, so she can make enough money to travel to other countries when she has time off from school. Rachael hopes to become an optometrist herself someday and travel to poorer countries to practice eye care. Rachael learned about a program called Doctors Without Borders, which provides humanitarian aid and medical assistance to countries that need it. The idea of helping people in need appeals to Rachael.

Rachael was surprised when she noticed that every country she visited uses the metric system. She is now becoming very adept at converting English measurements to metric and back again. She can even convert Celsius temperatures to Fahrenheit. Initially, Rachael was astonished at how many countries use the metric system. She did some research on the metric system by going to USMA.com and found that the only countries in the world that do not use the metric system are the United States, Liberia (in western Africa), and Myanmar (formerly Burma, in Southeast Asia). She also found that the metric system is a decimal-based system because it is based on multiples of 10. Just as English has become the global language for trade, the metric system has become the global language for measurement.

## CHAPTER 7 OUTLINE

# 7 pretest



Name ____________________

Section ________ Date ________

This pretest provides a preview of the types of exercises you will encounter in each section of this chapter. The answers for these exercises can be found in the back of the text. If you are working on your own, or ahead of the class, this pretest can help you identify the sections in which you should focus more of your time.

## Answers

1. ____________________
2. ____________________
3. ____________________
4. ____________________
5. ____________________
6. ____________________
7. ____________________
8. ____________________
9. ____________________
10. ____________________
11. ____________________

**7.1–7.2** Complete the statements.

**1.** 9 ft = ______ in.

**2.** 7 m = ______ cm

Simplify.

**3.** 5 min 90 s

Do the indicated operation.

**4.**
$$\begin{array}{r} 4 \text{ gal } 5 \text{ qt} \\ +\ 8 \text{ gal } 2 \text{ qt} \\ \hline \end{array}$$

**5.**
$$\begin{array}{r} 6 \text{ lb } 10 \text{ oz} \\ -\ 3 \text{ lb } 12 \text{ oz} \\ \hline \end{array}$$

Choose the reasonable measure.

**6.** The height of a doorway:

**(a)** 200 mm **(b)** 20 m **(c)** 2 m

Complete each statement using metric measurements.

**7.** 8 m = ________ mm

**8.** A fence is 2 ______ high.

**7.3** **9.** The correct dosage for a cough medicine is 40 ______.

**7.4** **10.** 77°F = ________ °C

**11.** 15°C = ________ °F

# 7.1 The Units of the English System

< 7.1 Objectives >

1 > Convert between two English units of measure

2 > Use a unit ratio to convert between English units of measure

3 > Simplify denominate numbers

4 > Perform operations with denominate numbers

Many arithmetic problems involve **units of measure.** When we measure an object, we give it a number and some unit. For instance, we might say a board is 6 feet long, a container holds 4 quarts, or a package weighs 5 pounds. Feet, quarts, and pounds are the units of measure.

The system you are probably most familiar with is called the **English system of measurement.** This system is used in the United States and a few other countries.

The following table lists the units of measurement you should be familiar with.

**NOTE**

Don't let the name mislead you. The English system is no longer used in England. England has converted to the metric system, which we will look at in Sections 7.2 and 7.3.

**English Units of Measure and Equivalents**

| Length | Weight |
|---|---|
| 1 foot (ft) = 12 inches (in.)<br>1 yard (yd) = 3 ft<br>1 mile (mi) = 5,280 ft | 1 pound (lb) = 16 ounces (oz)<br>1 ton = 2,000 lb |
| **Volume** | **Time** |
| 1 pint (pt) = 16 fluid ounces (fl oz)<br>1 quart (qt) = 2 pt<br>1 gallon (gal) = 4 qt | 1 minute (min) = 60 seconds (s)<br>1 hour (h) = 60 min<br>1 day = 24 h<br>1 week = 7 days |

You may want to use the equivalencies shown in the table to change from one unit to another. Here is one approach.

**Property**

**Converting Units in the English System**

To change from one unit to another, replace the unit of measure with the appropriate equivalent measure and multiply.

## Example 1 Converting Within the English System

< Objective 1 >

**NOTE**

We write 5 ft as 5(1 ft) and then change 1 ft to 12 in.

$5 \text{ ft} = 5(1 \text{ ft}) = 5(12 \text{ in.}) = 60 \text{ in.}$ Replace 1 ft with 12 in.

$3 \text{ lb} = 3(1 \text{ lb}) = 3(16 \text{ oz}) = 48 \text{ oz}$ Replace 1 lb with 16 oz.

$6 \text{ gal} = 6(1 \text{ gal}) = 6(4 \text{ qt}) = 24 \text{ qt}$ Replace 1 gal with 4 qt.

$48 \text{ in.} = 48(1 \text{ in.}) = 48\left(\frac{1}{12} \text{ ft}\right) = 4 \text{ ft}$ Because 12 in. = 1 ft, 1 in. = $\frac{1}{12}$ ft.

$180 \text{ min} = 180(1 \text{ min}) = 180\left(\frac{1}{60} \text{ h}\right) = 3 \text{ h}$ Because 60 min = 1 h, 1 min = $\frac{1}{60}$ h.

### Check Yourself 1

Complete each of the following statements.

**(a)** 4 ft = ______ in. **(b)** 12 qt = ______ pt
**(c)** 48 fl oz = ______ pt **(d)** 240 s = ______ min

**NOTE**

This is a variation on the method of units analysis discussed earlier in this text.

Here is another idea that may help you convert units. You can use a *unit ratio* to convert from one unit to another. A *unit ratio* is a fraction whose value is 1.

**Property**

### Using Unit Ratios

To decide which unit ratio to use, just choose one with the unit you *want* in the numerator (inches in Example 2) and the unit you *want to remove* in the denominator (feet in Example 2).

## Example 2 Using the Unit Ratio to Convert

< Objective 2 >

Convert 5 ft to inches.

To convert from feet to inches, you can multiply by the ratio $\frac{12 \text{ in.}}{1 \text{ ft}}$. So, to convert 5 ft to inches, write

$5 \text{ ft} = 5 \cancel{\text{ft}} \left(\frac{12 \text{ in.}}{1 \cancel{\text{ft}}}\right)$ We are multiplying by 1, and so the value of the expression is not changed.

$= 60 \text{ in.}$ Note that we can divide out units just as we do numbers.

**NOTE**

$\frac{12 \text{ in.}}{1 \text{ ft}}$ is a *unit ratio*. It can be reduced to 1.

### Check Yourself 2

Use a unit ratio to complete each of the following statements.

**(a)** 240 min = ______ h **(b)** 64 qt = ______ gal

**NOTE**

Historically, units were associated with various things. A foot was the length of a foot, of course. The yard was the distance from the end of a nose to the fingertips of an outstretched arm. Objects were weighed by comparing them with grains of barley.

You have now had a chance to use two different methods for converting from one unit of measurement to another. Use whichever approach seems easier for you.

From our work so far, it should be clear that one big disadvantage of the English system is that the relationships between units are all different. One foot is 12 in., 1 lb is 16 oz, and so on. We will see in Sections 7.2 and 7.3 that this problem does not exist in the metric system.

In our units analysis features, we discussed the difference between denominate numbers (those with units attached) and abstract numbers.

A denominate number may involve two or more different units. We regularly combine feet and inches, pounds and ounces, and so on. The measures 5 lb 6 oz and 4 ft 7 in. are examples. When simplifying a denominate number with multiple units, the *largest unit should include as much of the measure as possible.* For example, 3 ft 2 in. is simplified, whereas 2 ft 14 in. is not.

Example 3 shows the steps used to simplify a denominate number with multiple units.

## Example 3 Simplifying Denominate Numbers

< Objective 3 >

**NOTE**

18 in. is larger than 1 ft and can be simplified.

**(a)** Simplify 4 ft 18 in.

Write 18 in. as 1 ft 6 in. because 12 in. is 1 ft.

$$4 \text{ ft } 18 \text{ in.} = \underbrace{4 \text{ ft} + 1 \text{ ft}} + 6 \text{ in.} \quad (\overbrace{1 \text{ ft} + 6 \text{ in.}}^{18 \text{ in.}})$$

$$= 5 \text{ ft } 6 \text{ in.}$$

**(b)** Simplify 5 h 75 min.

Write 75 min as 1 h 15 min because 1 h is 60 min.

$$5 \text{ h } 75 \text{ min} = \underbrace{5 \text{ h} + 1 \text{ h}} + 15 \text{ min} \quad (\overbrace{1 \text{ h} + 15 \text{ min}}^{75 \text{ min}})$$

$$= 6 \text{ h } 15 \text{ min}$$

**Check Yourself 3**

**(a)** Simplify 5 lb 24 oz. **(b)** Simplify 7 ft 20 in.

Denominate numbers with the same units are called *like numbers.* We can always add or subtract denominate numbers according to the following rule.

**Step by Step**

**Adding Denominate Numbers**

**Step 1** Arrange the numbers so that the like units are in the same vertical column.

**Step 2** Add in each column.

**Step 3** Simplify if necessary.

Example 4 illustrates this rule for adding denominate numbers.

## Example 4 Adding Denominate Numbers

< Objective 4 >

Add 5 ft 4 in., 6 ft 7 in., and 7 ft 9 in.

**NOTES**

The columns here represent inches and feet.

Be sure to simplify the results.

| | | |
|---|---|---|
| 5 ft 4 in. | | Arrange in a vertical column. |
| 6 ft 7 in. | | |
| + 7 ft 9 in. | | |
| 18 ft 20 in. | | Add in each column. |
| = 19 ft 8 in. | | Simplify as before. |

### Check Yourself 4

Add 3 h 15 min, 5 h 50 min, and 2 h 40 min.

To subtract denominate numbers, we have a similar rule.

**Step by Step**

### Subtracting Denominate Numbers

**Step 1** Arrange the numbers so that the like units are in the same vertical column.

**Step 2** Subtract in each column. You may have to borrow from the larger unit at this point.

**Step 3** Simplify if necessary.

Consider the following example of subtracting denominate numbers.

## Example 5 Subtracting Denominate Numbers

Subtract 3 lb 6 oz from 8 lb 13 oz.

| | |
|---|---|
| 8 lb 13 oz | Arrange vertically. |
| − 3 lb 6 oz | |
| 5 lb 7 oz | Subtract in each column. |

### Check Yourself 5

Subtract 5 ft 9 in. from 10 ft 11 in.

As step 2 points out, subtracting denominate numbers may involve borrowing.

## Example 6 Subtracting Denominate Numbers

Subtract 5 ft 8 in. from 9 ft 3 in.

$$\begin{array}{r} 9 \text{ ft } 3 \text{ in.} \\ -\ 5 \text{ ft } 8 \text{ in.} \\ \hline \end{array}$$

Do you see the problem? We cannot subtract in the inches column.

To complete the subtraction, we borrow 1 ft and rename. The "borrowed" number will depend on the units involved.

$$\begin{array}{r} 9 \text{ ft } 3 \text{ in.} \\ -\ 5 \text{ ft } 8 \text{ in.} \\ \hline \end{array} \longrightarrow \begin{array}{r} 8 \text{ ft } 15 \text{ in.} \\ -\ 5 \text{ ft } \ 8 \text{ in.} \\ \hline 3 \text{ ft } \ 7 \text{ in.} \end{array}$$

**NOTES**

Borrowing with denominate numbers is not the same as in the place-value system, in which we always borrow a power of 10.

9 ft becomes 8 ft 12 in. Combine the 12 in. with the original 3 in.

### Check Yourself 6

Subtract 3 lb 9 oz from 8 lb 5 oz.

Certain types of problems involve multiplying or dividing denominate numbers by abstract numbers, that is, numbers without a unit of measure attached. The following rule is used.

**Step by Step**

**Multiplying or Dividing by Abstract Numbers**

**Step 1** Multiply or divide each part of the denominate number by the abstract number.

**Step 2** Simplify if necessary.

Examples 7 and 8 illustrate this procedure.

## Example 7 Multiplying Denominate Numbers

**(a)** Multiply $4 \times 5$ in.

$4 \times 5$ in. $= 20$ in. or 1 ft 8 in.

**(b)** Multiply $3 \times (2$ ft 7 in.).

$$\begin{array}{r} 2 \text{ ft } \ \ 7 \text{ in.} \\ \times \qquad\quad 3 \\ \hline 6 \text{ ft } 21 \text{ in.} \end{array}$$

Simplify. The product is 7 ft 9 in.

**NOTE**

Multiply each part of the denominate number by 3.

### Check Yourself 7

Multiply 5 lb 8 oz by 4.

Division is illustrated in Example 8.

## Example 8 Dividing Denominate Numbers

**NOTE**

Divide each part of the denominate number by 4.

Divide 8 lb 12 oz by 4.

$$\frac{8 \text{ lb } 12 \text{ oz}}{4} = 2 \text{ lb } 3 \text{ oz}$$

### Check Yourself 8

**Divide 9 ft 6 in. by 3.**

We encounter the need to make such calculations in many applications. Consider Example 9.

## Example 9 A Vocational Application

There were 482 lb 6 oz of steel in stock before a shipment of 219 lb 13 oz arrived. How much steel was in stock after the shipment?

We begin by lining up like numbers.

| 482 lb | 6 oz |
|---|---|
| 219 lb | 13 oz |
| 701 lb | 19 oz |

Then, since 19 oz = 1 lb 3 oz, we simplify our result to 702 lb 3 oz.

### Check Yourself 9

**There are 6 gal 1 qt of aluminum sealer on hand. A run of parts will require 1 gal 2 qt of sealer. How much sealer will be left after the run?**

### Check Yourself ANSWERS

1. **(a)** 48; **(b)** 24; **(c)** 3; **(d)** 4
2. **(a)** $\overset{4}{\cancel{240}} \cancel{\text{min}} \left(\frac{1 \text{ h}}{\underset{1}{\cancel{60}} \cancel{\text{min}}}\right) = 4$ h; **(b)** 16
3. **(a)** 6 lb 8 oz; **(b)** 8 ft 8 in.
4. 11 h 45 min
5. 5 ft 2 in.
6. Rename 8 lb 5 oz as 7 lb 21 oz. Then subtract for the result, 4 lb 12 oz.
7. 20 lb 32 oz, or 22 lb
8. 3 ft 2 in.
9. 4 gal 3 qt

## Reading Your Text

The following fill-in-the-blank exercises are designed to ensure that you understand some of the key vocabulary used in this section.

### SECTION 7.1

**(a)** To change from one ______________ to another, replace the unit of measure with the appropriate equivalent measure and multiply.

**(b)** ______________ numbers are those numbers that have units attached.

**(c)** Denominate numbers with the same units are called ______________ numbers.

**(d)** When simplifying a denominate number with multiple units, the ______________ unit should include as much of the measure as possible.

# 7.1 exercises

Basic Skills | Advanced Skills | Vocational-Technical Applications | Calculator/Computer | Above and Beyond

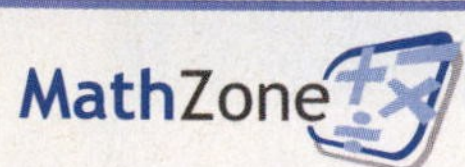


Boost your grade at mathzone.com!

- > Practice Problems
- > NetTutor
- > Self-Tests
- > e-Professors
- > Videos

Name ________________

Section ________ Date ________

## Answers

1. ________ 2. ________
3. ________ 4. ________
5. ________ 6. ________
7. ________ 8. ________
9. ________ 10. ________
11. ________ 12. ________
13. ________ 14. ________
15. ________ 16. ________
17. ________ 18. ________
19. ________ 20. ________
21. ________ 22. ________
23. ________ 24. ________
25. ________ 26. ________
27. ________ 28. ________
29. ________ 30. ________
31. ________ 32. ________
33. ________ 34. ________

< Objectives 1–2 >

*Complete the following statements.*

**1.** 8 ft = ________ in.

**2.** 9 gal = ________ qt

**3.** 3 lb = ________ oz

**4.** 300 s = ________ min

**5.** 360 min = ________ h

**6.** 5 pt = ________ fl oz

**7.** 4 days = ________ h

**8.** 6 h = ________ min

**9.** 16 qt = ________ gal

**10.** 11 min = ________ s

**11.** 10,000 lb = ________ tons

**12.** 5 mi = ________ ft

**13.** 30 pt = ________ qt

**14.** 64 fl oz = ________ pt

**15.** 64 oz = ________ lb

> Videos

**16.** 540 min = ________ h

**17.** 7 yd = ________ ft

**18.** 24 qt = ________ gal

**19.** 39 ft = ________ yd

**20.** 192 oz = ________ lb

**21.** 8 min = ________ s

**22.** 18 qt = ________ pt

**23.** 192 h = ________ days

**24.** 360 h = ________ days

**25.** 16 qt = ________ pt

**26.** 7 days = ________ h

**27.** $7\frac{1}{4}$ h = ________ min

**28.** 43 pt = ________ qt

**29.** 56 oz = ________ lb

**30.** 20 fl oz = ________ pt

**31.** 225 s = ________ min

**32.** 44 in. = ________ ft

**33.** 1.55 lb = ________ oz

**34.** 4.72 ft = ________ in.

< Objective 3 >

*Simplify.*

**35.** 4 ft 18 in.

**36.** 6 lb 20 oz

**37.** 7 qt 5 pt

**38.** 7 yd 50 in.

**39.** 5 gal 9 qt

**40.** 3 min 110 s

**41.** 9 min 75 s

**42.** 9 h 80 min

< Objective 4 >

*Add.*

**43.**
$$\begin{array}{r} 8 \text{ lb} \quad 7 \text{ oz} \\ +\ 6 \text{ lb}\ 15 \text{ oz} \\ \hline \end{array}$$

**44.**
$$\begin{array}{r} 9 \text{ ft} \quad 7 \text{ in.} \\ +\ 3 \text{ ft}\ 10 \text{ in.} \\ \hline \end{array}$$

**45.**
$$\begin{array}{r} 3 \text{ h}\ 20 \text{ min} \\ 4 \text{ h}\ 25 \text{ min} \\ +\ 5 \text{ h}\ 35 \text{ min} \\ \hline \end{array}$$

**46.**
$$\begin{array}{l} \quad 5 \text{ yd}\ 2 \text{ ft} \\ \quad 4 \text{ yd} \\ +\ 6 \text{ yd}\ 1 \text{ ft} \\ \hline \end{array}$$

**47.** 4 lb 7 oz, 3 lb 11 oz, and 5 lb 8 oz

**48.** 7 ft 8 in., 8 ft 5 in., and 9 ft 7 in.

*Subtract.*

**49.**
$$\begin{array}{r} 9 \text{ lb}\ 15 \text{ oz} \\ -\ 5 \text{ lb} \quad 8 \text{ oz} \\ \hline \end{array}$$

**50.**
$$\begin{array}{r} 7 \text{ ft}\ 11 \text{ in.} \\ -\ 4 \text{ ft} \quad 3 \text{ in.} \\ \hline \end{array}$$

**51.**
$$\begin{array}{r} 6 \text{ h}\ 30 \text{ min} \\ -\ 3 \text{ h}\ 50 \text{ min} \\ \hline \end{array}$$

**52.**
$$\begin{array}{r} 7 \text{ gal}\ 3 \text{ qt} \\ -\ 1 \text{ gal}\ 3 \text{ qt} \\ \hline \end{array}$$

**53.** Subtract 2 yd 2 ft from 5 yd 1 ft.

**54.** Subtract 2 h 30 min from 7 h 25 min.

*Multiply.*

**55.** 4 × 13 oz

**56.** 4 × 10 in.

**57.** 3 × (4 ft 5 in.)

**58.** 5 × (4 min 20 s)

## Answers

35. ________ 36. ________

37. ________

38. ________

39. ________ 40. ________

41. ________ 42. ________

43. ________ 44. ________

45. ________

46. ________

47. ________

48. ________

49. ________

50. ________

51. ________

52. ________

53. ________

54. ________

55. ________

56. ________

57. ________

58. ________

Answers

59. ____

60. ____

61. ____

62. ____

63. ____

64. ____

65. ____

66. ____

67. ____

68. ____

69. ____

70. ____

71. ____

72. ____

*Divide.*

**59.** $\frac{4 \text{ ft } 6 \text{ in.}}{2}$

**60.** $\frac{12 \text{ lb } 15 \text{ oz}}{3}$

**61.** $\frac{16 \text{ min } 28 \text{ s}}{4}$

**62.** $\frac{25 \text{ h } 40 \text{ min}}{5}$

*Solve the following applications.*

**63. Science and Medicine** In 1989, the United States emitted approximately 9 million tons of suspended particulates into the atmosphere. How many pounds of (suspended) particulates did the United States emit in 1989?

**64. Science and Medicine** In 1989, the United States emitted approximately 20 million tons of volatile organic compounds into the atmosphere. How many pounds of volatile organic compounds did the United States emit in 1989?

*Solve each of the following applications.*

**65. Construction** A railing for a deck requires pieces of cedar 4 ft 8 in., 11 ft 7 in., and 9 ft 3 in. long. What is the total length of material that is needed?

**66. Business and Finance** Ted worked 3 h 45 min on Monday, 5 h 30 min on Wednesday, and 4 h 15 min on Friday. How many hours did he work during the week?

**67. Crafts** A pattern requires a 2-ft 10-in. length of fabric. If a 2-yd length is used, what length remains?

**68. Crafts** You use 2 lb 8 oz of hamburger from a package that weighs 4 lb 5 oz. How much is left over?

**69. Crafts** A picture frame is to be 2 ft 6 in. long and 1 ft 8 in. wide. A 9-ft piece of molding is available for the frame. Will this be enough for the frame?

**70. Construction** A plumber needs two pieces of plastic pipe that are 6 ft 9 in. long and one piece that is 2 ft 11 in. long. He has a 16-ft piece of pipe. Is this enough for the job?

**71. Crafts** Mark uses 1 pt 9 fl oz and then 2 pt 10 fl oz from a container of film developer that holds 3 qt. How much of the developer remains?

**72. Business and Finance** Some flights limit passengers to 44 lb of checked-in luggage. Susan checks three pieces, weighing 20 lb 5 oz, 7 lb 8 oz, and 15 lb 7 oz. By how much was she under or over the limit?

73. **Business and Finance** Six packages weighing 2 lb 9 oz each are to be mailed. What is the total weight of the packages?

74. **Construction** A bookshelf requires four boards 3 ft 8 in. long and two boards 2 ft 10 in. long. How much lumber will be needed for the bookshelf?

75. **Business and Finance** You can buy three 12-oz cans of peanuts for $3 or one large can containing 2 lb 8 oz for the same price. Which is the better buy?

76. **Statistics and Mathematics** Rich, Susan, and Marc agree to share the driving on a 12-hour (12-h) trip. Rich has driven for 4 h 45 min, and Susan has driven for 3 h 30 min. How long must Marc drive to complete the trip?

*Determine whether each statement is* **true** *or* **false.**

77. A denominate number may have more than two units of measure attached.

78. It is impossible to divide a denominate number by an abstract number.

*In each of the following statements, fill in the blank with* **always, sometimes,** *or* **never.**

79. We can __________ add or subtract denominate numbers if we arrange the numbers so that the like units are in the same vertical column.

80. Subtraction of denominate numbers __________ requires borrowing.

Basic Skills | **Advanced Skills** | Vocational-Technical Applications | Calculator/Computer | Above and Beyond

*Solve each of the following.*

81.
```
   2 gal 3 qt 1 pt
 + 3 gal 2 qt 1 pt
```

82.
```
   7 weeks 3 days 15 hours
 + 3 weeks 9 days 10 hours
```

83.
```
   13 yd 15 ft 10 in.
 −  9 yd 16 ft 15 in.
```

84.
```
   8 gal 3 qt 2 pt
 − 5 gal 5 qt 3 pt
```

85.
```
 2 weeks 7 days 18 h 40 min
 ×                        2
```

86.
```
 4 gal 5 qt 3 pt 10 oz
 ×                   2
```

87. **Science and Medicine** The average number of times a human heart beats per minute is about 72. At what age has a person's heart beat 1 billion times?

## Answers

73. ____________

74. ____________

75. ____________

76. ____________

77. ____________

78. ____________

79. ____________

80. ____________

81. ____________

82. ____________

83. ____________

84. ____________

85. ____________

86. ____________

87. ____________

Answers

88. ______

89. ______

90. ______

91. ______

92. ______

93. ______

94. ______

95. ______

96. ______

97. ______

98. ______

**88. Science and Medicine** Each time a human heart beats, it carries about 1 ounce of blood. How many gallons of blood does the average human heart carry in a day? (See Exercise 87.)

**89.** A greyhound can run at an average rate of 37 mi/h for $\frac{5}{16}$ of a mile. How long does it take the greyhound to run $\frac{5}{16}$ of a mile?

Basic Skills | Advanced Skills | **Vocational-Technical Applications** | Calculator/Computer | Above and Beyond

**90. Mechanical Engineering** A piece of steel that is 13 ft 8 in long is to be sheared into four equal pieces. How long will each piece be?

**91. Manufacturing Technology** A part weighs 4 lb 11 oz. How much does a lot of 24 parts weigh?

**92. Automotive Technology** An engine block weighs 218 lb 12 oz. Each head weighs 36 lb 3 oz. There are two heads on the engine. What is the total weight of the block with the two heads?

**93. Automotive Technology** Each piston in an engine weighs 4 lb 13 oz. How much do the eight pistons add to the weight of an engine?

**94. Manufacturing Technology** A knee wall is to be constructed from 2 by 4 studs that are 4 ft 9 in. long. If the wall will use 13 studs, what is the total length of 2 by 4 required for the studs?

**95. Allied Health** A premature, newborn baby boy weighs 98 oz. Determine his weight in pounds and ounces.

Basic Skills | Advanced Skills | Vocational-Technical Applications | Calculator/Computer | **Above and Beyond**

**96. Science and Medicine**

**(a)** John is traveling at a speed of 60 mi/h. What is his speed in feet per second?

**(b)** Use the information in part **(a)** to develop a method to convert any speed from miles per hour to feet per second.

**97.** Refer to several sources and write a brief history of how the units that are currently used in the English system of measurement originated. Discuss some units that were previously used but are no longer in use today.

**98.** A unit of measurement used in surveying is the **chain.** There are 80 chains in a mile. If you measured the distance from your home to school, how many chains would you have traveled?

**99.** The average person takes about 17 breaths per minute. How many breaths have you taken in your lifetime?

**100.** **Science and Medicine** The average breath takes in about $1\frac{1}{2}$ pints of air. The air is about 20% oxygen. Of the oxygen that we breathe, about 25% makes its way into our bloodstream. How much oxygen does the average person take in every day? (See Exercise 99.)

**101.** What is your age in seconds? (Don't forget about the leap years!)

## Answers

99. ____________

100. ____________

101. ____________

## Answers

**1.** 96 **3.** 48 **5.** 6 **7.** 96 **9.** 4 **11.** 5 **13.** 15
**15.** 4 **17.** 21 **19.** 13 **21.** 480 **23.** 8 **25.** 32
**27.** 435 **29.** 3.5 **31.** 3.75 **33.** 24.8 **35.** 5 ft 6 in.
**37.** 9 qt 1 pt **39.** 7 gal 1 qt **41.** 10 min 15 s **43.** 15 lb 6 oz
**45.** 13 h 20 min **47.** 13 lb 10 oz **49.** 4 lb 7 oz **51.** 2 h 40 min
**53.** 2 yd 2 ft **55.** 3 lb 4 oz **57.** 13 ft 3 in. **59.** 2 ft 3 in.
**61.** 4 min 7 s **63.** 18 billion lb **65.** 25 ft 6 in. **67.** 3 ft 2 in.
**69.** Yes, 8 in. will remain **71.** 1 pt 13 fl oz **73.** 15 lb 6 oz
**75.** The 2-lb 8-oz can **77.** True **79.** always **81.** 6 gal 2 qt
**83.** 3 yd 1 ft 7 in. **85.** 6 weeks 1 day 13 h 20 min
**87.** 26 yr 4 mo 24 days **89.** 0.008 h or 30.4 s **91.** 112 lb 8 oz
**93.** 38 lb 8 oz **95.** 6 lb 2 oz **97.** Above and Beyond
**99.** Above and Beyond **101.** Above and Beyond

7.2

# Metric Units of Length

< 7.2 Objectives >

1 > Estimate metric units of length

2 > Convert between metric units of length

**NOTES**

Even in the United States, the metric system is used in science, medicine, the automotive industry, the food industry, and many other areas.

The basic unit of length in the metric system is also spelled *metre* (the British spelling).

The meter is one of the basic units of the International System of Units (abbreviated SI). This is a standardization of the metric system agreed to by scientists in 1960.

In Section 7.1 we studied the English system of measurement, which is used in the United States and a few other countries. Our work will now concentrate on the **metric system,** used throughout the rest of the world.

The metric system is based on one unit of length, the **meter (m).** In the eighteenth century the meter was defined to be one ten-millionth of the distance from the north pole to the equator. Today the meter is scientifically defined in terms of a wavelength in the spectrum of krypton-86 gas.

One big advantage of the metric system is that you can convert from one unit to another by simply multiplying or dividing by powers of 10. This advantage and the need for uniformity throughout the world have led to legislation that promotes the use of the metric system in the United States.

To see how the metric system works, we will start with measures of length and compare a basic English unit, the yard, with the meter.

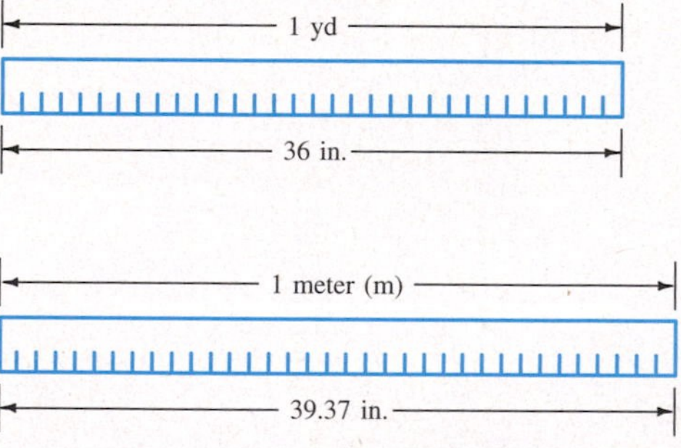


**NOTE**

There is a standard pattern of abbreviation in the metric system. We will introduce the abbreviation for each term as we go along. The abbreviation for meter is m (no period!).

As you can see, the meter is just slightly longer than the yard. It is used for measuring the same things you might measure in feet or yards. Look at Example 1 to get a feel for the size of the meter.

**Example 1** — **Estimating Metric Length**

< Objective 1 >

A room might be 6 meters (6 m) long.

A building lot could be 30 m wide.

A fence is 2 m tall.

### Check Yourself 1

Try to estimate the following lengths in meters.

(a) A traffic lane is ________ m wide.
(b) A small car is ________ m long.
(c) You are ________ m tall.

For other units of length, the meter is multiplied or divided by powers of 10. One commonly used unit is the **centimeter (cm).**

**Definition**

**Comparing Centimeters (cm) to Meters (m)**

$$1 \text{ centimeter (cm)} = \frac{1}{100} \text{ meter (m)}$$

**NOTE**

The prefix *centi* means one–hundredth. This should be no surprise. What is our cent? It is one hundredth of a dollar.

The following drawing relates the centimeter and the meter:

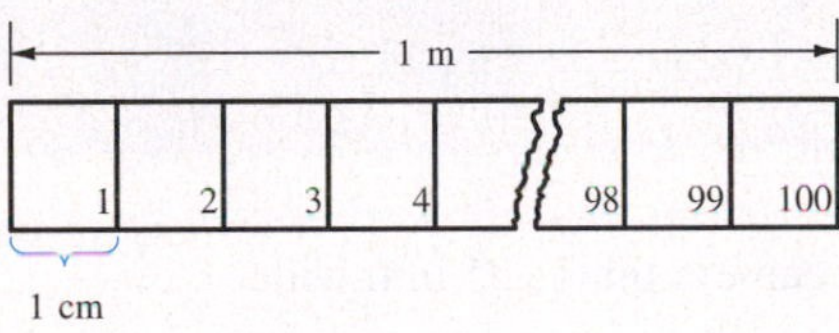

There are 100 cm in 1 m.

Just to give you an idea of the size of the centimeter, it is about the width of your little finger. There are about $2\frac{1}{2}$ cm to 1 in., and the unit is used to measure small objects. Look at Example 2 to get a feel for the length of a centimeter.

### Example 2 Estimating Metric Length

A small paperback book is 10 cm wide.

A playing card is 8 cm long.

A ballpoint pen is 16 cm long.

### Check Yourself 2

Try to estimate each of the following. Then use a metric ruler to check your guess.

(a) This page is ________ cm long.
(b) A dollar bill is ________ cm long.
(c) The seat of the chair you are on is ________ cm from the floor.

To measure *very* small things, the **millimeter (mm)** is used. To give you an idea of its size, a millimeter is about the thickness of a new dime.

**Definition**

**Comparing Millimeters (mm) to Meters (m)**

$$1 \text{ millimeter (mm)} = \frac{1}{1{,}000} \text{ m}$$

**NOTES**

The prefix *milli* means one–thousandth.

There are 10 mm to 1 cm.

The following diagram will help you see the relationships of the three units we have looked at.

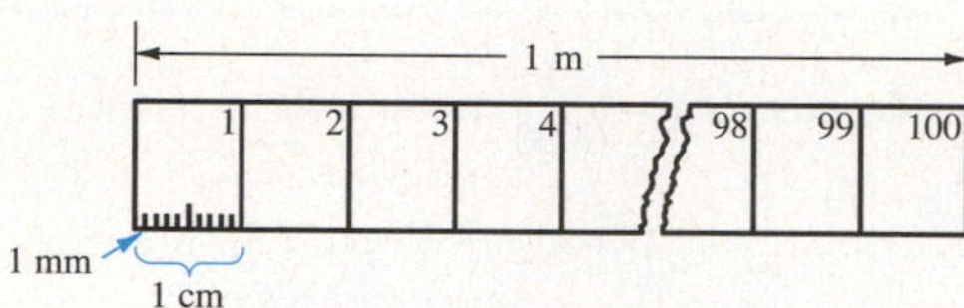


To get used to the millimeter, consider Example 3.

**Example 3** **Estimating Metric Length**

Standard camera film is 35 mm wide.

A small paper clip is 7 mm wide.

Some cigarettes are 100 mm long.

**Check Yourself 3**

Try to estimate each of the following. Then use a metric ruler to check your guess.

(a) Your pencil is ________ mm wide.
(b) The tabletop you are working on is ________ mm thick.

**NOTE**

The prefix *kilo* means 1,000. You are already familiar with this. For instance, 1 kilowatt (kW) = 1,000 watts (W).

The **kilometer (km)** is used to measure long distances. The kilometer is about $\frac{6}{10}$ of a mile.

**Definition**

**Comparing Kilometers (km) to Meters (m)**

1 kilometer (km) = 1,000 m

Example 4 shows how to get used to the kilometer.

## Example 4 Estimating Metric Length

The distance from New York to Boston is 338 km.

A popular distance for road races is 10 km.

Now that you have seen the four commonly used units of length in the metric system, you can review with the following Check Yourself exercise.

### Check Yourself 4

Choose the most reasonable measure in each of the following statements.

(a) The width of a doorway: 50 mm, 1 m, or 50 cm.
(b) The length of your pencil: 20 m, 20 mm, or 20 cm.
(c) The distance from your house to school: 500 km, 5 km, or 50 m.
(d) The height of a basketball center: 2.2 m, 22 m, or 22 cm.
(e) The width of a matchbook: 30 cm, 30 mm, or 3 mm.

As we said earlier, to convert units of measure within the metric system, we multiply or divide by the appropriate power of 10. To accomplish this, we move the decimal point to the right or left the required number of places. This is the big advantage of the metric system.

**Property**

**Converting Metric Measurements to Smaller Units**

To convert to a *smaller* unit of measure, we *multiply* by a power of 10, moving the decimal point *to the right.*

## Example 5 Converting Metric Length

< Objective 2 >

5.2 m = 520 cm — The *smaller* the unit, the *more* units it takes, so we multiply by 100 to convert from meters to centimeters.

8 km = 8,000 m — Multiply by 1,000.

6.5 m = 6,500 mm — Multiply by 1,000.

2.5 cm = 25 mm — Multiply by 10.

### Check Yourself 5

Complete the following by moving the decimal point the appropriate number of places.

(a) 3 km = ________ m
(b) 4.5 m = ________ cm
(c) 1.2 m = ________ mm
(d) 6.5 cm = ________ mm

Property

**Converting Metric Measurements to Larger Units**

To convert to a *larger* unit of measure, we *divide* by a power of 10, moving the decimal point *to the left.*

### Example 6 Converting Metric Length

**NOTE**

The *larger* the unit, the *fewer* units it takes, so *divide.*

43 mm = 4.3 cm   Divide by 10.

3,000 m = 3 km   Divide by 1,000.

450 cm = 4.5 m   Divide by 100.

**Check Yourself 6**

**Complete the following statements.**

**(a)** 750 cm = ________ m   **(b)** 5,000 m = ________ km
**(c)** 78 mm = ________ cm   **(d)** 3,500 mm = ________ m

We have introduced all the commonly used units of linear measure in the metric system. There are other prefixes that can be used to form other linear measures. The prefix *deci* means $\frac{1}{10}$, *deka* means 10, and *hecto* means 100. Their use is illustrated in the following table.

Definition

**Using Metric Prefixes**

1 *kilo*meter (km) = 1,000 m

1 *hecto*meter (hm) = 100 m

1 *deka*meter (dam) = 10 m

1 meter (m)

1 *deci*meter (dm) = $\frac{1}{10}$ m

1 *centi*meter (cm) = $\frac{1}{100}$ m

1 *milli*meter (mm) = $\frac{1}{1,000}$ m

You may find the following chart helpful when converting between metric units. Think of the chart as a set of stairs. Note that the largest unit is at the highest step on the stairs.

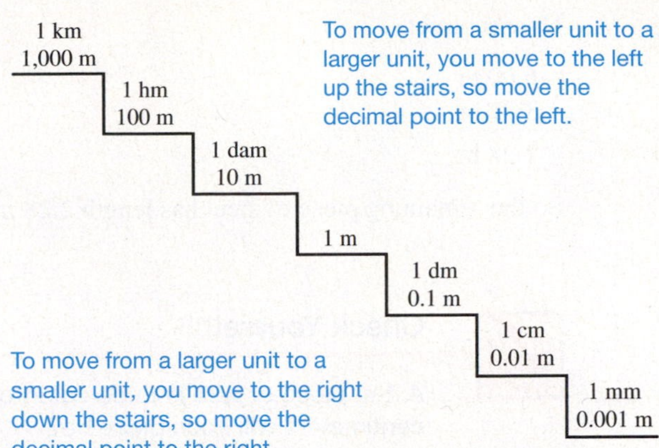


## Example 7 Converting Between Metric Lengths

**(a)** 800 cm = ? m

To convert from centimeters to meters, you can see from the chart that you must move the decimal point *two places to the left.*

800 cm = 8.00 m = 8 m

**(b)** 500 m = ? km

To convert from meters to kilometers, move the decimal point *three places to the left.*

500 m = .500 km = 0.5 km

**(c)** 6 m = ? mm

To convert from meters to millimeters, move the decimal point *three places to the right.*

6 m = 6000. mm = 6,000 mm

### Check Yourself 7

Complete each statement.

**(a)** 300 cm = ________ m **(b)** 370 mm = ________ m
**(c)** 4,500 m = ________ km

## Example 8 A Vocational Application

A piece of steel that is 6 m long has a piece 372 cm long removed. How much is left? Express your answer in meters.

First, we convert the length 372 cm to meters. Since we move the decimal point two places to the left, we see

372 cm = 3.72 m

Now subtract:

$$\begin{array}{r} 6.00 \text{ m} \\ -3.72 \text{ m} \\ \hline 2.28 \text{ m} \end{array}$$

So the remaining piece of steel has length 2.28 m.

### Check Yourself 8

A 2-m piece of wire is to be cut into five equal pieces. How long in centimeters will each piece be?

### Check Yourself ANSWERS

1. **(a)** About 3 m; **(b)** perhaps 5 m; **(c)** you are probably between 1.5 and 2 m tall.
2. **(a)** About 28 cm; **(b)** almost 16 cm; **(c)** about 45 cm
3. **(a)** About 8 mm; **(b)** probably between 25 and 30 mm
4. **(a)** 1 m; **(b)** 20 cm; **(c)** 5 km; **(d)** 2.2 m; **(e)** 30 mm
5. **(a)** 3,000 m; **(b)** 450 cm; **(c)** 1,200 mm; **(d)** 65 mm
6. **(a)** 7.5 m; **(b)** 5 km; **(c)** 7.8 cm; **(d)** 3.5 m
7. **(a)** 3 m; **(b)** 0.37 m; **(c)** 4.5 km **8.** 40 cm

## Reading Your Text

The following fill-in-the-blank exercises are designed to ensure that you understand some of the key vocabulary used in this section.

SECTION 7.2

**(a)** The ______________ system is based on one unit of length, the meter.

**(b)** The meter is just slightly longer than the ______________.

**(c)** The prefix ______________ means one–hundredth.

**(d)** The ______________ is used to measure long distances.

# 7.2 exercises

MathZone

Boost your grade at mathzone.com!
> Practice Problems
> NetTutor
> Self-Tests
> e-Professors
> Videos

Name ______________

Section ________ Date ________

< Objective 1 >

*Choose the most reasonable measure.*

1. The height of a ceiling
   (a) 25 m
   (b) 2.5 m
   (c) 25 cm

2. The diameter of a quarter
   (a) 24 mm
   (b) 2.4 mm
   (c) 24 cm

3. The height of a kitchen counter
   (a) 9 m
   (b) 9 cm
   (c) 90 cm

4. The diagonal measure of a television screen
   (a) 50 mm
   (b) 50 cm
   (c) 5 m

5. The height of a two-story building
   (a) 7 m
   (b) 70 m
   (c) 70 cm

6. An hour's drive on a freeway
   (a) 9 km
   (b) 90 m
   (c) 90 km

7. The width of a roll of cellophane tape
   (a) 1.27 mm
   (b) 12.7 mm
   (c) 12.7 cm

8. The width of a sheet of typing paper
   (a) 21.6 cm
   (b) 21.6 mm
   (c) 2.16 cm

9. The thickness of window glass
   (a) 5 mm
   (b) 5 cm
   (c) 50 mm

10. The height of a refrigerator
    (a) 16 m
    (b) 16 cm
    (c) 160 cm

11. The length of a ballpoint pen
    (a) 16 mm
    (b) 16 m
    (c) 16 cm

12. The width of a handheld calculator key
    (a) 1.2 mm
    (b) 12 mm
    (c) 12 cm

*Complete each statement, using a metric unit of length.*

13. A playing card is 6 __________ wide.

14. The diameter of a penny is 19 __________.

## Answers

1. ______________
2. ______________
3. ______________
4. ______________
5. ______________
6. ______________
7. ______________
8. ______________
9. ______________
10. ______________
11. ______________
12. ______________
13. ______________
14. ______________

## Answers

15. ______ 16. ______

17. ______ 18. ______

19. ______ 20. ______

21. ______ 22. ______

23. ______ 24. ______

25. ______

26. ______

27. ______

28. ______

29. ______

30. ______

31. ______

32. ______

33. ______

34. ______

35. ______

36. ______

37. ______

38. ______

39. ______

40. ______

**15.** A doorway is 2 __________ high.

**16.** A table knife is 22 __________ long.

**17.** A basketball court is 28 __________ long.

**18.** A commercial jet flies 800 __________ per hour.

**19.** The width of a nail file is 12 __________.

**20.** The distance from New York to Washington, D.C., is 387 __________.

**21.** A recreation room is 6 __________ long.

**22.** A ruler is 22 __________ wide.

**23.** A long-distance run is 35 __________.

**24.** A paperback book is 11 __________ wide.

< **Objective 2** >

*Complete each statement.*

**25.** 3,000 mm = __________ m

**26.** 150 cm = __________ m

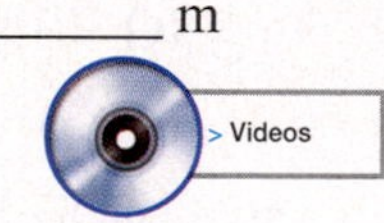


**27.** 8 m = __________ cm

**28.** 77 mm = __________ cm

**29.** 250 km = __________ cm

**30.** 500 cm = __________ m

> Videos

**31.** 25 cm = __________ mm

**32.** 150 mm = __________ m

**33.** 7,000 m = __________ km

**34.** 9 m = __________ cm

> Videos

**35.** 8 cm = __________ mm

**36.** 45 cm = __________ mm

**37.** 5 km = __________ m

**38.** 4,000 m = __________ km

**39.** 5 m = __________ mm

**40.** 7 km = __________ m

*Determine whether each statement is* **true** *or* **false.**

**41.** To convert to a larger measure in the metric system, we divide by a power of 10.

**42.** To convert kilometers to meters, move the decimal point three decimal places to the left.

*In each of the following statements, fill in the blank with* **always, sometimes,** *or* **never.**

**43.** When converting units of measure within the metric system, __________ multiply or divide by a power of 10.

**44.** If we are converting from a smaller unit to a larger unit, we ________ move the decimal point to the right.

*Use a metric ruler to measure the necessary dimensions and complete the statements.*

**45.** The perimeter of the parallelogram is __________ cm.

**46.** The perimeter of the triangle is __________ mm.

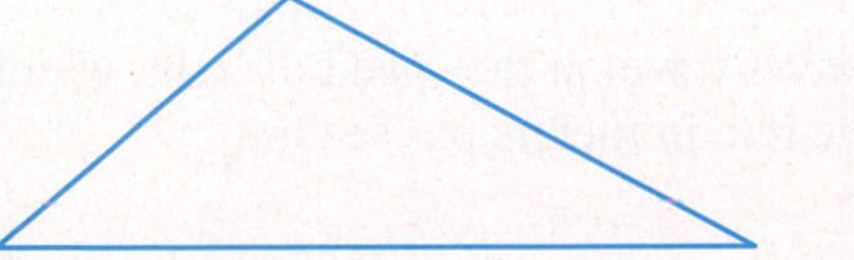

**47.** The perimeter of the rectangle is __________ cm.

**48.** Its area is __________ $cm^2$.

## Answers

41. ____________________

42. ____________________

43. ____________________

44. ____________________

45. ____________________

46. ____________________

47. ____________________

48. ____________________

## Answers

49. ______

50. ______

51. ______

52. ______

53. ______

54. ______

55. ______

56. ______

57. ______

49. The perimeter of the square is __________ mm.

50. The area of the square in exercise 49 is __________ $mm^2$.

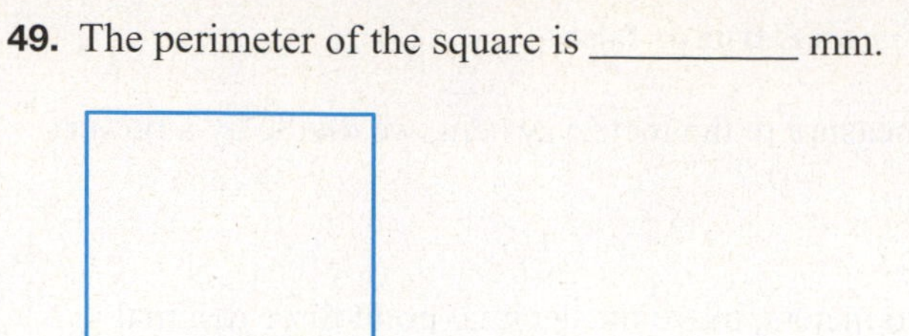


51. **Manufacturing Technology** A 2 by 6 plank is 3 m long. If lengths of 86 cm, 9.3 dm, and 29 cm are cut from the plank, how long is the remaining piece? chapter 7 > Make the Connection

52. **Mechanical Engineering** A piece of steel stock that is 2.5 m long has lengths of 82 cm, 2.4 dm, and 190 mm cut from it. How long is the remaining portion? chapter 7 > Make the Connection

53. **Agriculture** In a barn, there are 40 stalls in a distance of 51.2 m. How many centimeters wide is each stall?

54. **Electronics** A printed circuit board (PCB) measures 45 mm by 67 mm. What is the area of the board in *square centimeters?*

55. **Information Technology** Radio waves travel at the speed of light, which is 300,000 km per second. What is the rate in meters per second?

56. **Information Technology** Satellites are located in geostationary orbit at an altitude of approximately 35,786 km above the equator. What is this distance in meters?

57. **(a)** Determine the world record speed for both men and women in meters per second (m/s) for the following events: 100-, 400-, 1,500-, and 5,000-m run. The record times can be found at any one of several websites.

**(b)** Rank all the speeds obtained in order from fastest to slowest.

58. What units in the metric system would you use to measure each of the following quantities?

(a) Distance from Los Angeles to New York

(b) Your waist measurement

(c) Width of a hair

(d) Your height

Answers

58. ______

## Answers

1. (b) 3. (c) 5. (a) 7. (b) 9. (a) 11. (c) 13. cm 15. m 17. m 19. mm 21. m 23. km 25. 3 27. 800 29. 25,000,000 31. 250 33. 7 35. 80 37. 5,000 39. 5,000 41. True 43. always 45. 12 47. 14 49. 100 51. 92 cm 53. 128 cm 55. 300,000,000 m/s 57. Above and Beyond

# 7.3 Metric Units of Weight and Volume

< 7.3 Objectives >

1 > Use appropriate metric units of weight

2 > Convert metric units of weight

3 > Use appropriate metric units of volume

4 > Convert metric units of volume

**NOTE**

Technically, the gram is a unit of *mass* rather than weight. Weight is the force of gravity on an object. Thus, astronauts *weigh less* on the moon than on earth even though their masses are unchanged. For common use on earth, the terms *mass* and *weight* are still used interchangeably.

The basic unit of weight in the metric system is a very small unit called the **gram.** Think of a paper clip. It weighs roughly 1 gram (g). About 28 g make 1 oz in the English system. The gram is most often used to measure items that are fairly light. For heavier items, a more convenient unit of weight is the **kilogram (kg).** From the prefix *kilo* you should be able to deduce that a kilogram is equal to 1,000 grams.

**Definition**

**Comparing Kilograms (kg) to Grams (g)**

1 kilogram (kg) = 1,000 grams (g) (A kilogram is a bit more than 2 lb.)

### Example 1 Using Appropriate Units of Metric Weight

< Objective 1 >

The weight of a box of breakfast cereal is 320 g.

A woman might weigh 50 kg.

The weight of a nickel is 5 g.

**Check Yourself 1**

Choose the most reasonable measure.

(a) A penny: 30 g, 3 g, or 3 kg.
(b) A bar of soap: 120 g, 12 g, or 1.2 kg.
(c) A car: 5,000 kg, 1,000 kg, or 5,000 g.

Another metric unit of weight that you will encounter is the **milligram.**

**Definition**

**Comparing Milligrams (mg) to Grams (g)**

$$1 \text{ milligram (mg)} = \frac{1}{1{,}000} \text{ g}$$

**NOTES**

The prefix *milli* means one-thousandth.

kg, g, and mg are the units in common use.

A milligram is an extremely small unit. It is used, for example, in medicine for measuring drug amounts. Thus, a pill might contain 300 mg of aspirin.

Just as with units of length, converting metric units of weight is simply a matter of moving the decimal point. The following chart will help. Again, think of the chart as a set of stairs. Recall that the largest unit is at the highest step on the stairs.

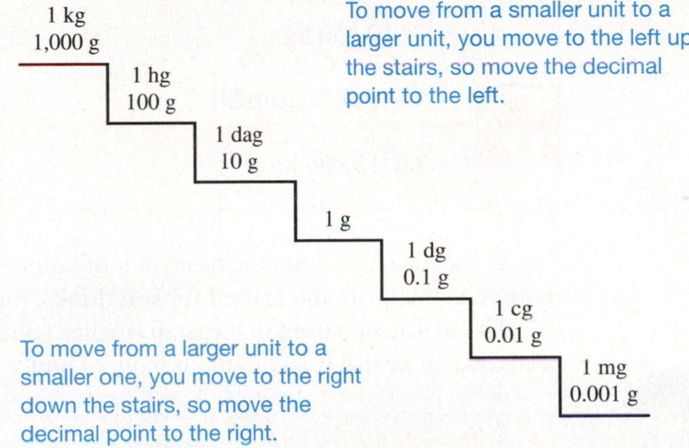


**Example 2** Converting Metric Weight

< Objective 2 >

Complete the following statements.

**(a)** 7 kg = ? g

7 kg = 7000$_\wedge$ g = 7,000 g

Move the decimal point *three places to the right* (to multiply by 1,000).

**NOTES**

We are converting to a *smaller* unit.

We are converting to a *larger* unit.

**(b)** 5,000 mg = ? g

5,000 mg = 5$_\wedge$000 g = 5 g

Move the decimal point *three places to the left* (to divide by 1,000).

**Check Yourself 2**

**(a)** 3,000 g = ________ kg **(b)** 500 cg = ________ g

For very heavy objects, the *metric ton* (t) is used.

**Definition**

**Metric Ton**

$$1 \text{ metric ton (t)} = 1{,}000 \text{ kg} \quad \text{so} \quad \frac{1 \text{ t}}{1{,}000 \text{ kg}} = 1 \quad \text{and} \quad \frac{1{,}000 \text{ kg}}{1 \text{ t}} = 1$$

We will use this measure in Example 3.

## Example 3 Converting to Metric Tons

Complete the following.

**(a)** 7,500 kg = ? t

$$7{,}500 \text{ kg} = \frac{7{,}500 \text{ kg}}{1} \times \frac{1 \text{ t}}{1{,}000 \text{ kg}} = 7.500 \text{ t}$$

Move the decimal point three places to the left to divide by 1,000.

**(b)** 12.25 t = ? kg

$$12.25 \text{ t} = \frac{12.25 \text{ t}}{1} \times \frac{1{,}000 \text{ kg}}{1 \text{ t}}$$
$$= 12{,}250 \text{ kg}$$

Move the decimal point three places to the right to multiply by 1,000.

**NOTE**

In both parts of Example 3, we are multiplying by 1.

### Check Yourself 3

**(a)** 13,400 kg = ________ t **(b)** 0.76 t = ________ kg

In the metric system, the basic unit of volume is the **liter (L).** The liter is slightly more than the quart and is used for soft drinks, milk, oil, gasoline, and so on.

The metric unit used to measure smaller volumes is the **milliliter (mL).** From the prefix we know that it is one-thousandth of a liter.

**Definition**

**Comparing Liters (L) to Milliliters (mL)**

1 liter (L) = 1,000 milliliters (mL)

**NOTE**

The liter is related to the meter. It is defined as the volume of a cube 10 cm on each edge, so

1 L = 1,000 $\text{cm}^3$

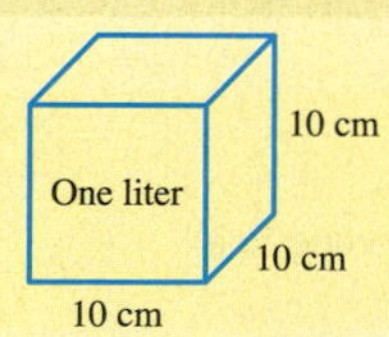


The milliliter is the volume of a cube 1 cm on each edge. So 1 mL is equal to 1 $\text{cm}^3$. These units can be used interchangeably.

**Note:** Scientists give measurements of volume in terms of cubic centimeters ($\text{cm}^3$). Example 4 will help you get used to the metric units of volume.

## Example 4 Using Appropriate Units of Metric Volume

< Objective 3 >

A teaspoon is about 5 mL or 5 $\text{cm}^3$.

A 6-oz cup of coffee is about 180 mL.

A quart of milk is 946 mL (just less than 1 L).

A gallon is just less than 4 L.

Now try these Check Yourself exercises.

**NOTE**

This unit of volume is also spelled *litre* (the British spelling).

## Check Yourself 4

**Choose the most reasonable measure.**

**(a)** A can of soup: 3 L, 30 mL, or 300 mL.
**(b)** A pint of cream: 4.73 L, 473 mL, or 47.3 mL.
**(c)** A home-heating oil tank: 100 L, 1,000 L, or 1,000 mL.
**(d)** A tablespoon: 150 mL, 1.5 L, or 15 mL.

Converting metric units of volume is again just a matter of moving the decimal point. A chart similar to the ones you saw earlier may be helpful.

**NOTE**

L, cL, and mL are the most commonly used units. We show the other units simply to indicate that the prefixes and abbreviations are used in a consistent fashion.

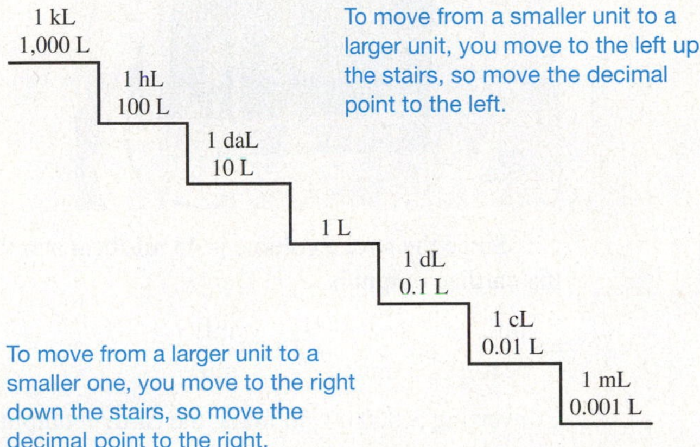


## Example 5 Converting Metric Volume

< Objective 4 >

**NOTES**

We are converting to a *smaller* unit.

We are converting to a *larger* unit.

We are converting to a *smaller* unit.

Complete the following statements.

**(a)** 4 L = ? mL

From the previous chart, we see that we should move the decimal point three places to the right (to multiply by 1,000).

4 L = 4.000‸ mL = 4,000 mL

**(b)** 3,500 mL = ? L

Move the decimal point three places to the left (to divide by 1,000).

3,500 mL = 3‸500 L = 3.5 L

**(c)** 30 cL = ? mL

Move the decimal point one place to the right (to multiply by 10).

30 cL = 30.0‸ mL = 300 mL

## Check Yourself 5

**Complete the following statements.**

**(a)** 5 L = ______ mL **(b)** 7,500 mL = ______ L
**(c)** 550 mL = ______ cL

Consider the following application from the field of health sciences.

## Example 6 A Vocational Application

Cardiac output, measured in liters per minute (L/min), is the product of a patient's stroke volume, in liters per beat, times the heart rate, in beats per minute (beats/min). Determine the cardiac output for a patient with a stroke volume of 45 milliliters per beat (mL/beat) and a heart rate of 80 beats/min.

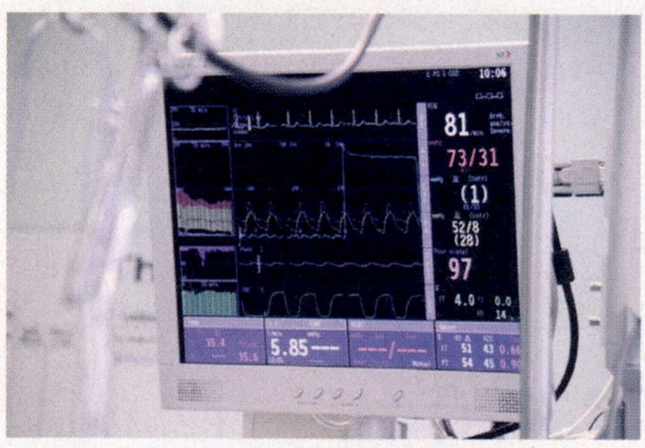

Since the stroke volume is 45 mL/beat and the patient's heart rate is 80 beats/min, the cardiac output is

$$45\frac{\text{mL}}{\text{beat}} \times 80\frac{\text{beats}}{\text{min}} = 3{,}600\frac{\text{mL}}{\text{min}}$$

Converting 3,600 mL to liters, the cardiac output is 3.6 L/min.

### Check Yourself 6

**Determine the cardiac output for a patient with a stroke volume of 68 mL/beat and a heart rate of 95 beats/min.**

### Check Yourself ANSWERS

**1.** **(a)** 3 g; **(b)** 120 g; **(c)** 1,000 kg **2.** **(a)** 3 kg; **(b)** 5 g
**3.** **(a)** 13.4 t; **(b)** 760 kg **4.** **(a)** 300 mL; **(b)** 473 mL; **(c)** 1,000 L; **(d)** 15 mL
**5.** **(a)** 5,000 mL; **(b)** 7.5 L; **(c)** 55 cL **6.** 6.46 L/min

## Reading Your Text

The following fill-in-the-blank exercises are designed to ensure that you understand some of the key vocabulary used in this section.

SECTION 7.3

**(a)** The basic unit of weight in the metric system is called the ___________.

**(b)** The ___________ is a small unit of weight that is used, for example, in measuring drug amounts.

**(c)** In the metric system, the basic unit of volume is the ___________.

**(d)** The ___________ is the volume of a cube 1 cm on each edge.

# 7.3 exercises

< Objective 1 >

*Choose the most reasonable measure of weight.*

**1.** A nickel
(a) 5 kg
(b) 5 g
(c) 50 g

**2.** A portable television set
(a) 8 g
(b) 8 kg
(c) 80 kg

**3.** A flashlight battery
(a) 8 g
(b) 8 kg
(c) 80 g

**4.** A 10-year-old boy
(a) 30 kg
(b) 3 kg
(c) 300 g

**5.** A Volkswagen Rabbit
(a) 100 kg
(b) 1,000 kg
(c) 1,000 g

**6.** A 10-lb bag of flour
(a) 45 kg
(b) 4.5 kg
(c) 45 g

**7.** A dinner fork
(a) 50 g
(b) 5 g
(c) 5 kg

**8.** A can of spices
(a) 3 g
(b) 300 g
(c) 30 g

**9.** A slice of bread
(a) 2 g
(b) 20 g
(c) 2 kg

**10.** A house paintbrush
(a) 120 g
(b) 12 kg
(c) 12 g

**11.** A sugar cube
(a) 2 mg
(b) 20 g
(c) 2 g

**12.** A salt shaker
(a) 10 g
(b) 100 g
(c) 1 g

*Complete each statement, using a metric unit of weight.*

**13.** A marshmallow weighs 5 _______.

**14.** A toaster weighs 2 _______.

**15.** 1 _______ is $\frac{1}{1,000}$ g.

**16.** A bag of peanuts weighs 100 _______.

**17.** An electric razor weighs 250 _______.

MathZone

Boost your grade at mathzone.com!
> Practice Problems
> NetTutor
> Self-Tests
> e-Professors
> Videos

Name _______________

Section _______ Date _______

## Answers

1. _______________
2. _______________
3. _______________
4. _______________
5. _______________
6. _______________
7. _______________
8. _______________
9. _______________
10. _______________
11. _______________
12. _______ 13. _______
14. _______ 15. _______
16. _______ 17. _______

## Answers

18. ______

19. ______

20. ______

21. ______

22. ______

23. ______

24. ______

25. ______

26. ______

27. ______

28. ______

29. ______

30. ______

31. ______

32. ______

33. ______

34. ______

35. ______

36. ______

37. ______

38. ______

**18.** A soupspoon weighs 50 _______.

**19.** A heavyweight boxer weighs 98 _______.

**20.** A vitamin C tablet weighs 500 _______.

**21.** A cigarette lighter weighs 30 _______.

**22.** A clock radio weighs 1.5 _______.

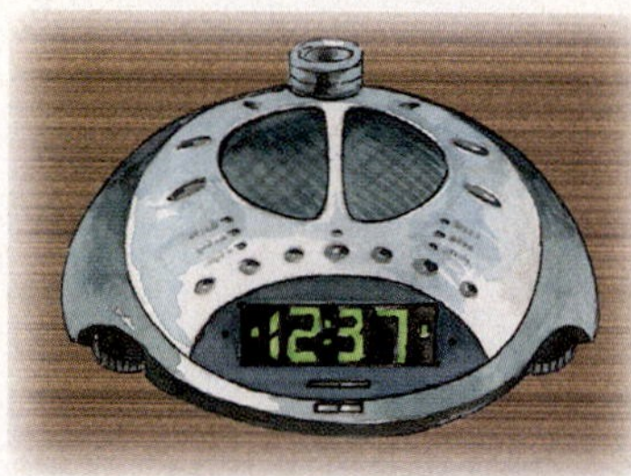


**23.** A household broom weighs 300 _______.

**24.** A 60-watt light bulb weighs 25 _______.

< Objective 2 >

*Complete each statement.*

**25.** 8 kg = ________ g

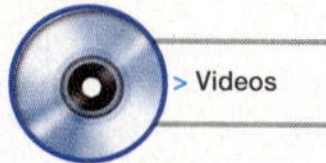


**26.** 5,000 mg = ________ g

**27.** 9,500 kg = ________ t

**28.** 3 kg = ________ g

**29.** 1.45 t = ________ kg

**30.** 12,500 kg = ________ t

**31.** 3 g = ________ mg

**32.** 2,000 g = ________ kg

< Objective 3 >

*Choose the most reasonable measure of volume.*

**33.** A bottle of wine

**(a)** 75 mL
**(b)** 7.5 L
**(c)** 750 mL

**34.** A gallon of gasoline

**(a)** 400 mL
**(b)** 4 L
**(c)** 40 L

**35.** A bottle of perfume

**(a)** 15 mL
**(b)** 150 mL
**(c)** 1.5 L

**36.** A can of frozen orange juice

**(a)** 1.5 L
**(b)** 150 mL
**(c)** 15 mL

**37.** A hot-water heater

**(a)** 200 mL
**(b)** 50 L
**(c)** 200 L

**38.** An oil drum

**(a)** 220 L
**(b)** 220 mL
**(c)** 22 L

39. A bottle of ink
(a) 60 $cm^3$
(b) 6 $cm^3$
(c) 600 $cm^3$

40. A cup of tea
(a) 18 mL
(b) 180 mL
(c) 18 L

41. A jar of mustard
(a) 150 mL
(b) 15 L
(c) 15 mL

42. A bottle of aftershave lotion
(a) 50 mL
(b) 5 L
(c) 5 mL

43. A cream pitcher
(a) 12 mL
(b) 120 mL
(c) 1.2 L

44. One tablespoon
(a) 1.5 mL
(b) 1.5 L
(c) 15 mL

*Complete each statement, using a metric unit of volume.*

45. A can of tomato soup is 300 _______.

46. 1 _______ is $\frac{1}{100}$ L.

47. A saucepan holds 1.5 _______.

48. A thermos bottle contains 500 _______ of liquid.

49. A coffee pot holds 720 _______.

50. A garbage can will hold 120 _______.

51. A car's engine capacity is 2,000 $cm^3$. It is advertised as a 2.0 _______ model.

52. A bottle of vanilla extract contains 60 _______.

53. 1 _______ is $\frac{1}{10}$ cL.

54. A can of soft drink is 35 _______.

55. A garden sprinkler delivers 8 _______ of water per minute.

56. 1 kL is 1,000 _______.

< Objective 4 >

*Complete each statement.*

57. 7 L = _______ mL

58. 4,000 $cm^3$ = _______ L

59. 4 hL = _______ L

60. 7 L = _______ cL

**Answers**

39. ____
40. ____
41. ____
42. ____
43. ____
44. ____
45. ____
46. ____
47. ____
48. ____
49. ____
50. ____
51. ____
52. ____
53. ____
54. ____
55. ____
56. ____
57. ____
58. ____
59. ____ 60. ____

## Answers

61. ______

62. ______

63. ______

64. ______

65. ______

66. ______

67. ______

68. ______

69. ______

70. ______

71. ______

72. ______

73. ______

74. ______

**61.** 8,000 mL = ________ L

**62.** 12 L = ________ mL

**63.** 5 L = ________ $cm^3$

**64.** 2 L = ________ cL

**65.** 75 cL = ________ mL

**66.** 5 kL = ________ L

**67.** 5 L = ________ cL

**68.** 400 mL = ________ cL

*Solve the following applications.*

**69. SCIENCE AND MEDICINE** The United States emitted 67.3 million metric tons (t) of carbon monoxide (CO) into the atmosphere in 1999. One metric ton equals 1,000 kg. How many kilograms of CO were emitted to the atmosphere in the United States during 1999?

**70. SCIENCE AND MEDICINE** The United States emitted 19.5 million t of nitrogen oxides (NO) into the atmosphere in 1987. One metric ton (1 t) equals 1,000 kg. How many kilograms of NO were emitted to the atmosphere in the United States during 1987?

*Determine whether each statement is* **true** *or* **false.**

**71.** A milliliter is the same as a cubic centimeter.

**72.** A kilogram is 1,000 times heavier than a milligram.

Basic Skills | Advanced Skills | **Vocational-Technical Applications** | Calculator/Computer | Above and Beyond

**73. ALLIED HEALTH** Many medications are expressed as percent solutions where the percent indicates how many grams of the active ingredient are dissolved in 100 milliliters (mL) of diluting element (usually water or saline). Consider a 0.6% solution of metaproterenol, an oral inhalation medication used to treat bronchospasm.

**(a)** Determine the number of milligrams (mg) of metaproterenol per milliliter of diluting element in a 0.6% solution.

**(b)** Determine the volume (in milliliters) of the 0.6% solution to be administered if a dose of 15 mg of metaproterenol is ordered.

**74. ALLIED HEALTH** Many medications are expressed as percent solutions where the percent indicates how many grams of the active ingredient are dissolved in 100 milliliters (mL) of diluting element (usually water or saline). Consider a 2.5% solution of Demerol, a potent pain medication.

**(a)** Determine the number of milligrams (mg) of Demerol per milliliter of diluting element in a 2.5% solution.

**(b)** Determine the volume (in milliliters) of the 2.5% solution to be administered if a dose of 120 mg of Demerol is ordered.

**75.** **Manufacturing Technology** There are 312.83 kg of steel in stock. A new part will use 1,600 mg. If 30 of these parts will be produced, how much steel will be left in stock?

**76.** **Manufacturing Technology** There are 8,370 mL of sealer in stock. A run of parts uses 2.4 L of sealer. How much is left after the run?

**77.** **Automotive Technology** In a 4-cylinder engine, the displacement of a single cylinder is 575 mL. What is the displacement of the engine?

**78.** **Manufacturing Technology** A green treated deck board weighs 4.19 kg. As it dries out, it loses 247 g of weight. How much does the board now weigh?

**79.** **Agriculture** A pallet containing 65 bags of fertilizer weighs 1.482 metric tons (1 metric ton = 1,000 kg). How many kilograms does each bag weigh?

Answers

75. ______
76. ______
77. ______
78. ______
79. ______
80. ______
81. ______
82. ______

Basic Skills | Advanced Skills | Vocational-Technical Applications | Calculator/Computer | **Above and Beyond**

**80.** Mass (weight) and volume are connected in the metric system. The weight of water in a cube 1 cm on a side is 1 g. Does such a relationship exist in the English system of measurement? If so, what is it? If not, why not?

**81.** **(a)** Determine how many liters of gasoline your car will hold.

**(b)** Using current prices, determine what a liter of gasoline should cost to make it competitive.

**(c)** How much would it cost to fill your car?

**82.** Do the following doses of medicine seem reasonable or unreasonable?

**(a)** Take 5 L of Kaopectate every morning.

**(b)** Soak your feet in 5 L of epsom salt bath every evening.

**(c)** Inject yourself with $\frac{3}{4}$ L of insulin every day.

## Answers

**1.** (b) **3.** (c) **5.** (b) **7.** (a) **9.** (b) **11.** (c) **13.** g **15.** mg **17.** g **19.** kg **21.** g **23.** g **25.** 8,000 **27.** 9.5 **29.** 1,450 **31.** 3,000 **33.** (c) **35.** (a) **37.** (c) **39.** (a) **41.** (a) **43.** (b) **45.** mL (or $cm^3$) **47.** L **49.** mL **51.** L **53.** mL **55.** L **57.** 7,000 **59.** 400 **61.** 8 **63.** 5,000 **65.** 750 **67.** 500 **69.** 67.3 billion kg **71.** True **73.** **(a)** 6 mg/mL; **(b)** 2.5 mL **75.** 312.782 kg **77.** 2.3 L **79.** 22.8 kg **81.** Above and Beyond

# 7.4 Converting Between the English and Metric Systems

< 7.4 Objectives >

1 > Convert between metric and English units of length

2 > Convert between metric and English units of weight

3 > Convert between metric and English units of volume

4 > Convert between metric and English units of temperature

Occasionally, it is necessary to convert between the English and metric systems. When such conversion is necessary, it is best to have a calculator and a conversion table. In this section, we provide conversion tables for length, weight, and volume. In addition, we provide formulas used for converting temperature. We begin by looking at length.

The following table will help you to convert from the metric to the English system for measuring length.

**Definition**

**Converting Between Metric and English Units (Length)**

1 m = 39.37 in.

1 cm = 0.394 in.

1 km = 0.62 mi

**Example 1** Converting Between Metric and English Units of Length

< Objective 1 >

Complete the following statements.

**(a)** 3 m = ? in.

$$3 \text{ m} = 3 \cancel{\text{m}} \left(\frac{39.37 \text{ in.}}{1 \cancel{\text{m}}}\right) = 118.11 \text{ in.}$$

This is a unit ratio equal to 1.

**(b)** 25 km = ? mi

$$25 \text{ km} = 25 \cancel{\text{km}} \left(\frac{0.62 \text{ mi}}{1 \cancel{\text{km}}}\right) = 15.5 \text{ mi}$$

**NOTE**

There are many ways to convert between the systems. You can just multiply or divide by the appropriate factor. We chose to use the unit ratio idea introduced in Section 7.1.

**Check Yourself 1**

Complete the following statements.

**(a)** 5 cm = ______ in. **(b)** 40 km = ______ mi

**Note:** Guidebooks for U.S. travelers give this tip for a quick conversion from kilometers to miles: Multiply the number of kilometers by 6 and drop the last digit. The result is the approximate number of miles. For example, a speed limit sign reads 90 km/h.

**NOTE**

This multiplies the number of kilometers by 0.6.

$90 \times 6 = 540$ ← Drop this digit. Multiply by 6 and drop the last digit, in this case 0.

90 km/h is approximately 54 mi/h.

You may also want to be able to convert units of the English system to those of the metric system. You can use the following conversion factors.

**Definition**

**Converting Between English and Metric Units of Length**

1 in. = 2.54 cm

1 yd = 0.914 m

1 mi = 1.6 km

**Example 2** **Converting Between English and Metric Units of Length**

Complete the following statements.

**(a)** 5 in. = ? cm

$$5 \text{ in.} = 5 \cancel{\text{in.}} \left(\frac{2.54 \text{ cm}}{1 \cancel{\text{in.}}}\right) \qquad \frac{2.54 \text{ cm}}{1 \text{ in.}} \text{ is a unit ratio equal to 1.}$$

$$= 12.7 \text{ cm}$$

**(b)** 12 mi = ? km

$$12 \text{ mi} = 12 \cancel{\text{mi}} \left(\frac{1.6 \text{ km}}{1 \cancel{\text{mi}}}\right) = 19.2 \text{ km}$$

**(c)** 5 yd = ? m

$$5 \text{ yd} = 5 \cancel{\text{yd}} \left(\frac{0.914 \text{ m}}{1 \cancel{\text{yd}}}\right) = 4.57 \text{ m}$$

**Check Yourself 2**

Complete the following statements.

**(a)** 8 in. = ______ cm **(b)** 20 mi = ______ km

To convert between units of weight in the two systems, use the following.

**Definition**

**Converting Between English and Metric Units of Weight**

1 kg = 2.2 lb

1 lb = 0.45 kg

1 oz = 28 g

### Example 3 Converting Between English and Metric Units of Weight

< Objective 2 >

**(a)** A roast beef weighs 3 kg. What is its weight in pounds?

$$3 \text{ kg} = 3 \cancel{\text{kg}} \left(\frac{2.2 \text{ lb}}{1 \cancel{\text{kg}}}\right) = 6.6 \text{ lb}$$

**(b)** A jar of spices weighs $1\frac{1}{2}$ oz. What is its weight in grams?

$$1\frac{1}{2} \text{ oz} = 1.5 \text{ oz} = 1.5 \cancel{\text{oz}} \left(\frac{28 \text{ g}}{1 \cancel{\text{oz}}}\right) = 42 \text{ g}$$

**(c)** A package weighs 5 lb. What is its weight in kilograms?

$$5 \text{ lb} = 5 \cancel{\text{lb}} \left(\frac{0.45 \text{ kg}}{1 \cancel{\text{lb}}}\right) = 2.25 \text{ kg}$$

**Check Yourself 3**

**(a)** A radio weighs 4 kg. What is its weight in pounds?

**(b)** A bag of peanuts weighs $3\frac{1}{2}$ oz. What is its weight in grams?

**(c)** If your cat weighs 8 lb, what is her weight in kilograms?

If you need to convert between units of volume in the English and metric systems, you can use the following table.

**Definition**

**Converting Between Metric and English Units of Volume**

1 L = 1.06 qt

1 qt = 0.95 L

1 fluid ounce (fl oz) = 30 mL

= 30 $\text{cm}^3$

Examples 4 and 5 show how to use these conversion factors.

### Example 4 Converting Between English and Metric Units of Volume

< Objective 3 >

A bottle's volume is 3 L. What is its capacity in quarts?

$$3 \text{ L} = 3 \cancel{\text{L}} \left(\frac{1.06 \text{ qt}}{1 \cancel{\text{L}}}\right) = 3.18 \text{ qt}$$

Note that $\frac{1.06 \text{ qt}}{1 \text{ L}}$ is a unit ratio equal to 1.

### Check Yourself 4

**A soft drink is bottled in a 2-L container. What does it contain in quarts?**

## Example 5 Converting Between English and Metric Units of Volume

A gasoline tank holds 16 gal. How many liters will it hold?

$$16 \text{ gal} = 16 \cancel{\text{gal}} \left(\frac{4 \text{ qt}}{1 \cancel{\text{gal}}}\right) = 64 \text{ qt}$$ First we convert to quarts.

$$= 64 \cancel{\text{qt}} \left(\frac{0.95 \text{ L}}{1 \cancel{\text{qt}}}\right) = 60.8 \text{ L}$$ Then we multiply by the unit ratio.

### Check Yourself 5

**A hot-water tank will hold 40 gal. What is its volume in liters?**

Temperature is expressed in **degrees Celsius** in the metric system, whereas **degrees Fahrenheit** is the unit of temperature used in the English system. The boiling point of water (at sea level) is 100 degrees Celsius, written 100°C, while it is 212 degrees Fahrenheit, written 212°F. The freezing point of water (at sea level) is 0°C, which corresponds to 32°F. The temperature on a hot day might be 30°C, corresponding to 86°F.

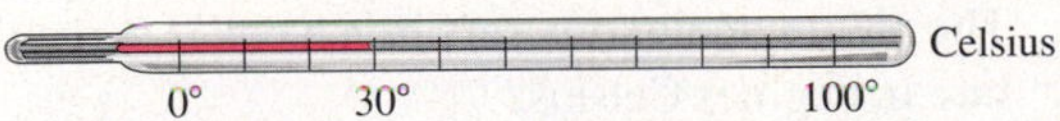


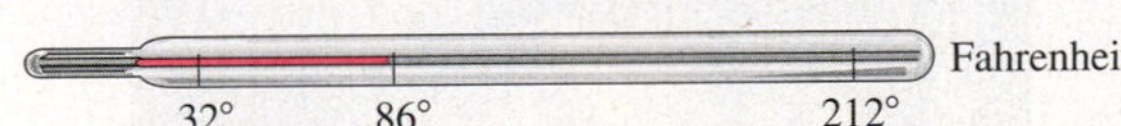


To convert between units of temperature, use the following formulas.

**Property**

### Converting Between English and Metric Units of Temperature

To convert from degrees Celsius (°C) to degrees Fahrenheit (°F), multiply by 9, divide by 5, and then add 32. A formula that describes this is

$$F = \frac{9C}{5} + 32$$

To convert from degrees Fahrenheit (°F) to degrees Celsius (°C), subtract 32, multiply by 5, and then divide by 9. A formula that describes this is

$$C = \frac{5(F - 32)}{9}$$

Example 6 shows the use of these formulas.

## Example 6 Converting Between English and Metric Units of Temperature

< Objective 4 >

To what temperature Fahrenheit does 18°C correspond?

Following the first formula given,

$$F = \frac{9(18)}{5} + 32 = \frac{162}{5} + 32 = 32.4 + 32 = 64.4$$

So the temperature is 64.4°F.

To what temperature Celsius does 77°F correspond?

Using the second rule,

$$C = \frac{5(77 - 32)}{9} = \frac{5(45)}{9} = \frac{225}{9} = 25$$

So the temperature is 25°C.

### Check Yourself 6

Complete the following:

**(a)** 12°C = ______ °F **(b)** 83°F = ______ °C

## Example 7 A Vocational Application

The melting point of a cast-iron block is approximately 2,300°F. What is this temperature in degrees Celsius?

Using the second formula given,

$$C = \frac{5(2{,}300 - 32)}{9} = \frac{5(2{,}268)}{9} = \frac{11{,}340}{9} = 1{,}260$$

So the corresponding temperature is 1,260°C.

### Check Yourself 7

The thermostat of a car opens at a temperature of 165°F. Convert this temperature to degrees Celsius.

If you would like to do a quick mental conversion from Celsius to Fahrenheit, the following estimation rule has been suggested: double the Celsius temperature and add 30. The result approximates the Fahrenheit temperature.

## Example 8 Estimating Fahrenheit Temperatures

Use the estimation rule described to convert 20°C to a Fahrenheit temperature. Then use the formula given prior to Example 6 to find the exact Fahrenheit temperature.

Doubling 20 gives 40, and adding 30 gives an estimate of 70°F.

Using the appropriate formula:

$$F = \frac{9C}{5} + 32 = \frac{9(20)}{5} + 32 = \frac{180}{5} + 32 = 36 + 32 = 68°\text{F}$$

### Check Yourself 8

**Use the estimation rule to convert 14°C to Fahrenheit. Then use the appropriate formula to find the exact Fahrenheit temperature.**

### Check Yourself ANSWERS

**1.** **(a)** 1.97 in.; **(b)** 24.8 mi **2.** **(a)** 20.32 cm; **(b)** 32 km
**3.** **(a)** 8.8 lb; **(b)** 98 g; **(c)** 3.6 kg **4.** 2.12 qt **5.** 152 L
**6.** **(a)** 53.6°F; **(b)** $28\frac{1}{3}°C$ **7.** Approx: 73.9°C
**8.** Estimate: 58°F; Exact: 57.2°F

### Reading Your Text

The following fill-in-the-blank exercises are designed to ensure that you understand some of the key vocabulary used in this section.

SECTION 7.4

**(a)** When conversion is necessary, it is best to have a ____________ and a conversion table.

**(b)** Guidebooks for U.S. travelers give a tip for quick conversion from ____________ to miles.

**(c)** Temperature is expressed in degrees ____________ in the metric system.

**(d)** The temperature on a ____________ day might be 30°C.

# 7.4 exercises

Basic Skills | Advanced Skills | Vocational-Technical Applications | Calculator/Computer | Above and Beyond

MathZone

Boost your grade at mathzone.com!

> Practice Problems
> NetTutor
> Self-Tests
> e-Professors
> Videos

Name ________________

Section ________ Date ________

## Answers

1. ________ 2. ________
3. ________ 4. ________
5. ________ 6. ________
7. ________ 8. ________
9. ________ 10. ________
11. ________ 12. ________
13. ________ 14. ________
15. ________ 16. ________
17. ________ 18. ________
19. ________ 20. ________
21. ________ 22. ________
23. ________ 24. ________
25. ________ 26. ________

< Objectives 1–4 >

*Complete each statement. (Round to the nearest hundredth.)*

**1.** 250 km = ________ mi

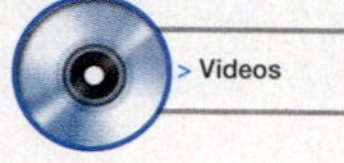


**2.** 9 cm = ________ in.

**3.** 150 mi = ________ km

**4.** 9 yd = ________ m

**5.** 2.6 m = ________ in.

**6.** 72 in. = ________ cm

**7.** 6 lb = ________ kg

**8.** 8 oz = ________ g

**9.** 0.25 kg = ________ oz

**10.** 5 lb = ________ g

**11.** 4 qt = ________ L

**12.** 7 L = ________ qt

**13.** 8 fl oz = ________ mL

**14.** 15.9 gal = ________ L

**15.** 760 mL = ________ qt

**16.** 15 L = ________ gal

**17.** 52°F = ________ °C

**18.** 6°C = ________ °F

**19.** 24°C = ________ °F

**20.** 95°F = ________ °C

**21.** 86°F = ________ °C

**22.** 10°C = ________ °F

**23.** 20°C = ________ °F

**24.** 72°F = ________ °C

**25.** 100°F = ________ °C

**26.** 27°C = ________ °F

*Solve the following applications.*

**27. Science and Medicine** A football team's fullback weighs 250 lb. How many kilograms does he weigh?

**28. Science and Medicine** Samantha's speedometer reads in kilometers per hour. If the legal speed limit is 55 mi/h, how fast can she drive?

*Determine whether each statement is* **true** *or* **false.**

**29.** A mile is shorter than a kilometer.

**30.** An inch is larger than a centimeter.

**31.** A quart is smaller then a liter.

**32.** A pound is more than half of a kilogram.

*In exercises 33 to 36, (a) use the estimation rule to convert the given Celsius temperature to Fahrenheit, and (b) find the exact Fahrenheit temperature.*

**33.** 15°C

**34.** 19°C

**35.** 8°C

**36.** 22°C

Basic Skills | Advanced Skills | **Vocational-Technical Applications** | Calculator/Computer | Above and Beyond

**37. Manufacturing Technology** The label for a primer states that it should not be applied to surfaces at less than 10°C. What is this temperature in degrees Fahrenheit?

**38. Manufacturing Technology** A construction adhesive should not be exposed to temperatures above 140°F. What is this temperature in degrees Celsius?

**39. Agriculture** In order for a corn seed to germinate, a soil temperature of at least 12.5°C is required. What is this temperature in degrees Fahrenheit?

**40. Manufacturing Technology** A copper-nickel alloy at 40% nickel becomes liquid at 1,280°C and becomes solid at 1,240°C. Convert these temperatures into degrees Fahrenheit.

**Answers**

27. ______

28. ______

29. ______

30. ______

31. ______

32. ______

33. ______

34. ______

35. ______

36. ______

37. ______

38. ______

39. ______

40. ______

Answers

41. ______

42. ______

43. ______

44. ______

45. ______

46. ______

47. ______

41. **Allied Health** An infant measures 51 cm long at birth. Determine the baby's length at birth in inches. Round to the nearest inch.

chapter 7 > Make the Connection

42. **Allied Health** An adult female patient is 5 ft 4 in. tall. Determine her height in centimeters. Round to the nearest cm.

chapter 7 > Make the Connection

43. **Manufacturing Technology** A machine is listed with a gross weight of 1,200 kg. The load limit of the trailer is listed at 2,500 lb. Can the machine be hauled on the trailer?

44. **Manufacturing Technology** A piece of steel rod stock is 5 m long. For a prototype, 6 ft 2 in. is used. How much is left (in meters)? Round to the nearest hundredth meter.

chapter 7 > Make the Connection

45. **Automotive Technology** The oil pan of a diesel engine calls for 2 gal of oil. How many liters of oil would this be?

46. **Manufacturing Technology** Which is larger: a metric ton of aluminum or a standard net ton of aluminum?

47. Complete the following puzzle.

**Across**

1. 6,000 milliliters
6. 1,760 yards
7. paradise
8. LV + LV
9. extraterrestrial
10. out of whack
13. tome
14. seven thousandths g

**Down**

1. 600 centimeters
2. top
3. ______ de France
4. ______ dm in 1 m
5. sixty thousand g
10. Presidential nickname
11. didn't lose
12. read-only memory

**MathWork Puzzle**

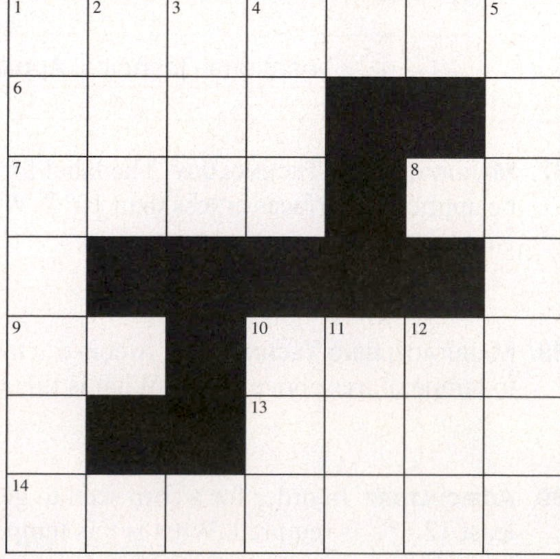

## Answers

**1.** 155 **3.** 240 **5.** 102.36 **7.** 2.7 **9.** 8.8 **11.** 3.8
**13.** 240 **15.** 0.81 **17.** 11.11 **19.** 75.2 **21.** 30 **23.** 68
**25.** 37.78 **27.** 112.5 **29.** False **31.** True
**33.** Approx: 60°F; Exact: 59°F **35.** Approx: 46°F; Exact: 46.4°F
**37.** 50°F **39.** 54.5°F **41.** 20 in. **43.** No; too heavy **45.** 7.6 L

**47.**

| | | | | | | |
|---|---|---|---|---|---|---|
| 6 | L | I | T | E | R | S |
| M | I | L | E | ■ | ■ | I |
| E | D | E | N | ■ | C | X |
| T | ■ | ■ | ■ | ■ | ■ | T |
| E | T | ■ | A | W | R | Y |
| R | ■ | ■ | B | O | O | K |
| S | E | V | E | N | M | G |

# Activity 19 :: Tool Sizes

Perhaps you work in construction, remodeling, or automotive repair. Or maybe you simply enjoy working on your own car or doing projects around the house. You have probably found the need to use both English and metric tool sizes.

The following drill bits came in a typical set. Convert each bit size to millimeters (mm), rounding to the nearest tenth of a millimeter.

| | | | | |
|---|---|---|---|---|
| $\frac{1}{16}$ in. | $\frac{5}{64}$ in. | $\frac{3}{32}$ in. | $\frac{7}{64}$ in. | $\frac{1}{8}$ in. |
| $\frac{9}{64}$ in. | $\frac{5}{32}$ in. | $\frac{3}{16}$ in. | $\frac{7}{32}$ in. | $\frac{1}{4}$ in. |

A set of wrenches with metric unit sizes consists of those listed in the following chart. Convert each size to a corresponding English unit wrench. In each case, find an English size accurate to the nearest $\frac{1}{32}$ of an inch.

| 8 mm | 10 mm | 12 mm | 13 mm | 14 mm | 17 mm |
|---|---|---|---|---|---|
| | | | | | |

Locate at least three other tool sizes in your home or apartment and make an appropriate conversion, either English to metric or metric to English.

| Definition/Procedure | Example | Reference |
|---|---|---|
| **The Units of the English System** | | Section 7.1 |
| The English system of measurement is in common use in the United States.<br>*English Units of Measure and Equivalents*<br>*Length*<br>1 foot (ft) = 12 inches (in.)<br>1 yard (yd) = 3 ft<br>1 mile (mi) = 5,280 ft<br>*Weight*<br>1 pound (lb) = 16 ounces (oz)<br>1 ton = 2,000 lb<br>*Volume*<br>1 pint (pt) = 16 fluid oz (fl oz)<br>1 quart (qt) = 2 pt<br>1 gallon (gal) = 4 qt<br>*Time*<br>1 minute (min) = 60 seconds (s)<br>1 hour (h) = 60 min<br>1 day = 24 h<br>1 week = 7 days | | *p.* 493 |
| **Unit Ratios** A fraction whose value is 1. Unit ratios can be used to convert units. | $\frac{12 \text{ in.}}{1 \text{ ft}}$ and $\frac{60 \text{ min}}{1 \text{ h}}$ are unit ratios. | *p.* 494 |
| *To Add Like Denominate Numbers* | | |
| **Step 1** Arrange the numbers so that the like units are in the same column.<br>**Step 2** Add in each column.<br>**Step 3** Simplify if necessary. | To add 4 ft 7 in. and 5 ft 10 in.:<br>4 ft 7 in.<br>+ 5 ft 10 in.<br>9 ft 17 in.<br>or 10 ft 5 in. | *p.* 495 |
| *To Multiply or Divide Denominate Numbers by Abstract Numbers* | | |
| **Step 1** Multiply or divide each part of the denominate number by the abstract number.<br>**Step 2** Simplify if necessary. | 2 × (3 yd 2 ft) = 6 yd 4 ft, or 7 yd 1 ft | *p.* 497 |
| **Metric Units of Length** | | Section 7.2 |
| **Metric units of length** are the meter (m), centimeter (cm), millimeter (mm), and kilometer (km). | | *pp.* 506–508 |
| *Basic Metric Prefixes*<br>*milli** means $\frac{1}{1,000}$<br>*centi** means $\frac{1}{100}$<br>*deci* means $\frac{1}{10}$<br>*kilo** means 1,000<br>*hecto* means 100<br>*deka* means 10 | *These are the most commonly used and should be memorized. | *pp.* 508–510 |

*Continued*

| Definition/Procedure | Example | Reference |
|---|---|---|

### *Converting Metric Units*

| Definition/Procedure | Example | Reference |
|---|---|---|
| You can use the following chart. | 500 cm = ? m<br>To convert from centimeters to meters, move the decimal point two places to the *left*.<br>500 cm = 5.00. cm = 5 m | p. 511 |

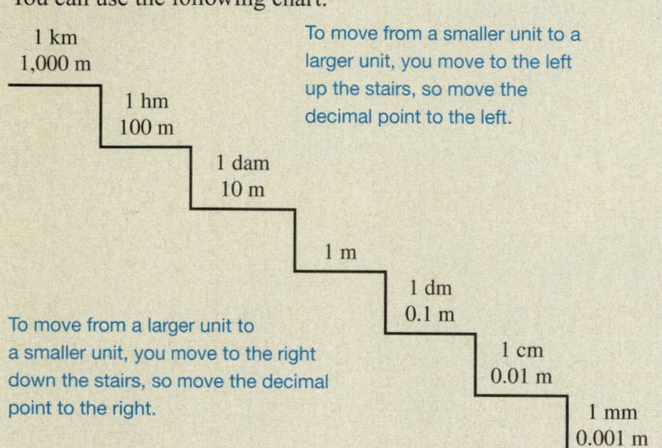


To convert between metric units, just move the decimal point the same number of places to the left or right as indicated by the chart.

### Metric Units of Weight and Volume

Section 7.3

| Definition/Procedure | Example | Reference |
|---|---|---|
| Conversions between units of volume (liters) or units of weight (grams) work in exactly the same fashion as those between units of length. | 3 L = ? mL<br>To convert from liters to milliliters, move the decimal point three places to the *right*.<br>3 L = 3.000. mL = 3,000 mL | p. 518 |

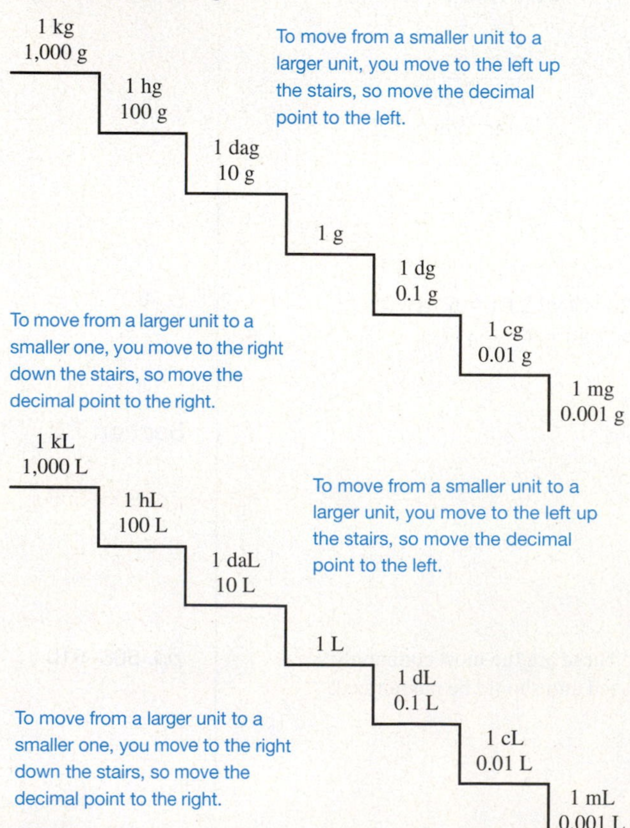

| Definition/Procedure | Example | Reference |
|---|---|---|
| **Converting Between the English and Metric Systems** | | Section 7.4 |
| ***Converting from Metric to English Units***<br>1 meter (m) ≈ 39.37 inches (in.)<br>1 centimeter (cm) ≈ 0.394 in.<br>1 kilometer (km) ≈ 0.62 mile (mi)<br>1 kilogram (kg) ≈ 2.2 pounds (lb)<br>1 liter (L) ≈ 1.06 quarts (qt) | 90 km ≈ 90(0.62) mi<br>= 55.8 mi | p. 528 |
| ***Converting from English to Metric Units***<br>1 inch (in.) ≈ 2.54 centimeters (cm)<br>1 yard (yd) ≈ 0.914 meter (m)<br>1 mile (mi) ≈ 1.6 kilometers (km)<br>1 quart (qt) ≈ 0.95 liter (L)<br>1 pound (lb) ≈ 0.45 kilogram (kg) | 20 lb ≈ 20(0.45) kg<br>= 9 kg | |
| ***Temperature Conversions***<br>From degrees Celsius (°C) to degrees Fahrenheit (°F):<br>$F = \frac{9C}{5} + 32$<br>From degrees Fahrenheit (°F) to degrees Celsius (°C):<br>$C = \frac{5(F - 32)}{9}$ | To convert 34°C,<br>$F = \frac{9(34)}{5} + 32 = 93.2$<br>To convert 75°F,<br>$C = \frac{5(75 - 32)}{9} \approx 23.9$ | p. 531 |

# summary exercises :: chapter 7

This summary exercise set is provided to give you practice with each of the objectives of this chapter. Each exercise is keyed to the appropriate chapter section. When you are finished, you can check your answers to the odd-numbered exercises against those presented in the back of the text. If you have difficulty with any of these questions, go back and reread the examples from that section. The answers to the even-numbered exercises appear in the *Instructor's Solutions Manual*. Your instructor will give you guidelines on how to best use these exercises in your instructional setting.

**7.1** *Complete each of the following statements.*

**1.** 11 ft = _______ in.

**2.** 72 h = _______ days

**3.** 6 gal = _______ qt

**4.** 80 fl oz = _______ pt

**5.** 4 lb = _______ oz

**6.** 5 mi = _______ ft

**7.** 8,000 lb = _______ tons

**8.** 16 pt = _______ qt

*Simplify.*

**9.** 3 ft 23 in.

**10.** 4 lb 20 oz

*Add.*

**11.**
```
   3 lb  9 oz
 + 5 lb 10 oz
```

**12.**
```
   5 h 20 min
   3 h 40 min
 + 2 h 20 min
```

*Subtract.*

**13.**
```
   7 ft 11 in.
 − 2 ft  4 in.
```

**14.**
```
   3 h 30 min
 − 1 h 50 min
```

*Multiply.*

**15.** 3 × (1 h 25 min)

*Divide.*

**16.** $\frac{10 \text{ lb } 12 \text{ oz}}{2}$

**17. BUSINESS AND FINANCE** John worked 6 h 15 min, 8 h, 5 h 50 min, 7 h 30 min, and 6 h during 1 week. What were the total hours worked?

**18. CONSTRUCTION** A room requires two pieces of floor molding 12 ft 8 in. long, one piece 6 ft 5 in. long, and one piece 10 ft long. Will 42 ft of molding be enough for the job?

**7.2** *Choose the most reasonable measure.*

**19.** A marathon race

**(a)** 40 km
**(b)** 400 km
**(c)** 400 m

**20.** The distance around your wrist

**(a)** 15 mm
**(b)** 15 cm
**(c)** 1.5 m

21. The diameter of a penny

(a) 19 cm

(b) 1.9 mm

(c) 19 mm

22. The width of a portable television screen

(a) 28 mm

(b) 28 cm

(c) 2.8 m

*Complete each statement, using a metric unit of length.*

23. A matchbook is 39 ____________ wide.

24. The distance from San Francisco to Los Angeles is 618 ____________.

25. A 1-lb coffee can has a diameter of 10 ____________.

*Complete each statement.*

26. 2 km = ____________ m

27. 3 cm = ____________ mm

28. 3,000 mm = ____________ m

29. 8 m = ____________ mm

30. 6 cm = ____________ m

31. 8 m = ____________ km

7.3 *Choose the most reasonable measure of weight.*

32. A quarter

(a) 6 g

(b) 6 kg

(c) 60 g

33. A tube of toothpaste

(a) 20 kg

(b) 200 g

(c) 20 g

34. A refrigerator

(a) 120 kg

(b) 1,200 kg

(c) 12 kg

35. A paperback book

(a) 1.2 kg

(b) 120 g

(c) 12 g

*Complete each statement, using a metric unit of weight.*

36. A loaf of bread weighs 500 _______.

37. A compact car weighs 900 _______.

38. A television set weighs 25 _______.

7.3 *Complete each statement.*

39. 5 kg = _______ g

40. 2,000 g = _______ kg

41. 5 t = _______ kg

42. 2,000 mg = _______ g

*Choose the most reasonable measure of volume.*

**43.** The gas tank of your car

**(a)** 500 mL

**(b)** 5 L

**(c)** 50 L

**44.** A bottle of eye drops

**(a)** 18 $cm^3$

**(b)** 180 $cm^3$

**(c)** 1.8 L

**45.** A can of soft drink

**(a)** 3.5 L

**(b)** 350 mL

**(c)** 35 mL

**46.** A punch bowl

**(a)** 200 L

**(b)** 20 L

**(c)** 200 mL

*Complete each statement, using a metric unit of volume.*

**47.** The crankcase of an automobile takes 5.5 _______ of oil.

**48.** The correct dosage for a cough medicine is 40 _______.

**49.** A bottle of iodine holds 20 _______.

**50.** A large mixing bowl holds 6 _______.

*Complete each statement.*

**51.** 5 L = _______ mL

**52.** 6,000 $cm^3$ = _______ L

**53.** 9 L = _______ $cm^3$

**54.** 10 mL = _______ L

**7.4** *Complete each statement. Round to the nearest hundredth.*

**55.** 8.3 m = _______ in.

**56.** 42 in. = _______ cm

**57.** 15 lb = _______ kg

**58.** 27.5 kg = _______ lb

**59.** 5.2 L = _______ qt

**60.** 18 gal = _______ L

**61.** 17°C = _______ °F

**62.** 41°F = _______ °C

**63.** 98.6°F = _______ °C

**64.** 6°C = _______ °F

**65.** 59°F = _______ °C

**66.** 30°C = _______ °F

**67.** 5°C = _______ °F

**68.** 35°F = _______ °C

# self-test 7

Name ______________

Section ________ Date ________

The purpose of this chapter test is to help you check your progress so that you can find sections and concepts that you need to review before the next exam. Allow yourself about an hour to take this test. At the end of that hour, check your answers against those given in the back of the text. If you missed any, note the section reference that accompanies the answer. Go back to that section and reread the examples until you have mastered that particular concept.

## Answers

1. ______________
2. ______________
3. ______________
4. ______________
5. ______________
6. ______________
7. ______________
8. ______________
9. ______________
10. ______________
11. ______________
12. ______________

*Complete the statements.*

**1.** 8 ft = _______ in.

**2.** 3 pt = _______ fl oz

*Simplify.*

**3.** 5 ft 21 in.

*Do the indicated operations.*

**4.**
```
   7 ft 9 in.
 + 3 ft 8 in.
```

**5.**
```
   7 lb  3 oz
 − 4 lb 10 oz
```

**6.** $4 \times (3 \text{ h } 50 \text{ min})$

**7.** $\dfrac{12 \text{ lb } 18 \text{ oz}}{3}$

**8.** **Construction** The Martins are fencing in a rectangular yard that is 110 ft long by 40 ft wide. If the fencing costs $3.50 per linear foot, what is the total cost of the fencing?

*Choose the most reasonable measure.*

**9.** The width of your hand

**(a)** 50 cm
**(b)** 10 cm
**(c)** 1 m

**10.** The speed limit on a freeway

**(a)** 10 km/h
**(b)** 100 km/h
**(c)** 100 m/h

**11.** A football player

**(a)** 12 kg
**(b)** 120 kg
**(c)** 120 g

**12.** A small can of tomato juice

**(a)** 4 L
**(b)** 400 mL
**(c)** 40 mL

**Answers**

**13.** ___________

**14.** ___________

**15.** ___________

**16.** ___________

**17.** ___________

**18.** ___________

**19.** ___________

**20.** ___________

**21.** ___________

**22.** ___________

*Complete the statements.*

**13.** 5 m = __________ mm

**14.** 3 kg = __________ g

**15.** 300 cL = __________ L

*Complete each statement, using multiplication of equivalents. Round to the nearest tenth.*

**16.** 40 in. = __________ cm

**17.** 5.2 km = __________ mi

**18.** 14.5 gal = __________ L

**19.** 150 lb = __________ kg

**20.** 12 in. = __________ cm

*Convert each temperature. Round to the nearest tenth.*

**21.** 58°F = __________ °C

**22.** 24°C = __________ °F

# cumulative review chapters 1-7

Name ____________

Section ______ Date ______

The following exercises are presented to help you review concepts from earlier chapters. This is meant as review material and not as a comprehensive exam. The answers are presented in the back of the text. Beside each answer is a section reference for the concept. If you have difficulty with any of these exercises, be certain to at least read through the summary related to that section.

**1.** A classroom is 7 yd wide by 8 yd long. If the room is to be recarpeted with material costing \$16 per square yard, find the cost of the carpeting.

**2.** Michael bought a washer-dryer combination that, with interest, cost \$959. He paid \$215 down and agreed to pay the balance in 12 monthly payments. Find the amount of each payment.

**3.** Evaluate the expression: $8 + 16 \div 4 \times 2$

**4.** Write the prime factorization for 168.

**5.** Find the greatest common factor of 12 and 20.

**6.** Arrange in order from smallest to largest: $\frac{5}{8}, \frac{3}{5}, \frac{2}{3}$

**7.** Multiply: $\frac{2}{3} \times 1\frac{4}{5} \times \frac{5}{8}$

**8.** Divide: $4\frac{1}{6} \div 10$

**9.** Find the least common multiple of 6, 15, and 20.

**10.** Add: $\frac{3}{5} + \frac{1}{6} + \frac{4}{15}$

**11.** Subtract: $7\frac{3}{8} - 3\frac{5}{6}$

**12.** You pay for purchases of \$14.95, \$18.50, \$11.25, and \$7 with a \$70 check. How much cash will you have left?

**13.** Find the area of a rectangle with length 6.4 centimeters (cm) and width 4.35 cm.

**14.** Find the perimeter of a rectangle with length 6.4 cm and width 4.35 cm.

**15.** Find the decimal equivalent of $\frac{9}{16}$.

**16.** Write the decimal form of $\frac{7}{13}$. Round to the nearest thousandth.

**Answers**

1. ____________
2. ____________
3. ____________
4. ____________
5. ____________
6. ____________
7. ____________
8. ____________
9. ____________
10. ____________
11. ____________
12. ____________
13. ____________
14. ____________
15. ____________
16. ____________

**Answers**

17. ______
18. ______
19. ______
20. ______
21. ______
22. ______
23. ______
24. ______
25. ______
26. ______
27. ______
28. ______
29. ______
30. ______
31. ______
32. ______
33. ______
34. ______ 35. ______
36. ______ 37. ______
38. ______ 39. ______

**17.** Solve for the unknown in the following proportion: $\frac{15}{x} = \frac{10}{16}$

**18.** If the scale on a map is $\frac{1}{4}$ in. equals 20 mi, how far apart are two towns that are 5 in. apart on the map?

**19.** Felipe traveled 342 mi, using 19 gal of gas. At this rate, how far can he travel on 25 gal?

**20.** Write as a percent: 0.375

**21.** Write as a simplified fraction: 12.5%

**22.** What is 43% of 8,200?

**23.** 315 is what percent of 140?

**24.** 120% of what number is 180?

**25.** A home that was purchased for $125,000 increased in value by 14% over a 3-year period. What was its value at the end of that period?

**26.** Complete the statement: 5 days = ________ hours

**27.** Subtract:

| | 4 min 10 s |
|---|---|
| − | 2 min 35 s |

**28.** Find the sum of 8 lb 14 oz and 12 lb 13 oz.

**29.** Find the difference between 7 ft 2 in. and 4 ft 5 in.

**30.** Multiply and simplify: $8 \times (2\text{ h } 40\text{ min})$

*Complete each statement.*

**31.** 43 cm = _______ m

**32.** 62 kg = _______ g

**33.** 740 mm = _______ cm

**34.** 14 L = _______ mL

**35.** 500 cm$^3$ = _______ L

*Complete each statement, using multiplication of equivalents. Round to the nearest tenth.*

**36.** 8.3 mi = _______ km

**37.** 68 kg = _______ lb

*Convert each temperature. Round to the nearest tenth.*

**38.** 85°F = _______ °C

**39.** 9°C = _______ °F

CHAPTER

8

# Geometry

## INTRODUCTION

Since the Norman conquests throughout Europe, Norman architecture has been extremely popular. In the eleventh century (ca. 1066), institutions such as churches, castles, and universities began including Norman windows in their design, as shown in the photograph. Later designs, such as Gothic architecture, grew out of the Norman designs.

Norman windows are one example of composite geometric figures. These are figures that incorporate several simpler geometric figures. Another composite figure, race tracks, are formed by attaching a semicircle to either end of a rectangle.

Whether determining the amount of fertilizer needed for the grass inside a racetrack, the amount of glass needed for a custom window, the amount of paint needed for the trim of a home design, or the amount of fluid that a bottle can hold, you will find that composite geometric figures are all around us.

The properties of geometric figures, such as perimeter and area, are critical for success in many trades. Architects, craftspeople, and contractors frequently work with two- and three-dimensional figures.

We examine the geometric properties of such figures throughout this chapter.

## CHAPTER 8 OUTLINE

Name ______________

Section ________ Date ________

This pretest provides a preview of the types of exercises you will encounter in each section of this chapter. The answers for these exercises can be found in the back of the text. If you are working on your own, or ahead of the class, this pretest can help you identify the sections in which you should focus more of your time.

**Answers**

1. ______________

2. ______________

3. ______________

4. ______________

**8.1** **1.** Find the circumference of the given circle. Use 3.14 for $\pi$.

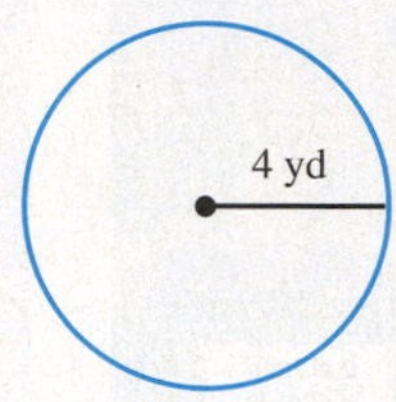


**2.** Find the area of the given figures. Use 3.14 for $\pi$.

**(a)**

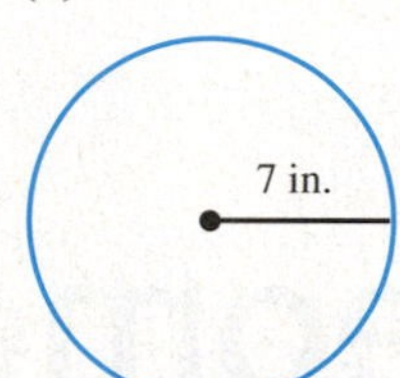


**(b)**

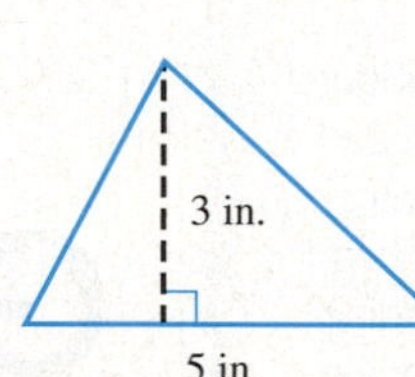


**(c)**

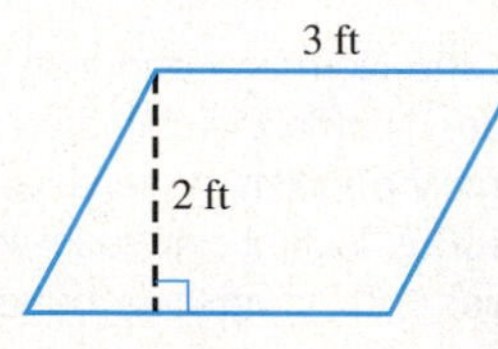


**8.2** **3.** Are the following lines parallel, perpendicular, or neither?

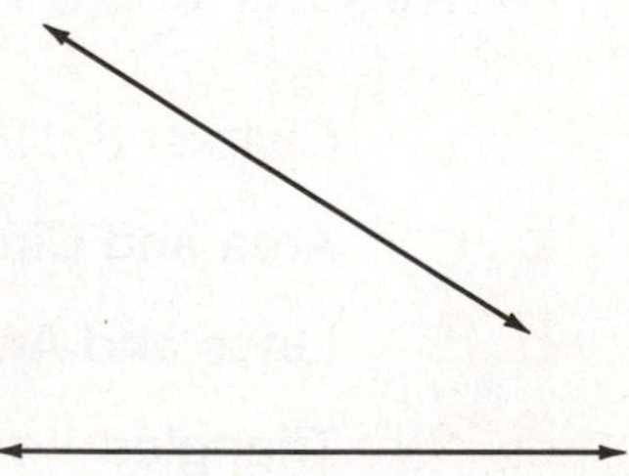

**4.** Use a protractor to measure $\angle BOA$. Label the triangle as acute, obtuse, or right.

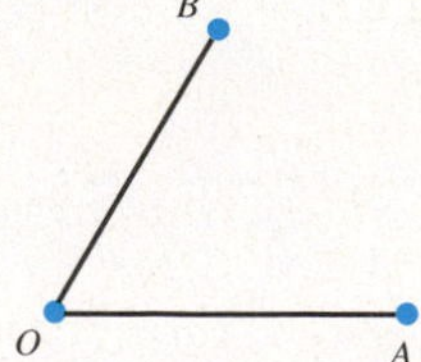

8.2 **5.** Find the measure of $\angle BCA$.

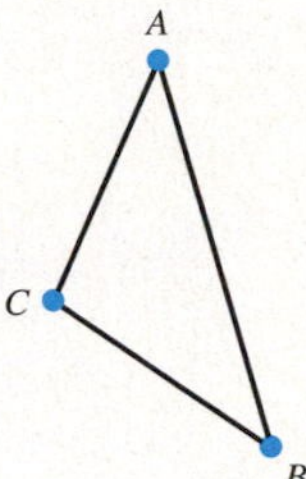


8.3 **6.** Label the triangle as equilateral, isosceles, or scalene.

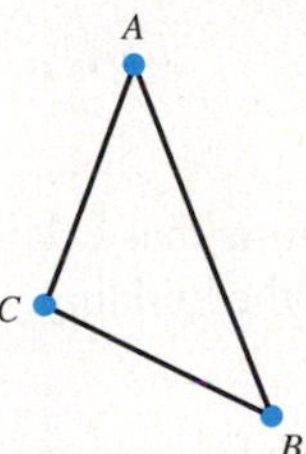


**7.** Label the triangle as acute, obtuse, or right.

**8.** Find the two triangles that are similar.

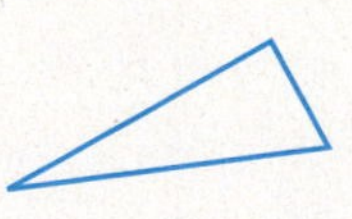
(*a*)

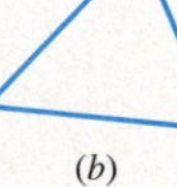
(*b*)

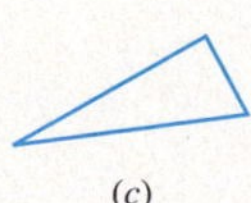
(*c*)

**9.** Find $\angle A$.

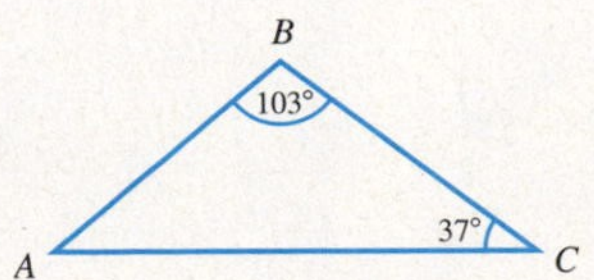


## Answers

5. ______

6. ______

7. ______

8. ______

9. ______

Answers

10. ______

11. ______

12. ______

13. ______

8.3 Find the length of the third leg in each pair of similar triangles.

10. 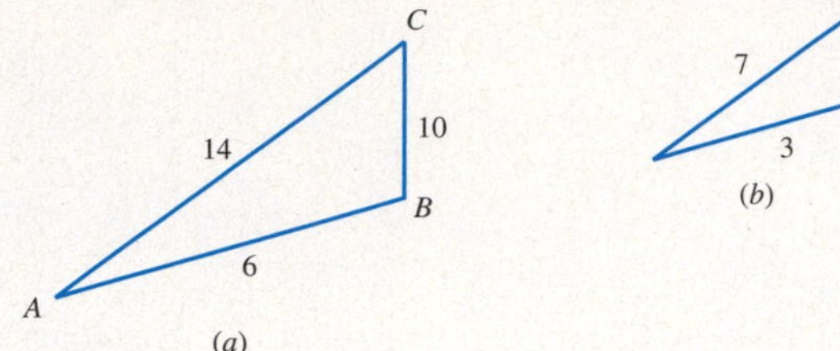


11. 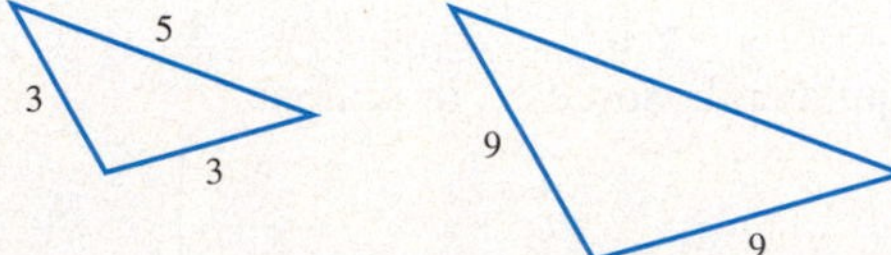


12. A building casts a shadow 99 m long at the same time that a flagpole 10 m high casts a shadow 18 m long. Find the height of the building.

8.4 13. Use the Pythagorean theorem to find the length of the hypotenuse for $\triangle ABC$.

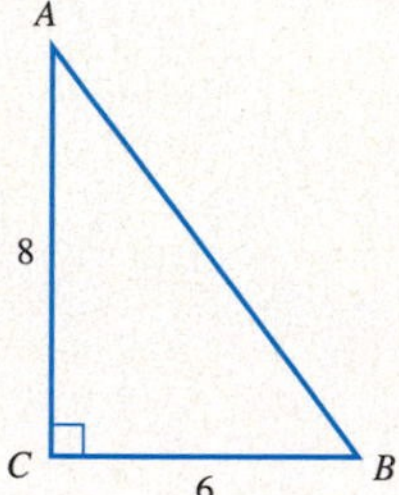

# 8.1 Area and Circumference

< 8.1 Objectives >

1 > Use $\pi$ to find the circumference of a circle

2 > Use $\pi$ to find the area of a circle

3 > Find the area of a parallelogram

4 > Find the area of a triangle

5 > Convert square units

We have previously seen how to find the perimeter and area of a rectangle. In this section we study these concepts as applied to circles, parallelograms, and triangles. The distance around the outside of a circle is closely related to the concept of perimeter. We call the perimeter of a circle the **circumference.**

**Definition**

**Circumference of a Circle**

The *circumference* of a circle is the distance around that circle.

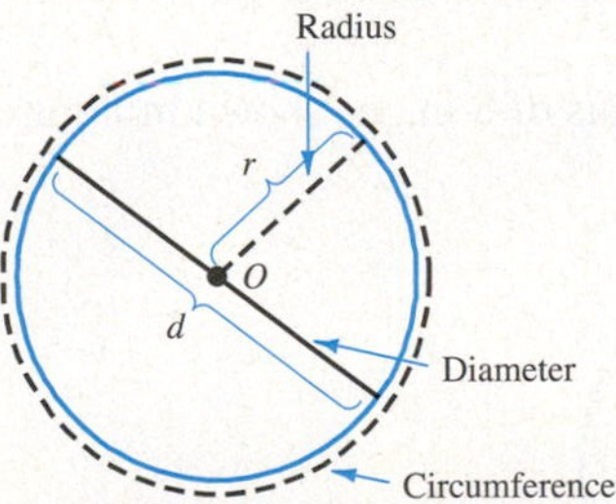


Figure 1

**NOTE**

The formula comes from the ratio

$\frac{C}{d} = \pi$

We begin by defining some terms. In the circle of Figure 1, $d$ represents the **diameter.** This is the distance across the circle through its center (labeled with the letter $O$, for **origin**). The **radius** $r$ is the distance from the center to a point on the circle. The diameter is always twice the radius.

It was discovered long ago that the ratio of the circumference of a circle to its diameter always stays the same. The ratio has a special name. It is named by the Greek letter $\pi$ (pi). Pi is approximately 3.14, rounded to two decimal places. We can write the following formula.

**Property**

**Formula for the Circumference of a Circle**

$C = \pi d$

## Example 1 Finding the Circumference of a Circle

< Objective 1 >

**NOTE**

Because 3.14 is an approximation for pi, we can only say that the circumference is approximately 14.1 ft. The symbol ≈ means "approximately."

A circle has a diameter of 4.5 ft, as shown in Figure 2. Find its circumference, using 3.14 for $\pi$. If your calculator has a $\boxed{\pi}$ key, use that key instead of a decimal approximation for $\pi$.

By the formula,

$$C = \pi d$$
$$\approx 3.14 \times 4.5 \text{ ft}$$
$$\approx 14.1 \text{ ft} \quad \text{(rounded to one decimal place)}$$

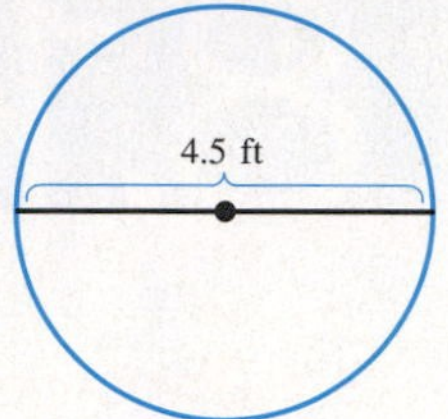


Figure 2

### Check Yourself 1

A circle has a diameter of $3\frac{1}{2}$ inches (in.). Find its circumference.

**Note:** In finding the circumference of a circle, you can use 3.14 as an approximation for pi. If you are using a calculator and want greater accuracy, use the $\boxed{\pi}$ key.

There is another useful formula for the circumference of a circle.

**Property**

**Formula for the Circumference of a Circle**

$C = 2\pi r$

## Example 2 Finding the Circumference of a Circle

**NOTE**

If you want to approximate $\pi$, you needn't worry about running out of decimal places. The value for pi has been calculated to over 51 billion decimal places on a computer (the printout would be over 150,000 pages long).

A circle has a radius of 8 in., as shown in Figure 3. Find its circumference, using 3.14 for $\pi$.

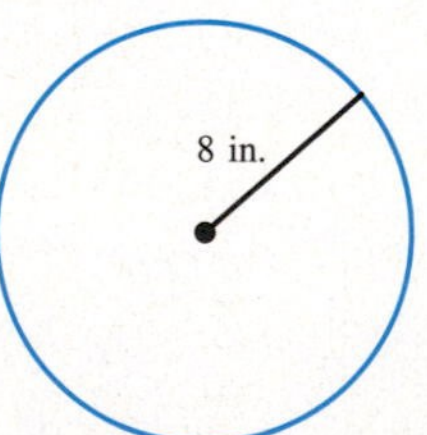


Figure 3

From the formula,

$$C = 2\pi r$$
$$\approx 2 \times 3.14 \times 8 \text{ in.}$$
$$\approx 50.2 \text{ in.} \quad \text{(rounded to one decimal place)}$$

**NOTE**

Because $d = 2r$ (the diameter is twice the radius) and $C = \pi d$, we have $C = \pi(2r)$, or $C = 2\pi r$.

### Check Yourself 2

Find the circumference of a circle with a radius of 2.5 in.

Sometimes we will want to combine the ideas of perimeter and circumference to solve a problem.

## Example 3 Finding Perimeter

> **NOTE**
>
> The distance around the semicircle is $\frac{1}{2}\pi d$.

We wish to build a wrought-iron frame gate according to the diagram in Figure 4. How many feet (ft) of material will be needed?

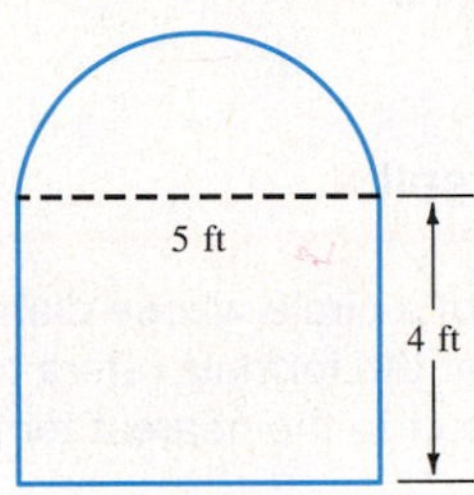


Figure 4

> **NOTE**
>
> Using a calculator with a $\boxed{\pi}$ key,
>
> 1 $\boxed{\div}$ 2 $\boxed{\times}$ $\boxed{\pi}$ $\boxed{\times}$ 5 $\boxed{=}$

The problem can be broken into two parts. The upper part of the frame is a semicircle (half a circle). The remaining part of the frame is just three sides of a rectangle.

$$\text{Circumference (upper part)} \approx \frac{1}{2} \times 3.14 \times 5 \text{ ft} \approx 7.9 \text{ ft}$$

$$\text{Perimeter (lower part)} = 4 + 5 + 4 = 13 \text{ ft}$$

Adding, we have

$$7.9 + 13 = 20.9 \text{ ft}$$

We will need approximately 20.9 ft of material.

### Check Yourself 3

**Find the perimeter of the following figure.**

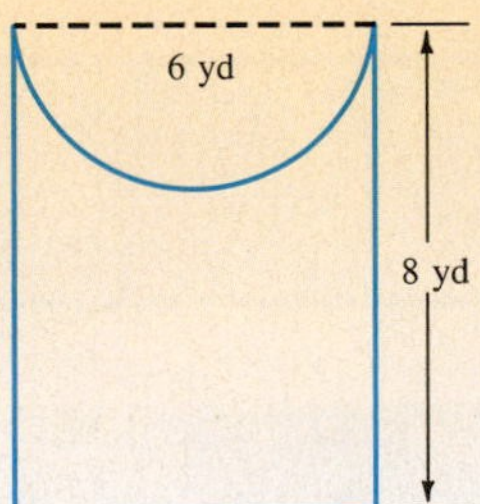


> **NOTE**
>
> This is read, "Area equals pi $r$ squared." You can multiply the radius by itself and then by pi.

The number pi ($\pi$), which we used to find circumference, is also used in finding the area of a circle. If $r$ is the radius of a circle, we have the following formula.

**Property**

**Formula for the Area of a Circle**

$$A = \pi r^2$$

## Example 4 Find the Area of a Circle

< Objective 2 >

A circle has a radius of 7 inches (in.) (see Figure 5). What is its area?

Use the area formula, using 3.14 for $\pi$ and $r = 7$ in.

$$A \approx 3.14 \times (7 \text{ in.})^2$$
$$\approx 153.86 \text{ in.}^2$$

Again, the area is an approximation because we use 3.14, an approximation for $\pi$.

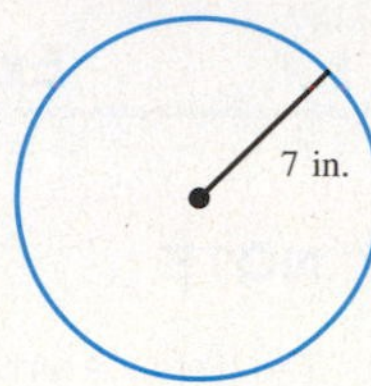

Figure 5

### Check Yourself 4

Find the area of a circle whose diameter is 4.8 centimeters (cm). Remember that the formula refers to the radius. Use 3.14 for $\pi$ and round your result to the nearest tenth of a square centimeter.

Two other important figures are parallelograms and triangles.

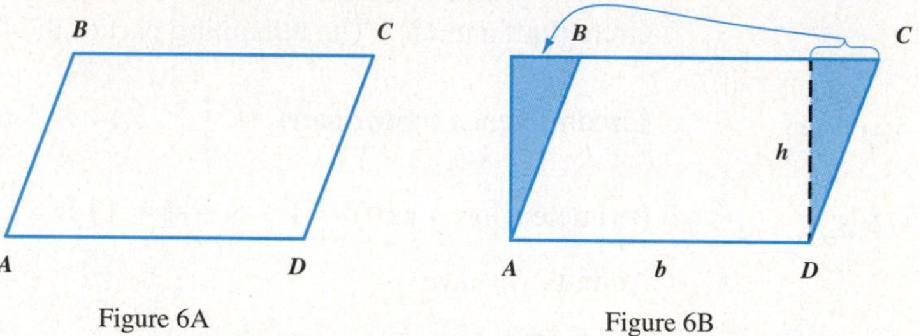

Figure 6A Figure 6B

In Figure 6 $ABCD$ is called a **parallelogram.** Its opposite sides are parallel and equal. In Figure 6B we have drawn a line from $D$ that forms a right angle with side $BC$. This cuts off one corner of the parallelogram. Now imagine that we move that corner over to the left side of the figure, as shown. This gives us a rectangle instead of a parallelogram. Because we did not change the area of the figure by moving the corner, the parallelogram has the same area as the rectangle, the product of the base and the height.

**Property**

**Formula for the Area of a Parallelogram** $A = b \cdot h$

## Example 5 Finding the Area of a Parallelogram

< Objective 3 >

A parallelogram has the dimensions shown in Figure 7. What is its area?

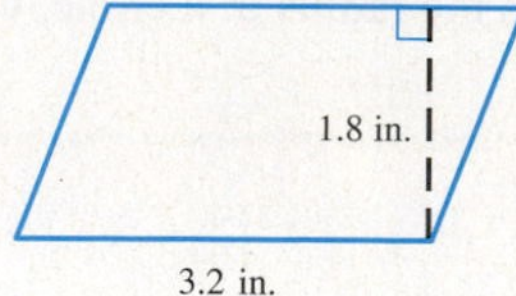

Figure 7

Use the formula, with $b = 3.2$ in. and $h = 1.8$ in.

$$A = b \cdot h$$
$$= 3.2 \text{ in.} \times 1.8 \text{ in.} = 5.76 \text{ in.}^2$$

**Check Yourself 5**

If the base of a parallelogram is $3\frac{1}{2}$ in. and its height is $1\frac{1}{2}$ in., what is its area?

Another common geometric figure is the **triangle.** It has three sides. An example is triangle $ABC$ in Figure 8.

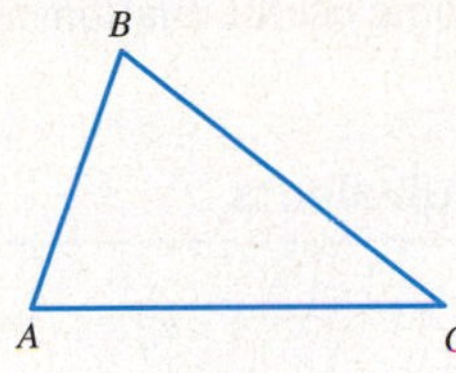

Figure 8A

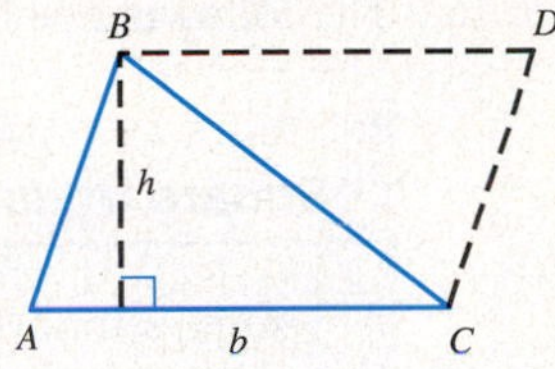

Figure 8B

$b$ is the base of the triangle. $h$ is the height, or the *altitude,* of the triangle.

Once we have a formula for the area of a parallelogram, it is not hard to find the area of a triangle. In Figure 8B we have drawn the dotted lines from $B$ to $D$ and from $C$ to $D$ parallel to the sides of the triangle, forming a parallelogram. The area of the triangle is then one-half the area of the parallelogram (which is $b \cdot h$ by our formula).

**Property**

**Formula for the Area of a Triangle**

$$A = \frac{1}{2} \cdot b \cdot h$$

## Example 6 Finding the Area of a Triangle

< Objective 4 >

A triangle has an altitude of 2.3 in., and its base is 3.4 in. (see Figure 9). What is its area?

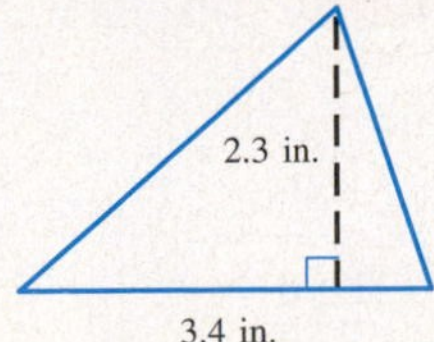

Figure 9

Use the triangle formula, with $b = 3.4$ in. and $h = 2.3$ in.

$$A = \frac{1}{2} \cdot b \cdot h$$
$$= \frac{1}{2} \times 3.4 \text{ in.} \times 2.3 \text{ in.} = 3.91 \text{ in.}^2$$

## Check Yourself 6

A triangle has a base of 10 feet (ft) and an altitude of 6 ft. Find its area.

Sometimes we will want to convert from one square unit to another. For instance, look at 1 $yd^2$ in Figure 10.

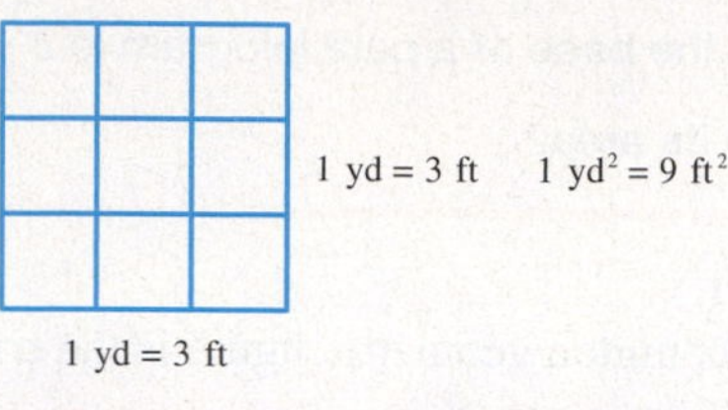


Figure 10

The following table gives some useful relationships.

**NOTE**

Originally the acre was the area that could be plowed by a team of oxen in a day!

**Square Units and Equivalents**

| | |
|---|---|
| 1 square foot ($ft^2$) | = 144 square inches ($in.^2$) |
| 1 square yard ($yd^2$) | = 9 $ft^2$ |
| 1 acre | = 4,840 $yd^2$ = 43,560 $ft^2$ |
| 1 square mile ($mi^2$) | = 640 acres |

## Example 7 Converting Between Feet and Yards in Finding Area

< Objective 5 >

A room has the dimensions 12 ft by 15 ft. How many square yards of linoleum will be needed to cover the floor?

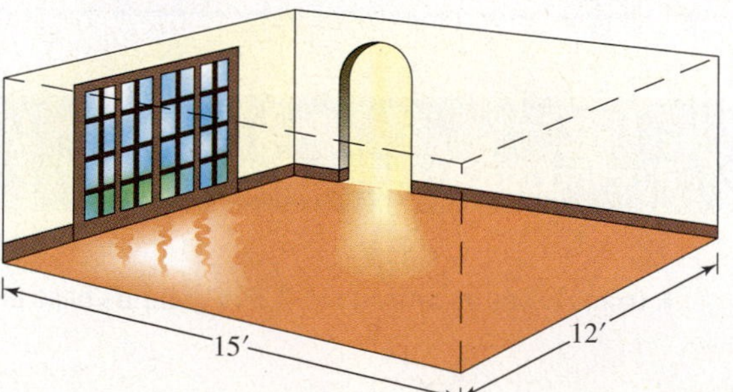


**NOTE**

We first found the area in square feet, and then converted to square yards. We could have first converted to yards (12 ft = 4 yd and 15 ft = 5 yd) and then multiplied.

$$A = 12 \text{ ft} \times 15 \text{ ft} = 180 \text{ ft}^2$$

$$= \overset{20}{\cancel{180}} \text{ ft}^2 \times \frac{1 \text{ yd}^2}{\underset{1}{\cancel{9}} \text{ ft}^2}$$

$$= 20 \text{ yd}^2$$

## Check Yourself 7

A hallway is 27 ft long and 4 ft wide. How many square yards of carpeting will be needed to carpet the hallway?

Example 8 illustrates the use of a common unit of area, the acre.

**Example 8** Converting Between Yards and Acres in Finding Area

A rectangular field is 220 yd long and 110 yd wide. Find its area in acres.

$A = 220 \text{ yd} \times 110 \text{ yd} = 24{,}200 \text{ yd}^2$

$= \overset{5}{\cancel{24{,}200}} \text{ yd}^2 \cdot \dfrac{1 \text{ acre}}{\underset{1}{\cancel{4{,}840}} \text{ yd}^2}$

$= 5 \text{ acres}$

**Check Yourself 8**

A proposed site for an elementary school is 220 yd long and 198 yd wide. Find its area in acres.

**Check Yourself ANSWERS**

1. $C \approx 11$ in. 2. $C \approx 15.7$ in. 3. $P \approx 31.4$ yd 4. $A \approx 18.1 \text{ cm}^2$

5. $A = 3\frac{1}{2} \text{ in.} \times 1\frac{1}{2} \text{ in.} = \frac{7}{2} \text{ in.} \times \frac{3}{2} \text{ in.} = 5\frac{1}{4} \text{ in.}^2$

6. $A = \frac{1}{2} \times 10 \text{ ft} \times 6 \text{ ft} = 30 \text{ ft}^2$

7. $12 \text{ yd}^2$ 8. 9 acres

## Reading Your Text

The following fill-in-the-blank exercises are designed to ensure that you understand some of the key vocabulary used in this section. Each sentence usually comes directly from the section. You will find the correct answers in the Answer Appendix.

SECTION 8.1

(a) We call the perimeter of a circle the ______________.

(b) The ______________ is the distance from the center to a point on the circle.

(c) The value for ______________ has been calculated to over 6 billion decimal places.

(d) Originally the ______________ was the area that could be plowed by a team of oxen in a day.

# 8.1 exercises

Basic Skills | Advanced Skills | Vocational-Technical Applications | Calculator/Computer | Above and Beyond



Name ______________

Section ________ Date ________

**Answers**

1. ______________
2. ______________
3. ______________
4. ______________
5. ______________
6. ______________
7. ______________
8. ______________
9. ______________
10. ______________

< Objective 1 >

*Find the circumference of each figure. Use 3.14 for π and round your answer to one decimal place.*

**1.**

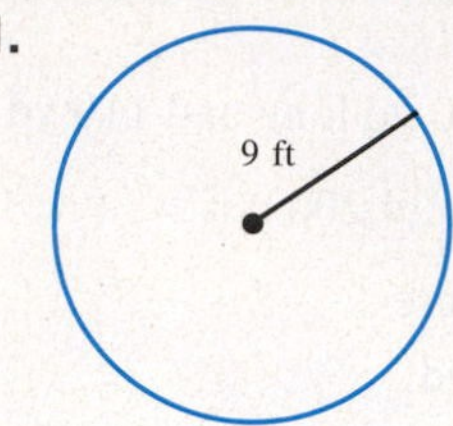


> Videos

**2.**

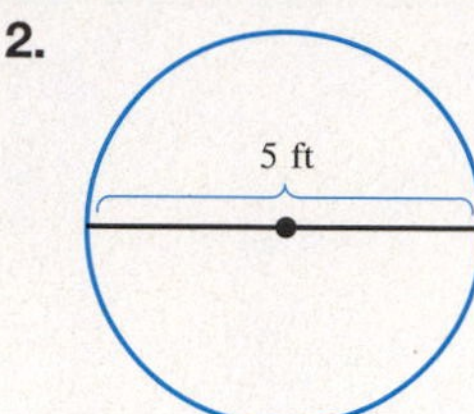


**3.**

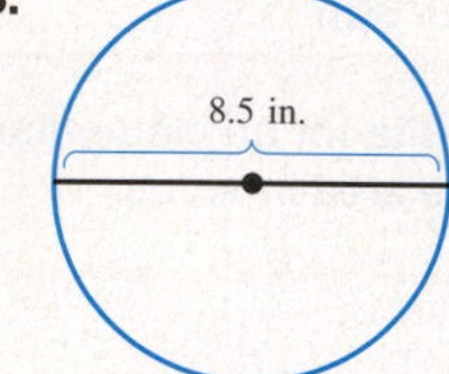


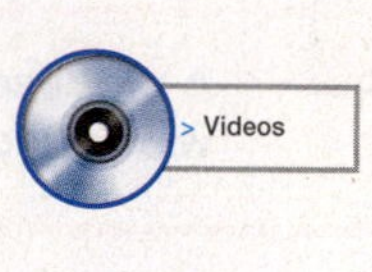


**4.**

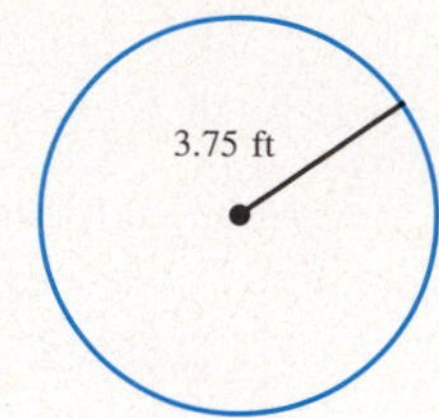


**5.**

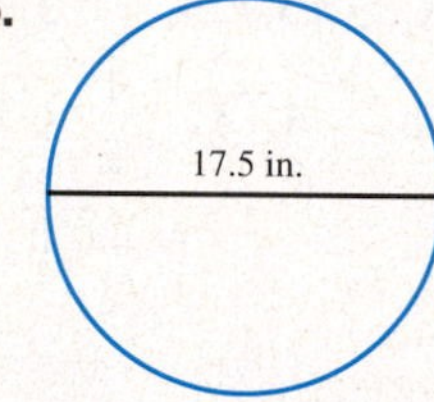


**6.**

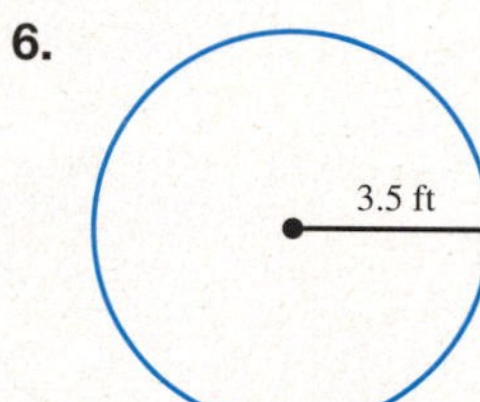


< Objective 2 >

*Find the area of each figure. Use 3.14 for π and round your answer to one decimal place.*

**7.**

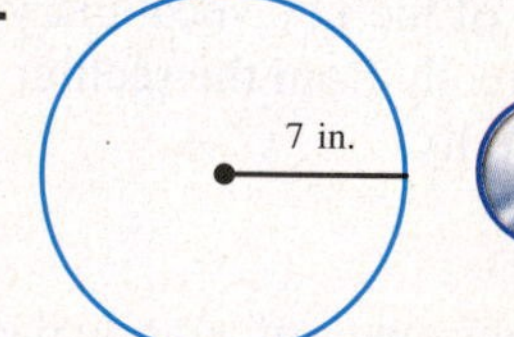


**8.**

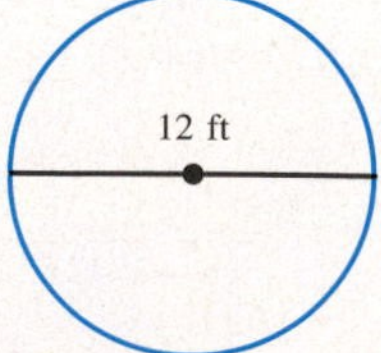


**9.**

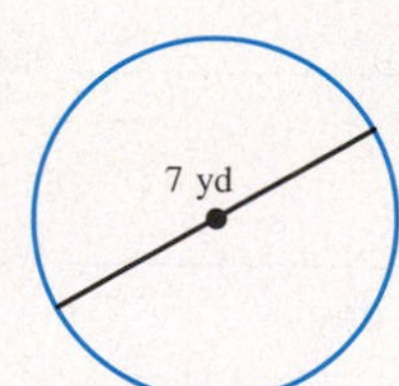


**10.**

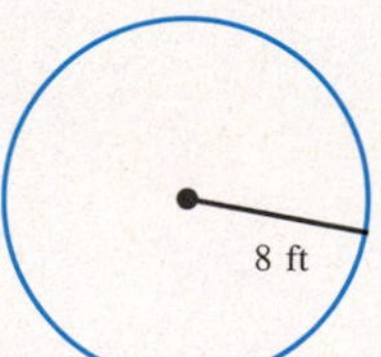

11.

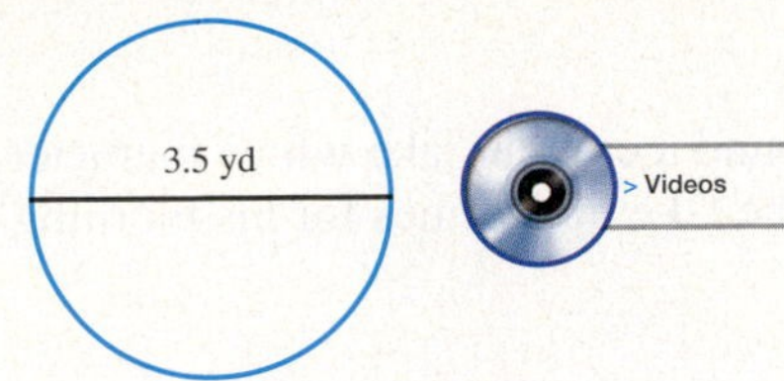


12.

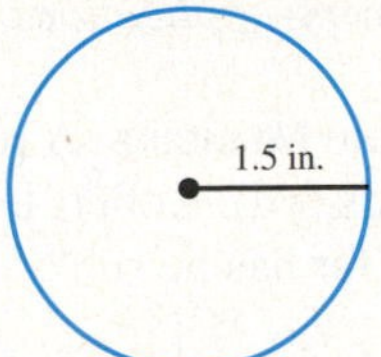


< Objective 3 >

*Find the area of each figure.*

13.

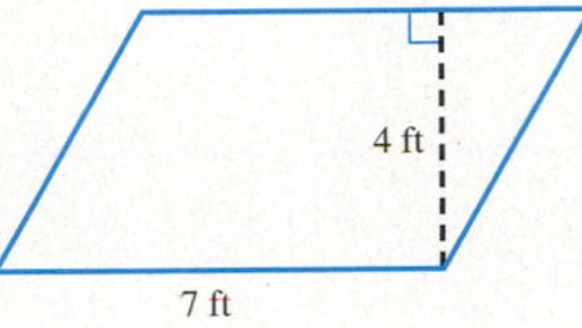


14.

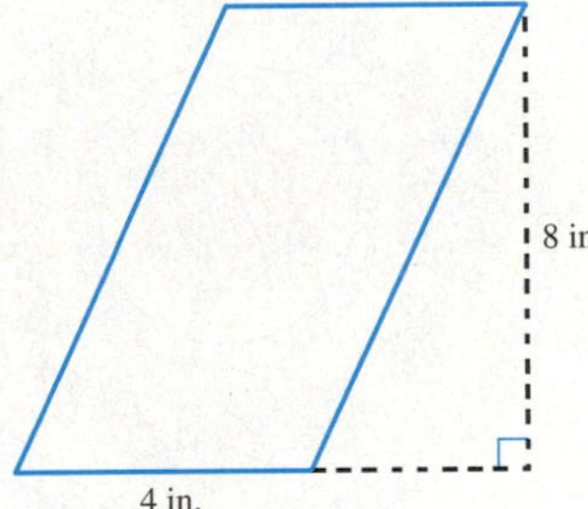


15.

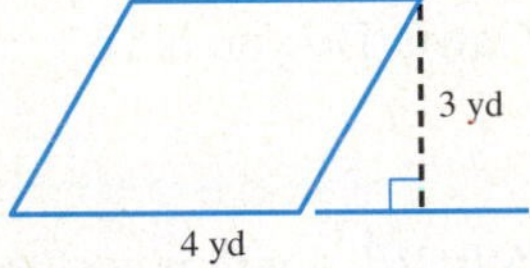


16.

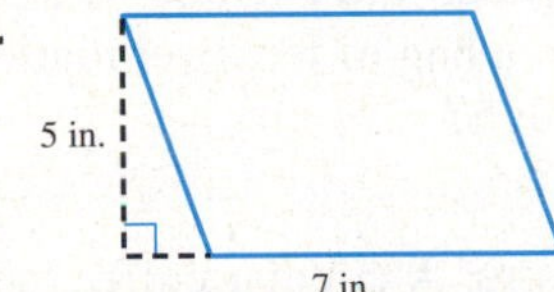


< Objective 4 >

17.

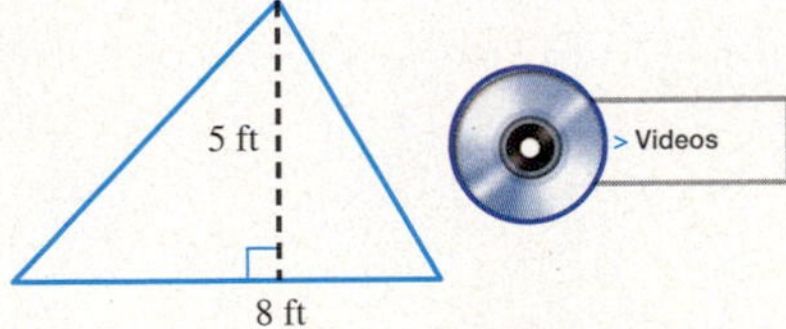


18.

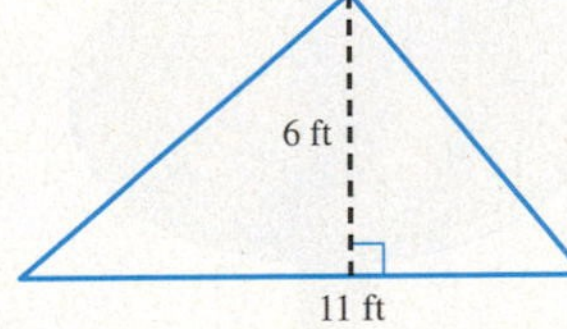


19.

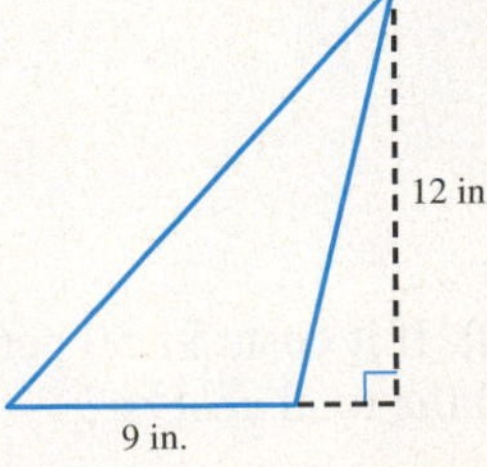


20.

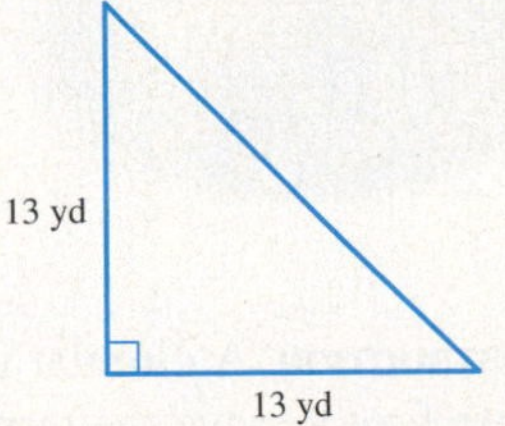


21.

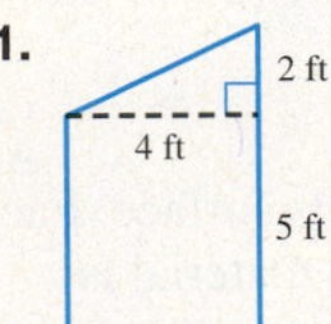

22.

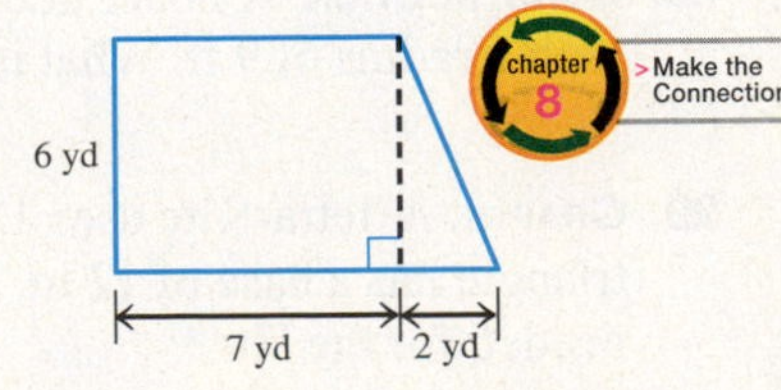


## Answers

11. ______

12. ______

13. ______

14. ______

15. ______

16. ______

17. ______

18. ______

19. ______

20. ______

21. ______

22. ______

Answers

23. ______

24. ______

25. ______

26. ______

27. ______

28. ______

29. ______

*Solve the following applications.*

23. **Science and Medicine** A path runs around a circular lake with a diameter of 1,000 yards (yd). Robert jogs around the lake three times for his morning run. How far has he run?

24. **Crafts** A circular rug is 6 feet (ft) in diameter. Binding for the edge costs \$1.50 per yard. What will it cost to bind the rug?

25. **Business and Finance** A circular piece of lawn has a radius of 28 ft. You have a bag of fertilizer that will cover 2,500 $ft^2$ of lawn. Do you have enough?

26. **Crafts** A circular coffee table has a diameter of 5 ft. What will it cost to have the top refinished if the company charges \$3 per square foot for the refinishing?

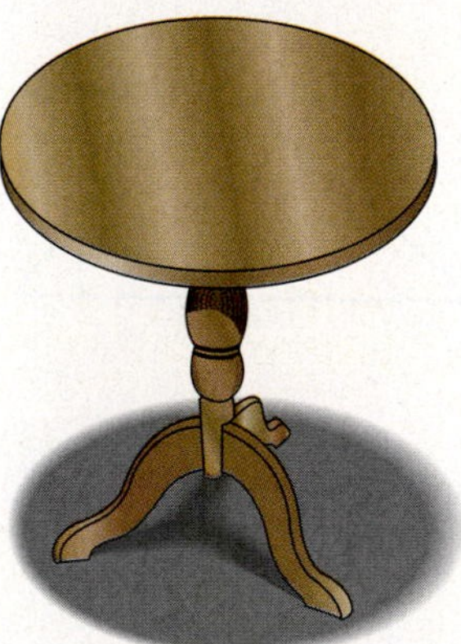

27. **Construction** A circular terrace has a radius of 6 ft. If it costs \$1.50 per square foot to pave the terrace with brick, what will the total cost be?

28. **Construction** A house addition is in the shape of a semicircle (a half-circle) with a radius of 9 ft. What is its area?

29. **Crafts** A Tetra-Kite uses 12 triangular pieces of plastic for its surface. Each triangle has a base of 12 in. and a height of 12 in. How much material is needed for the kite?

< Objective 5 >

**30.** **Construction** You buy a square lot that is 110 yd on each side. What is its size in acres?

**31.** **Crafts** You are making rectangular posters 12 in. by 15 in. How many square feet of material will you need for four posters?

**32.** **Crafts** Andy is carpeting a recreation room 18 ft long and 12 ft wide. If the carpeting costs $15 per square yard, what will be the total cost of the carpet?

**33.** **Business and Finance** A shopping center is rectangular, with dimensions of 550 yd by 440 yd. What is its size in acres?

**34.** **Construction** An A-frame cabin has a triangular front with a base of 30 ft and a height of 20 ft. If the front is to be glass that costs $3 per square foot, what will the glass cost?

*Find the perimeter of each figure. (The curves are semicircles.) Round answers to one decimal place.*

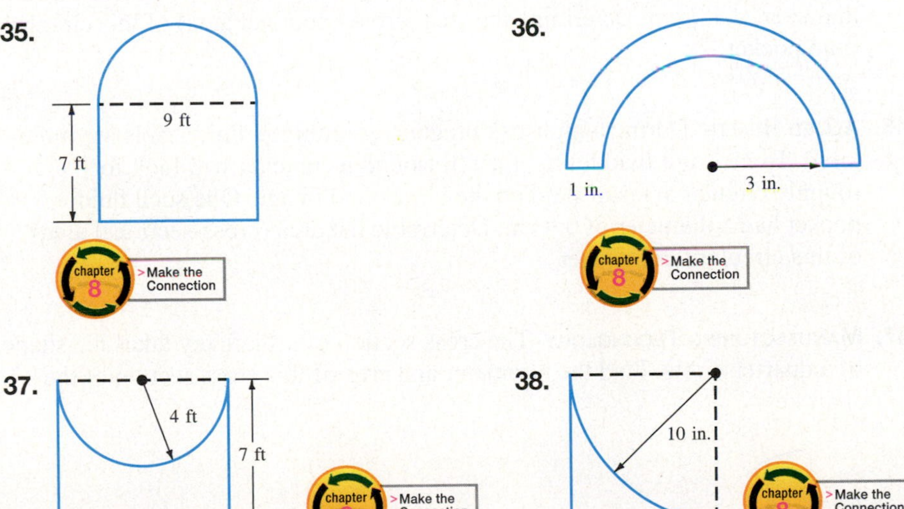


*Find the area of the shaded part in each figure. Use 3.14 for $\pi$. Round your answers to one decimal place.*

**39.**

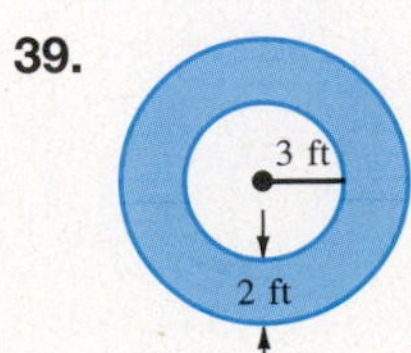


**40.**

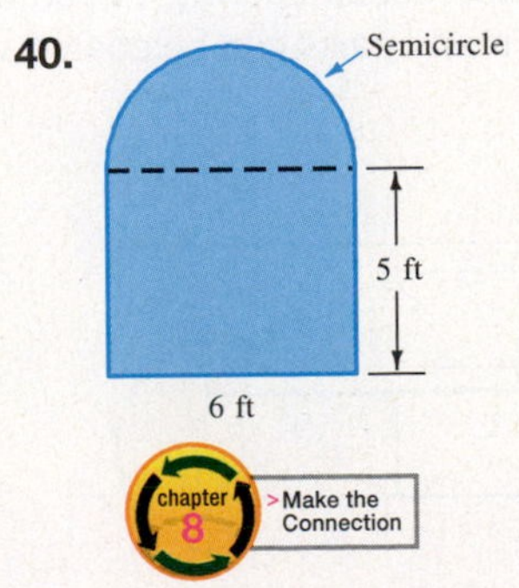


## Answers

30. ________

31. ________

32. ________

33. ________

34. ________

35. ________

36. ________

37. ________

38. ________

39. ________

40. ________

Answers

41. ____________________

42. ____________________

43. ____________________

44. ____________________

45. ____________________

46. ____________________

47. ____________________

48. ____________________

41.

42.

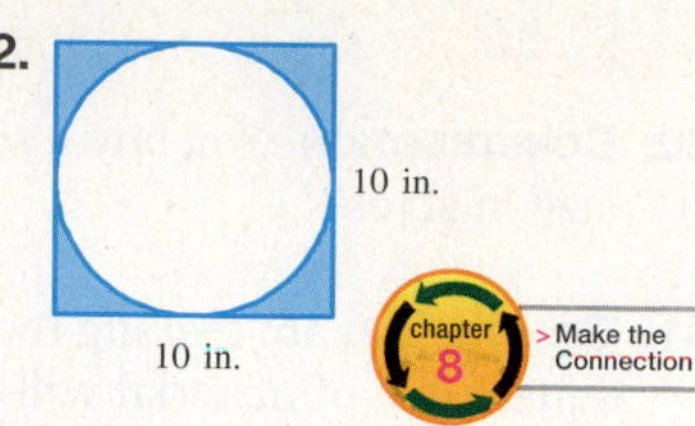


Basic Skills | Advanced Skills | **Vocational-Technical Applications** | Calculator/Computer | Above and Beyond

43. **Allied Health** An ultrasound technician measures the biparietal (or head) diameter of a 14-week-old fetus as 2.5 cm. Determine the head circumference of the fetus.

44. **Allied Health** An ultrasound technician measures the biparietal (or head) diameter of a 28-week-old fetus as 7.3 cm. Determine the head circumference of the fetus.

45. **Allied Health** During a high-risk pregnancy, amniotic fluid levels are monitored. To measure fluid levels, the ultrasound technician looks for roughly circular areas of fluid on the ultrasound image. One such fluid pocket had a diameter of 1.7 cm. Determine the area (cross-sectional area) of this circular fluid pocket.

46. **Allied Health** During a high-risk pregnancy, amniotic fluid levels are monitored. To measure fluid levels, the ultrasound technician will look for roughly circular areas of fluid on the ultrasound image. One such fluid pocket had a diameter of 0.9 cm. Determine the area (cross-sectional area) of this circular fluid pocket.

47. **Manufacturing Technology** The cross section of a shaft key takes the shape of a quarter-circle. Find the perimeter and area of this cross section of the shaft key.

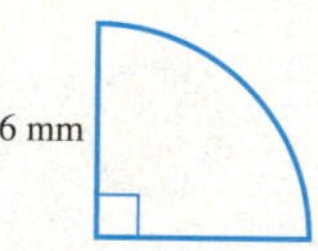


48. **Manufacturing Technology** The surface of this piece needs to be coated with a nonstick coating. A container of the coating can cover 28 square feet. How many parts can be coated with one can?

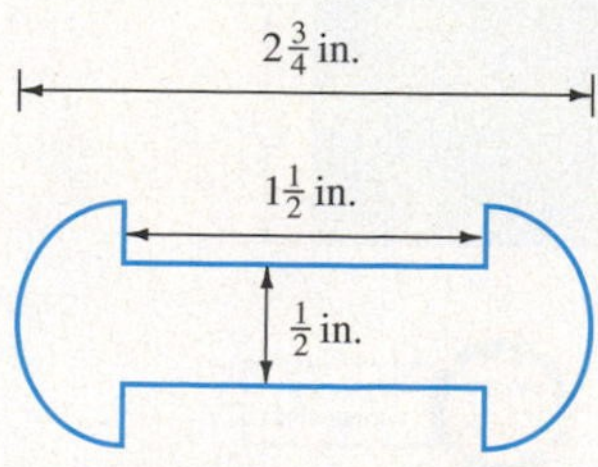

**49. MANUFACTURING TECHNOLOGY** How many square feet of decking are required to cover this deck?

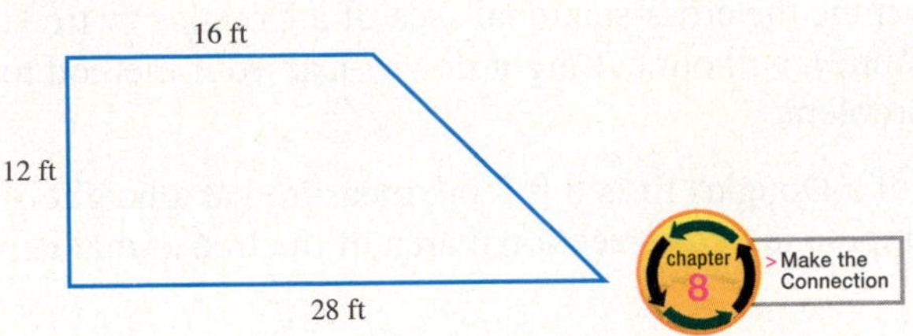


**50. AUTOMOTIVE TECHNOLOGY** The cross section of a cam shaft is shown.

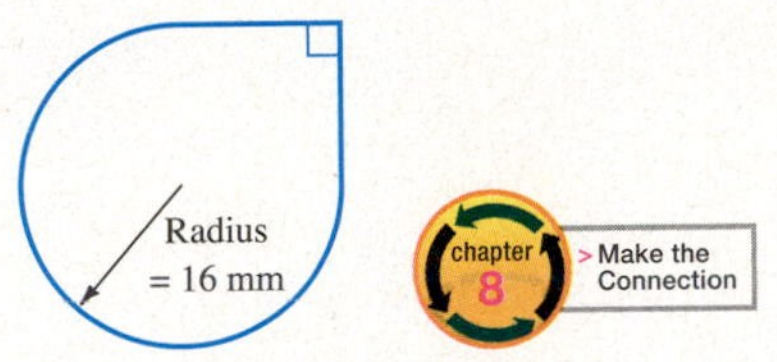


**(a)** Find the area of this cross section.

**(b)** Find the perimeter of the cross section.

Basic Skills | Advanced Skills | Vocational-Technical Applications | Calculator/Computer | Above and Beyond

**51.** Papa Doc's delivers pizza. The 8-in.-diameter pizza is \$8.99, and the price of a 16-in.-diameter pizza is \$17.98. Write a plan to determine which is the better buy.

**52.** The distance from Philadelphia to Sea Isle City is 100 mi. A car was driven this distance using tires with a radius of 14 in. How many revolutions of each tire occurred on the trip?

**53.** Find the area and the circumference (or perimeter) of each of the following:

**(a)** a penny, **(b)** a nickel, **(c)** a dime, **(d)** a quarter, **(e)** a half-dollar, **(f)** a silver dollar, **(g)** a Susan B. Anthony dollar, **(h)** a dollar bill, and **(i)** one face of the pyramid on the back of a \$1 bill.

## Answers

49. ______

50. ______

51. ______

52. ______

53. ______

## Answers

54. ______

55. ______

**54.** What is the effect on the area of a triangle if the base is doubled and the altitude is cut in half? Create some examples to demonstrate your ideas.

**55.** How would you determine the cross-sectional area of a Douglas fir tree (at, say, 3 ft above the ground), without cutting it down? Use your method to solve the following problem:

If the circumference of a Douglas fir is 6 ft 3 in., measured at a height of 3 ft above the ground, compute the cross-sectional area of the tree at that height.

## Answers

**1.** 56.5 ft **3.** 26.7 in. **5.** 55.0 in. **7.** 153.9 in.$^2$ **9.** 38.5 yd$^2$ **11.** 9.6 yd$^2$ **13.** 28 ft$^2$ **15.** 12 yd$^2$ **17.** 20 ft$^2$ **19.** 54 in.$^2$ **21.** 24 ft$^2$ **23.** $\approx$9,420 yd **25.** Yes; area $\approx$ 2,461.8 ft$^2$ **27.** $\approx$\$169.56 **29.** 864 in.$^2$ **31.** 5 ft$^2$ **33.** 50 acres **35.** 37.1 ft **37.** 34.6 ft **39.** 50.2 ft$^2$ **41.** 86 ft$^2$ **43.** $\approx$7.9 cm **45.** $\approx$2.3 cm$^2$ **47.** Perimeter $\approx$ 21.42 mm; Area $\approx$ 28.27 mm$^2$ **49.** 264 ft$^2$ **51.** Above and Beyond **53.** Above and Beyond **55.** $\approx$447.6 in$^2$.

# Activity 20 :: Exploring Circles

In this activity, we ask the question, How does multiplying the radius of a circle by a certain factor change the circumference? And how does this same action change the area? Begin by computing the following, using 3.14 for $\pi$. The first one is done for you.

| | Radius | Circumference | Factor | New Radius | New Circumference |
|---|---|---|---|---|---|
| Circle 1 | 10 ft | 62.8 ft | 2 | 20 ft | 125.6 ft |
| | 10 ft | | 3 | | |
| | 10 ft | | $\frac{1}{2}$ | | |
| Circle 2 | | | | | |
| | | | | | |
| | | | | | |

For circle 2, you and your group members choose values for the radius and for the multiplying factor.

How does the new circumference compare to the original circumference in each case? Try to describe your observations with a general statement.

Now check the effect on the area of each circle. Again using 3.14 for $\pi$, complete the following table.

| | Radius | Area | Factor | New Radius | New Area |
|---|---|---|---|---|---|
| Circle 1 | 10 ft | | 2 | | |
| | 10 ft | | 3 | | |
| | 10 ft | | $\frac{1}{2}$ | | |
| Circle 2 | | | | | |
| | | | | | |
| | | | | | |

How does the new area compare to the original area in each case? Try to describe your observations with a general statement.

Here is a challenge for your group: If you want to double the area of a circle, what should you multiply the radius by? (*Hint:* You will probably have to solve this by trial and error. And you will not find an exact answer, but you can determine an answer accurate to the nearest thousandth.)

# 8.2 Lines and Angles

< 8.2 Objectives >

1 > Recognize lines and line segments

2 > Recognize perpendicular and parallel lines

3 > Name an angle

4 > Determine whether an angle is right, acute, or obtuse

5 > Use a protractor to measure an angle

6 > Find the measures of angles using complements and supplements

7 > Find the measures of angles using vertical angles, alternate interior angles, and corresponding angles

**NOTE**

*Geo* means earth, just as it does in the words *geography* and *geology.*

Once the Egyptians and Babylonians had mastered the counting of their animals, they became interested in measuring their land. This is the foundation of geometry. Literally translated, *geometry* means earth measurement. Many of the topics we consider in geometry (topics such as angles, perimeter, and area) were first studied as part of surveying.

As is usually the case, we start the study of a new topic by learning some vocabulary. Most of the terms we will discuss are familiar to you. It is important that you understand what we mean when we use these words in the context of geometry.

We begin with the word *point.* A point is a location; it has no size and covers no area.

If we string points together forever, we create a *line.* In our studies we will consider only straight lines. We use arrowheads to indicate that a line goes on forever.

A piece of a line that has two endpoints is called a *line segment.*

## Example 1 — Recognizing Lines and Line Segments

< Objective 1 >

**NOTE**

The capital letters are labels for points.

Label each of the following as a line or a line segment.

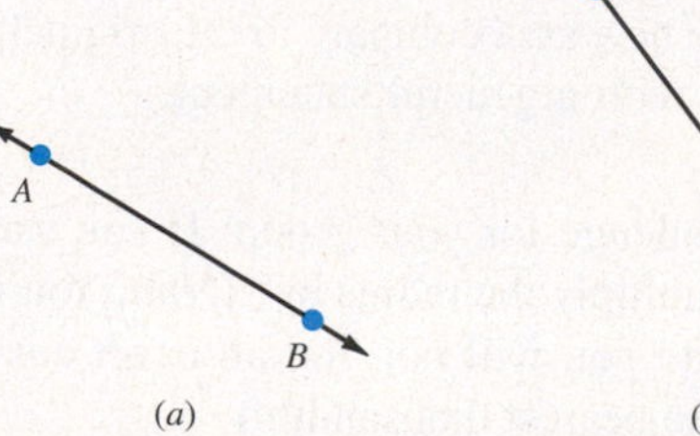


(a) (b)

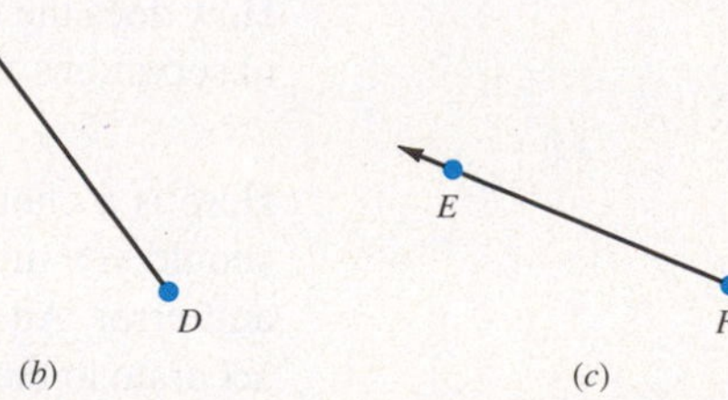


(c)

Both $(a)$ and $(c)$ continue forever in both directions. They are lines. Part $(b)$ has two endpoints. It is a line segment.

**Check Yourself 1**

Label each of the following as a line or a line segment.

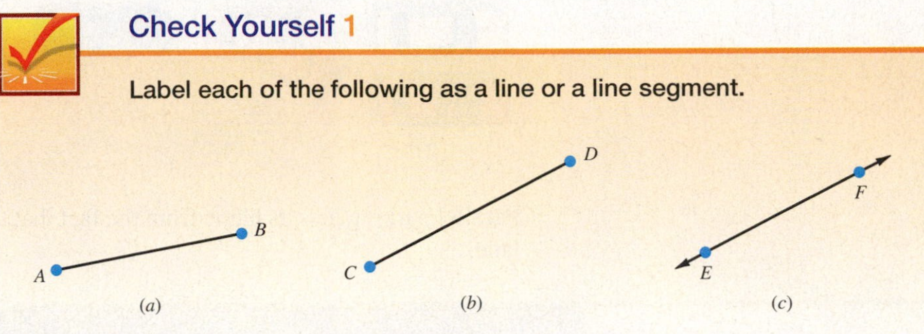


**Definition**

**Angle** — An **angle** is a geometric figure consisting of two line segments that share a common endpoint.

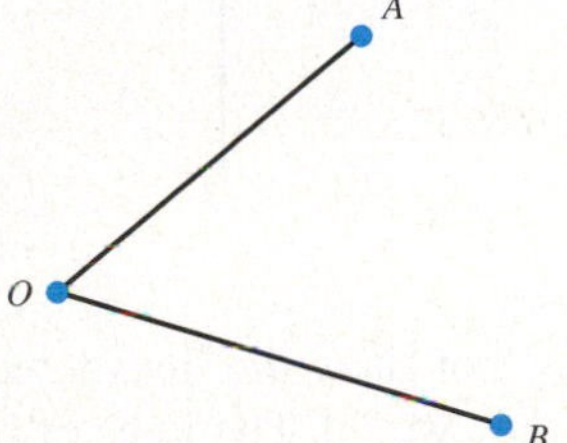


$OA$ and $OB$ are line segments. $O$ is the vertex of the angle.

Surveyors use an instrument called a *transit.* A transit allows surveyors to measure angles so that, from a mathematical description, they can determine exactly where a property line is.

**Definition**

**Perpendicular Lines** — When two lines cross (or intersect), they form four angles. If the lines intersect such that four equal angles are formed, we say that the two lines are **perpendicular.**

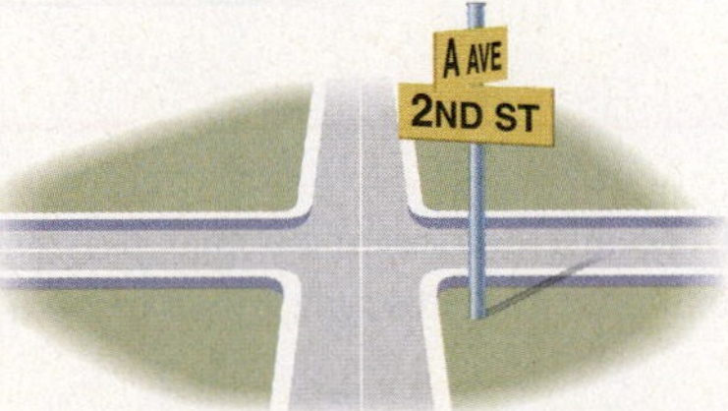


At most intersections, the two roads are perpendicular.

**Definition**

**Parallel Lines** — If two lines are drawn so that they never intersect (even if we extend the lines forever), we say that the two lines are **parallel.**

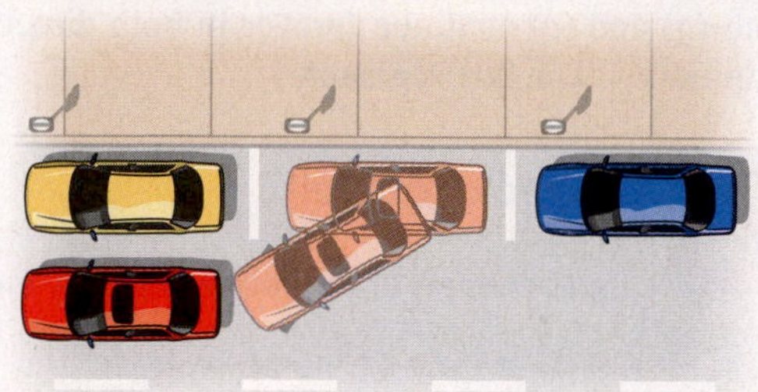

Parallel parking gets its name from the fact that the parking spot is parallel to the traffic lane.

**Example 2** **Recognizing Parallel and Perpendicular Lines**

< Objective 2 >

Label each pair of lines as parallel, perpendicular, or neither.

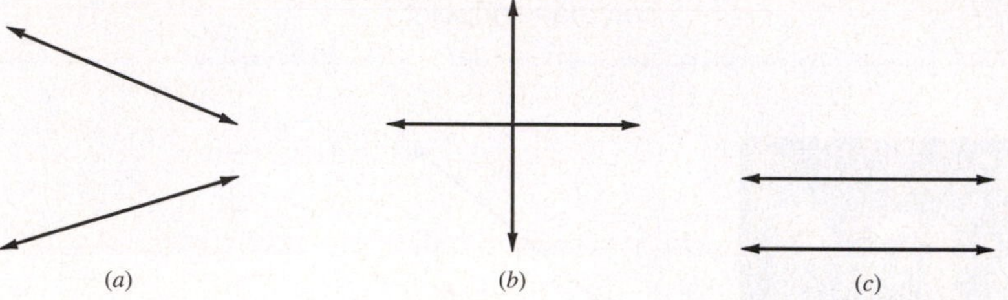


Although part (*a*) does not show the lines intersecting, if they were extended as the arrowheads indicate, they would. The lines of part (*b*) are perpendicular because the four angles formed are equal. Only the lines in part (*c*) are parallel.

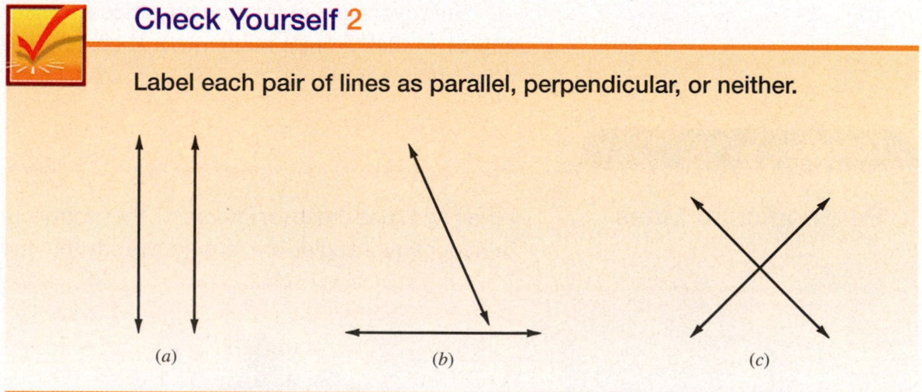

**Check Yourself 2**

Label each pair of lines as parallel, perpendicular, or neither.



We call the angle formed by two perpendicular lines or line segments a *right angle*. We designate a right angle by forming a small square.

We can refer to a specific angle by naming three points. The middle point is the vertex of the angle.

## Example 3 Naming an Angle

< Objective 3 >

Name the angle highlighted in blue.

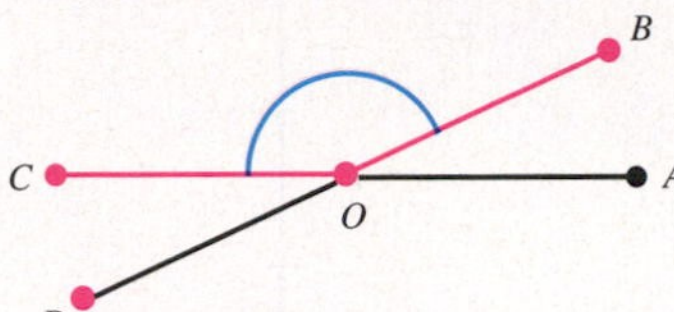


**NOTE**

We could also call this angle $\angle BOC$.

The vertex of the angle is $O$, and the angle begins at $C$ and ends at $B$, so we would name the angle $\angle COB$.

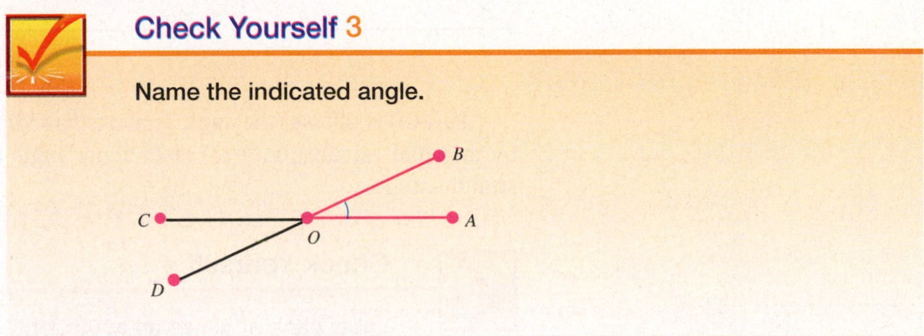

### Check Yourself 3

Name the indicated angle.



When there is no possibility of confusion, we may refer to an angle by naming only the label of the vertex, or perhaps using a symbol that appears in the angle. In the following figure, the angle shown may be named $\angle JKL$ or $\angle LKJ$ (as noted earlier), or by simply writing $\angle K$ or $\angle x$.

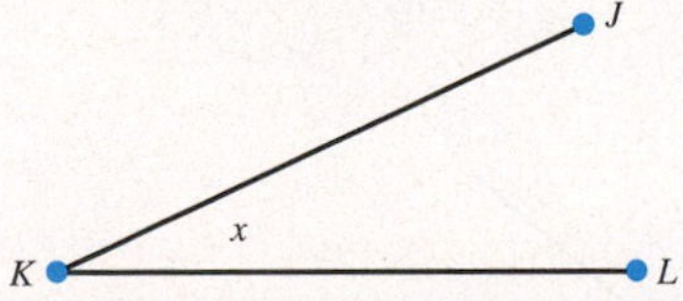


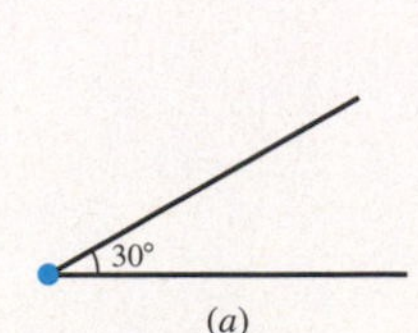

One way to measure an angle is to use a unit that we call a *degree.* There are 360 degrees (we write this as 360°) in a complete circle. Note in the picture on the left that there are four right angles in a circle. If we divide 360° by 4, we find that each right angle must measure 90°. Here are some other angles with their measurements.

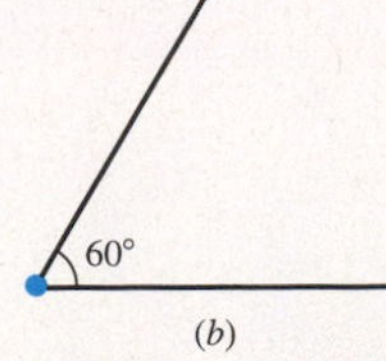


(a)



(b)

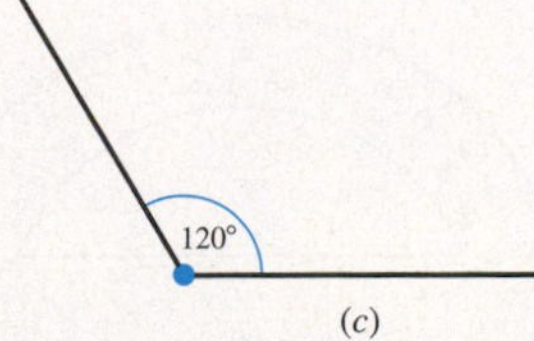


(c)

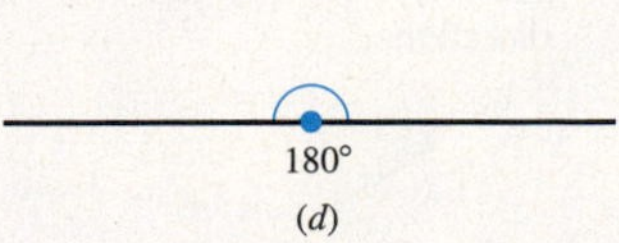


(d)

An *acute angle* measures between 0° and 90°. An *obtuse angle* measures between 90° and 180°. A *straight angle* measures 180°.

## Example 4 Labeling Types of Angles

< Objective 4 >

Label each of the following angles as an acute, obtuse, right, or straight angle.

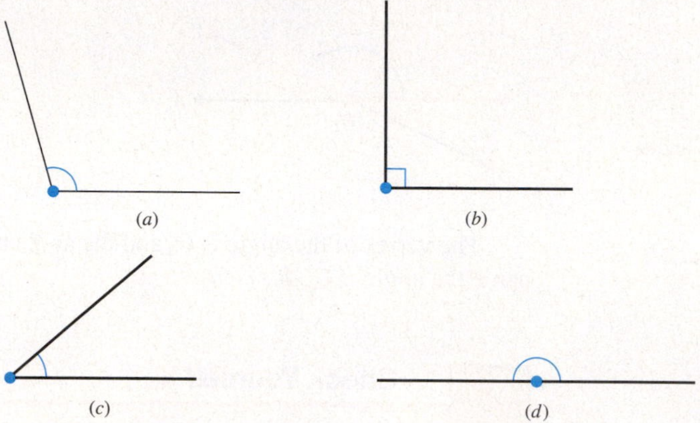


Part (*a*) is obtuse (the angle is more than 90°). Part (*b*) is a right angle (designated by the small square). Part (*c*) is an acute angle (it is less than 90°), and part (*d*) is a straight angle.

### Check Yourself 4

Label each angle as an acute, obtuse, right, or straight angle.

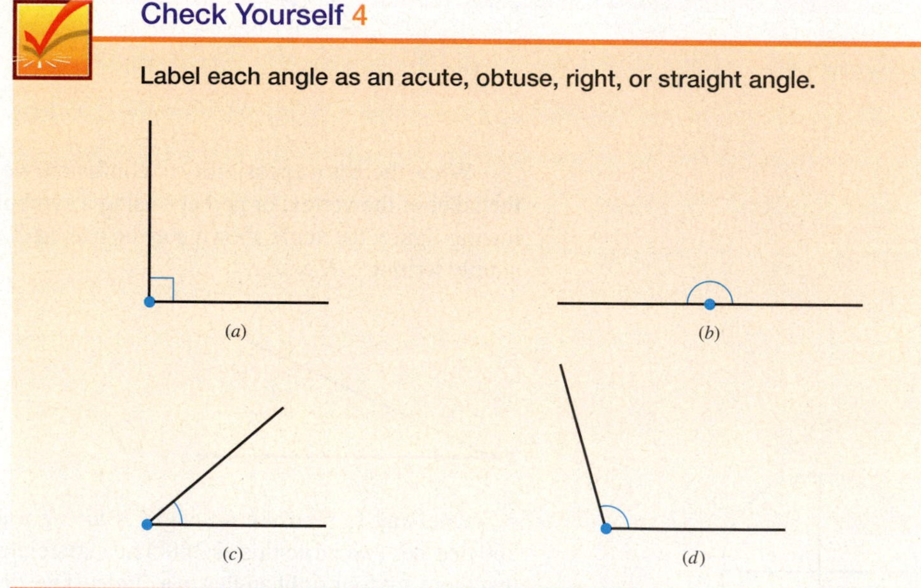


**NOTE**

Your protractor may show the degree measures in both directions.

When assigning a measurement to an angle, we usually use a tool called a *protractor.*

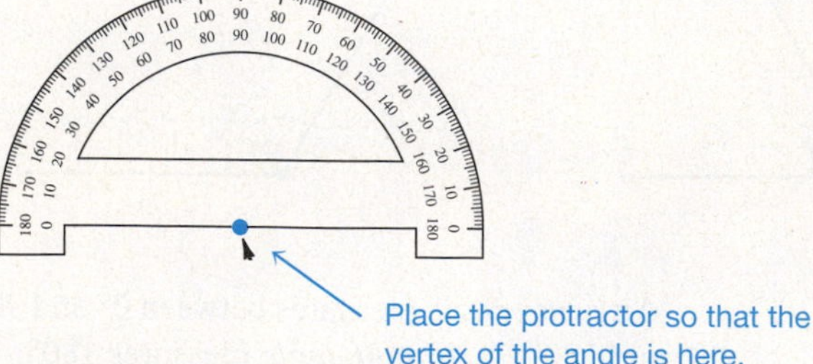

We read the protractor by placing one line segment of the angle at 0°. We then read the number that the other line segment passes through. This number represents the degree measurement of the angle. The point at the center of the protractor, the endpoint of the two line segments, is the vertex of the angle.

## Example 5 Measuring an Angle

< Objective 5 >

Use a protractor to estimate the measurement for each angle.

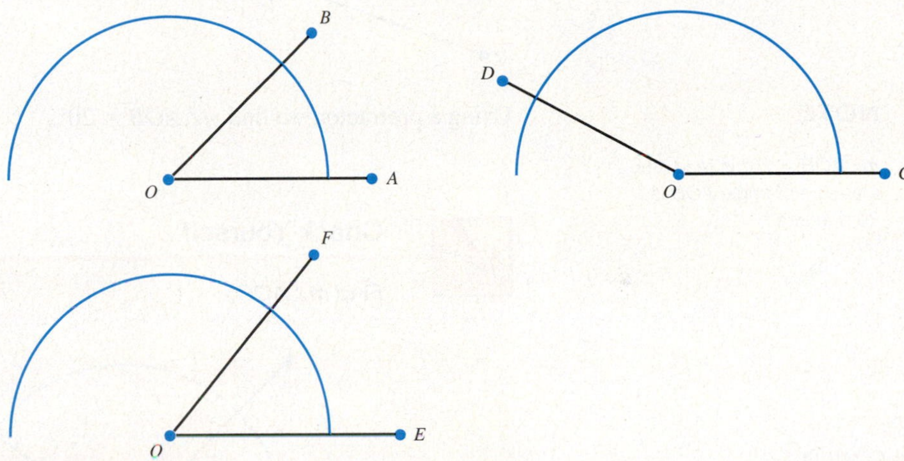


The measure of $\angle AOB$ is 45°. The measure of $\angle COD$ is 150°. The measure of $\angle EOF$ is between 50° and 55°. We could estimate that it is a 52° angle.

### Check Yourself 5

Use a protractor to estimate the measurement for each angle.

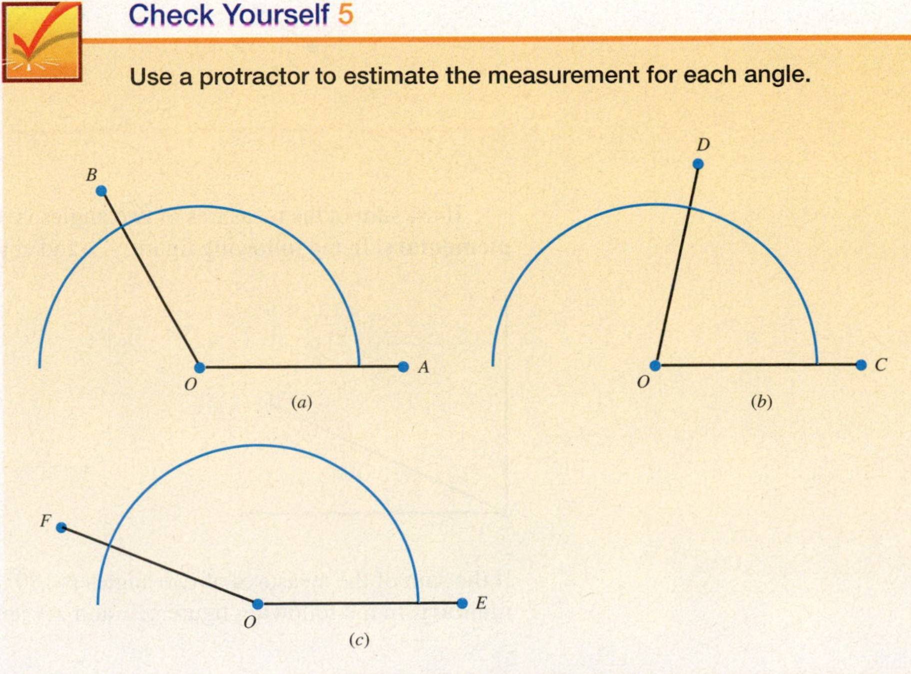


If we wish to refer to the degree measure of $\angle ABC$, we write $m\angle ABC$.

## Example 6 Measuring an Angle

Find $m\angle AOB$.

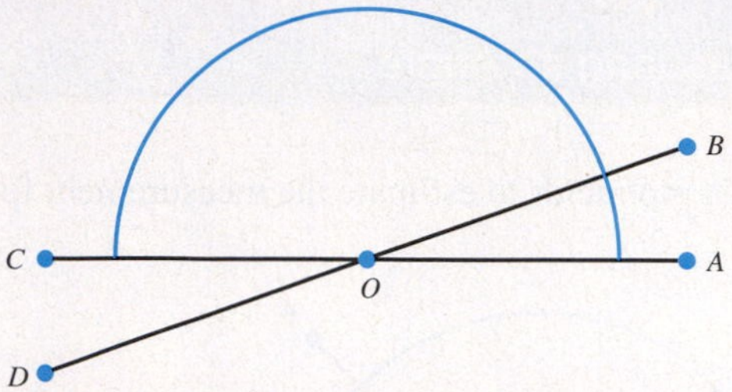


Using a protractor, we find $m\angle AOB = 20°$.

**NOTE**

$m\angle AOB = 20°$ is read "the measure of angle *AOB* is 20 degrees."

### Check Yourself 6

Find $m\angle AOC$.

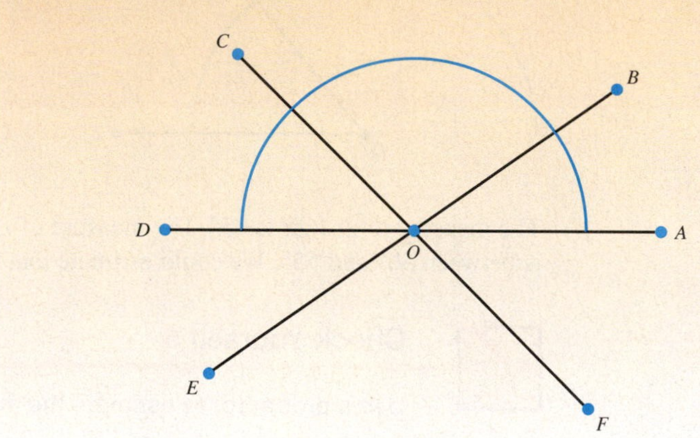


If the sum of the measures of two angles is 90°, the two angles are said to be **complementary.** In the following figure, $\angle x$ and $\angle y$ are complementary angles.

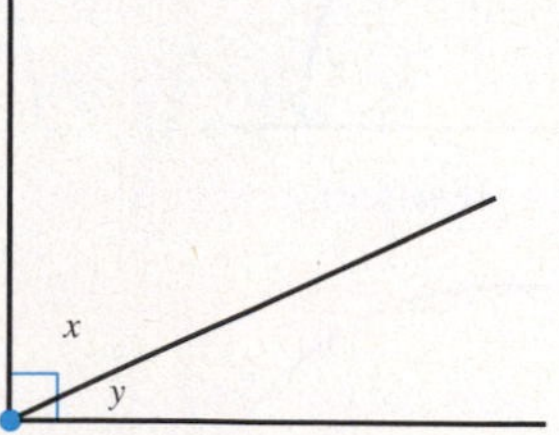


If the sum of the measures of two angles is 180°, the two angles are said to be **supplementary.** In the following figure, $\angle x$ and $\angle y$ are supplementary angles.

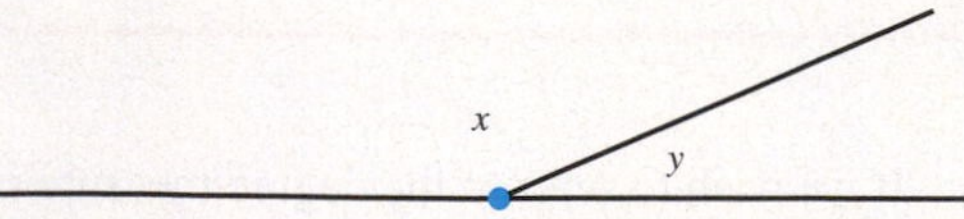

## Example 7 Finding the Measure of an Angle

< Objective 6 >

In each case, find the measure of angle $x$.

**(a)**

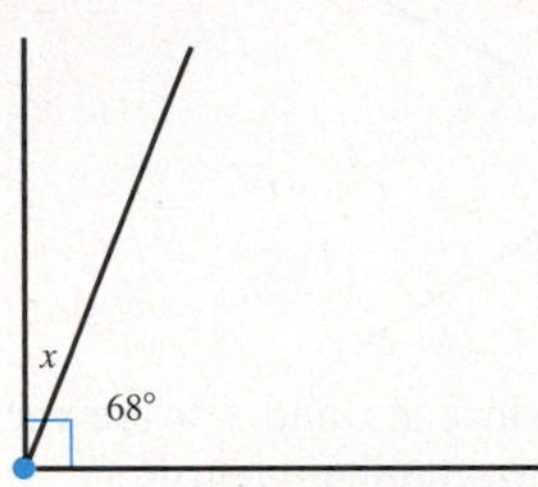


**(b)**

**(a)** Since $m\angle x + 68° = 90°$ (complementary angles), we must have

$$m\angle x = 90° - 68° = 22°$$

**(b)** Since $m\angle x + 37° = 180°$ (supplementary angles), we must have

$$m\angle x = 180° - 37° = 143°$$

### Check Yourself 7

In each case, find the measure of angle $x$.

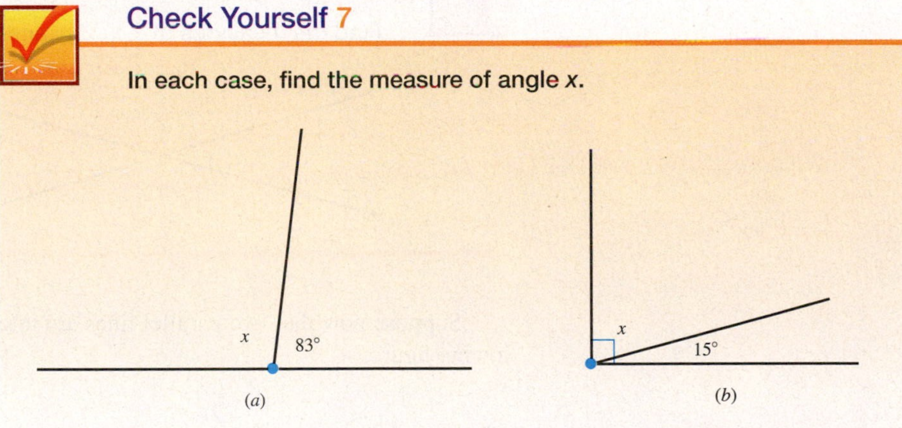


Suppose that two lines intersect, as shown in the following figure, forming four angles.

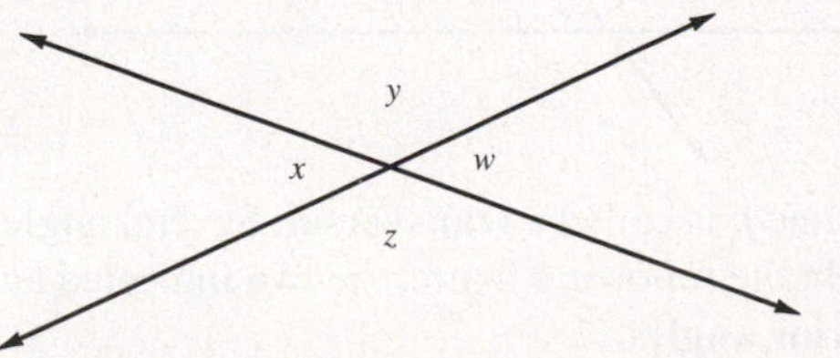


We say that $\angle x$ and $\angle w$ are **vertical angles.** Likewise, $\angle y$ and $\angle z$ are vertical angles. We have the following property:

**Property**

**Vertical Angles** Vertical angles have equal measure.

## Example 8 Finding the Measures of Angles

< Objective 7 >

Suppose $m\angle w = 59°$. Find the measures of $\angle x$, $\angle y$, and $\angle z$.

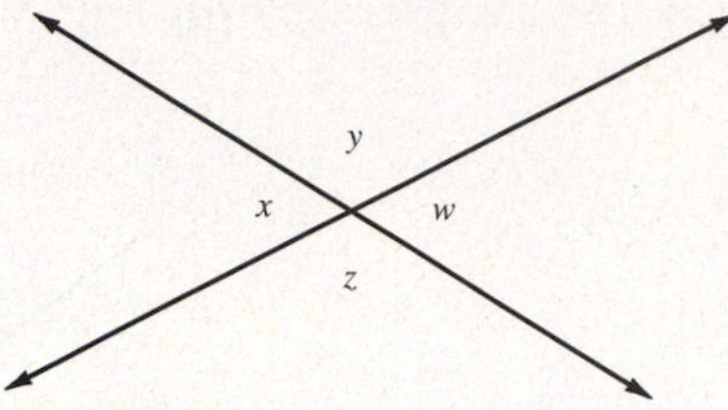


Note that $m\angle x = 59°$, since $\angle x$ and $\angle w$ are vertical angles.

Since $\angle x$ and $\angle y$ are supplementary,

$$\begin{aligned} m\angle y &= 180° - m\angle x \\ &= 180° - 59° \\ &= 121° \end{aligned}$$

We note that $\angle y$ and $\angle z$ are vertical angles, so $m\angle z = m\angle y$. $m\angle z = 121°$.

### Check Yourself 8

**Find the measures of $\angle x$, $\angle y$, and $\angle z$, if $m\angle w = 32°$.**

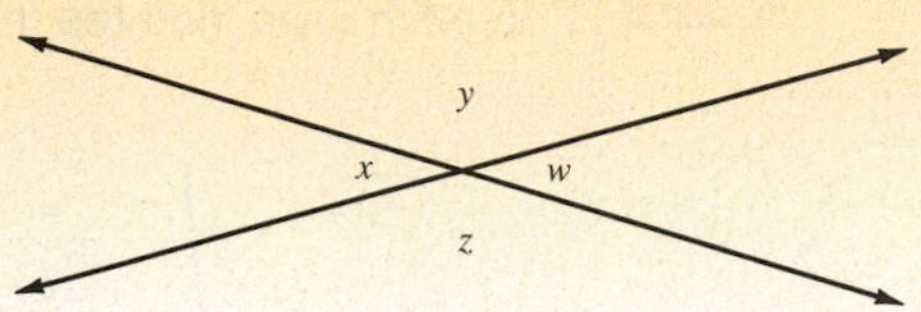


Suppose now that two parallel lines are intersected by a third line $p$ as in the following figure.

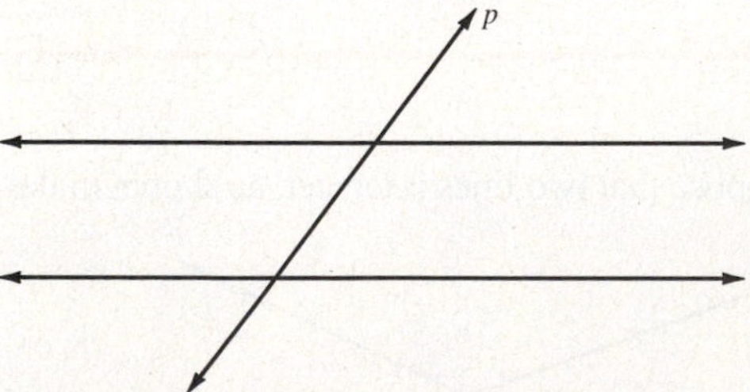


The line $p$ is called a **transversal.** Several angles are created in this situation.

In the following figure, the two indicated angles, $\angle x$ and $\angle y$, are called **alternate interior angles.**

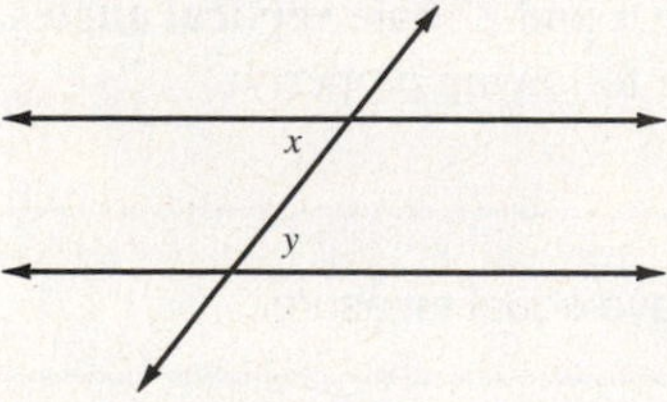

In this figure, $\angle a$ and $\angle b$ are called **corresponding angles.**

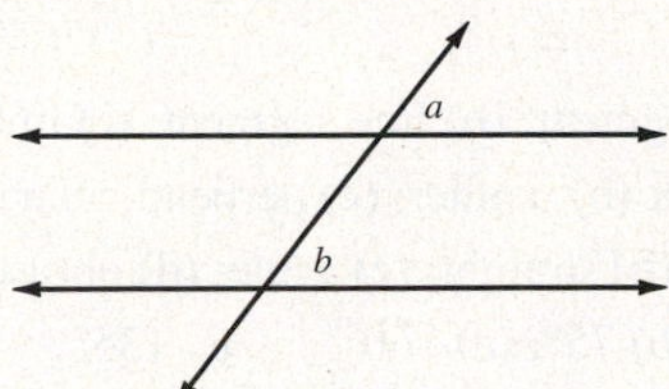


Property

**Parallel Lines and a Transversal**

When two parallel lines are intersected by a transversal,

1. Alternate interior angles have equal measure.
2. Corresponding angles have equal measure.

### Example 9 Finding the Measures of Angles

Given that $m\angle x = 125°$, find the measures of $\angle a$, $\angle b$, and $\angle c$.

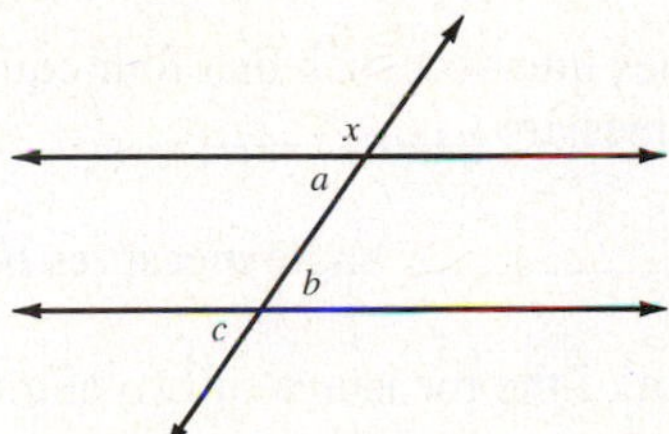


$m\angle x + m\angle a = 180°$ Supplementary angles

So $m\angle a = 180° - 125° = 55°$.

$m\angle a = m\angle b$ Alternate interior angles

So $m\angle b = 55°$.

$m\angle a = m\angle c$ Corresponding angles

So $m\angle c = 55°$.

### Check Yourself 9

Given that $m\angle x = 67°$, find $m\angle y$.

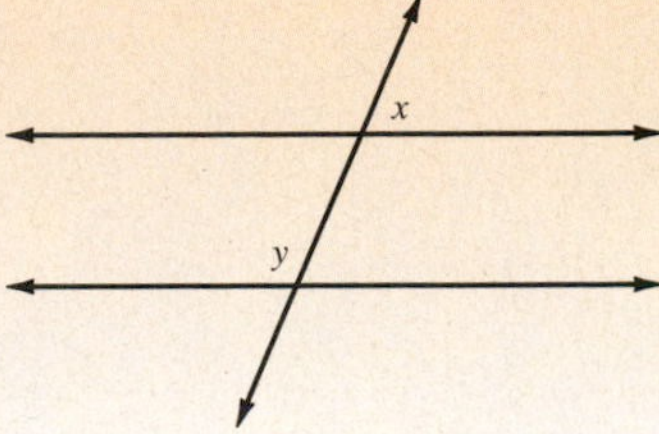

## Check Yourself ANSWERS

**1.** **(a)** Line segment; **(b)** line segment; **(c)** line
**2.** **(a)** Parallel; **(b)** neither; **(c)** perpendicular **3.** $\angle BOA$ or $\angle AOB$
**4.** **(a)** Right; **(b)** straight; **(c)** acute; **(d)** obtuse
**5.** **(a)** 120°; **(b)** 75°; **(c)** 160° **6.** 135° **7.** **(a)** 97°; **(b)** 75°
**8.** $m\angle x = 32°$; $m\angle y = 148°$; $m\angle z = 148°$ **9.** 113°

## Reading Your Text

The following fill-in-the-blank exercises are designed to ensure that you understand some of the key vocabulary used in this section. Each sentence usually comes directly from the section. You will find the correct answers in the Answer Appendix.

### SECTION 8.2

**(a)** *Geo* means _______________, just as it does in the words *geography* and *geology.*

**(b)** If two lines intersect such that four equal angles are formed, we say that the two lines are _______________.

**(c)** An _______________ angle measures between 90° and 180°.

**(d)** If the sum of the measures of two angles is 90°, the two angles are said to be _______________.

Basic Skills | Advanced Skills | Vocational-Technical Applications | Calculator/Computer | Above and Beyond

# 8.2 exercises

MathZone

Boost your grade at mathzone.com!
> Practice Problems
> NetTutor
> Self-Tests
> e-Professors
> Videos

Name ______________

Section ________ Date ________

< Objective 1 >

1. Draw line segment $AB$.

$A$ $B$

2. Draw line $EF$.

$E$ $F$

3. Draw line $AC$.

$A$ $C$

4. Draw line segment $BC$.

$B$ $C$

*Identify each object as a line or line segment.*

5.

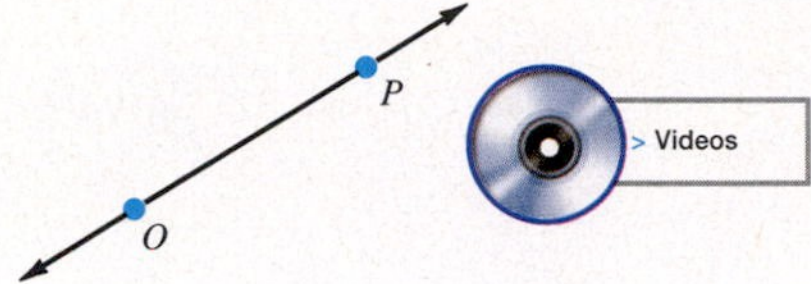


6.

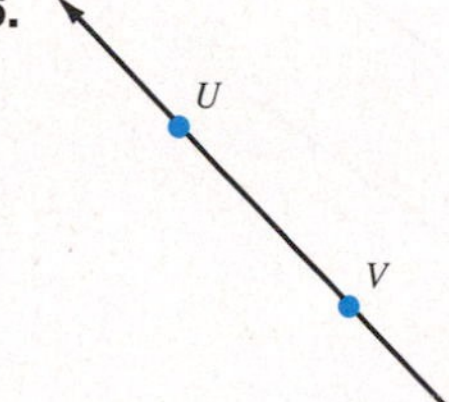


7.

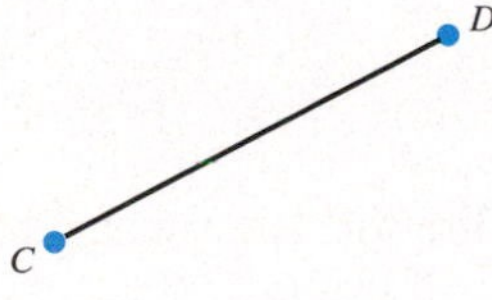


8.

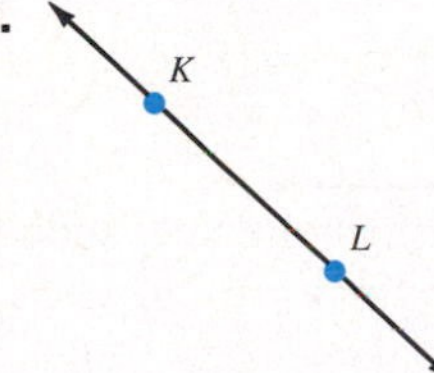


9.

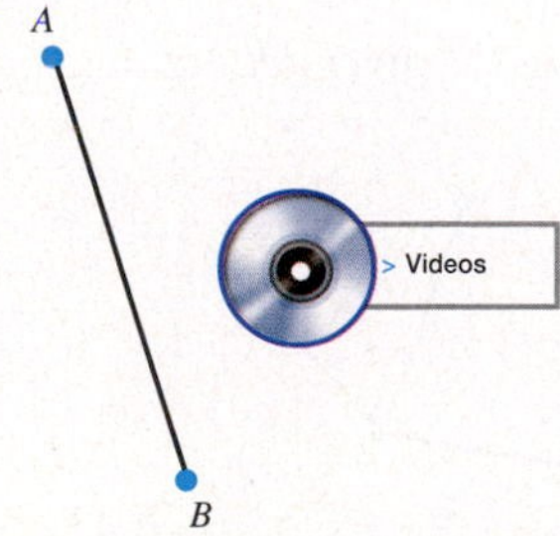


10.

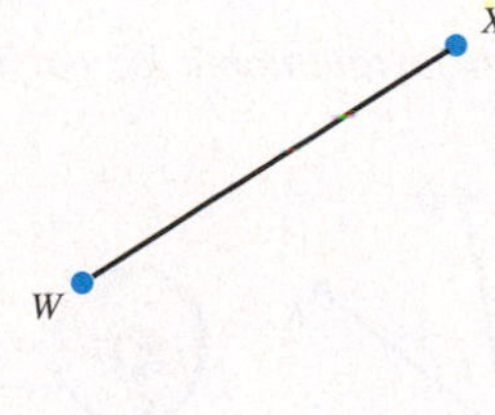


11.

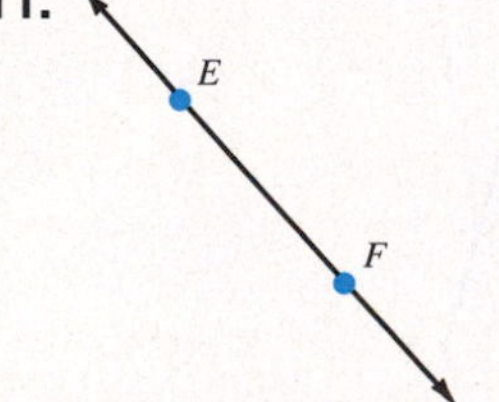


12.

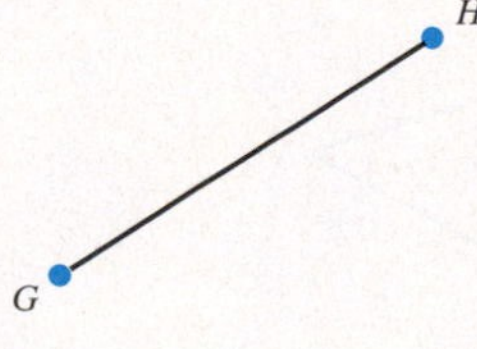


*Label exercises 13 to 18 as* **true** *or* **false.**

13. There are exactly two different line segments that can be drawn through two points.

14. There are exactly two different lines that can be drawn through two points.

## Answers

1. ______________
2. ______________
3. ______________
4. ______________
5. ______________
6. ______________
7. ______________
8. ______________
9. ______________
10. ______________
11. ______________
12. ______________
13. ______________
14. ______________

**Answers**

15. ______

16. ______

17. ______

18. ______

19. ______

20. ______

21. ______

22. ______

23. ______

24. ______

25. ______

26. ______

**15.** Two opposite sides of a square are parallel line segments.

**16.** Two adjacent sides of a square are perpendicular line segments.

**17.** $\angle ABC$ will always have the same measure as $\angle CAB$.

**18.** Two acute angles have the same measure.

< Objective 2 >

**19.** Are the following two lines parallel, perpendicular, or neither?

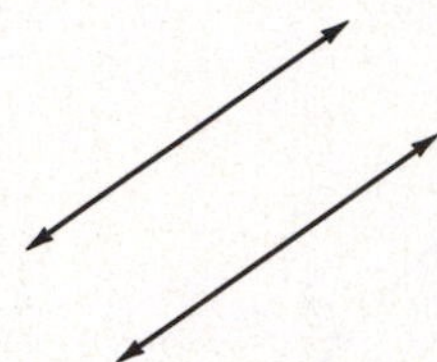

**20.** Are the following two lines parallel, perpendicular, or neither?

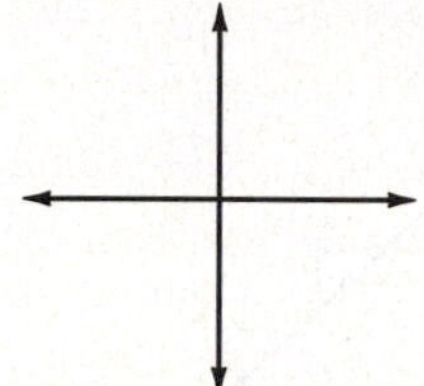

< Objective 3 >

*Give an appropriate name for each indicated angle, using the given letters.*

**21.**

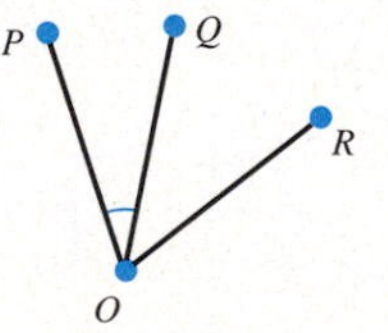


> Videos

**22.**

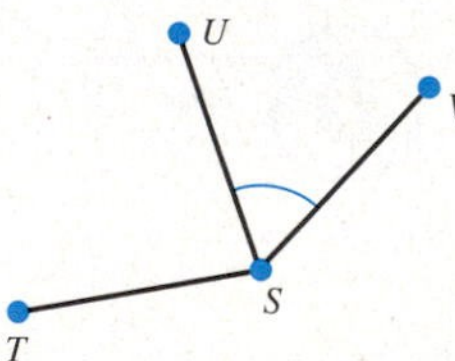


**23.**

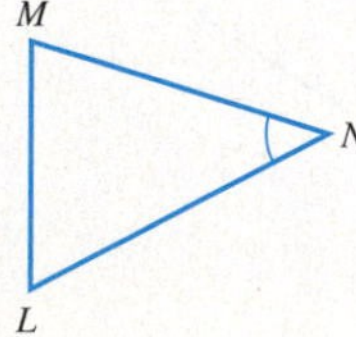


**24.**

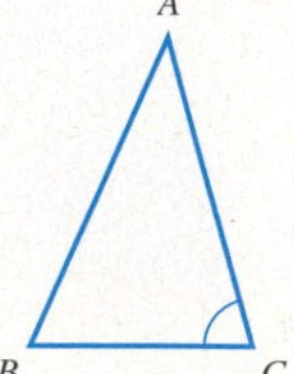


**25.**

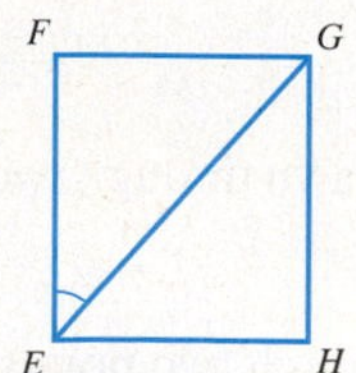


**26.**

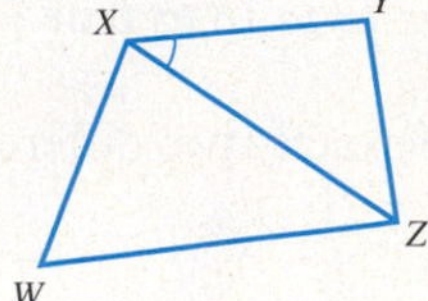

**27.**

**28.**

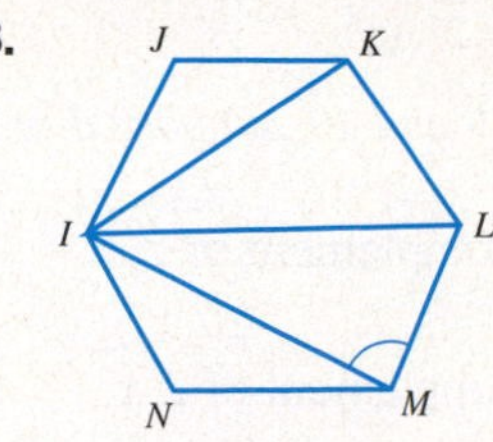


*For each angle described, give its measure in degrees. Sketch the angle.*

**29.** $\angle A$ represents $\frac{1}{6}$ of a complete circle.

**30.** $\angle B$ represents $\frac{1}{3}$ of a complete circle.

**31.** $\angle C$ represents $\frac{7}{12}$ of a complete circle.

**32.** $\angle D$ represents $\frac{11}{12}$ of a complete circle.

< **Objectives 4–5** >

*Measure each angle with a protractor. Identify the angle as acute, right, obtuse, or straight.*

**33.**

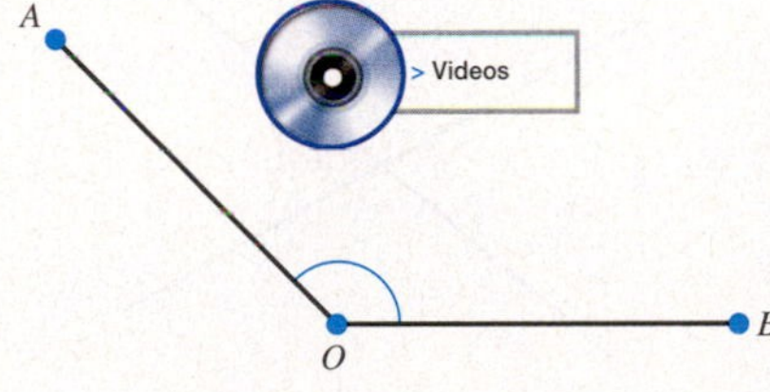


**34.**

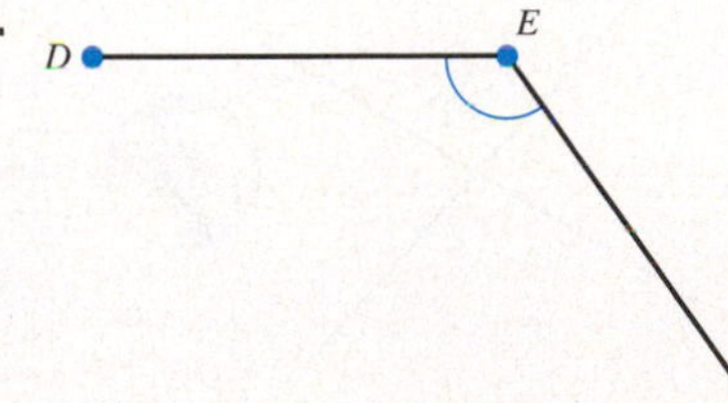


**35.**

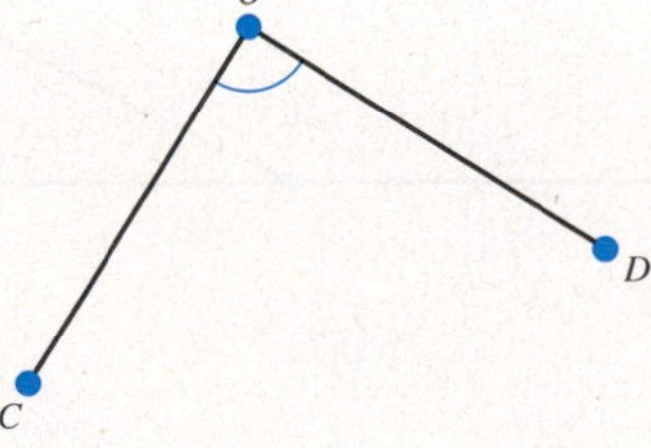


**36.**

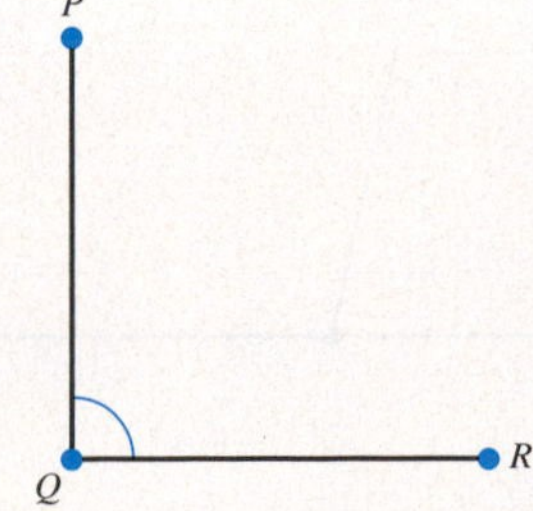


**37.**

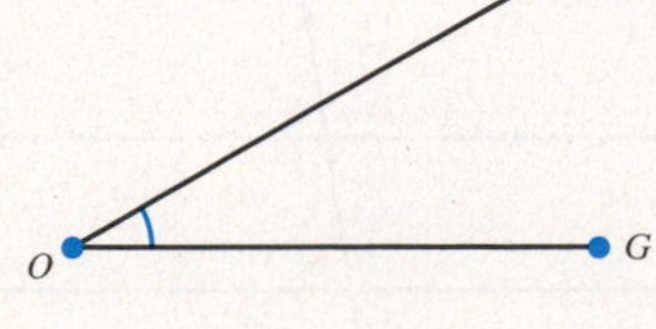


**38.**

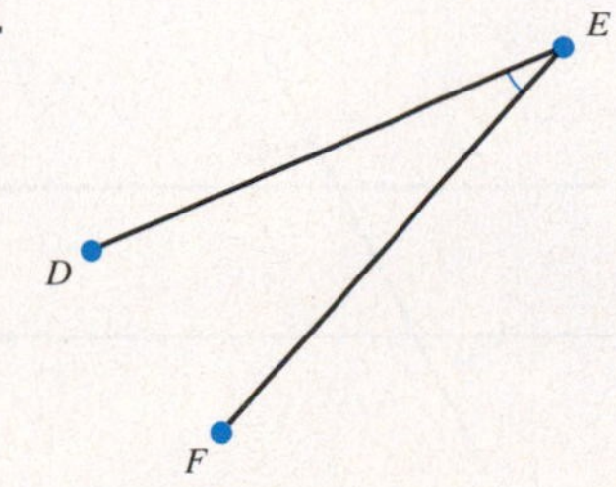


## Answers

27. ________

28. ________

29. ________

30. ________

31. ________

32. ________

33. ________

34. ________

35. ________

36. ________

37. ________

38. ________

## Answers

39. ____

40. ____

41. ____

42. ____

43. ____

44. ____

45. ____

46. ____

47. ____

48. ____

49. ____

50. ____

< Objectives 6–7 >

*For exercises 39 and 40, suppose that* $m\angle x = 29°$.

**39.** Find the complement of $\angle x$.

**40.** Find the supplement of $\angle x$.

*For exercises 41 and 42, suppose that* $m\angle y = 53°$.

**41.** Find the supplement of $\angle y$.

**42.** Find the complement of $\angle y$.

**43.** Find $m\angle x$.

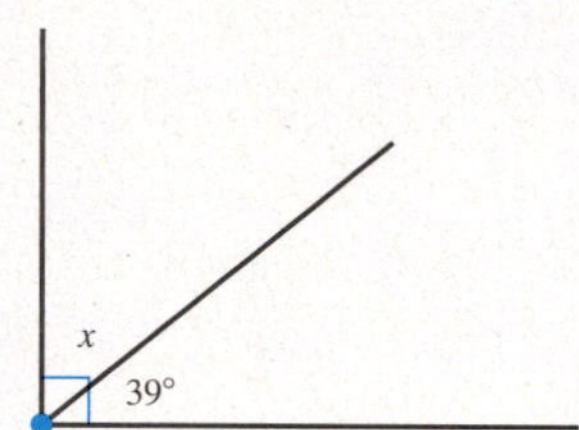


**44.** Find $m\angle y$.

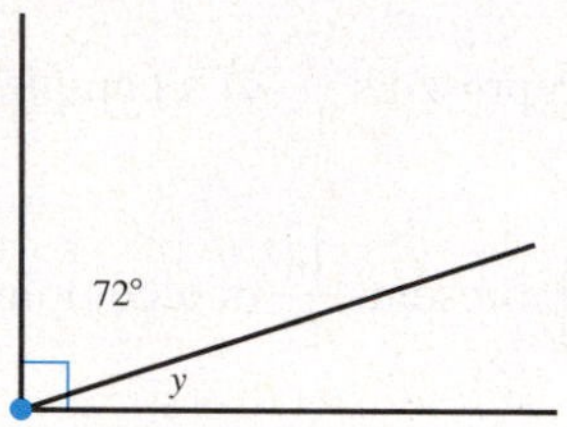


**45.** Find $m\angle x$ and $m\angle y$.

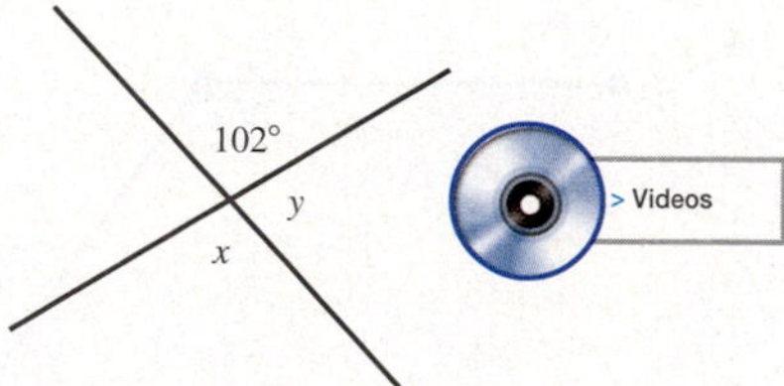


**46.** Find $m\angle a$ and $m\angle b$.

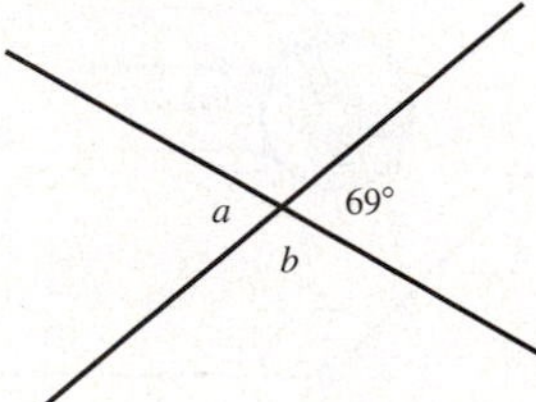


**47.** Find $m\angle w$.

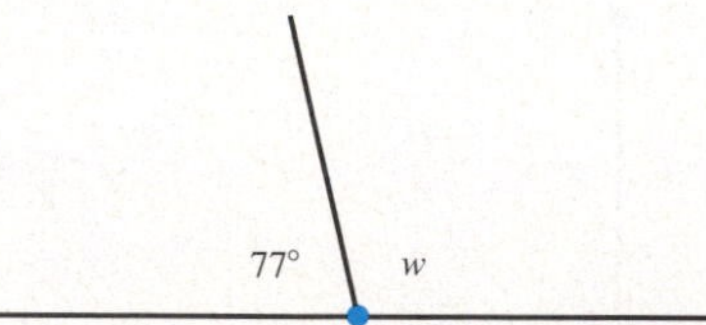


**48.** Find $m\angle z$.

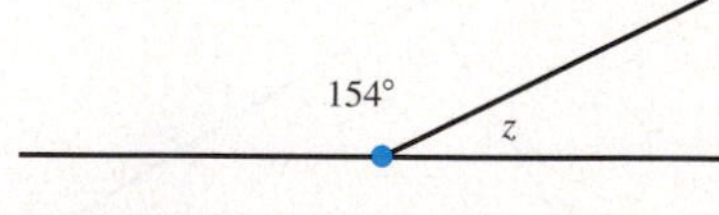


*For exercises 49 to 52, find* $m\angle a$, $m\angle b$, *and* $m\angle c$.

**49.**

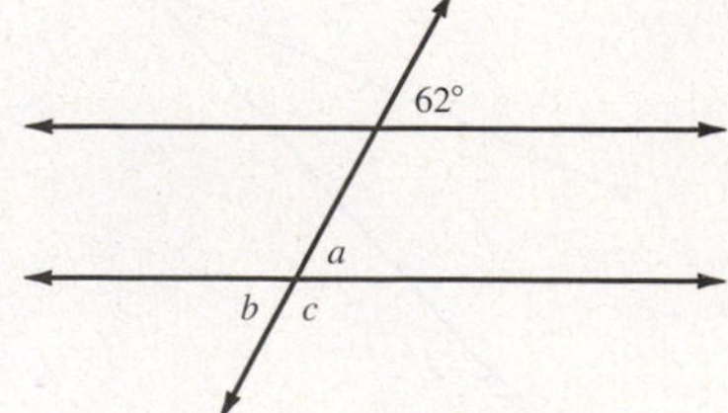


**50.**

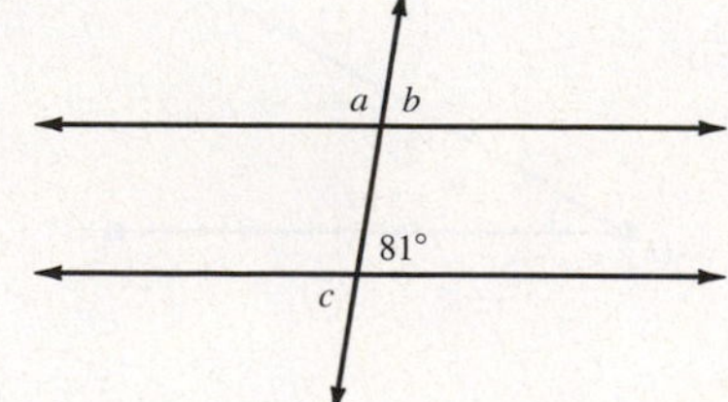

51.

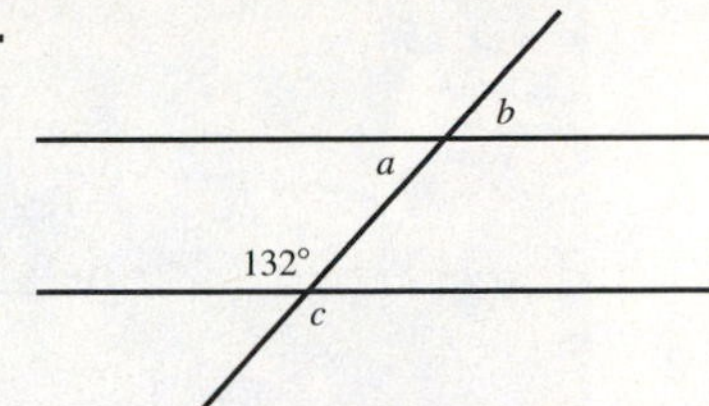


52.

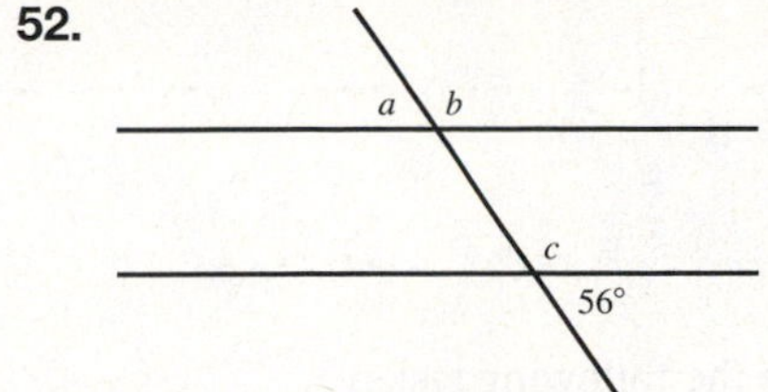


Answers

51. ______

52. ______

53. ______

54. ______

Basic Skills | Advanced Skills | **Vocational-Technical Applications** | Calculator/Computer | Above and Beyond

**53. Electronics** The following figure is a picture of an analog voltmeter. The needle rotates clockwise as the voltage increases. The total angular distance covered by the needle as it travels from 0 volts (V) to 10 V is 100°.

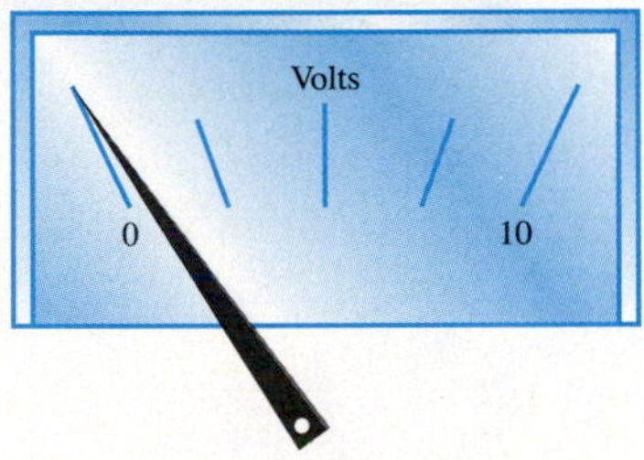


**(a)** Assuming that the angle is proportional to the voltage, what angular distance do you estimate the needle would travel from the initial 0-V position if 5 V are measured?

**(b)** If the angular distance traveled from the original 0-V location is 85°, estimate the voltage.

**54. Automotive Technology** In a 4-stroke engine, the crankshaft turns two complete turns for each cycle of a cylinder. How many degrees does the crankshaft turn for one cycle?

## Answers

**1.** A ———— B **3.**  **5.** Line

**7.** Line segment **9.** Line segment **11.** Line **13.** False **15.** True **17.** False **19.** Parallel **21.** $\angle POQ$ or $\angle QOP$ **23.** $\angle MNL$ or $\angle LNM$ **25.** $\angle FEG$ or $\angle GEF$ **27.** $\angle SVT$ or $\angle TVS$ **29.** 60°;

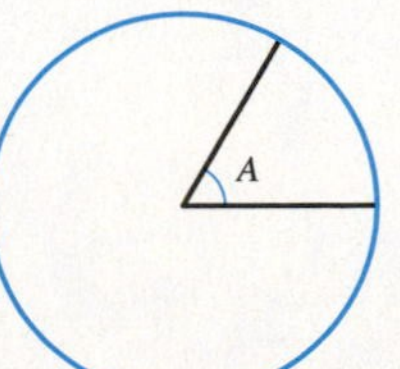


**31.** 210°;

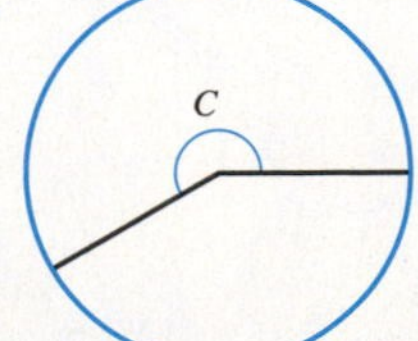


**33.** 135°; obtuse **35.** 90°; right **37.** 30°; acute **39.** 61° **41.** 127° **43.** 51° **45.** $m\angle x = 102°$; $m\angle y = 78°$ **47.** 103° **49.** $m\angle a = 62°$; $m\angle b = 62°$; $m\angle c = 118°$ **51.** $m\angle a = 48°$; $m\angle b = 48°$; $m\angle c = 132°$ **53.** **(a)** 50°; **(b)** 8.5 V

# Activity 21 :: Know the Angles

Your group members might wish to share the following tasks:

1. Draw any triangle, using a ruler. With your protractor, carefully measure the three interior angles, and find their sum.

   $\angle A =$ $\angle B =$ $\angle C =$ Sum =

   Do this again with another triangle of a different shape.

   $\angle A =$ $\angle B =$ $\angle C =$ Sum =

   What do you notice about the sums of the angles?

   Make a conjecture about the sum of the angles of *any* triangle. Then test your conjecture on another triangle.

2. A **quadrilateral** is a four-sided polygon. Draw any quadrilateral and measure the four interior angles with a protractor. Record these and find their sum.

   $\angle A =$ $\angle B =$ $\angle C =$ $\angle D =$ Sum =

   Make a conjecture concerning the sum of the interior angles of *any* quadrilateral. Then test your conjecture on another quadrilateral.

3. A **pentagon** is a five-sided polygon. Draw any pentagon and measure the five interior angles with a protractor. Record these and find their sum.

   $\angle A =$ $\angle B =$ $\angle C =$ $\angle D =$ $\angle E =$
   Sum =

   Make a conjecture concerning the sum of the interior angles of *any* pentagon. Then test your conjecture on another pentagon.

4. A **hexagon** is a six-sided polygon. Draw any hexagon and measure the six interior angles with a protractor. Record these and find their sum.

   $\angle A =$ $\angle B =$ $\angle C =$ $\angle D =$ $\angle E =$
   $\angle F =$ Sum =

   Make a conjecture concerning the sum of the interior angles of *any* hexagon. Then test your conjecture on another hexagon.

5. Now try to generalize. Suppose we have a polygon having $k$ sides. Give a formula for the sum of the interior angles. Sum =

# 8.3 Triangles

< **8.3 Objectives** >

1 > Classify triangles as acute, obtuse, or right

2 > Classify triangles as equilateral, isosceles, or scalene

3 > Find the measure of the third angle of a triangle

4 > Identify similar triangles

5 > Apply properties of similar triangles to find an unknown length

Now that you know something about angles, it is interesting to look again at triangles. Why is this shape called a triangle?

Literally, *triangle* means "three angles."

The same classifications we used for angles can be used for triangles. If a triangle has a right angle, we call it a *right triangle.*

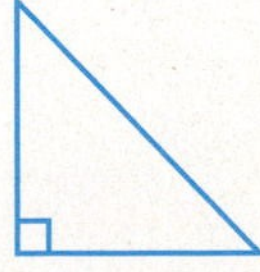

If it has three acute angles, it is called an *acute triangle.*

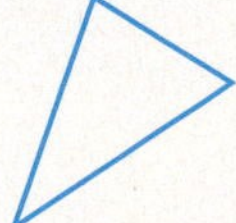

If it has an obtuse angle, it is called an *obtuse triangle.*

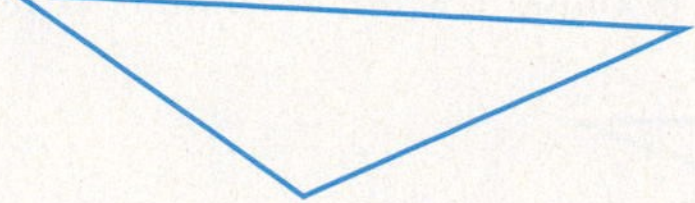


The Streeter/Hutchison Series in Mathematics Basic Mathematical Skills with Geometry

## Example 1 Identifying an Acute Triangle

< Objective 1 >

Which of the following triangles is acute?

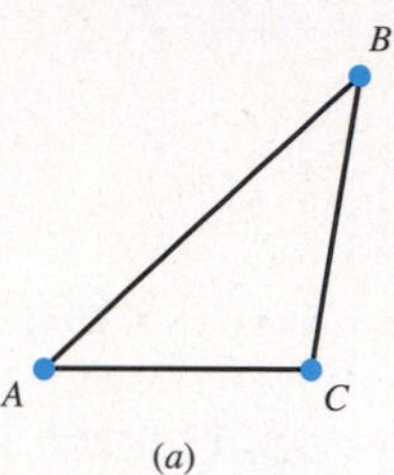

(a)

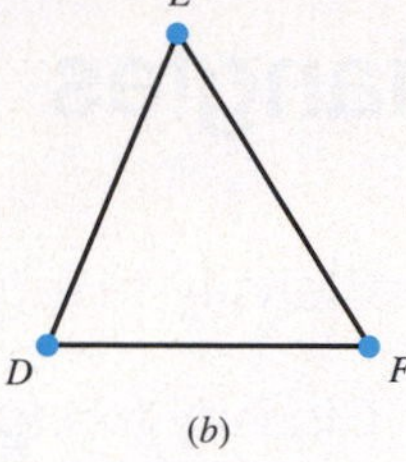

(b)

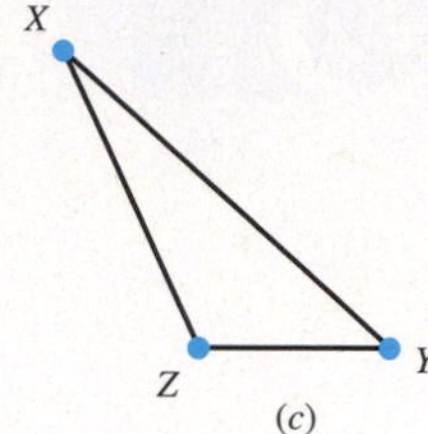

(c)

Only $\triangle DEF$ is an acute triangle. Both $\triangle ABC$ and $\triangle XYZ$ have one obtuse angle.

### Check Yourself 1

**Which of the following triangles is obtuse?**

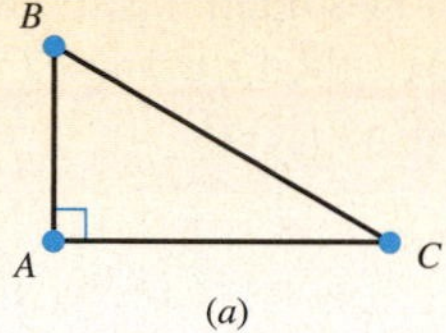

(a)

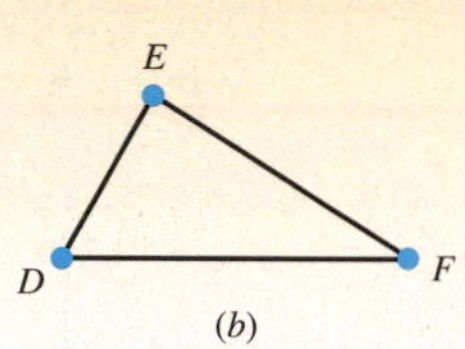

(b)

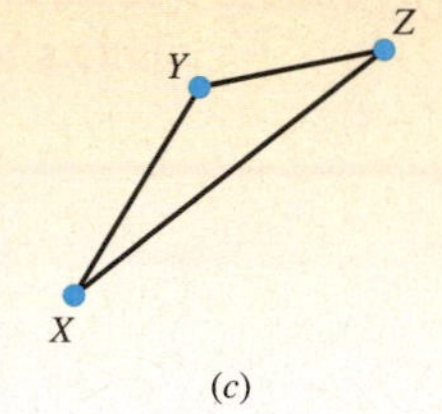

(c)

**NOTE**

All three sides of an equilateral triangle have the same length.

We can also classify triangles based on how many angles have the same measure.

A triangle is called an *equilateral triangle* if all three angles have the same measure.

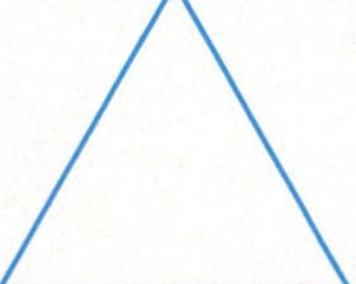

**NOTE**

Exactly two sides of an isosceles triangle have equal length.

A triangle is called an *isosceles triangle* if two angles have the same measure.

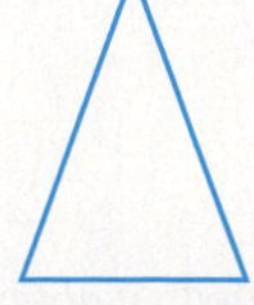

**NOTE**

All three sides of a scalene triangle have different lengths.

A triangle is called a *scalene triangle* if no two angles have the same measure.

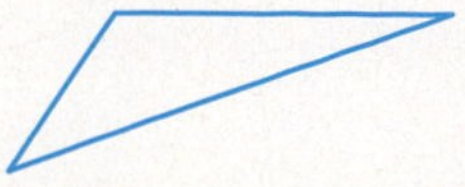

## Example 2 Labeling Types of Triangles

< Objective 2 >

Of the following triangles, which are equilateral? Isosceles? Scalene?

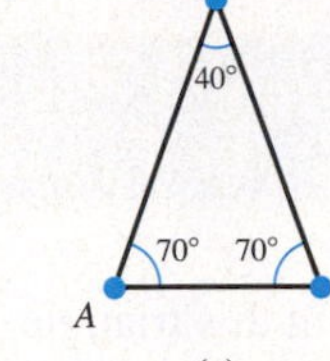

(a)

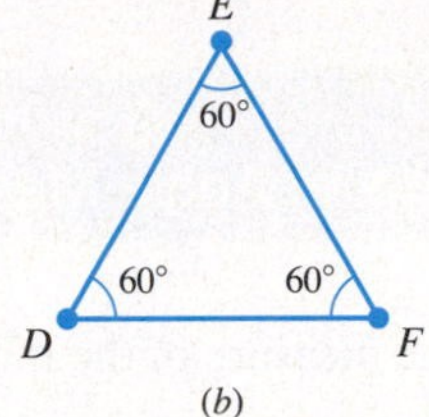

(b)

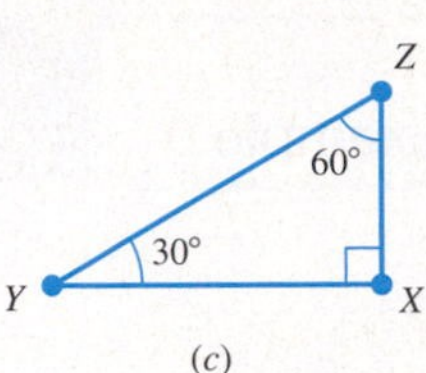

(c)

**NOTE**

Each of these triangles can be classified in different ways. *XYZ* is a right triangle, but it is also scalene.

$\triangle ABC$ is an isosceles triangle because two of the angles have the same measure. And $\triangle DEF$ is an equilateral triangle because all three angles have the same measure. $\triangle XYZ$ is a scalene triangle because no two angles have the same measure.

### Check Yourself 2

**Label each triangle as equilateral, isosceles, or scalene.**

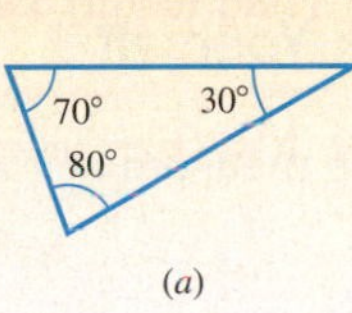

(a)

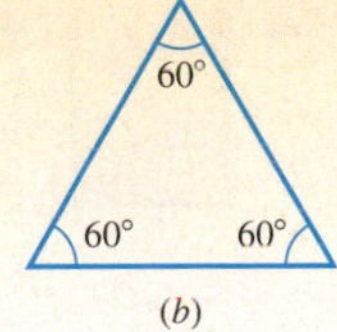

(b)

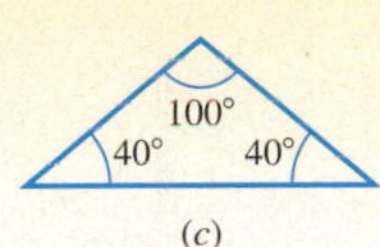

(c)

Go back and look at the sum of the angles inside each of the triangles in Example 2. You will note that they always add to 180°. No matter how we draw a triangle, the sum of the three angles inside the triangle will *always* be 180°.

Here is an experiment that might convince you that this is always the case.

1. Using a straight edge, draw any triangle on a sheet of paper.

2. Use scissors to cut out the triangle.
3. Rip the three vertices off of the triangle.

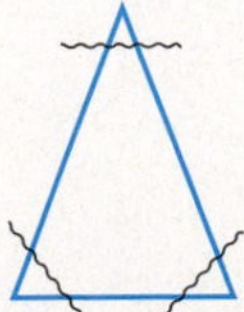

4. Lay the three vertices (with the points of the triangle touching) together. They will always form a straight angle, which we saw in Section 8.2 has a measure of 180°.

**Property**

## Angles of a Triangle

For any triangle *ABC*,

$m\angle A + m\angle B + m\angle C = 180°$

## Example 3 Finding an Angle Measure

< Objective 3 >

Find the measure of the third angle in this triangle.

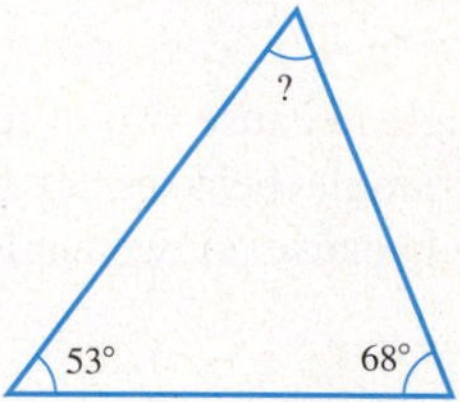


We need the three measurements to have a sum of 180°, so we add the two given measurements (53° + 68° = 121°). Then we subtract that from 180° (180° − 121° = 59°). This gives us the measure of the third angle, 59°.

### Check Yourself 3

Find the measure of $\angle ABC$.

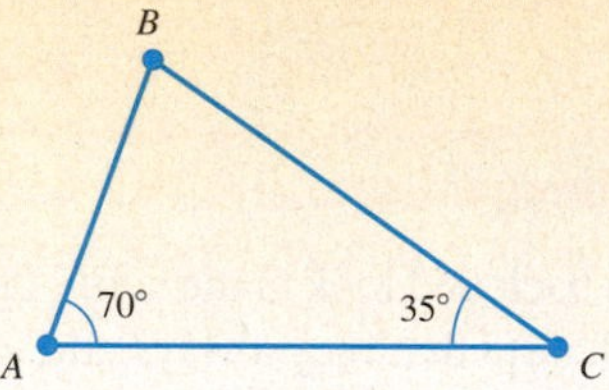


**Definition**

## Similar Triangles

If the measurements of the three angles in two different triangles are the same, we say the two triangles are **similar triangles.**

## Example 4 Identifying Similar Triangles

< Objective 4 >

Which two triangles are similar?

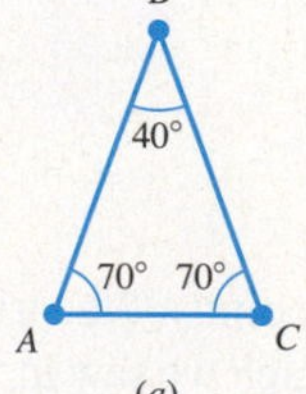

(*a*)

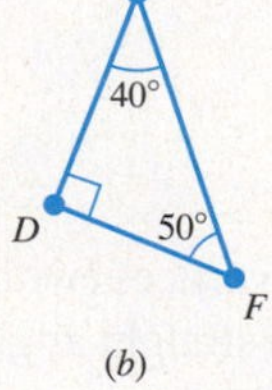

(*b*)

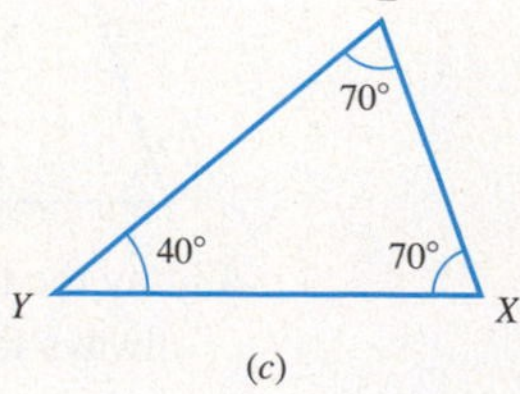

(*c*)

**NOTE**

This can be written $\triangle ABC \approx \triangle XYZ$

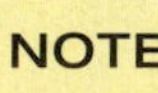

Although they are of different size, $\triangle ABC$ and $\triangle XYZ$ are similar because they have the same angle measurements.

## Check Yourself 4

Find the two triangles that are similar.

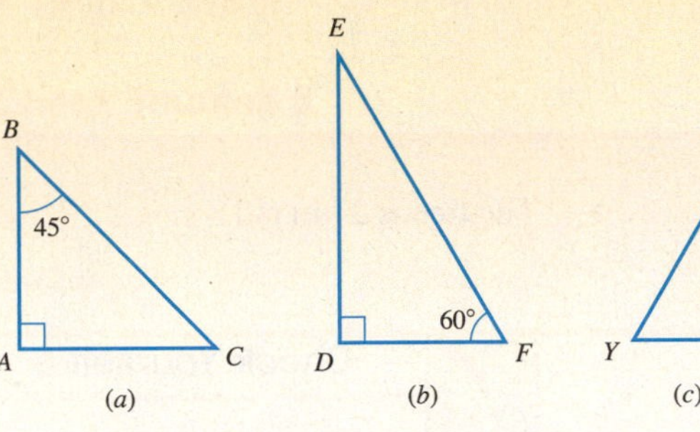


Finally, we return to an idea that we first saw in Chapter 5.

**Property**

**Similar Triangles** If two triangles are similar, their corresponding sides have the same ratio.

The most common use of this property of similar triangles occurs when you wish to find the height of a tall object. Example 5 illustrates this application of similar triangles.

## Example 5 Finding the Height of a Tree

< Objective 5 >

If a man who is 180 cm tall casts a shadow that is 60 cm long, how tall is a tree that casts a shadow that is 9 m long?

Note that, because of the sun, the man and his shadow form a similar triangle to the tree and its shadow. Because of this, we can use the common ratio to find the height of the tree.

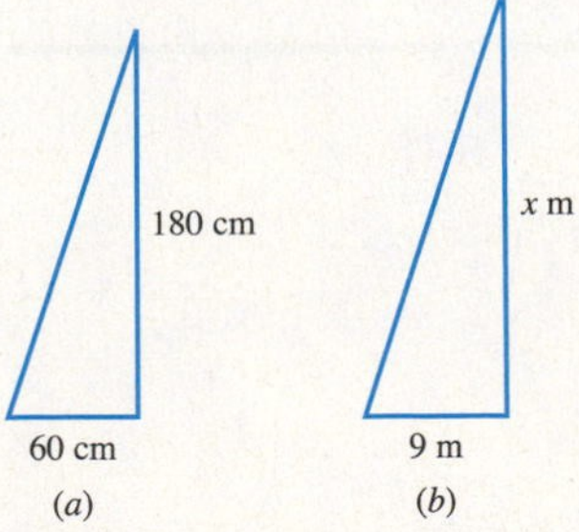

$$\frac{180 \text{ cm}}{60 \text{ cm}} = \frac{x \text{ m}}{9 \text{ m}}$$

$$x \cdot 60 = 180 \cdot 9$$

$$x = \frac{180 \cdot 9}{60}$$

$$x = 27$$

The tree is 27 m tall.

### Check Yourself 5

If a man who is 160 cm tall casts a shadow that is 120 cm long, how tall is a building that casts a shadow that is 60 m long?

### Check Yourself ANSWERS

**1.** $\triangle XYZ$ **2.** **(a)** Scalene; **(b)** equilateral; **(c)** isosceles **3.** 75°
**4.** $\triangle DEF$ and $\triangle XZY$ **5.** 80 m

## Reading Your Text

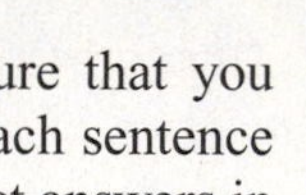

The following fill-in-the-blank exercises are designed to ensure that you understand some of the key vocabulary used in this section. Each sentence usually comes directly from the section. You will find the correct answers in the Answer Appendix.

### SECTION 8.3

**(a)** A triangle is called an ______________ triangle if all three angles have the same measure.

**(b)** A triangle is called an ______________ triangle if two angles have the same measure.

**(c)** If the measurements of the three angles in two different triangles are the same, we say that the two triangles are ______________ triangles.

**(d)** If two triangles are similar, their ______________ sides have the same ratio.

# 8.3 exercises

Boost your grade at mathzone.com!

- > Practice Problems
- > NetTutor
- > Self-Tests
- > e-Professors
- > Videos

Name ______________________

Section ________ Date ________

## Answers

1. ______________________
2. ______________________
3. ______________________
4. ______________________
5. ______________________
6. ______________________
7. ______________________
8. ______________________
9. ______________________
10. ______________________

< Objective 1 >

*Label the triangles as acute or obtuse.*

**1.**

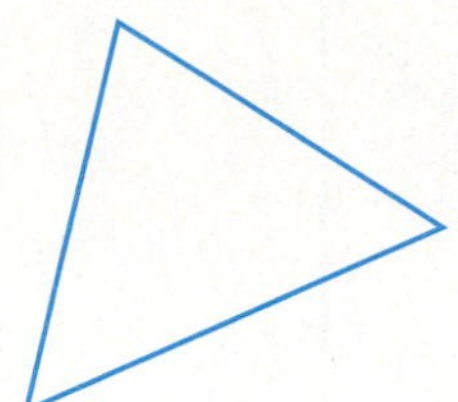

**2.**

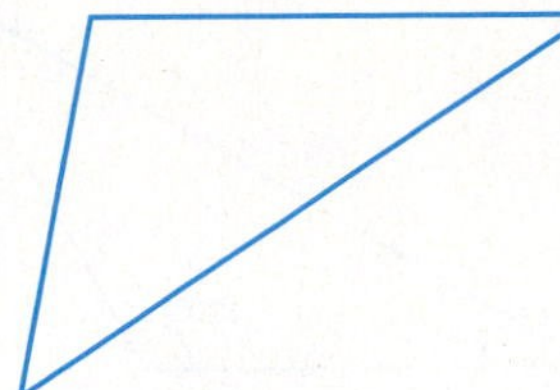

**3.**

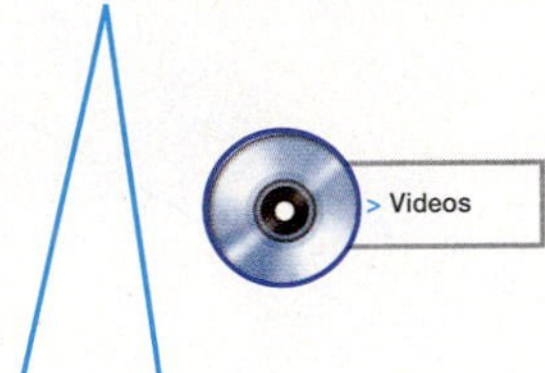


**4.**

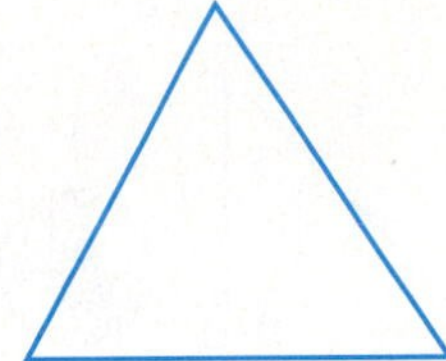

< Objective 2 >

*Label the triangles as equilateral, isosceles, or scalene.*

**5.**

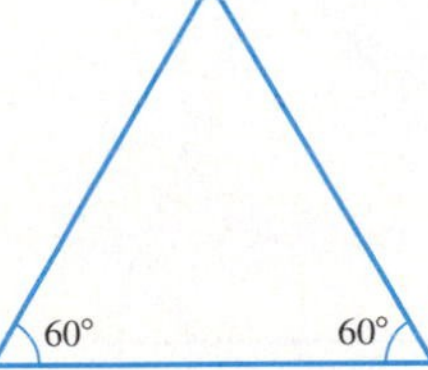


**6.**

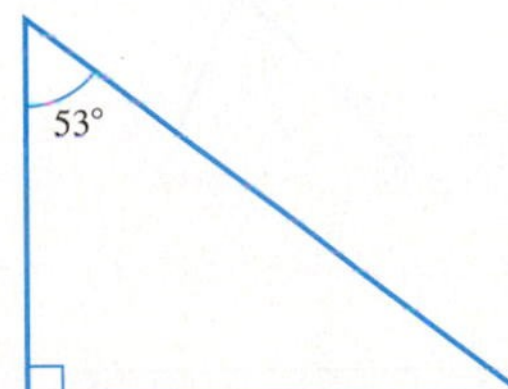


**7.**

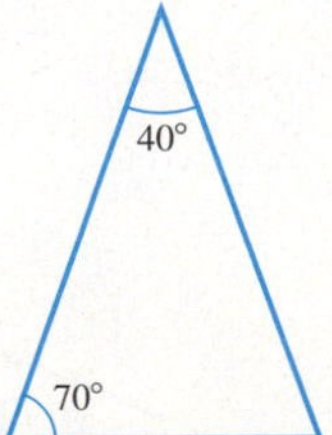


**8.**

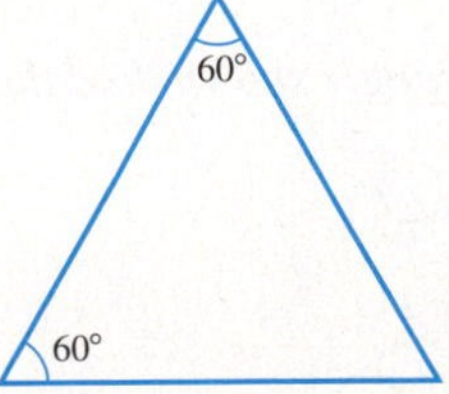


**9.**

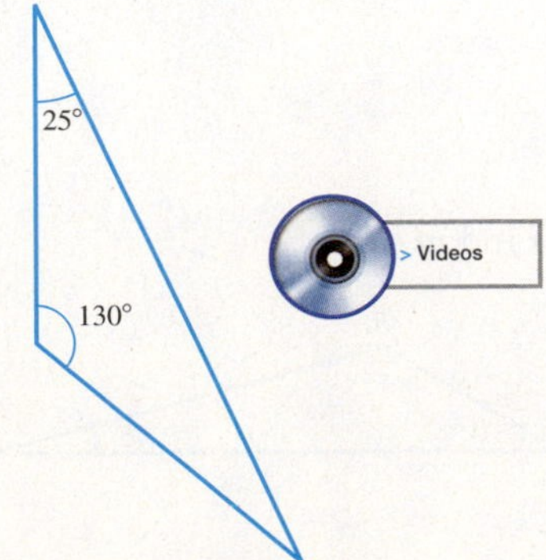


**10.**

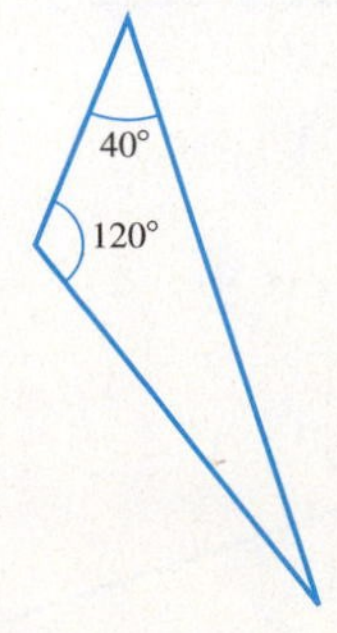

## Answers

11. ____________

12. ____________

13. ____________

14. ____________

15. ____________

16. ____________

17. ____________

18. ____________

19. ____________

20. ____________

< Objective 3 >

*Find the missing angle and then label the triangle as equilateral, isosceles, or scalene.*

**11.**

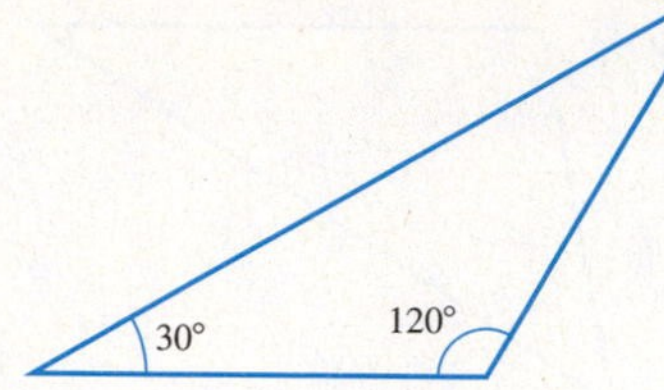


**12.**

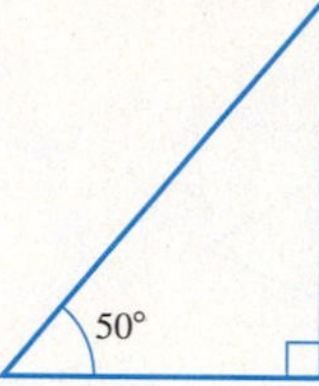


**13.**

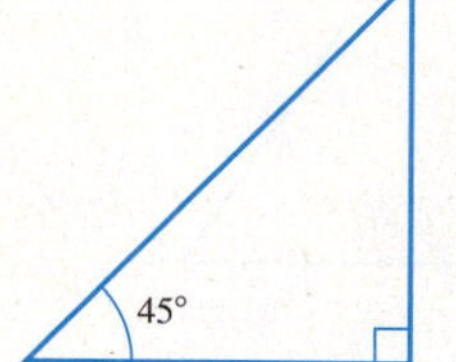


> Videos

**14.**

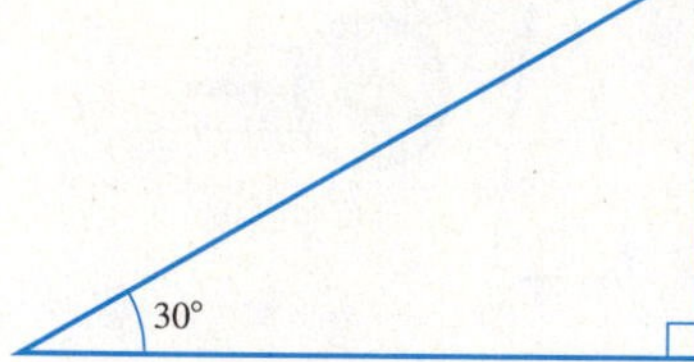


**15.**

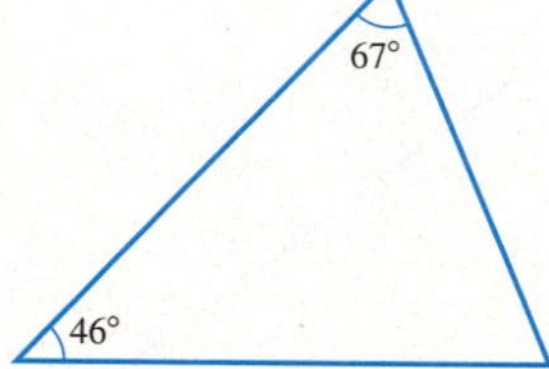


**16.**

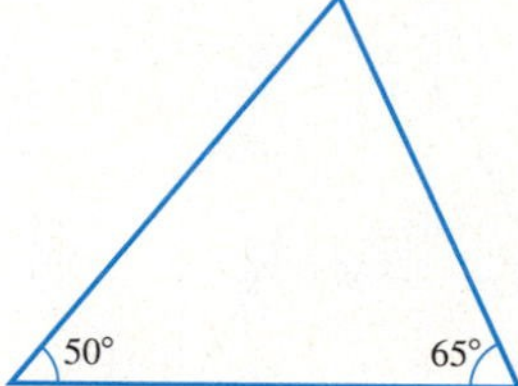


*For each triangle shown, find the indicated angle.*

**17.** Find $m\angle C$.

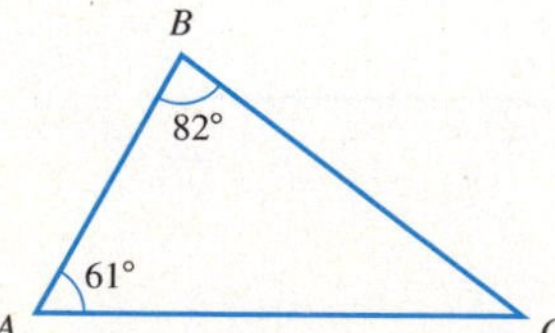


**18.** Find $m\angle B$.

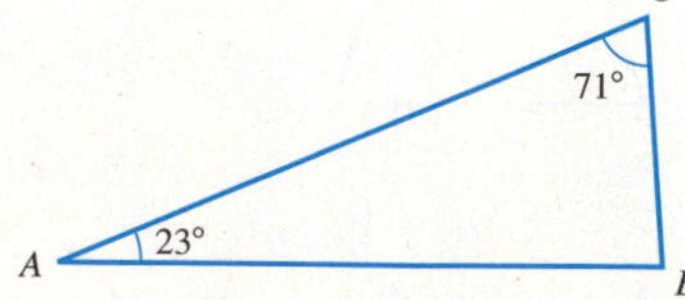


**19.** Find $m\angle A$.

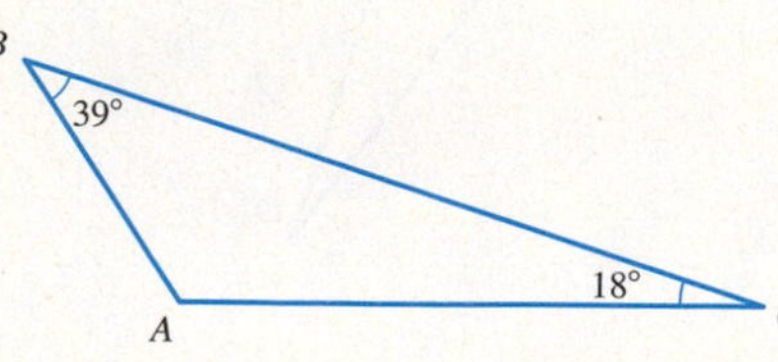


**20.** Find $m\angle B$.

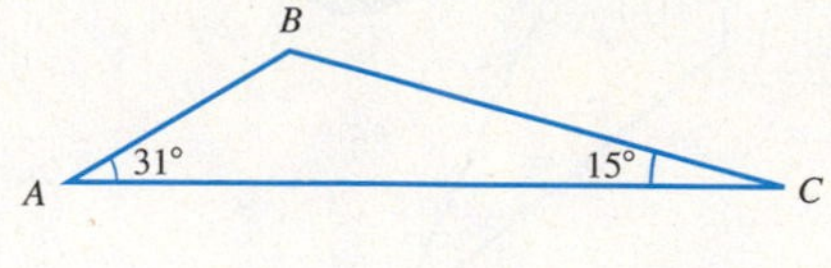

**21.** Find $m\angle B$.

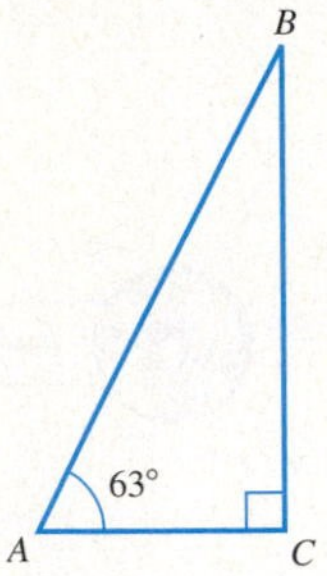

**22.** Find $m\angle A$.

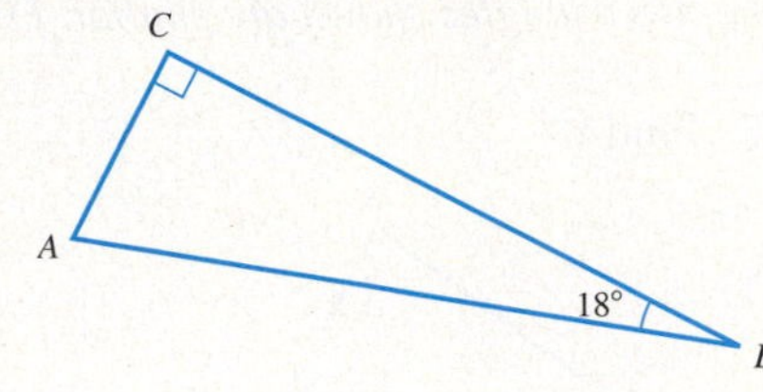


*Assume that the given triangle is isosceles.*

**23.** Find $m\angle A$ and $m\angle C$.

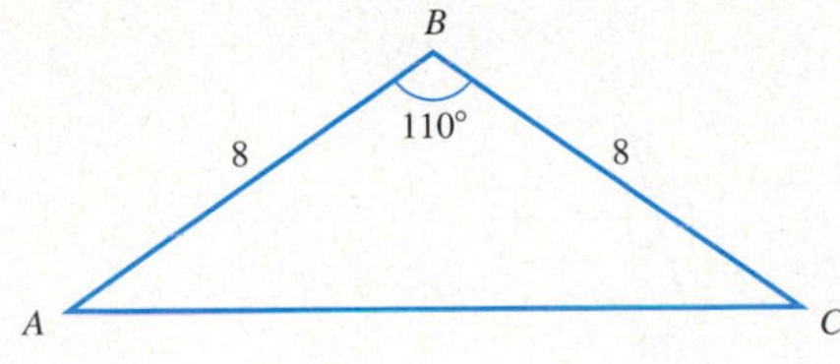


**24.** Find $m\angle D$ and $m\angle F$.

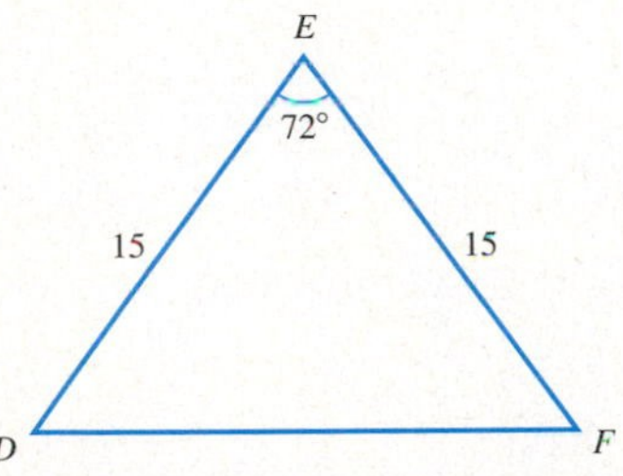


< Objective 4 >

**25.** Which two triangles are similar?

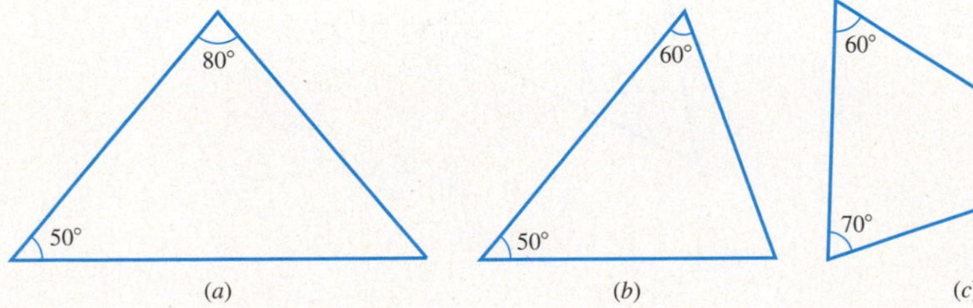


**26.** Which two triangles are similar?

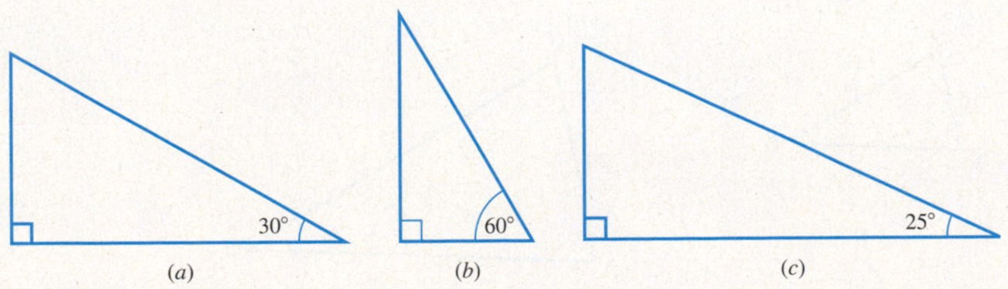


## Answers

21. __________

22. __________

23. __________

24. __________

25. __________

26. __________

## Answers

27. ______

28. ______

29. ______

30. ______

< Objective 5 >

*The two triangles shown are similar. Find the indicated side.*

**27.** Find *v*.

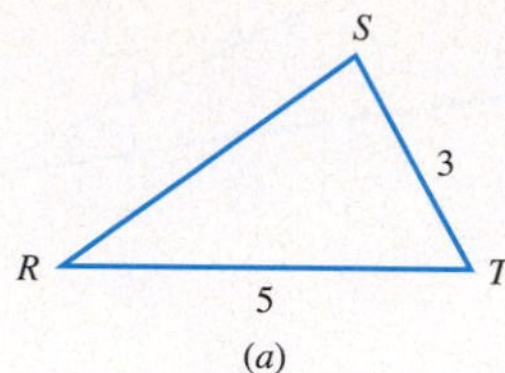


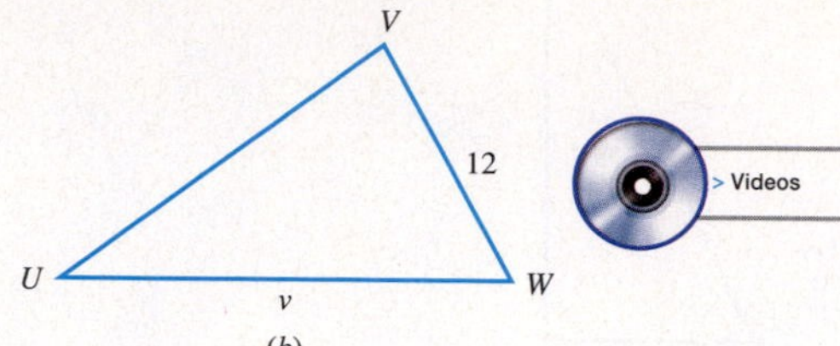


**28.** Find *f*.

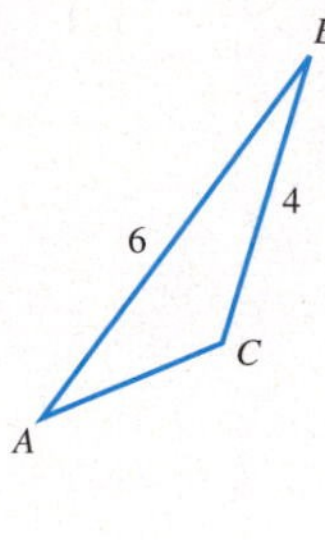


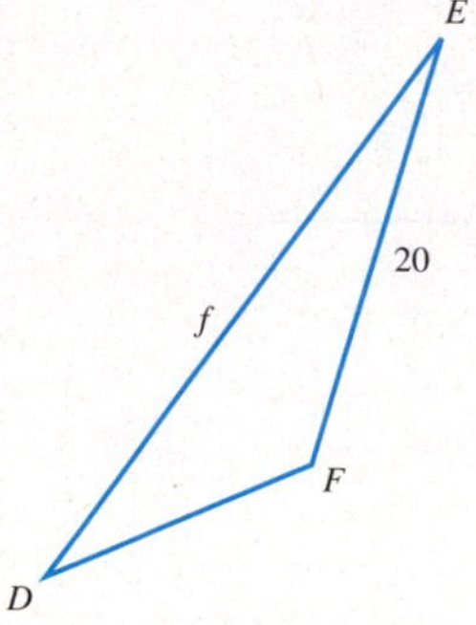


**29.** Find *g*.

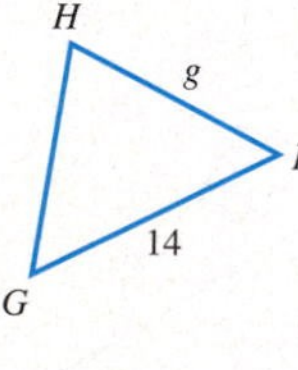


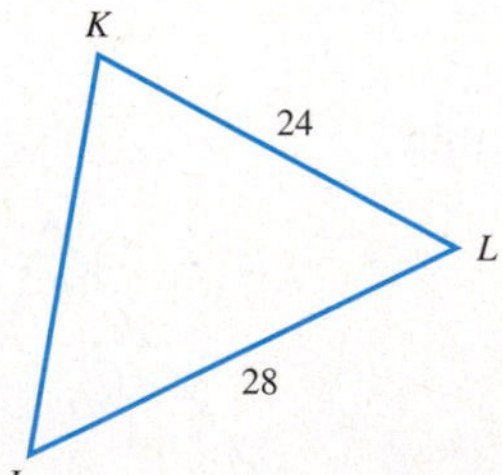


**30.** Find *m*.

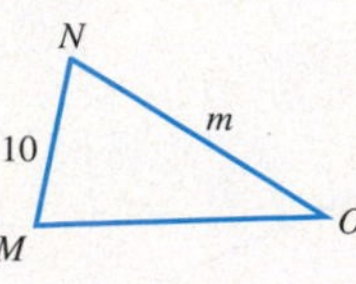


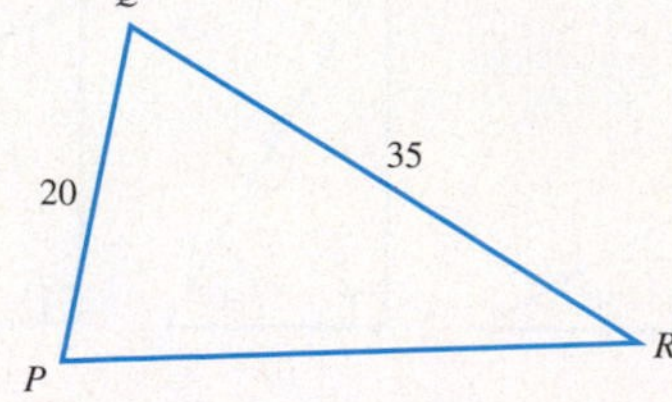

**31.** Find *t*.



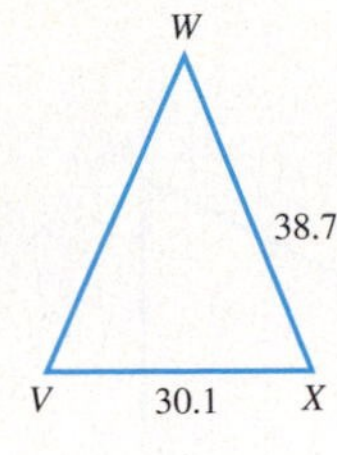


**32.** Find *e*.

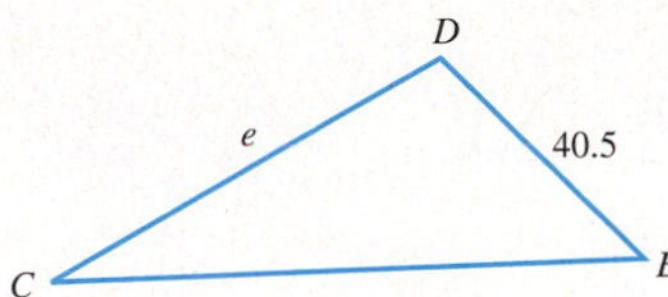


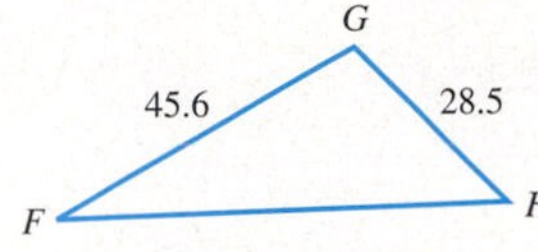


Basic Skills | **Advanced Skills** | Vocational-Technical Applications | Calculator/Computer | Above and Beyond

*First indicate which two triangles are similar. Then find the indicated side. Round your answer to the nearest hundredth.*

**33.** Find $\overline{KL}$.

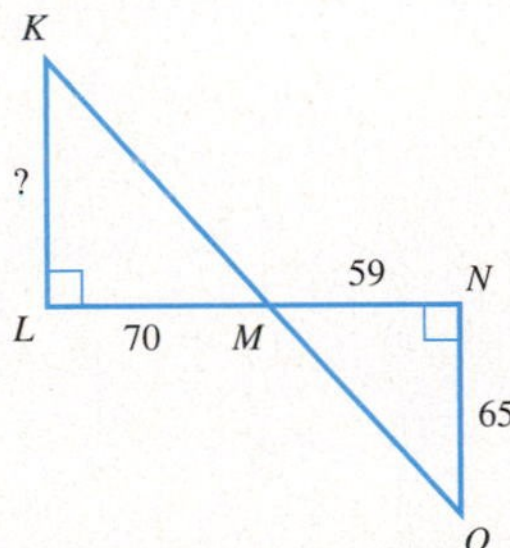


**34.** Find $\overline{PQ}$.

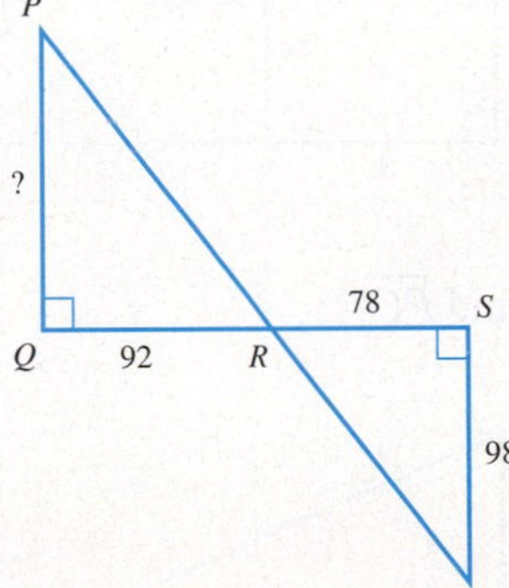


**35.** Find $\overline{VX}$.

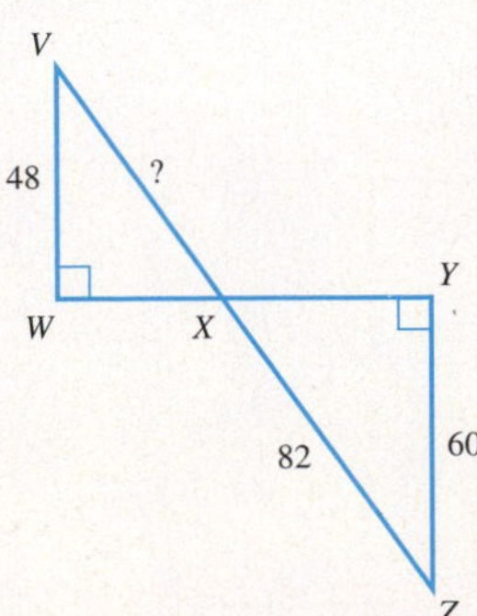


**36.** Find $\overline{AC}$.

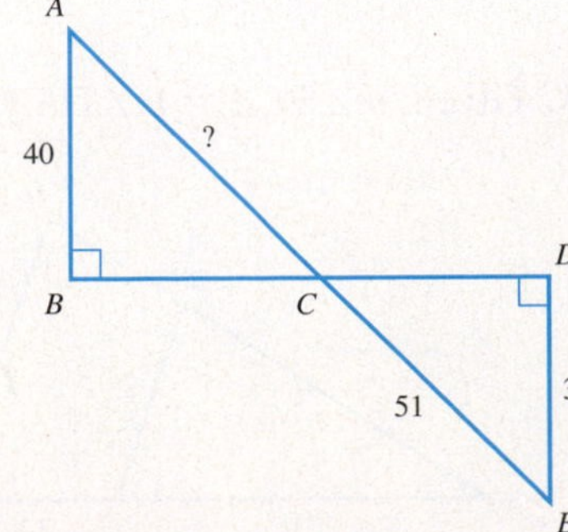


## Answers

31. ______

32. ______

33. ______

34. ______

35. ______

36. ______

Answers

37. ______

38. ______

39. ______

40. ______

41. ______

*Find the indicated side. If necessary, round to the nearest tenth of a unit.*

**37.** Find $\overline{DE}$.

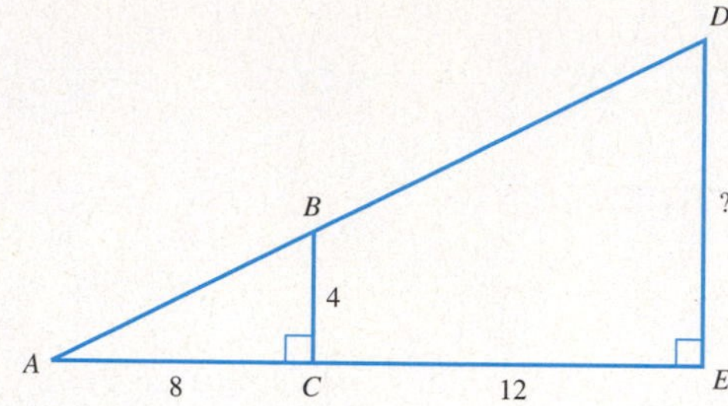


**38.** Find $\overline{IJ}$.

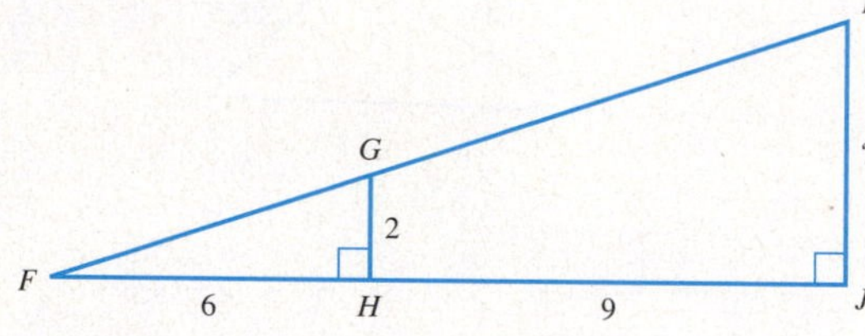


**39.** Find $\overline{KL}$.

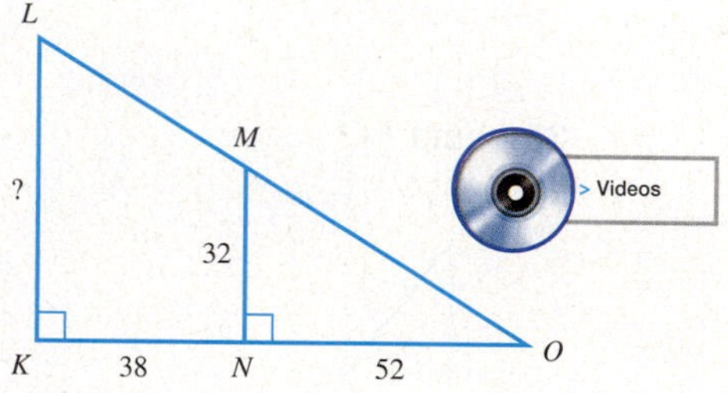


> Videos

**40.** Find $\overline{PQ}$.

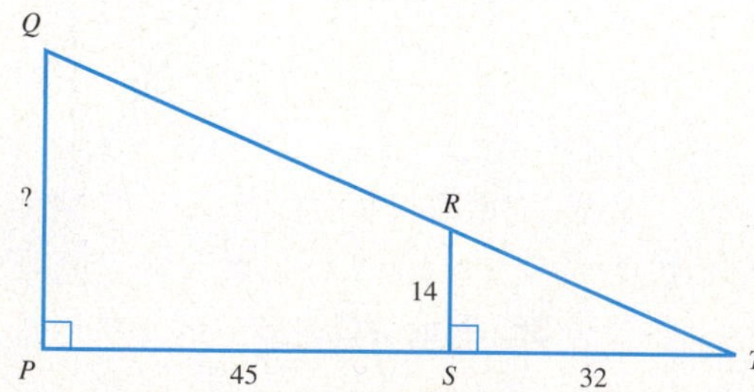


**41.** Given: $m\angle BCA = m\angle DEA$. Find $\overline{DE}$.

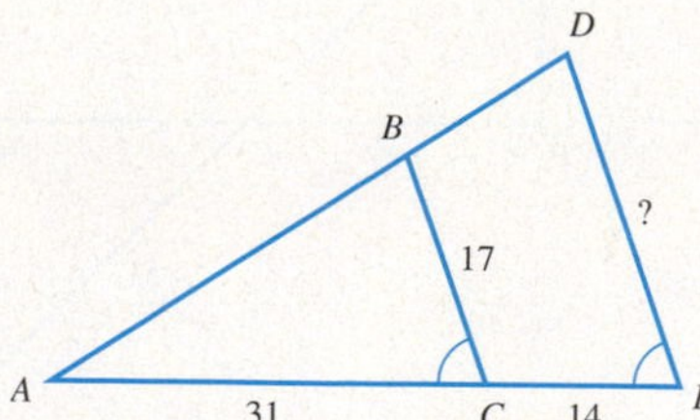

42. Given: $m\angle GHF = m\angle IJF$. Find $\overline{IJ}$.

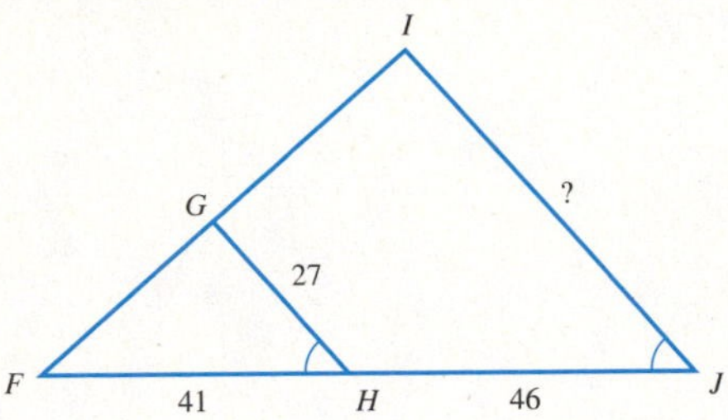


43. A light pole casts a shadow that measures 4 ft. At the same time, a yardstick casts a shadow that is 9 in. long. How tall is the pole?

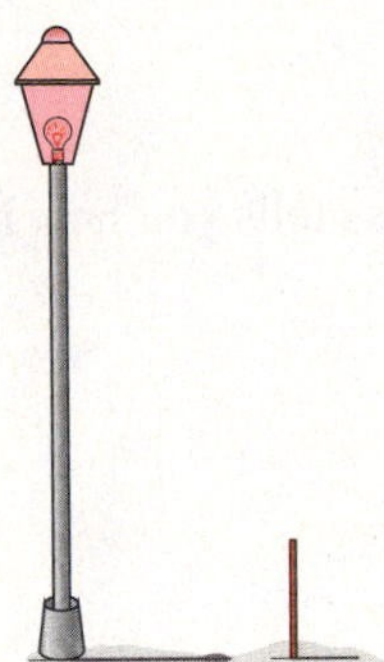

44. A tree casts a shadow that measures 5 m. At the same time, a meter stick casts a shadow that is 0.4 m long. How tall is the tree?

Basic Skills | Advanced Skills | **Vocational-Technical Applications** | Calculator/Computer | Above and Beyond

45. **Manufacturing Technology** In a common truss, the slope triangle tells you how high the truss is compared to the run. Find the height of this common truss. (*Hint:* The run is only half the span.)

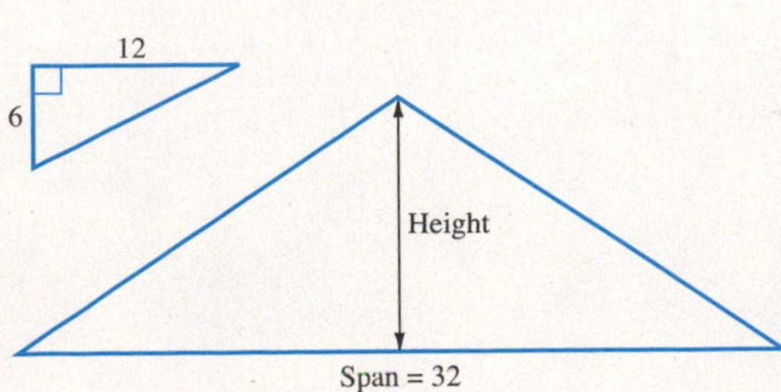


## Answers

42. ____________

43. ____________

44. ____________

45. ____________

**Answers**

46. ______

47. ______

48. ______

49. ______

50. ______

**46. Manufacturing Technology** A chipping hammer is shaped like a wedge with a tip angle of 25°.

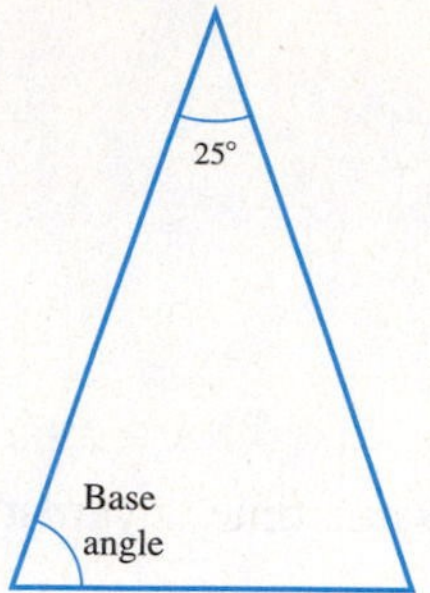


Find the angle at the base of the isosceles triangle.

**47. Manufacturing Technology** The slope triangle on a truss tells you how high the truss is compared to the run.

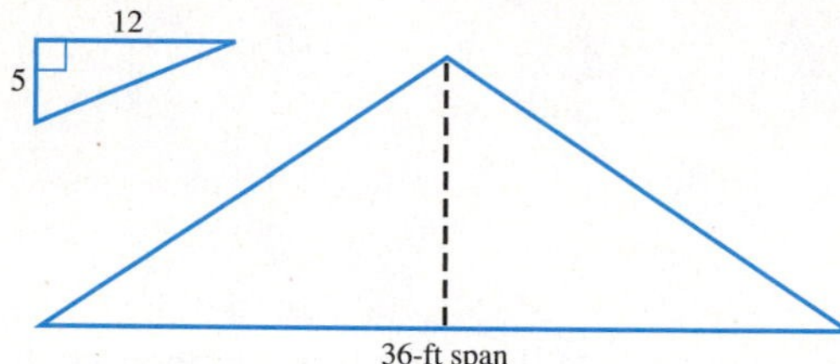


What is the height of this truss, given the $\frac{5}{12}$-slope triangle? Remember that the run is only half the span.

Basic Skills | Advanced Skills | Vocational-Technical Applications | Calculator/Computer | **Above and Beyond**

**48.** Use the ideas of similar triangles to determine the height of a pole or tree on your campus. Work with one or two partners.

*In exercises 49 to 51, one side of the triangle has been extended, forming what is called an* exterior angle. *In each case, find the measure of the indicated exterior angle.*

**49.**

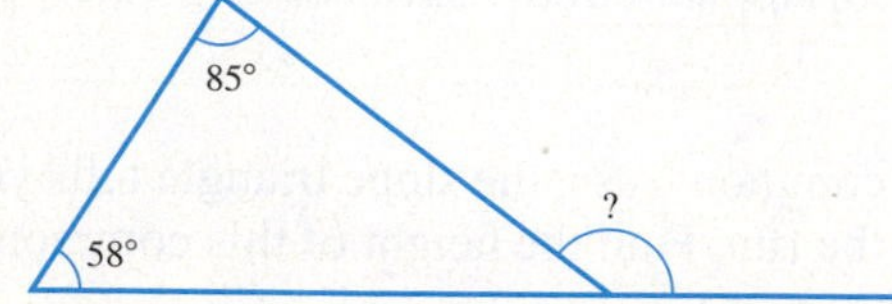


**50.**

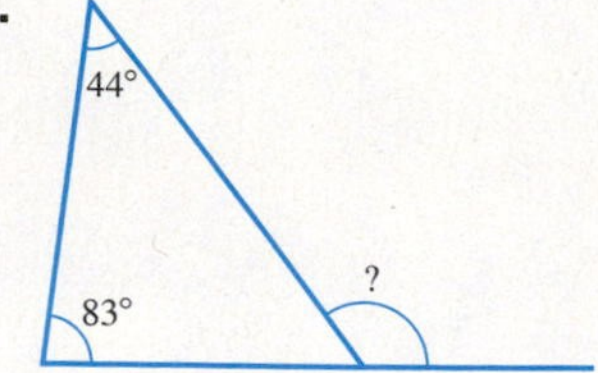

**51.** 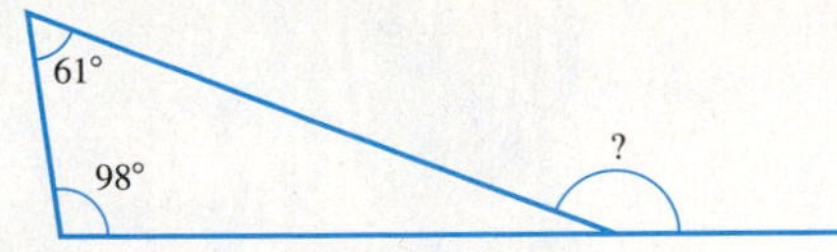


**52.** What do you observe from exercises 49 to 51? Write a general conjecture about an exterior angle of a triangle.

**53.** Write an argument to show that an equilateral triangle cannot have a right angle.

**54.** Argue that, given an equilateral triangle, the measure of each angle must be 60°.

**55.** Argue that a triangle cannot have more than one obtuse angle.

**56.** Is it possible to have an isosceles right triangle? If such a triangle exists, what can be said about the angles? Defend your statements.

**57.** Create an argument to support the following statement:

If $\triangle ABC$ is a right triangle, with $m\angle C = 90°$, then $\angle A$ and $\angle B$ must be acute and complementary.

## Answers

**1.** Acute **3.** Acute **5.** Equilateral **7.** Isosceles **9.** Isosceles **11.** 30°; isosceles **13.** 45°; isosceles **15.** 67°; isosceles **17.** 37° **19.** 123° **21.** 27° **23.** $m\angle A = 35°; m\angle C = 35°$ **25.** (*b*) and (*c*) **27.** 20 **29.** 12 **31.** 39.2 **33.** 77.12 **35.** 65.60 **37.** 10 **39.** 55.4 **41.** 24.7 **43.** 16 ft **45.** 8 ft **47.** 7.5 ft **49.** 143° **51.** 159° **53.** Above and Beyond **55.** Above and Beyond **57.** Above and Beyond

## Answers

51. ____
52. ____
53. ____
54. ____
55. ____
56. ____
57. ____

# Activity 22 ::
## Composite Geometric Figures

When first introduced to geometry in your math class, you worked with fairly straightforward figures such as squares, circles, and triangles. Most real-world objects cannot be described by such simple geometric figures. Consider the chair that you are using right now. The chair is probably made up of several shapes put together.

**Composite geometric figures** are figures formed by combining two or more simple geometric figures. One example of a composite geometric figure is the Norman window. Norman windows are windows constructed by combining a rectangle with a half-circle (see the figure). Given such a figure, there are several questions we could ask. We might ask an area question to find the amount of glass used, a perimeter question to find the amount of frame, or a combination question to determine the amount of wall space necessary to accommodate such a window.

Assume all measurements are in feet and round answers to two decimal places.

1. Find the area of the rectangular piece of glass in the Norman window pictured.
2. Find the area of the half-circle piece of glass.
3. Find the total area of the glass.
4. If the glass costs \$3 per square foot for a rectangular piece and \$4.75 per square foot for the circular piece, find the total cost of the glass.
5. Find the outer perimeter of the figure.
6. Find the length of framework needed for the Norman window (do not forget to include the strip of frame separating the rectangle and half-circle).
7. Find the cost of the framework if it costs \$1.25 per foot for straight pieces and \$3.25 per foot for curved pieces.
8. Use your answers to exercises 4 and 7 to determine the total cost of the Norman window pictured.
9. Find the dimensions of the smallest rectangle that would completely contain the Norman window.
10. Find the area of the "leftover" rectangle created by cutting the Norman window from the rectangle found in exercise 9.

# 8.4 Square Roots and the Pythagorean Theorem

< 8.4 Objectives >

1 > Find the square root of a perfect square

2 > Identify the hypotenuse of a right triangle

3 > Identify a perfect triple

4 > Use the Pythagorean theorem to find the length of a missing side of a right triangle

5 > Approximate the square root of a number

Some numbers can be written as the product of two identical factors, for example,

$9 = 3 \times 3$

Either factor is called a **square root** of the number. The symbol $\sqrt{\phantom{x}}$ (called a **radical sign**) is used to indicate a square root. Thus $\sqrt{9} = 3$ because $3 \times 3 = 9$.

**Example 1** Finding the Square Root

< Objective 1 >

**NOTE**

To use the $\boxed{\sqrt{\phantom{x}}}$ key with a scientific calculator, first enter the 49 and then press the key. With a graphing calculator, press the radical key first and then enter the 49 and a closing parenthesis.

Find the square root of 49 and of 16.

**(a)** $\sqrt{49} = 7$ Because $7 \times 7 = 49$

**(b)** $\sqrt{16} = 4$ Because $4 \times 4 = 16$

**Check Yourself 1**

Find the indicated square root.

**(a)** $\sqrt{121}$ **(b)** $\sqrt{36}$

The most frequently used theorem in geometry is undoubtedly the Pythagorean theorem. In this section you will use that theorem. You will also learn a little about the history of the theorem. It is a theorem that applies only to right triangles.

The side opposite the right angle of a right triangle is called the **hypotenuse.** The other two sides are called **legs.** Note that the legs are perpendicular to each other.

## Example 2 Identifying the Hypotenuse

< Objective 2 >

In the following right triangle, the side labeled $c$ is the hypotenuse.

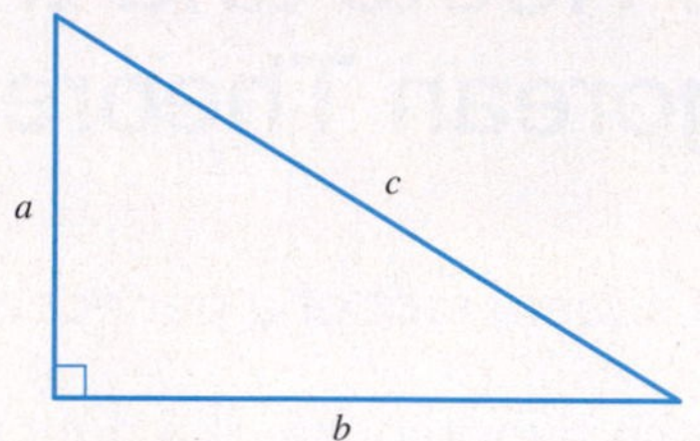

### Check Yourself 2

Which side represents the hypotenuse of the given right triangle?

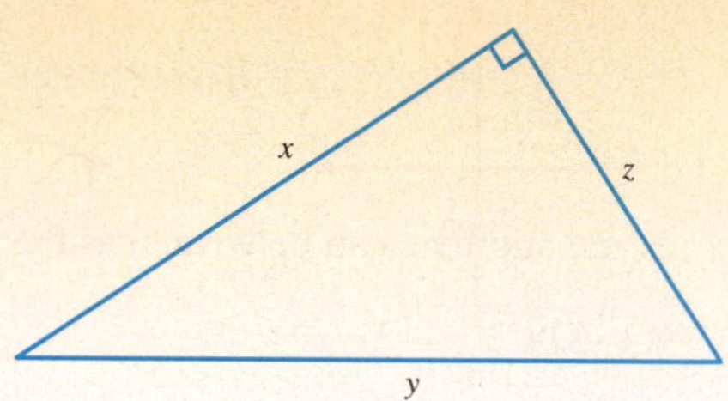


The numbers 3, 4, and 5 have a special relationship. Together they are called a **perfect triple,** which means that when you square all three numbers, the sum of the smaller squares equals the squared value of the largest number.

## Example 3 Identifying Perfect Triples

< Objective 3 >

Show that each of the following is a perfect triple.

**(a)** 3, 4, and 5

$3^2 = 9 \qquad 4^2 = 16 \qquad 5^2 = 25$

and $9 + 16 = 25$, so we can say that $3^2 + 4^2 = 5^2$.

**(b)** 7, 24, and 25

$7^2 = 49 \qquad 24^2 = 576 \qquad 25^2 = 625$

and $49 + 576 = 625$, so we can say that $7^2 + 24^2 = 25^2$.

### Check Yourself 3

Show that each of the following is a perfect triple.

**(a)** 5, 12, and 13 **(b)** 6, 8, and 10

All the triples that you have seen, and many more, were known by the Babylonians more than 4,000 years ago. Stone tablets that had dozens of perfect triples carved into them have been found. The basis of the Pythagorean theorem was understood long before the time of Pythagoras (ca. 540 B.C.). The Babylonians not only understood perfect triples but also knew how triples related to a right triangle.

**Property**

**The Pythagorean Theorem (Version 1)**

If the lengths of the three sides of a right triangle are all integers, they will form a perfect triple, with the hypotenuse as the longest side.

There are two other forms in which the Pythagorean theorem is regularly presented. It is important that you see the connection between the three forms.

**Property**

**The Pythagorean Theorem (Version 2)**

The square of the hypotenuse of a right triangle is equal to the sum of the squares of the other two sides.

**Property**

**The Pythagorean Theorem (Version 3)**

Given a right triangle with sides $a$ and $b$ and hypotenuse $c$, it is always true that

$c^2 = a^2 + b^2$

## Example 4 Finding the Length of a Side of a Right Triangle

< Objective 4 >

**NOTE**

This is the version that you will refer to in your algebra classes.

Find the missing integer length for each right triangle.

**(a)**

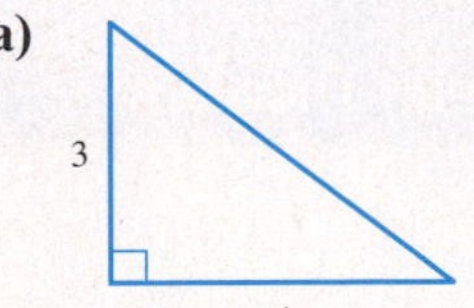


**(b)**

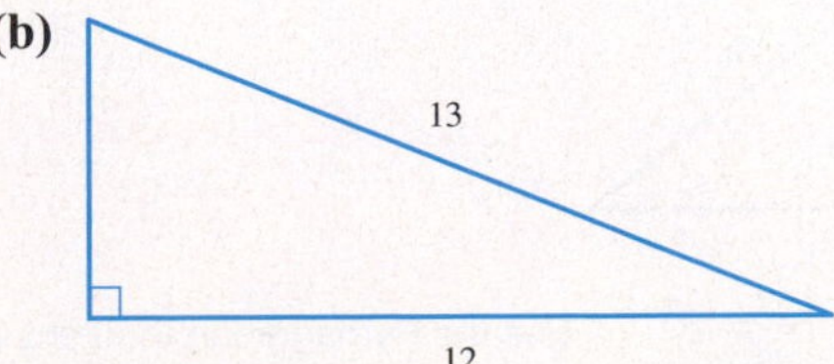


**(a)** A perfect triple will be formed if the hypotenuse is 5 units long, creating the triple 3, 4, 5. Note that $3^2 + 4^2 = 9 + 16 = 25 = 5^2$.

**(b)** The triple must be 5, 12, 13, which makes the missing length 5 units. Here, $5^2 + 12^2 = 25 + 144 = 169 = 13^2$.

### Check Yourself 4

Find the integer length of the unlabeled side for each right triangle.

(a)

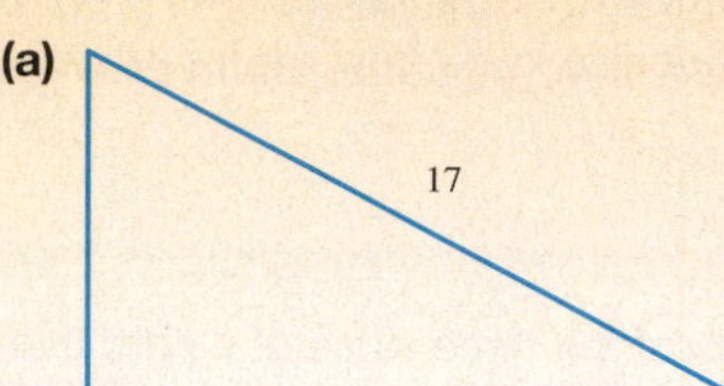


(b)

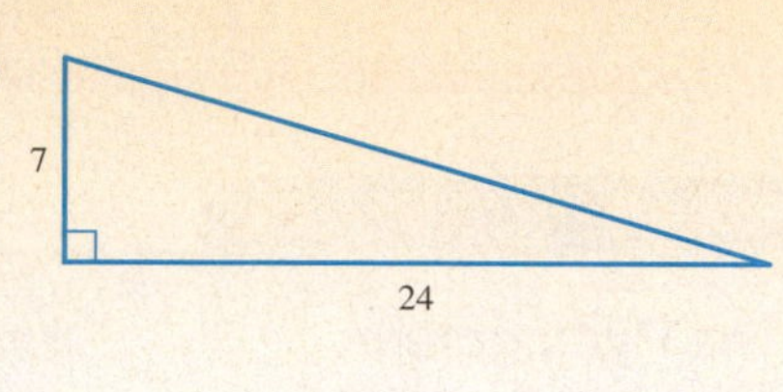


## Example 5 Using the Pythagorean Theorem

**NOTE**

The triangle has sides 6, 8, and 10.

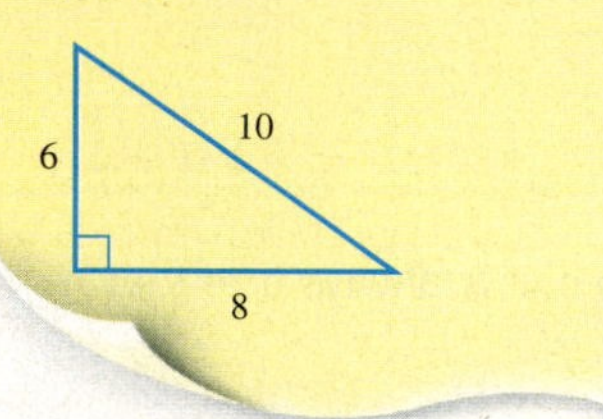


If the lengths of two sides of a right triangle are 6 and 8, find the length of the hypotenuse.

$c^2 = a^2 + b^2$ The value of the hypotenuse is found from the Pythagorean theorem with $a = 6$ and $b = 8$.

$c^2 = (6)^2 + (8)^2 = 36 + 64 = 100$

$c = \sqrt{100} = 10$ The length of the hypotenuse is 10 (because $10^2 = 100$).

### Check Yourself 5

Find the hypotenuse of a right triangle whose sides measure 9 and 12.

In some right triangles, the lengths of the hypotenuse and one side are given and we are asked to find the length of the missing side.

## Example 6 Using the Pythagorean Theorem

Find the missing length.

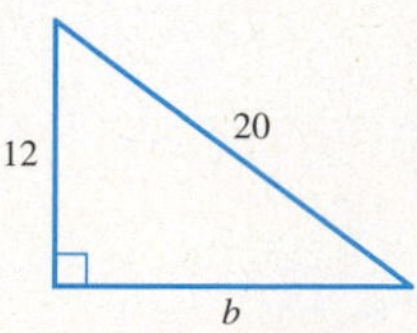


$a^2 + b^2 = c^2$ Use the Pythagorean theorem with $a = 12$ and $c = 20$.

$(12)^2 + b^2 = (20)^2$

$144 + b^2 = 400$

$b^2 = 400 - 144 = 256$

$b = \sqrt{256} = 16$ The missing side is 16.

### Check Yourself 6

Find the missing length for a right triangle with one leg measuring 8 centimeters (cm) and the hypotenuse measuring 10 cm.

Not every square root is a whole number. In fact, there are only 10 whole-number square roots for the numbers from 1 to 100. They are the square roots of 1, 4, 9, 16, 25, 36, 49, 64, 81, and 100. However, we can approximate square roots that are not whole numbers. For example, we know that the square root of 12 is not a whole number. We also know that its value must lie somewhere between the square root of 9 ($\sqrt{9} = 3$) and the square root of 16 ($\sqrt{16} = 4$). That is, $\sqrt{12}$ is between 3 and 4.

### Example 7 Approximating Square Roots

< Objective 5 >

Approximate $\sqrt{29}$.

$\sqrt{25} = 5$ and $\sqrt{36} = 6$, so $\sqrt{29}$ must be between 5 and 6.

### Check Yourself 7

$\sqrt{19}$ is between which of the following?

(a) 4 and 5 (b) 5 and 6 (c) 6 and 7

### Check Yourself ANSWERS

**1.** **(a)** 11; **(b)** 6 **2.** Side $y$ **3.** **(a)** $5^2 + 12^2 = 25 + 144 = 169$, $13^2 = 169$, so $5^2 + 12^2 = 13^2$; **(b)** $6^2 + 8^2 = 36 + 64 = 100$, $10^2 = 100$ so $6^2 + 8^2 = 10^2$ **4.** **(a)** 8; **(b)** 25 **5.** 15 **6.** 6 cm **7.** **(a)** 4 and 5

### Reading Your Text

The following fill-in-the-blank exercises are designed to ensure that you understand some of the key vocabulary used in this section. Each sentence usually comes directly from the section. You will find the correct answers in the Answer Appendix.

SECTION 8.4

**(a)** The symbol $\sqrt{\ }$ (called a ________________ sign) is used to indicate a square root.

**(b)** The side opposite the right angle of a right triangle is called the ________________.

**(c)** The ________________ theorem says that the square of the hypotenuse of a right triangle is equal to the sum of the squares of the other two sides.

**(d)** Not every square root is a ________________ number.

# 8.4 exercises

Basic Skills | Advanced Skills | Vocational-Technical Applications | Calculator/Computer | Above and Beyond

MathZone

Boost your grade at mathzone.com!

- Practice Problems
- NetTutor
- Self-Tests
- e-Professors
- Videos

Name ______________________

Section __________ Date __________

## Answers

1. ______________________
2. ______________________
3. ______________________
4. ______________________
5. ______________________
6. ______________________
7. ______________________
8. ______________________
9. ______________________
10. ______________________
11. ______________________
12. ______________________
13. ______________________
14. ______________________

< Objective 1 >

*Find the square root.*

**1.** $\sqrt{64}$

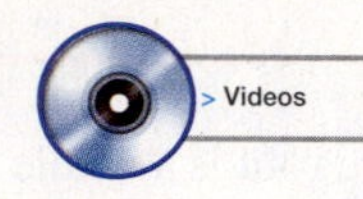


**2.** $\sqrt{121}$

**3.** $\sqrt{169}$

**4.** $\sqrt{196}$

< Objective 2 >

*Identify the hypotenuse of the given triangles by giving its letter.*

**5.**

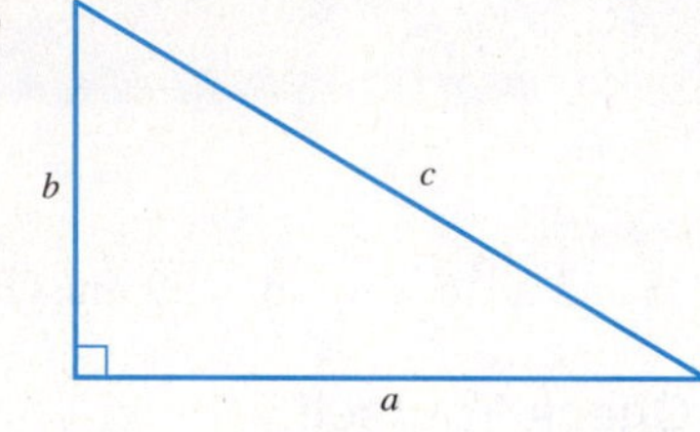


**6.**

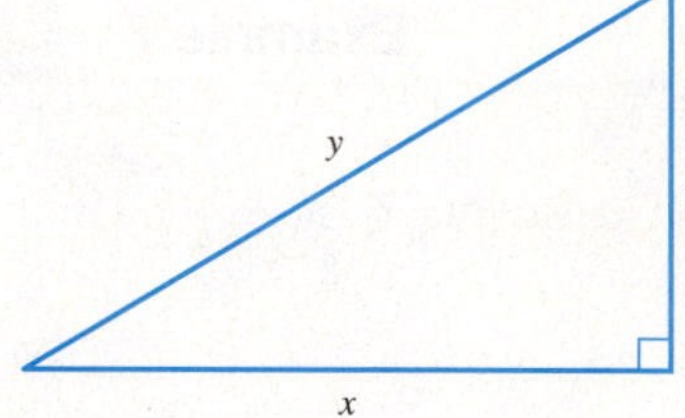


< Objective 3 >

*Identify which numbers are perfect triples.*

**7.** 3, 4, 5

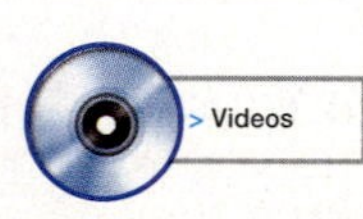


**8.** 4, 5, 6

**9.** 7, 12, 13

**10.** 5, 12, 13

**11.** 8, 15, 17

**12.** 9, 12, 15

< Objective 4 >

*Find the missing length for each right triangle.*

**13.**

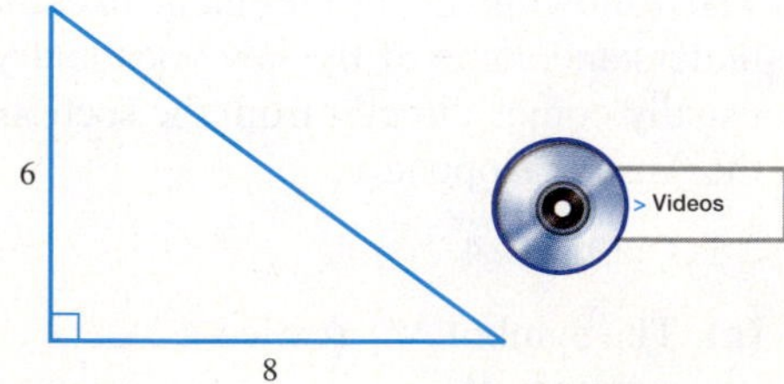


**14.**

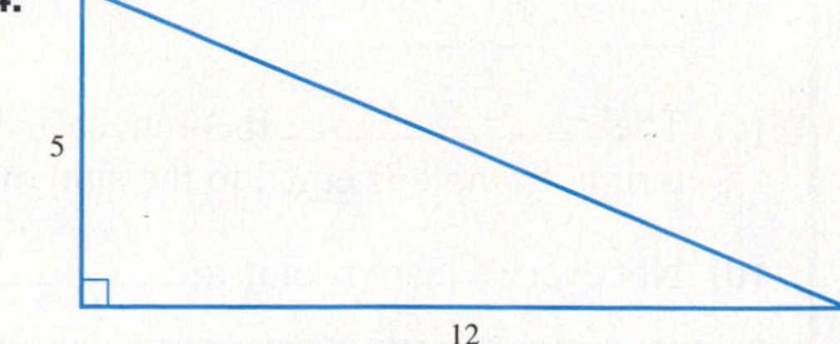




The Streeter/Hutchison Series in Mathematics Basic Mathematical Skills with Geometry

15.

8, 17

> Videos

16.

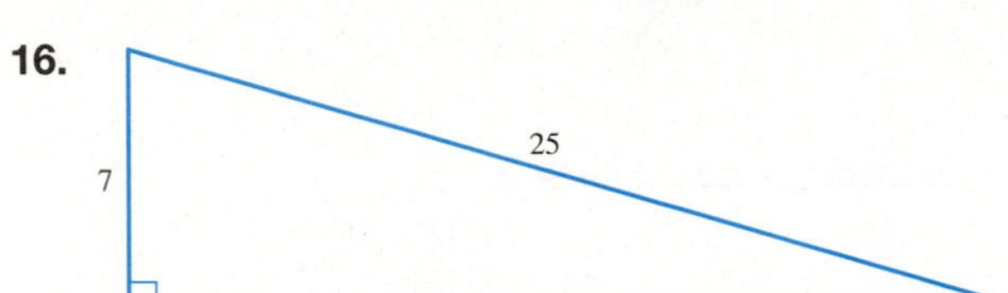


< Objective 5 >

*Select the correct approximation for each of the following.*

17. Is $\sqrt{23}$ between **(a)** 3 and 4, **(b)** 4 and 5, or **(c)** 5 and 6?

18. Is $\sqrt{15}$ between **(a)** 1 and 2, **(b)** 2 and 3, or **(c)** 3 and 4?

19. Is $\sqrt{44}$ between **(a)** 6 and 7, **(b)** 7 and 8, or **(c)** 8 and 9?

20. Is $\sqrt{31}$ between **(a)** 3 and 4, **(b)** 4 and 5, or **(c)** 5 and 6?

*Determine whether each statement is* **true** *or* **false.**

21. For any triangle with side lengths $a$, $b$, and $c$, it is true that $a^2 + b^2 = c^2$.

22. If we know the lengths of two of the sides of a right triangle, we can find the length of the third side.

*In each of the following statements, fill in the blank with* **always, sometimes,** *or* **never.**

23. The hypotenuse is __________ the longest side of a right triangle.

24. The square root of a whole number is ________ a whole number.

*Find the perimeter of each triangle shown.* (Hint: *First find the missing side.)*

25.

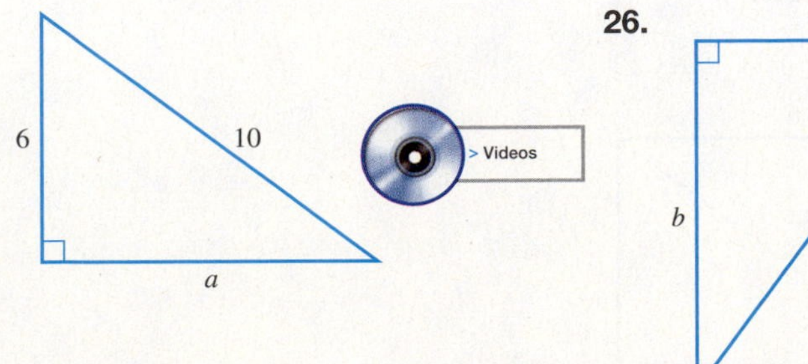


26.

9, b, 15

**Answers**

15. ________

16. ________

17. ________

18. ________

19. ________

20. ________

21. ________

22. ________

23. ________

24. ________

25. ________

26. ________

Answers

27. ______

28. ______

29. ______

30. ______

31. ______

32. ______

**27.**

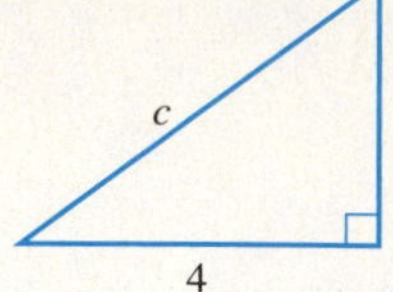


**28.**

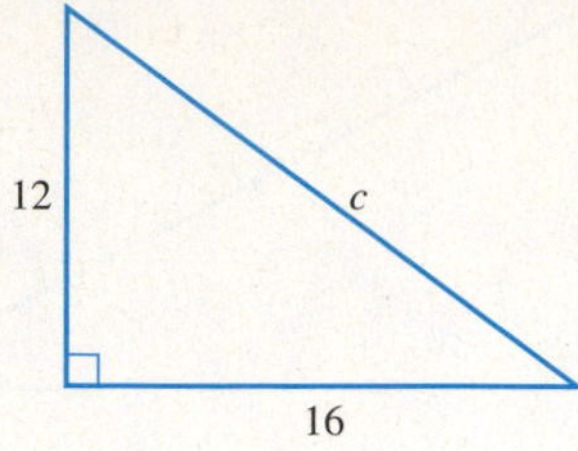


**29.** Find the altitude, $h$, of the isosceles triangle shown.

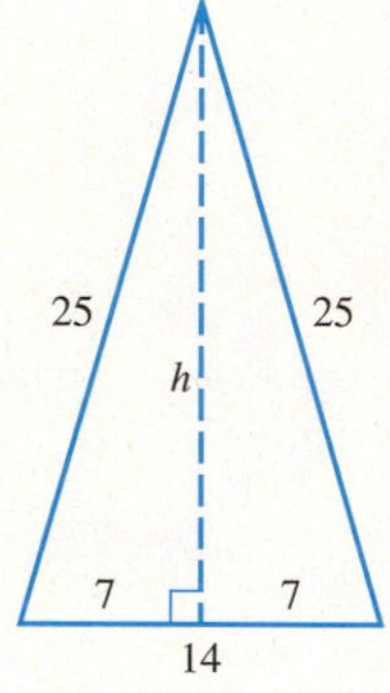


**30.** Find the altitude of the isosceles triangle shown. (*Hint:* The altitude shown bisects the base.)

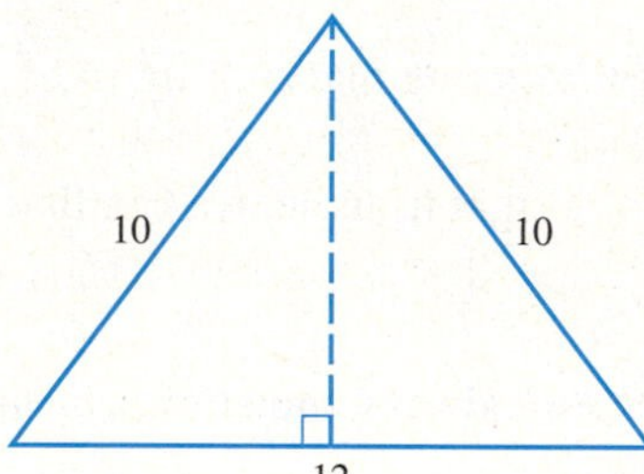


*Find the length of the diagonal of each rectangle.*

**31.**

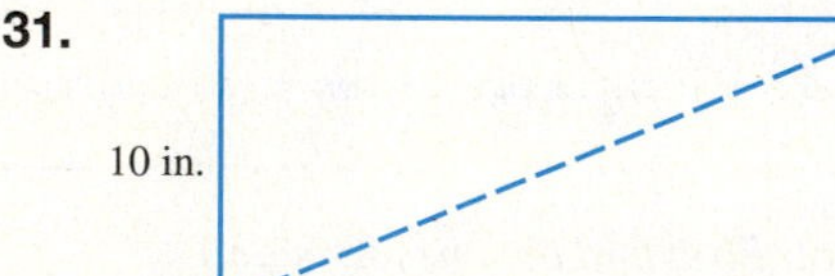


**32.**

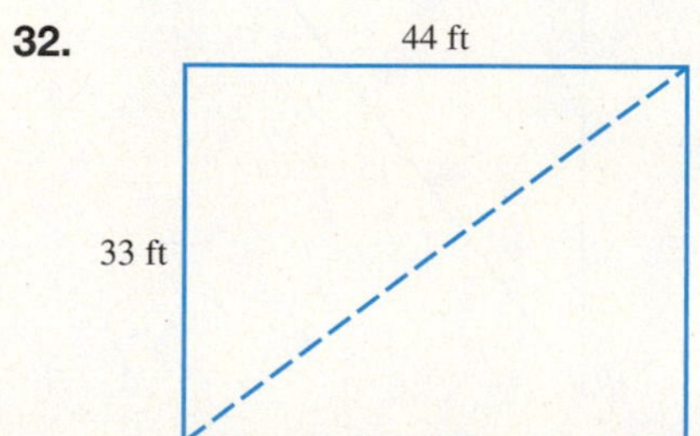

33. A castle wall, 24 feet high, is surrounded by a moat 7 feet across. Will a 26-foot ladder, placed at the edge of the moat, be long enough to reach the top of the wall?

Answers

33. ______

34. ______

35. ______

Basic Skills | Advanced Skills | **Vocational-Technical Applications** | Calculator/Computer | Above and Beyond

34. **Electronics** The following image is a small portion of a printed circuit board that is in the layout stage. The conductive trace being plotted between the solder pads is comprised of a vertical and a horizontal trace that meet at a right angle. The horizontal component is 0.86 in., and the vertical component is 0.92 in. If the trace could be run from point to point diagonally, its distance could be determined by finding $d = \sqrt{(0.86)^2 + (0.92)^2}$. How long would it be?

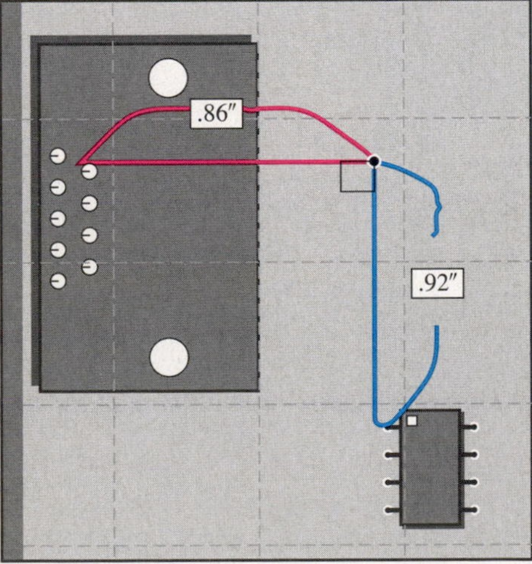


35. **Manufacturing Technology** Find the height of this truss.

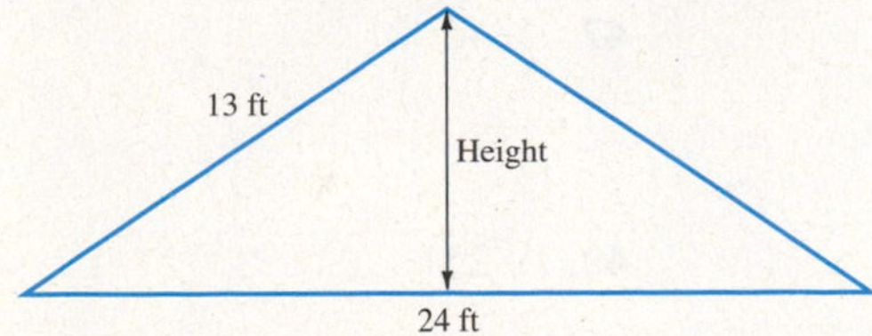

Basic Skills | Advanced Skills | Vocational-Technical Applications | **Calculator/Computer** | Above and Beyond

## Answers

36. ______
37. ______
38. ______
39. ______
40. ______
41. ______
42. ______
43. ______
44. ______
45. ______
46. ______
47. ______
48. ______
49. ______

To find a square root on your scientific calculator, you use the square root key. On some calculators, you simply enter the number and then press the square root key. With others, you must use the second function on the [$x^2$] (or [$y^x$]) key and specify the root you wish to find.

For example, to find the square root of 256 with a scientific calculator, you must enter

256 [$\sqrt{\ }$]

or perhaps

256 [2nd] [$\sqrt[x]{y}$ / $y^x$] 2 [=]

The display should show 16.

It is very likely that the square root of a number will not be "nice." Your calculator can give you the *approximate* square root in such a case. To find, for example, the square root of 29, enter 29 [$\sqrt{\ }$], and the display may show 5.385164807. This is an *approximation* of the square root. It is rounded to the nearest billionth. The calculator cannot display the exact answer because there is no end to the sequence of digits (and also no pattern). If you round to the nearest tenth, your approximation for the square root of 29 is 5.4.

*Use your calculator to find the square root of each of the following.*

**36.** 64

**37.** 144

**38.** 289

**39.** 1,024

**40.** 1,849

**41.** 784

**42.** 8,649

**43.** 5,329

*Use your calculator to approximate the following square roots. Round to the nearest tenth.*

**44.** $\sqrt{23}$

**45.** $\sqrt{31}$

**46.** $\sqrt{51}$

**47.** $\sqrt{42}$

**48.** $\sqrt{134}$

**49.** $\sqrt{251}$

**50.** A baseball diamond is the shape of a square that has sides of length 90 feet. Find the distance from home plate to second base. Round to the nearest tenth.

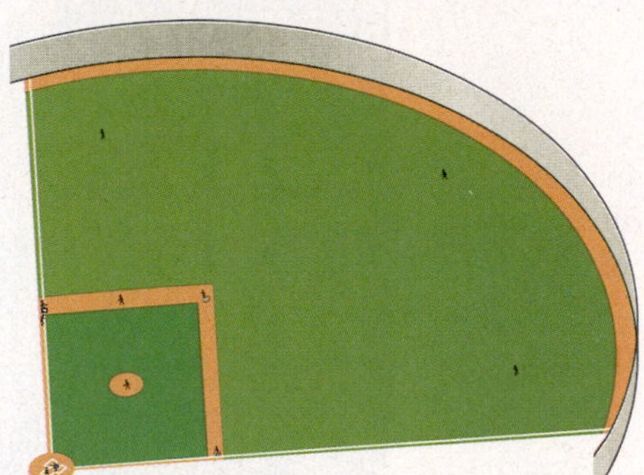

## Answers

50. ______

## Answers

**1.** 8 **3.** 13 **5.** *c* **7.** Yes **9.** No **11.** Yes **13.** 10 **15.** 15 **17.** (b) **19.** (a) **21.** False **23.** always **25.** 24 **27.** 12 **29.** 24 **31.** 26 in. **33.** Yes **35.** 5 ft **37.** 12 **39.** 32 **41.** 28 **43.** 73 **45.** 5.6 **47.** 6.5 **49.** 15.8

# Activity 23 ::
# The Pythagorean Theorem

In this activity, you and your group members will experiment with the Pythagorean theorem, one of the most famous results in all mathematics. This theorem is often used in construction to test that a manufactured (or built) right angle actually is a right angle.

For problems 1 and 2, we recommend that you work on graph paper to ensure that you have right angles.

**1.** Draw a right triangle of any size on your paper. Label the sides $a$, $b$, and $c$, where $c$ is the hypotenuse. With a metric ruler, measure and record the lengths of $a$, $b$, and $c$ to the nearest millimeter.

$a =$ $\quad$ $b =$ $\quad$ $c =$

Now compute: $\quad$ $a^2 =$ $\quad$ $b^2 =$ $\quad$ $c^2 =$

Compare the sum $a^2 + b^2$ with $c^2$. How close are they?

**2.** Repeat problem 1 with a completely new right triangle.

$a =$ $\quad$ $b =$ $\quad$ $c =$

Compute: $\quad$ $a^2 =$ $\quad$ $b^2 =$ $\quad$ $c^2 =$

How does the sum $a^2 + b^2$ compare to $c^2$?

**3.** Now draw a triangle that is *not* a right triangle. Again label the sides $a$, $b$, and $c$, where $c$ is the longest side. Measure and record as before:

$a =$ $\quad$ $b =$ $\quad$ $c =$

Compute: $\quad$ $a^2 =$ $\quad$ $b^2 =$ $\quad$ $c^2 =$

How does the sum $a^2 + b^2$ compare to $c^2$?

**4.** Locate an example of a right triangle on your campus and with a measuring tape find the lengths of the sides. Record and compute the following to see if it really is a right triangle.

$a =$ $\quad$ $b =$ $\quad$ $c =$

Compute: $\quad$ $a^2 =$ $\quad$ $b^2 =$ $\quad$ $c^2 =$

How does the sum $a^2 + b^2$ compare to $c^2$?

Do you conclude that the angle is in fact a right angle?

| Definition/Procedure | Example | Reference |
|---|---|---|
| **Area and Circumference** | | Section 8.1 |
| The **circumference** of a circle is the distance around that circle. The **radius** is the distance from the center to a point on the circle. The **diameter** is twice the radius.<br>The circumference is found by the following:<br>$C = \pi d = 2\pi r$<br>where $\pi$ is approximately 3.14.<br>The area of a circle is found by the following:<br>$A = \pi r^2$ | If $r = 4.5$ cm, then<br>$C = 2\pi r$<br>$\approx (2)(3.14)(4.5)$<br>$\approx 28.3$ cm (rounded)<br>If $r = 4.5$ cm, then<br>$A = \pi r^2$<br>$\approx (3.14)(4.5)^2$<br>$\approx 63.6\ \text{cm}^2$ (rounded) | p. 553 |
| The area of a parallelogram is found by the following:<br>$A = b \cdot h$<br>The area of a triangle is found by the following:<br>$A = \frac{1}{2} \cdot b \cdot h$<br>To convert from one unit of area to another, use these equivalents:<br>1 square foot = 144 square inches<br>1 square yard = 9 square feet<br>1 acre = 4,840 square yards = 43,560 $\text{ft}^2$<br>1 square mile = 640 acres | 1.8 ft<br>3.2 ft<br>$A = 3.2\ \text{ft} \times 1.8\ \text{ft}$<br>$= 5.76\ \text{ft}^2$<br>2.3 ft<br>3.4 ft<br>$A = \frac{1}{2} \times 3.4\ \text{ft} \times 2.3\ \text{ft}$<br>$= 3.91\ \text{ft}^2$<br>$180\ \text{ft}^2 = 180\ \text{ft}^2 \times \frac{1\ \text{yd}^2}{9\ \text{ft}^2}$<br>$= 20\ \text{yd}^2$ | p. 556 |
| **Lines and Angles** | | Section 8.2 |
| **Line** A series of points that goes on forever.<br>**Line segment** A piece of a line that has two endpoints. | A B<br>C D | p. 568 |
| **Angle** A geometric figure consisting of two line segments that share a common endpoint. | C<br>O D | p. 569 |

Continued

| Definition/Procedure | Example | Reference |
|---|---|---|
| **Perpendicular lines** Lines are *perpendicular* if they intersect to form four equal angles.<br>**Parallel lines** Lines are *parallel* if they never intersect. | These lines are perpendicular. 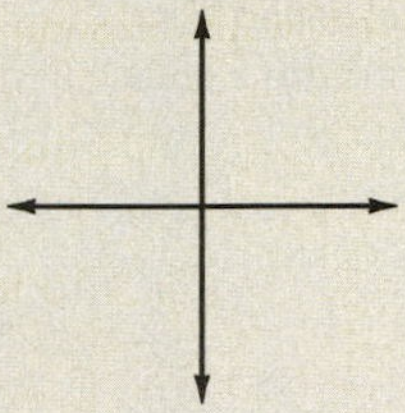 | *p.* 569 |
| **Acute angles** have a measure less than 90°.<br>**Obtuse angles** have a measure between 90° and 180°.<br>**Right angles** have a measure of 90°. | $\angle CEF$ is obtuse. 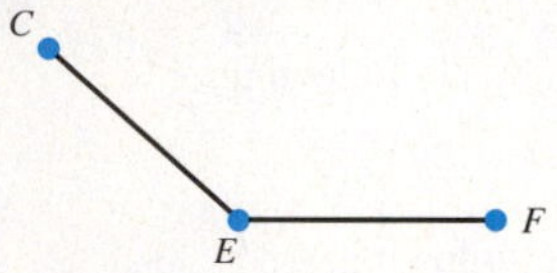<br>C E F | *pp.* 570–571 |
| **Straight angles** have a measure of 180°. |  A O B | *p.* 571 |
| **Complementary angles** Two angles are complementary if the sum of their measures is 90°. | 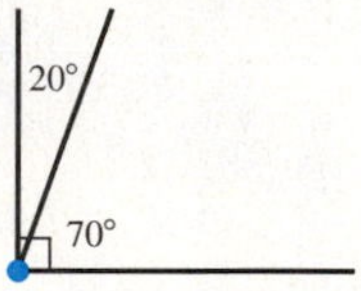 20° 70° | *p.* 574 |
| **Supplementary angles** Two angles are supplementary if the sum of their measures is 180°. | 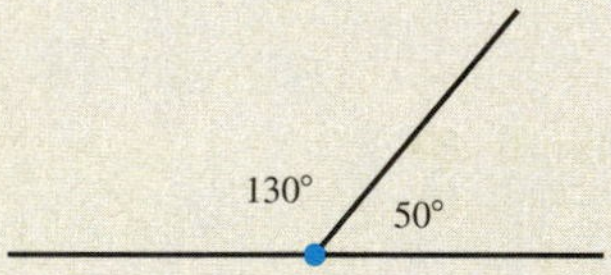 130° 50° | *p.* 574 |
| **Vertical angles** When two lines intersect, two pairs of vertical angles are formed. Vertical angles have equal measures. | 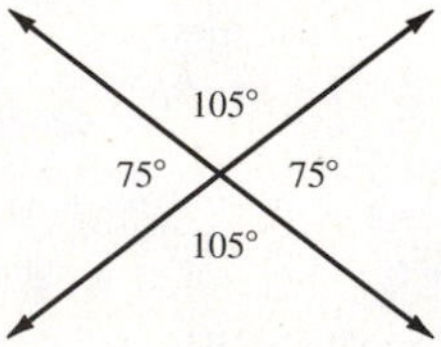 105° 75° 75° 105° | *p.* 575 |

| Definition/Procedure | Example | Reference |
|---|---|---|
| **Parallel lines and a transversal** When two parallel lines are intersected by a transversal,<br>**1.** Alternate interior angles have equal measures.<br>**2.** Corresponding angles have equal measures. | x, y, z<br>$m\angle x = m\angle y$<br>$m\angle y = m\angle z$ | p. 577 |
| **Triangles**<br>**Acute triangle** A triangle is *acute* if each angle measures less than 90°. | B, A, C | **Section 8.3**<br>p. 585 |
| **Obtuse triangle** A triangle is *obtuse* if an angle measures between 90° and 180°. | C, B, A | p. 585 |
| **Equilateral triangle** A triangle is *equilateral* if all three angles have the same measure. | B, A, C | p. 586 |
| **Isosceles triangle** A triangle is *isosceles* if two angles have the same measure. | H, F, G | p. 586 |

*Continued*

| Definition/Procedure | Example | Reference |
|---|---|---|
| **Scalene triangle** A triangle is *scalene* if no two angles have the same measure. | D, E, C | p. 586 |
| **Similar triangles** Two triangles are *similar* if the measures of the three angles in the two different triangles are the same. | C, 80°, A, 30°, 70°, B; E, 80°, D, 30°, 70°, F. $\Delta ACB$ and $\Delta DEF$ are similar triangles. | p. 588 |
| **Square Roots and the Pythagorean Theorem** | | Section 8.4 |
| The square root of a number is a value that, when squared, gives us that number. The length of the three sides of a right triangle will form a perfect triple. | 4, 5, 3 $3^2 + 4^2 = 5^2$ | p. 601 |



 

This summary exercise set is provided to give you practice with each of the objectives of this chapter. Each exercise is keyed to the appropriate chapter section. When you are finished, you can check your answers to the odd-numbered exercises against those presented in the back of the text. If you have difficulty with any of these questions, go back and reread the examples from that section. The answers to the even-numbered exercises appear in the *Instructor's Solutions Manual.* Your instructor will give you guidelines on how to best use these exercises in your instructional setting.

**8.1** *Find the perimeter or circumference of each figure. Use 3.14 for $\pi$. Round answers to the nearest tenth.*

**1.**

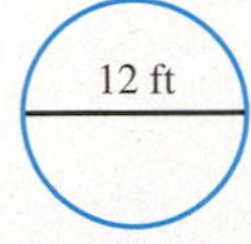


**2.**

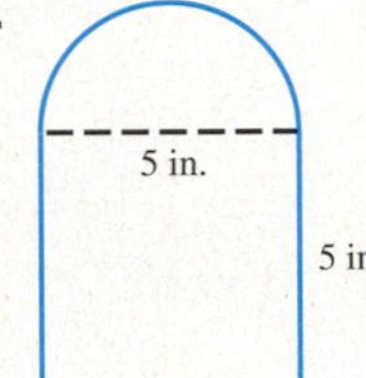


**3.** Find the area of a circle with radius 10 ft. Use 3.14 for $\pi$.

*Find the area of each figure.*

**4.**

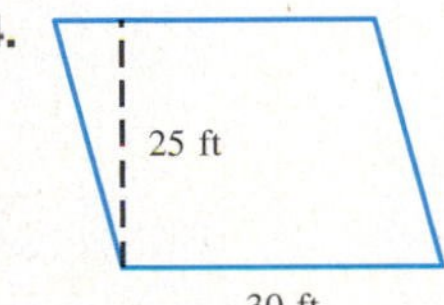


**5.**

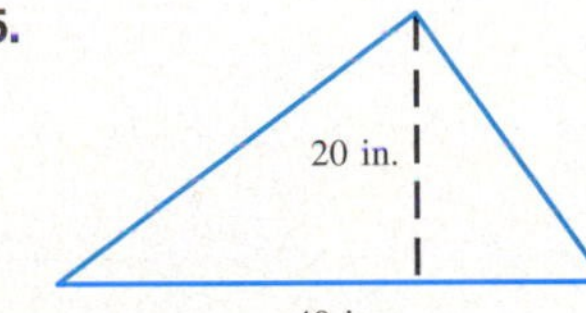


*Solve the applications.*

**6.** **Crafts** How many square feet of vinyl floor covering will be needed to cover the floor of a room that is 10 ft by 18 ft? How many square yards ($yd^2$) will be needed? (*Hint:* How many square feet are in a square yard?)

**7.** **Construction** A rectangular roof for a house addition measures 15 ft by 30 ft. A roofer will charge \$175 per "square" (100 $ft^2$). Find the cost of the roofing for the addition.

**8.2** *Name the angle; label it as acute, obtuse, right, or straight; then estimate its measure with a protractor.*

**8.**

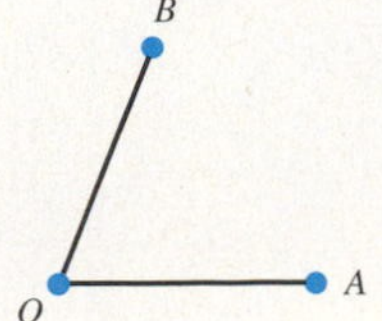


**9.**

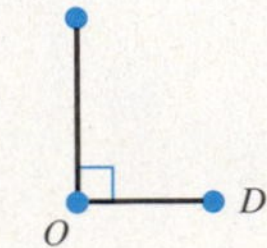

**10.**

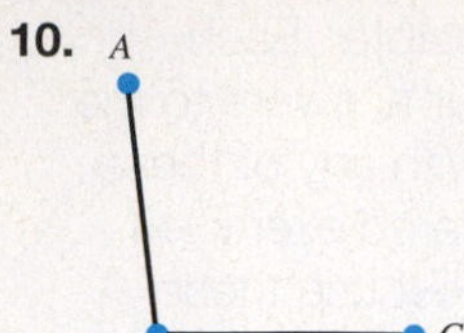


**11.**

R S T

**12.**

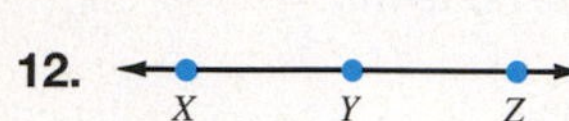


**13.**

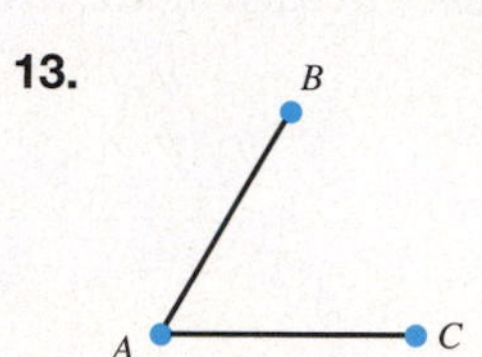


*Give the measure of each angle in degrees.*

**14.** $\angle A$ represents $\frac{3}{8}$ of a complete circle.

**15.** $\angle B$ represents $\frac{7}{10}$ of a complete circle.

**16.** If $m\angle x = 43°$, find the complement of $\angle x$.

**17.** If $m\angle y = 82°$, find the supplement of $\angle y$.

*Find $m\angle x$.*

**18.**

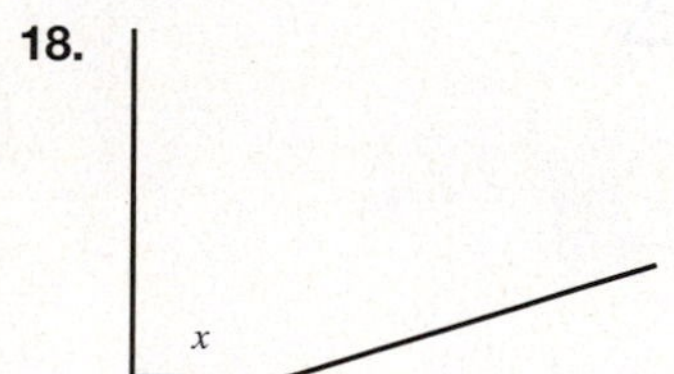
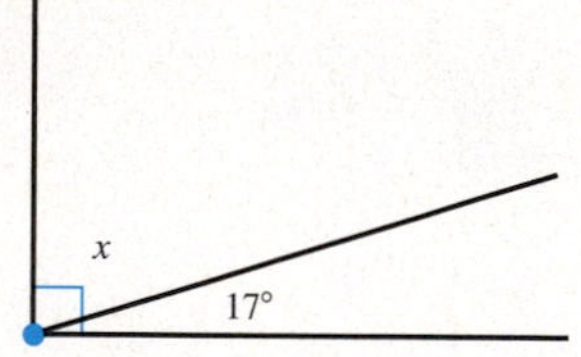


**19.**

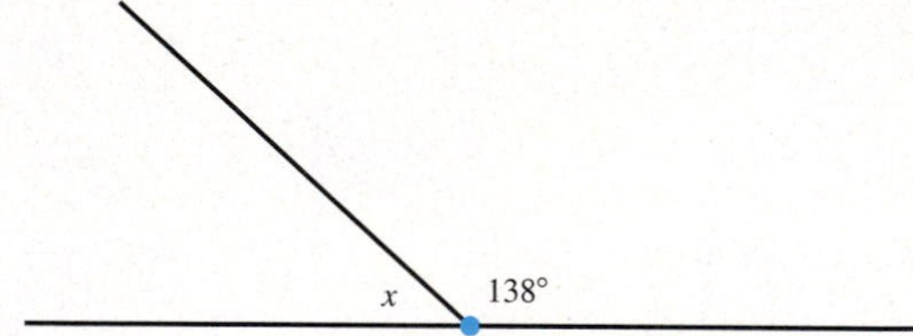


**20.**

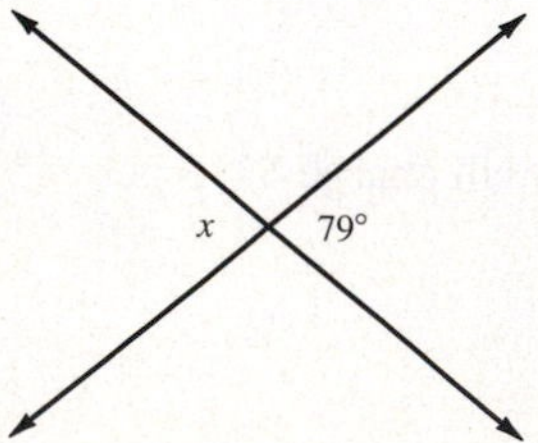


**21.**

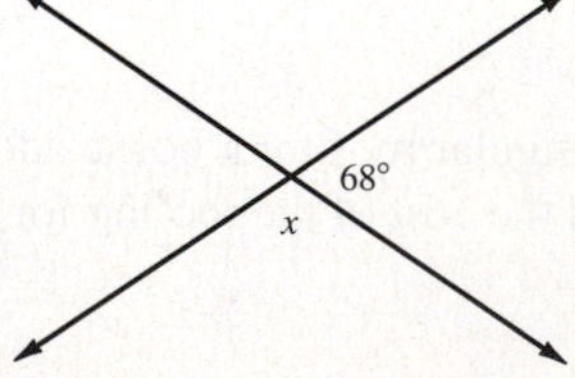


**22.**

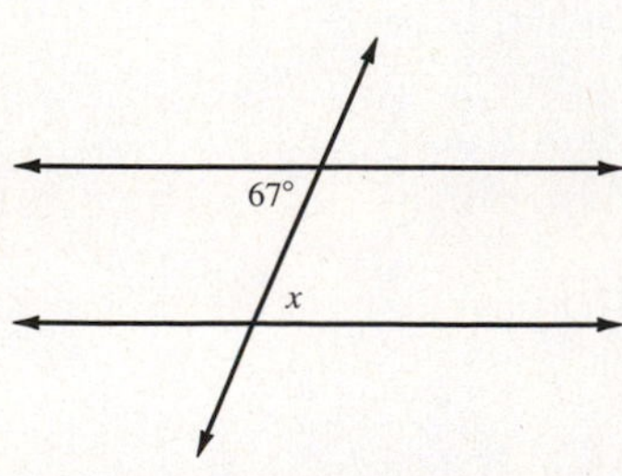


**23.**

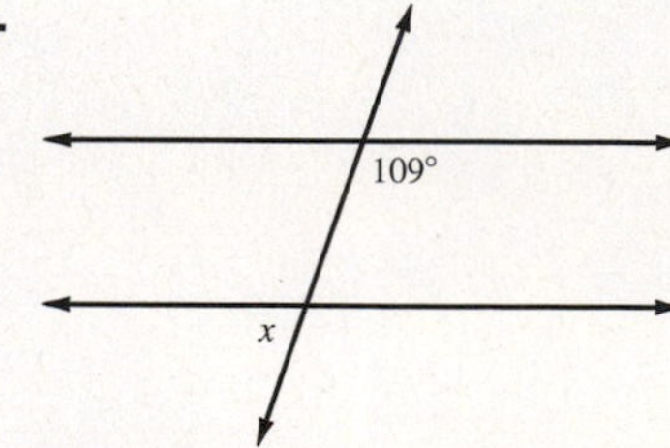

8.3 *Find the missing angle and then label the triangle as equilateral, isosceles, or scalene.*

**24.**

**25.**

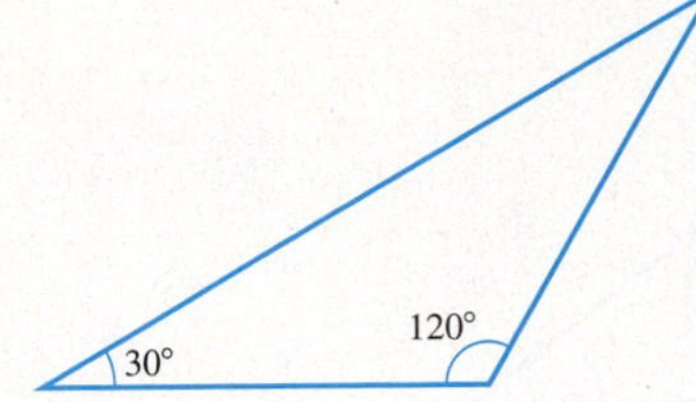


**26.**

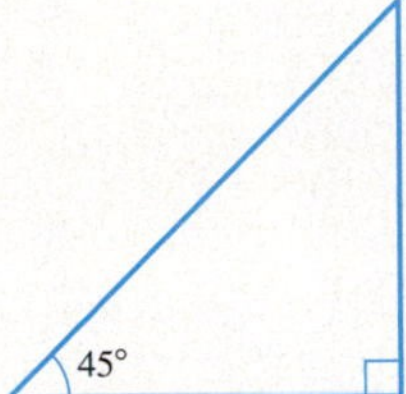


**27.**

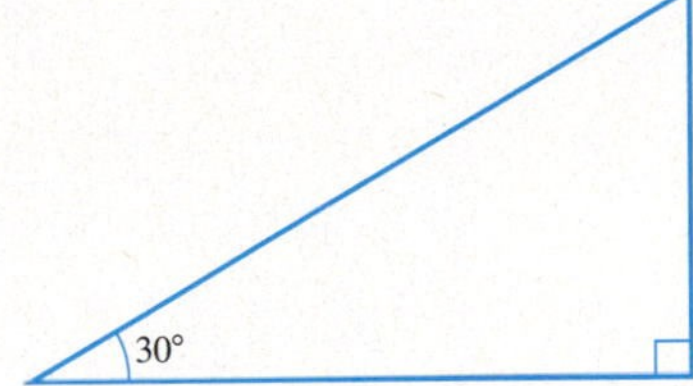


8.3 *Find the indicated side. Round results to the nearest tenth.*

**28.** Given: $\triangle MNO$ is similar to $\triangle PQR$. Find the length of $\overline{QR}$.

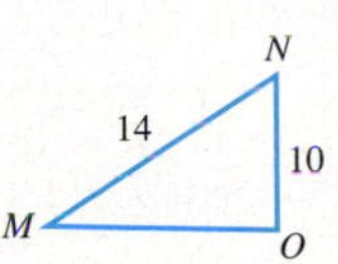


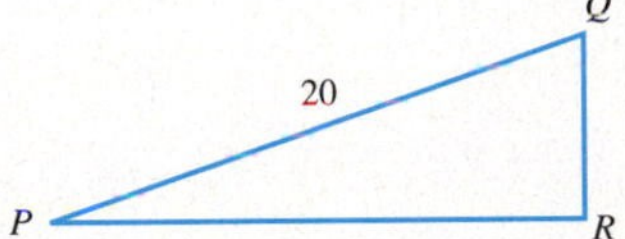


**29.** Given: $\triangle SUT$ is similar to $\triangle VWX$. Find the length of $\overline{WX}$.

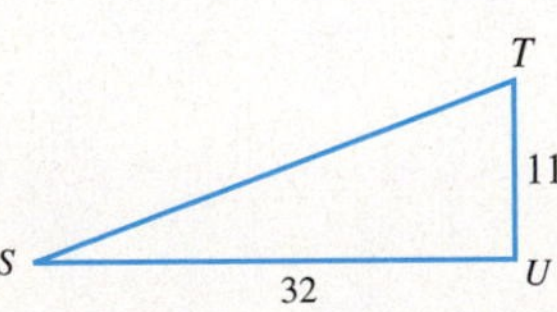


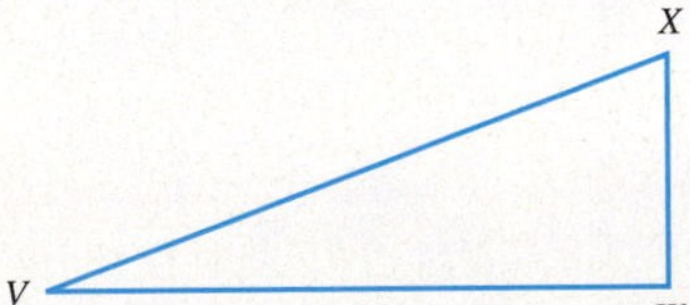


**30.** A tree casts a shadow that is 11.2 m long at the same time that a 4.0-m pole casts a shadow that is 1.4 m long. How tall is the tree?

8.4 *Solve each of the following square roots. Where necessary, round to the nearest hundredth.*

**31.** $\sqrt{324}$

**32.** $\sqrt{784}$

**33.** $\sqrt{189}$

**34.** $\sqrt{91}$

*Find the length of the unknown side.*

**35.**

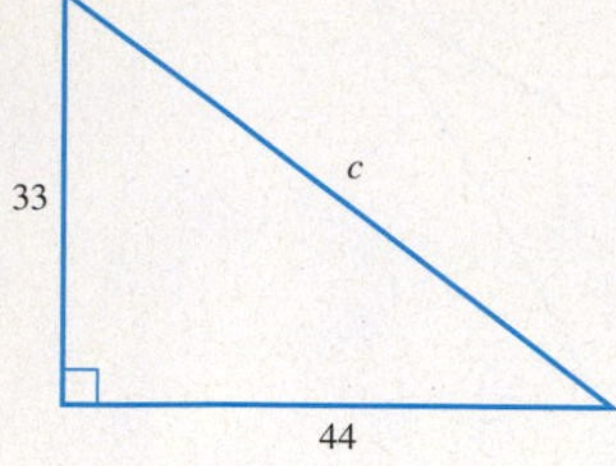


**36.**

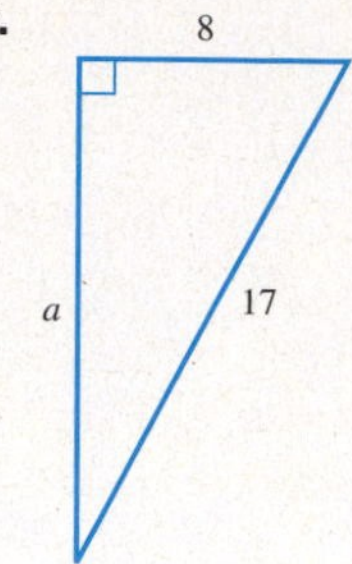

# self-test 8

Name ______________________

Section ________ Date ________

The purpose of this chapter test is to help you check your progress so that you can find sections and concepts that you need to review before the next exam. Allow yourself about an hour to take this test. At the end of that hour, check your answers against those given in the back of this text. If you missed any, note the section reference that accompanies the answer. Go back to that section and reread the examples until you have mastered that particular concept.

**Answers**

1. ______________________

2. ______________________

3. ______________________

4. ______________________

1. **Geometry** We find the circumference of a circle by multiplying the diameter of the circle by 3.14. If a circle has a diameter of 3.2 ft, find its circumference, to the nearest hundredth of a foot.

2. Label each pair of lines as parallel, perpendicular, or neither.

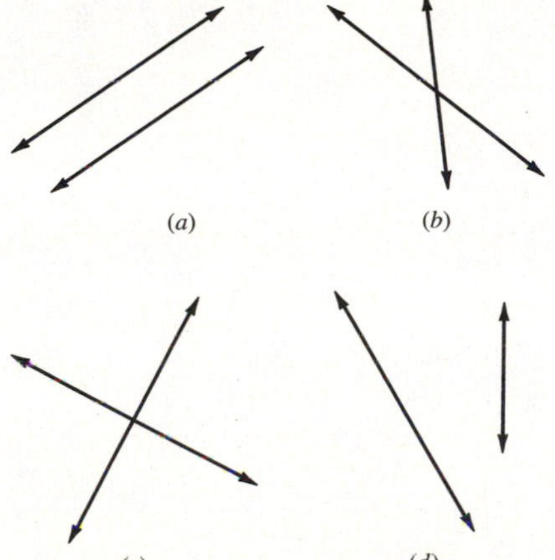


3. Label each of the angles as acute, obtuse, right, or straight.

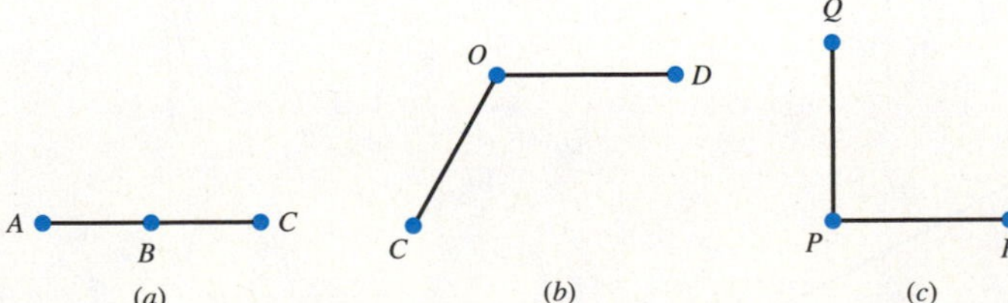


*Use a protractor to estimate the measurement of the angles.*

4.

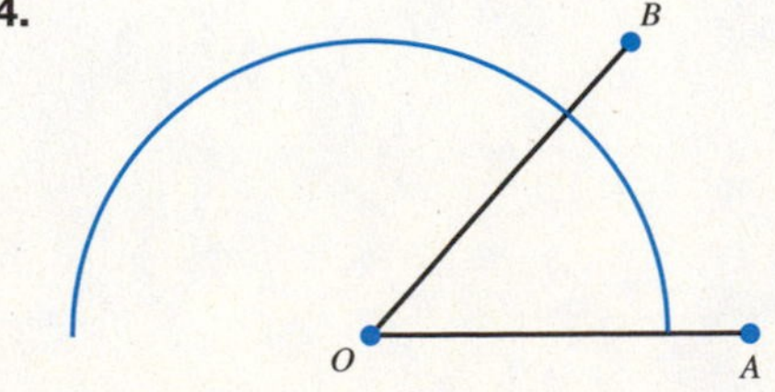

**Answers**

5. ____________

6. ____________

7. ____________

8. ____________

9. ____________

10. ____________

**5.**

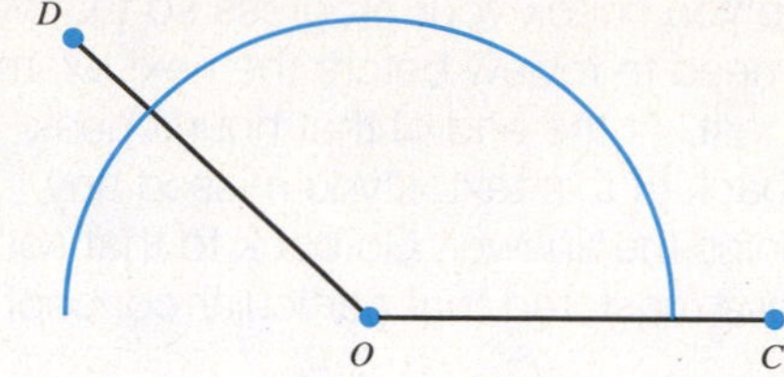


**6.** If $\angle A$ represents $\frac{5}{6}$ of a complete circle, give the measure of $\angle A$ in degrees.

*Find $m\angle x$.*

**7.**

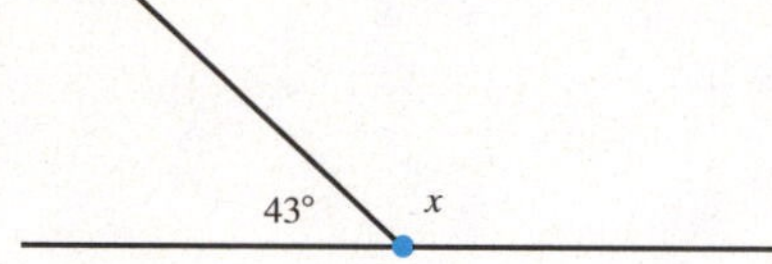


**8.**

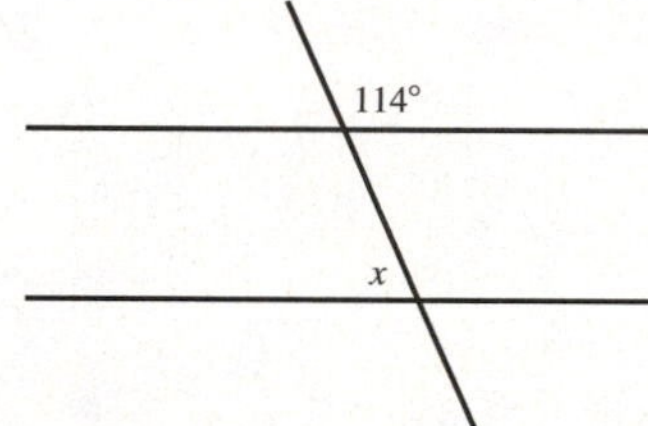


*Label the triangles as acute, obtuse, or right.*

**9.**

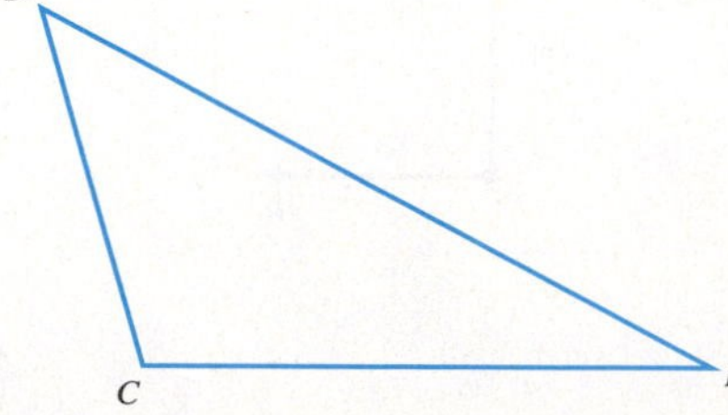


**10.**

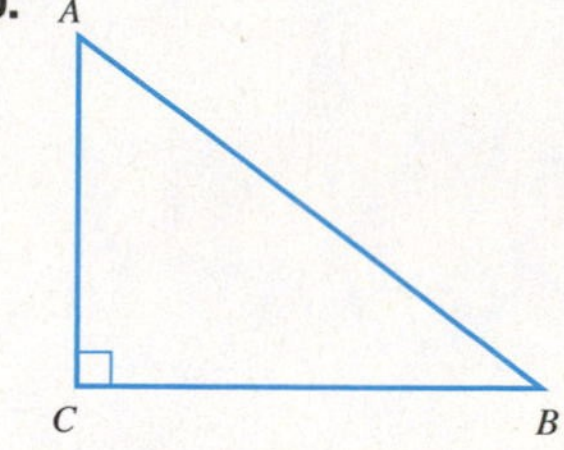

*Label the triangles as equilateral, isosceles, or scalene.*

**11.**

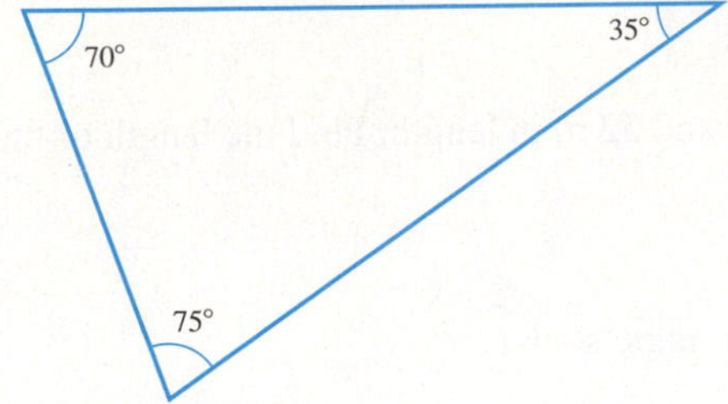


**12.**

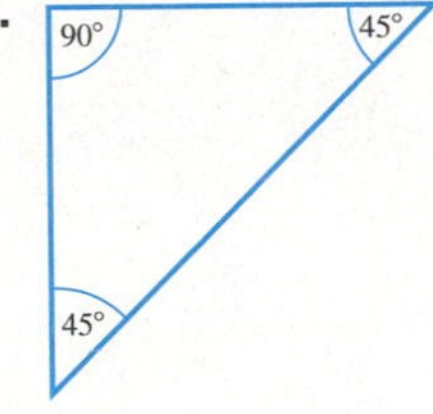


**13.** Find the two triangles that are similar.

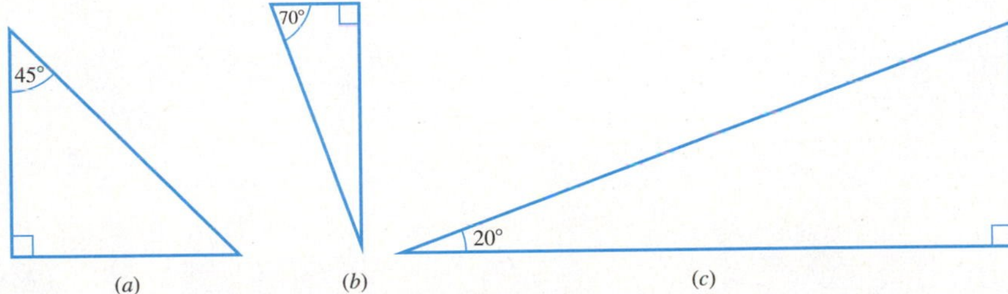


**14.** Find $m\angle A$.

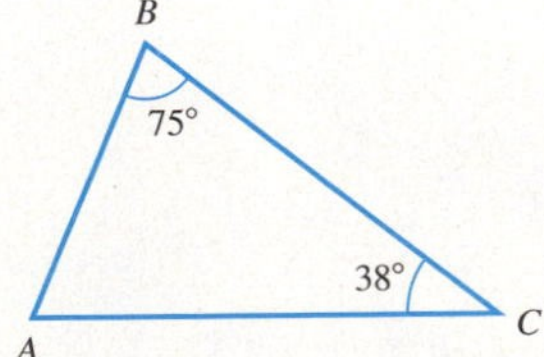


**15.** Given that $\triangle DEF$ is similar to $\triangle GHI$, find the length of $\overline{GH}$. Round to the nearest tenth.

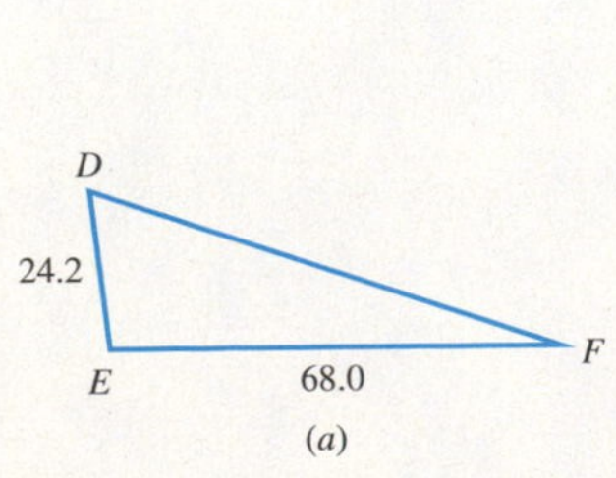


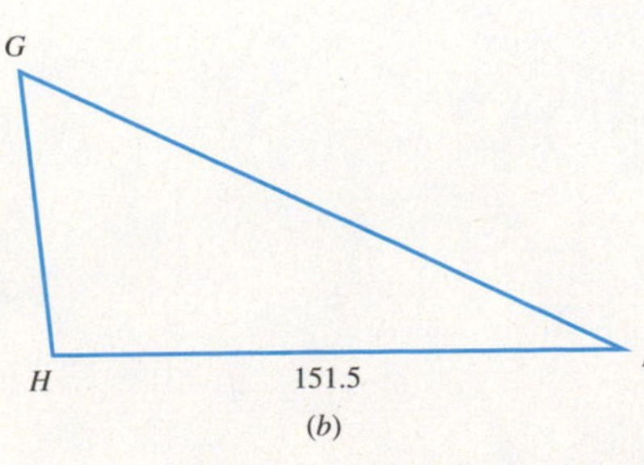


## Answers

11. ______________

12. ______________

13. ______________

14. ______________

15. ______________

Answers

16. ______________

17. ______________

18. ______________

**16.** Find the square root of 441.

**17.** The legs of a right triangle are 39 m and 52 m in length. Find the length of the hypotenuse.

**18.** Find the perimeter of the isosceles triangle shown.

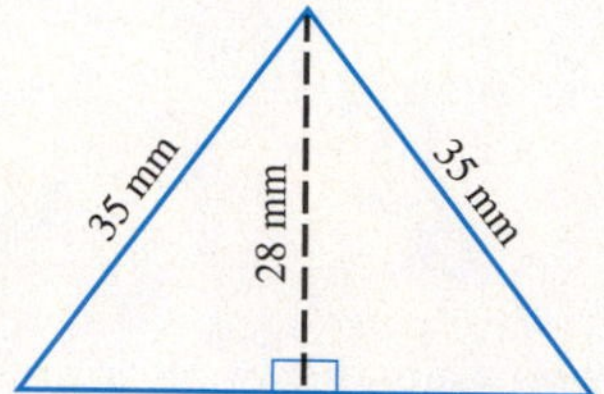

# cumulative review chapters 1-8

Name ________________

Section __________ Date __________

The following exercises are presented to help you review concepts from earlier chapters. This is meant as review material and not as a comprehensive exam. The answers are presented in the back of the text. Section references accompany the answers. If you have difficulty with any of these exercises, be certain to at least read through the summary related to those sections.

**1.** Give the place value of 6 in the number 4,865,201.

**2.** Evaluate: $82 - 2 \times 3^2$

**3.** Find the prime factorization for 630.

**4.** Find the greatest common factor (GCF) of 20, 24, and 32.

**5.** Find the least common multiple (LCM) of 20 and 24.

**6.** Multiply: $\frac{2}{3} \times 1\frac{4}{5} \times \frac{5}{8}$

**7.** Divide: $2\frac{5}{8} \div \frac{7}{12}$

**8.** Your living room measures $5\frac{2}{3}$ yd by $4\frac{1}{4}$ yd. If carpeting costs \$24 per square yard (yd$^2$), what will it cost to carpet the room?

**9.** If you drive 270 miles in $4\frac{1}{2}$ hours, what is your average speed?

**10.** Add: $\frac{3}{5} + \frac{1}{6} + \frac{2}{3}$

**11.** Subtract: $7\frac{3}{8} - 3\frac{5}{6}$

**12.** Adam's goal is to run 20 miles per week. So far he has run distances of $3\frac{1}{2}$ mi, $4\frac{2}{3}$ mi, and $5\frac{1}{4}$ mi. How much more must he do to reach his goal?

## Answers

1. ________________
2. ________________
3. ________________
4. ________________
5. ________________
6. ________________
7. ________________
8. ________________
9. ________________
10. ________________
11. ________________
12. ________________

**Answers**

13. ____

14. ____

15. ____

16. ____

17. ____

18. ____

19. ____

20. ____

21. ____

22. ____

23. ____

24. ____

*Find the indicated place values.*

**13.** 3 in 17.2396

**14.** 5 in 8.0915

**15.** Find the perimeter and area of the given figure.

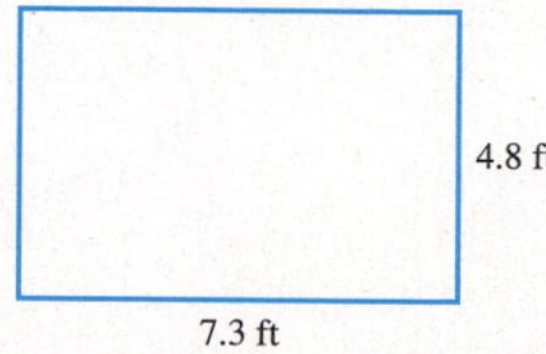


**16.** Write 0.125 as a simplified fraction.

**17.** Write the decimal equivalent of $\frac{5}{16}$.

**18.** Express the rate in simplest form: $\frac{81{,}000 \text{ dollars}}{4 \text{ years}}$

**19.** Solve for the unknown: $\frac{35}{14} = \frac{10}{w}$

**20.** If one gallon of paint covers 250 square feet ($ft^2$), how many square feet will $4\frac{1}{2}$ gallons cover?

**21.** Write 8.5% as a decimal.

**22.** Write 37.5% as a fraction.

**23.** Write $\frac{27}{40}$ as a percent.

**24.** Find 16% of 320.

25. 35% of what number is 525?

26. The number of students at a certain high school dropped by 6% since last year. There are now 1,269 students. How many were there last year?

*Complete each statement.*

27. 3 mi = ______________ yd

28. 250 mg = ______________ g

29. 5.8 km = ______________ m

30. Find the circumference of a circle whose radius is 7.9 ft. Use 3.14 for $\pi$ and round the result to the nearest tenth of a foot.

*Find the area for each figure. Use 3.14 for $\pi$.*

31.

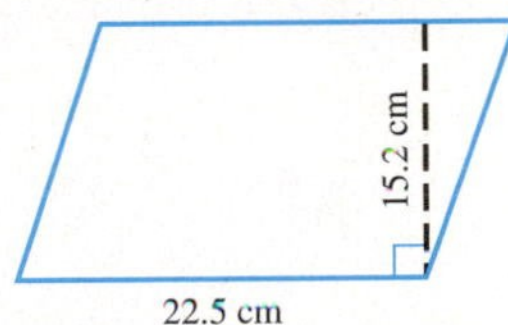


32.

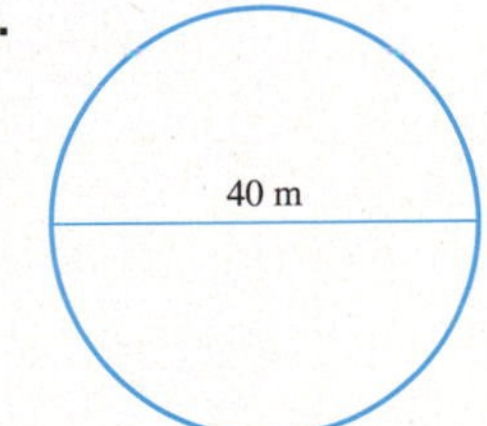


33.

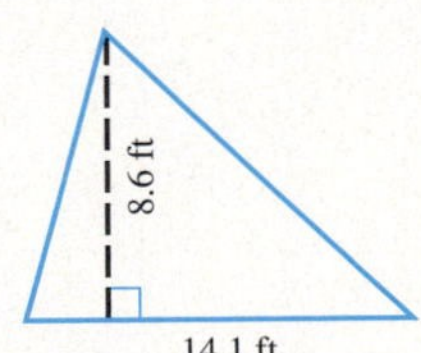


34. Use a protractor to find the measure of the given angle.

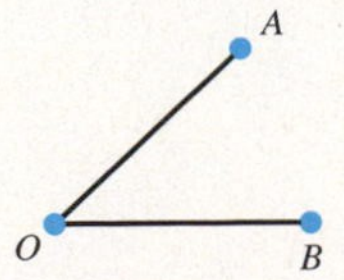


**Answers**

25. ______________

26. ______________

27. ______________

28. ______________

29. ______________

30. ______________

31. ______________

32. ______________

33. ______________

34. ______________

Answers

35. ______________________

36. ______________________

37. ______________________

**35.** Find the missing angle and identify the triangle as equilateral, isosceles, or scalene.

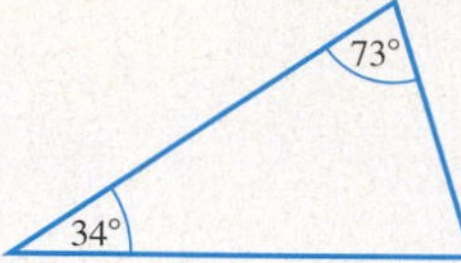


**36.** The given triangles are similar. Find $x$.

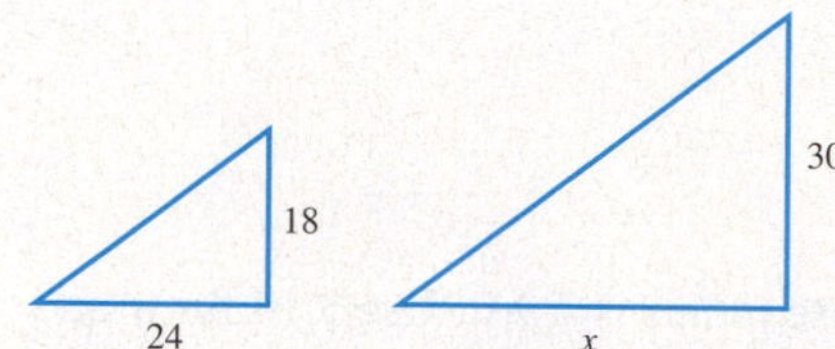


**37.** If two legs of a right triangle have lengths 8 ft and 15 ft, find the length of the hypotenuse.

# Data Analysis and Statistics

## INTRODUCTION

Alana is a statistical analyst for a small company in Woods Hole, Massachusetts. This particular company manufactures buoys that are used in the ocean to measure different properties of the water such as water temperature and how much salt is present. Alana tests the buoys for different properties such as strength and buoyancy. After all the research is compiled, she meets with the biotechnology team and they analyze and discuss ways to improve and streamline their models. Alana and the biotechnologists work together to create listings and data sets that will be used to recognize certain patterns. The team will also create graphs to assess the quality of the data and then perform statistical tests that will enable them to make certain assumptions about the buoys.

When all the research is analyzed and improvements are made, the buoys are ready to be sold and distributed to the marine biology institutions that will use them. The marine scientists will place them in various locations in the ocean, where the buoys will collect certain data and transmit the findings to satellites. The marine scientists collect this information to research and study their own data pertaining to the ocean. You will get a sense of how this is accomplished when you do Activity 26 on page 700.

## CHAPTER 9 OUTLINE

Name ____________________

Section ________ Date ________

This pretest provides a preview of the types of exercises you will encounter in each section of this chapter. The answers for these exercises can be found in the back of the text. If you are working on your own, or ahead of the class, this pretest can help you identify the sections in which you should focus more of your time.

**Answers**

1. ____________________
2. ____________________
3. ____________________
4. ____________________
5. ____________________
6. ____________________

9.1 **1.** Find the mean, median, and mode of the following set of numbers.

12, 16, 17, 18, 24, 24, 29, 33, 42, 42, 42

9.2 The total commercial fishery landings at the Cameron, LA, port (in millions of pounds, rounded) for each of the years 1997 to 2001 is shown in the following table. Use this table to work exercises 2 and 3.

**Commercial Fishery Landings; Cameron, LA**

| Year | 1997 | 1998 | 1999 | 2000 | 2001 | 2002 |
|---|---|---|---|---|---|---|
| Landings | 380 | 257 | 406 | 415 | 324 | 350 |

*Source:* National Marine Fisheries Service, Fisheries Statistics and Economics Division.

**2.** What was the increase in the commercial catch from 1997 to 2000?

**3.** What was the decrease in the commercial catch from 2000 to 2001?

9.3 Use the following graph to work exercises 4 to 6.

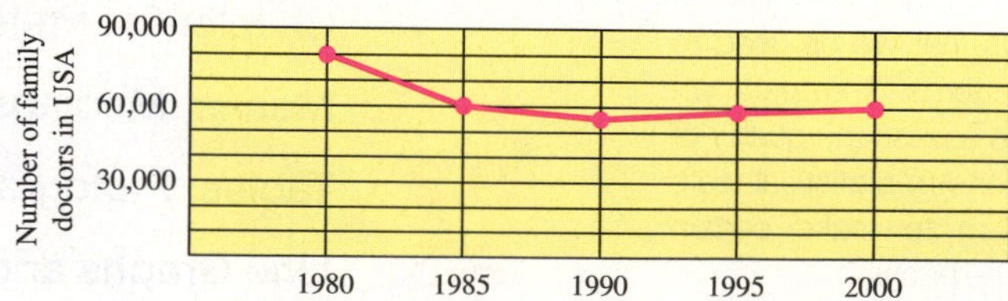


**4.** How many fewer family doctors were there in the United States in 1990 than in 1980?

**5.** What was the total change in the number of family doctors between 1980 and 2000?

**6.** In what 5-year period was the decrease in family doctors the greatest? What was the decrease?

**9.4** The following pie chart represents the portion of the worldwide $40,000,000,000 tourist industry that each country accounts for.

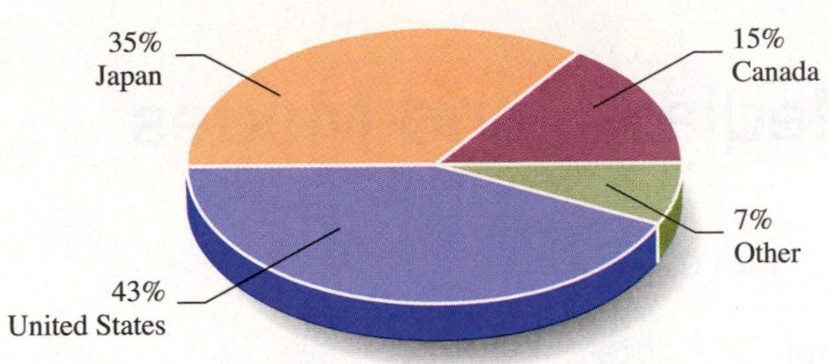


**7.** How many U.S. dollars are included?

**8.** How many non-U.S. dollars are included?

**9.5** The following table gives pump prices for medium-grade gas at a local station over a 2-week period (rounded up).

| Day | Mon. | Tues. | Wed. | Thu. | Fri. | Sat. | Sun. |
|---|---|---|---|---|---|---|---|
| **Week 1** | $1.80 | $1.83 | $1.85 | $1.85 | $1.88 | $1.92 | $1.88 |
| **Week 2** | $1.86 | $1.82 | $1.78 | $1.74 | $1.76 | $1.78 | $1.83 |

**9.** Give a five-number summary of the data shown in the table.

**10.** Construct a box-and-whiskers plot of the data given by the table.

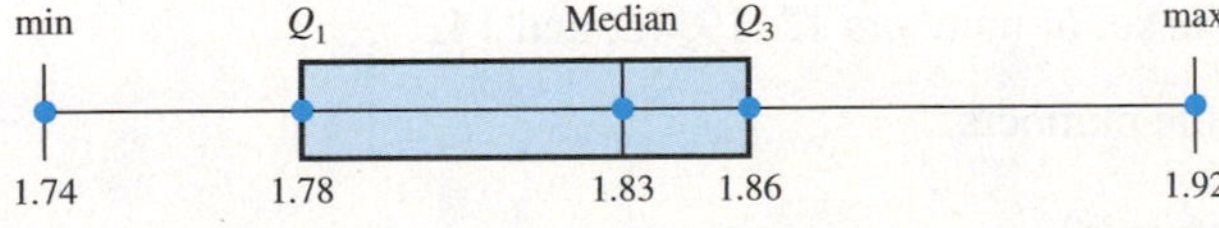


**Answers**

7. ______

8. ______

9. ______

10. ______

# 9.1 Means, Medians, and Modes

< 9.1 Objectives >

1 > Calculate the mean of a data set

2 > Find the median of a data set

3 > Compare the mean and median of a data set

4 > Find the mode of a data set

A very useful concept is the **average** of a group of numbers. An average is a number that is typical of a larger group of numbers. In mathematics we have several different kinds of averages that we can use to represent a larger group of numbers. The first of these is the **mean.**

Step by Step

**Finding the Mean**

To find the mean of a set of numbers:

Step 1 Add all the numbers in the set.

Step 2 Divide that sum by the number of items in the set.

**Example 1** Finding the Mean

< Objective 1 >

Find the mean of the set of numbers 12, 19, 15, and 14.

**Step 1** Add all the numbers.

$12 + 19 + 15 + 14 = 60$

**Step 2** Divide that sum by the number of items.

$60 \div 4 = 15$ There are four items in this group.

The mean of this set is 15.

**Check Yourself 1**

Find the mean of the set of numbers 17, 24, 19, and 20.

Next, we apply the concept of mean to a word problem.

## Example 2 Finding the Mean

The ticket prices (in dollars) for the nine concerts held at the Civic Arena this school year were

33, 31, 30, 59, 32, 35, 32, 36, 56

What was the mean price for these tickets?

**Step 1** Add all the numbers.

> **NOTE**
>
> We round to the nearest hundredth because we are dealing with money.

33 + 31 + 30 + 59 + 32 + 35 + 32 + 36 + 56 = 344

**Step 2** Divide by 9.

344 ÷ 9 = 38.22 Divide by 9 because there are 9 ticket prices.

The mean ticket price was $38.22.

### Check Yourself 2

The costs (in dollars) of the six textbooks that Aaron needs for the fall quarter are

75, 69, 57, 87, 76, 80

Find the mean cost of these books.

> **NOTE**
>
> The median is the middle value when the numbers are arranged in order.

Although the mean is probably the most common way to find an average for a set of numbers, it is not always the most representative. Another kind of average is called the **median.**

Step by Step

### Finding the Median

The *median* is the number for which there are as many instances that are above that number as there are instances below it. To find the median, follow these steps:

**Step 1** Rewrite the numbers in order from smallest to largest.

**Step 2** Count from both ends to find the number in the middle.

**Step 3** If there are two numbers in the middle, add them together and find their mean.

## Example 3 Finding the Median

< Objective 2 >

Find the median for the following sets of numbers.

**(a)** 35, 18, 27, 38, 19, 63, 22

**Step 1** Rewrite the numbers in order from smallest to largest.

18, 19, 22, 27, 35, 38, 63

**Step 2** Count from both ends to find the number in the middle.

Counting from both ends, we find that 27 is the median. There are three numbers above 27 and three numbers below it.

**(b)** 29, 88, 73, 81, 62, 37

**Step 1** Rewrite the numbers in order from smallest to largest.

29, 37, 62, 73, 81, 88

**Step 2** Count from both ends to find the number in the middle.

Counting from both ends, we find that there are two numbers in the middle, 62 and 73. We go on to step 3.

**Step 3** If there are two numbers in the middle, find their mean.

$$(62 + 73) \div 2 = 135 \div 2 = 67\frac{1}{2}$$

### Check Yourself 3

**Find the median for each set of numbers.**

**(a)** 8, 6, 19, 4, 21, 5, 27 **(b)** 43, 29, 13, 38, 29, 53

At times the median is a better representative of a set of numbers than the mean is. Example 4 illustrates such a case.

## Example 4 Comparing the Mean and the Median

< Objective 3 >

The following numbers represent the hourly wage of seven employees of a local chip manufacturing plant.

12, 11, 14, 16, 32, 13, 14

**(a)** Find the mean hourly wage.

**Step 1** Add all the numbers in the set.

$$12 + 11 + 14 + 16 + 32 + 13 + 14 = 112$$

**Step 2** Divide that sum by the number of items in the set.

$$112 \div 7 = 16$$

The mean wage is $16 an hour.

**(b)** Find the median wage for the seven workers.

**Step 1** Rewrite the numbers in order from smallest to largest.

11, 12, 13, 14, 14, 16, 32

**Step 2** Count from both ends to find the number in the middle.

The middle number is 14. There are three numbers equal to or above it and three numbers below it. The median salary is $14 per hour. Which salary do you think is more typical of the workers? Why?

**Check Yourself 4**

The following are Jessica's phone bills for each month of 2003.

26, 67, 31, 24, 15, 17, 41, 27, 17, 22, 26, 47

**(a)** Find the mean amount of her phone bills.
**(b)** Find the median amount of her phone bills.

**NOTE**

A set with two different modes is called **bimodal.**

Another measure used as an "average" is called the *mode.*

**Definition**

**Mode** The *mode* of a set of data is the item or number that appears most frequently.

### Example 5 Finding a Mode

< Objective 4 >

Find the mode for the set of numbers given.

22, 24, 24, 24, 24, 27, 28, 32, 32

The mode, 24, is the number that appears most frequently.

**Check Yourself 5**

Find the mode for the set of numbers given.

7, 7, 7, 9, 11, 13, 13, 15, 15, 15, 15, 21

One advantage of the mode is that it can be used with data that are not numbers.

### Example 6 Finding a Mode

**NOTE**

If each eye color appeared three times, there would be no mode! Not every data set has a mode.

Following are the eye colors from a class of 12 students. Which color is the mode?

blue, brown, hazel, blue, brown, brown, brown, brown, blue, brown, hazel, green

Because brown occurs most frequently, it is the mode.

### Check Yourself 6

**The following types of computers were available in the lab. Which type was the mode?**

**Apple, IBM, Compaq, Dell, Apple, IBM, Apple, Compaq, Dell, Apple, IBM, Apple, Dell, Apple, Compaq**

## Example 7 A Vocational Application

A pregnant, adult female patient tested positive for gestational diabetes during the last 3 months of her pregnancy. Blood glucose levels (in milligrams per 100 milliliters) were gathered and recorded on a regular basis. The results are tabulated here.

| | | | | | | |
|---|---|---|---|---|---|---|
| 78 | 104 | 103 | 101 | 78 | 120 | 103 |
| 97 | 75 | 128 | 90 | 98 | 80 | 128 |
| 119 | 106 | 83 | 99 | 101 | 108 | 127 |
| 118 | 96 | 125 | 78 | 123 | 124 | 92 |

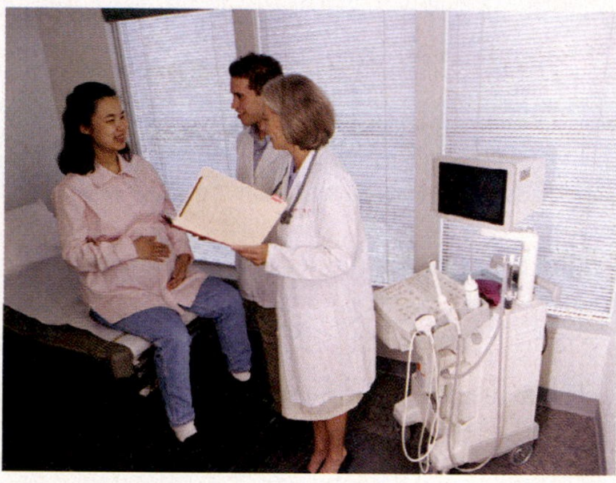

Compute the mean.

To do this, we must first find the sum of the numbers. In this case, a calculator (or computer) proves very useful, since there are 28 numbers to add. The sum of these is 2,882. Dividing by 28 produces 102.928 . . . . To the nearest whole number, the mean glucose level is 103.

### Check Yourself 7

**For the data given in Example 7, find the median.**

### Check Yourself ANSWERS

**1.** 20 **2.** $74 **3.** **(a)** 8; **(b)** $33\frac{1}{2}$ **4.** **(a)** $30; **(b)** $26 **5.** 15 **6.** Apple **7.** 102

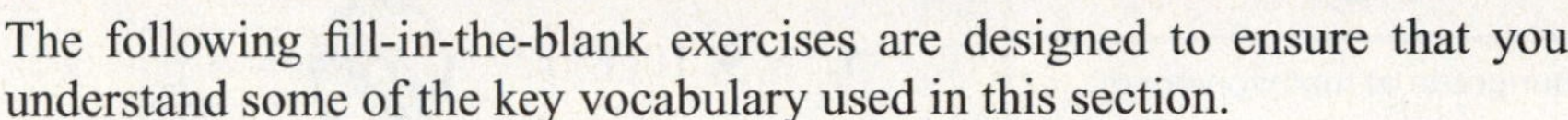

## Reading Your Text

The following fill-in-the-blank exercises are designed to ensure that you understand some of the key vocabulary used in this section.

SECTION 9.1

**(a)** To find the ______________ of a set of numbers, add the numbers in the set and divide by the number of items in the set.

**(b)** The ______________ is the middle value when the numbers are arranged in order.

**(c)** The ______________ of a set of data is the item or number that appears most frequently.

**(d)** A set with two different modes is called ______________.

# 9.1 exercises

Basic Skills | Advanced Skills | Vocational-Technical Applications | Calculator/Computer | Above and Beyond

MathZone

Boost your grade at mathzone.com!
- > Practice Problems
- > NetTutor
- > Self-Tests
- > e-Professors
- > Videos

Name ______________

Section ________ Date ________

## Answers

1. ______
2. ______
3. ______
4. ______
5. ______
6. ______ 7. ______
8. ______ 9. ______
10. ______ 11. ______
12. ______ 13. ______
14. ______ 15. ______
16. ______ 17. ______
18. ______ 19. ______
20. ______ 21. ______
22. ______ 23. ______
24. ______ 25. ______

< Objective 1 >

*Find the mean of each set of numbers.*

**1.** 6, 9, 10, 8, 12

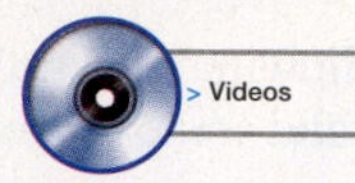


**2.** 13, 15, 17, 17, 18

**3.** 13, 15, 17, 19, 24, 25

**4.** 41, 43, 56, 67, 69, 72

**5.** 12, 14, 15, 16, 16, 16, 17, 22, 25, 27

**6.** 21, 25, 27, 32, 36, 37, 43, 43, 44, 51

**7.** 5, 8, 9, 11, 12

**8.** 7, 18, 11, 7, 12

**9.** 9, 8, 11, 14, 9

**10.** 21, 23, 25, 27, 22, 20

< Objective 2 >

*Find the median of each set of numbers.*

**11.** 2, 3, 5, 6, 10

**12.** 12, 13, 15, 17, 18

**13.** 23, 24, 27, 31, 36, 38

**14.** 1, 4, 9, 15, 25, 36

**15.** 46, 13, 47, 25, 68, 51

**16.** 26, 71, 33, 69, 71, 25, 75

< Objective 4 >

*Find the mode of each set of numbers.*

**17.** 17, 13, 16, 18, 17

**18.** 41, 43, 56, 67, 69, 72

**19.** 21, 44, 25, 27, 32, 36, 37, 44

**20.** 9, 8, 10, 9, 9, 10, 8

**21.** 12, 13, 7, 14, 4, 11, 9

**22.** 8, 2, 3, 3, 4, 9, 9, 3

**23.** The following are eye colors from a class of eight students. Which color is the mode?

hazel, green, brown, brown, blue, green, hazel, green

**24.** The weather in Philadelphia over the last 7 days was as follows:

rain, sunny, cloudy, rain, sunny, rain, rain

What type of weather was the mode?

*Solve the following applications.*

**25.** **Statistics** High temperatures of 86°, 91°, 92°, 103°, and 98°F were recorded for the first 5 days of July. What was the mean high temperature?

> Videos

chapter 9 > Make the Connection

26. **Statistics** A salesperson drove 238, 159, 87, 163, and 198 miles (mi) on a 5-day trip. What was the mean number of miles driven per day?

27. **Statistics** Highway mileage ratings for seven new diesel cars were 43, 29, 51, 36, 33, 42, and 32 miles per gallon (mi/gal). What was the mean rating?

28. **Statistics** The enrollments in the four elementary schools of a district are 278, 153, 215, and 198 students. What is the mean enrollment?

*Determine whether each statement is* **true** *or* **false.**

29. The mode can only be found for sets of numbers.

30. The mean can be larger than most of the data.

*In each of the following statements, fill in the blank with* **always, sometimes,** *or* **never.**

31. The mean and the median are __________ the same number.

32. To find the median, you must ________ put the numbers in order.

Basic Skills | **Advanced Skills** | Vocational-Technical Applications | Calculator/Computer | Above and Beyond

33. **Statistics** To get an A in history, you must have a mean of 90 on five tests. Your scores thus far are 83, 93, 88, and 91. How many points must you have on the final test to receive an A? (*Hint:* First find the total number of points you need to get an A.)

34. **Statistics** To pass biology, you must have a mean of 70 on six quizzes. So far your scores have been 65, 78, 72, 66, and 71. How many points must you have on the final quiz to pass biology?

35. **Statistics** Louis had scores of 87, 82, 93, 89, and 84 on five tests. Tamika had scores of 92, 83, 89, 94, and 87 on the same five tests. Who had the higher mean score? By how much?

## Answers

26. ____________
27. ____________
28. ____________
29. ____________
30. ____________
31. ____________
32. ____________
33. ____________
34. ____________
35. ____________

Answers

36. ______

37. ______

38. ______

39. ______

40. ______

41. ______

36. **Statistics** The Wong family had heating bills of \$105, \$110, \$90, and \$67 in the first 4 months of 2003. The bills for the same months of 2004 were \$110, \$95, \$75, and \$76. In which year was the mean monthly bill higher? By how much?

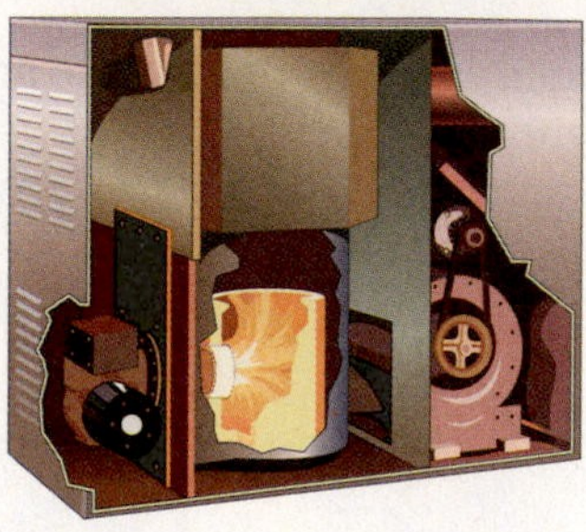

*Monthly energy use, in kilowatt-hours (kWh), by appliance type for four typical U.S. families is shown in the following table.*

| | Wong Family | McCarthy Family | Abramowitz Family | Gregg Family |
|---|---|---|---|---|
| Electric range | 97 | 115 | 80 | 96 |
| Electric heat | 1,200 | 1,086 | 1,103 | 975 |
| Water heater | 407 | 386 | 368 | 423 |
| Refrigerator | 127 | 154 | 98 | 121 |
| Lights | 75 | 99 | 108 | 94 |
| Air conditioner | 123 | 117 | 96 | 120 |
| Color TV | 39 | 45 | 21 | 47 |

37. **Science and Medicine** What is the mean number of kilowatt-hours used each month by the four families for heating their homes?

38. **Science and Medicine** What is the mean number of kilowatt-hours used each month by the four families for hot water?

39. **Science and Medicine** What is the mean number of kilowatt-hours used per appliance by the McCarthy family?

40. **Science and Medicine** What is the mean number of kilowatt-hours used per appliance by the Gregg family?

Basic Skills | Advanced Skills | **Vocational-Technical Applications** | Calculator/Computer | Above and Beyond

41. **Information Technology** Response times in milliseconds (ms) from your computer to a local router using ping are given by the following table.

| | |
|---|---|
| 2.2 | 2.5 |
| 2.3 | 2.4 |
| 1.9 | 2.2 |
| 2.0 | 2.4 |
| 2.5 | 2.5 |

**(a)** Compute the mean. **(b)** Compute the mode.
**(c)** Compute the median.

**42. Construction** The thicknesses (in mm) of several parts are as follows:

30.9, 30.7, 29, 30.6, 29.3, 31.2, 29.3

**(a)** Calculate the mean. Round to the nearest hundredth.
**(b)** Find the median.
**(c)** Find the mode.

**43. Construction** The following are the tensile strengths (MP) for several parts tested.

2,836, 2,861, 2,839, 2,849, 2,843, 2,812, 2,854

**(a)** Calculate the mean.
**(b)** Find the median.
**(c)** Find the mode.

**44. Automotive** Early in 2005, twenty gas stations from around the United States were surveyed to determine the price of regular unleaded gasoline. The raw data are as follows:

| | | | |
|---|---|---|---|
| 2.14 | 1.93 | 2.16 | 2.19 |
| 1.99 | 2.36 | 2.25 | 2.02 |
| 2.48 | 2.31 | 2.04 | 2.15 |
| 2.33 | 2.55 | 2.22 | 2.24 |
| 2.18 | 2.19 | 2.38 | 2.33 |

**(a)** Calculate the mean gas price.
**(b)** Calculate the median gas price.

**45. Welding** Throughout the day, welds are randomly chosen and tested for strength. The results are shown in the following table. [Tensile strength can be expressed in pounds per square inch (lb/in.$^2$).]

| | | | |
|---|---|---|---|
| 2,314 | 2,318 | 2,307 | 2,291 |
| 2,289 | 2,301 | 2,320 | 2,318 |
| 2,322 | 2,297 | 2,314 | 2,296 |
| 2,309 | 2,311 | 2,304 | 2,321 |

**(a)** Calculate the mean tensile strength.
**(b)** Calculate the median tensile strength.

## Answers

42. ______

43. ______

44. ______

45. ______

Basic Skills | Advanced Skills | Vocational-Technical Applications | **Calculator/Computer** | Above and Beyond

This explanation is followed by a set of exercises for which the calculator might be the preferred tool. As indicated by the placement of this explanation, you should refrain from using a calculator on the exercises that precede this.

Many calculators have built-in statistical functions that allow you to calculate the mean and median of a data set. Because these features vary widely between calculators, you will need to consult your instructor or the owner's manual for your calculator.

**Answers**

46. ________________

47. ________________

48. ________________

49. ________________

50. ________________

51. ________________

52. ________________

53. ________________

54. ________________

55. ________________

56. ________________

You can compute the mean of a data set without using built-in statistical functions as shown.

To compute the mean of the set: 2,253, 3,451, 2,157, 4,126, 967

**Step 1** Press the *clear* key. [CLEAR]

**Step 2** Press the *open parenthesis* key. [(]

**Step 3** Enter the numbers, separated by *plus* signs. 2253 [+] 3451 [+] 2157 [+] 4126 [+] 967

**Step 4** Press the *close parenthesis* key. [)]

**Step 5** Enter *division* and the number of items. [÷] 5

**Step 6** Press *enter* or equals [ENTER] or [=]

Your display should read 2590.8.

*Use your calculator to find the mean of each set of numbers.*

**46.** 48, 50, 51, 52, 49, 50

**47.** 20, 18, 17, 24, 22, 19

**48.** 346, 351, 353, 347, 341, 382, 373, 363

**49.** 1,560, 1,540, 1,570, 1,555, 1,565, 1,545, 1,557

**50.** 16,430, 15,487, 17,982, 11,290, 21,908, 16,545

**51.** 311,431, 286,356, 356,090, 292,007, 301,857, 299,005

**52.** 18, 21, 20, 22

**53.** 356, 371, 366, 373, 359, 363

**54.** 1,898, 1,913, 1,875, 1,937

**55.** 15,865, 16,270, 16,090, 15,904

**56.** **Business and Finance** The revenue for the leading apparel companies in the United States in 1997 is given in the following table.

| Company | Revenue (in millions) |
|---|---|
| Nike | $9,187 |
| Vanity Fair | 5,222 |
| Reebok | 3,637 |
| Liz Claiborne | 2,413 |
| Fruit of the Loom | 2,140 |
| Nine West | 1,865 |
| Kellwood | 1,521 |
| Warmaio | 1,437 |
| Jones Apparel | 1,387 |

What is the mean revenue taken in by these companies?

**57. Business and Finance**

| Unemployment in the United States (in thousands) | | |
|---|---|---|
| **Year** | **Employed** | **Unemployed** |
| 1993 | 120,259 | 8,940 |
| 1994 | 123,060 | 7,996 |
| 1995 | 124,900 | 7,404 |
| 1996 | 126,708 | 7,236 |
| 1997 | 129,558 | 6,739 |
| 1998 | 131,463 | 6,210 |
| 1999 | 133,488 | 5,880 |
| 2000 | 136,891 | 5,692 |
| 2001 | 136,933 | 6,801 |
| 2002 | 136,485 | 6,378 |

*Source:* U.S. Department of Labor, Bureau of Labor Statistics.

Find the mean number of employed and unemployed people per year from 1993 to 2002. Round to the nearest thousand.

*The work stoppages (strikes and lockouts) in the United States from 1995 to 2002 are given in the following table.*

| Year | No. of Stoppages | Work Days Idle |
|---|---|---|
| 1995 | 192 | 5,771 |
| 1996 | 273 | 4,889 |
| 1997 | 339 | 4,497 |
| 1998 | 387 | 5,116 |
| 1999 | 73 | 1,996 |
| 2000 | 394 | 20,419 |
| 2001 | 99 | 1,151 |
| 2002 | 46 | 6,596 |

*Source:* U.S. Department of Labor, Bureau of Labor Statistics.

**58.** Find the mean number of work stoppages per year from 1995 to 2002.

**59.** Find the mean number of work days idle from 1995 to 2002.

Basic Skills | Advanced Skills | Vocational-Technical Applications | Calculator/Computer | **Above and Beyond**

< Objective 3 >

**60. Business and Finance** Fred kept the following records of his utility bills for 12 months: \$53, \$51, \$43, \$37, \$32, \$29, \$34, \$41, \$58, \$55, \$49, and \$58.

**(a)** Find the mean of Fred's monthly utility bills.

**(b)** Find the median of Fred's monthly utility bills.

**(c)** Is the mean or median a more useful representative of Fred's monthly utility bills? Write a brief paragraph justifying your response.

## Answers

57. ______

58. ______

59. ______

60. ______

Answers

61. ______

62. ______

63. ______

64. ______

65. ______

66. ______

67. ______

**61.** **Business and Finance** The following scores were recorded on a 200-point final examination: 193, 185, 163, 186, 192, 135, 158, 174, 188, 172, 168, 183, 195, 165, 183.

**(a)** Find the mean final examination score.

**(b)** Find the median final examination score.

**(c)** Is the mean or median a more useful representative of the final examination scores? Write a brief paragraph justifying your response.

**62.** List the advantages and disadvantages of the mean, median, and mode.

**63.** In a certain math class, you take four tests and the final, which counts as two tests. Your grade is the average of the six tests. At the end of the course, you compute both the mean and the median.

**(a)** You want to convince the teacher to use the mean to compute your average. Write a note to your teacher explaining why this is a better choice. Choose numbers that make a convincing argument.

**(b)** You want to convince the teacher to use the median to compute your average. Write a note to your teacher explaining why this is a better choice. Choose numbers that make a convincing argument.

**64.** Create a set of five numbers such that the mean is equal to the median.

**65.** Create a set of five numbers such that the mean is greater than the median.

**66.** Create a set of five numbers such that the mean is less than the median.

**67.** Write a paragraph describing the conditions necessary for the mean of a data set to be greater than the median, less than the median, and equal to the median. How do you think the mode would compare to the mean and median in each of these situations?

## Answers

**1.** 9 **3.** $18\frac{5}{6}$ **5.** 18 **7.** 9 **9.** 10.2 **11.** 5 **13.** 29 **15.** $46\frac{1}{2}$ **17.** 17 **19.** 44 **21.** No mode **23.** Green **25.** 94°F **27.** 38 mi/gal **29.** False **31.** sometimes **33.** 95 points **35.** Louis's mean score was 87, Tamika's was 89. Tamika's average score was 2 points higher than Louis's **37.** 1,091 kWh **39.** 286 kWh **41.** **(a)** 2.29 ms; **(b)** 2.5 ms; **(c)** 2.35 ms **43.** **(a)** 2,842 MP; **(b)** 2,843 MP; **(c)** none **45.** **(a)** 2,308.25 lb/in.$^2$; **(b)** 2,310 lb/in.$^2$ **47.** 20 **49.** 1,556 **51.** 307,791 **53.** $364\frac{2}{3}$ **55.** 16,032.25 **57.** Employed: 129,975,000; unemployed: 6,928,000 **59.** 6,304 work days **61.** **(a)** 176 points; **(b)** 183 points; **(c)** Above and Beyond **63.** Above and Beyond **65.** Answer will vary **67.** Above and Beyond

# Activity 24 ::
## Car Color Preferences

While we tend to use the mean and median to describe the center of a data set, people who work in marketing and manufacturing use the mode at least as often. In many such applications, the mode of a data set is the natural way to describe the center.

### Out-of-Class Component

You should find a safe spot to observe cars. This could be an intersection, street, or even a parking lot. Record the color of the first 10 cars you see. Use broad color categories (such as *blue*), rather than more specific categories (such as *light blue* and *dark blue*). Make a second list of the next 25 cars you see.

### In-Class Component

You should have two data sets: one list of 10 colors and one list of 25 colors.

**1.** Find the mode of each of the two data sets.

**2.** **(a)** Do the modes differ?

**(b)** If so, which do you feel is more accurate, and why?

**3.** Create a data set of 35 colors by combining your two lists. Find the mode of this data set.

**4.** The method you used to gather your data is called **convenience sampling.** Briefly describe why the method might take on that name.

**5.** **(a)** Describe two benefits and two weaknesses of convenience sampling as a method of gathering data.

**(b)** Statisticians generally avoid convenience sampling, believing its weaknesses outweigh its strengths. Briefly describe how you might create a sample of car colors that more accurately mirrors the preferences of the population as a whole.

**6.** Go to www.mhhe.com/streeter to find the car color preferences for the U.S. population. Compare and contrast your data sets with that described on the website.

# 9.2 Tables, Pictographs, and Bar Graphs

< 9.2 Objectives >

1 > Read a table

2 > Interpret a table

3 > Create a pictograph

4 > Read a bar graph

**NOTE**

Rows read left to right.
Columns read top to bottom.

A **table** is a display of information in parallel rows or columns. Tables can be used anywhere that information is to be summarized.

The following is a table describing land area and world population. Each entry in the table is called a **cell.** This table will be used for Examples 1 and 2.

| Continent or Region | Land Area (1,000 mi$^2$) | % of Earth | Population 1900 | Population 1950 | Population 2000 |
|---|---|---|---|---|---|
| North America | 9,400 | 16.2 | 106,000,000 | 221,000,000 | 305,000,000 |
| South America | 6,900 | 11.9 | 38,000,000 | 111,000,000 | 515,000,000 |
| Europe | 3,800 | 6.6 | 400,000,000 | 392,000,000 | 510,000,000 |
| Asia (including Russia) | 17,400 | 30.1 | 932,000,000 | 1,591,000,000 | 4,028,000,000 |
| Africa | 11,700 | 20.2 | 118,000,000 | 229,000,000 | 889,000,000 |
| Oceana (including Australia) | 3,300 | 5.7 | 6,000,000 | 12,000,000 | 32,000,000 |
| Antarctica | 5,400 | 9.3 | Uninhabited | — | — |
| World total | 57,900 | | 1,600,000,000 | 2,556,000,000 | 6,279,000,000 |

*Source:* Bureau of the Census, U.S. Dept. of Commerce.

**Example 1** Reading a Table

< Objective 1 >

From the land area and world population table, find each of the following.

**(a)** What was the population of Africa in 1950?

Looking at the cell that is in the row labeled Africa and the column labeled 1950, we find a population of 229,000,000.

**(b)** What is the land area of Asia in square miles?

The cell in the row Asia and column labeled land area says 17,400. But note that the column is labeled "1,000 $mi^2$." The land area is 17,400 thousand square miles, or 17,400,000 $mi^2$.

### Check Yourself 1

Use the land area and world population table to answer each question.

**(a)** What was the population of South America in 1900?
**(b)** What is the land area of Africa as a percent of Earth's land area?

We can frequently use a table to find answers to questions that are not directly answered as part of the table. Example 2 will illustrate.

## Example 2 Interpreting a Table

< Objective 2 >

Use the world population and land area table to answer each of the following questions.

**(a)** To the nearest tenth of a percent, what percent of the world's population was in North America in the year 2000?

Note that 305,000,000 of the world's 6,279,000,000 people lived in North America.

$$\frac{305,000,000}{6,279,000,000} \approx 0.04857 \approx 4.9\%$$

Note that, although North America has more than 16% of the Earth's land area, it has less than 5% of the world's population.

**(b)** What percent of the Earth's habitable land is in Asia?

First, we must decide what is meant by "habitable land." We will assume anything outside of Antarctica is habitable. To find the amount of habitable land, we take the total of 57,900,000 and subtract Antarctica's 5,400,000. This leaves total habitable land of 52,500,000 $mi^2$.

$$\frac{17,400,000}{52,500,000} \approx 0.3314 \approx 33.1\%$$

**(c)** What was the mean population for the six populated regions in 1900?

Although we could add the six numbers, note that they have already been totaled. Using that total, we find the average.

$$\frac{1,600,000,000}{6} \approx 267,000,000$$

## Check Yourself 2

Use the world population and land area table to answer each of the following questions.

**(a)** To the nearest percent, what was the increase in the population of Africa between 1950 and 2000?

**(b)** Did world population increase by a greater percent between 1900 and 1950 or between 1950 and 2000?

If a table has only one or two columns of numeric information, it is often easier to interpret in picture form. A **graph** is a diagram that represents the connection between two or more things. A graph that uses pictures to represent quantities is called a **pictograph.**

## Example 3 Creating a Pictograph

< Objective 3 >

Create a pictograph that displays the information in the following table.

**Cars Registered in the United States**

| Year | Cars Registered |
|---|---|
| 1970 | 89,243,557 |
| 1975 | 106,705,934 |
| 1980 | 121,600,843 |
| 1985 | 131,664,029 |
| 1990 | 143,549,627 |
| 1995 | 136,066,045 |
| 2000 | 133,621,245 |

*Source:* Federal Highway Admin., U.S. Dept. of Transportation.

First, we must decide what to use as our picture unit. A car is the obvious choice in this case.

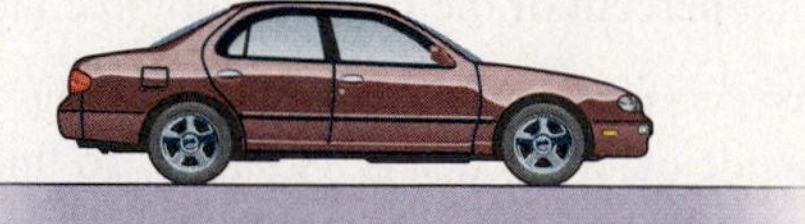

Second, we must decide the "value" of each car. Given the units in the table, we will let each car represent 10,000,000 registrations.

| Federal Highway Admin., U.S. Dept. of Transportation | |
|---|---|
| **Year** | **Cars Pictured** |
| 1970 | 9 |
| 1975 | $10\frac{2}{3}$ |
| 1980 | $12\frac{1}{6}$ |
| 1985 | $13\frac{1}{6}$ |
| 1990 | $14\frac{1}{3}$ |
| 1995 | $13\frac{2}{3}$ |
| 2000 | $13\frac{1}{3}$ |

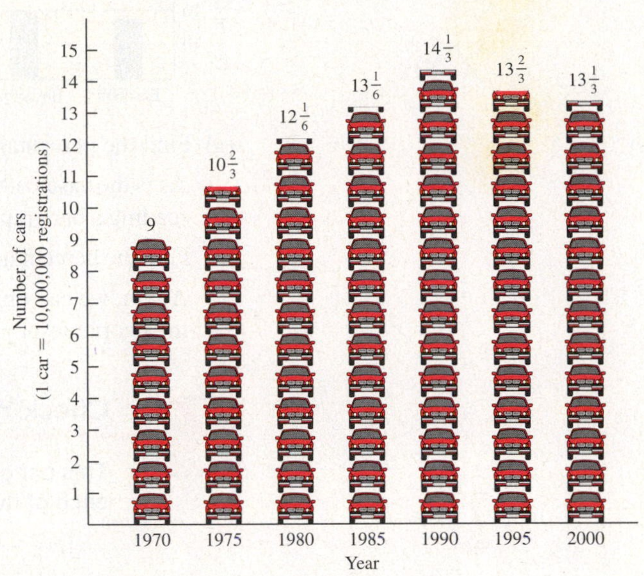

**Check Yourself 3**

Create a pictograph to represent the following table, which gives the percent of U.S. workers employed as farm workers.

| Economic Research Service, U.S. Dept. of Agriculture | |
|---|---|
| **Year** | **Percent of Workers in Farm Occupations** |
| 1800 | 73.2 |
| 1840 | 68.6 |
| 1880 | 57.1 |
| 1920 | 27.0 |
| 1960 | 6.1 |
| 2000 | 2.3 |

Other kinds of graphs also show the relationship of two sets of data. Perhaps the most common is the bar graph.

A bar graph is read in much the same manner as a pictograph.

**Example 4** **Reading a Bar Graph**

< Objective 4 >

The following bar graph represents the response to a 1995 Gallup poll that asked people what their favorite spectator sport was. In the graph, the information at the bottom describes the sport, and the information along the side describes the percentage of people surveyed. The height of the bar indicates the percentage of people who favor that particular sport.

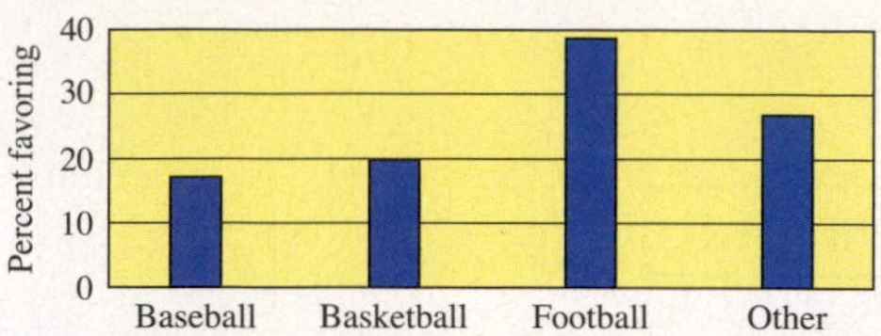


**(a)** Find the percentage of people for whom football is their favorite spectator sport.

As is the case with pictographs, we frequently have to estimate our answer when reading a bar graph. In this case, 38% would be a good estimate.

**(b)** Find the percentage of people for whom baseball is their favorite spectator sport.

Again, we can only estimate our answer. It appears to be approximately 17% of the people responding who favor baseball.

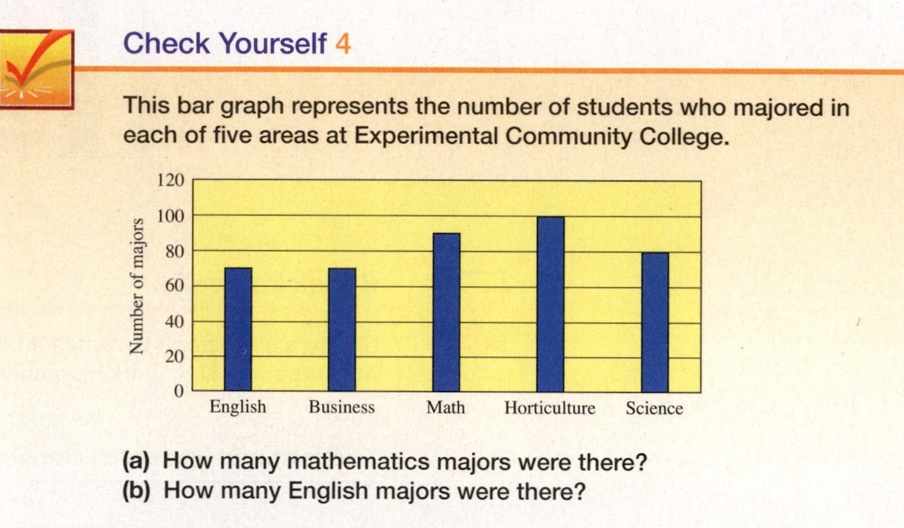


### Check Yourself 4

This bar graph represents the number of students who majored in each of five areas at Experimental Community College.

(a) How many mathematics majors were there?
(b) How many English majors were there?

Some bar graphs display additional information by using different colors or shading for different bars. With such graphs it is important to read the legend. The **legend** is the key that describes what each color or shade of the bar represents.

### Example 5 Reading the Legend of a Graph

The following bar graph represents the average student age at ECC.

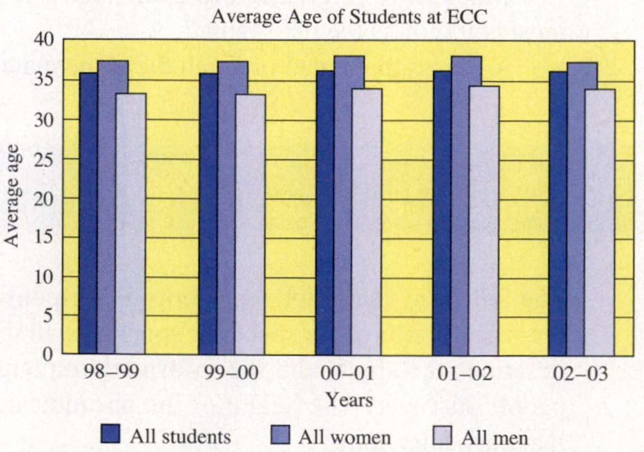

**(a)** What was the average age of female students in 2002–2003?

The legend tells us that the ages of all women are represented as the medium blue color. Looking at the height of the medium blue column for the year 2002–2003, we see the average age was about 37.

**(b)** Who tends to be older, male students or female students?

The medium blue bar is higher than the light blue bar in every year. Female students tend to be older than male students at ECC.

## Check Yourself 5

Use the graph in Example 5 to answer the following questions:

**(a)** Did the average age of female students increase or decrease between 2001–2002 and 2002–2003?
**(b)** What was the average age of male students in 2000–2001?

## Check Yourself ANSWERS

**1.** **(a)** 38,000,000; **(b)** 20.2% **2.** **(a)** 288%; **(b)** 1950–2000 (146% vs. 60%)

**3.**

**1 Worker = 10%**
**(Economic Research Service, U.S. Dept. of Agriculture)**

| Year | Workers Pictured |
|---|---|
| 1800 | $7\frac{1}{3}$ |
| 1840 | $6\frac{2}{3}$ |
| 1880 | $5\frac{2}{3}$ |
| 1920 | $2\frac{2}{3}$ |
| 1960 | $\frac{2}{3}$ |
| 2000 | $\frac{1}{4}$ |

(*continued*)

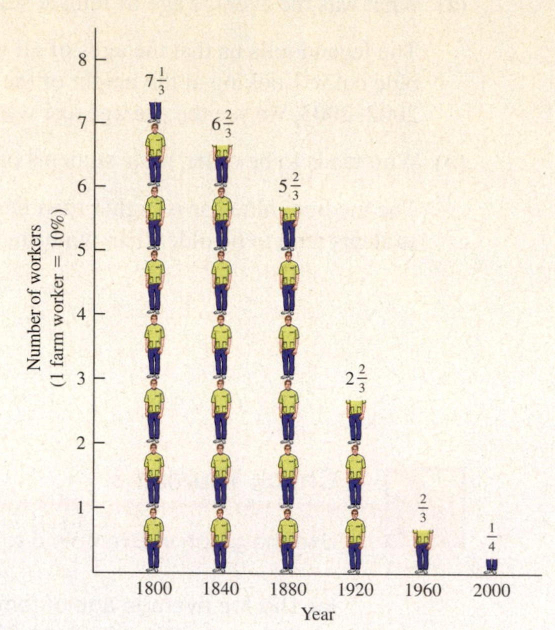


**4.** **(a)** 90; **(b)** 70 **5.** **(a)** It decreased; **(b)** 34

## Reading Your Text

The following fill-in-the-blank exercises are designed to ensure that you understand some of the key vocabulary used in this section.

### SECTION 9.2

**(a)** A table is a display of information in ______________ rows or columns.

**(b)** Each entry in the table is called a ______________.

**(c)** A graph that uses pictures to represent quantities is called a ___________.

**(d)** The ______________ is the key that describes what each color or shade of a bar (in a bar graph) represents.

**Basic Skills** | Advanced Skills | Vocational-Technical Applications | Calculator/Computer | Above and Beyond

# 9.2 exercises

< Objectives 1–2 >

*Use the world population and land area table reproduced here for exercises 1 to 10. Round answers to the nearest tenth or tenth of a percent.*

| Continent or Region | Land Area (1,000 mi²) | % of Earth | Population 1900 | Population 1950 | Population 2000 |
|---|---|---|---|---|---|
| North America | 9,400 | 16.2 | 106,000,000 | 221,000,000 | 305,000,000 |
| South America | 6,900 | 11.9 | 38,000,000 | 111,000,000 | 515,000,000 |
| Europe | 3,800 | 6.6 | 400,000,000 | 392,000,000 | 510,000,000 |
| Asia (including Russia) | 17,400 | 30.1 | 932,000,000 | 1,591,000,000 | 4,028,000,000 |
| Africa | 11,700 | 20.2 | 118,000,000 | 229,000,000 | 889,000,000 |
| Oceana (including Australia) | 3,300 | 5.7 | 6,000,000 | 12,000,000 | 32,000,000 |
| Antarctica | 5,400 | 9.3 | Uninhabited | — | — |
| World total | 57,900 | | 1,600,000,000 | 2,556,000,000 | 6,279,000,000 |

*Source:* Bureau of the Census, U.S. Dept. of Commerce.

**1.** **(a)** What was the population in North America in 1950?

**(b)** What is the total land area in North America as a percent of Earth?

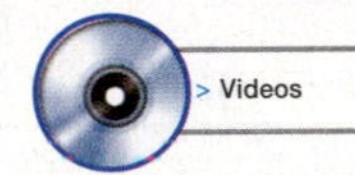


**2.** **(a)** What is the population of Europe in 2000?

**(b)** What is the total area of Europe?

**3.** **(a)** What was the percent increase in population in Asia from 1900 to 1950?

**(b)** What was the percent increase in population in Asia from 1950 to 2000?

**(c)** What was the population per square mile in Asia in 1950?

**(d)** What was the population per square mile in Asia in 2000?

**4.** Compare the population per square mile in Asia to the population per square mile in North America for the year 2000.

**5.** What was the mean population of all inhabited areas except Asia in 1950?

**6.** What is the percent increase in the population for all six inhabited continents, excluding Asia, from 1950 to 2000?

> Videos

**7.** **(a)** What percent of the Earth's inhabitable land is in North America?

**(b)** What percent of the world population in the year 2000 is in North America?

**8.** What was the percent increase in the population in South America from 1900 to 2000?

**MathZone**

Boost your grade at mathzone.com!

- > Practice Problems
- > NetTutor
- > Self-Tests
- > e-Professors
- > Videos

Name ______________

Section ________ Date ________

## Answers

1. ______________

2. ______________

3. ______________

4. ______________

5. ______________

6. ______________

7. ______________

8. ______________

Answers

9. ______

10. ______

9. (a) What was the number of people per square mile for the entire world in 1950?
   (b) What was the number of people per square mile for the entire world in 2000?
   (c) What was the percent increase in the number of people per square mile for the entire world from 1950 to 2000?

10. (a) What was the mean population of the six continents or land masses that were habitable in 2000?
   (b) What was the mean population in 1950?
   (c) What was the percent increase in the mean population from 1950 to 2000?

*In exercises 11 to 13, use the following table:*

**Gasoline Retail Prices, U.S. City Average, 1978–2002**

| | (cents per gallon, including taxes) | | | |
|---|---|---|---|---|
| **Average** | **Leaded Regular** | **Unleaded Regular** | **Unleaded Premium** | **All Types** |
| 1978 | 62.6 | 67.0 | NA | 65.2 |
| 1979 | 85.7 | 90.3 | NA | 88.2 |
| 1980 | 119.1 | 124.5 | NA | 122.1 |
| 1981 | 131.1 | 137.8 | 147.0 | 135.3 |
| 1982 | 122.2 | 129.6 | 141.5 | 128.1 |
| 1983 | 115.7 | 124.1 | 138.3 | 122.5 |
| 1984 | 112.9 | 121.2 | 136.6 | 119.8 |
| 1985 | 111.5 | 120.2 | 134.0 | 119.6 |
| 1986 | 85.7 | 92.7 | 108.5 | 93.1 |
| 1987 | 89.7 | 94.8 | 109.3 | 95.7 |
| 1988 | 89.9 | 94.6 | 110.7 | 96.3 |
| 1989 | 99.8 | 102.1 | 119.7 | 106.0 |
| 1990 | 114.9 | 116.4 | 134.9 | 121.7 |
| 1991 | NA | 114.0 | 132.1 | 119.6 |
| 1992 | NA | 112.7 | 131.6 | 119.0 |
| 1993 | NA | 110.8 | 130.2 | 117.3 |
| 1994 | NA | 111.2 | 130.5 | 117.4 |
| 1995 | NA | 114.7 | 133.6 | 120.5 |
| 1996 | NA | 123.1 | 141.3 | 128.8 |
| 1997 | NA | 123.4 | 141.6 | 129.1 |
| 1998 | NA | 105.9 | 125.0 | 111.5 |
| 1999 | NA | 116.5 | 135.7 | 122.1 |
| 2000 | NA | 151.0 | 169.3 | 156.3 |
| 2001 | NA | 146.1 | 165.7 | 153.1 |
| 2002 | NA | 135.5 | 157.8 | 144.1 |

*Source:* Energy Information Administration, U.S. Dept. of Energy, *Monthly Energy Review,* October 2003.

11. **(a)** What was the mean cost of unleaded regular gas in 1990?
**(b)** What was the mean cost of unleaded premium gas in 1997?

12. **(a)** What was the decrease in the price of unleaded regular gas from 1990 to 1998?
**(b)** What was the percent decrease in price of all types of gas from 1981 to 1998?

13. **(a)** What was the decrease in price of unleaded regular gas from 1982 to 1986?
**(b)** What was the decrease in price of unleaded premium gas from 1982 to 1986?

< Objective 3 >

14. Create a pictograph for the total world population in 1950, using the information given in the world population and land area table in the text.

**1 Individual ≈ 200 Million**

| Continent/Region | Population Individuals |
|---|---|
| North America | 1.1 |
| South America | 0.5 |
| Europe | 1.96 |
| Asia | 7.9 |
| Africa | 1.14 |
| Oceana | 0.06 |

15. Use the information in the table to create a pictograph.

**U.S. Car Sales**

| Year | Car Sales |
|---|---|
| 1970 | 8,403,000 |
| 1975 | 8,538,000 |
| 1980 | 8,979,000 |
| 1985 | 11,039,000 |
| 1990 | 9,484,000 |
| 1995 | 8,686,000 |
| 2000 | 8,847,000 |

**Answers**

11. ____________

12. ____________

13. ____________

14. ____________

15. ____________

Answers

16. ______

17. ______

18. ______

19. ______

20. ______

16. Use the following table to create a pictograph.

**World Motor Vehicle Production**

| Year | Vehicles |
|---|---|
| 1970 | 29,419,000 |
| 1980 | 38,565,000 |
| 1985 | 44,909,000 |
| 1990 | 48,554,000 |
| 1995 | 49,983,000 |
| 2000 | 57,528,000 |

< Objective 4 >

*Use the following graph, showing the total U.S. motor vehicle production for the years 1996 to 2002, to answer exercises 17 to 20.*

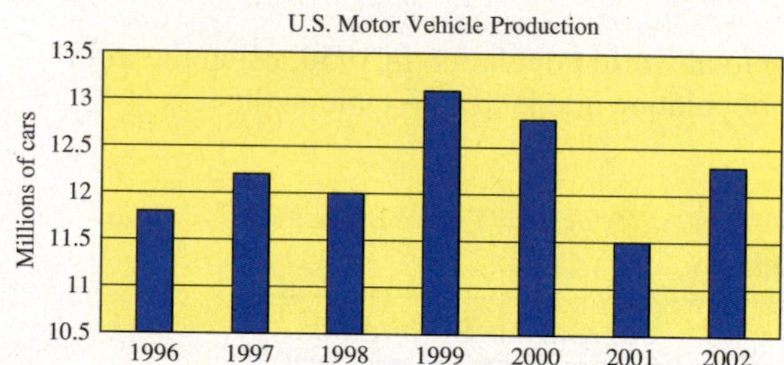


*Source:* Automotive News Market Data Book.

17. What was the production in 1998?

18. In what year did the greatest production occur?

19. Find the median number of cars produced in the 7 years.

20. In what year was the production decline the greatest compared to the previous year?

> Videos

*Use the following bar graph, showing the attendance at a circus for 7 days in August, to solve exercises 21 to 24.*

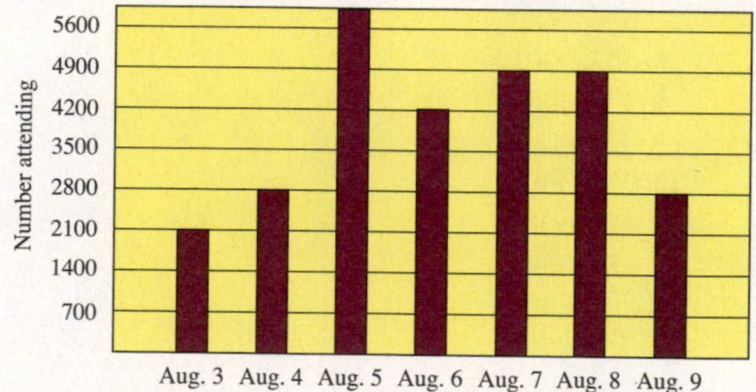

21. Find the attendance on August 4.

22. Which day had the greatest attendance?

23. Which day had the lowest attendance?

24. Find the median attendance over the 7 days.

*For exercises 25 to 28, use the following bar graph.*

### Sport Utility Vehicle Sales in the United States, 1993–2002

In 1993, 1,327,507 sport utility vehicles (SUVs) were sold in the United States, accounting for 26.3% of all sales of light vehicles (SUVs, minivans, vans, pickup trucks, and trucks under 14,000 lb). By 2002, sales of SUVs in the United States increased to 4,186,698, accounting for 48.3% of total light vehicle sales.

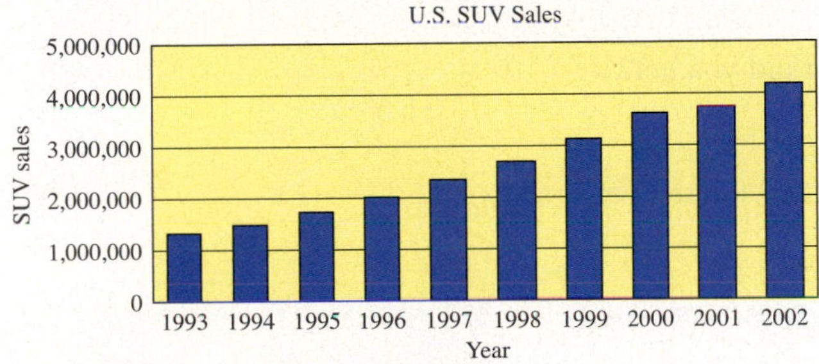


*Source:* Office of Transportation Technologies.

25. What were the sales of SUVs in 2000?

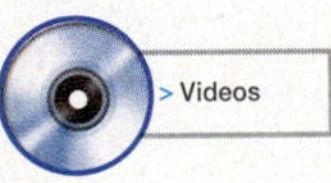


26. In what year did the greatest sales occur?

27. What was the percent increase in sales from 1993 to 2002?

28. In what year did the greatest increase in sales occur?

## Answers

21. ______

22. ______

23. ______

24. ______

25. ______

26. ______

27. ______

28. ______

Answers

29. ____

30. ____

31. ____

32. ____

33. ____

34. ____

35. ____

36. ____

37. ____

38. ____

*In exercises 29 to 32, use the nutritional facts given for Campbell's cream of mushroom soup. Assume you consumed 1 cup of soup.*

| Nutrition Facts | Amount / serving | %DV* | Amount / serving | %DV* |
|---|---|---|---|---|
| Serv. Size 1/2 cup (120mL) condensed soup | Total Fat 7g | 11% | Total Carb. 9g | 3% |
| Servings about 2.5 | Sat. Fat 2.5g | 13% | Fiber 1g | 4% |
| Calories 110 | Cholest. Less than 5mg | 1% | Sugars 1g | |
| Fat Cal. 60 | Sodium 870mg | 36% | Protein 2g | |
| *Percent Daily Values (DV) are based on a 2,000 calorie diet. | Vitamin A 0% • Vitamin C 0% • Calcium 2% • Iron 2% | | | |

Satisfaction guaranteed. For questions or comments, please call 1-800-257-8443. Please have code and date information on can end available. For recipes, information & more, visit Campbell's Community at www.campbellsoup.com 1261-56

**29.** How many calories have you consumed?

**30.** What percent of the daily value of saturated fat have you consumed?

**31.** What percent of fiber did you get?

**32.** How many grams of sodium did you get?

*In exercises 33 to 38, use the following table.*

| Soup | Calories | Fat | Total Protein | Sodium |
|---|---|---|---|---|
| Cream of mushroom | 110 | 7 g | 2 g | 870 mg |
| Cream of chicken | 130 | 8 g | 11 g | 890 mg |
| Split pea | 180 | 3.5 g | 10 g | 860 mg |
| Tomato | 100 | 0 g | 2 g | 760 mg |

**33.** Which soup has the least fat?

**34.** Which soup has the most sodium?

**35.** Which soup has the least sodium?

**36.** Which soup has the fewest calories?

**37.** Find the mean number of calories in the soups.

**38.** Find the mean number of milligrams of sodium in the soups.

Basic Skills | Advanced Skills | **Vocational-Technical Applications** | Calculator/Computer | Above and Beyond

Answers

39. ______

40. ______

**39. Mechanical Engineering** Use the following table.

| American Wire Gauge | Wire Diameter (in.) |
|---|---|
| 2 | 0.2576 |
| 3 | 0.2294 |
| 4 | 0.2043 |
| 6 | 0.1620 |
| 8 | 0.1285 |
| 10 | 0.1019 |
| 12 | 0.0808 |
| 14 | 0.0640 |
| 16 | 0.0508 |
| 18 | 0.0403 |
| 20 | 0.0319 |

**(a)** What is the diameter of a 12-gauge wire?

**(b)** What is the difference in diameter between 14-gauge and 10-gauge wire?

**40. Automotive** Use the following table.

| Temperature Protection (°F) | Required Percent of Ethylene Glycol |
|---|---|
| 15 | 22% |
| 10 | 26% |
| 5 | 29% |
| 0 | 34% |
| −10 | 39% |
| −20 | 43% |
| −30 | 48% |
| −40 | 53% |

**(a)** What percent ethylene glycol is required to provide protection down to −20°F?

**(b)** What is the temperature protection provided by a mixture of 29% ethylene glycol?

**(c)** If the percent of ethylene glycol is increased from 29% to 39%, how much does it change the temperature protection?

Answers

41. ____________

42. ____________

43. ____________

**41. Welding** Use the following information concerning properties of metals.

| Metal | Density (g/cm$^3$) | Melting Point (°C) |
|---|---|---|
| Iron | 7.87 | 1,538 |
| Aluminum | 2.699 | 660.4 |
| Copper | 8.93 | 1,084.9 |
| Tin | 5.765 | 231.9 |
| Titanium | 4.507 | 1,668 |

**(a)** What is the density of titanium?

**(b)** What is the difference in melting point between copper and iron?

**(c)** Which metal has the highest melting point?

Basic Skills | Advanced Skills | Vocational-Technical Applications | Calculator/Computer | **Above and Beyond**

**42.** Compare current gas prices in your area to those in the table given for exercises 11 to 13. Describe the differences.

**43.** Research national gas prices and compare them to the prices in the table given for exercises 11 to 13.

## Answers

**1.** **(a)** 221,000,000; **(b)** 16.2% **3.** **(a)** 70.7%; **(b)** 153.2%; **(c)** 91.4 people; **(d)** 231.5 people **5.** 193,000,000 people **7.** **(a)** 17.9%; **(b)** 4.9% **9.** **(a)** 44.1; **(b)** 108.4; **(c)** 145.8% **11.** **(a)** 116.4 ¢/gal; **(b)** 141.6 ¢/gal **13.** **(a)** 36.9 ¢/gal; **(b)** 33 ¢/gal **15.**

| Year | No. of Cars (in millions) |
|---|---|
| 1970 | 4.2 |
| 1975 | 4.3 |
| 1980 | 4.5 |
| 1985 | 5.5 |
| 1990 | 4.7 |
| 1995 | 4.3 |
| 2000 | 4.4 |

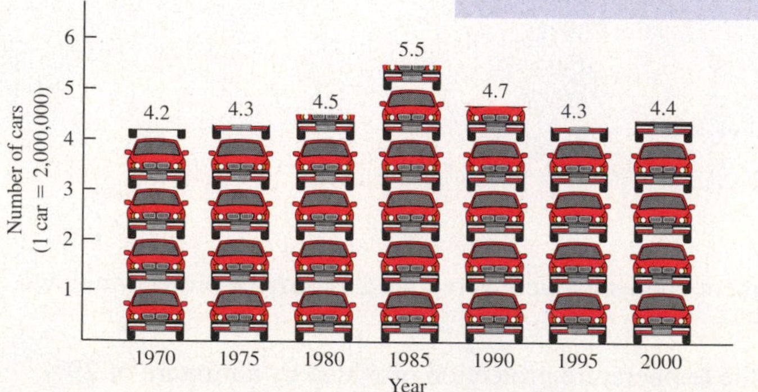


**17.** 12,000,000 cars **19.** 12,200,000 cars **21.** 2,800 people **23.** August 3 **25.** 3,600,000 SUVs **27.** 215.4% **29.** 220 **31.** 8% **33.** Tomato **35.** Tomato **37.** 130 cal **39.** **(a)** 0.0808 in.; **(b)** 0.0379 in. **41.** **(a)** 4.507 g/cm$^3$; **(b)** 453.1°C; **(c)** Titanium **43.** Above and Beyond

# 9.3 Line Graphs and Predictions

< 9.3 Objectives >

1 > Read a line graph

2 > Make a prediction from a line graph

We have seen that data can be represented graphically with a pictograph or a bar graph. Another useful type of graph is called a **line graph.** In a line graph, one of the types of information is almost always related to time (clock time, day, month, or year).

**Example 1** Reading a Line Graph

< Objective 1 >

This graph represents the number of regular season games won by the Dallas Cowboys each year of the 1990s. Note that the information across the bottom indicates the year and the information along the side indicates the number of victories.

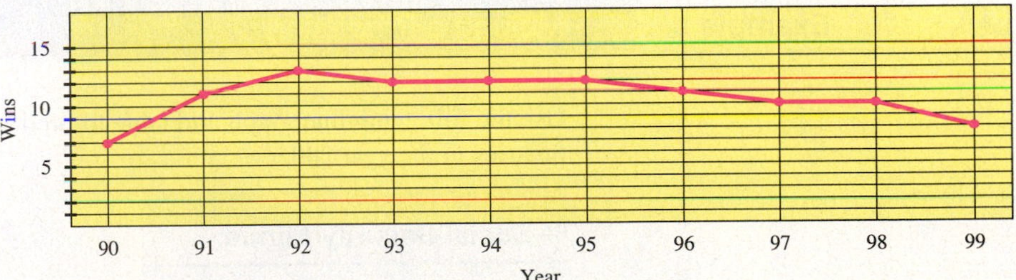


**(a)** How many games did they win in 1994?

We look across the bottom until we find 1994. We then look straight up until we see the line of the graph. Following across to the left, we see that they won 12 games in 1994.

**(b)** Find the mean number of games won by the Cowboys in the 1990s.

For each dot on the line, we look to the left to find how many victories it represents. We then add, using our definition of the mean, so

$$\text{Mean} = \frac{7 + 11 + 13 + 12 + 12 + 12 + 11 + 10 + 10 + 8}{10}$$

$$= \frac{106}{10}$$

$$= 10\frac{6}{10}$$

$$= 10\frac{3}{5}$$

The mean number of games won is $10\frac{3}{5}$.

**NOTE**

As a decimal, we could write this as 10.6.

### Check Yourself 1

The following graph indicates the high temperatures in Baltimore, Maryland, for a week in September.

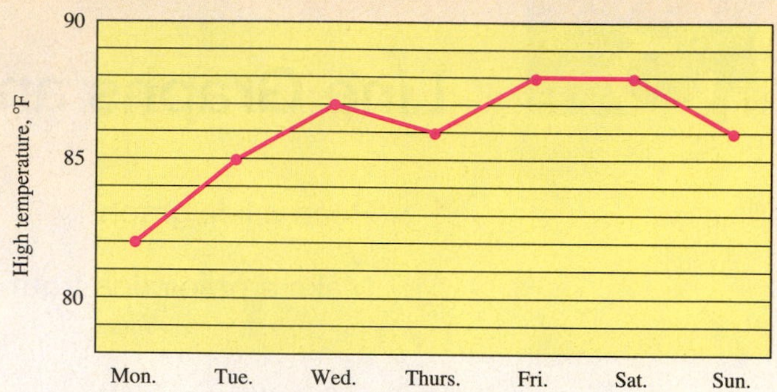


**(a)** What was the high temperature on Friday?
**(b)** Find the mean high temperature for that week.

It is often tempting, and sometimes useful, to use a line graph to predict a future value. Using an earlier trend to predict a future value is called **extrapolation.** This is something that statisticians warn us not to rely on, but it is done anyway. The key is not to predict very far from the data.

## Example 2 Making a Prediction

< Objective 2 >

Use the following line graph and table to predict the number of Social Security beneficiaries in the year 2005.

**Social Security Admin.**

| Year | Beneficiaries |
|---|---|
| 1955 | 6,000,000 |
| 1965 | 18,000,000 |
| 1975 | 28,000,000 |
| 1985 | 36,000,000 |
| 1995 | 41,000,000 |
| 2005 | ? |

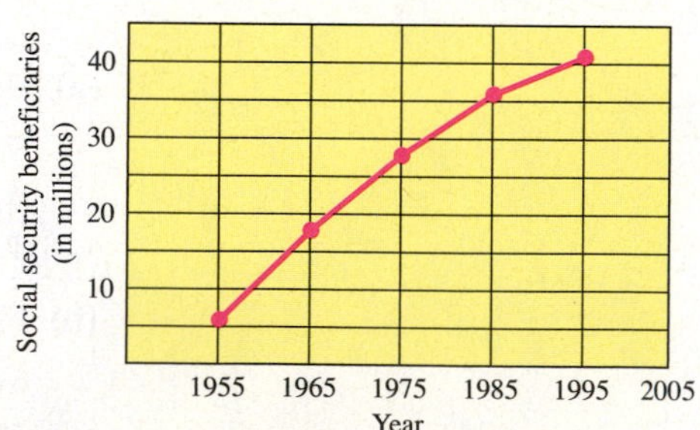


From the shape of the line graph, it would be reasonable to guess that the next point on the graph would continue on the same "curve."

**Social Security Admin.**

| Year | Beneficiaries |
|---|---|
| 2005 | 44,000,000 |

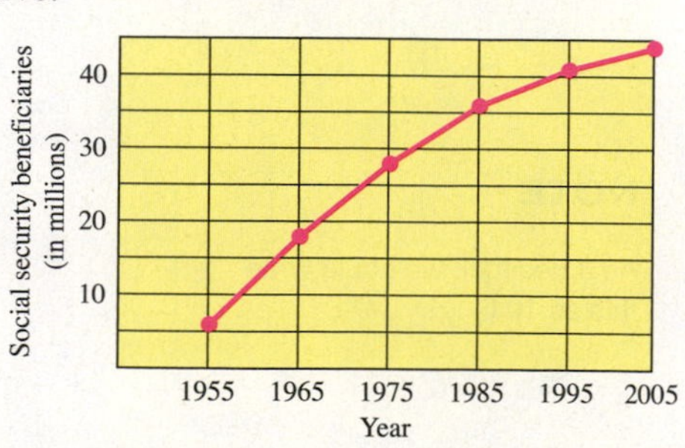

This point indicates that in 2005 we should expect about 44,000,000 Social Security beneficiaries. This number closely matches more sophisticated predictions for the number of beneficiaries in the year 2005.

## Check Yourself 2

**The following graph and table show the amount spent on health care (in billions of dollars) in the United States every 5 years from 1965 to 2000. Use that information to predict the amount spent in the year 2005.**

**National Center for Health Stats.**

| Year | Health Care Expenditures (in billions of $) |
|---|---|
| 1965 | 41 |
| 1970 | 73 |
| 1975 | 130 |
| 1980 | 247 |
| 1985 | 428 |
| 1990 | 700 |
| 1995 | 991 |
| 2000 | 1,285 |
| 2005 | ? |

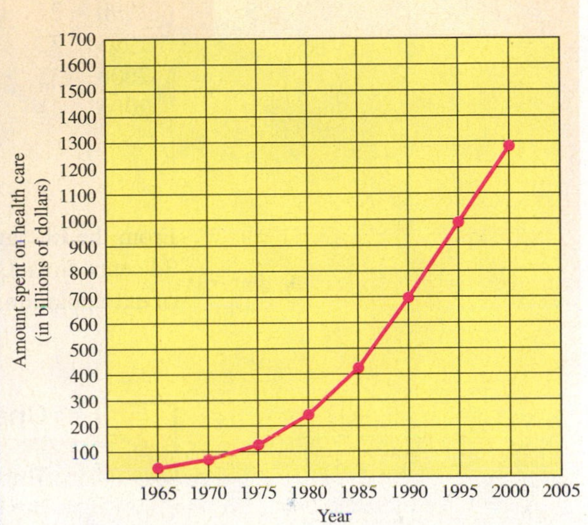


Here is an example yielding a result that is not quite as useful.

## Example 3 Making a Prediction

Use the following line graph and table to "predict" the cost of a first-class stamp on January 1, 2005.

| Year | Cost (¢) |
|---|---|
| 1960 | 4 |
| 1965 | 5 |
| 1970 | 6 |
| 1975 | 10 |
| 1980 | 15 |
| 1985 | 22 |
| 1990 | 25 |
| 1995 | 32 |
| 2000 | 33 |
| 2005 | ? |

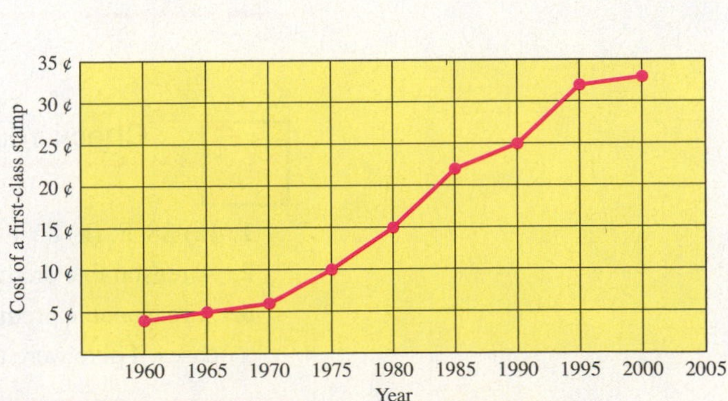

Place a straightedge so that it comes close to going through all the points.

| Year | Cost (¢) |
|---|---|
| 1960 | 4 |
| 1965 | 5 |
| 1970 | 6 |
| 1975 | 10 |
| 1980 | 15 |
| 1985 | 22 |
| 1990 | 25 |
| 1995 | 32 |
| 2000 | 33 |
| 2005 | 34 |

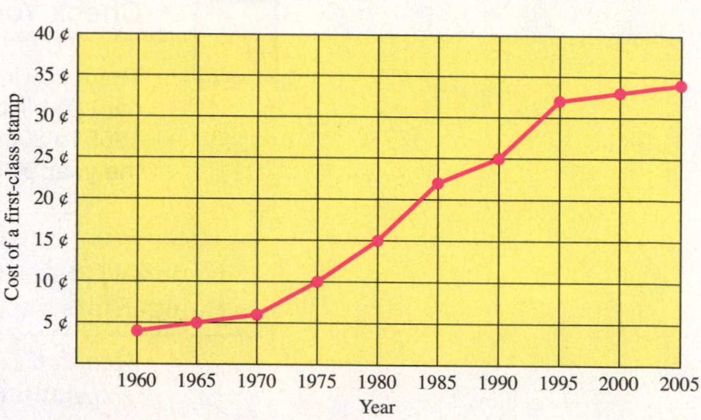


From the line graph, it would be reasonable to guess that in 2005 the cost would be 34 cents. In fact, on June 30, 2003, the cost was 37 cents. This is evidence of the danger of extrapolation, a danger we mentioned before Example 2.

## Check Yourself 3

**The following graph represents the number of larceny-theft cases in the United States every 5 years from 1980 to 1995. Use the graph to predict the number of cases in 2000.**

**FBI Uniform Crime Report**

| Year | Larceny-Theft Cases (in hundred thousands) |
|---|---|
| 1980 | 66 |
| 1985 | 73 |
| 1990 | 79 |
| 1995 | 80 |
| 2000 | ? |

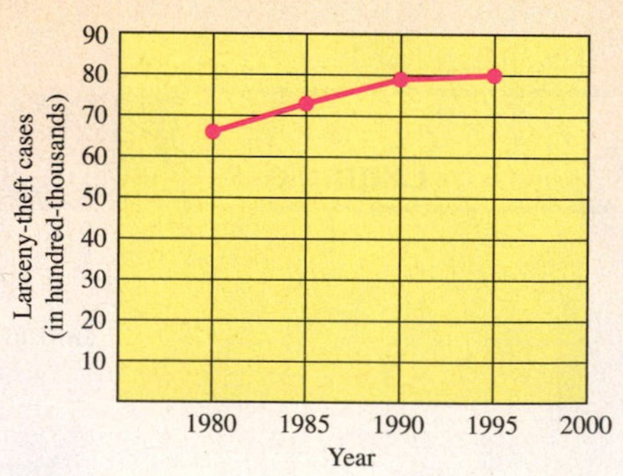

## Check Yourself ANSWERS

1. **(a)** 88°F; **(b)** 86°F
2. A reasonable prediction from the data would be about 1,600 billion dollars.
3. A reasonable prediction from the data would be about 82 hundred thousand cases. There were actually 70 hundred thousand.

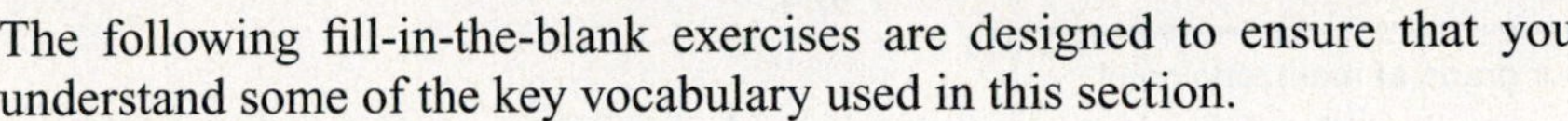

## Reading Your Text

The following fill-in-the-blank exercises are designed to ensure that you understand some of the key vocabulary used in this section.

### SECTION 9.3

**(a)** In a ____________ graph, one of the types of information is almost always related to time.

**(b)** It is often tempting to use a line graph to predict a ____________ value.

**(c)** Using an earlier trend to predict a future value is called ____________.

**(d)** The key, when using extrapolation, is not to ____________ very far from the data.

# 9.3 exercises

Basic Skills | Advanced Skills | Vocational-Technical Applications | Calculator/Computer | Above and Beyond

**MathZone**

Boost your grade at mathzone.com!

> Practice Problems
> NetTutor
> Self-Tests
> e-Professors
> Videos

Name ____________

Section ______ Date ______

## Answers

1. ____________
2. ____________
3. ____________
4. ____________
5. ____________
6. ____________
7. ____________
8. ____________

< Objective 1 >

*Use the following graph, showing the yearly utility costs of a family, for exercises 1 to 4.*

> Videos

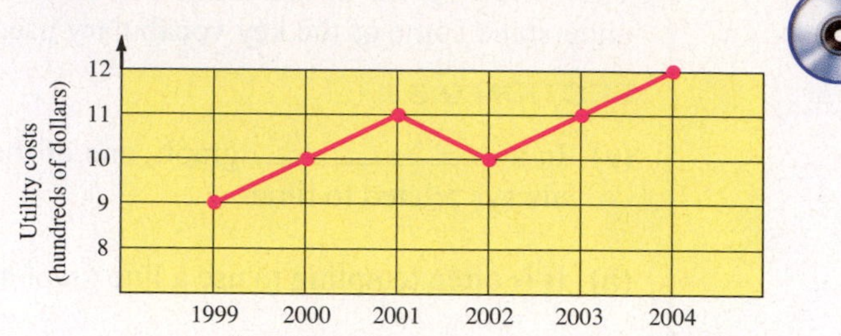


1. What was the cost in 2002?

2. What was the mean cost of utilities for this family in the 6 years from 1999 to 2004?

3. What was the decrease in the cost of utilities from 2001 to 2002?

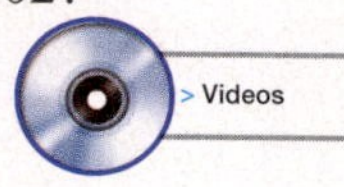


4. In what year was the cost of utilities the smallest?

*Use the following graph, showing the number of robberies in a town during the last 6 months of a year, for exercises 5 to 8.*

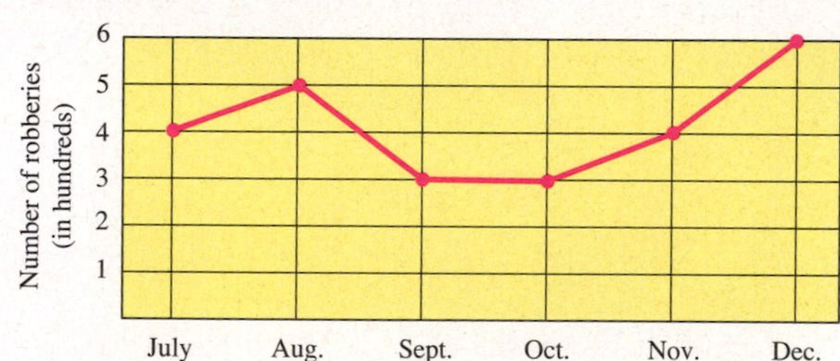


5. In which month did the greatest number of robberies occur?

6. How many robberies occurred in November?

7. Find the decrease in the number of robberies between August and September.

> Videos

8. What was the mean number of robberies per month over the last 6 months?

< Objective 2 >

**9.** The following graph and table show the income to the Hospital Insurance Trust Fund. Use this information to predict the income in the year 2005.

| Year | Total Income (in millions of $) |
|---|---|
| 1975 | 12,568 |
| 1980 | 25,415 |
| 1990 | 79,563 |
| 1995 | 114,847 |
| 2000 | 130,559 |

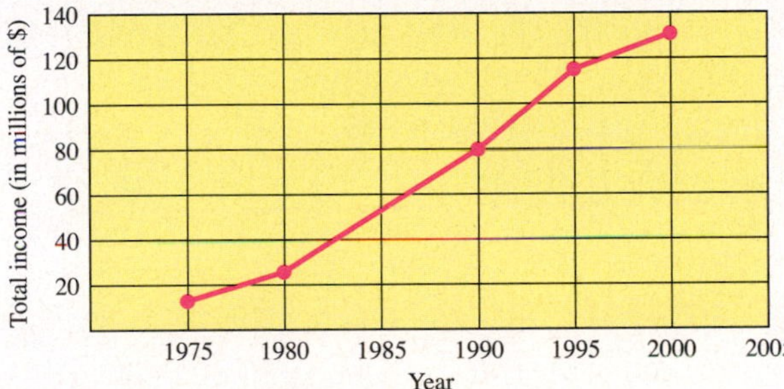


**10.** The following graph and table give the monthly principal and interest payments for a mortgage from 1999 to 2004. Use this information to predict the payment for 2005.

| Year | Payment |
|---|---|
| 1999 | $578 |
| 2000 | 613 |
| 2001 | 654 |
| 2002 | 675 |
| 2003 | 706 |
| 2004 | 730 |
| 2005 | |

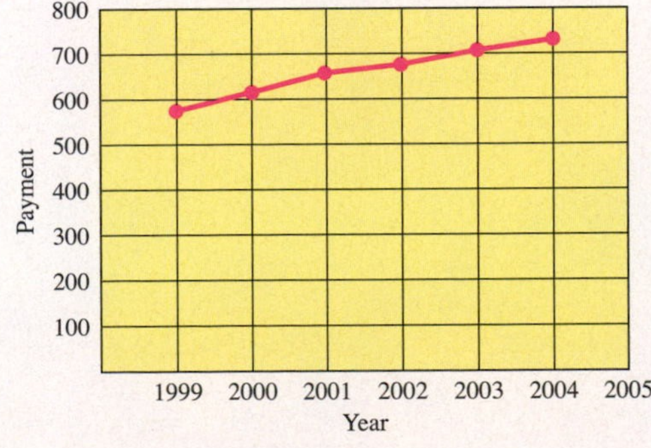


## Answers

9. ____________

10. ____________

Answers

11. ______

12. ______

Basic Skills | **Advanced Skills** | Vocational-Technical Applications | Calculator/Computer | Above and Beyond

**11.** The following information shows a relationship between years of formal education and typical income (in thousands of dollars) at age 30:

| Years | 8 | 10 | 12 | 14 | 16 |
|---|---|---|---|---|---|
| Income (×1,000) | 16 | 21 | 23 | 28 | 31 |

Use this information to make a line graph and then predict the yearly income associated with 18 years of formal education.

**12.** The following information shows a relationship between the amount spent per week on advertising by a small fast-food shop and the total sales per week:

| Amount spent | 10 | 20 | 30 | 40 |
|---|---|---|---|---|
| Sales | 200 | 380 | 625 | 790 |

Use this information to make a line graph and then predict the weekly sales associated with the expenditure of $50 (per week) on advertising.

13. The following information shows a relationship between the number of weeks on a special diet and the number of pounds lost during that time:

| Number of weeks | 2 | 4 | 6 | 8 |
|---|---|---|---|---|
| Number of pounds | 2 | 5 | 9 | 11 |

Use this information to make a line graph and then predict the number of pounds lost associated with 10 weeks on the special diet.

14. The following information shows a relationship between the speed of a certain car over a 100-mi test trip and the gas mileage obtained:

| Speed $\left(\text{in } \frac{\text{mi}}{\text{h}}\right)$ | 40 | 45 | 50 | 55 | 60 |
|---|---|---|---|---|---|
| Gas mileage | 28 | 25 | 21 | 18 | 16 |

Use this information to make a line graph and then predict the gas mileage associated with a speed of 65 mi/h.

Answers

13. ______

14. ______



## Answers

1. $1,000 3. $100 5. December 7. 200 9. $158,000,000,000
11. $35,000 13. 14 lb

# 9.4 Creating Bar Graphs and Pie Charts

< 9.4 Objectives >

1 > Use a table to create a bar graph

2 > Read a pie chart

3 > Create a pie chart from data

As we have seen, it is frequently easier to read information from a graph than it is from a table. In this section, we will look at two types of graphs that can be created from tables. We have already learned to read a bar graph. In Example 1, we will create one.

## Example 1 Creating a Bar Graph

< Objective 1 >

The following table represents the 2000 population of the six most populated urban areas in the world. Each population is the population of the city plus the population of all its suburbs. Create a bar graph from the information in the table.

**NOTE**

Mumbai was formerly listed as Bombay.

**Population of the World's Largest Urban Areas (U.N. Dept. for Economic and Social Info.)**

| City | 2000 Population |
|---|---|
| Tokyo, Japan | 26,400,000 |
| Mexico City, Mexico | 18,100,000 |
| Mumbai, India | 18,100,000 |
| Sao Paulo, Brazil | 17,800,000 |
| New York City, USA | 16,600,000 |
| Lagos, Nigeria | 13,400,000 |

*Source:* United Nations Human Settlements Programme.

We will let the vertical axis, the vertical line to the left of the graph, represent population. The six urban areas will be placed along the horizontal axis. To create a graph, we must decide on the scale for the vertical axis. The following steps will accomplish that.

1. Pick a number that is slightly larger than the biggest number we are to graph. 30,000,000 is slightly larger than 26,400,000.
2. Decide how long the axis will be. It is best if this length easily divides into the number of step 1. To accomplish this division, we choose 3 inches.

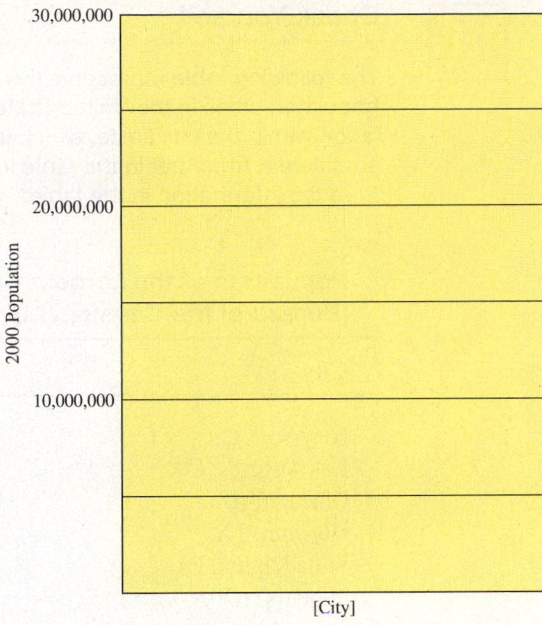


3. Scale the axis by dividing it with hashmarks. Label each hashmark with the appropriate number. In this graph, each inch will represent 10,000,000 people (the 30,000,000 divided by the 3 inches results in 10,000,000 people per inch).

Now, the height of each bar is determined by using the scale created for the axis. Remembering that we have 10,000,000 people per inch, we divide each population by 10,000,000. The result is the height of each bar. The height for Mexico City is 1.81 inches. Remember, all we can get from a bar graph is a rough approximation of the actual number.

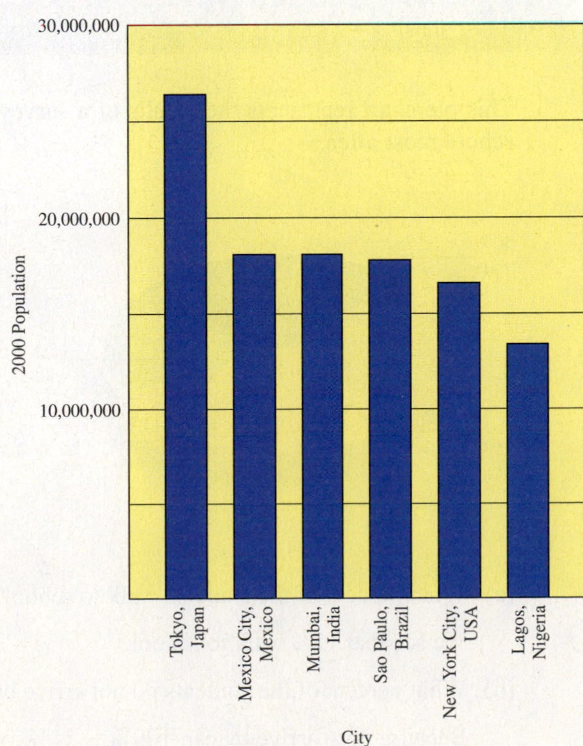

## Check Yourself 1

**The following table represents the 2000 population of the six most populated cities in the United States. Each population is the population within the city limits, which is why the New York population is so different from that in the table in Example 1. Create a bar graph from the information in the table.**

**Population of the Largest Cities in the United States (Bureau of the Census, U.S. Dept. of Commerce)**

| City | 2000 Population |
|---|---|
| New York City, NY | 8,008,000 |
| Los Angeles, CA | 3,695,000 |
| Chicago, IL | 2,896,000 |
| Houston, TX | 1,954,000 |
| Philadelphia, PA | 1,518,000 |
| Phoenix, AZ | 1,321,000 |

When a graph represents how some unit is divided, a *pie chart* is frequently used.

As you might expect, a **pie chart** is a circle. Wedges (or sectors) are drawn in the circle to show how much of the whole each part makes up.

## Example 2 Reading a Pie Chart

< Objective 2 >

This pie chart represents the results of a survey that asked students how they get to school most often.

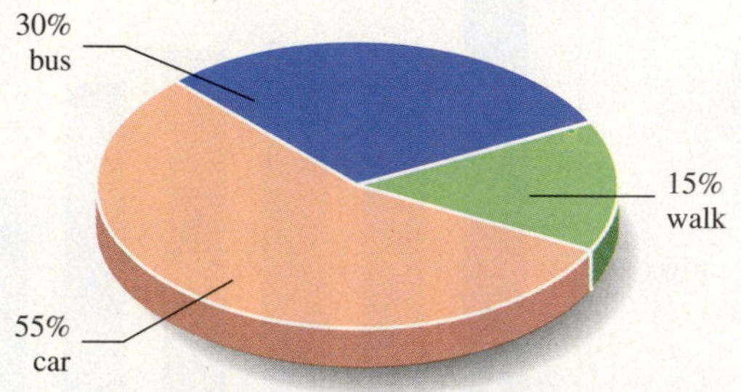


**(a)** What percent of the students walk to school?

We see that 15% walk to school.

**(b)** What percent of the students do not arrive by car?

Because 55% arrive by car, 100% − 55%, or 45%, do not.

## Check Yourself 2

**This pie chart represents the results of a survey that asked students whether they bought lunch, brought it, or skipped lunch altogether.**

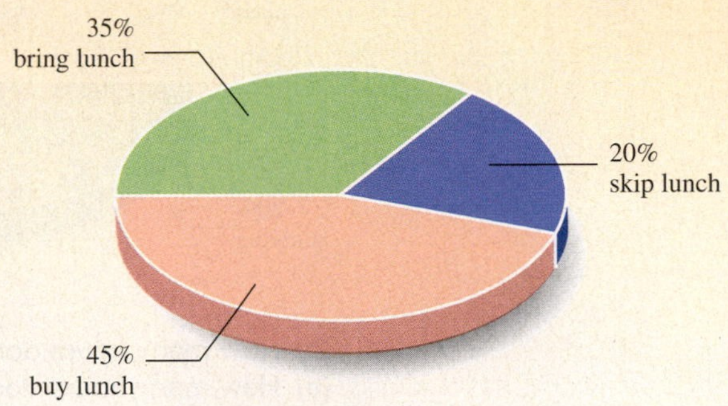


**(a) What percent of the students skipped lunch?**
**(b) What percent of the students did not buy lunch?**

If we know what the whole pie represents, we can also find out more about what each wedge represents. Example 3 illustrates this point.

## Example 3 Interpreting a Pie Chart

This pie chart shows how Sarah spent her \$12,000 college scholarship.

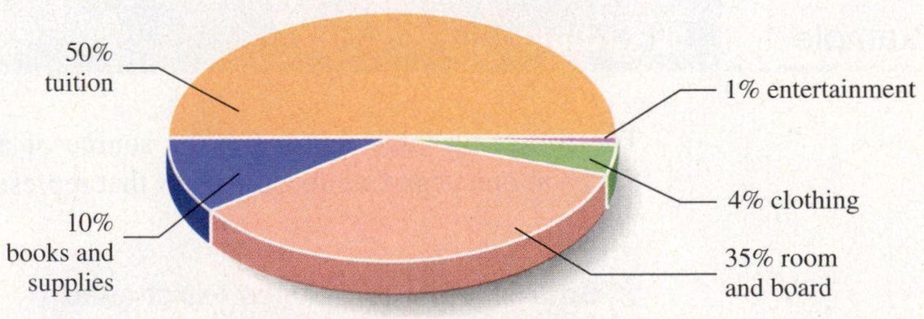


**(a)** How much did she spend on tuition?

50% of her \$12,000 scholarship, or \$6,000.

**(b)** How much did she spend on clothing and entertainment?

Together, 5% of the money was spent on clothing and entertainment, and $0.05 \times 12{,}000 = 600$. Therefore, \$600 was spent on clothing and entertainment.

## Check Yourself 3

This pie chart shows how Rebecca spends an average 24-h school day.

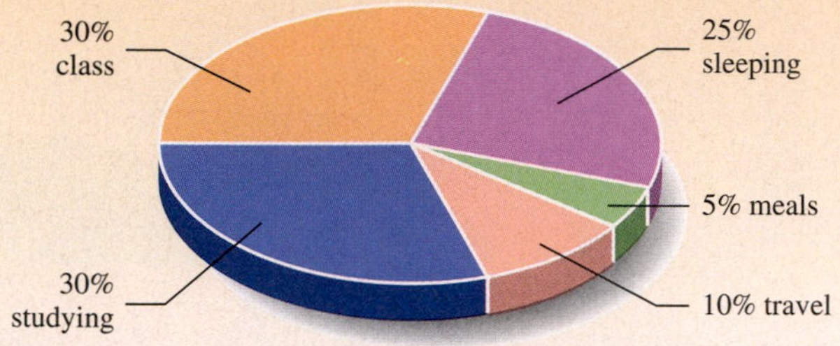


(a) How many hours does she spend sleeping each day?
(b) How many hours does she spend altogether studying and in class?

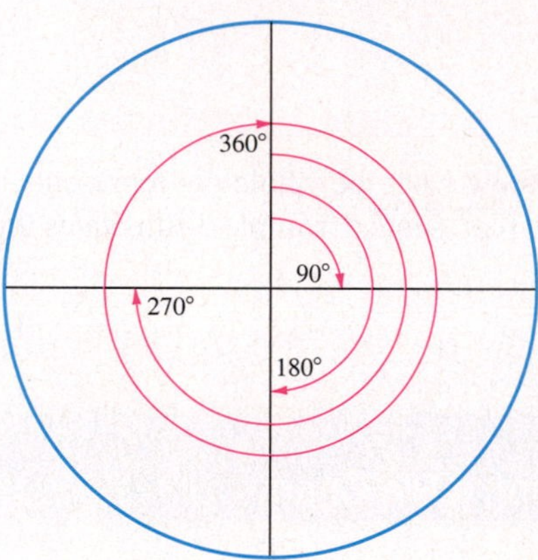


If we are creating a pie chart, how do we know how much of the circle to use for each piece? To make this decision requires that a scale be used for the circle. A standard scale has been established for all circles. As we saw in Chapter 8, each circle has 360°. That means that $\frac{1}{4}$ of the circle has $\frac{1}{4}$ of 360°, which is 90°.

With a protractor, we can now create our own pie chart.

## Example 4 Creating a Pie Chart

< Objective 3 >

The following table represents the source of automobiles purchased in the United States in one year. Create a pie chart that represents the same data.

**Source of Automobiles Purchased**

| Country of Origin | Number | % of Total |
|---|---|---|
| United States | 6,500,000 | 80 |
| Japan | 800,000 | 10 |
| Germany | 400,000 | 5 |
| All others | 400,000 | 5 |

*Source:* American Automotive Manufacturers' Association.

To find the size of the slice for each country, we take the given percent of 360°. We will create another table column to represent the degrees needed.

**Source of Automobiles Purchased**

| Country of Origin | Number | % of Total | Degrees |
|---|---|---|---|
| United States | 6,500,000 | 80 | 288 |
| Japan | 800,000 | 10 | 36 |
| Germany | 400,000 | 5 | 18 |
| All others | 400,000 | 5 | 18 |

*Source:* American Automotive Manufacturers' Association.

Using a protractor, we start with Japan and mark a section that is 36°.

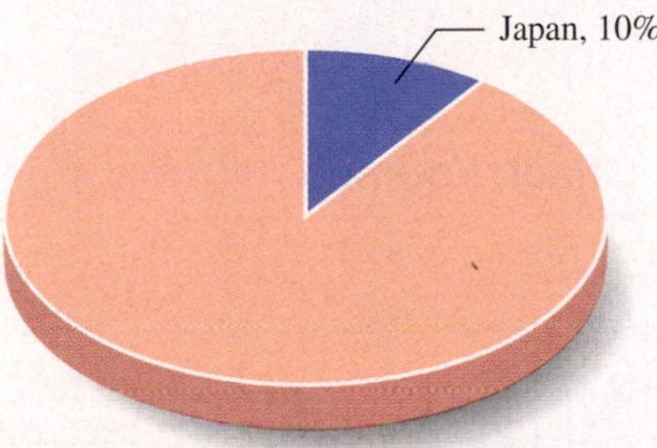


Again, using the protractor, we mark the 18° section for Germany and the 18° section for the other countries.

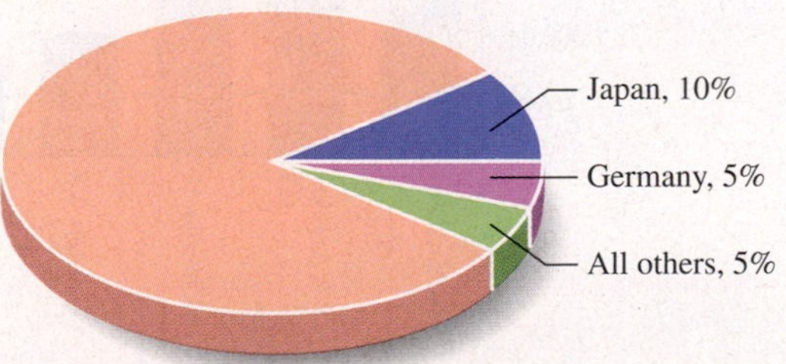


There is no need to measure the remainder of the pie. What is left is the 288° section for U.S.-made cars. Note that we saved the largest section for last. It is much easier to mark the smaller sections and leave the largest for last.

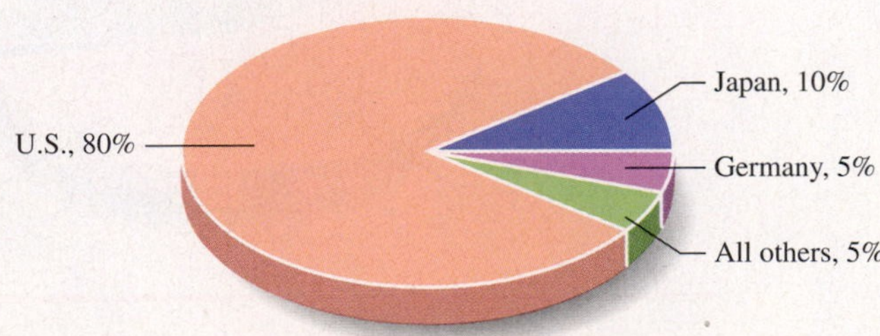

## Check Yourself 4

**Create a pie chart for the following table, which shows TV ownership for all U.S. homes.**

**TV Ownership**

| Number of TVs | % of U.S. Homes |
|---|---|
| 0 | 2 |
| 1 | 22 |
| 2 | 34 |
| 3 or more | 42 |

*Source:* Nielsen Media Research.

## Check Yourself ANSWERS

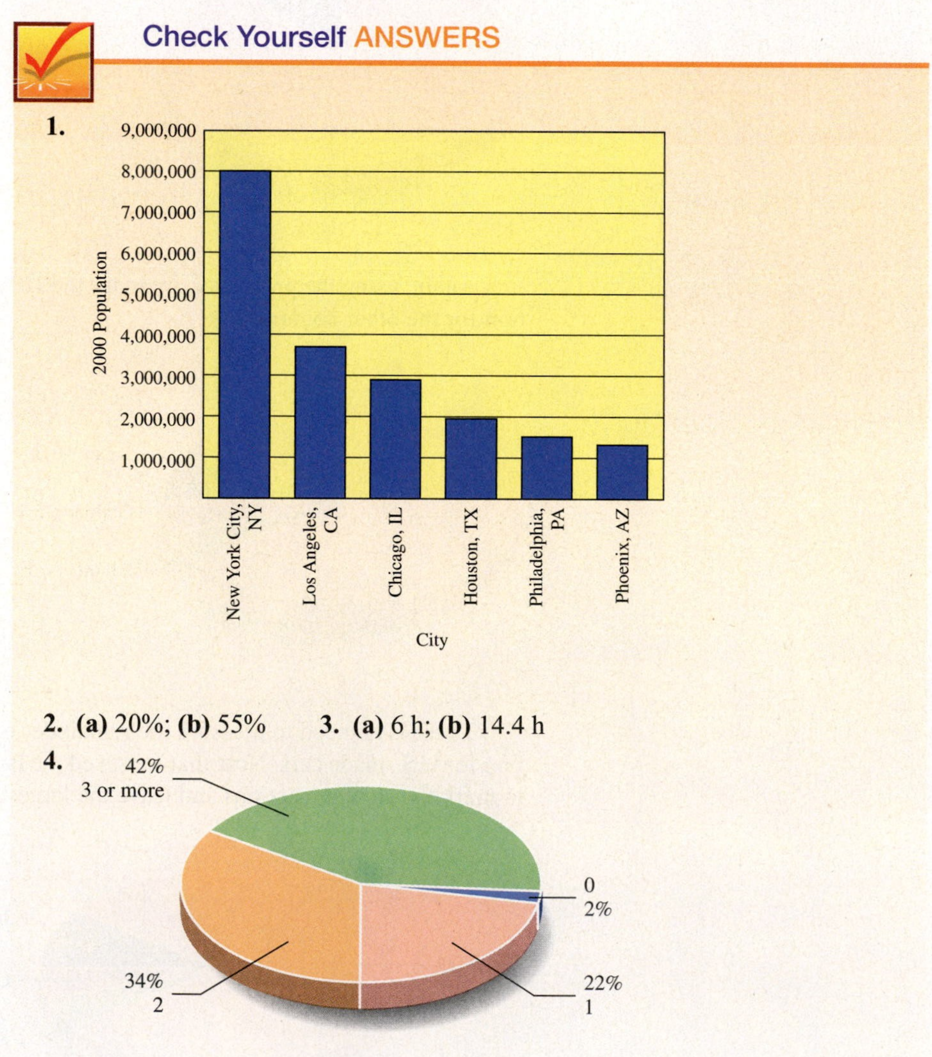


**2.** **(a)** 20%; **(b)** 55% **3.** **(a)** 6 h; **(b)** 14.4 h

## Reading Your Text

The following fill-in-the-blank exercises are designed to ensure that you understand some of the key vocabulary used in this section.

### SECTION 9.4

**(a)** It is frequently easier to read information from a graph than it is from a ______________.

**(b)** All we can get from a bar graph is a rough ______________ of the actual number.

**(c)** In a ______________ chart, wedges are drawn in a circle to show how much of the whole each part makes up.

**(d)** Each ______________ has 360°.

# 9.4 exercises

Basic Skills | Advanced Skills | Vocational-Technical Applications | Calculator/Computer | Above and Beyond

MathZone

Boost your grade at mathzone.com!
> Practice Problems
> NetTutor
> Self-Tests
> e-Professors
> Videos

Name ____________

Section ________ Date ________

**Answers**

1. ____________

2. ____________

3. ____________

4. ____________

5. ____________

6. ____________

7. ____________

< Objective 1 >

**1.** The following table gives the total U.S. population by age group, according to the 2000 census.

| Age | Population |
|---|---|
| 0–14 | 60,224,094 |
| 15–34 | 79,079,301 |
| 35–54 | 82,737,774 |
| 55–74 | 42,494,571 |
| 75+ | 16,603,839 |

*Source:* U.S. Census Bureau.

Construct a bar graph from this information.

**2.** The following table gives the median earnings of women aged 25 and older who work full-time, year round, by educational attainment, according to the 2000 census.

| Education | Earnings |
|---|---|
| HS diploma or GED | \$23,719 |
| Associate's degree | 30,178 |
| Bachelor's degree | 38,208 |
| Master's degree | 47,049 |
| Doctorate's degree | 55,620 |

*Source:* U.S. Census Bureau.

Create a bar graph from this information.

< Objective 2 >

*The following pie chart shows the budget for a local company. The total budget is \$600,000. Find the amount budgeted in each of the following categories.*

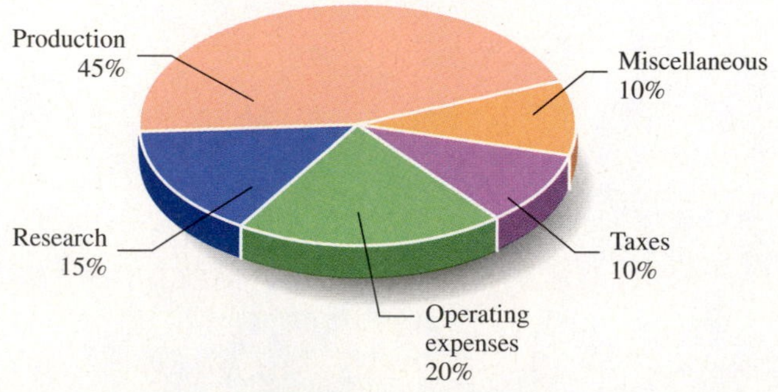


**3.** Production > Videos

**4.** Taxes

**5.** Research > Videos

**6.** Operating expenses

**7.** Miscellaneous

*The following pie chart shows the distribution of a person's total yearly income of $24,000. Find the amount budgeted for each category.*

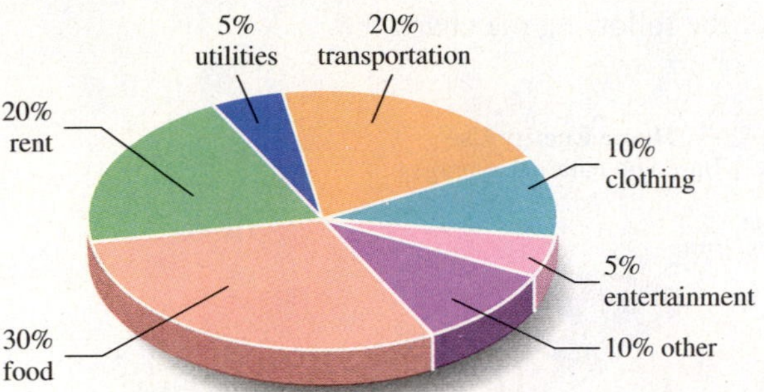


**8.** Food

**9.** Rent 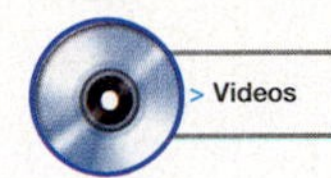


**10.** Utilities

**11.** Transportation

**12.** Clothing

**13.** Entertainment 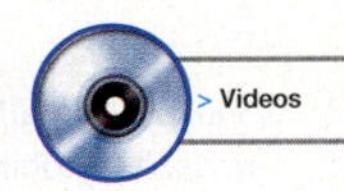


< Objective 3 >

**14.** The following table gives the number of Senate members with military service in the 106th U.S. Congress, by branch.

| Branch | Count |
|---|---|
| Army | 17 |
| Navy | 10 |
| Air Force | 4 |
| Marines | 6 |
| Coast Guard | 1 |
| National Guard | 2 |

Construct a pie chart from this information.

**15.** The following table gives the number of foreign-born residents of the United States by region of birth in the year 2000.

| Region | U.S. Population |
|---|---|
| Europe | 4,400,000 |
| Northern America | 700,000 |
| Latin America | 14,500,000 |
| Asia | 7,200,000 |
| Other areas | 1,600,000 |

*Source:* U.S. Census Bureau.

Construct a pie chart from the information.

**Answers**

8. ______

9. ______

10. ______

11. ______

12. ______

13. ______

14. ______

15. ______

Answers

16. ________

17. ________

Basic Skills | Advanced Skills | Vocational-Technical Applications | Calculator/Computer | Above and Beyond

16. **Manufacturing** Consider the following pie chart.

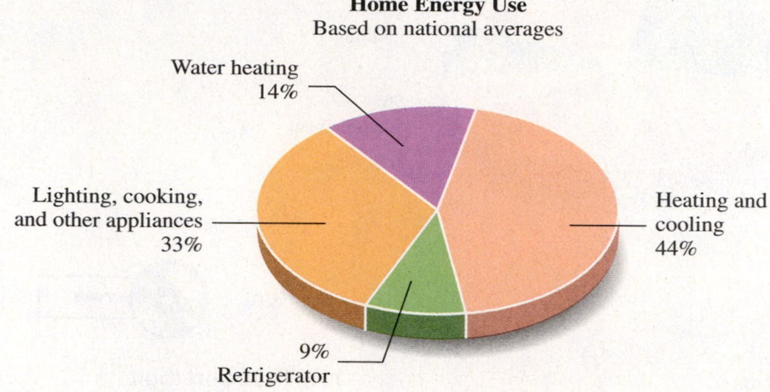


*Source:* "Small Electric Wind Systems: A U.S. Consumer's Guide," May 2001, revised October 2002. U.S. Department of Energy, Office of Energy Efficiency and Renewable Energy, Wind and Hydropower Technologies Program.

Assume that a household uses 10,600 kilowatt-hours (kWh) of electricity annually. Calculate the energy (in kWh) used annually for each category.

17. **Automotive** Create a circle graph for the following data concerning cars sold by various manufacturers.

| Manufacturer | Sales (1,000s of Vehicles) |
|---|---|
| Chevrolet | 43.4 |
| Dodge | 29.1 |
| Ford | 46.8 |
| Honda | 9.7 |
| Toyota | 28.1 |

**18. Health Sciences** The following table gives the number of live births, broken down by the race of the mother, in the United States for the year 2002.

| Race | Number |
|---|---|
| White | 3,174,760 |
| Black | 593,691 |
| American Indian | 42,368 |
| Asian or Pacific Islander | 210,907 |
| Total | 4,021,726 |

*Source:* National Vital Statistics Reports, Vol. 52, No. 10. December 17, 2003.

Construct a pie chart of this data.

**19. Automotive** Create a bar graph to display the data in the following mileage table.

| Model | Mileage |
|---|---|
| Taurus | 27 |
| Malibu | 29 |
| Stratus | 26 |
| Camry | 32 |

**20. Welding** Create a bar graph to display the melting points of the metals listed in the following table.

| Metal | Density (g/cm$^3$) | Melting Point (°C) |
|---|---|---|
| Iron | 7.87 | 1,538 |
| Aluminum | 2.699 | 660.4 |
| Copper | 8.93 | 1,084.9 |
| Tin | 5.765 | 231.9 |
| Titanium | 4.507 | 1,668 |

## Answers

18. ____________

19. ____________

20. ____________

## Answers

21. ______________

21. **Agriculture** Construct a pie chart for the harvest total shown in the following table.

| Crop | Harvest (1,000,000s of Tons) |
|---|---|
| Corn | 127 |
| Wheat | 93.2 |
| Barley | 32.8 |
| Soybeans | 87.4 |

## Answers

1.

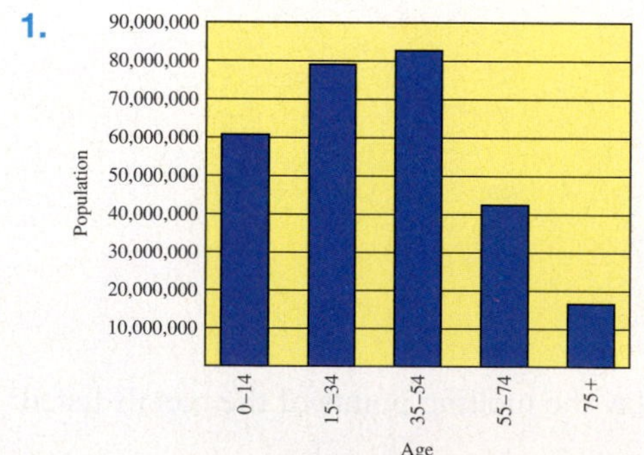


3. \$270,000 5. \$90,000
7. \$60,000 9. \$4,800
11. \$4,800 13. \$1,200

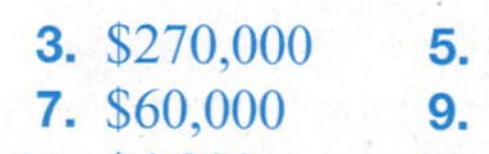

15.

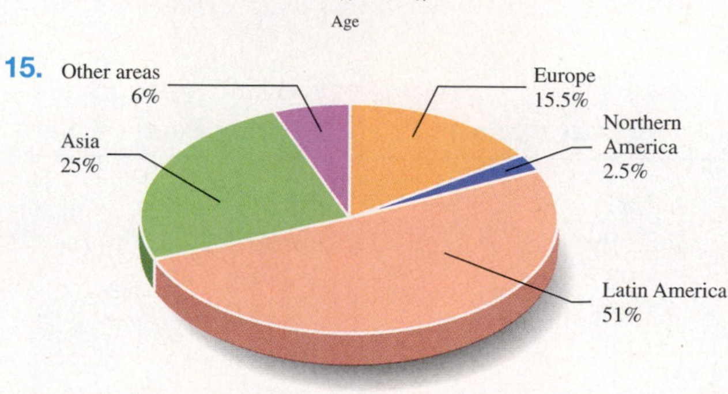

17.

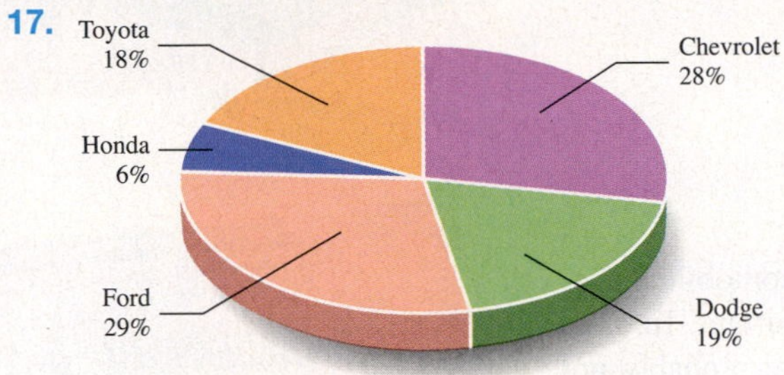
Toyota
18%
Chevrolet
28%
Honda
6%
Ford
29%
Dodge
19%

19.

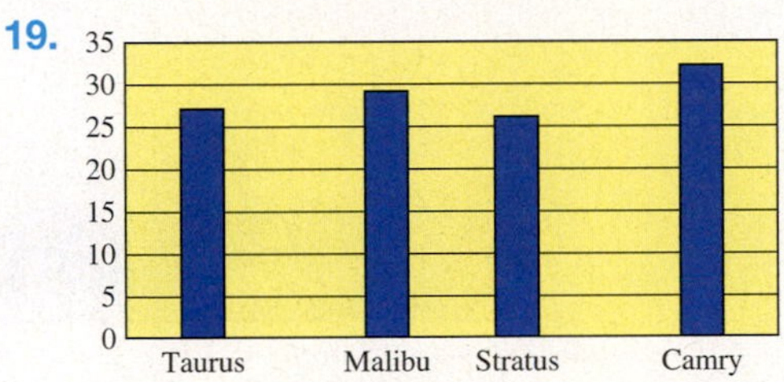
35
30
25
20
15
10
5
0
Taurus
Malibu
Stratus
Camry

21.

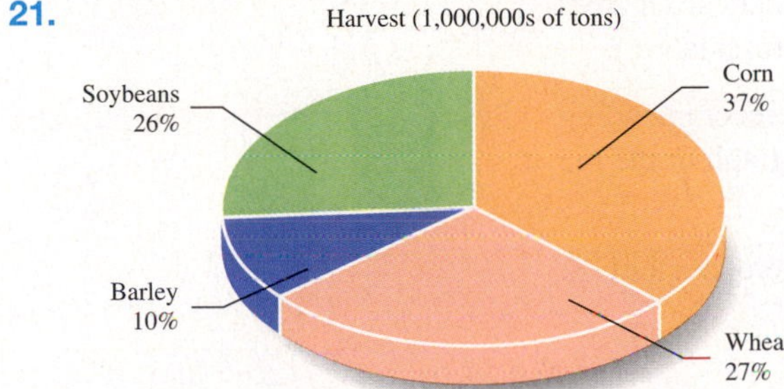
Harvest (1,000,000s of tons)
Soybeans
26%
Corn
37%
Barley
10%
Wheat
27%

# Activity 25 ::
## Graphing Car Color Data

Consider the data you gathered in Activity 24. If you worked in the automobile industry, information concerning customer color preferences would be important. To form conclusions, you need to present your data to other people (who are probably not statisticians). Bar graphs and pie charts are useful ways of presenting such data.

*Note:* If you did not complete Activity 24, a sample data set can be found on the website for this textbook.

**1.** Create a bar graph using the set of 35 car colors compiled in Activity 24.

**2.** Create a pie chart using the set of 35 car colors compiled in Activity 24.

**3.** **(a)** Briefly describe the view of your data given by the two graphs.

**(b)** In which graph is the mode more easily distinguished?

**(c)** In which is it easier to get a sense of the "whole" data set?

**4.** Compare your graphs to those of a classmate. Briefly describe how yours differ from your classmate's graphs (consider the differences in data sets and in presentation).

**5.** Write a short letter to the manager of an auto body and paint shop, recommending levels of inventory for different color paints. Include either the bar graph or the pie chart in your letter.

**6.** Go to www.mhhe.com/streeter to find graphs describing car color preferences for U.S. consumers.

# 9.5 Describing and Summarizing Data Sets

< 9.5 Objectives >

1 > Compute the quartiles of a data set

2 > Give the five-number summary of a data set

3 > Construct a box-and-whisker plot to describe a data set

4 > Interpret a box-and-whisker plot

In Section 9.1, we learned to describe a data set by using various measures of center. That is, we described the "average" member of a set, using the mean, median, and mode.

Often, though, this is not sufficient to provide a clear picture of a data set.

## Example 1 — Describing Data with Measures of Center

< Objective 1 >

Find the mean and median of each set of numbers.

**(a)** 1, 2, 4, 3, 5, 1, 1, 3

Recall from Section 9.1 that the mean is computed by adding the numbers together and dividing by the number of elements in the set.

$$\text{Mean} = \frac{1 + 2 + 4 + 3 + 5 + 1 + 1 + 3}{8} = \frac{20}{8} = 2.5$$

To compute the median, we list the numbers in increasing order and find the midpoint.

1, 1, 1, **2, 3,** 3, 4, 5

Because we have two elements in the "middle," the median is the mean of the pair of middle numbers.

$$\text{Median} = \frac{2 + 3}{2} = \frac{5}{2} = 2.5$$

**RECALL**

There will be two values in the "middle" when there are an even number of terms in the list.

**(b)** 2, 2, 2, **2, 3,** 3, 3, 3

$$\text{Mean} = \frac{2 + 2 + 2 + 2 + 3 + 3 + 3 + 3}{8} = \frac{20}{8} = 2.5$$

$$\text{Median} = \frac{2 + 3}{2} = \frac{5}{2} = 2.5$$

### Check Yourself 1

**Find the median and mode of each set of numbers.**

**(a)** 1, 2, 2, 7, 7, 7, 7, 10 **(b)** 24, 7, 7, 10, 6, 156, 7

**NOTES**

In fact, the median is the second quartile $Q_2$.

Just as a single number (the median) separates the data into two groups, it takes three numbers to separate data into four groups.

In Example 1, both sets had the same mean and the same median, yet they are very different lists of numbers. Clearly, neither the mean nor the median alone is sufficient to distinguish a list of numbers.

To assist us, we describe a data set using several numbers. One set of numbers we use is the **quartiles.** Just as the median divides a data set into halves, the quartiles divide the set into quarters (four equal parts). We use the notation $Q_1$, $Q_2$, and $Q_3$ to represent each of the three quartiles.

**Step by Step**

**Finding the Quartiles $Q_1$ and $Q_3$**

Step 1 Rewrite the given data set in ascending order.
Step 2 Find the median.
Step 3 Make a list of only those numbers (in order) that are to the left of the median.
Step 4 Find the median of the list created in step 3; this is the first quartile $Q_1$.
Step 5 Repeat steps 3 and 4 with those numbers that are to the right of the median; this gives the third quartile $Q_3$.

### Example 2 Finding Quartiles

Find the first and third quartiles, $Q_1$ and $Q_3$, of each data set.

**(a)** 1, 2, 4, 3, 5, 1, 1, 3

We sorted this list and found the median in Example 1.

1, 1, 1, 2, 3, 3, 4, 5

Median = 2.5

1, 1, 1, 2 are all to the left of the median. The median of this list is 1 (do you see why?), which is the first quartile.

$Q_1 = 1$

3, 3, 4, 5 are all to the right of the median. The median of this list is 3.5 (do you see why?), which is the third quartile.

$Q_3 = 3.5$

**NOTE**

We use a similar process to divide a set using deciles (tenths) and percentiles (hundredths).

**(b)** 2, 2, 2, 2, 3, 3, 3, 3

We found the median to be 2.5 in Example 1.

2, 2, 2, 2 are all to the left of the median. The median of this set gives the first quartile

$Q_1 = 2$

Similarly, the third quartile is 3.

**NOTE**

We refer to the smallest and largest members of a set as the **min** and **max**, respectively.

### Check Yourself 2

Find the first and third quartiles of each set of numbers.

**(a)** 1, 2, 2, 7, 7, 7, 7, 10 **(b)** 24, 7, 7, 10, 6, 156, 7

The final pair of numbers we will use is the smallest (minimum) and largest (maximum) elements of the data set.

We are now ready to give a **five-number summary** of a data set.

**Definition**

**Five-Number Summary**

The five-number summary associated with a data set is given by

Min, $Q_1$, Median, $Q_3$, Max

The five-number summary serves to distinguish between sets of data where a single number is not sufficient. We now find some five-number summaries.

### Example 3 Finding Five-Number Summaries

< Objective 2 >

Find the five-number summary of each data set.

**(a)** 1, 2, 4, 3, 5, 1, 1, 3

The smallest number is 1, so 1 is the min. The largest number is 5, so 5 is the max. Using our results from Examples 1 and 2, we have

1, 1, 2.5, 3.5, 5

**(b)** 2, 2, 2, 2, 3, 3, 3, 3

The five-number summary for this set is simply 2, 2, 2.5, 3, 3.

### Check Yourself 3

Give the five-number summary for each set of numbers.

**(a)** 1, 2, 2, 7, 7, 7, 7, 10 **(b)** 24, 7, 7, 10, 6, 156, 7

We can now see that our two data sets from Example 1 are distinct. We can use a picture to display these differences based on the five-number summary. The graph we create is called a **box-and-whisker plot.**

**Step by Step**

**Constructing Box-and-Whisker Plots**

**Step 1** Find the five-number summary of the given set of data.

**Step 2** Construct a horizontal number line from the min to the max values in the summary.

**Step 3** Mark off each of the numbers in the summary on the number line to scale (use small vertical lines).

**Step 4** Draw a box from $Q_1$ to $Q_3$ (so the number line is in the middle of the box).

## Example 4 Constructing Box-and-Whisker Plots

< Objective 3 >

Construct box-and-whisker plots for each data set.

**(a)** 1, 2, 4, 3, 5, 1, 1, 3

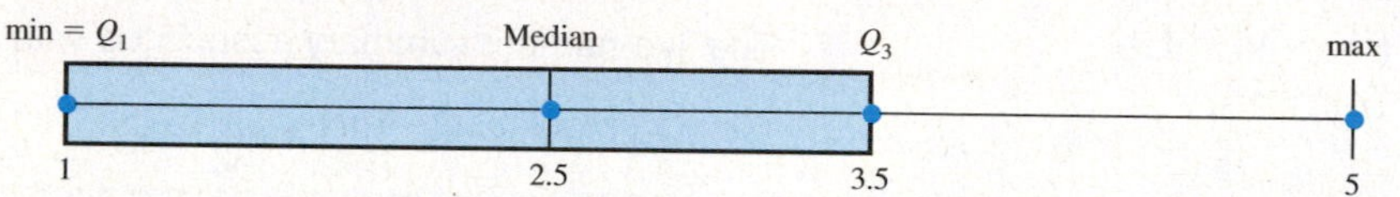


**(b)** 2, 2, 2, 2, 3, 3, 3, 3

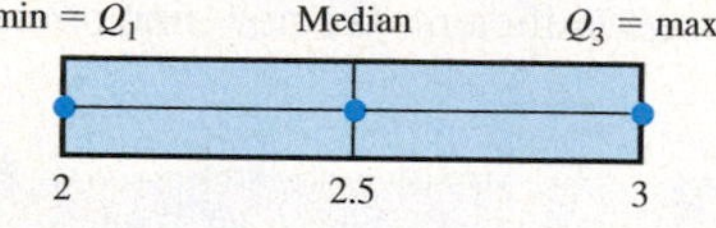


**NOTE**

It is also possible to construct a vertical box-and-whisker plot.

**Check Yourself 4**

Construct a box-and-whisker plot for each set of numbers.

**(a)** 1, 2, 2, 7, 7, 7, 7, 10 **(b)** 24, 7, 7, 10, 6, 56, 7

**NOTE**

Many graphing calculators will compute five-number summaries and even produce box-and-whisker plots.

Our box-and-whisker plots for each set of data look different, which reflects differences between the sets.

A common application of box-and-whisker plots involves stock prices.

## Example 5 Using Box-and-Whisker Plots

< Objective 4 >

Looking at investment options, we tracked the closing prices of two stocks (Microsoft—MSFT and Chevron-Texaco—CVX) on the New York Stock Exchange (NYSE) over a 2-week period. The following table shows closing prices for each stock.

| Stock | Mon. | Tues. | Wed. | Thu. | Fri. | Mon. | Tues. | Wed. | Thu. | Fri. |
|---|---|---|---|---|---|---|---|---|---|---|
| MSFT | 25.93 | 26.04 | 26.32 | 26.25 | 26.57 | 25.29 | 25.49 | 25.25 | 25.04 | 24.67 |
| CVX | 64.71 | 65.63 | 65.83 | 65.80 | 66.00 | 64.98 | 66.02 | 65.80 | 65.20 | 65.25 |

*Source:* Yahoo! Finance.

Construct box-and-whisker plots to compare the two stocks. Interpret your findings.

The five-number summaries are given for each stock.

MSFT: 24.67, 25.25, 25.71, 26.25, 26.57

CVX: 64.71, 65.20, 65.72, 65.83, 66.02

This gives us the following pair of box-and-whisker plots.

**NOTE**

We used the same scale for both plots.

MSFT

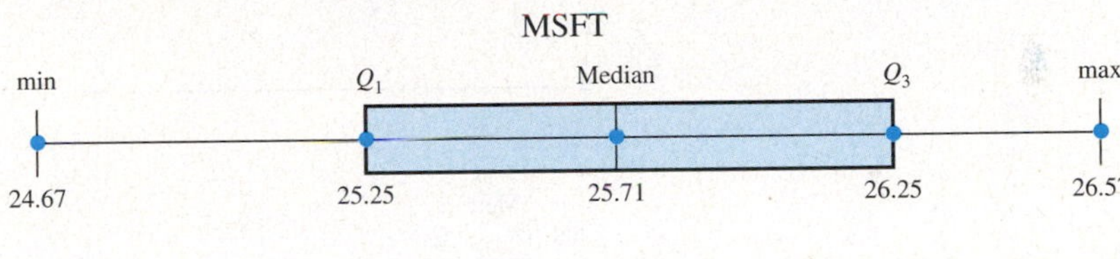


CVX

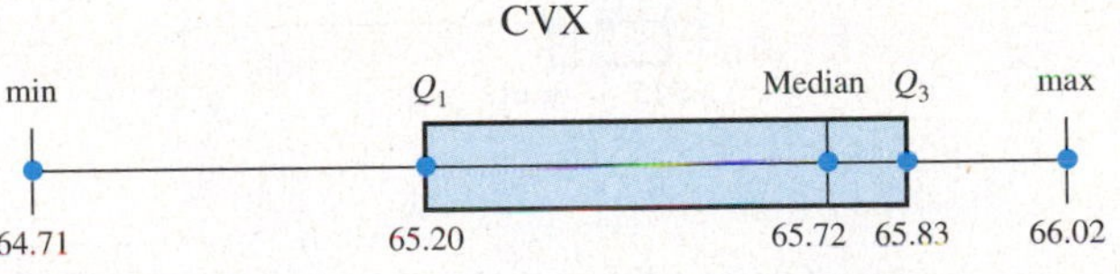


**NOTE**

See exercises 41 to 46 for more information concerning the range of values.

While the closing prices of CVX stock are higher than the closing prices of MSFT, there are other (more) interesting characteristics that the box-and-whisker plots allow us to see.

The MSFT plot is wider than the CVX plot. This means that there is a wider range of values in the MSFT summary. The MSFT stock exhibits greater *volatility*.

Also notice that the median, third quartile value, and max are close together in the CVX plot. This indicates that the numbers in the lower half of the CVX data are more spread out (show more variation) than the numbers in the upper half.

## Check Yourself 5

A different 2-week period is shown in the following table. Construct box-and-whisker plots to compare the results. Interpret your findings.

| Stock | Mon. | Tues. | Wed. | Thu. | Fri. |
|---|---|---|---|---|---|
| MSFT | 54.77 | 55.80 | 54.24 | 55.81 | 55.92 |
| CVX | 69.90 | 68.45 | 68.05 | 69.12 | 68.61 |
| MSFT | 56.39 | 56.97 | 56.27 | 55.35 | 51.46 |
| CVX | 68.18 | 68.50 | 68.15 | 68.32 | 68.11 |

## Check Yourself ANSWERS

**1.** **(a)** Median 7, mode 7; **(b)** median 7, mode 7
**2.** **(a)** $Q_1 = 2, Q_3 = 7$; **(b)** $Q_1 = 7, Q_3 = 24$
**3.** **(a)** 1, 2, 7, 7, 10; **(b)** 6, 7, 7, 24, 156
**4.** **(a)**

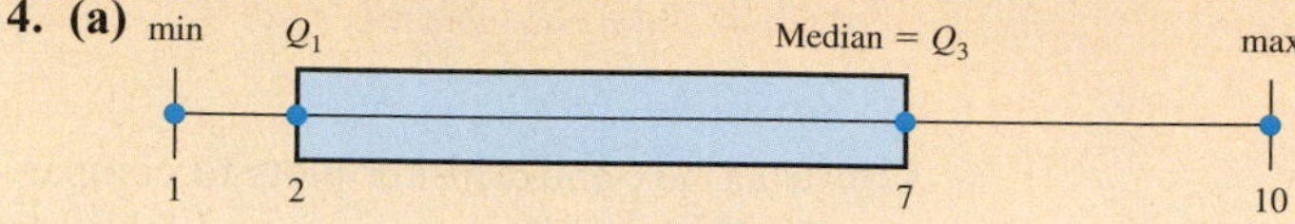


**(b)**

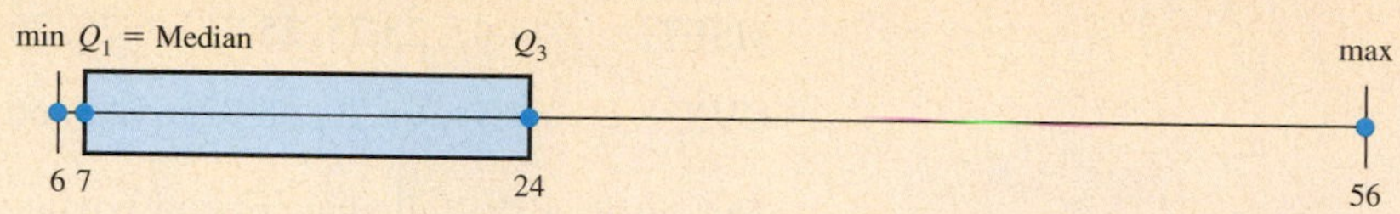


**5.**

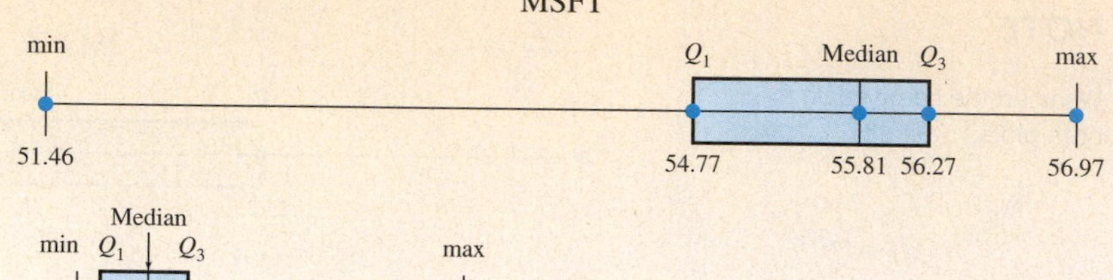


Median
min $Q_1$ $Q_3$ max
68.05 68.15 68.61 69.90
68.39

CVX

CVX still has higher prices than MSFT (though not nearly as much as in Example 5). MSFT exhibits greater volatility, but more of its values are at the high end.

## Reading Your Text

The following fill-in-the-blank exercises are designed to ensure that you understand some of the key vocabulary used in this section.

### SECTION 9.5

**(a)** The ______________ divide a data set into quarters (four equal parts).

**(b)** Just as a single number (the median) separates the data into two groups, it takes ______________ numbers to separate data into four groups.

**(c)** The five-number ______________ associated with a data set is given by min, $Q_1$, median, $Q_3$, max.

**(d)** A graph based on the five-number summary is called a ______________ plot.

| Basic Skills | Advanced Skills | Vocational-Technical Applications | Calculator/Computer | Above and Beyond |
|---|---|---|---|---|

# 9.5 exercises

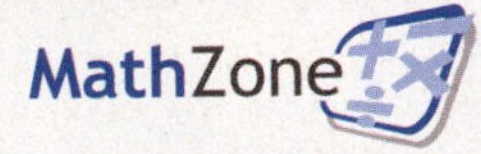

Boost your grade at mathzone.com!

> Practice Problems
> NetTutor
> Self-Tests
> e-Professors
> Videos

Name ____________

Section ________ Date ________

< Objective 1 >

*Find the median of each set of numbers.*

**1.** 2, 8, 5, 6, 9, 7, 4, 4, 5, 4, 3

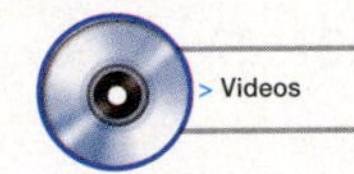


**2.** 7, 7, 5, 4, 1, 9, 8, 8, 8, 5, 2

**3.** 11, 12, 16, 14, 14, 14, 8, 12, 10, 18

**4.** 26, 30, 38, 67, 59, 21, 17, 85, 22, 22

**5.** 326, 245, 123, 222, 245, 300, 350, 602, 256

**6.** 0.10, 0.25, 0.24, 0.24, 0.30, 0.20, 0.18, 0.21, 0.28, 0.26

*Find the first and third quartiles, $Q_1$ and $Q_3$, of each set of numbers.*

**7.** 2, 8, 5, 6, 9, 7, 4, 4, 5, 4, 3

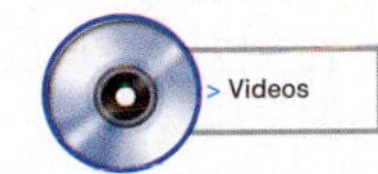


**8.** 7, 7, 5, 4, 1, 9, 8, 8, 8, 5, 2

**9.** 11, 12, 16, 14, 14, 14, 8, 12, 10, 18

**10.** 26, 30, 38, 67, 59, 21, 17, 85, 22, 22

**11.** 326, 245, 123, 222, 245, 300, 350, 602, 256

**12.** 0.10, 0.25, 0.24, 0.24, 0.30, 0.20, 0.18, 0.21, 0.28, 0.26

*Find the min and max of each set of numbers.*

**13.** 2, 8, 5, 6, 9, 7, 4, 4, 5, 4, 3

**14.** 7, 7, 5, 4, 1, 9, 8, 8, 8, 5, 2

## Answers

1. ____________
2. ____________
3. ____________
4. ____________
5. ____________
6. ____________
7. ____________
8. ____________
9. ____________
10. ____________
11. ____________
12. ____________
13. ____________
14. ____________

## Answers

15. ______

16. ______

17. ______

18. ______

19. ______

20. ______

21. ______

22. ______

23. ______

24. ______

25. ______

26. ______

**15.** 11, 12, 16, 14, 14, 14, 8, 12, 10, 18

**16.** 26, 30, 38, 67, 59, 21, 17, 85, 22, 22

**17.** 326, 245, 123, 222, 245, 300, 350, 602, 256

**18.** 0.10, 0.25, 0.24, 0.24, 0.30, 0.20, 0.18, 0.21, 0.28, 0.26

< Objective 2 >

*Give the five-number summary of each set of numbers.*

**19.** 2, 8, 5, 6, 9, 7, 4, 4, 5, 4, 3

**20.** 7, 7, 5, 4, 1, 9, 8, 8, 8, 5, 2

**21.** 11, 12, 16, 14, 14, 14, 8, 12, 10, 18

**22.** 26, 30, 38, 67, 59, 21, 17, 85, 22, 22

**23.** 326, 245, 123, 222, 245, 300, 350, 602, 256

**24.** 0.10, 0.25, 0.24, 0.24, 0.30, 0.20, 0.18, 0.21, 0.28, 0.26

< Objective 3 >

*Construct a box-and-whisker plot for each set of numbers.*

**25.** 2, 8, 5, 6, 9, 7, 4, 4, 5, 4, 3

**26.** 7, 7, 5, 4, 1, 9, 8, 8, 8, 5, 2

27. 11, 12, 16, 14, 14, 14, 8, 12, 10, 18

28. 26, 30, 38, 67, 59, 21, 17, 85, 22, 22

29. 326, 245, 123, 222, 245, 300, 350, 602, 256

30. 0.10, 0.25, 0.24, 0.24, 0.30, 0.20, 0.18, 0.21, 0.28, 0.26

*Determine whether each statement is* **true** *or* **false.**

31. The quartiles divide the data set into three equal parts.

32. The median is equal to the second quartile.

*In each of the following statements, fill in the blank with* **always, sometimes,** *or* **never.**

33. Before finding the quartiles, we must ______________ sort the data.

34. The third quartile is ______________ smaller than the median.

## Answers

27. ______________

28. ______________

29. ______________

30. ______________

31. ______________

32. ______________

33. ______________

34. ______________

Basic Skills | **Advanced Skills** | Vocational-Technical Applications | Calculator/Computer | Above and Beyond

Answers

35. ______________________

< Objective 4 >

*The following table gives the maximum depth and total bottom time for 25 recreational scuba dives. Use this table for exercises 35 and 36.*

**Diving Data**

| Dive no. | Max Depth (ft) | Bottom Time (s) |
|---|---|---|
| 1 | 24 | 55 |
| 2 | 24 | 10 |
| 3 | 26 | 22 |
| 4 | 30 | 35 |
| 5 | 45 | 31 |
| 6 | 58 | 45 |
| 7 | 109 | 25 |
| 8 | 40 | 35 |
| 9 | 42 | 30 |
| 10 | 42 | 26 |
| 11 | 48 | 29 |
| 12 | 64 | 31 |
| 13 | 50 | 32 |
| 14 | 72 | 24 |
| 15 | 42 | 35 |
| 16 | 55 | 33 |
| 17 | 64 | 24 |
| 18 | 71 | 32 |
| 19 | 63 | 27 |
| 20 | 45 | 30 |
| 21 | 43 | 30 |
| 22 | 59 | 29 |
| 23 | 59 | 26 |
| 24 | 51 | 20 |
| 25 | 78 | 23 |

**35. Science and Medicine**

**(a)** Give the five-number summary of the depth data.

**(b)** Construct a box-and-whisker plot for the depth data.

**(c)** Describe the depth data based on the box-and-whisker plot.

**36. Science and Medicine**

**(a)** Give the five-number summary of the bottom time data.

**(b)** Construct a box-and-whisker plot for the bottom time data.

**(c)** Describe the bottom time data based on the box-and-whisker plot.

**37. Business and Finance** Closing prices for Adobe Systems stock are shown during a 2-week period in 2003. Construct and interpret a box-and-whisker plot for these data.

**ADBE (Adobe Systems)**

| **Date** | 2/3 | 2/4 | 2/5 | 2/6 | 2/7 | 2/10 | 2/11 | 2/12 | 2/13 | 2/14 |
|---|---|---|---|---|---|---|---|---|---|---|
| **Price** | 27.06 | 26.43 | 26.52 | 26.52 | 26.14 | 26.63 | 26.70 | 26.78 | 26.93 | 27.43 |

*Source:* Yahoo! Finance.

**38. Business and Finance** The closing prices for Kellogg Company stock are shown during the same 2-week period in 2003 as the Adobe Systems data in exercise 37.

**K (Kellogg Co.)**

| **Date** | 2/3 | 2/4 | 2/5 | 2/6 | 2/7 | 2/10 | 2/11 | 2/12 | 2/13 | 2/14 |
|---|---|---|---|---|---|---|---|---|---|---|
| **Price** | 32.32 | 32.53 | 32.33 | 32.24 | 32.10 | 32.12 | 31.50 | 31.30 | 31.21 | 31.65 |

*Source:* Yahoo! Finance.

**(a)** Construct a box-and-whisker plot for the Kellogg Company closing stock price data.

**(b)** Describe any distinctive features shown by the plot.

**(c)** Compare your plot with the one constructed in exercise 37.

**Answers**

36. ______

37. ______

38. ______

Answers

39. 

40. 

41. 

42. 

43. 

**39. Statistics** The following table gives the mean temperature (in degrees Fahrenheit) for the month of July over a 20-year period in Roanoke, VA.

| **Year** | 1981 | 1982 | 1983 | 1984 | 1985 | 1986 | 1987 | 1988 | 1989 | 1990 |
|---|---|---|---|---|---|---|---|---|---|---|
| **Temp.** | 76.1° | 75.5° | 77.0° | 73.2° | 76.5° | 78.8° | 79.2° | 77.0° | 75.9° | 76.8° |
| **Year** | 1991 | 1992 | 1993 | 1994 | 1995 | 1996 | 1997 | 1998 | 1999 | 2000 |
| **Temp.** | 77.9° | 76.9° | 80.2° | 77.5° | 76.9° | 74.4° | 76.1° | 77.5° | 79.1° | 73.3° |

*Source:* NOAA; NCDC.

(a) Construct a box-and-whisker plot based on the data.

(b) Discuss any significant features of the plot.

**40. Statistics** The following table gives the mean temperature (in degrees Fahrenheit) for the month of January over a 20-year period in Dickinson, ND.

| **Year** | 1981 | 1982 | 1983 | 1984 | 1985 | 1986 | 1987 | 1988 | 1989 | 1990 |
|---|---|---|---|---|---|---|---|---|---|---|
| **Temp.** | 26.6° | −0.2° | 27.3° | 22.5° | 13.1° | 23.1° | 24.7° | 13.4° | 18.5° | 24.5° |
| **Year** | 1991 | 1992 | 1993 | 1994 | 1995 | 1996 | 1997 | 1998 | 1999 | 2000 |
| **Temp.** | 11.6° | 27.7° | 9.6° | 6.1° | 16.5° | 6.2° | 8.5° | 17.1° | 14.7° | 19.3° |

*Source:* NOAA; NCDC.

(a) Construct a box-and-whisker plot based on the data.

(b) Discuss any significant features of the plot.

(c) Compare the box-and-whisker plot constructed in exercise 39 with the one constructed in this exercise.

One measure used to describe a data set is the **range.** The range of a data set is given by the difference between the max and the min of the set. The range describes the variability of the data (that is, how much do the numbers vary).

Range = max − min

*Find the range of each set of numbers.*

**41.** 2, 8, 5, 6, 9, 7, 4, 4, 5, 4, 3

**42.** 7, 7, 5, 4, 1, 9, 8, 8, 8, 5, 2

**43.** 11, 12, 16, 14, 14, 14, 8, 12, 10, 18

44. 26, 30, 38, 67, 59, 21, 17, 85, 22, 22

45. 326, 245, 123, 222, 245, 300, 350, 602, 256

46. 0.10, 0.25, 0.24, 0.24, 0.30, 0.20, 0.18, 0.21, 0.28, 0.26

Another measure that we use is the **interquartile range** (IQR). The IQR is given by the difference between the third quartile and the first quartile. The IQR measures how large an interval is needed to contain the middle 50% of the data. It is used to measure variability and to assist in determining if there are any *outliers* in the data.

$\text{IQR} = Q_3 - Q_1$

*Find the IQR of each set of numbers.*

47. 2, 8, 5, 6, 9, 7, 4, 4, 5, 4, 3

48. 7, 7, 5, 4, 1, 9, 8, 8, 8, 5, 2

49. 11, 12, 16, 14, 14, 14, 8, 12, 10, 18

50. 26, 30, 38, 67, 59, 21, 17, 85, 22, 22

51. 326, 245, 123, 222, 245, 300, 350, 602, 256

52. 0.10, 0.25, 0.24, 0.24, 0.30, 0.20, 0.18, 0.21, 0.28, 0.26

One characteristic we look for when describing and analyzing a data set is the presence of **outliers.** An outlier of a data set is a number that is "far away" from most of the other numbers in the set. We use the IQR to determine whether a data set has any outliers.

A number from a set is an outlier if it is more than 1.5 IQRs from either the first quartile or the third quartile. That is, we compute the following "boundaries" for outliers.

Lower boundary: $Q_1 - 1.5 \times \text{IQR}$

Upper boundary: $Q_3 + 1.5 \times \text{IQR}$

Any number in the data set that is less than the lower boundary or greater than the upper boundary is considered an outlier.

*Find any outliers in each set of numbers.*

53. 2, 8, 5, 6, 9, 7, 4, 4, 5, 4, 3

54. 7, 7, 5, 4, 1, 9, 8, 8, 8, 5, 2

55. 11, 12, 16, 14, 14, 14, 8, 12, 10, 18

56. 26, 30, 38, 67, 59, 21, 17, 85, 22, 22

## Answers

44. ______

45. ______

46. ______

47. ______

48. ______

49. ______

50. ______

51. ______

52. ______

53. ______

54. ______

55. ______

56. ______

57. 326, 245, 123, 222, 245, 300, 350, 602, 256

58. 0.10, 0.25, 0.24, 0.24, 0.30, 0.20, 0.18, 0.21, 0.28, 0.26

Some outliers are so far from the rest of the data that we call them **extreme outliers.** An extreme outlier is more than 3 IQRs from one of the quartiles. Outliers that are not extreme outliers are called **mild outliers.**

To determine the extreme outliers, again we set boundaries.

Lower boundary: $Q_1 - 3 \times \text{IQR}$

Upper boundary: $Q_3 + 3 \times \text{IQR}$

*Classify any outliers of each data set as extreme or mild.*

59. 2, 8, 5, 6, 9, 7, 4, 4, 5, 4, 3

60. 7, 7, 5, 19, 4, 1, 9, 8, 8, 8, 5, 2

61. 11, 12, 16, 14, 14, 14, 8, 12, 10, 18

62. 26, 30, 38, 67, 59, 21, 17, 85, 22, 22

63. 326, 245, 61, 222, 245, 300, 350, 842, 256

64. 0.10, 0.25, 0.24, 0.24, 0.92, 0.20, 0.18, 0.21, 0.28, 0.26

Basic Skills | Advanced Skills | Vocational-Technical Applications | Calculator/Computer | **Above and Beyond**

65. Research Microsoft Corporation's (MSFT) closing stock prices for the most recent 2-week period (see Example 5). Give the five-number summary of these data, construct a box-and-whisker plot for the data, and interpret your display.

66. Research Chevron-Texaco's (CVX) closing stock prices for the most recent 2-week period (see Example 5). Give the five-number summary of these data, construct a box-and-whisker plot for the data, and interpret your display.

## Answers

57. ______
58. ______
59. ______
60. ______
61. ______
62. ______
63. ______
64. ______
65. ______
66. ______

## Answers

**1.** 5 **3.** 13 **5.** 256 **7.** 4; 7 **9.** 11; 14 **11.** 233.5; 338
**13.** 2; 9 **15.** 8; 18 **17.** 123; 602 **19.** 2, 4, 5, 7, 9
**21.** 8, 11, 13, 14, 18 **23.** 123, 233.5, 256, 338, 602
**25.**

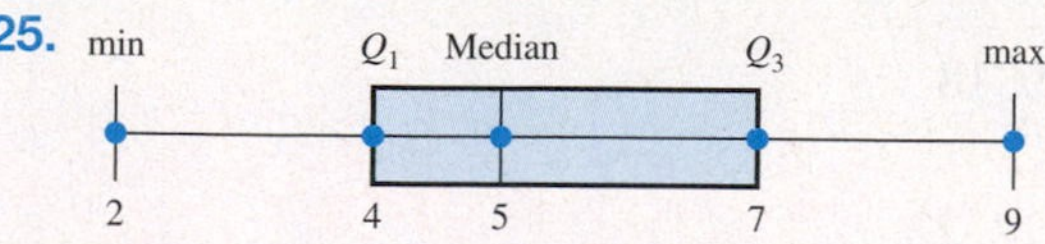

27. 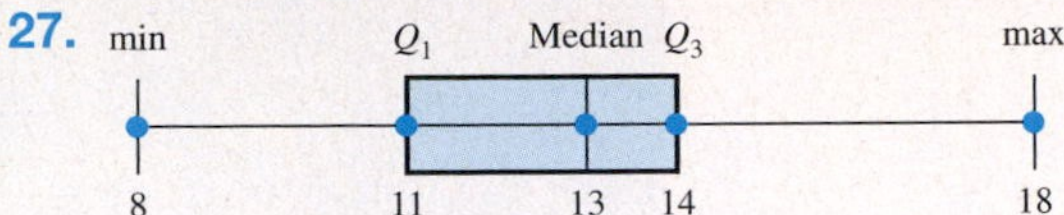


29. 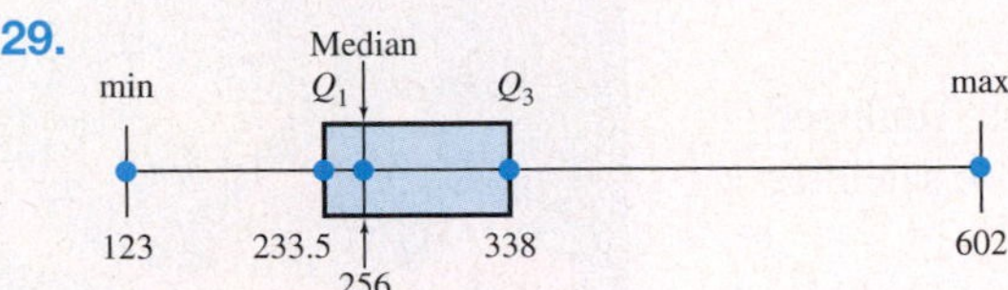


31. False 33. always

35. (a) 24, 42, 50, 63.5, 109

(b) 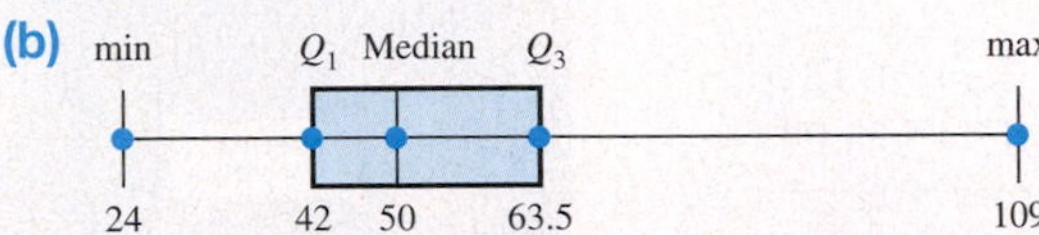


(c) Above and Beyond

37. 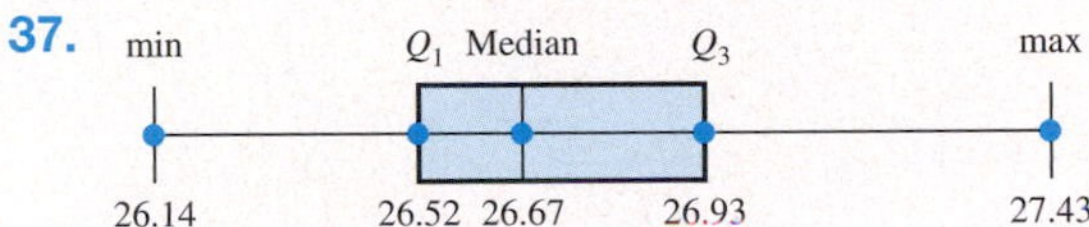


39. (a) 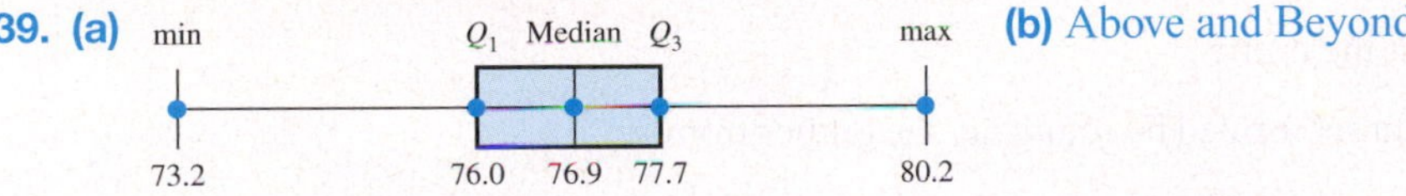


(b) Above and Beyond

41. 7 43. 10 45. 479 47. 3 49. 3 51. 104.5

53. None 55. None 57. 602 59. None 61. None

63. Mild: 61; extreme: 842 65. Above and Beyond

# Activity 26 ::
# Outliers in Scientific Data

Marine scientists have placed numerous buoys in the oceans. These buoys are equipped to measure various properties of the water and to relay these measurements to satellites at regular intervals. The data are then collected and analyzed.

The table shows midday temperature data (in degrees Fahrenheit) collected off the coast of Cape Charles, VA, over a 12-day period in July 2003.

| **Date** | 7/1 | 7/2 | 7/3 | 7/4 | 7/5 | 7/6 | 7/7 | 7/8 | 7/9 | 7/10 | 7/11 | 7/12 |
|---|---|---|---|---|---|---|---|---|---|---|---|---|
| **Temp.** | 72.0° | 73.2° | 71.8° | 72.6° | 73.3° | 74.1° | 74.8° | 76.0° | 57.7° | 77.5° | 77.1° | 78.2° |

*Source:* NODC.

1. Find the mean temperature for the period shown.
2. Give the five-number summary for these data.
3. Compute the interquartile range of these data.
4. Identify any outliers in the data set.
5. Compute the mean of the data set without the outlier.
6. Give the five-number summary for the data set formed by removing the outlier from the table.

   Investigation showed that the buoy responsible for the given data was nonterminally damaged sometime between 7/8 and 7/10. It is believed that a small, personal watercraft struck the buoy about when it sent the July 9 data.

7. Describe how you might use this information to analyze salinity data measured by the same buoy.
8. Go to the text website to access more weather data.

| Definition/Procedure | Example | Reference |
|---|---|---|

## Means, Medians, and Modes

Section 9.1

### Finding the Mean

To find the *mean* for a set of numbers, follow these two steps:

**Step 1** Add all the numbers in the set.

**Step 2** Divide that sum by the number of items in the set.

Given the numbers 4, 8, 17, 23,

$4 + 8 + 17 + 23 = 52$

$\text{Mean} = \frac{52}{4} = 13$

*p.* 632

### Finding the Median

The *median* is the number for which there are as many instances that are above that number as there are instances below it. To find the median follow these steps:

**Step 1** Rewrite the numbers in order from smallest to largest.

**Step 2** Count from both ends to find the number in the middle.

**Step 3** If there are two numbers in the middle, add them together and find their mean.

Given the numbers 9, 2, 5, 13, 7, 3, rewrite them as

2, 3, 5, 7, 9, 13

The middle numbers are 5 and 7

$\frac{5 + 7}{2} = \frac{12}{2} = 6$

Median $= 6$

*p.* 633

### Finding the Mode

The *mode* is the number that occurs most frequently in a set of numbers.

Given the numbers 2, 3, 3, 3, 5, 5, 7, 7, 9, 11.
3 is the mode.

*p.* 635

## Tables, Pictographs, and Bar Graphs

Section 9.2

A **table** is a display of information in parallel columns or rows.

*p.* 646

**Federal Highway Admin., U.S. Dept. of Transportation**

| Year | Cars Registered |
|---|---|
| 1970 | 89,243,557 |
| 1975 | 106,705,934 |
| 1980 | 121,600,843 |
| 1985 | 131,664,029 |
| 1990 | 143,549,627 |
| 1995 | 136,066,045 |
| 2000 | 133,621,245 |

A **graph** is a diagram that relates two different pieces of information. One of the most common graphs is the **bar graph.**

*p.* 648

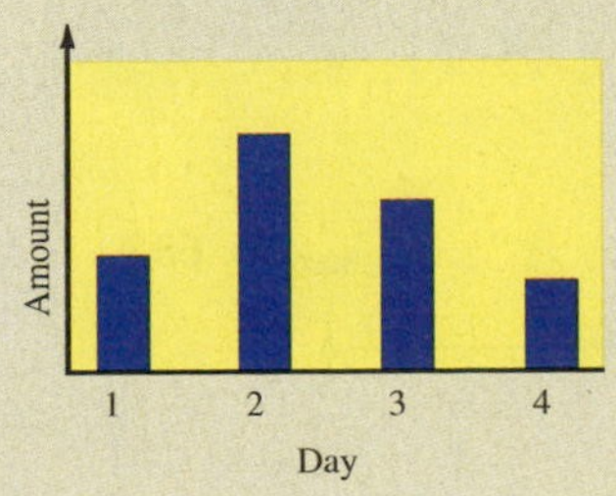

Amount

1 2 3 4

Day

 

| Definition/Procedure | Example | Reference |
|---|---|---|
| **Line Graphs and Predictions** | | Section 9.3 |
| **Line graph** In line graphs, one of the axes is usually related to time. | 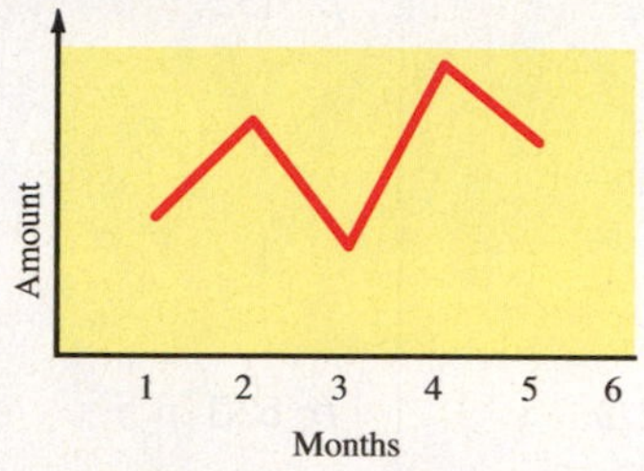 | p. 661 |
| **Creating Bar Graphs and Pie Charts** | | Section 9.4 |
| **Pie chart** Pie charts are graphs that show the component parts of a whole. | Each percent is shown as the percent of a 360° circle.<br>30% of 360° =<br>$0.30 \times 360 = 108°$<br>20% of 360° =<br>$0.20 \times 360 = 72°$ | p. 672 |
| **Describing and Summarizing Data Sets** | | Section 9.5 |
| *Finding Quartiles*<br>To find the first quartile, find the median of the set of numbers to the left of the median.<br>The third quartile is the median of the numbers to the right of the median. | Given 2, 3, 5, 7, 9, 13, we found the median to be 6.<br>The list to the left of 6 is 2, 3, 5 and has a median of 3.<br>Therefore, $Q_1 = 3$.<br>Similarly, $Q_3 = 9$. | p. 686 |
| *Finding a Five-Number Summary*<br>The five-number summary is given by the list<br>min, $Q_1$, median, $Q_3$, max | The five-number summary of the preceding list is<br>2, 3, 6, 9, 13 | p. 687 |
| *Box-and-Whisker Plots*<br>Mark off the five-number summary on a number line from min to max and draw a rectangle between the quartiles. | 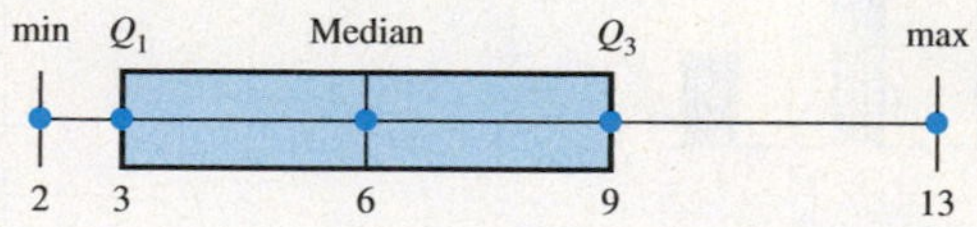 | p. 688 |

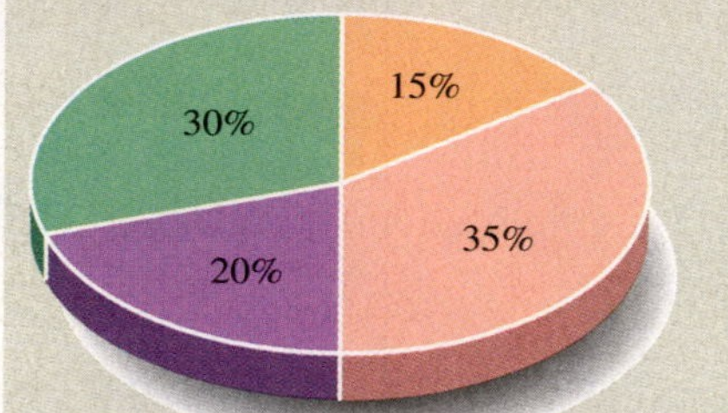

This summary exercise set is provided to give you practice with each of the objectives of this chapter. Each exercise is keyed to the appropriate chapter section. When you are finished, you can check your answers to the odd-numbered exercises against those presented in the back of the text. If you have difficulty with any of these exercises, go back and reread the examples from that section. The answers to the even-numbered exercises appear in the *Instructor's Solutions Manual.* Your instructor will give you guidelines on how to best use these exercises in your instructional setting.

**9.1** *Find the mean for each set of numbers.*

**1.** 8, 6, 7, 4, 5

**2.** 12, 14, 17, 19, 13

**3.** 117, 121, 122, 118, 115, 125, 123, 119

**4.** 134, 126, 128, 129, 133, 125, 122, 127

**5.** Elmer had test scores of 89, 71, 93, and 87 on his four math tests. What was his mean score?

**6.** The costs (in dollars) of the seven textbooks that Jacob needs for the spring semester are 77, 66, 55, 49, 85, 80, and 78. Find the mean cost of these books.

*Find the median and the mode for each set of data.*

**7.** 16, 20, 20, 19, 18

**8.** 8, 9, 9, 11, 11, 8, 7, 11, 12, 14, 10

**9.** 26, 31, 28, 35, 27, 28, 31, 30, 28, 30

**10.** 15, 18, 21, 23, 17, 19, 30, 35, 15, 32

**11.** Anita's first four test scores in her mathematics class were 88, 91, 86, and 93. What score must she get on her next test to have a mean of 90?

**12.** The weekly sales of a small company for 3 weeks were \$2,400, \$2,800, and \$3,300. How much do sales need to be in the fourth week to achieve a mean of \$3,000?

**9.2** *Use the following table for exercises 13 to 20.*

**World Motor Vehicle Production, 1950–2000**

| | (in thousands) | | | | | |
|---|---|---|---|---|---|---|
| **Year** | **United States** | **Canada** | **Europe** | **Japan** | **Other** | **World Total** |
| 2000 | 12,778 | 2,966 | 15,176 | 10,145 | 15,978 | 57,334 |
| 1999 | 13,025 | 3,057 | 15,395 | 9,895 | 14,388 | 55,760 |
| 1998 | 12,003 | 2,570 | 15,467 | 10,050 | 13,258 | 53,031 |
| 1995 | 11,985 | 2,408 | 17,045 | 10,196 | 8,349 | 49,983 |
| 1990 | 9,783 | 1,928 | 18,866 | 13,487 | 4,496 | 48,554 |
| 1985 | 11,653 | 1,933 | 16,113 | 12,271 | 2,939 | 44,909 |
| 1980 | 8,010 | 1,324 | 15,496 | 11,043 | 2,692 | 38,565 |
| 1970 | 8,284 | 1,160 | 13,049 | 5,289 | 1,637 | 29,419 |
| 1960 | 7,905 | 398 | 6,837 | 482 | 866 | 16,488 |
| 1950 | 8,006 | 388 | 1,991 | 32 | 160 | 10,577 |

*Note:* As far as can be determined, production refers to vehicles locally manufactured.
*Source:* American Automobile Manufacturers Assn.

13. What was the motor vehicle production in Japan in 1950? 2000?

14. What was the motor vehicle production in countries outside the United States in 1950? 2000?

15. What was the percent increase in motor vehicle production in the United States from 1950 to 2000?

16. What was the percent increase in motor vehicle production in countries outside the United States from 1950 to 2000?

17. What percent of world motor vehicle production occurred in Japan in 2000?

18. What percent of world motor vehicle production occurred in the United States in 2000?

19. What percent of world motor vehicle production occurred outside the United States and Japan in 2000?

20. Between 1950 and 2000, did the production of motor vehicles increase by a greater percent in Canada or Europe?

21. The following table represents the 10 most expensive areas for home prices in the United States in 2000. Create a pictograph to represent these data.

| City | Avg. Price |
|---|---|
| Palo Alto, CA | $974,237 |
| Beverly Hills, CA | 936,250 |
| San Mateo, CA | 873,250 |
| La Jolla, CA | 767,500 |
| San Francisco, CA | 759,250 |
| Beverly Hills-South, CA | 743,375 |
| Greenwich, CT | 706,500 |
| Hollywood Hills, CA | 663,375 |
| Wellesley, MA | 634,783 |

*Source:* Home Price Comparison Index.

22. The following table represents the 10 most affordable areas in which to live in the United States in 2000. Create a pictograph to represent these data.

| City | Avg. Price |
|---|---|
| Mt. Pleasant, MI | $103,640 |
| Sioux City, IA | 108,000 |
| Eau Claire, WI | 109,654 |
| Hastings, NE | 117,000 |
| Minot, ND | 124,400 |
| Yankton, SD | 124,750 |
| Stroudsburg, PA | 125,450 |
| Helena, MT | 126,475 |
| Fort Wayne, IN | 126,725 |
| Tulsa, OK | 129,950 |

*Source:* Home Price Comparison Index.

*Use the following bar graph for exercises 23 and 24.*

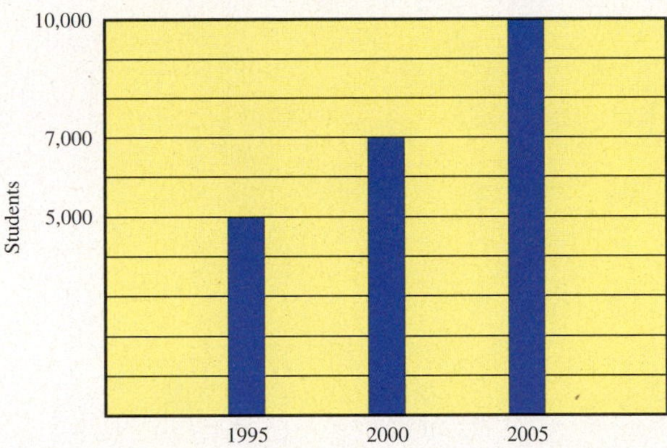


**23.** How many more students were enrolled in 2005 than in 1995?

**24.** What was the percent increase from 1995 to 2000?

**9.3** *Use the following line graph for exercises 25 to 27.*

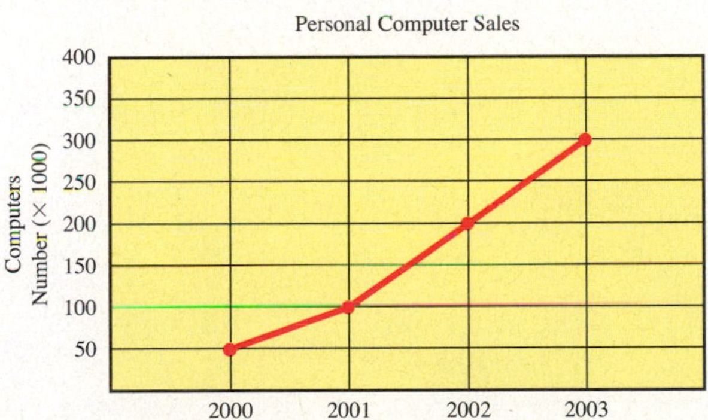


**25.** How many more personal computers were sold in 2003 than in 2000?

**26.** What was the percent increase in sales from 2000 to 2003?

**27.** Predict the sales of personal computers in the year 2004.

**9.4** *The following table shows the U.S. motor vehicle production in thousands, by source, in 2002.*

| Source | Number (in thousands) | Percent |
|---|---|---|
| General Motors | 4,093 | 33% |
| Ford | 3,413 | 28% |
| Chrysler | 1,837 | 15% |
| Foreign-based domestics | 2,985 | 24% |

*Source:* Automotive News Market Data Books.

**28.** Create a bar graph from the motor vehicle production table.

**29.** Create a pie chart from the motor vehicle production table.

*The pie charts show the production of foreign-based domestic automobiles, by manufacturer, in the United States in* 1992 *and* 2002.

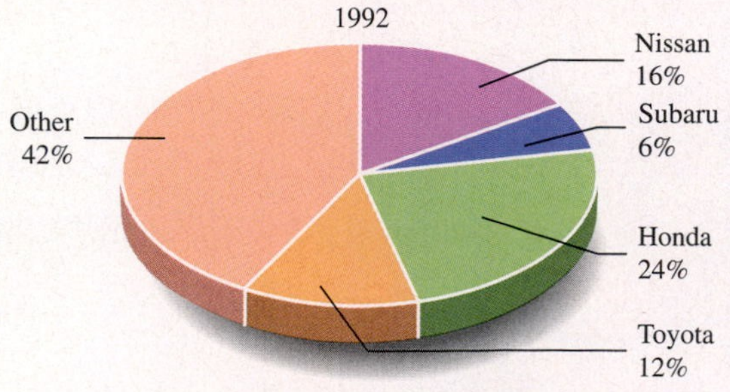


2002
Other
35%
Nissan
14%
Subaru
4%
Toyota
22%
Honda
25%

*Source:* Automotive News Market Data Books.

30. What was the percent of Nissan and Subaru automobiles produced in 2002?

31. What was the percent of Honda and Toyota automobiles produced in 2002?

32. Which company's production increased the most between 1992 and 2002?

## 9.5

33. Give the five-number summary for the given set of numbers.

30, 32, 21, 35, 28, 28, 24, 23, 26, 30

34. Construct a box-and-whiskers plot of the given set of numbers.

30, 32, 21, 35, 28, 28, 24, 23, 26, 30

35. The scores on the first examination for an algebra class are as follows:

93, 79, 84, 62, 66, 94, 90, 87, 74, 76, 77, 72, 68, 62, 74, 85, 98, 69, 97, 78, 71

Construct a box-and-whisker plot for the examination grades and describe the results.

# self-test 9



Name ______________________

Section __________ Date __________

The purpose of this self-test is to help you check your progress so that you can find sections and concepts that you need to review before the next exam. Allow yourself about an hour to take this test. At the end of that hour, check your answers against those given in the back of the text. If you missed any, note the section reference that accompanies the answer. Go back to that section and reread the examples until you have mastered that particular concept.

## Answers

1. ______________________

2. ______________________

3. ______________________

4. ______________________

5. ______________________

6. ______________________

7. ______________________

**1.** Find the mean of the numbers 12, 19, 15, 20, 11, and 13.

**2.** Find the median of the numbers 8, 9, 15, 3, 1.

**3.** Find the median of the numbers 12, 18, 9, 10, 16, 6.

**4.** Find the mode of the numbers 6, 2, 3, 6, 2, 9, 2, 6, 6.

**5.** **Construction** A bus carried 234 passengers on the first day of a newly scheduled route. The next 4 days there were 197, 172, 203, and 214 passengers. What was the mean number of riders per day?

**6.** **Statistics** To earn an A in biology, you must have a mean of 90 on four tests. Your scores thus far are 87, 89, and 91. How many points must you have on the final test to earn the A?

**7.** The following hair colors are from a class of seven students. What color is the mode?

brown, black, red, blonde, brown, brown, blue, gray

*In exercises 8 to 11, use the following table, which describes technology in the U.S. public schools from* 1995 *to* 1998.

**Technology in U.S. Public Schools, 1995–98**

| Technology | Number of Schools | | | |
|---|---|---|---|---|
| | 1995 | 1996 | 1997 | 1998 |
| **Schools with modems**[1] | **30,768** | **37,889** | **40,876** | **61,930** |
| Elementary | 16,010 | 20,250 | 22,234 | 35,066 |
| Junior high | 5,652 | 6,929 | 7,417 | 10,996 |
| Senior high[2] | 8,790 | 10,277 | 10,781 | 14,540 |
| **Schools with networks**[1] | **24,604** | **29,875** | **32,299** | **49,178** |
| Elementary | 11,693 | 14,868 | 16,441 | 26,422 |
| Junior high | 4,599 | 5,590 | 6,035 | 9,003 |
| Senior high[2] | 8,159 | 9,166 | 9,565 | 12,853 |
| **Schools with CD-ROMs**[1] | **34,480** | **43,499** | **46,388** | **64,200** |
| Elementary | 18,343 | 24,353 | 26,377 | 37,908 |
| Junior high | 6,510 | 7,952 | 8,410 | 11,023 |
| Senior high[2] | 9,327 | 10,756 | 11,140 | 13,985 |
| **Schools with Internet access**[1] | **NA** | **14,211** | **35,762** | **60,224** |
| Elementary | NA | 7,608 | 21,026 | 34,195 |
| Junior high | NA | 2,707 | 5,752 | 10,888 |
| Senior high[2] | NA | 3,736 | 8,984 | 13,829 |

NA = Not applicable. (1) Includes schools for special and adult education, not shown separate with grade spans of K-3, K-5, K-6, K-8, and K-12. (2) Includes schools with grade spans of technical and alternative high schools and schools with grade spans of 7–12, 9–12, and 10.

*Source:* Quality Education Data, Inc., Denver, CO.

**8.** What is the increase in schools with modems from 1995 to 1998?

**9.** How many senior high schools had modems, networks, or CD-ROMs in 1998?

**10.** What is the percent increase in public schools with Internet access from 1996 to 1998?

**11.** What is the percent increase in elementary schools that have modems, networks, or Internet access from 1996 to 1998?

**12.** Use the information in the following table to create a pictograph.

| Year | Population of United States (in millions) |
|---|---|
| 1950 | 151 |
| 1960 | 179 |
| 1970 | 203 |
| 1980 | 227 |
| 1990 | 248 |
| 2000 | 281 |

Let each figure represent 30,000,000 people.

## Answers

8. ______

9. ______

10. ______

11. ______

12. ______

Answers

13. ______

14. ______

15. ______

16. ______

17. ______

18. ______

19. ______

20. ______

*The following bar graph represents the number of bankruptcy filings during a recent 5-year period.*

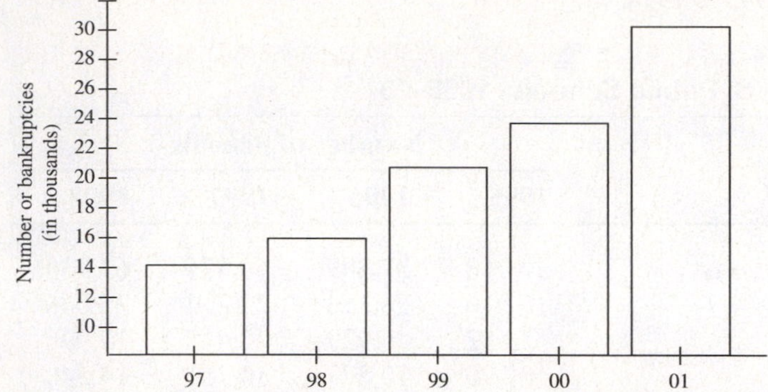


**13.** How many people filed for bankruptcy in 1998?

**14.** How many people filed for bankruptcy in 2001?

**15.** What was the increase in filings from 1999 to 2001?

**16.** What was the increase in filings from 1997 to 2001?

**17.** Which year had the greatest increase in filings?

*The following graph shows ticket sales for the last 6 months of the year.*

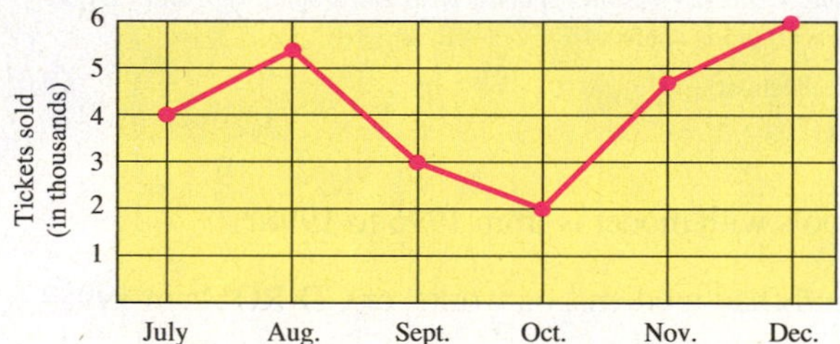


**18.** What month had the greatest number of ticket sales?

**19.** Between what two months did the greatest decrease in ticket sales occur?

**20.** The following information shows a relationship between the number of workers absent from the assembly line and the number of defects coming off the line. Use this information to create a line graph and then predict the number of defects coming off the line if five workers are absent.

| Number of Workers Absent | Number of Defects |
|---|---|
| 0 | 9 |
| 1 | 10 |
| 2 | 12 |
| 3 | 16 |
| 4 | 18 |

Answers

21. The following information represents a relationship between age and college education. Create a bar graph from this information.

| Age Group | Percent with 4 Years of College |
|---|---|
| 25–34 | 24 |
| 35–44 | 27 |
| 45–54 | 21 |
| 55–64 | 15 |
| 65+ | 11 |

21. ______

22. The following table represents the injuries suffered by students in classes involving participation.

| Type of Injury | Number |
|---|---|
| Knee | 24 |
| Ankle | 14 |
| Elbow | 4 |
| Wrist | 10 |
| Others | 8 |

Create a pie chart to show the distribution of the injury types.

22. ______

23. ______

24. ______

*The pie chart represents the way a new company ships its goods.*

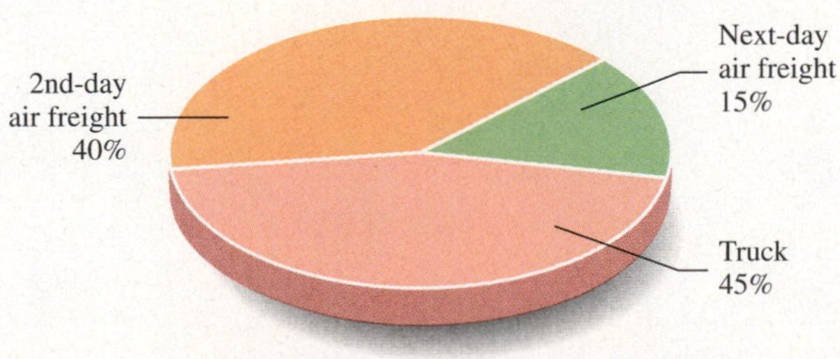


23. What percentage was shipped by truck?

24. What percentage was shipped by truck or second-day air freight?

 The Streeter/Hutchison Series in Mathematics Basic Mathematical Skills with Geometry

Answers

25. ______

**25.** November precipitation levels over a 20-year period in Amarillo, TX, are given by the following table.

**November Precipitation Levels; Amarillo, TX**

| **Year** | 1981 | 1982 | 1983 | 1984 | 1985 | 1986 | 1987 | 1988 | 1989 | 1990 |
|---|---|---|---|---|---|---|---|---|---|---|
| **Precipitation** | 1.50 | 0.76 | 0.36 | 1.10 | 0.42 | 1.83 | 0.44 | 0.30 | 0.00 | 0.52 |
| **Year** | 1991 | 1992 | 1993 | 1994 | 1995 | 1996 | 1997 | 1998 | 1999 | 2000 |
| **Precipitation** | 0.68 | 0.81 | 0.52 | 0.60 | 0.06 | 1.32 | 1.17 | 0.34 | 0.00 | 0.96 |

*Source:* NOAA; NCDC.

Construct a box-and-whisker plot for the precipitation data.

Name ______

Section ______ Date ______

The following exercises are presented to help you review concepts from earlier chapters. This is meant as review material and not as a comprehensive exam. The answers are presented in the back of the text. Beside each answer is a section reference for the concept. If you have difficulty with any of these exercises, be certain to at least read through the summary related to that section.

**1.** What is the place value of 6 in the numeral 126,489?

*Perform the indicated operation.*

**2.** $\begin{array}{r} 5{,}306 \\ 389 \\ +\ 26{,}583 \\ \hline \end{array}$

**3.** $\begin{array}{r} 74{,}983 \\ -\ 35{,}695 \\ \hline \end{array}$

**4.** $86 \times 305$

**5.** $27\overline{)8{,}322}$

**6.** $86{,}135 - 37{,}547$

**7.** $2.45 \times 30.7$

**8.** $\frac{4}{7} \times \frac{28}{24}$

**9.** $\frac{11}{15} \div \frac{121}{90}$

**10.** $3\frac{2}{3} + 5\frac{5}{6} - 2\frac{5}{12}$

*Solve for the unknown.*

**11.** $\frac{4}{7} = \frac{8}{x}$

**12.** $\frac{3}{5} = \frac{x}{15}$

**13.** Write 18% as a decimal and fraction.

**14.** Write $\frac{17}{40}$ as a decimal and percent.

*Do the indicated operations.*

**15.** $\begin{array}{r} 7\text{ lb}\ \ 9\text{ oz} \\ +\ 3\text{ lb}\ 12\text{ oz} \\ \hline \end{array}$

**16.** $\begin{array}{r} 4\text{ min}\ 10\text{ s} \\ -\ 2\text{ min}\ 35\text{ s} \\ \hline \end{array}$

*Complete each statement.*

**17.** 8 km = ________ m

**18.** 3,000 mg = ________ g

**19.** 500 cm = ________ m

**20.** 25 cL = ________ mL

**21.** **Statistics** According to the line graph, between what two years was the increase in benefits the greatest?

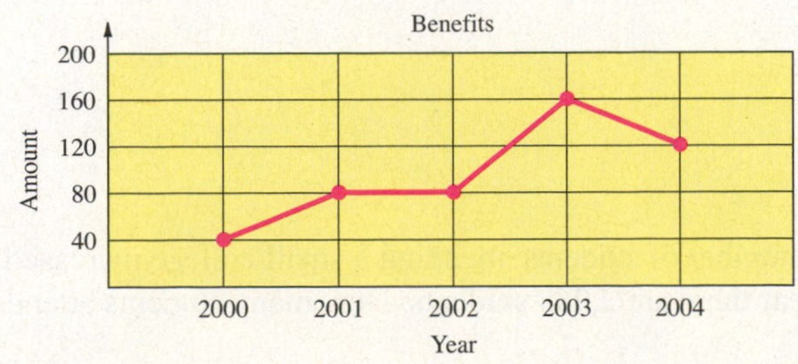


**Answers**

1. ______

2. ______

3. ______

4. ______

5. ______

6. ______

7. ______

8. ______

9. ______

10. ______

11. ______

12. ______

13. ______

14. ______

15. ______ 16. ______

17. ______ 18. ______

19. ______ 20. ______

21. ______

Answers

22. ______
23. ______
24. ______
25. ______
26. ______
27. ______
28. ______
29. ______
30. ______
31. ______
32. ______
33. ______
34. ______
35. ______

22. **Business and Finance** Construct a bar graph to represent the following data.

| Type of Stock | Number of Stocks |
|---|---|
| Industrial capital gains | 110 |
| Industrial consumer's goods | 184 |
| Public utilities | 60 |
| Railroads | 15 |
| Banks | 25 |
| Property liability insurance | 16 |

23. Calculate the mean, median, and mode for the following data.

11, 9, 3, 6, 7, 9, 8, 11, 12, 13, 11, 11, 4, 8, 12

24. If a boat uses 14 gal of gas to go 102 mi, how many gallons would be needed to go 510 mi?

*Express each of the following as a simplified rate.*

25. $\frac{1{,}760 \text{ ft}}{20 \text{ s}}$

26. $\frac{133 \text{ pitches}}{7 \text{ innings}}$

27. **Geometry** The floor of a room that is 12 ft by 18 ft is to be carpeted. If the price of the carpet is $17 per square yard, what will the carpet cost?

28. If you drive 152 miles in $3\frac{1}{6}$ hours, what is your average speed?

29. **Geometry** A rectangle has length $8\frac{3}{5}$ cm and width $5\frac{7}{10}$ cm. Find the perimeter.

30. **Geometry** The sides of a square each measure $13\frac{5}{6}$ ft. Find the perimeter of the square.

31. What is $9\frac{1}{2}\%$ of 1,400?

32. 15 is what percent of 7,500?

33. 111 is 60% of what number?

34. Find $\frac{2}{3}$ of $6\frac{1}{2}$.

35. **Social Science** The number of students attending a small college increased 6% since last year. This year there are 2,968 students. How many students attended last year?

CHAPTER

10

# The Real Number System

## INTRODUCTION

As he does the books for a medium-sized firm, Trinh reflects on how his knowledge of negative numbers helps him. Money comes in as revenue and goes out as costs. He also has lists of assets and debts to contend with.

Signed number arithmetic is used to model many situations that come up in business. Scientists, sports statisticians and fans, and even families considering their own personal finances find value in both positive and negative numbers.

In this chapter, we develop the skills necessary to perform arithmetic with real numbers. We also learn to model applications with real numbers; this leads into Chapter 11, which provides an introduction to algebra.

The real numbers are a tool that you will apply to several real-world applications in the activities, such as Activity 27 on p. 740, which requires you to gather data on the weather where you live.

## CHAPTER 10 OUTLINE

Name ____________

Section ________ Date ________

This pretest provides a preview of the types of exercises you will encounter in each section of this chapter. The answers for these exercises can be found in the back of the text. If you are working on your own, or ahead of the class, this pretest can help you identify the sections in which you should focus more of your time.

**10.1**

**1.** Place the following set in ascending order.

$$4, -7, 1, \frac{1}{2}, 9, -3, -\frac{3}{4}$$

**2.** Determine the maximum and minimum values of the following.

$-5, 4, 2, -4, 3, 0, 7, -3$

Evaluate the following.

**3.** $|-4|$

**4.** $-|6|$

**10.2–10.3**

**5.** **(a)** $12 + (-5)$
**(b)** $-9 + (-3)$

**6.** **(a)** $-7 - 15$
**(b)** $14 - (-3)$

**10.4–10.5**

**7.** **(a)** $(-3)(12)$
**(b)** $(-6)(-9)$

**8.** **(a)** $\frac{-24}{-8}$
**(b)** $40 \div (-8)$

**9.** $-4^2 \div (-2) + 8 \cdot 2$

**10.** $4 - 2^3 \div 2 \cdot (-3) + 4$

## Answers

1. ____________
2. ____________
3. ____________
4. ____________
5. ____________
6. ____________
7. ____________
8. ____________
9. ____________
10. ____________

# 10.1 Real Numbers and Order

< 10.1 Objectives >

1 > Represent an integer on a number line

2 > Order a set of real numbers

3 > Identify extreme values

4 > Simplify absolute value expressions

The numbers used to count things—1, 2, 3, 4, 5, and so on—are called the **natural** (or **counting) numbers.** The **whole numbers** consist of the natural numbers and zero—0, 1, 2, 3, 4, 5, and so on. They can be represented on a number line like the one shown. Zero (0) is considered the origin.

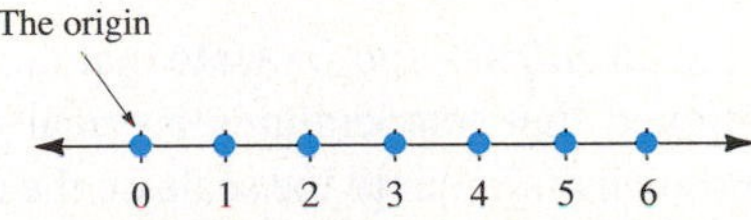


The number line continues indefinitely in both directions.

When numbers are used to represent physical quantities (such as altitude, temperature, and amount of money), it may be necessary to distinguish between *positive* and *negative* quantities. It is convenient to represent these quantities with plus (+) or minus (−) signs. For instance,

The Empire State building is 1,250 feet tall (+1,250).

The altitude of Death Valley is 282 ft *below* sea level (−282).

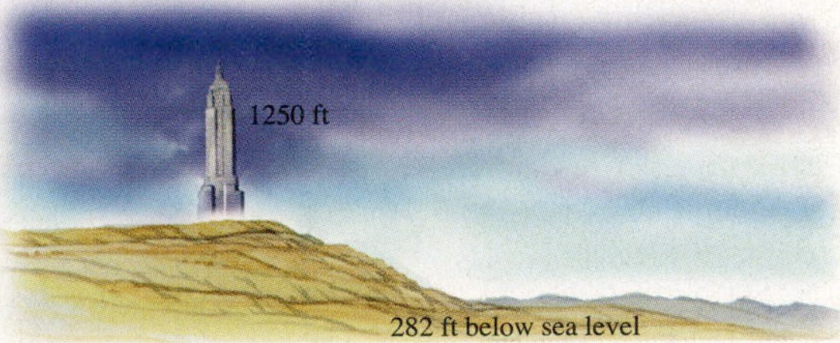


The temperature in Chicago is 10° *below* zero (−10°).

An account could show a *gain* of $100 (+100), or a *loss* of $100 (−100).

These numbers suggest the need to extend the whole numbers to include both positive numbers (such as +100) and negative numbers (such as −282).

To represent the negative numbers, we extend the number line to the *left* of zero and name equally spaced points.

Numbers used to name points to the right of zero are positive numbers. They are written with a positive $(+)$ sign or with no sign at all.

$+6$ and 9 are positive numbers

Numbers used to name points to the left of zero are negative numbers. They are always written with a negative $(-)$ sign.

$-3$ and $-20$ are negative numbers

Read "negative 3."

Positive and negative numbers considered together (along with zero) are **real numbers.**

Here is the number line extended to include both positive and negative numbers.

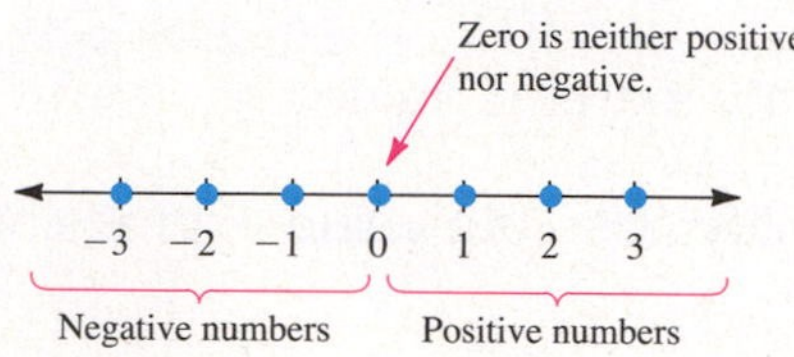


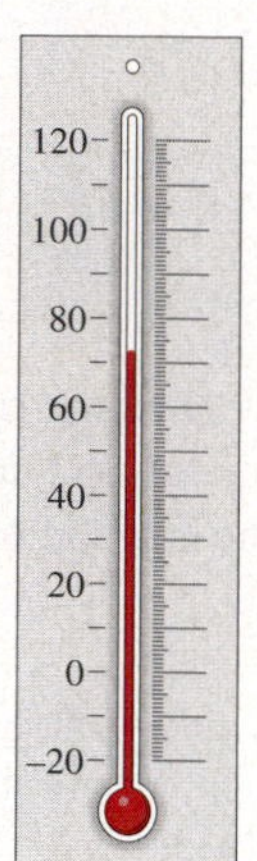


The numbers used to name the points shown on the preceding number line are called the **integers.** The integers consist of the natural numbers, their negatives, and the number 0. We can write

$$\ldots, -3, -2, -1, 0, 1, 2, 3, \ldots$$

The dots are called *ellipses* and indicate that the pattern continues.

We mentioned that temperatures are real numbers. Note that the scale on the thermometer shown is similar to the scale on the number line.

## Example 1 Representing Integers on the Number Line

< Objective 1 >

Represent the following integers on the number line shown.

$-3, -12, 8, 15, -7$

−12 −7 −3 8 15

−15 −10 −5 0 5 10 15

### Check Yourself 1

**Represent the following integers on a number line.**

**$-1, -9, 4, -11, 8, 20$**

−15 −10 −5 0 5 10 15 20

The set of numbers on the number line is *ordered.* The numbers get smaller as you move to the left on the number line and larger as you move to the right. When a set of numbers is written from smallest to largest, the numbers are said to be in *ascending order.*

## Example 2 Ordering Real Numbers

< Objective 2 >

Place each set of numbers in ascending order.

**(a)** 9, −5, −8, 3, 7

From smallest to largest, the numbers are

−8, −5, 3, 7, 9

Note that this is the order in which the numbers appear on a number line.

**(b)** 3, −2, 18, −20, −13

From smallest to largest, the numbers are

−20, −13, −2, 3, 18

### Check Yourself 2

Place each set of numbers in ascending order.

**(a)** 12, −13, 15, 2, −8, −3 **(b)** 3, 6, −9, −3, 8

**RECALL**

In Section 9.5, we called these values the **min** and **max**.

The least and greatest numbers in a set are called the **extreme values.** The least element is called the **minimum,** and the greatest element is called the **maximum.**

## Example 3 Labeling Extreme Values

< Objective 3 >

Determine the minimum and maximum values of each set of numbers.

**(a)** 9, −5, −8, 3, 7

From our previous ordering of these numbers, we see that −8, the least element, is the minimum, and 9, the greatest element, is the maximum.

**(b)** 3, −2, 18, −20, −13

−20 is the minimum and 18 is the maximum.

### Check Yourself 3

Determine the minimum and maximum values of each set of numbers.

**(a)** 12, −13, 15, 2, −8, −3 **(b)** 3, 6, −9, −3, 8

Integers are not the only kind of real numbers. Decimals and fractions are also real numbers.

## Example 4 Identifying Real Numbers as Integers

Which of the following real numbers are also integers?

**(a)** 145 is an integer. **(b)** $-28$ is an integer.

**(c)** 0.35 is not an integer. **(d)** $-\frac{2}{3}$ is not an integer.

### Check Yourself 4

Which of the following real numbers are also integers?

$-23 \quad 1{,}054 \quad -0.23 \quad 0 \quad -500 \quad -\frac{4}{5}$

**RECALL**

We call zero (0) the origin.

An important idea for our work in this chapter is the **absolute value** of a number. This represents the distance of the point named by the number from the origin on the number line.

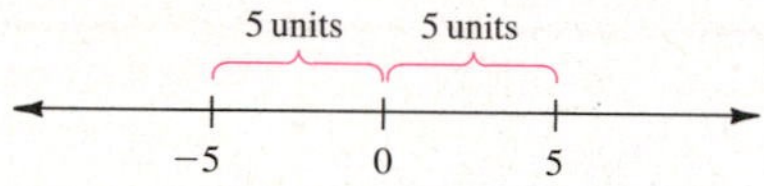


The absolute value of 5 is 5. The absolute value of $-5$ is also 5.

In symbols we write

$|5| = 5$ and $|-5| = 5$

Read "the absolute value of 5." Read "the absolute value of negative 5."

The absolute value of a number does *not* depend on whether the number is to the right or to the left of the origin, but on its *distance* from the origin.

## Example 5 Simplifying Absolute Value Expressions

< Objective 4 >

**(a)** $|7| = 7$

**(b)** $|-7| = 7$

**(c)** $-|-7| = -7$

This is the *negative,* or opposite, of the absolute value of negative 7.

**(d)** $|-10| + |10| = 10 + 10 = 20$

**(e)** $|8 - 3| = |5| = 5$ Absolute value bars serve as another set of grouping symbols, so do the operation inside first.

**(f)** $|8| - |3| = 8 - 3 = 5$

Here, evaluate the absolute values and then subtract.

## Check Yourself 5

Evaluate.

(a) $|8|$ (b) $|-8|$ (c) $-|-8|$
(d) $|-9| + |4|$ (e) $|9 - 4|$ (f) $|9| - |4|$

The language of real numbers can be applied to many situations. Example 6 gives you an idea of the variety of problems that are best modeled with real numbers.

## Example 6 Describing Applications with Real Numbers

Represent each quantity with an integer.

**(a)** A sick infant's temperature drops by 4 degrees Fahrenheit (°F) after being given acetaminophen.

After taking the medication, the infant's temperature went down, so we need a negative number.

$-4$°F

**RECALL**

You should include units in your answer to an application.

**(b)** A project is $1,500 over budget.

The cost of the project went up, so we use a positive number.

+$1,500 (or simply $1,500)

## Check Yourself 6

Represent each quantity with an integer.

(a) After an hour, it is noticed that an intravenous solution (IV) is running 30 milliliters (mL) ahead of schedule.
(b) The bridge is on schedule to be completed 7 months early.

## Check Yourself ANSWERS

**1.** Number line with points marked at $-11$, $-9$, $-1$, 4, 8, 20 (scale: $-20$, $-15$, $-10$, $-5$, 0, 5, 10, 15, 20)

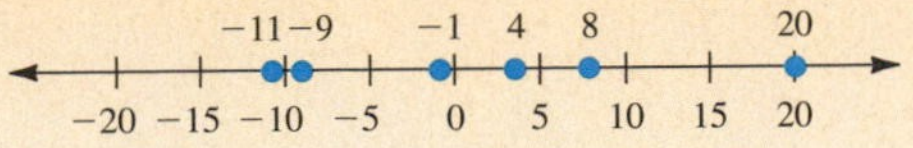

**2.** **(a)** $-13, -8, -3, 2, 12, 15$; **(b)** $-9, -3, 3, 6, 8$
**3.** **(a)** Minimum is $-13$; maximum is 15; **(b)** minimum is $-9$; maximum is 8
**4.** $-23$, 1,054, 0, and $-500$ **5.** **(a)** 8; **(b)** 8; **(c)** $-8$; **(d)** 13; **(e)** 5; **(f)** 5
**6.** **(a)** +30 mL; **(b)** –7 months

## Reading Your Text

The following fill-in-the-blank exercises are designed to ensure that you understand some of the key vocabulary used in this section.

### SECTION 10.1

**(a)** The whole numbers consist of the natural numbers and ____________.

**(b)** ____________ numbers are used to describe below-zero temperatures.

**(c)** When a set of numbers is written from smallest to largest, the numbers are said to be in ____________ order.

**(d)** The ____________ of a number is given by its distance from the origin on the number line.

Basic Skills | Advanced Skills | Vocational-Technical Applications | Calculator/Computer | Above and Beyond

# 10.1 exercises

MathZone
Boost your grade at mathzone.com!
> Practice Problems
> NetTutor
> Self-Tests
> e-Professors
> Videos

Name ______________

Section ________ Date ________

*Represent each quantity with a real number.*

**1.** An altitude of 400 feet (ft) above sea level

**2.** An altitude of 80 ft below sea level

**3.** A loss of \$200

**4.** A profit of \$400

**5.** A decrease in population of 25,000

**6.** An increase in population of 12,500

< Objective 1 >

*Represent the integers on the number lines shown.*

**7.** 5, −15, 18, −8, 3

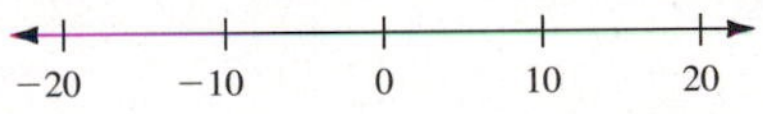
−20
−10
0
10
20

**8.** −18, 4, −5, 13, 9

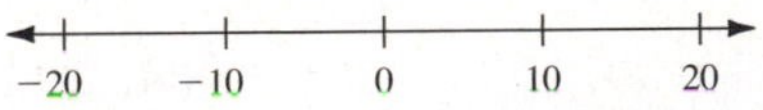
−20
−10
0
10
20

*Which of the following numbers are integers?*

**9.** $5, -\frac{2}{9}, 175, -234, -0.64$

> Videos

**10.** $-45, 0.35, \frac{3}{5}, 700, -26$

< Objective 2 >

*Place each of the following sets in ascending order.*

**11.** 3, −5, 2, 0, −7, −1, 8

> Videos

**12.** −2, 7, 1, −8, 6, −1, 0

**13.** 9, −2, −11, 4, −6, 1, 5

**14.** 23, −18, −5, −11, −15, 14, 20

< Objective 3 >

*For each set, determine the maximum and minimum values.*

**15.** 5, −6, 0, 10, −3, 15, 1, 8

> Videos

**16.** 9, −1, 3, 11, −4, 2, 5, −2

**17.** 21, −15, 0, 7, −9, 16, −3, 11

**18.** −22, 0, 22, −31, 18, −5, 3

## Answers

1. ______ 2. ______

3. ______ 4. ______

5. ______

6. ______

7. ______

8. ______

9. ______

10. ______

11. ______

12. ______

13. ______

14. ______

15. ______

16. ______

17. ______

18. ______

© The McGraw-Hill Companies. All Rights Reserved. The Streeter/Hutchison Series in Mathematics Basic Mathematical Skills with Geometry

## Answers

19. ______ 20. ______
21. ______ 22. ______
23. ______ 24. ______
25. ______ 26. ______
27. ______ 28. ______
29. ______ 30. ______
31. ______ 32. ______
33. ______ 34. ______
35. ______
36. ______
37. ______
38. ______
39. ______
40. ______
41. ______
42. ______
43. ______
44. ______
45. ______
46. ______

< Objective 4 >

*Evaluate.*

**19.** $|17|$ 

**20.** $|28|$

**21.** $|-10|$ 

**22.** $|-7|$

**23.** $-|3|$

**24.** $-|5|$

**25.** $-|-8|$

**26.** $-|-13|$

**27.** $|-2| + |3|$

**28.** $|4| + |-3|$

**29.** $|-9| + |9|$ 

**30.** $|11| + |-11|$

**31.** $|4| - |-4|$

**32.** $|5| - |-5|$

**33.** $|15| - |8|$

**34.** $|11| - |3|$

**35.** $|15 - 8|$

**36.** $|11 - 3|$

**37.** $|-9| + |2|$

**38.** $|-7| + |4|$

**39.** $|-8| - |-7|$

**40.** $|-9| - |-4|$

*Represent each quantity with a real number.*

**41.** **Business and Finance** The withdrawal of $50 from a checking account.

**42.** **Business and Finance** The deposit of $200 into a savings account.

**43.** **Science and Medicine** A temperature decrease of 10°F in 1 hour.

**44.** **Social Science** An increase of 25,000 in a city's population.

**45.** **Business and Finance** An increase of 75 points in the Dow-Jones average.

**46.** **Statistics** An eight-game losing streak by the local baseball team.

47. **Business and Finance** A country exported $90,000,000 more than it imported, creating a positive trade balance.

48. **Business and Finance** A stock lost 8.5% of its value.

*Label each statement as* **true** *or* **false.**

49. All whole numbers are integers.

50. All integers are real numbers.

51. All integers are whole numbers.

52. All real numbers are integers.

*Fill in each blank with $>$, $<$, or $=$ to make a true statement.*

53. $-9$ ___ $-6$

54. $|-9|$ ___ $|-6|$

55. $|-9|$ ___ $6$

56. $9$ ___ $|-6|$

*Represent the real numbers on the number line shown.*

57. $-\frac{3}{4}, 2\frac{1}{2}, -\frac{1}{2}, -3\frac{1}{4}, \frac{2}{3}$

−5 −4 −3 −2 −1 0 1 2 3 4 5

58. $3.2, -1.4, 0.7, -0.9, -2.1$

−5 −4 −3 −2 −1 0 1 2 3 4 5

*Place each of the following sets in ascending order.*

59. $-\frac{1}{2}, \frac{3}{4}, -\frac{5}{6}, \frac{2}{3}, -\frac{1}{3}$

60. $\frac{3}{7}, -\frac{6}{7}, \frac{1}{7}, -\frac{1}{2}, \frac{2}{7}$

61. $-6.1, -5.9, 6.1, 5.9, -6.0$

62. $3.5, -5.3, -3.5, 5.3, 4$

**Answers**

47. ____________
48. ____________
49. ____________
50. ____________
51. ____________
52. ____________
53. ____________
54. ____________
55. ____________
56. ____________
57. ____________
58. ____________
59. ____________
60. ____________
61. ____________
62. ____________

For each set, determine the maximum and minimum values.

**63.** $3, 0, \frac{1}{2}, -\frac{2}{3}, 5, \frac{3}{4}, -\frac{1}{6}$

**64.** $-3, 2, \frac{7}{12}, -\frac{3}{4}, \frac{5}{6}, -\frac{10}{3}, \frac{5}{2}$

**65.** $-3.3, 4\frac{1}{2}, -3, -2.8, 4.3, 4.8$

**66.** $-11, 4\frac{1}{2}, \frac{15}{4}, -10.9, -11.1, 0$

Place absolute value bars in the proper location on the left side of the equation in order to make it true.

**67.** $6 + (-2) = 4$

**68.** $8 + (-3) = 5$

**69.** $6 + (-2) = 8$

**70.** $8 + (-3) = 11$

Basic Skills | Advanced Skills | **Vocational-Technical Applications** | Calculator/Computer | Above and Beyond

Represent each quantity with a real number.

**71. Agricultural Technology** The erosion of 5 cm of topsoil from an Iowa cornfield.

**72. Agricultural Technology** The formation of 2.5 cm of new topsoil on the African savanna.

**73. Construction Technology** The elevations, in inches, of several points on a jobsite are as follows:

$-18, 27, -84, 37, 59, -13, 4, 92, 49, 66, -45$

Arrange the elevations in ascending order.

**74. Electronics** Several 12-volt (V) batteries were tested using a voltmeter. The voltage values were entered into a table indicating their value in reference to 12 V. Determine the maximum and minimum voltage measurements taken.

| Battery | Variance from 12 V (in V) |
|---|---|
| Cell 1 | +1 |
| Cell 2 | 0 |
| Cell 3 | −1 |
| Cell 4 | −3 |
| Cell 5 | +2 |

## Answers

63. ____

64. ____

65. ____

66. ____

67. ____

68. ____

69. ____

70. ____

71. ____

72. ____

73. ____

74. ____

**75.** **Electrical Engineering** Several resistors were tested using an ohmmeter. The resistance values were entered into a table indicating their value in reference to 10,000 ohms (10 kΩ). List the resistors in ascending order according to their measured resistance.

| Battery | Variance from 10,000 Ω (in Ω) |
|---|---|
| Resistor 1 | +175 |
| Resistor 2 | −60 |
| Resistor 3 | −188 |
| Resistor 4 | +10 |
| Resistor 5 | +218 |
| Resistor 6 | −65 |
| Resistor 7 | −302 |

**76.** **Electrical Engineering** Which of the resistors in exercise 75 had a measured value furthest from 10,000 Ω?

## Answers

**1.** 400 ft or (+400 ft) **3.** −$200 **5.** −25,000 people

**7.** 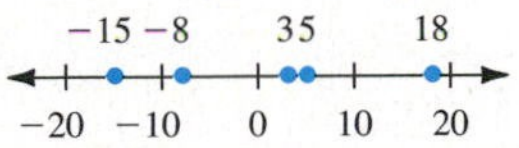


**9.** 5, 175, −234 **11.** −7, −5, −1, 0, 2, 3, 8

**13.** −11, −6, −2, 1, 4, 5, 9 **15.** Min: −6; max: 15

**17.** Min: −15; max: 21 **19.** 17 **21.** 10 **23.** −3 **25.** −8

**27.** 5 **29.** 18 **31.** 0 **33.** 7 **35.** 7 **37.** 11 **39.** 1

**41.** −$50 **43.** −10°F **45.** +75 points **47.** +$90,000,000

**49.** True **51.** False **53.** < **55.** >

**57.** 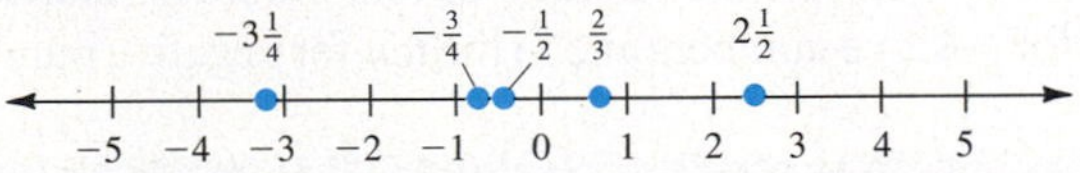


**59.** $-\frac{5}{6}, -\frac{1}{2}, -\frac{1}{3}, \frac{2}{3}, \frac{3}{4}$ **61.** −6.1, −6.0, −5.9, 5.9, 6.1

**63.** Min: $-\frac{2}{3}$; max: 5 **65.** Min: −3.3; max: 4.8

**67.** $6 + (-2) = 4$ or $|6 + (-2)| = 4$ **69.** $6 + |(-2)| = 8$ or $|6| + |(-2)| = 8$

**71.** −5 cm **73.** −84 in., −45 in., −18 in., −13 in., 4 in., 27 in., 37 in., 49 in., 59 in., 66 in., 92 in. **75.** 9,698 Ω, 9,812 Ω, 9,935 Ω, 9,940 Ω, 10,010 Ω, 10,175 Ω, 10,218 Ω

Answers

75. ____________

76. ____________

# 10.2 Adding Real Numbers

< 10.2 Objectives >

1 > Add two numbers with the same sign

2 > Add two numbers with opposite signs

In Section 10.1, we introduced the idea of real numbers. Now we will examine the four arithmetic operations (addition, subtraction, multiplication, and division) and see how those operations are performed when real numbers are involved. We start by considering addition.

An application may help. As before, we represent a gain of money as a positive number and a loss as a negative number.

| Beginning Balance | +$300 |
| --- | --- |
| Adjustments | +$400 |
| Ending Balance | $700 |

| Beginning Balance | -$300 |
| --- | --- |
| Adjustments | -$400 |
| Ending Balance | -$700 |

| Beginning Balance | +$300 |
| --- | --- |
| Adjustments | -$400 |
| Ending Balance | -$100 |

| Beginning Balance | -$300 |
| --- | --- |
| Adjustments | +$400 |
| Ending Balance | $100 |

If you gain \$300 and then gain \$400, the result is a gain of \$700:

$300 + 400 = 700$

If you lose \$300 and then lose \$400, the result is a loss of \$700:

$-300 + (-400) = -700$

If you gain \$300 and then lose \$400, the result is a loss of \$100:

$300 + (-400) = -100$

If you lose \$300 and then gain \$400, the result is a gain of \$100:

$-300 + 400 = 100$

The number line can be used to illustrate the addition of real numbers. Starting at the origin, we move to the *right* for positive numbers and to the *left* for negative numbers.

**Example 1** Adding Real Numbers on the Number Line

< Objective 1 >

**(a)** Add $3 + 2$.

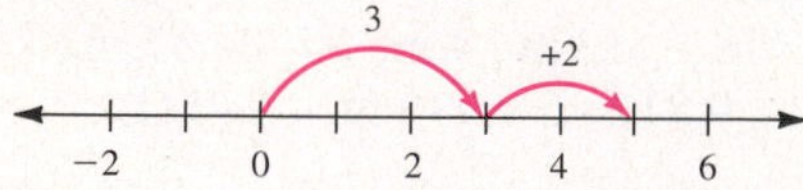


Start at the origin and move 3 units to the right. Then, move 2 units more to the right to find the sum.

$3 + 2 = 5$

**(b)** Add $\frac{4}{3} + \frac{2}{3}$.

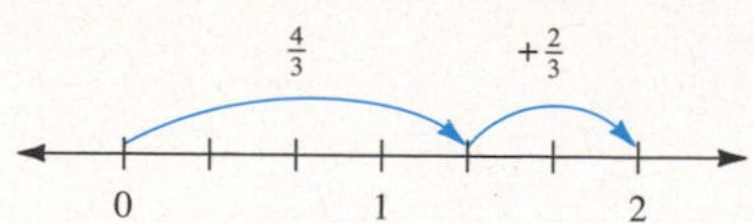

Start at the origin and move $\frac{4}{3}$ units to the right. Then move $\frac{2}{3}$ more to the right to find the sum. So we have

$$\frac{4}{3} + \frac{2}{3} = \frac{6}{3} = 2$$

### Check Yourself 1

Add.

(a) $5 + 6$ (b) $\frac{5}{4} + \frac{7}{4}$

The number line also helps you visualize the sum of two negative numbers. Remember to move left for negative numbers.

### Example 2 Adding Real Numbers with the Same Sign

**(a)** Add $-3 + (-4)$.

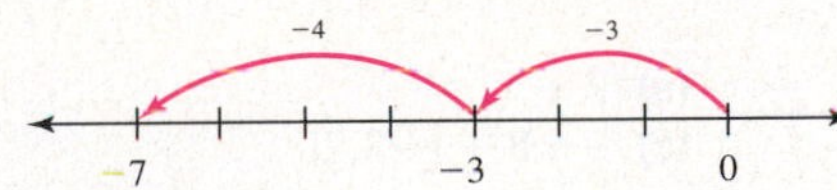


Start at the origin and move 3 units to the left. Then move 4 more units to the left to find the sum. From the graph we see that the sum is

$$-3 + (-4) = -7$$

**(b)** Add $-\frac{3}{2} + \left(-\frac{1}{2}\right)$.

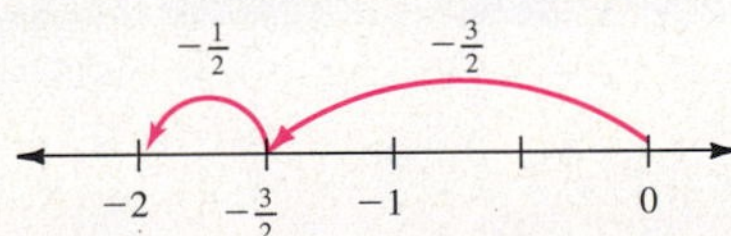


As before, we start at the origin. From that point move $\frac{3}{2}$ units left. Then move another $\frac{1}{2}$ unit left to find the sum. In this case

$$-\frac{3}{2} + \left(-\frac{1}{2}\right) = -2$$

### Check Yourself 2

Add.

(a) $-4 + (-5)$ (b) $-3 + (-7)$

(c) $-5 + (-15)$ (d) $-\frac{5}{2} + \left(-\frac{3}{2}\right)$

You have probably noticed some helpful patterns in Examples 1 and 2. These patterns will allow you to do the work mentally without having to use the number line. Look at the following rule.

**Property**

**Adding Real Numbers with the Same Sign**

If two numbers have the same sign, add their absolute values. Give the sum the sign of the original numbers.

## Example 3 Adding Real Numbers

**NOTE**

The sum of two positive numbers is positive, and the sum of two negative numbers is negative.

**(a)** $-8 + (-5) = -13$ — Add the absolute values $(8 + 5 = 13)$ and give the sum the sign $(-)$ of the original numbers.

**(b)** $[-3 + (-4)] + (-6)$ — Add inside the brackets as your first step.

$= -7 + (-6) = -13$

**Check Yourself 3**

Add mentally.

**(a)** $7 + 9$

**(b)** $-7 + (-9)$

**(c)** $-5.8 + (-3.2)$

**(d)** $[-5 + (-2)] + (-3)$

We can also use the number line to illustrate the addition of real numbers with *different* signs.

## Example 4 Adding Real Numbers with Opposite Signs

< Objective 2 >

**(a)** Add $3 + (-6)$.

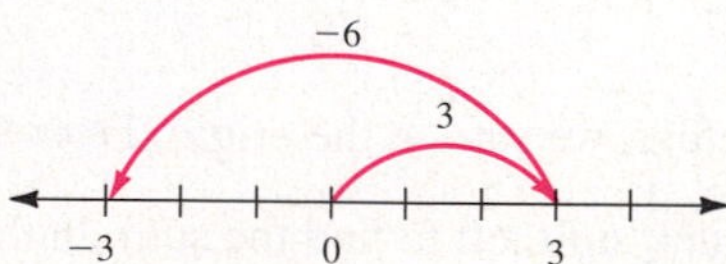


First move 3 units to the right of the origin. Then move 6 units to the left.

$3 + (-6) = -3$

**(b)** Add $-4 + 7$.

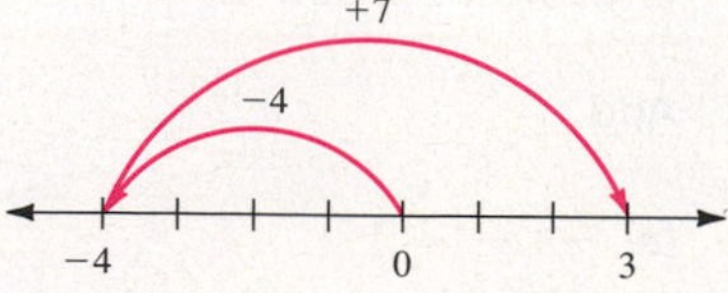


This time move 4 units to the left of the origin as the first step. Then move 7 units to the right.

$-4 + 7 = 3$

### Check Yourself 4

Add.

(a) $7 + (-5)$
(b) $4 + (-8)$
(c) $-4 + 9$
(d) $-7 + 3$

You have no doubt noticed that, in adding a positive number and a negative number, sometimes the sum is positive and sometimes it is negative. The result depends on which of the numbers has the larger absolute value. This leads us to the second part of our addition rule.

**Property**

**Adding Real Numbers with Different Signs**

If two numbers have different signs, subtract their absolute values, the smaller from the larger. Give the result the sign of the number with the larger absolute value.

### Example 5 Adding Real Numbers

Add.

(a) $7 + (-19) = -12$

Because the two numbers have different signs, subtract their absolute values ($19 - 7 = 12$). The sum has the sign ($-$) of the number with the larger absolute value, $-19$.

(b) $-13 + 7 = -6$

Subtract the absolute values ($13 - 7 = 6$). The sum has the sign ($-$) of the number with the larger absolute value, $-13$.

(c) $-4.5 + 8.2 = 3.7$

Subtract the absolute values ($8.2 - 4.5 = 3.7$). The sum has the sign ($+$) of the number with the larger absolute value, 8.2.

**RECALL**

Real numbers can be fractions and decimals as well as integers.

### Check Yourself 5

Add mentally.

(a) $5 + (-14)$
(b) $-7 + (-8)$
(c) $-8 + 15$
(d) $7 + (-8)$
(e) $-\frac{2}{3} + \left(-\frac{7}{3}\right)$
(f) $5.3 + (-2.3)$

There are two properties of addition that we should mention before concluding this section. First, the sum of any number and 0 is always that number. In symbols,

**Property**

**Additive Identity Property**

For any number $a$,

$a + 0 = 0 + a = a$

## Example 6 Adding Zero

**NOTE**

Numbers do not change their value after addition with 0. Zero is called the **additive identity.**

Add.

**(a)** $9 + 0 = 9$

**(b)** $0 + (-8) = -8$

**(c)** $-25 + 0 = -25$

### Check Yourself 6

Add.

**(a)** $8 + 0$ **(b)** $0 + (-7)$ **(c)** $-36 + 0$

We need one further definition to state our second property. Every number has an *opposite.* It corresponds to a point that is the same distance from the origin as the given number, but in the opposite direction.

**NOTES**

The opposite of a number is also called its **additive inverse.**

3 and $-3$ are opposites.

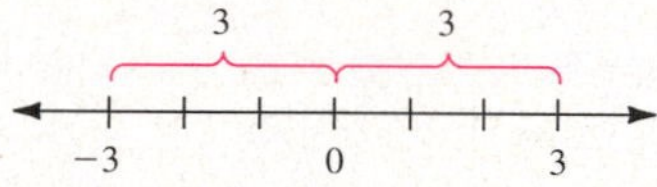


The opposite of 9 is $-9$.

The opposite of $-15$ is 15.

Our second property states that the sum of any number and its opposite is 0.

**Property**

**Additive Inverse Property**

For any number $a$, there exists a number $-a$ such that

$$a + (-a) = -a + a = 0$$

In words: The sum of any number and its opposite, or additive inverse, is 0.

## Example 7 Adding Inverses

Add.

**(a)** $9 + (-9) = 0$

**(b)** $-15 + 15 = 0$

**(c)** $-2.3 + 2.3 = 0$

**(d)** $\frac{4}{5} + \left(-\frac{4}{5}\right) = 0$

### Check Yourself 7

Add.

(a) $-17 + 17$ (b) $12 + (-12)$

(c) $\frac{1}{3} + \left(-\frac{1}{3}\right)$ (d) $-1.6 + 1.6$

**NOTE**

All properties of addition from Section 1.2 apply when negative numbers are involved.

We can now use the associative and commutative properties of addition, first introduced in Section 1.2, to find the sum when more than two real numbers are involved. Example 8 illustrates these properties.

### Example 8 Adding Real Numbers

**NOTE**

We use the commutative property to reverse the order of addition for $-3$ and 5. We then group $-5$ and 5. Do you see why?

$$\begin{aligned} -5 + (-3) + 5 &= -5 + 5 + (-3) \\ &= [-5 + 5] + (-3) \\ &= 0 + (-3) \\ &= -3 \end{aligned}$$

### Check Yourself 8

Add.

(a) $-4 + 5 + (-3)$ (b) $-8 + 4 + 8$

Real numbers appear in many situations.

### Example 9 An Application of Real Numbers

A vendor earned profits of $-\$86.75$, \$111.50, and \$123 one weekend (Friday through Sunday). What was the vendor's total weekend profit?

We add the vendor's daily profits to get the weekend profit.

$$-86.75 + 111.50 + 123 = 147.75$$

So the vendor earned a weekend profit of \$147.75.

### Check Yourself 9

A softball team scored 3 runs in one inning, gave up 4 runs in another inning, scored 2 runs after that, and gave up 3 runs in the final inning. How far was the team ahead at the end of the game?

### Check Yourself ANSWERS

**1.** **(a)** 11; **(b)** 3 **2.** **(a)** −9; **(b)** −10; **(c)** −20; **(d)** −4
**3.** **(a)** 16; **(b)** −16; **(c)** −9; **(d)** −10 **4.** **(a)** 2; **(b)** −4; **(c)** 5; **(d)** −4
**5.** **(a)** −9; **(b)** −15; **(c)** 7; **(d)** −1; **(e)** −3; **(f)** 3 **6.** **(a)** 8; **(b)** −7; **(c)** −36
**7.** **(a)** 0; **(b)** 0; **(c)** 0; **(d)** 0 **8.** **(a)** −2; **(b)** 4 **9.** −2 (they lost by 2)

## Reading Your Text

The following fill-in-the-blank exercises are designed to ensure that you understand some of the key vocabulary used in this section.

SECTION 10.2

**(a)** The sum of two negative numbers is always ______________.

**(b)** Adding a ______________ number can be illustrated on a number line by moving to the left.

**(c)** When adding numbers with different signs, the result has the same sign as the number with the larger ______________ value.

**(d)** The sum of any number and its opposite, or additive inverse, is ______________.

Basic Skills | Advanced Skills | Vocational-Technical Applications | Calculator/Computer | Above and Beyond

# 10.2 exercises

MathZone

Boost your grade at mathzone.com!
> Practice Problems
> NetTutor
> Self-Tests
> e-Professors
> Videos

Name ______________

Section ________ Date ________

< Objectives 1–2 >

*Perform the indicated operation.*

**1.** $3 + 6$

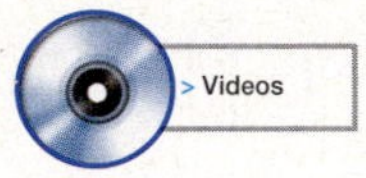


**2.** $5 + 9$

**3.** $11 + 5$

**4.** $8 + 7$

**5.** $-2 + (-3)$

**6.** $-1 + (-9)$

**7.** $-13 + (-24)$

**8.** $-1{,}234 + (-887)$

**9.** $9 + (-3)$

**10.** $10 + (-4)$

**11.** $8 + (-14)$

**12.** $7 + (-11)$

**13.** $-4 + 17$

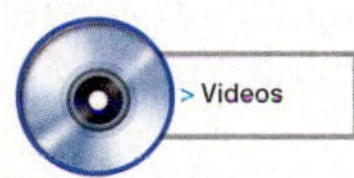


**14.** $-87 + 23$

**15.** $-732 + 1{,}104$

**16.** $2{,}417 + (-7{,}332)$

**17.** $\frac{3}{4} + \frac{5}{4}$

**18.** $\frac{1}{2} + \frac{4}{5}$

**19.** $-\frac{3}{5} + \left(-\frac{7}{5}\right)$

**20.** $-\frac{1}{8} + \left(-\frac{3}{8}\right)$

**21.** $-\frac{2}{3} + \left(-\frac{1}{4}\right)$

**22.** $-\frac{3}{4} + \left(-\frac{5}{12}\right)$

**23.** $-2\frac{2}{3} + \left(-1\frac{1}{2}\right)$

**24.** $-6\frac{1}{2} + \left(-\frac{3}{5}\right)$

## Answers

1. ______
2. ______
3. ______
4. ______
5. ______ 6. ______
7. ______ 8. ______
9. ______ 10. ______
11. ______ 12. ______
13. ______ 14. ______
15. ______ 16. ______
17. ______ 18. ______
19. ______ 20. ______
21. ______ 22. ______
23. ______ 24. ______

**Answers**

25. ______
26. ______
27. ______
28. ______
29. ______
30. ______
31. ______
32. ______
33. ______
34. ______
35. ______
36. ______
37. ______
38. ______
39. ______
40. ______
41. ______
42. ______
43. ______ 44. ______
45. ______ 46. ______

**25.** $\frac{3}{4} + \left(-\frac{1}{4}\right)$

**26.** $-\frac{7}{12} + \frac{1}{3}$

**27.** $-\frac{2}{5} + \frac{13}{20}$

**28.** $\frac{2}{3} + \left(-\frac{5}{6}\right)$

**29.** $5\frac{1}{3} + \left(-4\frac{4}{5}\right)$

**30.** $-17\frac{3}{4} + 21\frac{1}{3}$

**31.** $-1.6 + (-2.3)$

**32.** $-3.5 + (-2.6)$

**33.** $-3.6 + 7.6$

**34.** $13.4 + (-11.4)$

**35.** $-9 + 0$

**36.** $0 + (-15)$

**37.** $14 + (-14)$

**38.** $-5 + 5$

**39.** **STATISTICS** Beach Channel High School's football team scored one field goal (3 points) and gave up a touchdown (7 points) in the first quarter. In the third quarter, the team scored another field goal and gave up a safety (2 points). The team gave up a final field goal in the fourth quarter. By how much did the team lose?

**40.** **BUSINESS AND FINANCE** Jean deposited a check for \$625, wrote two for \$68.74 and \$29.95, and used her debit card to pay for a purchase of \$57.65. What is her new account balance?

**41.** **SCIENCE AND MEDICINE** The temperature dropped by 23°F from a high of 8°F. What was the low temperature?

chapter 10 > Make the Connection

**42.** **SCIENCE AND MEDICINE** The overnight low temperature was listed as −14°C. The temperature rose 19°C by noon. What was the noontime temperature?

*Label each of the following statements as* **true** *or* **false.**

**43.** $-10 + 6 = 6 + (-10)$

**44.** $5 + (-9) = -9 + 5$

**45.** $|-3| + |2| = |-3 + 2|$

**46.** $|-8| + |3| = |-8 + 3|$

Basic Skills | **Advanced Skills** | Vocational-Technical Applications | Calculator/Computer | Above and Beyond

*Compute, as indicated.*

**47.** $-9 + (-17) + 9$

**48.** $15 + (-3) + (-15)$

**49.** $2 + 5 + (-11) + 4$

**50.** $7 + (-9) + (-5) + 6$

**51.** $1 + (-2) + 3 + (-4)$

**52.** $(-9) + 0 + (-2) + 12$

**53.** $\frac{5}{3} + \left(-\frac{4}{3}\right) + \frac{5}{3}$

**54.** $-\frac{6}{5} + \left(-\frac{13}{5}\right) + \frac{4}{5}$

**55.** $-\frac{3}{2} + \left(-\frac{7}{4}\right) + \frac{3}{4}$

**56.** $\frac{2}{3} + \left(-\frac{5}{6}\right) + \left(-\frac{1}{2}\right)$

**57.** $2.8 + (-5.5) + (-2.9)$

**58.** $-5.4 + (-2.1) + (-3.5)$

**59.** $|3 + (-4)|$

**60.** $|-11 + 9|$

**61.** $|-17 + 8|$

**62.** $|-27 + 14|$

**63.** $|-5 + (-6)|$

**64.** $|-17 + (-14)|$

**65.** $|-3 + 2 + (-4)|$

**66.** $|-2 + 7 + (-5)|$

**67.** $|2 + (-3)| + |-3 + 2|$

**68.** $|8 + (-10)| + |-12 + 14|$

*Evaluate and round each result to the nearest tenth.*

**69.** $-4.1967 + 5.2943 + (-3.1698)$

**70.** $5.3297 + 4.1897 + (-3.2869)$

**71.** $-7.19863 + 4.8629 + 3.2689 + (-5.7936)$

**72.** $3.6829 + 4.5687 + 7.28967 + (-5.1623)$

## Answers

47. ________

48. ________

49. ________

50. ________

51. ________

52. ________

53. ________

54. ________

55. ________

56. ________

57. ________

58. ________

59. ________

60. ________

61. ________ 62. ________

63. ________ 64. ________

65. ________ 66. ________

67. ________ 68. ________

69. ________ 70. ________

71. ________ 72. ________

Basic Skills | Advanced Skills | **Vocational-Technical Applications** | Calculator/Computer | Above and Beyond

## Answers

73. ____________

74. ____________

75. ____________

76. ____________

77. ____________

**73. Information Technology** Amir, a network administrator, has a budget of \$50,000 at the beginning of April. He has the following entries to enter in the budget for the month of April: \$1,000 for travel, \$9,550 for technology, \$542 for miscellaneous expenses, \$443 received from returns, \$123 for supplies, and \$150 for subscriptions. How much money does Amir have left in his budget? Make sure to write an integer expression that represents the change in the budget.

**74. Information Technology** Fred has been hired to redesign a database that is having performance issues. He finds that one table in the database called CUSTOMERS has field sizes of FIRST NAME and LAST NAME to be 100 bytes. He knows from experience that first names average around 30 bytes and last names average around 45 bytes. A character is a byte. Fred knows that wasting space on a very large database can cause performance issues. Write an integer expression that represents the change in the field sizes. By how much does Fred need to modify the field sizes?

**75. Mechanical Engineering** A pneumatic actuator is operated by a pressurized air reservoir. At the beginning of the operator's shift, the pressure in the reservoir was 126 pounds per square inch (lb/in.$^2$). At the end of each hour, the operator records the change in pressure of the reservoir. The values (in lb/in.$^2$) recorded for this shift were a drop of 12, a drop of 7, a rise of 32, a drop of 17, a drop of 15, a rise of 31, a drop of 4, and a drop of 14. What is the pressure in the tank at the end of the shift?

**76. Mechanical Engineering** A diesel engine for an industrial shredder has an 18-quart (qt) oil capacity. When the maintenance technician checked the oil, it was 7 qt low. Later that day, she added 4 qt to the engine. What was the oil level after the 4 qt were added?

**Electrical Engineering** Dry cells or batteries have a positive and negative terminal. When correctly connected in series (positive to negative), the voltage of each cell can be added together. If a cell is connected and its terminals are reversed, the current will flow in the opposite direction.

For example, if three 3-volt cells are connected in series and one cell is inserted backwards, the resulting voltage is 3 V.

$3\text{ V} + 3\text{ V} + (-3)\text{ V} = 3\text{ V}$

The voltages are added together because the cells are in series, but you must pay attention to the current flow.

*Now solve exercises 77 and 78.*

**77.** Assume you have a 24-V cell and a 12-V cell with their negative terminals connected. What would the resulting voltage be if measured from the positive terminals?

| + 24 V − | − 12 V + |
|---|---|

**78.** If a 24-V cell, an 18-V cell, and 12-V cell are supposed to be connected in series and the 18-V cell is accidentally reversed, what would the total voltage be?

+ 24 V −    − 18 V +    + 12 V −

Basic Skills | Advanced Skills | Vocational-Technical Applications | Calculator/Computer | **Above and Beyond**

*Place absolute value bars in the proper location on the left side of the equation in order to make the statement true.*

**79.** $-3 + 7 = 10$

**80.** $-5 + 9 = 14$

**81.** $-6 + 7 + (-4) = 3$

**82.** $-10 + 15 + (-9) = 4$

## Answers

78. ______

79. ______

80. ______

81. ______

82. ______

## Answers

**1.** 9 **3.** 16 **5.** $-5$ **7.** $-37$ **9.** 6 **11.** $-6$ **13.** 13 **15.** 372 **17.** 2 **19.** $-2$ **21.** $-\frac{11}{12}$ **23.** $-4\frac{1}{6}$ **25.** $\frac{1}{2}$ **27.** $\frac{1}{4}$ **29.** $\frac{8}{15}$ **31.** $-3.9$ **33.** 4 **35.** $-9$ **37.** 0 **39.** 6 points **41.** $-15°F$ **43.** True **45.** False **47.** $-17$ **49.** 0 **51.** $-2$ **53.** 2 **55.** $-\frac{5}{2}$ **57.** $-5.6$ **59.** 1 **61.** 9 **63.** 11 **65.** 5 **67.** 2 **69.** $-2.1$ **71.** $-4.9$ **73.** \$39,078 **75.** 120 lb/in.$^2$ **77.** 12 V **79.** $|-3| + 7 = 10$ **81.** $|-6 + 7 + (-4)| = 3$

# Activity 27 :: Hometown Weather

The local weather provides us with many interesting applications of real numbers and data gathering.

1. Collect the daily high and low temperatures in your locale for the previous week.
2. Compute the mean high and mean low temperatures for the week.
3. List each day's high and low temperatures as their distance from the mean for the week. For example, if the weekly mean high temperature was 65°F, and Tuesday's high temperature was 62°F, then list it as $-3$°F.
4. Consider the differences from the weekly mean high temperature listed in exercise 3. What is the mean of this set of numbers?
5. Consider the differences from the weekly mean low temperature listed in exercise 3. What is the mean of this set of numbers?
6. Explain your answers to exercises 4 and 5.
7. Find the average annual high temperature for your locale.
8. Use real numbers to describe the difference between each of the high and low temperatures for the previous week and the annual averages.
9. Construct a line graph of the high temperatures for the previous week.
10. Add a second line graph to the graph constructed in exercise 9 for the low temperatures.
11. Add horizontal lines to your graph, one for the annual average high temperature and one for the annual average low temperature.
12. Describe the relation between the answers to exercise 8 and the graph.
13. Go to www.mhhe.com/streeter for more links concerning weather data.

# 10.3 Subtracting Real Numbers

< 10.3 Objective >

## 1 > Find the difference of two real numbers

To begin our discussion of subtraction when real numbers are involved, we look back at a problem using natural numbers. Of course, we know that

$$8 - 5 = 3$$

From our work in adding real numbers in Section 10.2, we know that it is also true that

$$8 + (-5) = 3$$

Comparing these equations, we see that the results are the same. This leads us to an important pattern. Any subtraction problem can be written as a problem in addition. Subtracting 5 is the same as adding the opposite of 5, or $-5$. We can write this fact as follows:

$$8 - 5 = 8 + (-5) = 3$$

This leads us to the following rule for subtracting real numbers.

**Step by Step**

**Subtracting Real Numbers**

**Step 1** Rewrite the subtraction problem as an addition problem by

**a.** Changing the subtraction symbol ($-$) to an addition symbol ($+$)

**b.** Replacing the number being subtracted with its opposite

**Step 2** Add the resulting real numbers as before.

In symbols,

$$a - b = a + (-b)$$

Example 1 illustrates the use of this process for subtracting.

### Example 1 Subtracting Real Numbers

< Objective 1 >

**NOTE**

Each subtraction is rewritten as an addition.

(a) $15 - 7 = 15 + (-7)$

$= 8$

Change the subtraction symbol ($-$) to an addition symbol ($+$).

Replace 7 with its opposite, $-7$.

(b) $9 - 12 = 9 + (-12) = -3$

(c) $-6 - 7 = -6 + (-7) = -13$

**(d)** $-\frac{3}{5} - \frac{7}{5} = -\frac{3}{5} + \left(-\frac{7}{5}\right) = -\frac{10}{5} = -2$

**(e)** $2.1 - 3.4 = 2.1 + (-3.4) = -1.3$

**(f)** Subtract 5 from $-2$.

We write the statement as $-2 - 5$ and proceed as before:

$-2 - 5 = -2 + (-5) = -7$

### Check Yourself 1

Subtract.

(a) $18 - 7$ (b) $5 - 13$ (c) $-7 - 9$

(d) $-\frac{5}{6} - \frac{7}{6}$ (e) $-2 - 7$ (f) $5.6 - 7.8$

The subtraction rule is used in the same way when the number being subtracted is negative. Change the subtraction to addition. Replace the negative number being subtracted with its opposite, which is positive. Example 2 illustrates this principle.

## Example 2 Subtracting Real Numbers

Subtract.

**(a)** $5 - (-2) = 5 + (+2) = 5 + 2 = 7$

Change the subtraction to addition.

Replace $-2$ with its opposite, $+2$ or 2.

**(b)** $7 - (-8) = 7 + (+8) = 7 + 8 = 15$

**(c)** $-9 - (-5) = -9 + 5 = -4$

**(d)** $-12.7 - (-3.7) = -12.7 + 3.7 = -9$

**(e)** $-\frac{3}{4} - \left(-\frac{7}{4}\right) = -\frac{3}{4} + \left(+\frac{7}{4}\right) = \frac{4}{4} = 1$

**(f)** Subtract $-4$ from $-5$. We write

$-5 - (-4) = -5 + 4 = -1$

### Check Yourself 2

Subtract.

(a) $8 - (-2)$ (b) $3 - (-10)$ (c) $-7 - (-2)$

(d) $-9.8 - (-5.8)$ (e) $7 - (-7)$

We are now ready to describe negative mixed numbers, such as $-2\frac{1}{2}$.

You should recall that we define a positive mixed number as the sum of the whole number part and the fraction part. For instance,

$$5\frac{1}{4} = 5 + \frac{1}{4}$$

This, of course, agrees with the number line approach, in which $5\frac{1}{4}$ is $\frac{1}{4}$ of a unit to the right of 5 on a number line.

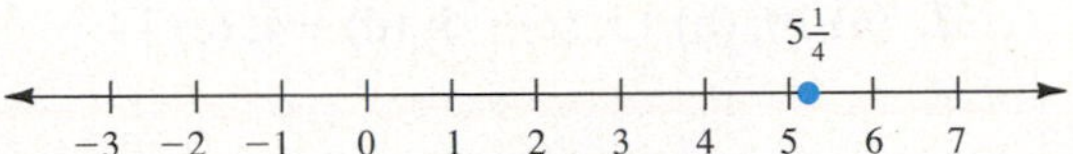


Looking again at a number line, we can also locate $-2\frac{1}{2}$, which is $\frac{1}{2}$ of a unit to the left of −2 on the number line.

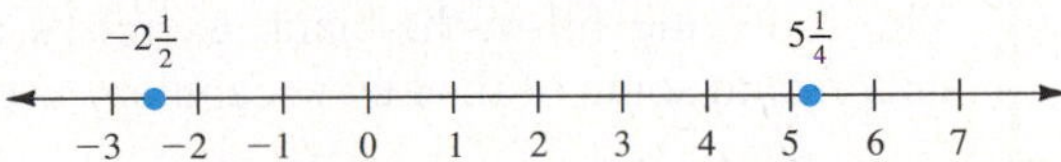


This suggests that $-2\frac{1}{2} = -2 - \frac{1}{2}$ or $-2 + \left(-\frac{1}{2}\right)$.

## Example 3 An Application of Subtraction

From a seaside cliff 1,700 feet above sea level in the Cayman Islands, Nicole looks south to where the Cayman Trench is located. The Cayman Trench is the deepest part of the Caribbean, reaching a depth of 24,576 feet.

How far above the bottom of the Cayman Trench is Nicole standing?

To gauge the distance, we subtract, treating the depth of the trench as a negative number because the trench is below sea level.

$$1{,}700 - (-24{,}576) = 1{,}700 + (+24{,}576)$$
$$= +26{,}276$$

Nicole is standing 26,276 feet above the bottom of the Cayman Trench.

### Check Yourself 3

**The high temperature one year for a midwestern town was 88°F. During the winter, the temperature dipped as low as −14°F. What was the temperature range experienced by the town that year?**

### Check Yourself ANSWERS

**1.** **(a)** 11; **(b)** −8; **(c)** −16; **(d)** −2; **(e)** −9; **(f)** −2.2
**2.** **(a)** 10; **(b)** 13; **(c)** −5; **(d)** −4; **(e)** 14 **3.** 102°F

### Reading Your Text

The following fill-in-the-blank exercises are designed to ensure that you understand some of the key vocabulary used in this section.

**SECTION 10.3**

**(a)** Any subtraction problem can be written as a problem in ____________.

**(b)** We define the ______________ of real numbers by $a - b = a + (-b)$.

**(c)** To subtract real numbers, change the operation to addition and replace the second number with its ______________.

**(d)** The opposite of a negative number is a ____________ number.

Basic Skills | Advanced Skills | Vocational-Technical Applications | Calculator/Computer | Above and Beyond

# 10.3 exercises

< Objective 1 >

*Perform the indicated operation.*

**1.** $21 - 13$

**2.** $36 - 22$

**3.** $82 - 45$

**4.** $103 - 56$

**5.** $8 - 10$

**6.** $14 - 19$

**7.** $24 - 45$

**8.** $136 - 352$

**9.** $-5 - 3$

**10.** $-15 - 8$

**11.** $-9 - 14$

**12.** $-8 - 12$

**13.** $3 - (-4)$

**14.** $6 - (-8)$

**15.** $5 - (-11)$

**16.** $7 - (-5)$

**17.** $7 - (-12)$

**18.** $3 - (-10)$

**19.** $-36 - (-24)$

**20.** $-28 - (-11)$

**21.** $-19 - (-27)$

**22.** $-11 - (-16)$

**23.** $-11 - (-11)$

**24.** $-15 - (-15)$

**25.** $0 - (-8)$

**26.** $0 - (-11)$

**27.** **Statistics** On April 2, 2003, the high temperature in the United States was recorded as 94°F in Wink, Texas. The low temperature for the day was −23°F, recorded in both Deadhorse and Northway, Alaska (at −4°F, Presque Isle, Maine, recorded the lowest temperature in the contiguous states). What was the temperature range in the United States on that day?

chapter 10 > Make the Connection

**28.** **Statistics** What was the temperature range in the contiguous 48 states on April 2, 2003 (from exercise 27)?

chapter 10 > Make the Connection

MathZone

Boost your grade at mathzone.com!

- > Practice Problems
- > NetTutor
- > Self-Tests
- > e-Professors
- > Videos

Name ____________

Section ________ Date ________

## Answers

1. ________ 2. ________

3. ________ 4. ________

5. ________ 6. ________

7. ________ 8. ________

9. ________ 10. ________

11. ________ 12. ________

13. ________ 14. ________

15. ________ 16. ________

17. ________ 18. ________

19. ________ 20. ________

21. ________ 22. ________

23. ________ 24. ________

25. ________ 26. ________

27. ________ 28. ________

**Answers**

29. ______

30. ______

31. ______

32. ______

33. ______

34. ______

35. ______

36. ______

37. ______

38. ______

39. ______

40. ______

41. ______

42. ______

43. ______

44. ______

45. ______

46. ______

**29. STATISTICS** The lowest temperature ever recorded in the state of Oregon was $-54°F$ (in Seneca, on February 10, 1933). The state's record high temperature occurred in Pendleton on August 10, 1898, when it reached 119°F. What is the historical temperature range in the state of Oregon?

chapter 10 > Make the Connection

**30. BUSINESS AND FINANCE** A government agency is operating despite a \$2.3 million budget deficit. Congress authorizes a \$3.5 million allocation in order for the agency to meet an additional \$1.9 million in payroll costs. How much is in the agency's budget after these transactions?

*Fill in each blank with* **always, sometimes,** *or* **never.**

**31.** The difference between two negative numbers is ______ negative.

**32.** A positive number subtracted from a negative number is ______ negative.

**33.** The difference between two positive numbers is ______ negative.

**34.** A negative number subtracted from a positive number is ______ negative.

Basic Skills | **Advanced Skills** | Vocational-Technical Applications | Calculator/Computer | Above and Beyond

*Compute, as indicated.*

**35.** $\frac{15}{7} - \frac{8}{7}$

**36.** $\frac{17}{8} - \frac{9}{8}$

**37.** $\frac{7}{6} - \frac{19}{6}$

**38.** $\frac{5}{9} - \frac{32}{9}$

**39.** $-\frac{2}{5} - \frac{7}{10}$

**40.** $-\frac{5}{9} - \frac{7}{18}$

**41.** $\frac{3}{4} - \left(-\frac{3}{2}\right)$

**42.** $\frac{5}{6} - \left(-\frac{7}{6}\right)$

**43.** $\frac{6}{7} - \left(-\frac{5}{14}\right)$

**44.** $\frac{11}{16} - \left(-\frac{7}{8}\right)$

**45.** $-\frac{3}{4} - \left(-\frac{11}{4}\right)$

**46.** $-\frac{1}{2} - \left(-\frac{5}{8}\right)$

47. $-\frac{2}{3} - \left(-\frac{3}{4}\right)$

48. $\frac{3}{10} - \frac{2}{3}$

49. $2\frac{3}{8} - 5\frac{3}{4}$

50. $11\frac{3}{4} - 6\frac{4}{5}$

51. $6\frac{1}{2} - \left(-5\frac{4}{5}\right)$

52. $3\frac{1}{6} - \left(-4\frac{2}{3}\right)$

53. $-2\frac{5}{6} - 3\frac{1}{2}$

54. $-4\frac{7}{10} - 3\frac{1}{3}$

55. $-8\frac{2}{3} - \left(-1\frac{1}{5}\right)$

56. $-3\frac{1}{2} - \left(-4\frac{1}{4}\right)$

57. $7.9 - 5.4$

58. $11.7 - 4.5$

59. $7.8 - 11.6$

60. $14.3 - 25.5$

61. $-3.4 - 4.7$

62. $-8.1 - 7.6$

63. $8.3 - (-5.7)$

64. $6.5 - (-4.3)$

65. $8.9 - (-11.7)$

66. $14.5 - (-24.6)$

67. $-12.7 - (-5.7)$

68. $-5.6 - (-2.6)$

69. $-6.9 - (-10.1)$

70. $-3.4 - (-7.6)$

Basic Skills | Advanced Skills | **Vocational-Technical Applications** | Calculator/Computer | Above and Beyond

**Construction** *The elevation of the reference point used to set up a job is 362 inches (in.). Find the difference in elevation at the given points (use a negative sign for elevations below the reference point).*

71. 311 in.

72. 491 in.

## Answers

47. ________
48. ________
49. ________
50. ________
51. ________
52. ________
53. ________
54. ________
55. ________
56. ________
57. ________
58. ________
59. ________
60. ________
61. ____ 62. ____
63. ____ 64. ____
65. ____ 66. ____
67. ____ 68. ____
69. ____ 70. ____
71. ____ 72. ____

Answers

73. ____________

74. ____________

75. ____________

76. ____________

77. ____________

78. ____________

79. ____________

80. ____________

**Manufacturing Technology** *At the beginning of the week, there were 2,489 pounds (lb) of steel in inventory. Report the weekly change in steel inventory for the given end-of-week inventories.*

**73.** 2,581 lb

**74.** 2,111 lb

**75. Electrical Engineering** A certain electric motor spins at 5,400 rotations per minute (rpm) when unloaded. When a load is applied, the motor spins at 4,250 rpm. What is the change in rpm after loading?

**76. Electrical Engineering** A cooling fan used to help dissipate heat from an electronic device has three modes of operation: off, low speed, and high speed. Low speed moves air at 34 cubic feet per minute ($ft^3$/min). High speed moves air at 52 $ft^3$/min. What is the difference in the volumes of air moved by the low and high speeds?

Basic Skills | Advanced Skills | Vocational-Technical Applications | **Calculator/Computer** | Above and Beyond

We now present a set of exercises for which the calculator might be the preferred tool. As indicated by the placement of this explanation, you should refrain from using a calculator on the exercises that precede this.

Your scientific or graphing calculator has a key that makes a number negative. This key is different from the *minus* key that we use for subtraction. The *negative* key is marked [(−)] or [+/−]. With a scientific calculator, the negative key must be pressed *after* the number is entered. We assume you are using a graphing calculator.

To evaluate the expression: 132 + 547 + (−234) − 112 − (−327)

**Step 1** Press the *clear* key. [CLEAR]

**Step 2** Enter the numbers as written. You do not need to use parentheses. Instead, use the negative key.

132 [+]
547 [+]
[(−)] 234 [−]
112 [−]
[(−)] 327

**Step 3** Press *enter* or equals. [ENTER] or [=]

Your display should read 660.

*Use your calculator to evaluate each expression.*

**77.** $8 + 4 - (-3) - 2$

**78.** $-27 - 43 - (-29) + 13$

**79.** $145 - (-547) + (-92) - 234$

**80.** $10{,}945 - (-2{,}347) + (-7{,}687) + 41$

## Answers

**1.** 8 **3.** 37 **5.** $-2$ **7.** $-21$ **9.** $-8$ **11.** $-23$ **13.** 7
**15.** 16 **17.** 19 **19.** $-12$ **21.** 8 **23.** 0 **25.** 8
**27.** 117°F **29.** 173°F **31.** sometimes **33.** sometimes **35.** 1
**37.** $-2$ **39.** $-\frac{11}{10}$ or $-1\frac{1}{10}$ **41.** $\frac{9}{4}$ or $2\frac{1}{4}$ **43.** $\frac{17}{14}$ or $1\frac{3}{14}$ **45.** 2
**47.** $\frac{1}{12}$ **49.** $-3\frac{3}{8}$ **51.** $12\frac{3}{10}$ **53.** $-6\frac{1}{3}$ **55.** $-7\frac{7}{15}$
**57.** 2.5 **59.** $-3.8$ **61.** $-8.1$ **63.** 14 **65.** 20.6 **67.** $-7$
**69.** 3.2 **71.** $-51$ in. **73.** $+92$ lb **75.** $-1{,}150$ rpm **77.** 13
**79.** 366

# Activity 28 ::
## Plus/Minus Ratings in Hockey

The plus/minus statistic in professional hockey provides an application of real number arithmetic. A player is awarded $+1$ point in this category if he is on the ice when his team scores a goal regardless of who is given credit for scoring. He is awarded $-1$ point if the opposing team scores while he is on the ice.

For the 2002–2003 hockey season, Peter Forsberg and Milan Hejduk (both with the Colorado Avalanche) shared top honors in the plus/minus category. They each had a plus/minus rating of +52 for the season.

1. Midway through the 2002–2003 season, Peter Forsberg entered a game against the Dallas Stars with a plus/minus rating of +36. In the game, which Colorado won 5 to 3, Forsberg was on the ice for two of his own team's goals and one goal by Dallas. What was his plus/minus for the game?
2. What was Forsberg's season total plus/minus rating after the game against Dallas?
3. In the same game, Dallas defenseman Philippe Boucher was on the ice for two Dallas goals and three Colorado goals. What was Boucher's plus/minus rating for the game?
4. If Boucher entered the game with a plus/minus rating of $+21$, what was his rating after the game?
5. Pittsburgh Penguins forward Mario Lemieux finished the 2002–2003 season as the league's eighth leading scorer. However, his plus/minus rating for the season was $-25$.
   **(a)** Lemieux entered a pair of games against the Boston Bruins with a $-12$ plus/minus rating. Boston won the first game (in Boston) 6 to 3. Lemieux was on the ice for one Pittsburgh goal and three Boston goals. Boston won the second game (in Pittsburgh) 2 to 1. Lemieux was on the ice for all three goals. What was his plus/minus rating for the pair of games?
   **(b)** What was his season total after the two-game series?
6. Go to www.mhhe.com/streeter for more NHL statistics.

*Source:* ESPN.

# 10.4 Multiplying Real Numbers

< 10.4 Objectives >

1> Find the product of two or more real numbers

2> Find the reciprocal of a real number

3> Evaluate expressions involving real numbers

When you first considered multiplication in arithmetic, you thought of it as repeated addition. Our work with the addition of real numbers can tell us about multiplication when real numbers are involved. For example,

$$3 \cdot 4 = \underbrace{4 + 4 + 4}_{} = 12$$

We interpret multiplication as repeated addition to find the product, 12.

Now, consider the product $3(-4)$:

$$3(-4) = (-4) + (-4) + (-4) = -12$$

Because multiplication is *commutative,* we know that the order of the factors does not matter. Therefore,

$$-4 \cdot 3 = 3 \cdot (-4) = -12$$

Looking at these products suggests the first portion of our rule for multiplying real numbers. The product of a positive number and a negative number is negative.

**Property**

**Multiplying Real Numbers with Different Signs**

The product of two numbers with different signs is negative.

To use this rule when multiplying two numbers with different signs, multiply their absolute values and attach a negative sign.

**Example 1** — **Multiplying Real Numbers**

< Objective 1 >

Multiply.

**(a)** $5(-6) = -30$

The product is negative.

**NOTE**

Multiply numerators together, and then multiply denominators together. Simplify the result.

**(b)** $-10(10) = -100$

**(c)** $8(-12) = -96$

**(d)** $-\frac{3}{4}\left(\frac{2}{5}\right) = -\frac{3}{10}$

## Check Yourself 1

Multiply.

**(a)** $-7(5)$ **(b)** $-12(9)$ **(c)** $-15(8)$ **(d)** $\left(-\frac{5}{7}\right)\left(\frac{4}{5}\right)$

The product of two negative numbers is harder to visualize. The following pattern may help you see how we can determine the sign of the product.

**NOTES**

$-1(-2)$ is the opposite of $-2$.

We present a more detailed explanation at the end of the section.

This number is decreasing by 1.

$$3(-2) = -6$$
$$2(-2) = -4$$
$$1(-2) = -2$$
$$0(-2) = 0$$
$$-1(-2) = 2$$

Do you see that the product is *increasing* by 2 each time?

What should the product $-2(-2)$ be? Continuing the pattern shown, we see that

$$-2(-2) = 4$$

This suggests that the product of two negative numbers is positive, which is the case.

**Property**

**Multiplying Real Numbers with the Same Sign**

The product of two numbers with the same sign is positive.

## Example 2 Multiplying Real Numbers

Multiply.

**(a)** $9 \cdot 7 = 63$ — The product of two positive numbers (same sign, +) is positive.

**(b)** $-8(-5) = 40$ — The product of two negative numbers (same sign, −) is positive.

**(c)** $-\frac{1}{2}\left(-\frac{1}{3}\right) = \frac{1}{6}$

## Check Yourself 2

Multiply.

(a) $10 \cdot 12$ (b) $-8(-9)$ (c) $-\frac{2}{3}\left(-\frac{6}{7}\right)$

Two numbers, 0 and 1, have special properties in multiplication.

**Property**

**Multiplicative Identity Property**

The product of 1 and any number is that number. The number 1 is called the **multiplicative identity.** In symbols,

$a \cdot 1 = 1 \cdot a = a$

**Property**

**Multiplicative Property of Zero**

The product of 0 and any number is 0. In symbols,

$a \cdot 0 = 0 \cdot a = 0$

## Example 3 Multiplying Real Numbers

Find each product.

**(a)** $1(-7) = -7$

**(b)** $15(1) = 15$

**(c)** $-7(0) = 0$

**(d)** $0 \cdot 12 = 0$

**(e)** $-\frac{4}{5}(0) = 0$

## Check Yourself 3

Multiply.

(a) $-10(1)$ (b) $0(-17)$ (c) $\frac{5}{7}(1)$ (d) $0\left(\frac{3}{4}\right)$

To complete our discussion of the properties of multiplication, we state the following.

**Property**

**Multiplicative Inverse Property**

For any nonzero number $a$, there is a number $\frac{1}{a}$ such that

$a \cdot \frac{1}{a} = 1$

$\frac{1}{a}$ is called the **multiplicative inverse,** or the **reciprocal,** of $a$. The product of any nonzero number and its reciprocal is 1.

## Example 4 Finding a Reciprocal

< Objective 2 >

**RECALL**

The product of two negative numbers is a positive number.

**(a)** The reciprocal of 3 is $\frac{1}{3}$ because $3 \cdot \frac{1}{3} = 1$.

**(b)** The reciprocal of $-5$ is $\frac{1}{-5}$ or $-\frac{1}{5}$ because $-5\left(-\frac{1}{5}\right) = 1$.

**(c)** The reciprocal of $\frac{2}{3}$ is $\frac{1}{\frac{2}{3}}$ or $\frac{3}{2}$ because $\frac{2}{3} \cdot \frac{3}{2} = 1$.

### Check Yourself 4

Find the multiplicative inverse (or the reciprocal) of each of the following numbers.

(a) 6 (b) $-4$ (c) $\frac{1}{4}$ (d) $-\frac{3}{5}$

In addition to the properties just mentioned, we can extend the commutative and associative properties for multiplication to real numbers. Example 5 is an application of the associative property of multiplication.

## Example 5 Multiplying Real Numbers

< Objective 3 >

Find the following product.

$-3(2)(-7)$

Applying the associative property, we can group the first two factors to write

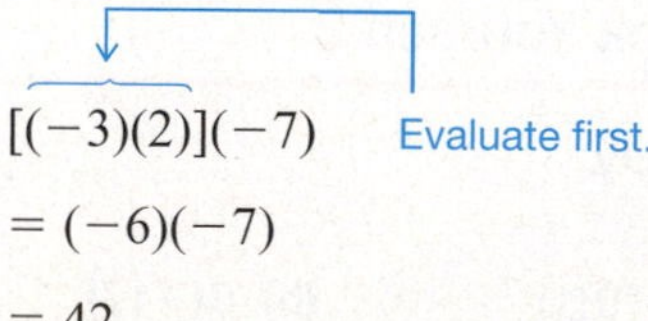

$[(-3)(2)](-7)$ Evaluate first.

$= (-6)(-7)$

$= 42$

**NOTE**

This "grouping" can be done mentally.

### Check Yourself 5

Find the product.

$-5(-8)(-2)$

We will take a closer look at how the order of operations comes into play when negative numbers are involved in Section 10.5. To do that, we need to learn some basic skills involving negative numbers.

The first such skill involves quantities with multiple negative signs. Our approach takes into account two ways of looking at positive and negative numbers.

First, a negative sign indicates the opposite of the number which follows. For instance, we have already said that the opposite of 5 is $-5$, whereas the opposite of $-5$ is 5. This last instance can be translated as $-(-5) = 5$.

Second, any number must correlate to some point on the number line. That is, any nonzero number is either positive or negative. No matter how many negative signs a quantity has, you can always simplify it so that it is represented by a negative or a positive number (one negative sign or none).

## Example 6 Simplifying Real Numbers with Negative Signs

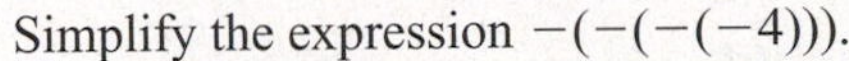

**NOTE**

You should see a pattern emerge. An even number of negative signs gives a positive number, whereas an odd number of negative signs produces a negative number.

Simplify the expression $-(-(-(-4)))$.

The opposite of $-4$ is 4, so $-(-4) = 4$.

The opposite of 4 is $-4$, so $-(-(-4)) = -4$. The opposite of this last number, $-4$, is 4, so

$$-(-(-(-4))) = 4$$

**Check Yourself 6**

Simplify the expression $-(-(-(-(-(-(-12))))))$.

We should also learn to evaluate expressions that contain both an exponent and a negative sign. Example 7 provides us with the opportunity to do just that.

## Example 7 Evaluating Expressions

Evaluate each expression.

**(a)** $(-5)^2$

This means $(-5) \cdot (-5)$, which is 25.

**> CAUTION**

Many students make careless errors when evaluating these types of expressions. Remember that $(-5)^2 \neq -5^2$.

**(b)** $-5^2$

This is not the same as the example in part **(a).** One way to look at this is to say that $-5^2$ is the *opposite* of $5^2 = 25$ so that

$$-5^2 = -25$$

Alternatively, we can look at $-5$ as shorthand notation for $-1 \cdot 5$. In which case, the order of operations requires that we compute the exponent prior to performing the multiplication. Therefore,

$$-5^2 = -(5^2) = -(25) = -25$$

**(c)** $(-5)^3$

This means $(-5) \cdot (-5) \cdot (-5)$.

**RECALL**

In Example 6, we saw that multiplying an odd number of negative signs yields a negative number.

From our earlier work in Example 5 of this section, we know we can use the associative property.

$$\begin{aligned}(-5)^3 &= (-5)\cdot(-5)\cdot(-5)\\ &= [(-5)\cdot(-5)]\cdot(-5)\\ &= 25\cdot(-5)\\ &= -125\end{aligned}$$

**Check Yourself 7**

Evaluate each expression.

(a) $-7^3$ (b) $(-7)^3$ (c) $(-3)^4$

(d) $-3^4$ (e) $\left(-\frac{2}{3}\right)^2$ (f) $-\frac{2^2}{3}$

Of course, there are many applications that involve both positive and negative numbers.

## Example 8 An Application of Real Numbers

The manager responsible for worker productivity at the TarCo manufacturing plant conducts a study on the amount of time workers do not engage in productive work. The manager finds that the average worker begins working 10 minutes after the shift begins, leaves for lunch 5 minutes early, and returns 10 minutes late. The manager also finds that the average employee works 15 minutes after the shift is over. Finally, the manager finds that employees spend an average of 22 minutes engaged in their 15-minute coffee break. If the plant employs 230 people, how much productivity does the plant lose each day?

We can compute the productivity lost in a day by the average worker as follows:

$$-10 + (-5) + (-10) + 15 + (15 - 22) = -17$$

Because the plant employs 230 people, the total lost productivity can be found by multiplication.

$$230(-17) = -3{,}910$$

The plant loses a total of 3,910 minutes of productivity per day (or about 65.2 hours).

**Check Yourself 8**

A math professor grades an exam as follows. Each correct answer is worth 4 points, each question left blank is worth 0 points, and each incorrect answer is worth −2 points. If a student answered 21 questions correctly, leaves 1 question blank, and answers 3 questions incorrectly, what score did the student earn on the exam?

**Property**

### The Product of Two Negative Numbers

This is a detailed explanation of why the product of two negative numbers is positive.

From our earlier work, we know that the sum of a number and its opposite is 0:

$5 + (-5) = 0$

Multiply both sides of the equation by $-3$:

$(-3)[5 + (-5)] = (-3)(0)$

Because the product of 0 and any number is 0, on the right we have 0.

$(-3)[5 + (-5)] = 0$

We use the distributive property on the left.

$(-3)(5) + (-3)(-5) = 0$

We know that $(-3)(5) = -15$, so the equation becomes

$-15 + (-3)(-5) = 0$

We now have a statement of the form

$-15 + \square = 0$

in which $\square$ is the value of $(-3)(-5)$. We also know that $\square$ is the number that must be added to $-15$ to get 0, so $\square$ is the opposite of $-15$, or 15. This means that

$(-3)(-5) = 15$ The product is positive!

Regardless of which numbers we use in this argument the resulting product of two negative numbers will always be positive.

## Check Yourself ANSWERS

**1.** **(a)** $-35$; **(b)** $-108$; **(c)** $-120$; **(d)** $-\frac{4}{7}$ **2.** **(a)** 120; **(b)** 72; **(c)** $\frac{4}{7}$
**3.** **(a)** $-10$; **(b)** 0; **(c)** $\frac{5}{7}$; **(d)** 0 **4.** **(a)** $\frac{1}{6}$; **(b)** $-\frac{1}{4}$; **(c)** 4; **(d)** $-\frac{5}{3}$ **5.** $-80$
**6.** $-12$ **7.** **(a)** $-343$; **(b)** $-343$; **(c)** 81; **(d)** $-81$; **(e)** $\frac{4}{9}$; **(f)** $-\frac{4}{3}$ **8.** 78

## Reading Your Text

The following fill-in-the-blank exercises are designed to ensure that you understand some of the key vocabulary used in this section.

SECTION 10.4

**(a)** The product of two numbers with different signs is ______________.

**(b)** The product of two negative numbers is ______________.

**(c)** The number 1 is called the multiplicative ______________.

**(d)** $\frac{1}{a}$ is called the multiplicative inverse or ______________ of $a$.

# 10.4 exercises

**Basic Skills** | Advanced Skills | Vocational-Technical Applications | Calculator/Computer | Above and Beyond

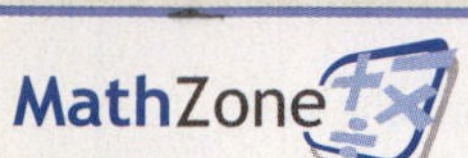

Boost your grade at mathzone.com!

> Practice Problems
> NetTutor
> Self-Tests
> e-Professors
> Videos

Name ______

Section ______ Date ______

**Answers**

1. ______ 2. ______
3. ______ 4. ______
5. ______ 6. ______
7. ______ 8. ______
9. ______ 10. ______
11. ______ 12. ______
13. ______ 14. ______
15. ______ 16. ______
17. ______ 18. ______
19. ______ 20. ______
21. ______ 22. ______
23. ______ 24. ______
25. ______ 26. ______
27. ______ 28. ______

< Objective 1 >

*Perform the indicated operation.*

**1.** $4 \cdot 10$

**2.** $3 \cdot 14$

**3.** $5(-12)$ > Videos

**4.** $10(-2)$

**5.** $8(-10)$

**6.** $13(-7)$

**7.** $-8(9)$

**8.** $-12(3)$

**9.** $-11(12)$

**10.** $-17(5)$

**11.** $-8(-7)$ > Videos

**12.** $-9(-8)$

**13.** $-5(-12)$

**14.** $-7(-3)$

< Objective 2 >

**15.** $1(-18)$

**16.** $-3(1)$

**17.** $\frac{3}{4} \cdot \frac{4}{3}$

**18.** $-\frac{5}{3}\left(-\frac{3}{5}\right)$

**19.** $-5\left(-\frac{1}{5}\right)$ > Videos

**20.** $7 \cdot \frac{1}{7}$

**21.** $5\left(-\frac{1}{5}\right)$

**22.** $-7\left(\frac{1}{7}\right)$

**23.** $-5(0)$

**24.** $0\left(-\frac{2}{3}\right)$

< Objective 3 >

**25.** $-5(3)(-2)$

**26.** $-4(2)(-3)$

**27.** $8(-3)(7)$

**28.** $13(-2)(6)$

29. $-3(-5)(-2)$

30. $-6(4)(-3)$

31. $-5(-4)(2)$

32. $-5(2)(-6)$

33. $-9(-12)(0)$

34. $-13(0)(-7)$

35. $-(-3)$

36. $-(-9)$

37. $-(-(-1))$

38. $-(-(-11))$

39. $-(-(-(-(-123))))$

40. $-(-(-(-(-(-80)))))$

41. $-6^3$

42. $(-6)^3$

43. $-6^2$

44. $(-6)^2$

45. $(-8)^2$ 

46. $-8^2$ 

**Statistics** *A professor grades an exam by awarding 5 points for each correct answer and subtracting 2 points for each incorrect answer. Points are neither added nor subtracted for answers left blank. What is the exam score of each of the following students?*

47. A student answers 14 questions correctly and 4 incorrectly while leaving 2 questions blank.

48. A student answers 16 questions correctly and 3 incorrectly while leaving 1 question blank.

49. **Social Science** A gambler lost \$45 per hour at a slot machine over a 4-hour period. How much money did the gambler lose?

50. **Social Science** A poker player loses \$325 per hour at a poker table over a 3-hour period. After a break, she wins \$145 per hour for two hours. How did she do, overall?

*Determine whether each statement is* **true** *or* **false.**

51. The square of a negative number is negative.

52. The opposite of the square of a number is negative.

## Answers

29. ______ 30. ______

31. ______ 32. ______

33. ______ 34. ______

35. ______ 36. ______

37. ______ 38. ______

39. ______

40. ______

41. ______

42. ______

43. ______

44. ______

45. ______

46. ______

47. ______

48. ______

49. ______

50. ______

51. ______

52. ______

53. The opposite of the opposite of a negative number is negative.

54. The cube of a negative number is negative.

Basic Skills | **Advanced Skills** | Vocational-Technical Applications | Calculator/Computer | Above and Beyond

*Compute, as indicated.*

55. $4\left(-\frac{3}{2}\right)$

56. $9\left(-\frac{2}{3}\right)$

57. $-\frac{1}{4}(8)$

58. $-\frac{3}{2}(4)$

59. $-9\left(-\frac{2}{3}\right)$

60. $-6\left(-\frac{3}{2}\right)$

61. $\frac{4}{5}\left(-\frac{3}{8}\right)$

62. $-\frac{2}{3}\left(-\frac{6}{7}\right)$

63. $-\frac{3}{4}\left(-\frac{10}{21}\right)$

64. $-\frac{1}{5}\left(\frac{3}{4}\right)$

65. $-\frac{1}{3}\left(\frac{6}{5}\right)(-10)$

66. $-\frac{1}{2}\left(\frac{4}{3}\right)(-6)$

67. $\left(-2\frac{5}{6}\right)\left(3\frac{1}{2}\right)$

68. $\left(-4\frac{7}{10}\right)\left(3\frac{1}{3}\right)$

69. $\left(-8\frac{2}{3}\right)\left(-1\frac{1}{5}\right)$

70. $\left(-3\frac{1}{2}\right)\left(-4\frac{1}{4}\right)$

71. $3.25(-4)$

72. $5.4(-5)$

73. $-1.25(-12)$

74. $-1.5(-20)$

## Answers

53. ______ 54. ______

55. ______ 56. ______

57. ______ 58. ______

59. ______

60. ______

61. ______

62. ______

63. ______

64. ______

65. ______

66. ______

67. ______

68. ______

69. ______

70. ______

71. ______

72. ______

73. ______

74. ______

Basic Skills | Advanced Skills | **Vocational-Technical Applications** | Calculator/Computer | Above and Beyond

**75. Manufacturing Technology** Companies will occasionally sell products at a loss in order to draw in customers or as a reward to good customers. The theory is that customers will buy other products along with the discounted product and the net result will be a profit.

Beguhn Industries sells five different products. For each unit of product sold, the company makes or loses the following: product A, makes \$18; product B, loses \$4; product C, makes \$11; product D, makes \$38; and product E, loses \$15. During the previous month, Beguhn Industries sold 127 units of product A, 273 units of product B, 201 units of product C, 377 units of product D, and 43 units of product E.

Calculate the profit or loss for the month.

**76. Mechanical Engineering** The bending moment created by a center support on a steel beam is approximated by the formula $-\frac{1}{4}PL^3$, in which $P$ is the load on each side of the center support and $L$ is the length of the beam on each side of the center support (assuming a symmetrical beam and load).

If the total length of the beam is 24 ft (12 ft on each side of the center) and the total load is 4,124 lb (2,062 lb on each side of the center), what is the bending moment (in ft-lb) at the center support?

Basic Skills | Advanced Skills | Vocational-Technical Applications | **Calculator/Computer** | Above and Beyond

**77.** Use a calculator to complete the following table.

| | |
|---|---|
| $4 \cdot 3$ | 12 |
| $4 \cdot 2$ | 8 |
| $4 \cdot 1$ | |
| $4 \cdot 0$ | |
| $4(-1)$ | |
| $4(-2)$ | |
| $4(-3)$ | |
| $4(-4)$ | |

**78.** Use a calculator to complete the following table.

| | |
|---|---|
| $-4 \cdot 3$ | $-12$ |
| $-4 \cdot 2$ | $-8$ |
| $-4 \cdot 1$ | |
| $-4 \cdot 0$ | |
| $-4(-1)$ | |
| $-4(-2)$ | |
| $-4(-3)$ | |
| $-4(-4)$ | |

## Answers

75. ______

76. ______

77. ______

78. ______

## Answers

**1.** 40 **3.** $-60$ **5.** $-80$ **7.** $-72$ **9.** $-132$ **11.** 56 **13.** 60
**15.** $-18$ **17.** 1 **19.** 1 **21.** $-1$ **23.** 0 **25.** 30 **27.** $-168$
**29.** $-30$ **31.** 40 **33.** 0 **35.** 3 **37.** $-1$ **39.** $-123$
**41.** $-216$ **43.** $-36$ **45.** 64 **47.** 62 **49.** \$180 **51.** False
**53.** True **55.** $-6$ **57.** $-2$ **59.** 6 **61.** $-\frac{3}{10}$ **63.** $\frac{5}{14}$ **65.** 4
**67.** $-9\frac{11}{12}$ **69.** $10\frac{2}{5}$ **71.** $-13$ **73.** 15 **75.** +\$17,086

**77.**

| | |
|---|---|
| $4 \cdot 3$ | 12 |
| $4 \cdot 2$ | 8 |
| $4 \cdot 1$ | 4 |
| $4 \cdot 0$ | 0 |
| $4(-1)$ | $-4$ |
| $4(-2)$ | $-8$ |
| $4(-3)$ | $-12$ |
| $4(-4)$ | $-16$ |

# 10.5 Dividing Real Numbers and the Order of Operations

< 10.5 Objectives >

1> Find the quotient of two real numbers

2> Recognize that division by zero is undefined

3> Use the order of operations to evaluate expressions involving real numbers

You know from your work in arithmetic that multiplication and division are related operations. We can use this fact, and our work from Section 10.4, to determine rules for the division of real numbers. Every division problem can be stated as an equivalent multiplication problem. For instance,

$$\frac{15}{5} = 3 \qquad \text{because} \qquad 15 = 5 \cdot 3$$

$$\frac{-24}{6} = -4 \qquad \text{because} \quad -24 = (6)(-4)$$

$$\frac{-30}{-5} = 6 \qquad \text{because} \quad -30 = (-5)(6)$$

These examples illustrate that because the two operations are related, the rule of signs that we stated in Section 10.4 for multiplication is also true for division.

**Property**

**Dividing Real Numbers**

The quotient of two numbers with different signs is negative.
The quotient of two numbers with the same sign is positive.

To divide two real numbers, divide their absolute values. Then attach the proper sign according to the preceding rule.

**Example 1** **Dividing Real Numbers**

< Objective 1 >

Divide.

(a) $\frac{28}{7} = 28 \div 7 = 4$ ← Positive

(28: Positive; 7: Positive)

(b) $\frac{-36}{-4} = -36 \div (-4) = 9$ ← Positive

(−36: Negative; −4: Negative)

(c) $\frac{-42}{7} = -42 \div 7 = -6$



(d) $\frac{75}{-3} = 75 \div (-3) = -25$



(e) $\frac{15.2}{-3.8} = -4$

## Check Yourself 1

Divide.

(a) $\frac{-55}{11}$ (b) $\frac{80}{20}$ (c) $\frac{-48}{-8}$ (d) $\frac{144}{-12}$ (e) $\frac{-13.5}{-2.7}$

You should be very careful when 0 is involved in a division problem. Remember that 0 divided by any nonzero number is just 0. Recall that

$$\frac{0}{-7} = 0 \quad \text{because} \quad 0 = (-7)(0)$$

However, if zero is the *divisor,* we have a special problem. Consider

$$\frac{9}{0} = ?$$

This means that $9 = 0 \cdot ?$.

Can 0 times a number ever be 9? No, so there is no solution.

Because $\frac{9}{0}$ cannot be replaced by any number, we agree that *division by* 0 *is not allowed.*

**Property**

**Division by Zero** Division by 0 is undefined.

## Example 2 Division and Zero

< Objective 2 >

Divide, if possible.

(a) $\frac{7}{0}$ is undefined.

(b) $\frac{-9}{0}$ is undefined.

**(c)** $\frac{0}{5} = 0$

**(d)** $\frac{0}{-8} = 0$

**Note:** The expression $\frac{0}{0}$ is called an **indeterminate form.** You will learn more about this in later mathematics classes.

## Check Yourself 2

Divide if possible.

(a) $\frac{0}{3}$ (b) $\frac{5}{0}$ (c) $\frac{-7}{0}$ (d) $\frac{0}{-9}$

In Section 10.4, we began to develop the skills to evaluate more complex expressions involving negative numbers. We continue that work in examples 3–7.

We begin with the reminder that any number must correlate to some point on the number line. That is, any nonzero number is either positive or negative.

With fractions, this fact is most easily seen as follows:

$$-\frac{3}{5} = \frac{-3}{5} = \frac{3}{-5}$$

All three quantities represent the same point on the number line

$-\frac{3}{5}$

−3 −2 −1 0 1 2 3

In this text, we generally choose to write negative fractions with the minus sign outside the fraction, such as $-\frac{3}{5}$.

In Section 10.4, we used these facts about multiple negative signs to help us evaluate integers. Here, we evaluate fractions with multiple negative signs.

## Example 3 Simplifying Fractions with Negative Signs

Simplify each expression.

**(a)** $-\frac{-3}{4}$

This is the opposite of $\frac{-3}{4}$ which is $\frac{3}{4}$, a positive number.

**(b)** $-\frac{-3}{-4}$

The fraction part represents a negative number divided by another negative number, which is positive.

$$\frac{-3}{-4} = \frac{3}{4}$$

Therefore,

$$-\frac{-3}{-4} = -\left(\frac{-3}{-4}\right) = -\frac{3}{4}$$

### Check Yourself 3

Simplify each expression.

(a) $-\frac{7}{-10}$ (b) $-\frac{-2}{-3}$

When symbols of grouping, or more than one operator, are involved in an expression, we must always remember to follow the rules for the order of operations.

**Property**

### Order of Operations

The order of operations was first presented in Section 1.7.

**"Please Excuse My Dear Aunt Sally."**

1. Perform all operations inside grouping symbols. Grouping symbols include parentheses, brackets, absolute value signs, fraction bars, and radical signs.
2. Apply all exponents.
3. Perform all multiplication and division operations, from left to right.
4. Perform all addition and subtraction operations, from left to right.

### Example 4 Multiplying Real Numbers

< Objective 3 >

Evaluate each expression.

(a) $7(-9 + 12)$ Evaluate inside the parentheses first.

$= 7(3) = 21$

(b) $-8(-7) - 40$ Multiply first, then subtract.

$= 56 - 40$

$= 16$

(c) $(-5)^2 - 3$ Evaluate the power first.

$= (-5)(-5) - 3$ Note that $(-5)^2 = (-5)(-5) = 25$

$= 25 - 3$

$= 22$

(d) $-5^2 - 3$ Note that $-5^2 = -25$. The power applies *only* to the 5.

$= -25 - 3$

$= -28$

### Check Yourself 4

Evaluate each expression.

(a) $8(-9 + 7)$ (b) $-3(-5) + 7$
(c) $(-4)^2 - (-4)$ (d) $-4^2 - (-4)$

Because the fraction bar also serves as a grouping symbol, all operations in the numerator or denominator should each be done first, as illustrated in Example 5.

## Example 5 Dividing Real Numbers

Evaluate each expression.

(a) $\frac{-6(-7)}{3} = \frac{42}{3} = 14$ Multiply in the numerator and then divide.

(b) $\frac{3 + (-12)}{3} = \frac{-9}{3} = -3$ Add in the numerator and then divide.

(c) $\frac{-4 + 2(-6)}{-6 - 2} = \frac{-4 + (-12)}{-6 - 2}$ Multiply in the numerator. Then add in the numerator and subtract in the denominator.

$= \frac{-16}{-8} = 2$ Divide as the last step.

### Check Yourself 5

Evaluate each expression.

(a) $\frac{-4 + (-8)}{6}$ (b) $\frac{3 - 2(-6)}{-5}$ (c) $\frac{-2(-4) - (-6)(-5)}{(-4)(11)}$

Many students have difficulty applying the distributive property when negative numbers are involved. Just remember that the sign of a number "travels" with that number.

## Example 6 Applying the Distributive Property with Negative Numbers

**RECALL**

We usually enclose negative numbers in parentheses in the middle of an expression to avoid careless errors.

Evaluate each expression.

(a) $-7(3 + 6) = -7 \cdot 3 + (-7) \cdot 6$ Apply the distributive property.

$= -21 + (-42)$ Multiply first and then add.

$= -63$

(b) $-3(5 - 6) = -3[5 + (-6)]$ First, change the subtraction to addition.

$= -3 \cdot 5 + (-3)(-6)$ Distribute the $-3$.

$= -15 + 18$ Multiply first and then add.

$= 3$

**RECALL**

We use brackets rather than nesting parentheses to avoid careless errors.

**(c)** $5(-2 - 6) = 5[-2 + (-6)]$
$= 5 \cdot (-2) + 5 \cdot (-6)$
$= -10 + -30$
$= -40$ The sum of two negative numbers is negative.

### Check Yourself 6

Evaluate each expression.

**(a)** $-2(-3 + 5)$ **(b)** $4(-3 + 6)$ **(c)** $-7(-3 - 8)$

Combining the elements from Examples 3 to 6 requires that we carefully apply the order of operations. You must remain vigilant with any negative signs.

### Example 7 Evaluating Expressions

Evaluate each expression.

**(a)** $4 + 2 \cdot (5 - 7)^2$
$= 4 + 2 \cdot (-2)^2$ Evaluate inside parentheses first.
$= 4 + 2 \cdot 4$ Apply the exponent.
$= 4 + 8$ Multiply.
$= 12$ Add.

**RECALL**

Fraction bars are grouping symbols. Evaluate the numerator and denominator separately.

**(b)** $\dfrac{3 - (-2)^3}{-7 + 3}$

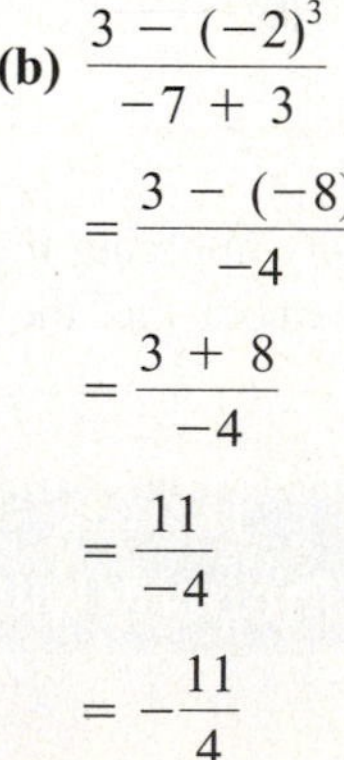

$= \dfrac{3 - (-8)}{-4}$

$= \dfrac{3 + 8}{-4}$

$= \dfrac{11}{-4}$

$= -\dfrac{11}{4}$

**NOTE**

In future courses you will find that we rarely change an improper fraction to a mixed number.

### Check Yourself 7

Evaluate each expression.

**(a)** $-35 - (3 - 7)^3$ **(b)** $\dfrac{2 - 3 \cdot (1 - 5)^2}{(-3)^3 + (-2)^4}$

## Check Yourself ANSWERS

**1.** **(a)** $-5$; **(b)** 4; **(c)** 6; **(d)** $-12$; **(e)** 5 **2.** **(a)** 0; **(b)** undefined; **(c)** undefined; **(d)** 0 **3.** **(a)** $\frac{7}{10}$; **(b)** $-\frac{2}{3}$ **4.** **(a)** $-16$; **(b)** 22; **(c)** 20; **(d)** $-12$ **5.** **(a)** $-2$; **(b)** $-3$; **(c)** $\frac{1}{2}$ **6.** **(a)** $-4$; **(b)** 12; **(c)** 77 **7.** **(a)** 29; **(b)** $\frac{46}{11}$

## Reading Your Text

The following fill-in-the-blank exercises are designed to ensure that you understand some of the key vocabulary used in this section.

### SECTION 10.5

**(a)** The quotient of two numbers with different signs is ______________.

**(b)** The quotient of two negative numbers is ______________.

**(c)** ______________ by 0 is undefined.

**(d)** Every nonzero number is either ______________ or negative.

# 10.5 exercises

Basic Skills | Advanced Skills | Vocational-Technical Applications | Calculator/Computer | Above and Beyond

MathZone

Boost your grade at mathzone.com!

> Practice Problems
> NetTutor
> Self-Tests
> e-Professors
> Videos

Name ______________

Section ________ Date ________

## Answers

1. ______ 2. ______
3. ______ 4. ______
5. ______ 6. ______
7. ______ 8. ______
9. ______ 10. ______
11. ______ 12. ______
13. ______ 14. ______
15. ______ 16. ______
17. ______ 18. ______
19. ______ 20. ______
21. ______ 22. ______
23. ______ 24. ______
25. ______ 26. ______

< Objectives 1–3 >

*Perform the indicated operations.*

1. $\frac{70}{14}$

2. $\frac{48}{6}$

3. $\frac{-20}{-4}$

4. $\frac{-75}{-3}$

5. $\frac{-24}{8}$

6. $\frac{56}{-7}$

7. $\frac{50}{-5}$

8. $\frac{-52}{4}$

9. $\frac{0}{-8}$

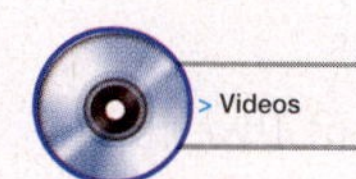


10. $\frac{-9}{-1}$

11. $\frac{-17}{1}$

12. $\frac{18}{0}$

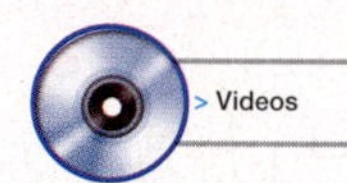


13. $\frac{-27}{-1}$

14. $\frac{0}{8}$

15. $\frac{-10}{0}$

16. $\frac{-32}{1}$

17. $\frac{-8}{32}$

18. $\frac{-6}{-30}$

19. $\frac{24}{-16}$

20. $\frac{-25}{10}$

21. $\frac{-28}{-42}$

22. $\frac{-125}{-75}$

23. $-\frac{-13}{52}$

24. $-\frac{-52}{-13}$

25. $-\frac{-12}{-15}$

26. $-\frac{-91}{-7}$

27. $5(7 - 2)$

28. $7(8 - 5)$

29. $2(5 - 8)$

30. $6(14 - 16)$

31. $(-3)(9 - 7)$

32. $(-6)(12 - 9)$

33. $(-3)(-2 - 5)$

34. $(-2)(-7 - 3)$

35. $\dfrac{-6(-3)}{2}$

36. $\dfrac{-9(5)}{-3}$

37. $\dfrac{24}{-4 - 8}$

38. $\dfrac{36}{-7 + 3}$

39. $\dfrac{55 - 19}{-12 - 6}$

40. $\dfrac{-11 - 7}{-14 + 8}$

41. $\dfrac{7 - 5}{2 - 2}$

42. $\dfrac{-11 - (-3)}{-4 - (-4)}$

43. $\dfrac{-9(-6) - 10}{18 - (-4)}$

44. $\dfrac{4 - 2(-6)}{-14 - (-6)}$

45. **Business and Finance** Michelle deposits \$1,000 in her checking account each month. She writes a check for \$100 for car insurance and \$200 for her car payment. She also makes a \$55 payment on her student loan. How much money is left for her to use each week (assume there are 4 weeks in the month)?

46. **Business and Finance** An advertising agency lost \$42,000 in revenue last year when a major client left for another agency. What was the agency's monthly loss in revenue?

47. **Science and Medicine** At noon, the temperature was 70°F. It dropped at a constant rate until 5 P.M., when it was 58°F. What was the hourly change in temperature?

chapter 10 > Make the Connection

48. **Science and Medicine** A chemist has 84 ounces (oz) of a solution, which she pours into test tubes. If the chemist pours $\frac{2}{3}$ oz in each test tube, how many can she fill?

*Fill in each blank with* **always, sometimes,** *or* **never.**

49. The sum of a positive number and the square of a negative number is ______________ negative.

50. The product of three negative numbers is ______________ negative.

## Answers

27. ______
28. ______
29. ______
30. ______
31. ______
32. ______
33. ______
34. ______
35. ______
36. ______
37. ______
38. ______
39. ______
40. ______
41. ______
42. ______ 43. ______
44. ______ 45. ______
46. ______ 47. ______
48. ______
49. ______ 50. ______

## Answers

51. ______ 52. ______

53. ______ 54. ______

55. ______ 56. ______

57. ______ 58. ______

59. ______ 60. ______

61. ______ 62. ______

63. ______ 64. ______

65. ______ 66. ______

67. ______ 68. ______

69. ______ 70. ______

71. ______

72. ______

73. ______

74. ______

75. ______

76. ______

77. ______

78. ______

79. ______

80. ______

**51.** The sum of a negative number and the product of negative numbers is ______ negative.

**52.** A negative number subtracted from the square of a negative number is ______ negative.

Basic Skills | **Advanced Skills** | Vocational-Technical Applications | Calculator/Computer | Above and Beyond

*Compute, as indicated.*

**53.** $(-2)(-7) + (2)(-3)$

**54.** $(-3)(-6) + (4)(-2)$

**55.** $(-7)(3) - (-2)(-8)$

**56.** $(-5)(2) - (3)(-4)$

**57.** $\dfrac{3(6) - (-4)(8)}{6 - (-4)}$

**58.** $\dfrac{5(-2) - (-4)(-5)}{-4 - 2}$

**59.** $\dfrac{2(-5) + 4(6 - 8)}{3(-4 + 2)}$

**60.** $\dfrac{-3(-5) - 3(5 - 8)}{4(-8 + 6)}$

**61.** $(-7)^2 - 17$

**62.** $(-6)^2 - 20$

**63.** $-7^2 - 17$

**64.** $-6^2 - 20$

**65.** $(-4)^2 - (-2)(-5)$

**66.** $-4^2 - (-2)(-5)$

**67.** $(-6)^2 - (-3)^2$

**68.** $-6^2 - 3^2$

**69.** $(-8)^2 - 8^2$

**70.** $-11^2 - (-11)^2$

**71.** $5 + 3(4 - 6)^2$

**72.** $8 - 2(3 - 6)^3$

**73.** $-20 \div 2 + 10 \cdot 2$

**74.** $-6 \cdot 3 \div 2 - 9$

**75.** $\dfrac{4 - (-3)^2}{-7(4) + 3}$

**76.** $\dfrac{4 - (-2)^3}{5^2 - (-2)^2}$

**77.** $-60 \div (-3)(4) - 2^3 + 4$

**78.** $\dfrac{16 \div (-2)(-4)}{2^3 + 4}$

**79.** $\dfrac{7 - (-1)}{3^2 - 9 \div (-3)}$

**80.** $\dfrac{4 - (-3) + 1}{2 - 2 \div (-2)}$

Basic Skills | Advanced Skills | **Vocational-Technical Applications** | Calculator/Computer | Above and Beyond

**81.** **Manufacturing Technology** Peer's Pipe Fitters started the month of July with 1,789 gallons (gal) of liquified petrolium gas (LP) in their tank. After 21 working days, there were 676 gal left in the tank. If the same amount was used each day, how much LP was consumed each day?

**82.** **Business and Finance** Three friends bought equal shares in an investment for a total of \$21,000. They sold it later for \$17,232. How much did each person profit?

Basic Skills | Advanced Skills | Vocational-Technical Applications | **Calculator/Computer** | Above and Beyond

We now present a set of exercises for which the calculator might be the preferred tool. As indicated by the placement of this explanation, you should refrain from using a calculator on the exercises that precede this.

Recall from Section 10.3 that your calculator has a *negative* key. Using this makes multiplication and division of real numbers a relatively straightforward process. Remember that you need to press the negative key *after* entering the number on a scientific calculator, but press it *before* entering the number on a graphing calculator.

To evaluate the expression: $457(-734)$

**Step 1** Press the *clear* key. [CLEAR]

**Step 2** Enter the numbers as written. Use the negative key, as necessary, and the proper operation. 457 [×] [(−)] 734

**Step 3** Press *enter* or equals. [ENTER] or [=]

Your display should read −335438.

If we replace [×] with [÷] in step 2, we get $457 \div (-734) \approx -0.623$.

You can also use a calculator to raise real numbers to a power. You should be able to find a *power* key, either a [^] (called a *caret*) or a [$y^x$]. Use this key to separate the base from the power.

To evaluate the expression: $(-3)^6$

**Step 1** Press the *clear* key. [CLEAR]

**Step 2** Enter the negative key followed by the base number. [(] [(−)] 3 [)]
Enter the *power* key and the exponent. [^] 6 or [$y^x$] 6

**Step 3** Press *enter* or equals. [ENTER] or [=]

Your display should read 729.

*For exercises 83 to 94, use your calculator to evaluate each expression. Round your answer to two decimal places.*

**83.** $25(-21)$

**84.** $15(-45)$

## Answers

81. ______

82. ______

83. ______

84. ______

## Answers

85. ______
86. ______
87. ______
88. ______
89. ______
90. ______
91. ______
92. ______
93. ______
94. ______

**85.** $-34(-28)$ **86.** $-71(-19)$

**87.** $345 \div (-25)$ **88.** $128 \div (-28)$

**89.** $-564 \div 36$ **90.** $-232 \div 52$

**91.** $-28 \div (-14)$ **92.** $-456 \div (-124)$

**93.** $(-4)^5$ **94.** $(-5)^4$

## Answers

**1.** 5 **3.** 5 **5.** $-3$ **7.** $-10$ **9.** 0 **11.** $-17$ **13.** 27 **15.** Undefined **17.** $-\frac{1}{4}$ **19.** $-\frac{3}{2}$ **21.** $\frac{2}{3}$ **23.** $\frac{1}{4}$ **25.** $-\frac{4}{5}$ **27.** 25 **29.** $-6$ **31.** $-6$ **33.** 21 **35.** 9 **37.** $-2$ **39.** $-2$ **41.** Undefined **43.** 2 **45.** \$161.25 **47.** $-2.4$°F **49.** never **51.** sometimes **53.** 8 **55.** $-37$ **57.** 5 59. 3 **61.** 32 **63.** $-66$ **65.** 6 **67.** 27 **69.** 0 **71.** 17 **73.** 10 **75.** $\frac{1}{5}$ **77.** 76 **79.** $\frac{2}{3}$ **81.** $53 \frac{\text{gal}}{\text{day}}$ **83.** $-525$ **85.** 952 **87.** $-13.8$ **89.** $-15.67$ **91.** 2 **93.** $-1{,}024$

# Activity 29 :: Building Molecules

Every atom has an associated **valence number.** The valence of an atom is the number of electrons in its outermost electron-energy level.

From the valence number, we can determine the number of electrons that an atom must exchange in a *covalent bond* in order to form a stable molecule. We refer to this number as the **covalent bonding number.**

If an atom's valence is less than 4, then its covalent bonding number is the same as its valence number. In this case, the atom needs to give up electrons to another atom in order to form a covalent bond. The number of electrons that such an atom must give up is equal to its covalent bonding number.

Note that it is common, in chemistry, to include a "+" symbol before the covalent bonding number of an atom when it is a positive number.

If an atom's valence is greater than 4, then its covalent bonding number is equal to its valence minus 8 (this will always be zero or a negative number). These atoms gain electrons when forming covalent bonds. Specifically, they must gain the same number of electrons as the absolute value of their covalent bonding number in order to form a stable molecule.

An atom whose valence is exactly 4 can either gain or give up 4 electrons when forming a stable molecule. These atoms have a covalent bonding number of either $+4$ or $-4$, as the situation requires.

A stable molecule is formed when the sum of all the covalent bonding numbers of the atoms in the molecule equals 0.

The possible covalent bonding numbers for atoms are

$-4, -3, -2, -1, 0, 1, 2, 3, 4$

1. Find the covalent bonding number for each of the atoms given in the following table.

| Atom | Valence Number | Covalent Bonding Number |
|---|---|---|
| Boron | 3 | |
| Calcium | 2 | |
| Carbon | 4 | |
| Chlorine | 7 | |
| Hydrogen | 1 | |
| Nitrogen | 5 | |
| Phosphorus | 5 | |
| Sulfur | 6 | |

2. If the covalent bonding numbers of the atoms in a molecule add up to $-2$, and only one more atom is to be added, what covalent bonding number should it have? What valence number should it have?

3. A molecule contains 2 boron atoms. How many sulfur atoms would need to be added to make a stable molecule?

**4.** When combining hydrogen and chlorine, what is the fewest number of each atom that can be used to create a stable molecule?

**5.** When combining hydrogen and nitrogen, what is the fewest number of each atom that can be used to create a stable molecule?

**6.** When combining phosphorus and calcium, what is the fewest number of each atom that can be used to create a stable molecule?

**7.** When combining carbon and boron, what is the fewest number of each atom that can be used to create a stable molecule?

**8.** When combining carbon and nitrogen, what is the fewest number of each atom that can be used to create a stable molecule?

| Definition/Procedure | Example | Reference |
|---|---|---|
| **Real Numbers and Order** | | **Section 10.1** |
| **Positive numbers** Numbers used to name points to the right of 0 on the number line.<br>**Negative numbers** Numbers used to name points to the left of 0 on the number line.<br>**Real numbers** The set containing all of the numbers corresponding to points on the number line. | Negative numbers Positive numbers<br>−3 −2 −1 0 1 2 3<br>Zero is neither positive nor negative. | *p. 718* |
| **Integers** The set consisting of the natural numbers, their opposites, and 0. | The integers are<br>$\{\ldots, -3, -2, -1, 0, 1, 2, 3, \ldots\}$ | |
| **Absolute value** The distance on the number line between the point named by a number and 0.<br>The absolute value of a number is always positive or 0. | The absolute value of a number $a$ is written $\|a\|$.<br>$\|7\| = 7 \qquad \|-8\| = 8$ | *p. 720* |
| **Adding Real Numbers** | | **Section 10.2** |
| ***To Add Real Numbers*** | | |
| **1.** If two numbers have the same sign, add their absolute values. Give the sum the sign of the original numbers.<br>**2.** If two numbers have different signs, subtract the smaller absolute value from the larger. Give the result the sign of the number with the larger absolute value. | $5 + 8 = 13$<br>$-3 + (-7) = -10$<br>$5 + (-3) = 2$<br>$7 + (-9) = -2$ | *p. 730* |
| **Opposites** Two numbers are opposites if the points name the same distance from 0 on the number line, but in opposite directions.<br>The opposite of a positive number is negative.<br>The opposite of a negative number is positive.<br>0 is its own opposite. | 5 units 5 units<br>−5 0 5<br>The opposite of 5 is −5.<br>3 units 3 units<br>−3 0 3<br>The opposite of −3 is 3. | *p. 732* |
| **Subtracting Real Numbers** | | **Section 10.3** |
| ***To Subtract Real Numbers*** | | |
| To subtract real numbers, add the first number and the opposite of the number being subtracted. | $4 - (-2) = 4 + 2 = 6$<br>Replace −2 with its opposite, 2 | *p. 741* |

*Continued*

| Definition/Procedure | Example | Reference |
|---|---|---|
| **Multiplying Real Numbers** | | Section 10.4 |
| *To Multiply Real Numbers* | | |
| To multiply real numbers, multiply the absolute values of the numbers. Then attach a sign to the product according to the following rules:<br>**1.** If the numbers have different signs, the product is negative.<br>**2.** If the numbers have the same sign, the product is positive. | $5 \cdot 7 = 35$<br>$(-4)(-6) = 24$<br>$(8)(-7) = -56$ | pp. 751–752 |
| **Dividing Real Numbers and the Order of Operations** | | Section 10.5 |
| *To Divide Real Numbers* | | |
| To divide real numbers, divide the absolute values of the numbers. Then attach a sign to the quotient according to the following rules:<br>**1.** If the numbers have the same sign, the quotient is positive.<br>**2.** If the numbers have different signs, the quotient is negative. | $\frac{-8}{-2} = 4$<br>$27 \div (-3) = -9$<br>$\frac{-16}{8} = -2$ | p. 763 |
| Always follow the proper **order of operations** when evaluating an expression.<br>**1.** Perform all operations inside grouping symbols.<br>**2.** Apply any exponents.<br>**3.** Perform all multiplication and division operations, from left to right.<br>**4.** Perform all addition and subtraction operations, from left to right. | $4 + 2(5 - 7)^2$<br>$= 4 + 2(-2)^2$ **Please**<br>$= 4 + 2 \cdot (4)$ **Excuse**<br>$= 4 + 8$ **My Dear**<br>$= 12$ **Aunt Sally** | p. 766 |

This summary exercise set is provided to give you practice with each of the objectives of this chapter. Each exercise is keyed to the appropriate chapter section. When you are finished, you can check your answers to the odd-numbered exercises against those presented in the back of the text. If you have difficulty with any of these questions, go back and reread the examples from that section. The answers to the even-numbered exercises appear in the *Instructor's Solutions Manual.* Your instructor will give you guidelines on how to best use these exercises in your instructional setting.

**10.1** *Represent the integers on the number line shown.*

**1.** $6, -18, -3, 2, 15, -9$

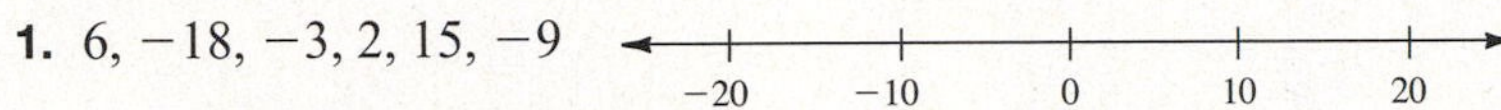


*Place each of the following sets in ascending order.*

**2.** $4, -3, 6, -7, 0, 1, -2$

**3.** $-\frac{1}{2}, -\frac{2}{3}, \frac{3}{5}, -\frac{4}{5}, \frac{5}{6}, \frac{7}{10}$

*Determine the maximum and minimum values for each set of numbers.*

**4.** $4, -2, 5, 1, -6, 3, -4$

**5.** $-4, 2, 5, -9, 8, 1, -6$

*Evaluate.*

**6.** $|9|$

**7.** $|-9|$

**8.** $-|9|$

**9.** $-|-9|$

**10.** $|12 - 8|$

**11.** $|8| - |12|$

**12.** $-|8 - 12|$

**13.** $|-8| - |-12|$

**10.2** *Add.*

**14.** $-3 + (-8)$

**15.** $10 + (-4)$

**16.** $6 + (-6)$

**17.** $-16 + (-16)$

**18.** $-18 + 0$

**19.** $\frac{3}{8} + \left(-\frac{11}{8}\right)$

**20.** $5.7 + (-9.7)$

**21.** $-18 + 7 + (-3)$

**10.3** *Subtract.*

**22.** $8 - 13$

**23.** $-7 - 10$

**24.** $10 - (-7)$

**25.** $-5 - (-1)$

**26.** $-9 - (-9)$

**27.** $0 - (-2)$

**28.** $-\frac{5}{4} - \left(-\frac{17}{4}\right)$

**29.** $7.9 - (-8.1)$

*Perform the indicated operations.*

**30.** $|4 - 8|$

**31.** $|4| - |8|$

**32.** $|-4 - 8|$

**33.** $|-4| - |-8|$

**34.** $-6 - (-2) + 3$

**35.** $-5 - (5 - 8)$

**36.** $7 - (3 - 7) + 4$

**37.** Subtract $-7$ from $-8$.

**38.** Subtract $-9$ from the sum of 6 and $-2$.

**10.4** *Multiply.*

**39.** $10(-7)$

**40.** $-8(-5)$

**41.** $-3(-15)$

**42.** $1(-15)$

**43.** $0(-8)$

**44.** $\frac{2}{3}\left(-\frac{3}{2}\right)$

**45.** $-4\left(\frac{3}{8}\right)$

**46.** $-\frac{5}{4}(-1)$

**47.** $-8(-2)(5)$

**48.** $-4(-3)(2)$

**49.** $\frac{2}{5}(-10)\left(-\frac{5}{2}\right)$

**50.** $\frac{4}{3}(-6)\left(-\frac{3}{4}\right)$

*Perform the indicated operations.*

**51.** $2(-4 + 3)$

**52.** $2(-3) - (-5)(-3)$

**53.** $(2 - 8)(2 + 8)$

**10.5** *Divide.*

**54.** $\frac{80}{16}$

**55.** $\frac{-63}{7}$

**56.** $\frac{-81}{-9}$

**57.** $\frac{0}{-5}$

**58.** $\frac{32}{-8}$

**59.** $\frac{-7}{0}$

*Perform the indicated operations.*

**60.** $\frac{-8 + 6}{-8 - (-10)}$

**61.** $\frac{2(-3) - 1}{5 - (-2)}$

**62.** $\frac{(-5)^2 - (-2)^2}{-5 - (-2)}$

# self-test 10

Name ______

Section ______ Date ______

The purpose of this self-test is to help you check your progress so that you can find sections and concepts that you need to review before the next exam. Allow yourself about an hour to take this test. At the end of that hour, check your answers against those given in the back of this text. If you missed any, note the section reference that accompanies the answer. Go back to that section and reread the examples until you have mastered that particular concept.

## Answers

1. ______
2. ______
3. ______
4. ______
5. ______
6. ______
7. ______
8. ______
9. ______
10. ______
11. ______
12. ______
13. ______
14. ______
15. ______
16. ______
17. ______
18. ______
19. ______

*Represent the integers on the number line shown.*

**1.** $5, -12, 4, -7, 18, -17$

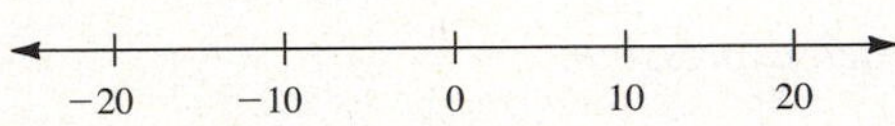


**2.** Place the following data set in ascending order: $4, -3, -6, 5, 0, \frac{3}{4}, \frac{1}{2}, 2, -2$

**3.** Determine the maximum and minimum of the following data set: $3, 2, -5, 6, 1, -2$

*Evaluate.*

**4.** $|7|$

**5.** $|-7|$

**6.** $|18 - 7|$

**7.** $|18| - |-7|$

*Add.*

**8.** $-8 + (-5)$

**9.** $6 + (-9)$

**10.** $-9 + (-12)$

**11.** $-\frac{5}{3} + \frac{8}{3}$

*Subtract.*

**12.** $9 - 15$

**13.** $-9 - 15$

**14.** $5 - (-4)$

**15.** $-7 - (-7)$

*Multiply.*

**16.** $(-8)(5)$

**17.** $(-9)(-7)$

**18.** $(4.5)(-6)$

**19.** $(-2)(-3)(-4)$

## Answers

20. ________

21. ________

22. ________

23. ________

24. ________

25. ________

*Divide.*

**20.** $\frac{75}{-3}$

**21.** $\frac{-27}{-9}$

**22.** $\frac{-45}{9}$

**23.** $\frac{9}{0}$

*Evaluate each expression.*

**24.** $-8 - (-3 + 7)^2$

**25.** $\frac{5 + (-9) - 6}{(-3)^2 + (-2)^3}$

# cumulative review chapters 1-10

Name ______________________

Section __________ Date __________

**Answers**

1. ______________________
2. ______________________
3. ______________________
4. ______________________
5. ______________________
6. ______________________
7. ______________________
8. ______________________
9. ______________________
10. ______________________
11. ______________________
12. ______________________
13. ______________________
14. ______________________
15. ______________________
16. ______________________
17. ______________________
18. ______________________
19. ______________________
20. ______________________
21. ______________________

The following exercises are presented to help you review concepts from earlier chapters. This is meant as review material and not as a comprehensive exam. The answers are presented in the back of the text. Beside each answer is a section reference for the concept. If you have difficulty with any of these exercises, be certain to at least read through the summary related to that section.

**1.** What is the place value of 9 in the numeral 4,593,657?

*Perform the indicated operation.*

**2.** $\begin{array}{r} 7{,}623 \\ 3{,}006 \\ +\ 131{,}602 \\ \hline \end{array}$

**3.** $\begin{array}{r} 125{,}678 \\ -\ 96{,}105 \\ \hline \end{array}$

**4.** $105 \times 509$

**5.** $56\overline{)22{,}540}$

**6.** $103.456 - 89.769$

**7.** $30.45 \times 60.34$

**8.** $\frac{8}{11} \times \frac{33}{56}$

**9.** $\frac{7}{15} \div \frac{21}{45}$

**10.** $6\frac{1}{9} - 3\frac{4}{27} + 5\frac{7}{18}$

**11.** Solve for the unknown: $\frac{6}{3} = \frac{8}{x}$.

**12.** Write 58% as a decimal and fraction.

**13.** Write $\frac{12}{25}$ as a decimal and percent.

*Simplify.*

**14.** 7 ft 22 in.

**15.** 8 lb 20 oz

*Do the indicated operations.*

**16.** $\begin{array}{r} 5\text{ ft}\ \ 8\text{ in.} \\ +\ 6\text{ ft}\ 10\text{ in.} \\ \hline \end{array}$

**17.** $\begin{array}{r} 5\text{ lb}\ \ 8\text{ oz} \\ -\ 2\text{ lb}\ 10\text{ oz} \\ \hline \end{array}$

**18.** $3 \times (3\text{ h }30\text{ min})$

**19.** $\frac{10\text{ min }45\text{ s}}{5}$

*Solve the applications.*

**20.** **Crafts** A plan for a bookcase requires three pieces of lumber 2 ft 8 in. long and two pieces 3 ft 4 in. long. What is the total length of material that is needed?

**21.** **Business and Finance** You can buy three bottles of dishwashing liquid, each containing 1 pt 6 fl oz, on sale for $2.40. For the same price you can buy a large container holding 2 qt. Which is the better buy?

Answers

22. ________
23. ________
24. ________
25. ________
26. ________
27. ________
28. ________
29. ________
30. ________
31. ________
32. ________
33. ________
34. ________
35. ________
36. ________
37. ________
38. ________
39. ________
40. ________

**22. GEOMETRY** A rectangle has length 8.6 cm and width 5.7 cm. Find the area of the rectangle.

**23. GEOMETRY** The sides of a square each measure $8\frac{1}{2}$ in. Find the area of the square.

**24. GEOMETRY** Find the circumference of a circle whose diameter is 8.2 ft. Use 3.14 for $\pi$, and round the result to the nearest tenth.

**25.** 116 is 145% of what number?

**26. BUSINESS AND FINANCE** The sales tax on an item costing \$136 is \$10.20. What is the sales tax rate?

*Complete each statement.*

**27.** 17 g = ________ kg

**28.** 82 cm = ________ mm

**29.** Use a protractor to find the measure of the given angle.

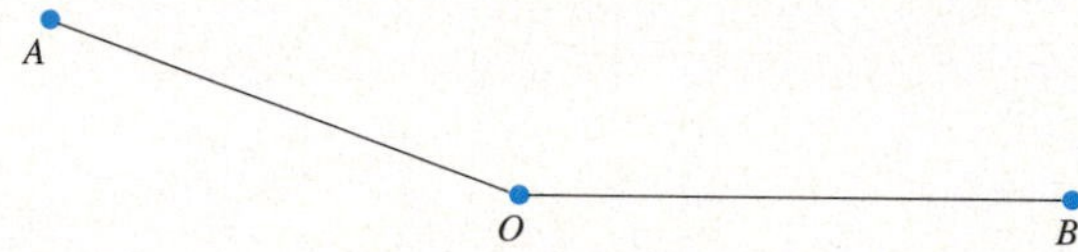


**30. SOCIAL SCIENCE** According to the following bar graph, how many more students were enrolled in the university in 2004 than in 1995?

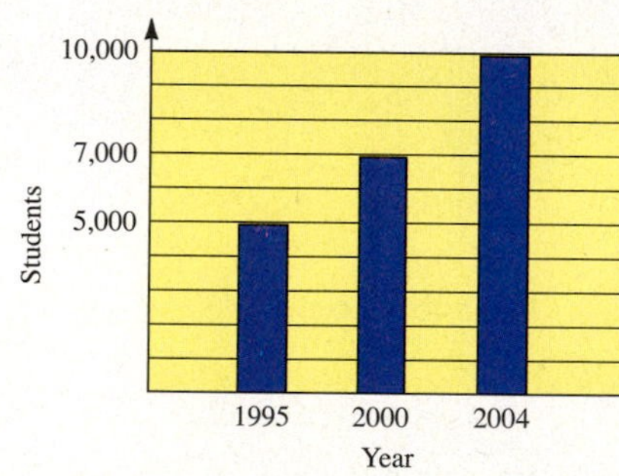


**31.** Calculate the mean, median, and mode for the following data.

15, 16, 18, 13, 17, 19, 17, 21

*Evaluate the expressions.*

**32.** $-9 + 13$

**33.** $17 - (-3)$

**34.** $-9 - 23$

**35.** $|-8| - |-23|$

**36.** $|-8 - 23|$

**37.** $(-9)(-12)$

**38.** $\frac{-36}{-9}$

**39.** $(-7)^2 - (16 \div 2)$

**40.** $45 \div 5 \times 2^3$

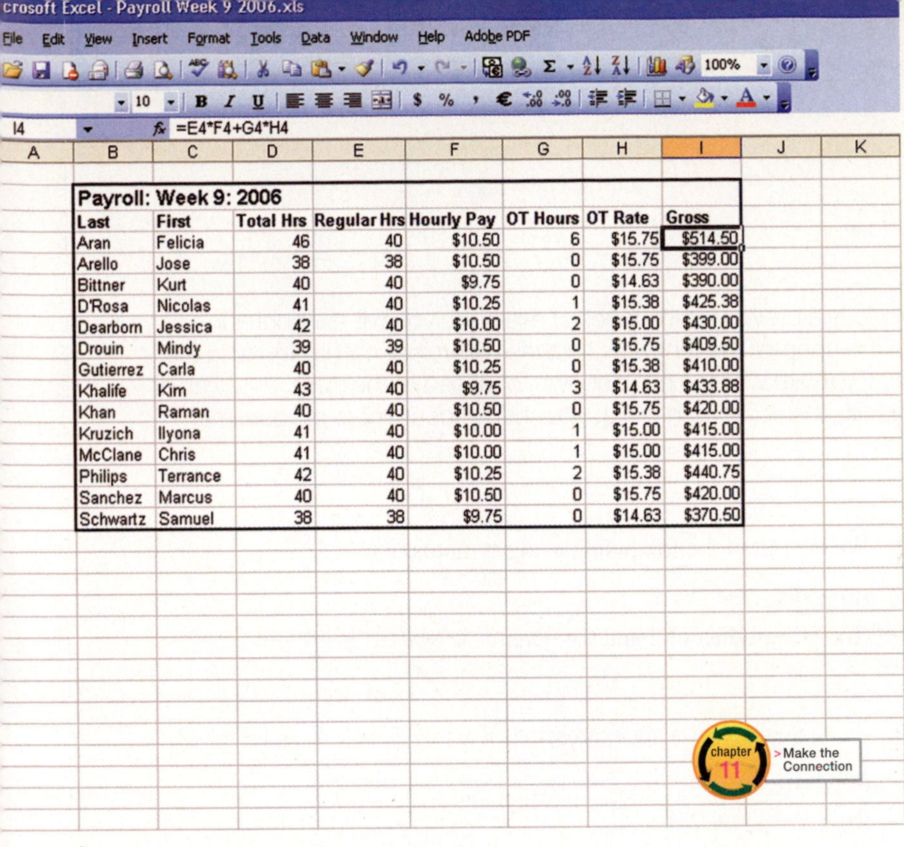

crosoft Excel - Payroll Week 9 2006.xls

I4 $f_x$ =E4*F4+G4*H4

| Payroll: Week 9: 2006 | | | | | | | |
|---|---|---|---|---|---|---|---|
| Last | First | Total Hrs | Regular Hrs | Hourly Pay | OT Hours | OT Rate | Gross |
| Aran | Felicia | 46 | 40 | $10.50 | 6 | $15.75 | $514.50 |
| Arello | Jose | 38 | 38 | $10.50 | 0 | $15.75 | $399.00 |
| Bittner | Kurt | 40 | 40 | $9.75 | 0 | $14.63 | $390.00 |
| D'Rosa | Nicolas | 41 | 40 | $10.25 | 1 | $15.38 | $425.38 |
| Dearborn | Jessica | 42 | 40 | $10.00 | 2 | $15.00 | $430.00 |
| Drouin | Mindy | 39 | 39 | $10.50 | 0 | $15.75 | $409.50 |
| Gutierrez | Carla | 40 | 40 | $10.25 | 0 | $15.38 | $410.00 |
| Khalife | Kim | 43 | 40 | $9.75 | 3 | $14.63 | $433.88 |
| Khan | Raman | 40 | 40 | $10.50 | 0 | $15.75 | $420.00 |
| Kruzich | Ilyona | 41 | 40 | $10.00 | 1 | $15.00 | $415.00 |
| McClane | Chris | 41 | 40 | $10.00 | 1 | $15.00 | $415.00 |
| Philips | Terrance | 42 | 40 | $10.25 | 2 | $15.38 | $440.75 |
| Sanchez | Marcus | 40 | 40 | $10.50 | 0 | $15.75 | $420.00 |
| Schwartz | Samuel | 38 | 38 | $9.75 | 0 | $14.63 | $370.50 |



# CHAPTER 11

# An Introduction to Algebra

## INTRODUCTION

Small businesses often use spreadsheet software to keep track of things. This is especially true of the many home-based businesses that have cropped up since computers became so common.

Spreadsheets, such as Microsoft Excel, became popular because of how easy they are to use. Most people can learn to use spreadsheet software by taking a single course at a local community college, or even by working through a tutorial on their own. While larger firms need more complex software, requiring extensive training, small businesses find that spreadsheets (and perhaps database software, such as Access) can help them with most of their bookkeeping needs.

While many people have come to rely on spreadsheets, those who can realize their full power have a strong background in algebra. This is because spreadsheets can be thought of as multidimensional algebra machines. Consider a typical payroll spreadsheet.

While it may seem that each "cell" is simply typed in, the truth is that only the name, total hours, and hourly rate are manually entered on each line. The other fields, such as the gross pay, are determined by formulas. You can see such a formula, which refers to other cells in the spreadsheet, in the formula line. This allows someone to copy the formula to every employee, without having to retype it each time.

## CHAPTER 11 OUTLINE

# 11 pretest



Name ____________

Section ________ Date ________

This pretest provides a preview of the types of exercises you will encounter in each section of this chapter. The answers for these exercises can be found in the back of the text. If you are working on your own, or ahead of the class, this pretest can help you identify the sections in which you should focus more of your time.

**Answers**

1. ____________
2. ____________
3. ____________
4. ____________
5. ____________
6. ____________
7. ____________
8. ____________
9. ____________
10. ____________

**11.1**

1. Write each of the following, using symbols.

   **(a)** $y$ decreased by 1

   **(b)** The product of 2 and the quantity $a + b$

**11.2**

2. Evaluate the following expressions if $x = -3$, $y = 4$, and $z = -2$.

   **(a)** $4x^2 - 3y + z$

   **(b)** $(y - x)^2 + 2z$

**11.3** Combine like terms.

3. $8m^2 - 2m + 3m^2$

4. $\frac{17}{6}a - 3 - \frac{5}{6}a + 7$

**11.4–11.6** Solve the following equations.

5. $x + 6 = 2$

6. $8x - 4 = 7x + 6$

7. $14x - 6x = 48$

8. $-\frac{2}{5}x = 8$

9. $8x - 2 = 4x + 10$

10. $8(3x + 2) - 10x = 12x - 2$

# 11.1 From Arithmetic to Algebra

< 11.1 Objectives >

1 > Use the symbols and language of algebra

2 > Identify algebraic expressions

In arithmetic, you learned how to do calculations with numbers by using the basic operations of addition, subtraction, multiplication, and division.

In algebra, we still use numbers and the same four operations. However, we also use letters to represent numbers. Letters such as $x$, $y$, $L$, and $W$ are called **variables** when they represent numerical values.

You are familiar with the four symbols ($+$, $-$, $\times$, $\div$) used to indicate the fundamental operations of arithmetic.

To see how these operations are indicated in algebra, we begin with addition.

**Definition**

**Addition** $x + y$ means the *sum* of $x$ and $y$ or $x$ *plus* $y$.

**Example 1** Writing Expressions That Indicate Addition

< Objective 1 >

**(a)** The *sum* of $a$ and 3 is written as $a + 3$.

**(b)** $L$ *plus* $W$ is written as $L + W$.

**(c)** 5 *more than* $m$ is written as $m + 5$.

**(d)** $x$ *increased by* 7 is written as $x + 7$.

**NOTE**

Some other words that tell you to add are *more than* and *increased by*.

**Check Yourself 1**

Write, using symbols.

**(a)** The sum of $y$ and 4 **(b)** $a$ plus $b$

**(c)** 3 more than $x$ **(d)** $n$ increased by 6

Next, we look at how subtraction is indicated in algebra.

**Definition**

**Subtraction** $x - y$ means the *difference* of $x$ and $y$ or $x$ *minus* $y$. Subtracting $y$ is the same as adding its opposite, so $x - y = x + (-y)$.

## Example 2 Writing Expressions That Indicate Subtraction

**NOTE**

Some other words that indicate subtraction are *decreased by* and *less than.*

**(a)** *r minus s* is written as $r - s$.

**(b)** The *difference* of $m$ and 5 is written as $m - 5$.

**(c)** *x decreased by* 8 is written as $x - 8$.

**(d)** 4 *less than a* is written as $a - 4$.

### Check Yourself 2

Write, using symbols.

**(a)** $w$ minus $z$
**(b)** The difference of $a$ and 7
**(c)** $y$ decreased by 3
**(d)** 5 less than $b$

You have seen that the operations of addition and subtraction are written exactly the same way in algebra as in arithmetic. This is not true in multiplication because the sign × looks like the letter $x$. So in algebra we use other symbols to show multiplication to avoid any confusion. Here are some ways to write multiplication.

**Definition**

**Multiplication**

| | | |
|---|---|---|
| A centered dot | $x \cdot y$ | These all indicate the *product* of $x$ and $y$ or $x$ times $y$. $x$ and $y$ are called the **factors** of the product $xy$. |
| Parentheses | $(x)(y)$ | |
| Writing the letters next to each other | $xy$ | |

## Example 3 Writing Expressions That Indicate Multiplication

**NOTE**

You can place letters next to each other or numbers and letters next to each other to show multiplication. But you *cannot* place numbers side by side to show multiplication: 37 means the number "thirty-seven," not 3 times 7.

**(a)** The product of 5 and $a$ is written as $5 \cdot a$, $(5)(a)$, or $5a$. The last expression, $5a$, is the shortest and the most common way of writing the product.

**(b)** 3 times 7 can be written as $3 \cdot 7$ or $(3)(7)$.

**(c)** Twice $z$ is written as $2z$.

**(d)** The product of 2, $s$, and $t$ is written as $2st$.

**(e)** 4 more than the product of 6 and $x$ is written as $6x + 4$.

### Check Yourself 3

Write, using symbols.

**(a)** $m$ times $n$
**(b)** The product of $h$ and $b$
**(c)** The product of 8 and 9
**(d)** The product of 5, $w$, and $y$
**(e)** 3 more than the product of 8 and $a$

Before moving on to division, look at how we combine the symbols learned so far.

**Definition**

**Expression** An **expression** is a meaningful collection of numbers, variables, and signs of operation.

## Example 4 Identifying Expressions

< Objective 2 >

**NOTE**

Not every collection of symbols is an expression.

**(a)** $2m + 3$ is an expression. It means that we multiply 2 and $m$, and then add 3.

**(b)** $x + \cdot + 3$ is not an expression. The three operations in a row have no meaning.

**(c)** $y = 2x - 1$ is not an expression. The equal sign is not an operation sign.

**(d)** $3a + 5b - 4c$ is an expression. Its meaning is clear.

### Check Yourself 4

Identify which are expressions and which are not.

**(a)** $7 - \cdot x$ **(b)** $6 + y = 9$
**(c)** $a + b - c$ **(d)** $3x - 5yz$

To write more complicated products in algebra, we need some "punctuation marks." Parentheses ( ) mean that an expression is to be thought of as a single quantity. Brackets [ ] are used in exactly the same way as parentheses in algebra. Example 5 shows the use of these signs of grouping.

## Example 5 Expressions with More Than One Operation

**NOTES**

This can be read as "3 times the quantity $a$ plus $b$."

No parentheses are needed in (b) because the 3 multiplies *only* the $a$.

**(a)** 3 times the sum of $a$ and $b$ is written as

$$3(a + b)$$

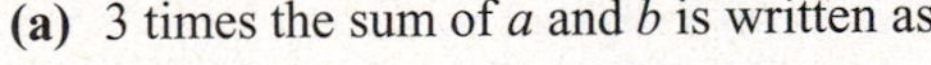

The sum of $a$ and $b$ is a single quantity, so it is enclosed in parentheses.

**(b)** The sum of 3 times $a$ and $b$ is written as $3a + b$.

**(c)** 2 times the difference of $m$ and $n$ is written as $2(m - n)$.

**(d)** The product of $s$ plus $t$ and $s$ minus $t$ is written as $(s + t)(s - t)$.

**(e)** The product of $b$ and 3 less than $b$ is written as $b(b - 3)$.

### Check Yourself 5

Write, using symbols.

**(a)** Twice the sum of $p$ and $q$
**(b)** The sum of twice $p$ and $q$
**(c)** The product of $a$ and the quantity $b - c$
**(d)** The product of $x$ plus 2 and $x$ minus 2
**(e)** The product of $x$ and 4 more than $x$

**NOTE**

In algebra the fraction form is usually used.

Now look at the operation of division. In arithmetic, you use the division sign ÷, the long division symbol $\overline{)\quad}$, and fraction notation. For example, to indicate the quotient when 9 is divided by 3, you could write

$$9 \div 3 \quad \text{or} \quad 3\overline{)9} \quad \text{or} \quad \frac{9}{3}$$

**Definition**

**Division**

$\frac{x}{y}$ means $x$ *divided by* $y$ or the *quotient* of $x$ and $y$.

### Example 6 Writing Expressions That Indicate Division

Write, using symbols.

**(a)** $m$ divided by 3 is written as $\frac{m}{3}$.

**(b)** The quotient of $a$ plus $b$ and 5 is written as $\frac{a + b}{5}$.

**(c)** The sum of $p$ and $q$ divided by the difference of $p$ and $q$ is written as $\frac{p + q}{p - q}$.

**Check Yourself 6**

Write, using symbols.

**(a)** $r$ divided by $s$
**(b)** The quotient when $x$ minus $y$ is divided by 7
**(c)** The difference of $a$ and 2 divided by the sum of $a$ and 2

Of everything we have studied so far, algebra lends itself to the greatest variety of applications. In Example 7, we model one such application.

### Example 7 Modeling Applications with Algebra

**NOTE**

When choosing a letter to use as a variable, it is often a good idea to choose one that reminds us of what it represents.

Carla earns $10.25 per hour in her job. Write an equation that sets her weekly gross pay equal to an expression involving the number of hours she works per week.

We need to name the variables we are using.

Let $P$ be Carla's weekly gross pay and $h$ be the number of hours she works in a week.

Then the equation $P = 10.25h$ gives Carla's weekly gross pay based on the number of hours she works.

**NOTE**

The word *twice* indicates multiplication by 2.

## Check Yourself 7

Specs for an engine cylinder call for the stroke length to be two more than twice the diameter of the cylinder. Write an equation relating stroke length to cylinder diameter.

We close this section by listing many of the common words used to indicate arithmetic operations.

**Table: Words Indicating Operations**

The operations listed are usually indicated by the words shown.

| | |
|---|---|
| **Addition (+)** | Plus, and, more than, increased by, sum |
| **Subtraction (−)** | Minus, from, less than, decreased by, difference |
| **Multiplication (·)** | Times, of, by, product |
| **Division (÷)** | Divided, into, per, quotient |

## Check Yourself ANSWERS

**1.** **(a)** $y + 4$; **(b)** $a + b$; **(c)** $x + 3$; **(d)** $n + 6$
**2.** **(a)** $w - z$; **(b)** $a - 7$; **(c)** $y - 3$; **(d)** $b - 5$
**3.** **(a)** $mn$; **(b)** $hb$; **(c)** $8 \cdot 9$ or $(8)(9)$; **(d)** $5wy$; **(e)** $8a + 3$
**4.** **(a)** Not an expression; **(b)** not an expression; **(c)** an expression; **(d)** an expression
**5.** **(a)** $2(p + q)$; **(b)** $2p + q$; **(c)** $a(b - c)$; **(d)** $(x + 2)(x - 2)$; **(e)** $x(x + 4)$
**6.** **(a)** $\frac{r}{s}$; **(b)** $\frac{x - y}{7}$; **(c)** $\frac{a - 2}{a + 2}$ **7.** $S = 2d + 2$

## Reading Your Text

The following fill-in-the-blank exercises are designed to ensure that you understand some of the key vocabulary used in this section.

SECTION 11.1

**(a)** In algebra, we often use letters, called ______________, to represent numerical values that can vary depending upon the application.

**(b)** $x - y$ means the ______________ of $x$ and $y$.

**(c)** $x \cdot y$, $(x)(y)$, and $xy$ are all ways of indicating ______________ in algebra.

**(d)** When choosing a ______________ to use as a variable, it is often a good idea to choose one that reminds us of what it represents.

# 11.1 exercises

Basic Skills | Advanced Skills | Vocational-Technical Applications | Calculator/Computer | Above and Beyond

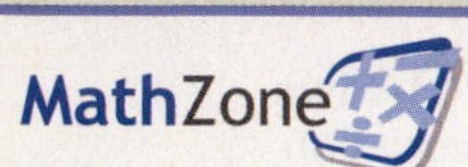

Boost your grade at mathzone.com!
> Practice Problems
> NetTutor
> Self-Tests
> e-Professors
> Videos

Name ______________

Section ________ Date ________

**Answers**

1. ______ 2. ______
3. ______ 4. ______
5. ______ 6. ______
7. ______ 8. ______
9. ______ 10. ______
11. ______ 12. ______
13. ______ 14. ______
15. ______ 16. ______
17. ______ 18. ______
19. ______ 20. ______
21. ______ 22. ______
23. ______ 24. ______
25. ______ 26. ______
27. ______ 28. ______

< Objective 1 >

*Write each of the following phrases, using symbols.*

1. The sum of $c$ and $d$

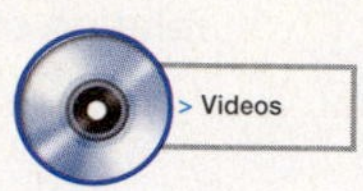


2. $a$ plus 7
3. $w$ plus $z$
4. The sum of $m$ and $n$
5. $x$ increased by 2
6. 3 more than $b$
7. 10 more than $y$
8. $m$ increased by 4
9. $a$ minus $b$
10. 5 less than $s$
11. $b$ decreased by 7
12. $r$ minus 3
13. 6 less than $r$

14. $x$ decreased by 3
15. $w$ times $z$
16. The product of 3 and $c$
17. The product of 5 and $t$

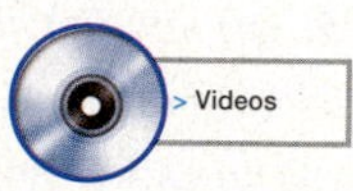


18. 8 times $a$
19. The product of 8, $m$, and $n$
20. The product of 7, $r$, and $s$
21. The product of 3 and the quantity $p$ plus $q$
22. The product of 5 and the sum of $a$ and $b$
23. Twice the sum of $x$ and $y$

24. 3 times the sum of $m$ and $n$
25. The sum of twice $x$ and $y$
26. The sum of 3 times $m$ and $n$
27. Twice the difference of $x$ and $y$
28. 3 times the difference of $c$ and $d$

29. The quantity $a$ plus $b$ times the quantity $a$ minus $b$

30. The product of $x$ plus $y$ and $x$ minus $y$

31. The product of $m$ and 3 less than $m$

32. The product of $a$ and 7 more than $a$

33. $x$ divided by 5 

34. The quotient when $b$ is divided by 8

35. The quotient of $a$ plus $b$, and 7

36. The difference $x$ minus $y$, divided by 9

37. The difference of $p$ and $q$, divided by 4

38. The sum of $a$ and 5, divided by 9

39. The sum of $a$ and 3, divided by the difference of $a$ and 3

40. The difference of $m$ and $n$, divided by the sum of $m$ and $n$

< Objective 2 >

*Identify which are expressions and which are not.*

41. $2(x + 5)$

42. $4 + (x - 3)$

43. $4 + \div m$

44. $6 + a = 7$

45. $2b = 6$

46. $x(y + 3)$

47. $2a + 5b$

48. $4x + \cdot 7$

49. **Social Science** The Earth's population has doubled in the last 40 years. If we let $x$ represent the Earth's population 40 years ago, what is the population today, in terms of $x$?

50. **Science and Medicine** It is estimated that the Earth is losing 4,000 species of plants and animals each year to extinction. Let $S$ represent the number of species living last year and represent the number of species alive this year, in terms of $S$.

## Answers

29. ______
30. ______
31. ______
32. ______
33. ______
34. ______
35. ______ 36. ______
37. ______ 38. ______
39. ______ 40. ______
41. ______
42. ______
43. ______
44. ______
45. ______
46. ______
47. ______
48. ______
49. ______
50. ______

## Answers

51. ______
52. ______
53. ______
54. ______
55. ______
56. ______
57. ______
58. ______
59. ______
60. ______
61. ______
62. ______
63. ______
64. ______
65. ______
66. ______
67. ______
68. ______
69. ______
70. ______

**51. Business and Finance** The simple interest earned when a principal $P$ is invested at a rate $r$ for a time period $t$ is given by the product of the principal, the rate, and the time. Write an expression for the simple interest earned.

**52. Science and Medicine** The kinetic energy of a particle of mass $m$ is found by taking one-half of the product of the mass and the square of the velocity $v$. Write an expression for the kinetic energy of a particle.

*Match each phrase with the proper expression.*

**53.** 8 decreased by $x$ **(a)** $x - 8$

**54.** 8 less than $x$ **(b)** $8 - x$

**55.** The difference between 8 and $x$

**56.** 8 from $x$

Basic Skills | **Advanced Skills** | Vocational-Technical Applications | Calculator/Computer | Above and Beyond

*Write each of the following phrases, using symbols. Use the variable $x$ to represent the number in each case.*

**57.** 5 more than a number

**58.** A number increased by 8

**59.** 7 less than a number

**60.** A number decreased by 10

**61.** 9 times a number

**62.** Twice a number

**63.** 6 more than 3 times a number

**64.** 5 times a number, decreased by 10

**65.** Twice the sum of a number and 5

**66.** 3 times the difference of a number and 4

**67.** The product of 2 more than a number and 2 less than that same number

**68.** The product of 5 less than a number and 5 more than that same number

**69.** The quotient of a number and 7

**70.** A number divided by 3

71. The sum of a number and 5, divided by 8

72. The quotient when 7 less than a number is divided by 3

73. 6 more than a number divided by 6 less than that same number

74. The quotient when 3 less than a number is divided by 3 more than that same number

Basic Skills | Advanced Skills | **Vocational-Technical Applications** | Calculator/Computer | Above and Beyond

75. **Allied Health** The standard dosage given to a patient is equal to the product of the desired dose $D$ and the available quantity $Q$ divided by the available dose $H$. Write the standard dosage calculation formula.

76. **Information Technology** Mindy is the manager of the help desk at a large cable company. She notices that, on average, her staff can handle 50 calls per hour. Last week, during a thunderstorm, the call volume increased from 65 to 150 calls per hour. To figure out the average number of customers in the system, she needs to take the quotient of the average rate of customer arrivals (the call volume) $a$ and the average rate at which customers are served $h$ minus the average rate of customer arrivals $a$. Write a formula for the average number of customers in the system.

77. **Construction Technology** K Jones Manufacturing produces hex bolts and carriage bolts. It sold 284 more hex bolts than carriage bolts last month. Write a formula that describes the number of carriage bolts it sold last month. Let $H$ be the number of hex bolts sold last month.

78. **Electrical Engineering** Electrical power $P$ is the product of voltage $V$ and current $I$. Express this relationship algebraically.

## Answers

71. ______
72. ______
73. ______
74. ______
75. ______
76. ______
77. ______
78. ______

## Answers

1. $c + d$ 3. $w + z$ 5. $x + 2$ 7. $y + 10$ 9. $a - b$ 11. $b - 7$ 13. $r - 6$ 15. $wz$ 17. $5t$ 19. $8mn$ 21. $3(p + q)$ 23. $2(x + y)$ 25. $2x + y$ 27. $2(x - y)$ 29. $(a + b)(a - b)$ 31. $m(m - 3)$ 33. $\frac{x}{5}$ 35. $\frac{a + b}{7}$ 37. $\frac{p - q}{4}$ 39. $\frac{a + 3}{a - 3}$ 41. Expression 43. Not an expression 45. Not an expression 47. Expression 49. $2x$ 51. $Prt$ 53. (b) 55. (b) 57. $x + 5$ 59. $x - 7$ 61. $9x$ 63. $3x + 6$ 65. $2(x + 5)$ 67. $(x + 2)(x - 2)$ 69. $\frac{x}{7}$ 71. $\frac{x + 5}{8}$ 73. $\frac{x + 6}{x - 6}$ 75. $\frac{DQ}{H}$ 77. $H - 284$

# 11.2 Evaluating Algebraic Expressions

< 11.2 Objectives >

1 > Substitute real numbers for the variables in an expression

2 > Use the order of operations to evaluate an expression

When using algebra to solve problems, we often want to find the value of an algebraic expression, given particular values for the variables. Finding the value of an expression is called *evaluating the expression* and uses the following steps.

**Step by Step**

**To Evaluate an Algebraic Expression**

**Step 1** Replace each variable with the given number value.

**Step 2** Do the necessary arithmetic operations, following the rules for the order of operations.

**Example 1** Evaluating Algebraic Expressions

< Objective 1 >

**NOTE**

We use parentheses when we make the initial substitution. This helps us to avoid careless errors.

Suppose that $a = 5$ and $b = 7$.

**(a)** To evaluate $a + b$, we replace $a$ with 5 and $b$ with 7.

$a + b = (5) + (7) = 12$

**(b)** To evaluate $3ab$, we again replace $a$ with 5 and $b$ with 7.

$3ab = 3(5)(7) = 105$

**Check Yourself 1**

If $x = 6$ and $y = 7$, evaluate.

**(a)** $y - x$ **(b)** $5xy$

We are now ready to evaluate algebraic expressions that require following the rules for the order of operations.

**Example 2** Evaluating Algebraic Expressions

< Objective 2 >

Evaluate the following expressions if $a = 2$, $b = 3$, $c = 4$, and $d = 5$.

**(a)** $5a + 7b = 5(2) + 7(3)$ Multiply first.

$= 10 + 21 = 31$ Then add.

> **CAUTION**
>
> The expression in part (b) is different from
>
> $(3c)^2 = [3(4)]^2$
>
> $= 12^2 = 144$

**(b)** $3c^2 = 3(4)^2$ Apply the exponent.

$= 3 \cdot 16 = 48$ Then multiply.

**(c)** $7(c + d) = 7[(4) + (5)]$ Add inside the parentheses.

$= 7 \cdot 9 = 63$

**(d)** $5a^4 - 2d^2 = 5(2)^4 - 2(5)^2$ Apply the exponents.

$= 5 \cdot 16 - 2 \cdot 25$ Multiply.

$= 80 - 50 = 30$ Subtract.

### Check Yourself 2

If $x = 3$, $y = 2$, $z = 4$, and $w = 5$, evaluate the following expressions.

(a) $4x^2 + 2$ (b) $5(z + w)$ (c) $7(z^2 - y^2)$

To evaluate algebraic expressions when a fraction bar is used, start by doing all the work in the numerator, and then do the work in the denominator. Divide the numerator by the denominator as the last step.

## Example 3 Evaluating Algebraic Expressions

If $p = 2$, $q = 3$, and $r = 4$, evaluate:

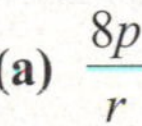

**(a)** $\frac{8p}{r}$

> **RECALL**
>
> The fraction bar, like parentheses, is a grouping symbol. Work first in the numerator and then in the denominator.

Replace $p$ with 2 and $r$ with 4.

$$\frac{8p}{r} = \frac{8(2)}{(4)} = \frac{16}{4} = 4$$ Divide as the last step.

**(b)** $\frac{7q + r}{p + q} = \frac{7(3) + (4)}{(2) + (3)}$ Now evaluate the top and bottom separately.

$$= \frac{21 + 4}{2 + 3} = \frac{25}{5} = 5$$

### Check Yourself 3

Evaluate the following if $c = 5$, $d = 8$, and $e = 3$.

(a) $\frac{6c}{e}$ (b) $\frac{4d + e}{c}$ (c) $\frac{10d - e}{d + e}$

When the algebraic expression contains parentheses or other grouping symbols, remember to follow the proper order of operations.

## Example 4 Evaluating Expressions

**NOTE**

We can interchange parentheses and brackets as convenient.

Evaluate each of the following expressions if $x = 3$, $y = -4$, and $z = 1$.

**(a)** $3x(y + 2z)$

Replace $x$ with 3, $y$ with $-4$, and $z$ with 1 and evaluate properly.

$$\begin{aligned} 3x(y + 2z) &= 3(3)[(-4) + 2(1)] \\ &= 3(3)(-4 + 2) \\ &= 3(3)(-2) \\ &= -18 \end{aligned}$$

**(b)** $(1 - 2x)(z - y)$

$$\begin{aligned} (1 - 2x)(z - y) &= [1 - 2(3)][(1) - (-4)] \\ &= (-5)(5) \\ &= -25 \end{aligned}$$

### Check Yourself 4

Evaluate each of the expressions if $a = 3$, $b = 0$, and $c = -6$.

**(a)** $3c(c - a)$ **(b)** $\dfrac{ab}{c - a}$

## Example 5 Evaluating Expressions

**RECALL**

The rules for the order of operations require that we multiply first and then add.

Evaluate $5a + 4b$ if $a = -2$ and $b = 3$.

Replace $a$ with $-2$ and $b$ with 3.

$$\begin{aligned} 5a + 4b &= 5(-2) + 4(3) \\ &= -10 + 12 \\ &= 2 \end{aligned}$$

### Check Yourself 5

Evaluate $3x + 5y$ if $x = -2$ and $y = -5$.

We follow the same rules no matter how many variables are in the expression.

## Example 6 Evaluating Expressions

Evaluate the following expressions if $a = -4$, $b = 2$, $c = -5$, and $d = 6$.

This becomes $-(-20)$, or $+20$.

**(a)** 
$$\begin{aligned} 7a - 4c &= 7(-4) - 4(-5) \\ &= -28 + 20 \\ &= -8 \end{aligned}$$

> **CAUTION**

When a squared variable is replaced by a negative number, square the negative.

$(-5)^2 = (-5)(-5) = 25$

The exponent applies to −5!

$-5^2 = -(5 \cdot 5) = -25$

The exponent applies only to 5!

Apply the exponent first, and then multiply by 7.

**(b)** $7c^2 = 7(-5)^2 = 7 \cdot 25$
$= 175$

**(c)** $b^2 - 4ac = (2)^2 - 4(-4)(-5)$
$= 4 - 4(-4)(-5)$
$= 4 - 80$
$= -76$

Add inside the parentheses first.

**(d)** $b(a + d) = (2)[(-4) + (6)]$
$= 2(2)$
$= 4$

## Check Yourself 6

Evaluate if $p = -4$, $q = 3$, and $r = -2$.

**(a)** $5p - 3r$ **(b)** $2p^2 + q$ **(c)** $p(q + r)$
**(d)** $-q^2$ **(e)** $(-q)^2$

If an expression involves a fraction, remember that the fraction bar is a grouping symbol. This means that you should do the required operations first in the numerator and then in the denominator separately. Divide as the last step.

## Example 7 Evaluating Expressions

Evaluate the following expressions if $x = 4$, $y = -5$, $z = 2$, and $w = -3$.

**(a)** $$\frac{z - 2y}{x} = \frac{(2) - 2(-5)}{(4)} = \frac{2 + 10}{4}$$
$$= \frac{12}{4} = 3$$

**(b)** $$\frac{3x - w}{2x + w} = \frac{3(4) - (-3)}{2(4) + (-3)} = \frac{12 + 3}{8 + (-3)}$$
$$= \frac{15}{5} = 3$$

## Check Yourself 7

Evaluate if $m = -6$, $n = 4$, and $p = -3$.

**(a)** $\dfrac{m + 3n}{p}$ **(b)** $\dfrac{4m + n}{m + 4n}$

When a calculator is used to evaluate an expression, the same order of operations that we introduced in Section 1.7 is followed.

| | Algebraic Notation | Calculator Notation |
|---|---|---|
| Addition | $6 + 2$ | 6 [+] 2 |
| Subtraction | $4 - 8$ | 4 [−] 8 |
| Multiplication | $(3)(-5)$ | 3 [×] ([(−)] 5) or 3 [×] 5 [+/−] |
| Division | $\frac{8}{6}$ | 8 [÷] 6 |
| Exponential | $3^4$ | 3 [^] 4 or 3 [$y^x$] 4 |
| | $(-3)^4$ | [(] [(−)] 3 [)] [^] 4 or [(] 3 [+/−] [)] [$y^x$] 4 |

The use of this notation is illustrated in Example 8.

## Example 8 Evaluating Expressions

Evaluate each of the following expressions if $A = 2.3$, $B = 8.4$, and $C = 4.5$. Round your answer to the nearest tenth.

**(a)** $A + B(-C)$

Letting $A$, $B$, and $C$ take on the given values, we have

2.3 [+] 8.4 [×] ([(−)]4.5) = −35.5

**(b)** $-B + (-A)C^2$

Substituting the given values, we have

[(−)]8.4 [+] ([(−)]2.3) [×] (4.5)[^]2 = −54.975

Rounding to the nearest tenth gives us −55.0.

### Check Yourself 8

Evaluate each of the following expressions when $A = -2$, $B = 3$, and $C = 5$.

**(a)** $A + B(-C)$ **(b)** $C + BA^3$ **(c)** $4(B - C)/(2A)$

Many applications require us to evaluate algebraic expressions.

## Example 9 Applying Algebra

The lens formula (in the study of *optics*) states that the *focal length* (the distance between a *lens* and the *focal point*) of a lens is given by the formula

$$\frac{d_o d_i}{d_o + d_i}$$

in which $d_o$ is the distance of an object from a thin lens and $d_i$ is the distance of the object's image from the lens. Find the focal length of a lens if an object 24 inches (in.) from a lens produces an image 1 in. from the lens.

We substitute as before.

$$\frac{d_o d_i}{d_o + d_i} = \frac{(24)(1)}{(24) + (1)}$$

$$= \frac{24}{25}$$

So the focal length is $\frac{24}{25}$ in. (or 0.96 in.).

## Check Yourself 9

In an electric circuit with electromotive force of $E$ volts and resistance $R$ ohms, the rate of change in the current with respect to resistance is given by

$$-\frac{E}{R^2} \quad \text{amperes per ohm}$$

Find the rate of change in the current with respect to resistance if $E = 100$ volts and $R = 12$ ohms.

## Check Yourself ANSWERS

**1.** **(a)** 1; **(b)** 210 **2.** **(a)** 38; **(b)** 45; **(c)** 84 **3.** **(a)** 10; **(b)** 7; **(c)** 7
**4.** **(a)** 162; **(b)** 0 **5.** −31 **6.** **(a)** −14; **(b)** 35; **(c)** −4; **(d)** −9; **(e)** 9
**7.** **(a)** −2; **(b)** −2 **8.** **(a)** −17; **(b)** −19; **(c)** 2
**9.** $-\frac{25}{36}$ amperes per ohm (approximately −0.69)

## Reading Your Text

The following fill-in-the-blank exercises are designed to ensure that you understand some of the key vocabulary used in this section.

### SECTION 11.2

**(a)** To evaluate an algebraic expression, first replace each ______________ by the given numerical value.

**(b)** When evaluating an algebraic expression that includes a __________ bar, you should do all the work in the numerator and denominator separately.

**(c)** Scientific and graphing calculators follow the familiar order of ______________.

**(d)** When a squared variable is replaced by a negative number and evaluated, the result is a ______________ number.

# 11.2 exercises

Basic Skills | Advanced Skills | Vocational-Technical Applications | Calculator/Computer | Above and Beyond

**MathZone**

Boost your grade at mathzone.com!

- Practice Problems
- NetTutor
- Self-Tests
- e-Professors
- Videos

Name ____________

Section ________ Date ________

**Answers**

1. ____________
2. ____________
3. ____________
4. ____________
5. ________ 6. ________
7. ________ 8. ________
9. ________ 10. ________
11. ________ 12. ________
13. ________ 14. ________
15. ________ 16. ________
17. ________ 18. ________
19. ________ 20. ________
21. ________ 22. ________
23. ________ 24. ________
25. ________ 26. ________

< Objectives 1–2 >

*Evaluate each of the expressions if $a = -2$, $b = 5$, $c = -4$, and $d = 6$.*

**1.** $3c - 2b$

**2.** $4c - 2b$

**3.** $8b + 2c$

**4.** $7a - 2c$

**5.** $-b^2 + b$

**6.** $(-b)^2 + b$

**7.** $3a^2$

**8.** $6c^2$

**9.** $c^2 - 2d$

**10.** $3a^2 + 4c$

**11.** $2a^2 + 3b^2$

**12.** $4b^2 - 2c^2$

**13.** $2(a + b)$

**14.** $5(b - c)$

**15.** $4(2a - d)$

**16.** $6(3c - d)$

**17.** $a(b + 3c)$

**18.** $c(3a - d)$

**19.** $\frac{6d}{c}$

**20.** $\frac{8b}{5c}$

**21.** $\frac{3d + 2c}{b}$

**22.** $\frac{2b + 3d}{2a}$

**23.** $\frac{2b - 3a}{c + 2d}$

**24.** $\frac{3d - 2b}{5a + d}$

**25.** $d^2 - b^2$

**26.** $c^2 - a^2$

27. $(d - b)^2$

28. $(c - a)^2$

29. $(d - b)(d + b)$

30. $(c - a)(c + a)$

31. $d^3 - b^3$

32. $c^3 + a^3$

33. $(d - b)^3$

34. $(c + a)^3$

35. $(d - b)(d^2 + db + b^2)$

36. $(c + a)(c^2 - ac + a^2)$

37. $b^2 + a^2$

38. $d^2 - a^2$

39. $(b + a)^2$

40. $(d - a)^2$

41. $a^2 + 2ad + d^2$

42. $b^2 - 2bc + c^2$

43. **Geometry** The formula for the area of a triangle is given by $A = \frac{1}{2}bh$. Find the area of a triangle if $b = 4$ cm and $h = 8$ cm.

44. **Geometry** The perimeter of a rectangle with length $L$ and width $W$ is given by the formula $P = 2L + 2W$. Find the perimeter of a rectangle if its length is 10 in. and its width is 5 in.

45. **Business and Finance** The simple interest $I$ on a principal $P$ dollars at interest rate $r$ for time $t$ is given by $I = Prt$. Find the simple interest earned on a principal of \$6,000 at 3% for 2 years. *Hint:* 3% = 0.03.

46. **Business and Finance** Use the simple interest formula in exercise 45 to find the interest earned on a principal of \$12,500 at 4.5% for 3 years.

47. **Science and Medicine** A formula that relates Celsius and Fahrenheit temperatures is $F = \frac{9}{5}C + 32$. If the low temperature is $-10°C$ one day, what was the Fahrenheit equivalent?

48. **Geometry** The area of a circle with radius $r$ is $A = \pi r^2$. Use $\pi \approx 3.14$ to approximate the area of a circle if the radius is 3 ft.

**Answers**

27. ______
28. ______
29. ______
30. ______
31. ______
32. ______
33. ______
34. ______
35. ______
36. ______
37. ______
38. ______
39. ______
40. ______
41. ______
42. ______
43. ______
44. ______
45. ______
46. ______
47. ______
48. ______

Basic Skills | **Advanced Skills** | Vocational-Technical Applications | Calculator/Computer | Above and Beyond

## Answers

49. ______

50. ______

51. ______

52. ______

53. ______

54. ______

55. ______

56. ______

57. ______

58. ______

59. ______

60. ______

*Evaluate each expression if* $x = -3, y = 5,$ *and* $z = \frac{2}{3}$.

**49.** $x^2 - y$

**50.** $\frac{y - x}{z}$

**51.** $z - y^2$

**52.** $z - \frac{z + x}{y - x}$

*In each of the following problems, decide if the given values for the variables make the statement* **true** *or* **false.**

**53.** $x - 7 = 2y + 5; x = 22, y = 5$

**54.** $3(x - y) = 6; x = 5, y = -3$

**55.** $2(x + y) = 2x + y; x = -4, y = -2$

**56.** $x^2 - y^2 = x - y; x = 4, y = -3$

Basic Skills | Advanced Skills | **Vocational-Technical Applications** | Calculator/Computer | Above and Beyond

**57. Allied Health** The concentration, in micrograms per milliliter ($\mu$g/mL), of an antihistamine in a patient's bloodstream can be approximated using the formula $-2t^2 + 13t + 1$, in which $t$ is the number of hours since the drug was administered. Approximate the concentration of the antihistamine 1 hour after it has been administered.

**58. Allied Health** Use the formula given in exercise 57 to approximate the concentration of the antihistamine 3 hours after it has been administered.

**59. Electrical Engineering** Evaluate $\frac{rT}{5{,}252}$ for $r = 1{,}180$ and $T = 3$ (round to the nearest thousandth).

**60. Mechanical Engineering** The kinetic energy (in joules) of a particle is given by $\frac{1}{2}mv^2$. Find the kinetic energy of a particle if its mass is 60 kg and its velocity is 6 m/s.

Basic Skills | Advanced Skills | Vocational-Technical Applications | **Calculator/Computer** | Above and Beyond

*Use your calculator to evaluate each expression if* $x = -2.34$, $y = -3.14$, *and* $z = 4.12$. *Round your results to the nearest tenth.*

**61.** $x + yz$

**62.** $y - 2z$

**63.** $x^2 - z^2$

**64.** $x^2 + y^2$

**65.** $\dfrac{xy}{z - x}$

**66.** $\dfrac{y^2}{zy}$

**67.** $\dfrac{2x + y}{2x + z}$

**68.** $\dfrac{x^2y^2}{xz}$

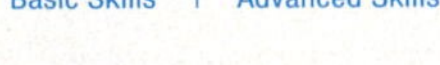

Basic Skills | Advanced Skills | Vocational-Technical Applications | Calculator/Computer | **Above and Beyond**

**69.** Write an English interpretation for each of the following algebraic expressions.

**(a)** $(2x^2 - y)^3$ **(b)** $3n - \dfrac{n - 1}{2}$ **(c)** $(2n + 3)(n - 4)$

**70.** Is $a^n + b^n = (a + b)^n$? Try a few numbers and decide if you think this is true for all numbers, true for some numbers, or never true. Write an explanation of your findings and give examples.

## Answers

**1.** −22 **3.** 32 **5.** −20 **7.** 12 **9.** 4 **11.** 83 **13.** 6
**15.** −40 **17.** 14 **19.** −9 **21.** 2 **23.** 2 **25.** 11 **27.** 1
**29.** 11 **31.** 91 **33.** 1 **35.** 91 **37.** 29 **39.** 9 **41.** 16
**43.** 16 cm$^2$ **45.** \$360 **47.** 14°F **49.** 4 **51.** $-\dfrac{73}{3}$ **53.** True
**55.** False **57.** 12 μg/mL **59.** 0.674 **61.** −15.3 **63.** −11.5
**65.** 1.1 **67.** 14 **69.** Above and Beyond

## Answers

61. ______

62. ______

63. ______

64. ______

65. ______

66. ______

67. ______

68. ______

69. ______

70. ______

# Activity 30 :: Evaluating Net Pay

Many people are paid based on the number of hours they work. However, while a person may earn a fixed number of dollars per hour, the person's actual paycheck differs from this straightforward multiplication.

The **gross pay** of an hourly employee is determined by multiplying the number of hours worked by the amount paid per hour. More generally, gross pay is the amount earned before any money is deducted. A person's **net pay** is the amount the person actually receives, after all deductions.

1. Ilyona earns \$12.50 per hour working at her local library. Find her gross pay if she works a 35-hour week.
2. The federal government deducts 6% of her gross pay for taxes and an additional 7% for FICA. The state also deducts 5% of her gross for state taxes. How much do the federal and state governments deduct from her pay?
3. Ilyona contributes \$25 each week to her benefits package, and \$8 each week is paid as city employees' union dues. Find Ilyona's net weekly pay.
4. Find her yearly gross and net earnings. Assume she is paid for 52 weeks.

Obviously, it is not efficient for a large company to compute steps 1 to 4 manually, one at a time, for each employee. By creating and using formulas, the process can be made more efficient.

5. Create an expression using $r$ for hourly pay and $t$ for the number of hours worked that describes a person's gross pay.
6. Create an expression that describes a person's net pay. Assume the deductions stated in part 2 apply.
7. Ilyona's supervisor is paid \$15.75 per hour and works 40 hours per week. Use the expression found in exercise 6 to find the supervisor's net pay.
8. Go to www.mhhe.com/streeter to view spreadsheets describing payroll for a larger company.



 

# 11.3 Adding and Subtracting Algebraic Expressions

< 11.3 Objectives >

1 > Identify terms and coefficients

2 > Identify like terms

3 > Combine like terms

To find the perimeter of a rectangle, we add 2 times the length and 2 times the width. In the language of algebra, this can be written as

**RECALL**

The perimeter of a figure is the distance around that figure.

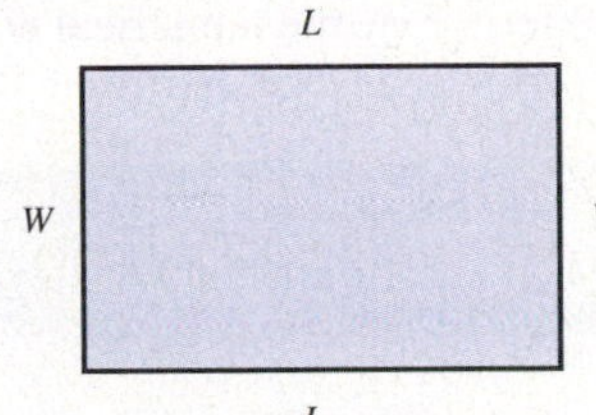


Perimeter $= 2L + 2W$

We call $2L + 2W$ an **algebraic expression,** or more simply an **expression.** Recall from Section 11.1 that an expression is a mathematical idea written in symbols. It can be thought of as a meaningful collection of letters, numbers, and operation signs.

Some expressions are

1. $5x^2$
2. $3a + 2b$
3. $4x^3 - 2y + 1$
4. $3(x^2 + y^2)$

In algebraic expressions, the addition and subtraction signs break the expressions into smaller parts called *terms*.

**Definition**

| **Term** | A **term** is a number, or the product of a number and one or more variables, raised to a power. |
|---|---|

In an expression, each sign (+ or −) is a part of the term that follows the sign.

## Example 1 Identifying Terms

< Objective 1 >

**RECALL**

Each term "owns" the sign that precedes it.

**(a)** $5x^2$ has one term.

**(b)** $3a + 2b$ has two terms: $3a$ and $2b$.
Term Term

**(c)** $4x^3 - 2y + 1$ has three terms: $4x^3$, $-2y$, and 1.
Term Term Term

### Check Yourself 1

List the terms of each expression.

(a) $2b^4$ (b) $5m + 3n$ (c) $2s^2 - 3t - 6$

A term in an expression may have any number of factors. For instance, $5xy$ is a term. It has factors of 5, $x$, and $y$. The number factor of a term is called the **numerical coefficient.** So for the term $5xy$, the numerical coefficient is 5.

## Example 2 Identifying the Numerical Coefficient

**(a)** $4a$ has the numerical coefficient 4.

**(b)** $6a^3b^4c^2$ has the numerical coefficient 6.

**(c)** $-7m^2n^3$ has the numerical coefficient $-7$.

**(d)** Because $1 \cdot x = x$, the numerical coefficient of $x$ is understood to be 1.

### Check Yourself 2

Give the numerical coefficient for each of the following terms.

(a) $8a^2b$ (b) $-5m^3n^4$ (c) $y$

If terms contain exactly the *same letters* (or variables) raised to the *same powers,* they are called **like terms.**

## Example 3 Identifying Like Terms

< Objective 2 >

**(a)** The following are like terms.

$6a$ and $7a$

$5b^2$ and $b^2$

$10x^2y^3z$ and $-6x^2y^3z$

$-3m^2$ and $m^2$

Each pair of terms has the same letters, with each letter raised to the same power—the numerical coefficients do not need to be the same.

**(b)** The following are *not* like terms.

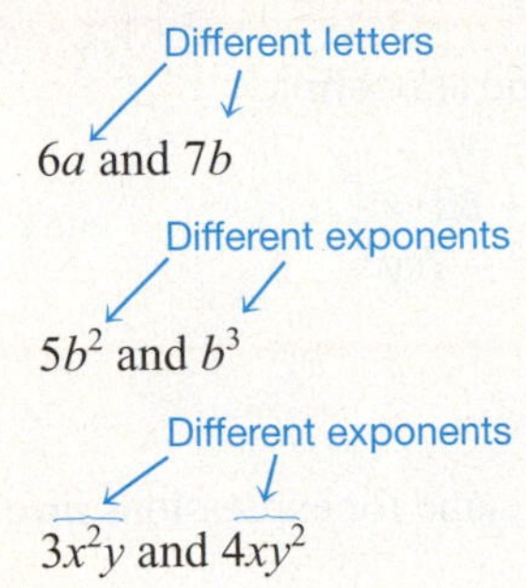


$6a$ and $7b$

$5b^2$ and $b^3$

$3x^2y$ and $4xy^2$

### Check Yourself 3

Circle the like terms.

$5a^2b$ $ab^2$ $a^2b$ $-3a^2$ $4ab$ $3b^2$ $-7a^2b$

Like terms of an expression can always be combined into a single term. Look at the following:

$$\underbrace{x + x}_{2x} + \underbrace{x + x + x + x + x}_{5x} = \underbrace{x + x + x + x + x + x + x}_{7x}$$

**RECALL**

Here we use the distributive property from Section 1.5.

Rather than having to write out all those $x$'s, try

$2x + 5x = (2 + 5)x = 7x$

In the same way,

$9b + 6b = (9 + 6)b = 15b$

and

$10a - 4a = (10 - 4)a = 6a$

This leads us to the following rule.

**Step by Step**

**Combining Like Terms**

To combine like terms, use the following steps.

**Step 1** Add or subtract the numerical coefficients.

**Step 2** Attach the common variables.

### Example 4 Combining Like Terms

< Objective 3 >

**RECALL**

When any factor is multiplied by 0, the product is 0.

Combine like terms.

**(a)** $8m + 5m = (8 + 5)m = 13m$

**(b)** $5pq^3 - 4pq^3 = 1pq^3 = pq^3$ Multiplication by 1 is understood.

**(c)** $7a^3b^2 - 7a^3b^2 = 0a^3b^2 = 0$

### Check Yourself 4

Combine like terms.

(a) $6b + 8b$
(b) $12x^2 - 3x^2$
(c) $8xy^3 - 7xy^3$
(d) $9a^2b^4 - 9a^2b^4$

The idea is the same for expressions involving more than two terms.

## Example 5 Combining Like Terms

**NOTE**

The distributive property can be used over any number of like terms.

Combine like terms.

**(a)** $4xy - xy + 2xy$

$= (4 - 1 + 2)xy$

$= 5xy$

Only like terms can be combined.

**(b)** $\overbrace{8x - 2x} + 5y$

$= 6x + 5y$

**NOTE**

With practice you will do these steps mentally instead of writing them out.

Like terms    Like terms

**(c)** $5m + 8n + 4m - 3n$

$= (5m + 4m) + (8n - 3n)$

$= 9m + 5n$

We use the associative and commutative properties.

**(d)** $4x^2 + 2x - 3x^2 + x$

$= (4x^2 - 3x^2) + (2x + x)$

$= x^2 + 3x$

As these examples illustrate, combining like terms often means changing the grouping and the order in which the terms are written. Again, all this is possible because of the properties of addition that we introduced in Section 1.2.

### Check Yourself 5

Combine like terms.

(a) $4m^2 - 3m^2 + 8m^2$
(b) $9ab + 3a - 5ab$
(c) $4p + 7q + 5p - 3q$

You may not realize it, but adding and subtracting algebraic expressions occurs all the time in the world. You have probably combined like terms successfully many times before ever taking this course.

## Example 6 An Application of Algebra

In anticipation of a holiday rush, a produce market receives 6 cases of apples and 4 cases of oranges from their supplier. The market already had 2 cases of apples and 2 cases of oranges in stock. How many cases of each does the market have after the delivery?

We add

6 apples + 4 oranges + 2 apples + 2 oranges

and combine like terms using the *commutative* property.

(6 apples + 2 apples) + (4 oranges + 2 oranges)
= 8 apples + 6 oranges

**NOTE**

We cannot add apples and oranges.

Therefore, the market begins the day with 8 cases of apples and 6 cases of oranges.

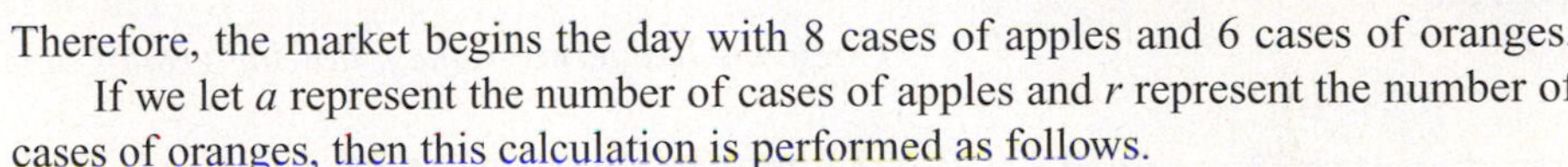

If we let $a$ represent the number of cases of apples and $r$ represent the number of cases of oranges, then this calculation is performed as follows.

$$\begin{aligned}6a + 4r + 2a + 2r &= (6a + 2a) + (4r + 2r)\\ &= 8a + 6r\end{aligned}$$

### Check Yourself 6

An electronics store has 8 two-packs and 20 ten-packs of 3.5-in. floppy disks in stock. They receive a shipment of 48 two-packs and 24 ten-packs. Algebraically represent the number and type of packages of disks that the store has after the shipment arrives.

### Check Yourself ANSWERS

**1. (a)** $2b^4$; **(b)** $5m$, $3n$; **(c)** $2s^2$, $-3t$, $-6$ **2. (a)** 8; **(b)** $-5$; **(c)** 1
**3.** The like terms are $5a^2b$, $a^2b$, and $-7a^2b$ **4. (a)** $14b$; **(b)** $9x^2$; **(c)** $xy^3$; **(d)** 0
**5. (a)** $9m^2$; **(b)** $4ab + 3a$; **(c)** $9p + 4q$ **6.** $56x + 44y$

### Reading Your Text

The following fill-in-the-blank exercises are designed to ensure that you understand some of the key vocabulary used in this section.

SECTION 11.3

**(a)** The product of a number and a variable is called a ______________.

**(b)** The ______________ factor of a term is called a coefficient.

**(c)** Terms that contain exactly the same variables raised to the same powers are called ______________ terms.

**(d)** The ______________ property enables us to combine like terms into a single term.

# 11.3 exercises

Basic Skills | Advanced Skills | Vocational-Technical Applications | Calculator/Computer | Above and Beyond

MathZone

Boost your grade at mathzone.com!
- Practice Problems
- NetTutor
- Self-Tests
- e-Professors
- Videos

Name ______________

Section ________ Date ________

## Answers

1. ______________
2. ______________
3. ________ 4. ________
5. ______________
6. ______________
7. ________ 8. ________
9. ______________
10. ______________
11. ________ 12. ________
13. ________ 14. ________
15. ________ 16. ________
17. ________ 18. ________
19. ________ 20. ________
21. ________ 22. ________
23. ________ 24. ________

< Objective 1 >

*List the terms of the following expressions.*

**1.** $5a + 2$

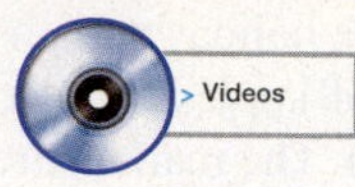


**2.** $7a - 4b$

**3.** $4x^3$

**4.** $3x^2$

**5.** $3x^2 + 3x - 7$

**6.** $2a^3 - a^2 + a$

< Objective 2 >

*Circle the like terms in the following groups of terms.*

**7.** $5ab, 3b, 3a, 4ab$

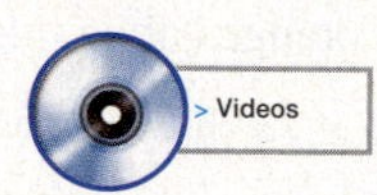


**8.** $9m^2, 8mn, 5m^2, 7m$

**9.** $4xy^2, 2x^2y, 5x^2, -3x^2y, 5y, 6x^2y$

**10.** $8a^2b, 4a^2, 3ab^2, -5a^2b, 3ab, 5a^2b$

< Objective 3 >

*Combine the like terms.*

**11.** $3m + 7m$

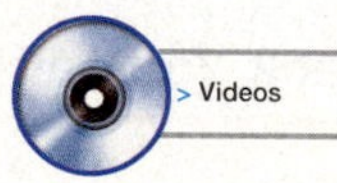


**12.** $6a^2 + 8a^2$

**13.** $7b^3 + 10b^3$

**14.** $7rs + 13rs$

**15.** $21xyz + 7xyz$

**16.** $4mn^2 + 15mn^2$

**17.** $9z^2 - 3z^2$

**18.** $7m - 6m$

**19.** $5a^3 - 5a^3$

**20.** $13xy - 9xy$

**21.** $19n^2 - 18n^2$

**22.** $7cd - 7cd$

**23.** $21p^2q - 6p^2q$

**24.** $17r^3s^2 - 8r^3s^2$

**25.** $10x^2 - 7x^2 + 3x^2$

**26.** $13uv + 5uv - 12uv$

**27.** $9a - 7a + 4b$

**28.** $5m^2 - 3m + 6m^2$

**29.** $7x + 5y - 4x - 4y$

**30.** $6a^2 + 11a + 7a^2 - 9a$

**31.** $4a + 7b + 3 - 2a + 3b - 2$

**32.** $5p^2 + 2p + 8 + 4p^2 + 5p - 6$

**33.** $\frac{2}{3}m + 3 + \frac{4}{3}m$

**34.** $\frac{1}{5}a - 2 + \frac{4}{5}a$

**35.** $\frac{13}{5}x + 2 - \frac{3}{5}x + 5$

**36.** $\frac{17}{12}y + 7 + \frac{7}{12}y - 3$

**37.** $2.3a + 7 + 4.7a + 3$

**38.** $5.8m + 4 - 2.8m + 11$

*Perform the indicated operations.*

**39.** Find the sum of $5a^4$ and $8a^4$.

**40.** What is the sum of $9p^2$ and $12p^2$?

**41.** Subtract $12a^3$ from $15a^3$.

**42.** Subtract $5m^3$ from $18m^3$.

**43.** Subtract $4x$ from the sum of $8x$ and $3x$.

**44.** Subtract $8ab$ from the sum of $7ab$ and $5ab$.

**45.** Subtract $3mn^2$ from the sum of $9mn^2$ and $5mn^2$.

**46.** Subtract $4x^2y$ from the sum of $6x^2y$ and $12x^2y$.

## Answers

25. ______
26. ______
27. ______
28. ______
29. ______
30. ______
31. ______
32. ______
33. ______
34. ______
35. ______
36. ______
37. ______
38. ______
39. ______
40. ______
41. ______
42. ______
43. ______ 44. ______
45. ______ 46. ______

Answers

47. ______

48. ______

49. ______

50. ______

51. ______

52. ______

53. ______

54. ______

55. ______

56. ______

57. ______

58. ______

59. ______

60. ______

61. ______

62. ______

63. ______

64. ______

47. **Geometry** A rectangle has sides that measure $8x + 9$ and $6x - 7$. Find the simplified expression that represents its perimeter.

48. **Geometry** A triangle has sides measuring $3x + 7$, $4x - 9$, and $5x + 6$. Find the simplified expression that represents its perimeter.

Basic Skills | **Advanced Skills** | Vocational-Technical Applications | Calculator/Computer | Above and Beyond

49. **Business and Finance** The cost of producing $x$ units of an item is $150 + 25x$. The revenue from selling $x$ units is $90x - x^2$. The profit is given by the revenue minus the cost. Find the simplified expression that represents the profit.

chapter 11 > Make the Connection

50. **Business and Finance** The revenue from selling $y$ units is $3y^2 - 2y + 5$ and the cost of producing $y$ units is $y^2 + y - 3$. Find the simplified expression that represents the profit.

chapter 11 > Make the Connection

*Use the distributive property to remove the parentheses in each expression. Then, simplify each expression by combining like terms.*

51. $2(3x + 2) + 4$

52. $3(4z + 5) - 9$

53. $5(6a - 2) + 12a$

54. $7(4w - 3) - 25w$

55. $4s + 2(s + 4) + 4$

56. $5p + 4(p + 3) - 8$

*Evaluate each of the following expressions if $a = 2$, $b = 3$, and $c = 5$. Be sure to combine like terms, when possible, as the first step.*

57. $7a^2 + 3a$

58. $11b^2 - 9b$

59. $3c^2 + 5c^2$

60. $9b^3 - 5b^3$

61. $5b + 3a - 2b$

62. $7c - 2b + 3c$

63. $5ac^2 - 2ac^2$

64. $5a^3b - 2a^3b$

Basic Skills | Advanced Skills | **Vocational-Technical Applications** | Calculator/Computer | Above and Beyond

**Answers**

65. ____________

66. ____________

67. ____________

68. ____________

69. ____________

70. ____________

71. ____________

72. ____________

73. ____________

**65.** **ALLIED HEALTH** The ideal body weight, in pounds, for a woman can be approximated by substituting her height, in inches, into the formula $105 + 5(h - 60)$. Use the distributive property to simplify the expression.

**66.** **ALLIED HEALTH** Use exercise 65 to approximate the ideal body weight for a woman who stands 5 ft 4 in. tall.

**67.** **MECHANICAL ENGINEERING** A primary beam can support a load of $54p$. A second beam is added that can support a load of $32p$. What is the total load that the two beams can support?

**68.** **MECHANICAL ENGINEERING** Two objects are spinning on the same axis. The moment of inertia of the first object is $\frac{6^3}{12}b$. The moment of inertia of the second object is given by $\frac{30^3}{36}b$. The total moment of inertia is given by the sum of the moments of inertia of the two objects. Write a simplified expression for the total moment of inertia for the two objects described.

Basic Skills | Advanced Skills | Vocational-Technical Applications | **Calculator/Computer** | Above and Beyond

*Use your calculator to evaluate each expression for the given values of the variables. Round your results to the nearest tenth.*

**69.** $7x^2 - 5y^3$; $x = 7.1695$, $y = 3.128$

**70.** $2x^2 + 3y + 5x$; $x = 3.61$, $y = 7.91$

**71.** $(4x^2y)(2xy^2) - 5x^3y$; $x = 1.29$, $y = 2.56$

**72.** $3x^3y - 4xy + 2x^2y^2$; $x = 3.26$, $y = 1.68$

Basic Skills | Advanced Skills | Vocational-Technical Applications | Calculator/Computer | **Above and Beyond**

**73.** A toy store begins the day with four Frisbees and eight basketballs in stock. During the morning shift, two Frisbees and one basketball are sold. In the afternoon, a shipment containing six Frisbees arrived. The afternoon shift sells three Frisbees and two basketballs.

Algebraically represent the number of Frisbees and of basketballs that are left at the end of the day (use $f$ to represent the number of Frisbees and $b$ to represent the number of basketballs).

Answers

74. ______

75. ______

76. ______

77. ______

78. ______

79. ______

80. ______

**74.** Determine the number of pounds of each type of coffee that a retailer has at the end of the day, given the following information.

A retailer begins the day with 24 pounds (lb) of Kona coffee, 17 lb of Italian roast, and 12 lb of Sumatran roast.

The retailer sells 8 lb of the Kona variety, 11 lb of the Italian, and 7 lb of the Sumatran. A delivery brings 4 lb of Kona and 16 lb of Sumatran coffees.

Express your answer algebraically, using $K$, $I$, and $S$ to represent the number of pounds of Kona, Italian, and Sumatran coffees, respectively.

**75.** Write a paragraph explaining the difference between $n^2$ and $2n$.

**76.** Complete the explanation: "$x^3$ and $3x$ are not the same because . . ."

**77.** Complete the statement: "$x + 2$ and $2x$ are different because . . ."

**78.** Write an English phrase for each algebraic expression.

**(a)** $2x^3 + 5x$ **(b)** $(2x + 5)^3$ **(c)** $6(n + 4)^2$

**79.** Work with another student to complete this exercise. Place >, <, or = in the blank in these statements.

$1^2$ ______ $2^1$

$2^3$ ______ $3^2$

$3^4$ ______ $4^3$

$4^5$ ______ $5^4$

What happens as the table of numbers is extended? Try more examples.

What sign seems to occur the most in your table: >, <, or =?

Write an algebraic statement for the pattern of signs in this table. Do you think this is a pattern that continues? Add more lines to the table and extend the pattern to the general case by writing the pattern in algebraic notation. Write a short paragraph stating your conjecture.

**80.** Work with other students on this exercise.

**Part 1:** Evaluate the three expressions $\frac{n^2 - 1}{2}$, $n$, and $\frac{n^2 + 1}{2}$, using odd values of $n$: 1, 3, 5, 7, etc. Make a chart like the following one and complete it.

| $n$ | $a = \frac{n^2 - 1}{2}$ | $b = n$ | $c = \frac{n^2 + 1}{2}$ | $a^2$ | $b^2$ | $c^2$ |
|---|---|---|---|---|---|---|
| 1 | | | | | | |
| 3 | | | | | | |
| 5 | | | | | | |
| 7 | | | | | | |
| 9 | | | | | | |
| 11 | | | | | | |
| 13 | | | | | | |
| 15 | | | | | | |

**Part 2:** The numbers, $a$, $b$, and $c$ that you get in each row have a surprising relationship to each other. Complete the last three columns and work together to discover this relationship. You may want to find out more about the history of this famous number pattern.

## Answers

**1.** $5a, 2$ **3.** $4x^3$ **5.** $3x^2, 3x, -7$ **7.** $5ab, 4ab$
**9.** $2x^2y, -3x^2y, 6x^2y$ **11.** $10m$ **13.** $17b^3$ **15.** $28xyz$ **17.** $6z^2$
**19.** $0$ **21.** $n^2$ **23.** $15p^2q$ **25.** $6x^2$ **27.** $2a + 4b$ **29.** $3x + y$
**31.** $2a + 10b + 1$ **33.** $2m + 3$ **35.** $2x + 7$ **37.** $7a + 10$
**39.** $13a^4$ **41.** $3a^3$ **43.** $7x$ **45.** $11mn^2$ **47.** $28x + 4$
**49.** $P = -x^2 + 65x - 150$ **51.** $6x + 8$ **53.** $42a - 10$ **55.** $6s + 12$
**57.** 34 **59.** 200 **61.** 15 **63.** 150 **65.** $5h - 195$ **67.** $86p$
**69.** 206.8 **71.** 260.6 **73.** $5f + 5b$ **75.** Above and Beyond
**77.** Above and Beyond **79.** Above and Beyond

# Activity 31 :: Writing Equations

In Section 11.1, you learned to translate phrases to algebraic expressions. In most applications, you need more than an expression; you need an equation.

**1.** Write an algebraic equation for the statement "Three more than a number is 9."

**2.** Write an algebraic equation describing "an employee's gross pay is the hourly pay times the number of hours worked."

**3.** Use the equation in exercise 2 to determine the gross pay of someone who works 40 hours, earning $9.75 per hour.

**4.** Create an equation that determines the net pay if the employee in exercise 3 pays a total of 16% of the gross pay to the federal and state governments.

**5.** Determine the net pay for the employee in exercise 3.

**6.** Write a paragraph describing some reasons why, or situations in which, forming an equation to compute net pay might be useful.

**7.** **(a)** Describe another situation in which constructing an equation would be useful.
**(b)** Construct an equation for the situation described in part **(a).**

**8.** Go to www.mhhe.com/streeter to view more situations and the equations used to describe them.

# 11.4 Using the Addition Property to Solve an Equation

< 11.4 Objectives >

1 > Determine whether a given number is a solution for an equation

2 > Use the addition property to solve an equation

In this section we begin working with one of the most important tools of mathematics, the equation. The ability to recognize and solve various types of equations is probably the most useful algebraic skill you will learn. To begin with, we define the word *equation*.

**Definition**

**Equation** An **equation** is a mathematical statement that two expressions are equal.

Some examples are $3 + 4 = 7$, $x + 3 = 5$, $P = 2L + 2W$.

As you can see, an equal sign (=) separates the two equal expressions. We call these expressions the *left side* and the *right side* of the equation.

$x + 3 = 5$

Left side   Equals   Right side

**NOTE**

An equation such as

$x + 3 = 5$

is called a **conditional equation** because it can be either true or false depending on the value given to the variable.

An equation may be either true or false. For instance, $3 + 4 = 7$ is true because both sides name the same number. What about an equation such as $x + 3 = 5$ that has a letter or variable on one side? Any number can replace $x$ in the equation. However, only one number will make this equation a true statement.

$$x + 3 = 5$$

If $x = 1$: $(1) + 3 = 5$ is false

If $x = 2$: $(2) + 3 = 5$ is true

If $x = 3$: $(3) + 3 = 5$ is false

The number 2 is called a **solution** (or *root*) of the equation $x + 3 = 5$ because substituting 2 for $x$ gives a true statement.

**Definition**

**Solution** A **solution** for an equation is any value for the variable that makes the equation a true statement.

**Example 1** Verifying a Solution

< Objective 1 >

**(a)** Is 3 a solution for the equation $2x + 4 = 10$?

To find out, replace $x$ with 3 and evaluate $2x + 4$ on the left.

| Left Side | | Right Side |
|---|---|---|
| $2(3) + 4$ | $\stackrel{?}{=}$ | 10 |
| $6 + 4$ | $\stackrel{?}{=}$ | 10 |
| 10 | $=$ | 10 |

Because $10 = 10$ is a true statement, 3 is a solution of the equation.

**(b)** Is 5 a solution for the equation $3x - 2 = 2x + 1$?

To find out, replace $x$ with 5 and evaluate each side separately.

**RECALL**

The rules for the order of operation require that we multiply first; then add or subtract.

| Left Side | | Right Side |
|---|---|---|
| $3(5) - 2$ | $\stackrel{?}{=}$ | $2(5) + 1$ |
| $15 - 2$ | $\stackrel{?}{=}$ | $10 + 1$ |
| 13 | $\stackrel{?}{=}$ | 11 |

Because the two sides do not name the same number, we do not have a true statement, and 5 is not a solution.

**Check Yourself 1**

For the equation

$2x - 1 = x + 5$

**(a)** Is 4 a solution? **(b)** Is 6 a solution?

You may be wondering whether an equation can have more than one solution. It certainly can. For instance,

$x^2 = 9$

has two solutions. They are 3 and $-3$ because

$3^2 = 9$ and $(-3)^2 = 9$

In this chapter, however, we work with *linear equations in one variable.* These are equations that can be put into the form

$ax + b = 0$

in which $x$ is the variable, $a$ and $b$ are any numbers, and $a$ is not equal to 0. In a linear equation, the variable can appear only to the first power. No other power ($x^2$, $x^3$, etc.) can appear. Linear equations are also called **first-degree equations.** The degree of an equation in one variable is the highest power to which the variable appears.

**Property**

**Linear Equations**

Linear equations in one variable that can be written in the form

$ax + b = 0 \qquad a \neq 0$

have exactly one solution.

## Example 2 Identifying Expressions and Equations

Label each of the following as an expression, a linear equation, or an equation that is not linear.

**(a)** $4x + 5$ is an expression.

**(b)** $2x + 8 = 0$ is a linear equation.

**(c)** $3x^2 - 9 = 0$ is an equation that is not linear.

**(d)** $5x = 15$ is a linear equation.

### Check Yourself 2

Label each as an expression, a linear equation, or an equation that is nonlinear.

**(a)** $2x^2 = 8$ **(b)** $2x - 3 = 0$

**(c)** $5x - 10$ **(d)** $2x + 1 = 7$

It is not difficult to find the solution for an equation such as $x + 3 = 8$ by guessing the answer to the question, What plus 3 is 8? Here the answer to the question is 5, which is also the solution for the equation. But for more complicated equations you are going to need something more than guesswork. A better method is to transform the given equation to an *equivalent equation* whose solution can be found by inspection. Here is a definition.

**Definition**

**Equivalent Equations**

Equations that have exactly the same solution(s) are called **equivalent equations.**

**NOTE**

In some cases we write the equation in the form

$\square = x$

The number will be our solution when the equation has the variable isolated on either side.

The following are all equivalent equations:

$2x + 3 = 5 \qquad 2x = 2 \qquad$ and $\qquad x = 1$

They all have the same solution, 1. We say that a linear equation is *solved* when it is transformed to an equivalent equation of the form

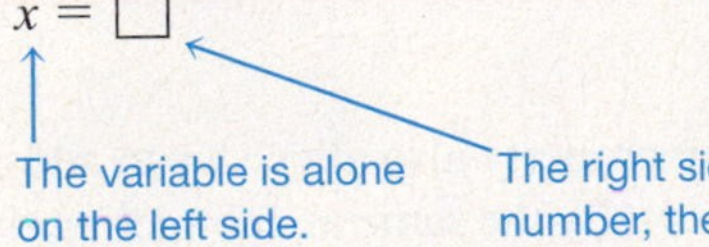

The addition property of equality is the first property you need in order to transform an equation to an equivalent form.

**Property**

### The Addition Property of Equality

If $\quad a = b$

then $\quad a + c = b + c$

In words, adding the same quantity to both sides of an equation gives an equivalent equation.

> **NOTE**
>
> An equation is a statement that the two sides are equal. Adding the same quantity to both sides does not change the equality or "balance."

Here is an example of applying this property to solve an equation.

### Example 3 Using the Addition Property to Solve an Equation

< Objective 2 >

Solve

$x - 3 = 9$

Remember that our goal is to isolate $x$ on one side of the equation. Because 3 is being subtracted from $x$, we can add 3 to remove it. We must use the addition property to add 3 to both sides of the equation.

$$\begin{array}{rcl} x - 3 & = & 9 \\ +3 & & +3 \\ \hline x & = & 12 \end{array}$$

Adding 3 leaves $x$ alone on the left.

Because 12 is the solution for the equivalent equation $x = 12$, it is the solution for our original equation.

> **NOTE**
>
> To check, replace $x$ with 12 in the original equation:
>
> $x - 3 = 9$
>
> $(12) - 3 \stackrel{?}{=} 9$
>
> $9 = 9$
>
> Because we have a true statement, 12 is the solution.

**Check Yourself 3**

Solve and check.

$x - 5 = 4$

The addition property also allows us to add a negative number to both sides of an equation. This is really the same as subtracting the same quantity from both sides.

### Example 4 Using the Addition Property to Solve an Equation

**RECALL**

Earlier we stated that we could write an equation in the equivalent form $x = \square$ or $\square = x$, in which $\square$ represents some number. Suppose we have an equation like

$12 = x + 7$

Subtracting 7 isolates $x$ *on the right*

$$\begin{array}{rcl} 12 &=& x + 7 \\ -7 & & -7 \\ \hline 5 &=& x \end{array}$$

The solution is 5.

Solve

$x + 5 = 9$

In this case, 5 is *added* to $x$ on the left. We can use the addition property to subtract 5 from both sides. This leaves the variable $x$ alone on one side of the equation.

$$\begin{array}{rcr} x + 5 &=& 9 \\ -5 & & -5 \\ \hline x &=& 4 \end{array}$$

The solution is 4. To check, replace $x$ with 4 in the original equation.

$(4) + 5 = 9$ (True)

**Check Yourself 4**

Solve and check.

$x + 6 = 13$

What if the equation has a variable term on both sides? Then we use the addition property to add a term involving the variable to get the desired result.

### Example 5 Using the Addition Property to Solve an Equation

Solve

$5x = 4x + 7$

We start by subtracting $4x$ from both sides of the equation. Do you see why? Remember that an equation is solved when we have an equivalent equation of the form $x = \square$.

$$\begin{array}{rcr} 5x &=& 4x + 7 \\ -4x & & -4x \\ \hline x &=& 7 \end{array}$$

Adding $-4x$ to both sides *removes* $4x$ from the right.

To check: Because 7 is a solution for the equivalent equation $x = 7$, it should be a solution for the original equation. To find out, replace $x$ with 7:

$5(7) \stackrel{?}{=} 4(7) + 7$

$35 \stackrel{?}{=} 28 + 7$

$35 = 35$ (True)

**Check Yourself 5**

Solve and check.

$7x = 6x + 3$

You may have to apply the addition property more than once to solve an equation. Look at Example 6.

## Example 6 Using the Addition Property to Solve an Equation

**NOTE**

This equation could also be written as

$7x + (-8) = 6x$

Solve

$7x - 8 = 6x$

We want all variables on *one* side of the equation. If we choose the left, we subtract $6x$ from both sides of the equation. This removes $6x$ from the right:

$$\begin{array}{rcr} 7x - 8 & = & 6x \\ -6x & & -6x \\ \hline x - 8 & = & 0 \end{array}$$

We want the variable alone, so we add 8 to both sides. This isolates $x$ on the left.

$$\begin{array}{rcr} x - 8 & = & 0 \\ +8 & & +8 \\ \hline x & = & 8 \end{array}$$

The solution is 8. We leave it to you to check this result.

### Check Yourself 6

Solve and check.

$9x + 3 = 8x$

Often an equation has more than one variable term *and* more than one number. You have to apply the addition property twice when solving these equations.

## Example 7 Using the Addition Property to Solve an Equation

Solve

$5x - 7 = 4x + 3$

We would like the variable terms on the left, so we start by subtracting $4x$ from both sides of the equation to remove that term from the right side of the equation:

**NOTE**

You could just as easily have added 7 to both sides and *then* subtracted $4x$. The result is the same. In fact, some students prefer to combine the two steps.

$$\begin{array}{rcr} 5x - 7 & = & 4x + 3 \\ -4x & & -4x \\ \hline x - 7 & = & 3 \end{array}$$

Now, to isolate the variable, we add 7 to both sides.

$$\begin{array}{rcr} x - 7 & = & 3 \\ +7 & & +7 \\ \hline x & = & 10 \end{array}$$

The solution is 10. To check, replace $x$ with 10 in the original equation:

$$5(10) - 7 \stackrel{?}{=} 4(10) + 3$$
$$43 = 43 \quad \text{(True)}$$

## Check Yourself 7

Solve and check.

**(a)** $4x - 5 = 3x + 2$ **(b)** $6x + 2 = 5x - 4$

**RECALL**

By *simplify* we mean to combine all like terms.

In solving an equation, you should always simplify each side as much as possible before using the addition property.

## Example 8 Combining Like Terms to Solve an Equation

Solve

Like terms ↓ ↘ Like terms ↙ ↘

$$5 + 8x - 2 = 2x - 3 + 5x$$

Because like terms appear on each side of the equation, we start by combining the numbers on the left (5 and −2). Then we combine the like terms ($2x$ and $5x$) on the right. We have

$$3 + 8x = 7x - 3$$

Now we apply the addition property, as before:

$$\begin{array}{rcll} 3 + 8x &=& 7x - 3 & \\ -7x &=& -7x & \text{Subtract } 7x. \\ \hline 3 + x &=& -3 & \text{Subtract 3.} \\ -3 & & -3 & \\ \hline x &=& -6 & \text{Isolate } x. \end{array}$$

The solution is −6. To check, always return to the original equation. That catches any possible errors in simplifying. Replacing $x$ with −6 gives

$$5 + 8(-6) - 2 \stackrel{?}{=} 2(-6) - 3 + 5(-6)$$
$$5 - 48 - 2 \stackrel{?}{=} -12 - 3 - 30$$
$$-45 = -45 \quad \text{(True)}$$

## Check Yourself 8

Solve and check.

**(a)** $3 + 6x + 4 = 8x - 3 - 3x$ **(b)** $5x + 21 + 3x = 20 + 7x - 2$

We may need to apply some of the properties discussed in Chapter 1 in solving equations. Example 9 illustrates the use of the distributive property to clear an equation of parentheses.

## Example 9 Using the Distributive Property and Solving Equations

**NOTE**

$2(3x + 4)$
$= 2(3x) + 2(4)$
$= 6x + 8$

Solve

$$2(3x + 4) = 5x - 6$$

Applying the distributive property on the left, we have

$$6x + 8 = 5x - 6$$

We can then proceed as before:

$$\begin{array}{rcll} 6x + 8 &=& 5x - 6 & \\ -5x & & -5x & \text{Subtract } 5x. \\ \hline x + 8 &=& -6 & \\ -8 & & -8 & \text{Subtract 8.} \\ \hline x &=& -14 & \end{array}$$

The solution is $-14$. We leave it to you to check this result.

**Remember:** Always return to the original equation to check.

### Check Yourself 9

Solve and check each of the following equations.

**(a)** $4(5x + 2) = 19x + 20$ **(b)** $3(5x + 1) = 2(7x + 3) - 16$

Of course, there are many applications that require us to use the addition property to solve an equation. Consider the consumer application in Example 10.

## Example 10 A Consumer Application

An appliance store is having a sale on washers and dryers. It charges $999 for a washer and dryer combination. If the washer sells for $649, how much is someone paying for the dryer as part of the combination?

Let $d$ be the cost of the dryer and solve the equation $d + 649 = 999$ to answer the question.

**RECALL**

You should always answer an application problem with a full sentence.

$$\begin{array}{rcll} d + 649 &=& 999 & \\ -649 & & -649 & \text{Subtract 649 from both sides.} \\ \hline d &=& 350 & \end{array}$$

The dryer adds $350 to the price.

### Check Yourself 10

Of 18,540 votes cast in the school board election, 11,320 went to Carla. How many votes did her opponent Marco receive? Who won the election?

Let $m$ be the number of votes Marco received and solve the equation $11{,}320 + m = 18{,}540$ in order to answer the questions.

## Check Yourself ANSWERS

1. **(a)** 4 is not a solution; **(b)** 6 is a solution
2. **(a)** Nonlinear equation; **(b)** linear equation; **(c)** expression; **(d)** linear equation
3. $x = 9$ **4.** $x = 7$ **5.** $x = 3$ **6.** $x = -3$ **7.** **(a)** $x = 7$; **(b)** $x = -6$
8. **(a)** $x = -10$; **(b)** $x = -3$ **9.** **(a)** $x = 12$; **(b)** $x = -13$
10. Marco received 7,220 votes; Carla won the election.

## Reading Your Text

The following fill-in-the-blank exercises are designed to ensure that you understand some of the key vocabulary used in this section.

SECTION 11.4

**(a)** An ______________ is a mathematical statement that two expressions are equal.

**(b)** A solution for an equation is a value for the variable that makes the equation a ______________ statement.

**(c)** Equivalent equations have the same ______________.

**(d)** The answer to an application should always be given using a full ______________.

# 11.4 exercises

Basic Skills | Advanced Skills | Vocational-Technical Applications | Calculator/Computer | Above and Beyond

MathZone

Boost your grade at mathzone.com!

> Practice Problems
> NetTutor
> Self-Tests
> e-Professors
> Videos

Name ____________

Section ________ Date ________

## Answers

1. ________ 2. ________
3. ________ 4. ________
5. ________ 6. ________
7. ________ 8. ________
9. ________ 10. ________
11. ________ 12. ________
13. ________ 14. ________
15. ________ 16. ________
17. ________ 18. ________
19. ________ 20. ________
21. ________ 22. ________
23. ________
24. ________
25. ________
26. ________

< Objective 1 >

*Is the number shown in parentheses a solution for the given equation?*

**1.** $x + 4 = 9$ (5)

**2.** $x + 2 = 11$ (8)

**3.** $x - 15 = 6$ $(-21)$

**4.** $x - 11 = 5$ (16)

**5.** $5 - x = 2$ (4)

**6.** $10 - x = 7$ (3)

**7.** $4 - x = 6$ $(-2)$

**8.** $5 - x = 6$ $(-3)$

**9.** $3x + 4 = 13$ (8)

**10.** $5x + 6 = 31$ (5)

**11.** $4x - 5 = 7$ (2)

**12.** $2x - 5 = 1$ (3)

**13.** $5 - 2x = 7$ $(-1)$

**14.** $4 - 5x = 9$ $(-2)$

**15.** $4x - 5 = 2x + 3$ (4)

**16.** $5x + 4 = 2x + 10$ (4)

**17.** $x + 3 + 2x = 5 + x + 8$ (5)

**18.** $5x - 3 + 2x = 3 + x - 12$ $(-2)$

**19.** $\frac{3}{4}x = 18$ (20)

**20.** $\frac{3}{5}x = 24$ (40)

**21.** $\frac{3}{5}x + 5 = 11$ (10)

**22.** $\frac{2}{3}x + 8 = -12$ $(-6)$

*Label each of the following as an expression or a linear equation.*

**23.** $2x + 1 = 9$

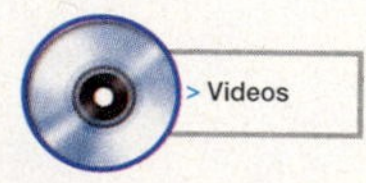


**24.** $7x + 14$

**25.** $2x - 8$

**26.** $5x - 3 = 12$

27. $7x + 2x + 8 - 3$

28. $x + 5 = 13$

29. $2x - 8 = 3$

30. $12x - 5x + 2 + 5$

< Objective 2 >

*Solve and check the following equations.*

31. $x + 9 = 11$

32. $x - 4 = 6$

33. $x - 8 = 3$

34. $x + 11 = 15$

35. $x - 8 = -10$

36. $x + 5 = 2$

37. $x + 4 = -3$

38. $x - 5 = -4$

39. $11 = x + 5$

40. $x + 7 = 0$

41. $4x = 3x + 4$

42. $7x = 6x - 8$

43. $11x = 10x - 10$

44. $9x = 8x + 5$

45. $6x + 3 = 5x$

46. $12x - 6 = 11x$

47. $8x - 4 = 7x$

48. $9x - 7 = 8x$

49. $2x + 3 = x + 5$

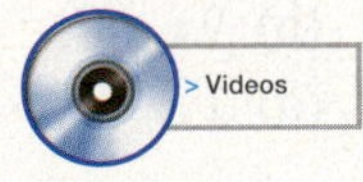


50. $3x - 2 = 2x + 1$

51. $5x - 7 = 4x - 3$

52. $8x + 5 = 7x - 2$

53. $7x - 2 = 6x + 4$

54. $10x - 3 = 9x - 6$

55. $3 + 6x + 2 = 3x + 11 + 2x$

56. $6x - 3 + 2x = 7x + 8$

## Answers

27. ________

28. ________

29. ________

30. ________

31. ____ 32. ____

33. ____ 34. ____

35. ____ 36. ____

37. ____ 38. ____

39. ____ 40. ____

41. ____ 42. ____

43. ________

44. ________

45. ________

46. ________

47. ________

48. ________

49. ________

50. ________

51. ________

52. ________

53. ________

54. ________

55. ________

56. ________

**Answers**

57. ______

58. ______

59. ______

60. ______

61. ______

62. ______

63. ______

64. ______

65. ______

66. ______

67. ______

68. ______

**57.** $4x + 7 + 3x = 5x + 13 + x$

**58.** $5x + 9 + 4x = 9 + 8x - 7$

**59.** $3x - 5 + 2x - 7 + x = 5x + 2$

**60.** $5x + 8 + 3x - x + 5 = 6x - 3$

**61.** **Crafts** Jeremiah had found 50 bones for a Halloween costume. In order to complete his 62-bone costume, how many more does he need?

Let $b$ be the number of bones he needs, and use the equation $b + 50 = 62$ to solve the problem.

**62.** **Business and Finance** Four hundred tickets to the opening of an art exhibit were sold. General admission tickets cost \$5.50, whereas students were only required to pay \$4.50 for tickets. If total ticket sales were \$1,950, how many of each type of ticket were sold?

Let $x$ be the number of general admission tickets sold and $400 - x$ be the number of student tickets sold. Use the equation $5.5x + 4.5(400 - x) = 1{,}950$ to solve the problem.

**63.** **Business and Finance** A shop pays \$2.25 for each copy of a magazine and sells the magazines for \$3.25 each. If the fixed costs associated with the sale of these magazines are \$50 per month, how many must the shop sell in order to realize \$175 in profit from the magazines?

Let $m$ be the number of magazines the shop must sell, and use the equation $3.25m - 2.25m - 50 = 175$ to solve the problem.

chapter 11 > Make the Connection

**64.** **Number Problem** The sum of a number and 15 is 22. Find the number.

Let $x$ be the number and solve the equation $x + 15 = 22$ to find the number.

**65.** Which of the following is equivalent to the equation $8x + 5 = 9x - 4$?

**(a)** $17x = -9$ **(b)** $x = -9$
**(c)** $8x + 9 = 9x$ **(d)** $9 = 17x$

**66.** Which of the following is equivalent to the equation $5x - 7 = 4x - 12$?

**(a)** $9x = 19$ **(b)** $9x - 7 = -12$
**(c)** $x = -18$ **(d)** $x - 7 = -12$

**67.** Which of the following is equivalent to the equation $12x - 6 = 8x + 14$?

**(a)** $4x - 6 = 14$ **(b)** $x = 20$
**(c)** $20x = 20$ **(d)** $4x = 8$

**68.** Which of the following is equivalent to the equation $7x + 5 = 12x - 10$?

**(a)** $5x = -15$ **(b)** $7x - 5 = 12x$
**(c)** $-5 = 5x$ **(d)** $7x + 15 = 12x$

**True** *or* **false?**

**69.** Every linear equation with one variable has exactly one solution.

**70.** Isolating the variable on the right side of the equation results in a negative solution.

Basic Skills | **Advanced Skills** | Vocational-Technical Applications | Calculator/Computer | Above and Beyond

*Solve and check the following equations.*

**71.** $4(3x + 4) = 11x - 2$

**72.** $2(5x - 3) = 9x + 7$

**73.** $3(7x + 2) = 5(4x + 1) + 17$

**74.** $5(5x + 3) = 3(8x - 2) + 4$

**75.** $\frac{5}{4}x - 1 = \frac{1}{4}x + 7$

**76.** $\frac{7}{5}x + 3 = \frac{2}{5}x - 8$

**77.** $\frac{9}{2}x - \frac{3}{4} = \frac{7}{2}x + \frac{5}{4}$

**78.** $\frac{11}{3}x + \frac{1}{6} = \frac{8}{3}x + \frac{19}{6}$

## Answers

**1.** Yes **3.** No **5.** No **7.** Yes **9.** No **11.** No **13.** Yes **15.** Yes **17.** Yes **19.** No **21.** Yes **23.** Linear equation **25.** Expression **27.** Expression **29.** Linear equation **31.** 2 **33.** $x = 11$ **35.** $x = -2$ **37.** $x = -7$ **39.** $x = 6$ **41.** $x = 4$ **43.** $x = -10$ **45.** $x = -3$ **47.** $x = 4$ **49.** $x = 2$ **51.** $x = 4$ **53.** $x = 6$ **55.** $x = 6$ **57.** $x = 6$ **59.** $x = 14$ **61.** 12 **63.** 225 **65.** (c) **67.** (a) **69.** True **71.** $x = -18$ **73.** $x = 16$ **75.** $x = 8$ **77.** $x = 2$

## Answers

69. ______

70. ______

71. ______

72. ______

73. ______

74. ______

75. ______

76. ______

77. ______

78. ______

# Activity 32 :: Graphing Solutions

You have now solved many equations using algebra. Often, it is convenient to present a picture of the solutions for an equation instead of giving a set of numbers. One method to present a picture uses the familiar number line.

**1.** Plot the points $\{-2, 0.5, 3\}$ on a number line.

**2.** Solve the equation $x + 5 = 8$.

**3.** Plot the solution to the equation in exercise 2 on a number line.

We often use a number line to present sets of numbers. For example, the set of numbers greater than 2 is written algebraically as $\{x \mid x > 2\}$, and is shown on a number line as

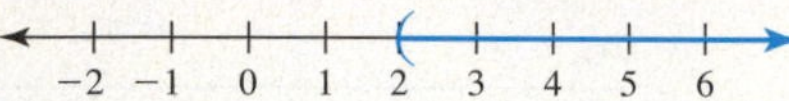



Whereas the set of numbers less than or equal to 3 is written as $\{x \mid x \leq 3\}$ and shown on a number line as

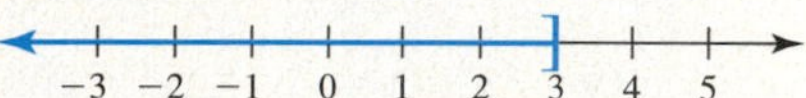



**4.** Graph the set of numbers $\{x \mid x < 3\}$ on a number line.

**5.** Graph the set of numbers $\{x \mid x \geq -1\}$ on a number line.

**6.** Solve the inequality $x + 2 > 0$.

**7.** Graph every solution to exercise 6 on a number line.

**NOTE**

We read $\{x \mid x > 2\}$ as "the set of every value of $x$ for which $x$ is greater than 2."

# 11.5 Using the Multiplication Property to Solve an Equation

< 11.5 Objectives >

1 > Use the multiplication property to solve an equation

2 > Combine like terms before solving an equation

3 > Use algebra to solve percent problems

**Property**

**The Multiplication Property of Equality**

If $a = b$ then $ac = bc$ when $c \neq 0$

In words, multiplying both sides of an equation by the same nonzero number gives an equivalent equation.

**NOTES**

Do you see why the number cannot be 0? Multiplying by 0 gives $0 = 0$. We have lost the variable!

Again, as long as you do the *same* thing to *both* sides of an equation, the "balance" is maintained.

We work through some examples using this rule.

**Example 1** Solving Equations by Using the Multiplication Property

< Objective 1 >

**RECALL**

Multiplying both sides by $\frac{1}{6}$ is equivalent to dividing both sides by 6.

Solve

$6x = 18$

Here the variable $x$ is multiplied by 6. So we apply the multiplication property and multiply both sides by $\frac{1}{6}$. Keep in mind that we want an equation of the form

$x = \square$

$\frac{1}{6}(6x) = \frac{1}{6}(18)$

**NOTE**

$\frac{1}{6}(6x) = \frac{\not{6}x}{\not{6}} = x$

We then have $x$ alone on the left, which is what we want.

$$\frac{1}{6}(6x) = \frac{1}{6}(18)$$

$$\frac{1}{6} \cdot \frac{6x}{1} = \frac{1}{6} \cdot \frac{18}{1}$$

$$\frac{\not{6}x}{\not{6}} = \frac{\not{18}^{3}}{\not{6}_{1}} \qquad 18 \div 6 = 3.$$

$$x = 3$$

The solution is 3. To check, replace $x$ with 3:

$$6(3) \stackrel{?}{=} 18$$

$$18 = 18 \qquad \text{(True)}$$

**Check Yourself 1**

Solve and check.

$8x = 32$

**Example 2** Solving Equations by Using the Multiplication Property

**NOTES**

Because division is defined in terms of multiplication, we can also divide both sides of an equation by the same nonzero number.

In this case, $x$ is multiplied by $-9$, so we divide both sides by $-9$ to isolate $x$ on the left.

Solve

$$-9x = 54$$

$$\frac{-9x}{-9} = \frac{54}{-9}$$

$$x = -6$$

The solution is $-6$. To check:

$$(-9)(-6) \stackrel{?}{=} 54$$

$$54 = 54 \qquad \text{(True)}$$

**Check Yourself 2**

Solve and check.

$-10x = -60$

Example 3 illustrates the use of the multiplication property when fractions appear in an equation.

**Example 3** Solving Equations by Using the Multiplication Property

**(a)** Solve

$$\frac{x}{3} = 6$$

**NOTE**

$\frac{x}{3} = \frac{1}{3}x$

Here $x$ is *divided* by 3. We use multiplication to isolate $x$.

$$3\left(\frac{x}{3}\right) = 3 \cdot 6$$

This leaves $x$ alone on the left because $3\left(\frac{x}{3}\right) = \frac{3}{1} \cdot \frac{x}{3} = \frac{x}{1} = x$

$$x = 18$$

To check:

$$\frac{(18)}{3} \stackrel{?}{=} 6$$

$$6 = 6 \quad \text{(True)}$$

**NOTE**

$\frac{x}{5} = \frac{1}{5}x$

**(b)** Solve

$$\frac{x}{5} = -9$$

$$5\left(\frac{x}{5}\right) = 5(-9)$$

Because $x$ is divided by 5, multiply both sides by 5.

$$x = -45$$

The solution is $-45$. To check, we replace $x$ with $-45$:

$$\frac{(-45)}{5} \stackrel{?}{=} -9$$

$$-9 = -9 \quad \text{(True)}$$

The solution is verified.

### Check Yourself 3

**Solve and check.**

**(a)** $\frac{x}{7} = 3$ **(b)** $\frac{x}{4} = -8$

When the variable is multiplied by a fraction with a numerator other than 1, there are two approaches to finding the solution.

## Example 4 Solving Equations by Using Reciprocals

Solve

$$\frac{3}{5}x = 9$$

One approach is to multiply by 5 as the first step.

$$5\left(\frac{3}{5}x\right) = 5 \cdot 9$$

$$3x = 45$$

Now we divide by 3.

$$\frac{3x}{3} = \frac{45}{3}$$

$$x = 15$$

To check:

$$\frac{3}{5}(15) \stackrel{?}{=} 9$$

$$9 = 9 \quad \text{(True)}$$

A second approach combines the multiplication and division steps and is generally more efficient. We multiply by $\frac{5}{3}$.

$$\frac{5}{3}\left(\frac{3}{5}x\right) = \frac{5}{3} \cdot 9$$

$$x = \frac{5}{\cancel{3}_1} \cdot \frac{\cancel{9}^3}{1} = 15$$

So $x = 15$, as before.

**NOTE**

Recall that $\frac{5}{3}$ is the *reciprocal* of $\frac{3}{5}$, and the product of a number and its reciprocal is just 1! So $\left(\frac{5}{3}\right)\left(\frac{3}{5}\right) = 1$

**Check Yourself 4**

**Solve and check.**

$$\frac{2}{3}x = 18$$

You may sometimes need to simplify an equation before applying the methods of this section. Example 5 illustrates this situation.

**Example 5** Combining Like Terms and Solving Equations

< Objective 2 >

Solve and check:

$$3x + 5x = 40$$

Using the distributive property, we can combine the like terms on the left to write

$$8x = 40$$

We can now proceed as before.

$$\frac{8x}{8} = \frac{40}{8} \quad \text{Divide by 8.}$$

$$x = 5$$

The solution is 5. To check, we return to the original equation. Substituting 5 for $x$ yields

$$3(5) + 5(5) \stackrel{?}{=} 40$$
$$15 + 25 \stackrel{?}{=} 40$$
$$40 = 40 \quad \text{(True)}$$

### Check Yourself 5

Solve and check.

$7x + 4x = -66$

As with the addition property, many applications require the multiplication property. One of the most useful set of applications involves percent problems.

In Section 6.4, you learned to use the percent relationship

$$\frac{A}{B} = R$$

to solve percent problems. We did this by writing the percent relationship as a proportion in which $R = \frac{r}{100}$.

$$\frac{A}{B} = \frac{r}{100}$$

We then used the proportion rule to rewrite the equation.

$$100A = rB$$

So, for instance, if the question asked us to find 45% of 80, we would identify $r = 45$ and $B = 80$.

$$100A = (45)(80)$$
$$100A = 3{,}600$$

The next step was to divide both sides by the coefficient of the variable, just as we have been doing throughout this section.

$$\frac{100A}{100} = \frac{3{,}600}{100}$$
$$A = 36$$

So, 45% of 80 is 36.

We can simplify this process by using algebra. First, we can rewrite the percent relationship by multiplying both sides by the base.

$$\frac{A}{B} = R$$
$$B\left(\frac{A}{B}\right) = R \cdot B$$
$$A = RB$$

This form is especially useful if we write the rate as a decimal rather than a fraction.

**RECALL**

In percent problems, $A$ is the amount, $B$ is the base, and $R$ is the rate.

**RECALL**

In Section 6.3, you learned to identify the base, rate, and amount in a percent problem.

**NOTE**

This says that the amount is equal to the product of the base and the rate.

## Example 6 Using Algebra to Solve a Percent Problem

< Objective 3 >

**(a)** What is 75% of 360?

The rate is given as 75%, which we write as a decimal, $R = 0.75$. Translating the question, we have

$$A = RB$$
$$A = (0.75)(360)$$
$$A = 270$$

75% of 360 is 270.

Alternatively, we can translate the question directly into an algebraic equation.

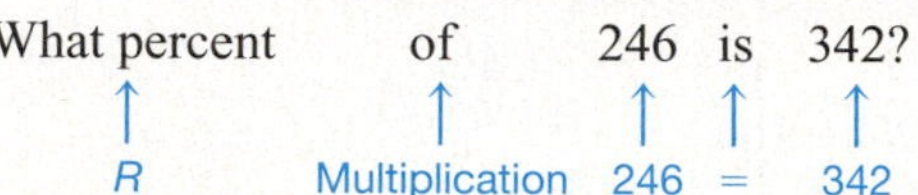


Which gives,

$$A = 0.75 \cdot 360$$
$$= 270$$

**(b)** What percent of 246 is 342?

This time, we begin by translating the question directly into an algebraic equation.

What percent of 246 is 342?
↑ R ↑ Multiplication ↑ 246 ↑ = ↑ 342

**NOTE**

The base is 246 and the amount is 342.

This gives us

$$R \cdot 246 = 342$$

or $246R = 342$

We can solve this equation using the methods of this section.

**RECALL**

Because the amount is larger than the base, the rate is greater than 100%.

$$\frac{\cancel{246}R}{\cancel{246}} = \frac{342}{246}$$

$$R = \frac{342}{246}$$

$$= \frac{57}{41}$$ The GCF of 246 and 342 is 6.

$$\approx 1.39$$

$$= 139\%$$ Move the decimal two places to the right and attach the percent symbol.

Therefore, 342 is about 139% of 246.

### Check Yourself 6

Use algebra to solve each percent application.

**(a)** 240 is what percent of 400?
**(b)** 57 is 30% of what number?

Of course, we can use algebra to solve applications that involve percents, as well.

## Example 7 Solving a Percent Application with Algebra

A saleswoman earns a 5% commission on her sales. If she wants to earn \$1,800 in commissions in one month, how much does she need to sell?

The question we are being asked is, "\$1,800 is 5% of what number?" We can translate the question into an algebraic equation by writing the rate in decimal form, $5\% = 0.05$.

$$1{,}800 = 0.05x$$

$$\frac{1{,}800}{0.05} = \frac{0.05x}{0.05}$$

$$36{,}000 = x$$

She must sell \$36,000 in order to earn \$1,800 in commissions.

### Check Yourself 7

**Patrick pays \$525 interest for a 1-year loan at 10.5%. What was the amount of his loan?**

We can solve many types of problems with algebra besides percent problems.

## Example 8 An Application Involving the Multiplication Property

On his first day on the job in a photography lab, Samuel processed all the film given to him. The following day, his boss gave him four times as much film to process. Over the two days, he processed 60 rolls of film. How many rolls did he process on the first day?

Let $x$ be the number of rolls Samuel processed on his first day and solve the equation $x + 4x = 60$ to answer the question.

$$x + 4x = 60$$

$$5x = 60 \quad \text{Combine like terms first.}$$

$$\frac{1}{5}(5x) = \frac{1}{5}(60) \quad \text{Multiply by } \frac{1}{5}\text{, to isolate the variable.}$$

$$x = 12$$

Samuel processed 12 rolls of film on his first day.

**RECALL**

You should always use a sentence to give the answer to an application.

**NOTE**

The yen (¥) is the monetary unit of Japan.

### Check Yourself 8

**On a recent trip to Japan, Marilyn exchanged \$1,200 and received 139,812 yen. What exchange rate did she receive?**

**Let $x$ be the exchange rate and solve the equation $1{,}200x = 139{,}812$ to answer the question (to the nearest hundredth).**

## Check Yourself ANSWERS

**1.** $x = 4$ **2.** $x = 6$ **3. (a)** $x = 21$; **(b)** $x = -32$ **4.** $x = 27$
**5.** $x = -6$ **6. (a)** 60%; **(b)** 190 **7.** $5,000
**8.** Marilyn received 116.51 yen for each dollar.

## Reading Your Text

The following fill-in-the-blank exercises are designed to ensure that you understand some of the key vocabulary used in this section.

### SECTION 11.5

**(a)** Multiplying both sides of an equation by the same nonzero number yields an ______________ equation.

**(b)** Division is defined in terms of ______________.

**(c)** Multiplying an equation by 5 is the same as ______________ by $\frac{1}{5}$.

**(d)** The product of a nonzero number and its ______________ is 1.

Basic Skills | Advanced Skills | Vocational-Technical Applications | Calculator/Computer | Above and Beyond

# 11.5 exercises

< Objective 1 >

*Solve for x and check your result.*

1. $5x = 20$ 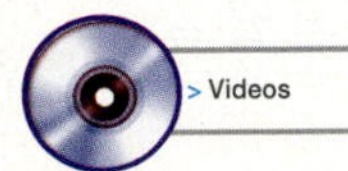

2. $6x = 30$

3. $9x = 54$

4. $6x = -42$

5. $63 = 9x$

6. $66 = 6x$

7. $4x = -16$

8. $-3x = 27$

9. $-9x = 72$

10. $10x = -100$

11. $6x = -54$

12. $-7x = 49$

13. $-4x = -12$ 

14. $52 = -4x$

15. $-42 = 6x$

16. $-7x = -35$

17. $-6x = -54$

18. $-4x = -24$

19. $\frac{x}{2} = 4$ 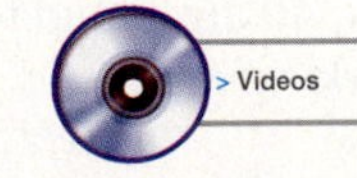

20. $\frac{x}{3} = 2$

21. $\frac{x}{5} = 3$

22. $\frac{x}{8} = 5$

23. $6 = \frac{x}{7}$

24. $6 = \frac{x}{3}$

25. $\frac{x}{5} = -4$

26. $\frac{x}{7} = -5$

27. $-\frac{x}{3} = 8$

28. $-\frac{x}{8} = -3$

29. $\frac{2}{3}x = 6$ 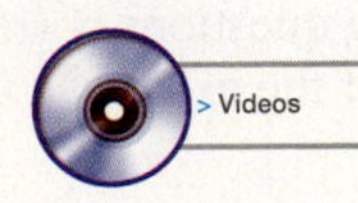

30. $\frac{4}{5}x = 8$

31. $\frac{3}{4}x = -15$

32. $\frac{7}{8}x = -21$

33. $-\frac{2}{5}x = 10$ > Videos

34. $-\frac{5}{6}x = -15$

Boost your grade at mathzone.com!
- > Practice Problems
- > NetTutor
- > Self-Tests
- > e-Professors
- > Videos

Name ______________

Section ________ Date ________

## Answers

| 1. ______ | 2. ______ |
|---|---|
| 3. ______ | 4. ______ |
| 5. ______ | 6. ______ |
| 7. ______ | 8. ______ |
| 9. ______ | 10. ______ |
| 11. ______ | 12. ______ |
| 13. ______ | 14. ______ |
| 15. ______ | 16. ______ |
| 17. ______ | 18. ______ |
| 19. ______ | 20. ______ |
| 21. ______ | 22. ______ |
| 23. ______ | 24. ______ |
| 25. ______ | 26. ______ |
| 27. ______ | 28. ______ |
| 29. ______ | 30. ______ |
| 31. ______ | 32. ______ |
| 33. ______ | 34. ______ |

**Answers**

35. ______ 36. ______
37. ______ 38. ______
39. ______ 40. ______
41. ______ 42. ______
43. ______
44. ______
45. ______
46. ______
47. ______
48. ______
49. ______
50. ______
51. ______
52. ______
53. ______
54. ______
55. ______
56. ______
57. ______
58. ______

< Objective 2 >

**35.** $5x + 4x = 36$

**36.** $8x - 3x = -50$

**37.** $16x - 9x = -42$

**38.** $5x + 7x = 60$

**39.** $4x - 2x + 7x = 36$

**40.** $6x + 7x - 5x = -48$

< Objective 3 >

*Solve each of the following percent problems.*

**41.** What is 65% of 300?

**42.** 15% of 140 is what number?

**43.** Find 80% of 80.

**44.** What is 6% of 550?

**45.** What percent of 220 is 66?

**46.** 104 is what percent of 260?

**47.** 102 is what percent of 85?

**48.** What percent of 130 is 299?

**49.** 15 is 4% of what number?

**50.** 16% of what number is 24?

**51.** Find the base if 240% of the base is 36.

**52.** Find the base if 375% of the base is 600.

*Solve each of the following applications.*

**53.** **Business and Finance** Roberto has 26% of his pay withheld for deductions. If he earns \$550 per week, what amount is withheld? chapter 11 > Make the Connection

**54.** **Business and Finance** A real estate agent's commission rate is 6%. What is the amount of commission on the sale of a \$185,000 home? chapter 11 > Make the Connection

**55.** **Social Science** Of the 60 people who started a training program, 45 were successful. What is the dropout rate?

**56.** **Business and Finance** In a shipment of 250 parts, 40 are found to be defective. What percent of the shipment is in good working order?

**57.** **Science and Medicine** There are 117 mL of acid in 900 mL of a solution (acid and water). What percent of the solution is water?

**58.** **Statistics** Marla needs to answer 70% of the questions correctly on her final exam in order to receive a C for the course. If the exam has 120 questions, how many can she miss?

**59. Crafts** Returning from Mexico City, Sung-A exchanged her remaining 450 pesos for \$41.70. What exchange rate did she receive?

Use the equation $450x = 41.70$ to solve this problem (round to the nearest thousandth).

**60. Business and Finance** Upon arrival in Portugal, Nicolas exchanged \$500 and received 417.35 euros (€). What exchange rate did he receive?

Use the equation $500x = 417.35$ to solve this problem (round to the nearest hundredth).

**61. Science and Technology** On Tuesday, there were twice as many patients in the clinic as on Monday. Over the two-day period, 48 patients were treated. How many patients were treated on Monday?

Let $p$ be the number of patients that came in on Monday and use the equation $p + 2p = 48$ to answer the question.

**62. Number Problem** Two-thirds of a number is 46. Find the number.

Use the equation $\frac{2}{3}x = 46$ to solve the problem.

Basic Skills | **Advanced Skills** | Vocational-Technical Applications | Calculator/Computer | Above and Beyond

*Certain equations involving decimals can be solved by the methods of this section. For instance, to solve $2.3x = 6.9$, we use the multiplication property to divide both sides of the equation by 2.3. This isolates x on the left, as desired. Use this idea to solve each of the following equations for x.*

**63.** $3.2x = 12.8$

**64.** $5.1x = -15.3$

**65.** $-4.5x = 13.5$

**66.** $-8.2x = -32.8$

**67.** $1.3x + 2.8x = 12.3$

**68.** $2.7x + 5.4x = -16.2$

**69.** $9.3x - 6.2x = 12.4$

**70.** $12.5x - 7.2x = -21.2$

**71. Information Technology** A 50-Gbyte-capacity hard drive contains 30 Gbytes of used space. What percent of the hard drive is full?

**72. Information Technology** A compression program reduces the size of files and folders by 36%. If a folder contains 17.5 Mbytes, how large will it be after it is compressed?

## Answers

59. ____________

60. ____________

61. ____________

62. ____________

63. ____________

64. ____________

65. ____________

66. ____________

67. ____________

68. ____________

69. ____________

70. ____________

71. ____________

72. ____________

**Answers**

73. ______

74. ______

75. ______

76. ______

77. ______

78. ______

79. ______

80. ______

81. ______

82. ______

83. ______

84. ______

**73.** **Automotive Technology** It is estimated that 8% of rebuilt alternators do not last through the 90-day warranty period. If a parts store had six bad alternators returned during the year, how many did it sell?

chapter 11 > Make the Connection

**74.** **Agricultural Technology** A farmer sold 2,200 bushels of barley on the futures market. Because of a poor harvest, he was only able to make 94% of his bid. How many bushels did he actually harvest?

chapter 11 > Make the Connection

Basic Skills | Advanced Skills | Vocational-Technical Applications | **Calculator/Computer** | Above and Beyond

*Use your calculator to solve each equation. Round your answers to the nearest hundredth.*

**75.** $230x = 157$

**76.** $31x = -15$

**77.** $-29x = 432$

**78.** $-141x = -3{,}467$

**79.** $23.12x = 94.6$

**80.** $46.1x = -1$

Basic Skills | Advanced Skills | Vocational-Technical Applications | Calculator/Computer | **Above and Beyond**

**81.** Describe the difference between the multiplication property and the addition property for solving equations. Give examples of when to use each property.

**82.** Describe when you should add a quantity to or subtract a quantity from both sides of an equation as opposed to when you should multiply or divide both sides by the same quantity.

*Motors, Windings, and More! sells every motor, regardless of type, for* \$2.50. *This vendor also has a deal in which customers can choose whether to receive a markdown or free shipping. Shipping costs are* \$1.00 *per item. If you do not choose the free shipping option, you can deduct* 17.5% *from your total order (but not the cost of shipping).*

**83.** If you buy six motors, calculate the total cost for each of the two options. Which option is cheaper?

**84.** Is one option *always* cheaper than the other? Justify your result.

## Answers

**1.** $x = 4$ **3.** $x = 6$ **5.** $x = 7$ **7.** $x = -4$ **9.** $x = -8$ **11.** $x = -9$ **13.** $x = 3$ **15.** $x = -7$ **17.** $x = 9$ **19.** $x = 8$ **21.** $x = 15$ **23.** $x = 42$ **25.** $x = -20$ **27.** $x = -24$ **29.** $x = 9$ **31.** $x = -20$ **33.** $x = -25$ **35.** $x = 4$ **37.** $x = -6$ **39.** $x = 4$ **41.** 195 **43.** 64 **45.** 30% **47.** 120% **49.** 375 **51.** 15 **53.** \$143 **55.** 25% **57.** 87% **59.** 0.093 dollars per peso **61.** 16 patients **63.** $x = 4$ **65.** $x = -3$ **67.** $x = 3$ **69.** $x = 4$ **71.** 60% **73.** 75 alternatives **75.** 0.68 **77.** −14.9 **79.** 4.09 **81.** Above and Beyond **83.** Above and Beyond

# 11.6 Combining the Properties to Solve Equations

< 11.6 Objectives >

1> Solve an equation using both the addition and multiplication properties

2> Combine like terms first, and then solve an equation using both properties

In our examples in Sections 11.4 and 11.5, either the addition property or the multiplication property was used in solving an equation. Often, finding a solution requires the use of both properties.

**Example 1** Solving Equations

< Objective 1 >

**(a)** Solve

$$4x - 5 = 7$$

Here $x$ is *multiplied* by 4. The result, $4x$, then has 5 subtracted from it on the left side of the equation. These two operations mean that both properties must be applied in solving the equation.

Because there is only one variable term, we start by adding 5 to both sides:

$4x - 5 + 5 = 7 + 5$ or $4x = 12$ The first step is to *isolate* the variable term on one side of the equation.

We now divide both sides by 4:

$$\frac{4x}{4} = \frac{12}{4}$$ Next, *isolate* the variable.

$$x = 3$$

The solution is 3. To check, replace $x$ with 3 in the original equation. Be careful to follow the rules for the order of operations.

$$4(3) - 5 \stackrel{?}{=} 7$$
$$12 - 5 \stackrel{?}{=} 7$$
$$7 = 7 \quad \text{(True)}$$

**(b)** Solve

$$3x + 8 = -4$$
$$3x + 8 - 8 = -4 - 8$$ Subtract 8 from both sides.
$$3x = -12$$

> CAUTION

Use the addition property before applying the multiplication property. That is, do not divide by 4 until after you have added 5!

**NOTE**

Isolate the variable term.

NOTE

Isolate the variable.

Now divide both sides by 3 to isolate $x$ on the left.

$$\frac{3x}{3} = \frac{-12}{3}$$

$$x = -4$$

The solution is $-4$. We leave it to you to check this result.

**Check Yourself 1**

Solve and check.

(a) $6x + 9 = -15$ (b) $5x - 8 = 7$

The variable may appear in any position in an equation. Just apply the rules carefully as you try to write an equivalent equation, and you will find the solution. Example 2 illustrates this property.

## Example 2 Solving Equations

Solve

$$3 - 2x = 9$$

$$3 - 3 - 2x = 9 - 3$$ First subtract 3 from both sides.

$$-2x = 6$$

NOTE

$\frac{-2}{-2} = 1$, so we divide by $-2$ to isolate $x$.

Now divide both sides by $-2$. This leaves $x$ alone on the left.

$$\frac{-2x}{-2} = \frac{6}{-2}$$

$$x = -3$$

The solution is $-3$. We leave it to you to check this result.

**Check Yourself 2**

Solve and check.

$10 - 3x = 1$

You may also have to combine multiplication with addition or subtraction to solve an equation. Consider Example 3.

## Example 3 Solving Equations

**(a)** Solve

$$\frac{x}{5} - 3 = 4$$

To get the $x$ term alone, we first add 3 to both sides.

$$\frac{x}{5} - 3 + 3 = 4 + 3$$

$$\frac{x}{5} = 7$$

To undo the division, multiply both sides of the equation by 5.

$$5\left(\frac{x}{5}\right) = 5 \cdot 7$$

$$x = 35$$

The solution is 35. Just return to the original equation to check the result.

$$\frac{(35)}{5} - 3 \stackrel{?}{=} 4$$

$$7 - 3 \stackrel{?}{=} 4$$

$$4 = 4 \quad \text{(True)}$$

**(b)** Solve

$$\frac{2}{3}x + 5 = 13$$

$$\frac{2}{3}x + 5 - 5 = 13 - 5 \quad \text{First subtract 5 from both sides.}$$

$$\frac{2}{3}x = 8$$

Now multiply both sides by $\frac{3}{2}$, the reciprocal of $\frac{2}{3}$.

$$\left(\frac{3}{2}\right)\left(\frac{2}{3}x\right) = \left(\frac{3}{2}\right)8$$

or

$$x = 12$$

The solution is 12. We leave it to you to check this result.

### Check Yourself 3

Solve and check.

(a) $\frac{x}{6} + 5 = 3$ (b) $\frac{3}{4}x - 8 = 10$

In Section 11.4, you learned how to solve certain equations when the variable appeared on both sides. Example 4 shows you how to extend that work by using the multiplication property of equality.

### Example 4 Combining Properties to Solve an Equation

Solve

$$6x - 4 = 3x - 2$$

We begin by bringing all the variable terms to one side. To do this, we subtract $3x$ from both sides. This eliminates the variable term from the right side.

$$6x - 4 = 3x - 2$$
$$6x - 4 - 3x = 3x - 2 - 3x$$
$$3x - 4 = -2$$

We now isolate the variable term by adding 4 to both sides.

$$3x - 4 = -2$$
$$3x - 4 + 4 = -2 + 4$$
$$3x = 2$$

Finally, divide by 3.

$$\frac{3x}{3} = \frac{2}{3}$$
$$x = \frac{2}{3}$$

Check:

$$6\left(\frac{2}{3}\right) - 4 \stackrel{?}{=} 3\left(\frac{2}{3}\right) - 2$$
$$4 - 4 \stackrel{?}{=} 2 - 2$$
$$0 = 0 \qquad \text{(True)}$$

The basic idea is to use our two properties to form an equivalent equation with the $x$ isolated. Here we subtracted $3x$ and then added 4. You can do these steps in either order. Try it for yourself the other way. In either case, the multiplication property is then used as the *last step* in finding the solution.

**Check Yourself 4**

Solve and check.

$7x - 5 = 3x + 5$

Here are two approaches to solving equations in which the coefficient on the right side is greater than the coefficient on the left side.

## Example 5 Combining Properties to Solve an Equation (Two Methods)

Solve $4x - 8 = 7x + 7$.

**Method 1**

$4x - 8 - 7x = 7x + 7 - 7x$ — Bring the variable terms to the same (left) side.

$-3x - 8 = 7$

$-3x - 8 + 8 = 7 + 8$ — Isolate the variable term.

$-3x = 15$

$\frac{-3x}{-3} = \frac{15}{-3}$ — Isolate the variable.

$x = -5$

We let you check this result.

To avoid a negative coefficient (−3, in this example), some students prefer a different approach.

This time we work toward having the number on the *left* and the $x$ term on the *right,* or $\square = x$.

**Method 2**

$$4x - 8 = 7x + 7$$

$$4x - 8 - 4x = 7x + 7 - 4x \quad \text{Bring the variable terms to the same (right) side.}$$

$$-8 = 3x + 7$$

$$-8 - 7 = 3x + 7 - 7 \quad \text{Isolate the variable term.}$$

$$-15 = 3x$$

$$\frac{-15}{3} = \frac{3x}{3} \quad \text{Isolate the variable.}$$

$$-5 = x$$

**NOTE**

It is usually easier to isolate the variable term on the side that results in a positive coefficient.

Because $-5 = x$ and $x = -5$ are equivalent equations, it really makes no difference; the solution is still −5! You may use whichever approach you prefer.

**Check Yourself 5**

Solve $5x + 3 = 9x - 21$ by finding equivalent equations of the form $x = \square$ and $\square = x$ to compare the two methods of finding the solution.

When possible, we start by combining like terms on each side of the equation.

## Example 6 Combining Terms to Solve an Equation

< Objective 2 >

Solve.

$$7x - 3 + 5x + 4 = 6x + 25 \quad \text{Start by combining like terms.}$$

$$12x + 1 = 6x + 25$$

$$12x + 1 - 6x = 6x + 25 - 6x \quad \text{Bring the variables to one side.}$$

$$6x + 1 = 25$$

$$6x + 1 - 1 = 25 - 1 \quad \text{Isolate the variable term.}$$

$$6x = 24$$

$$\frac{6x}{6} = \frac{24}{6} \quad \text{Isolate the variable.}$$

$$x = 4$$

The solution is 4. We leave the check to you.

**Check Yourself 6**

Solve and check.

$$9x - 6 - 3x + 1 = 2x + 15$$

It may also be necessary to remove grouping symbols to solve an equation. Example 7 illustrates this property.

## Example 7 Solving Equations That Contain Parentheses

Solve and check.

$5(x - 3) - 2x = x + 7$ Apply the distributive property.

$5x - 15 - 2x = x + 7$ Combine like terms.

$3x - 15 = x + 7$

**NOTE**

$5(x - 3)$
$= 5[x + (-3)]$
$= 5x + 5(-3)$
$= 5x + (-15)$
$= 5x - 15$

We now have an equation that we can solve by the usual methods. First, bring the variable terms to one side, then isolate the variable term, and finally, isolate the variable.

$3x - 15 - x = x + 7 - x$ Subtract $x$ to bring the variable terms to the same side.

$2x - 15 = 7$

$2x - 15 + 15 = 7 + 15$ Add 15 to isolate the variable term.

$2x = 22$

$\frac{2x}{2} = \frac{22}{2}$ Divide by 2 to isolate the variable.

$x = 11$

The solution is 11. To check, substitute 11 for $x$ in the original equation. Again note the use of our rules for the order of operations.

$5[(11) - 3] - 2(11) \stackrel{?}{=} (11) + 7$ Simplify terms in parentheses.

$5 \cdot 8 - 2 \cdot 11 \stackrel{?}{=} 11 + 7$ Multiply.

$40 - 22 \stackrel{?}{=} 11 + 7$ Add and subtract.

$18 = 18$ A true statement.

### Check Yourself 7

Solve and check.

$7(x + 5) - 3x = x - 7$

We say that an equation is "solved" when we have an equivalent equation of the form

$x = \square$ or $\square = x$ in which $\square$ is some number

The steps of solving a linear equation are as follows:

**Step by Step**

**To Solve a Linear Equation**

**Step 1** Use the distributive property to remove any grouping symbols.
**Step 2** Combine like terms on each side of the equation.
**Step 3** Add or subtract variable terms to bring the variable term to one side of the equation.
**Step 4** Add or subtract numbers to isolate the variable term.
**Step 5** Multiply by the reciprocal of the coefficient to isolate the variable.

There are a host of applications involving linear equations.

## Example 8 Applying Algebra

In an election, the winning candidate had 160 more votes than the loser did. If the total number of votes cast was 3,260, how many votes did each candidate receive?

We first set up the problem. Let $x$ represent the number of votes received by the loser. Then the winner received $x + 160$ votes.

We can set up an equation by adding the number of votes the candidates received. This must total 3,260.

$$x + (x + 160) = 3{,}260$$ Combine like terms.

$$2x + 160 = 3{,}260$$ Subtract 160 from both sides.

$$2x = 3{,}100$$ Divide both sides by 2.

$$x = 1{,}550$$

The loser received 1,550 votes. Therefore, the winner received $x + 160 = 1{,}550 + 160 = 1{,}710$ votes.

### Check Yourself 8

The Randolphs used 12 more gallons (gal) of fuel oil in October than in September and twice as much oil in November as in September. If they used 132 gal for the 3 months, how much was used each month?

### Check Yourself ANSWERS

**1.** **(a)** $x = -4$; **(b)** $x = 3$ **2.** $x = 3$ **3.** **(a)** $x = -12$; **(b)** $x = 24$ **4.** $x = \frac{5}{2}$ **5.** $x = 6$ **6.** $x = 5$ **7.** $x = -14$ **8.** 30 gal in September, 42 gal in October, 60 gal in November

## Reading Your Text

The following fill-in-the-blank exercises are designed to ensure that you understand some of the key vocabulary used in this section.

**SECTION 11.6**

**(a)** The first goal for solving an equation is to ______________ the variable term on one side of the equation.

**(b)** Apply the ______________ property before applying the multiplication property.

**(c)** Always return to the ______________ equation to check your result.

**(d)** An equation in the form $x = \square$ or $\square = x$ has been ______________.

# 11.6 exercises

Basic Skills | Advanced Skills | Vocational-Technical Applications | Calculator/Computer | Above and Beyond

**MathZone**

Boost your grade at mathzone.com!

> Practice Problems
> NetTutor
> Self-Tests
> e-Professors
> Videos

Name ____________

Section ________ Date ________

**Answers**

1. ______ 2. ______
3. ______ 4. ______
5. ______ 6. ______
7. ______ 8. ______
9. ______ 10. ______
11. ______ 12. ______
13. ______ 14. ______
15. ______ 16. ______
17. ______ 18. ______
19. ______ 20. ______
21. ______ 22. ______
23. ______ 24. ______
25. ______ 26. ______
27. ______ 28. ______
29. ______ 30. ______
31. ______ 32. ______
33. ______ 34. ______

< Objective 1 >

*Solve for x and check your result.*

**1.** $2x + 1 = 9$

**2.** $3x - 1 = 17$

**3.** $3x - 2 = 7$

**4.** $5x + 3 = 23$

**5.** $4x + 7 = 35$

**6.** $7x - 8 = 13$

**7.** $2x + 9 = 5$

**8.** $6x + 25 = -5$

**9.** $4 - 7x = 18$

**10.** $8 - 5x = -7$

**11.** $3 - 4x = -9$

**12.** $5 - 4x = 25$

**13.** $\frac{x}{2} + 1 = 5$

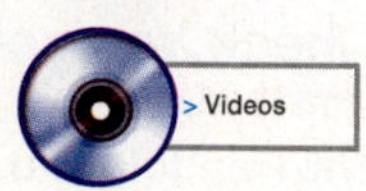


**14.** $\frac{x}{3} - 2 = 3$

**15.** $\frac{x}{4} - 5 = 3$

**16.** $\frac{x}{5} + 3 = 8$

**17.** $\frac{2}{3}x + 5 = 17$

**18.** $\frac{3}{4}x - 5 = 4$

**19.** $\frac{4}{5}x - 3 = 13$

**20.** $\frac{5}{7}x + 4 = 14$

**21.** $5x = 2x + 9$

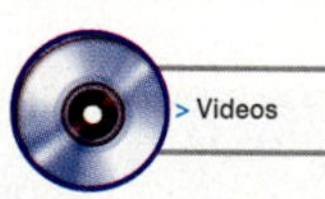


**22.** $7x = 18 - 2x$

**23.** $3x = 10 - 2x$

**24.** $11x = 7x + 20$

**25.** $9x + 2 = 3x + 38$

**26.** $8x - 3 = 4x + 17$

**27.** $4x - 8 = x - 14$

**28.** $6x - 5 = 3x - 29$

**29.** $5x + 7 = 2x - 3$

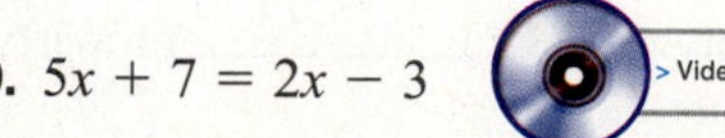


**30.** $9x + 7 = 5x - 3$

**31.** $7x - 3 = 9x + 5$

**32.** $5x - 2 = 8x - 11$

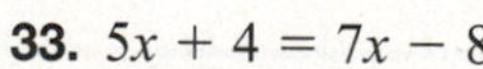

**33.** $5x + 4 = 7x - 8$

**34.** $2x + 23 = 6x - 5$

< Objective 2 >

**35.** $2x - 3 + 5x = 7 + 4x + 2$

**36.** $8x - 7 - 2x = 2 + 4x - 5$

**37.** $6x + 7 - 4x = 8 + 7x - 26$

**38.** $7x - 2 - 3x = 5 + 8x + 13$

**39.** $9x - 2 + 7x + 13 = 10x - 13$

**40.** $5x + 3 + 6x - 11 = 8x + 25$

**41.** $8x - 7 + 5x - 10 = 10x - 12$

**42.** $10x - 9 + 2x - 3 = 8x - 18$

**43.** **Social Science** There were 55 more yes votes than no votes on an election measure. If 735 votes were cast in all, how many yes votes were there?

**44.** **Business and Finance** Juan worked twice as many hours as Jerry. Marcia worked 3 more hours than Jerry. If they worked a total of 31 hours, how many hours did each employee work?

**45.** **Business and Finance** Francine earns $120 per month more than Rob. If they earn a total of $2,680 per month, how much does Francine earn each month?

**46.** **Science and Medicine** To determine the upper limit for a person's heart rate during aerobic training, subtract the person's age from 220, and then multiply the result by $\frac{9}{10}$. Determine the age of a person if the person's upper limit heart rate is 153.

Basic Skills | **Advanced Skills** | Vocational-Technical Applications | Calculator/Computer | Above and Beyond

*Solve each of the following equations.*

**47.** $7(2x - 1) - 5x = x + 25$

**48.** $9(3x + 2) - 10x = 12x - 7$

**49.** $3x + 2(4x - 3) = 6x - 9$

**50.** $7x + 3(2x + 5) = 10x + 17$

**51.** $\frac{8}{3}x - 3 = \frac{2}{3}x + 15$

**52.** $\frac{12}{5}x + 7 = 31 - \frac{3}{5}x$

**53.** $\frac{2x}{5} - 5 = \frac{12x}{5} + 8$

**54.** $\frac{3x}{7} - 5 = \frac{24x}{7} + 7$

**55.** $5.3x - 7 = 2.3x + 5$

**56.** $9.8x + 2 = 3.8x + 20$

Basic Skills | Advanced Skills | **Vocational-Technical Applications** | Calculator/Computer | Above and Beyond

**57.** **Agricultural Technology** The estimated yield $Y$ of a field of corn (in bushels per acre) can be found by multiplying the rainfall $r$, in inches, during the growing season by 16 and then subtracting 15. This relationship can be modeled by the formula

$Y = 16r - 15$

If a farmer wants a yield of 159 bushels per acre, then we can write the equation shown to determine the amount of rainfall required.

$159 = 16r - 15$

## Answers

35. ______ 36. ______

37. ______ 38. ______

39. ______ 40. ______

41. ______

42. ______

43. ______

44. ______

45. ______

46. ______

47. ______

48. ______

49. ______

50. ______

51. ______

52. ______

53. ______

54. ______

55. ______

56. ______

57. ______

Answers

58. ______

59. ______

60. ______

61. ______

62. ______

63. ______

64. ______

How much rainfall is necessary to achieve a yield of 159 bushels of corn per acre?

**58. Construction Technology** The number of studs $s$ required to build a wall (with studs spaced 16 inches on center) is equal to the one more than $\frac{3}{4}$ times the length of the wall $w$, in feet. We model this with the formula

$$s = \frac{3}{4}w + 1$$

If a contractor uses 22 studs to build a wall, how long is the wall?

**59. Allied Health** The internal diameter $D$ [in millimeters (mm)] of an endotracheal tube for a child is calculated using the formula

$$D = \frac{t + 16}{4}$$

in which $t$ is the child's age (in years).

How old is a child who requires an endotracheal tube with an internal diameter of 7 mm?

**60. Mechanical Engineering** The number of BTUs required to heat a house is $2\frac{3}{4}$ times the volume of the air in the house (in cubic feet). What is the maximum air volume that can be heated with a 90,000-BTU furnace?

Basic Skills | Advanced Skills | Vocational-Technical Applications | Calculator/Computer | **Above and Beyond**

**61.** Create an equation of the form $ax + b = c$ that has 2 as a solution.

**62.** Create an equation of the form $ax + b = c$ that has $-6$ as a solution.

**63.** The equation $3x = 3x + 5$ has no solution, whereas the equation $7x + 8 = 8$ has zero as a solution. Explain the difference between an equation that has zero as a solution and an equation that has no solution.

**64.** Construct an equation for which every real number is a solution.

## Answers

**1.** $x = 4$ **3.** $x = 3$ **5.** $x = 7$ **7.** $x = -2$ **9.** $x = -2$ **11.** $x = 3$ **13.** $x = 8$ **15.** $x = 32$ **17.** $x = 18$ **19.** $x = 20$ **21.** $x = 3$ **23.** $x = 2$ **25.** $x = 6$ **27.** $x = -2$ **29.** $x = -\frac{10}{3}$ **31.** $x = -4$ **33.** $x = 6$ **35.** $x = 4$ **37.** $x = 5$ **39.** $x = -4$ **41.** $x = \frac{5}{3}$ **43.** 395 votes **45.** \$1,400 **47.** $x = 4$ **49.** $x = -\frac{3}{5}$ **51.** $x = 9$ **53.** $x = -\frac{13}{2}$ **55.** $x = 4$ **57.** $10\frac{7}{8}$ in. **59.** 12 yr **61.** Above and Beyond **63.** Above and Beyond

| Definition/Procedure | Example | Reference |
|---|---|---|
| **From Arithmetic to Algebra** | | Section 11.1 |
| **Addition** $x + y$ means the **sum** of $x$ **and** $y$, or $x$ **plus** $y$. Some other words indicating addition are *more than* and *increased by.* | The sum of $x$ and 5 is $x + 5$.<br>7 more than $a$ is $a + 7$.<br>$b$ increased by 3 is $b + 3$. | *pp.* 787–790 |
| **Subtraction** $x - y$ means the **difference** of $x$ **and** $y$, or $x$ **minus** $y$. Some other words indicating subtraction are *less than* and *decreased by.* | The difference of $x$ and 3 is $x - 3$.<br>5 less than $p$ is $p - 5$.<br>$a$ decreased by 4 is $a - 4$. | |
| **Multiplication**<br>$\left.\begin{matrix} x \cdot y \\ (x)(y) \\ xy \end{matrix}\right\}$ These all mean the **product** of $x$ **and** $y$, or $x$ **times** $y$. | The product of $m$ and $n$ is $mn$.<br>The product of 2 and the sum of $a$ and $b$ is $2(a + b)$. | |
| **Division** $\frac{x}{y}$ means $x$ **divided by** $y$, or the **quotient** when $x$ is divided by $y$. | $n$ divided by 5 is $\frac{n}{5}$. The sum of $a$ and $b$, divided by 3, is $\frac{a + b}{3}$. | |
| **Evaluating Algebraic Expressions** | | Section 11.2 |
| **Step 1** Replace each variable with the given number value.<br>**Step 2** Do the necessary arithmetic operations, following the rules for the order of operations. | Evaluate<br>$\frac{4a - b}{2c}$<br>if $a = -6$, $b = 8$, and $c = -4$.<br>$\frac{4a - b}{2c} = \frac{4(-6) - (8)}{2(-4)} = \frac{-24 - 8}{-8} = \frac{-32}{-8} = 4$ | *p.* 796 |
| **Adding and Subtracting Algebraic Expressions** | | Section 11.3 |
| **Term** A number, or the product of a number and one or more variables, raised to a power. | $4a^2$ and $3a^2$ are like terms. | *pp.* 807–808 |
| **Like terms** Terms that contain exactly the same variables raised to the same powers. | $5x^2$ and $2xy^2$ are not like terms. | |
| ***Combining Like Terms*** | | |
| **Step 1** Add or subtract the numerical coefficients.<br>**Step 2** Attach the common variables. | $5a + 3a = 8a$<br>$7xy - 3xy = 4xy$ | *p.* 809 |

*Continued*

| Definition/Procedure | Example | Reference |
|---|---|---|
| **Using the Addition Property to Solve an Equation** | | **Section 11.4** |
| **Equation** A statement that two expressions are *equal*.<br>**Solution** Any value for the variable that makes an equation a true statement. | $3x - 5 = 7$ is an equation.<br>4 is a solution to the equation because<br>$3(4) - 5 \stackrel{?}{=} 7$<br>$12 - 5 \stackrel{?}{=} 7$<br>$7 = 7$ (True) | p. 819 |
| **Equivalent equations** Equations that have exactly the same set of solutions. | $3x - 5 = 7$ and $x = 4$ are equivalent equations. | p. 821 |
| **The addition property** If $a = b$, then $a + c = b + c$. Adding (or subtracting) the same quantity to both sides of an equation yields an equivalent equation. | $x - 5 = 7$<br>$+5 \quad +5$<br>$x = 12$ | p. 822 |
| **Using the Multiplication Property to Solve an Equation** | | **Section 11.5** |
| **The multiplication property** If $a = b$ and $c \neq 0$, then $ac = bc$. Multiplying (or dividing) both sides of an equation by the same nonzero number yields an equivalent equation. | $5x = 20$<br>$\frac{5x}{5} = \frac{20}{5}$<br>$x = 4$ | p. 833 |
| *To solve a percent problem algebraically,* translate the problem into algebra (writing the rate as a decimal) and use the multiplication rule to solve. | 30% of what number is 45?<br>$0.3x = 45$<br>$\frac{0.3x}{0.3} = \frac{45}{0.3}$<br>$x = 150$ | p. 838 |
| **Combining the Properties to Solve Equations** | | **Section 11.6** |
| **Solving linear equations** We say that an equation is solved when we have an equivalent equation of the form $x = \square$ or $\square = x$ in which $\square$ is some number.<br>The steps for solving a linear equation follow.<br>Step 1 Use the distributive property to remove any grouping symbols.<br>Step 2 Combine like terms on each side of the equation.<br>Step 3 Add or subtract variable terms to bring the variable term to one side of the equation.<br>Step 4 Add or subtract numbers to isolate the variable term.<br>Step 5 Multiply by the reciprocal of the coefficient to isolate the variable. | Solve:<br>$3x - 6 + 4x = 3x + 14$<br>$7x - 6 = 3x + 14$<br>$7x - 6 - 3x = 3x + 14 - 3x$<br>$4x - 6 = 14$<br>$4x - 6 + 6 = 14 + 6$<br>$4x = 20$<br>$\frac{4x}{4} = \frac{20}{4}$<br>$x = 5$ | p. 850 |

# summary exercises :: chapter 11

This summary exercise set is provided to give you practice with each of the objectives of this chapter. Each exercise is keyed to the appropriate chapter section. When you are finished, you can check your answers to the odd-numbered exercises against those presented in the back of the text. If you have difficulty with any of these questions, go back and reread the examples from that section. The answers to the even-numbered exercises appear in the *Instructor's Solutions Manual*. Your instructor will give you guidelines on how to best use these exercises in your instructional setting.

**11.1** *Write, using symbols.*

**1.** 5 more than $y$

**2.** $c$ decreased by 10

**3.** The product of 8 and $a$

**4.** The quotient when $y$ is divided by 3

**5.** 5 times the product of $m$ and $n$

**6.** The product of $a$ and 5 less than $a$

**7.** 3 more than the product of 17 and $x$

**8.** The quotient when $a$ plus 2 is divided by $a$ minus 2

**11.2** *Evaluate the expressions if* $x = -3$, $y = 6$, $z = -4$, *and* $w = 2$.

**9.** $3x + w$

**10.** $5y - 4z$

**11.** $x + y - 3z$

**12.** $5z^2$

**13.** $3x^2 - 2w^2$

**14.** $3x^3$

**15.** $5(x^2 - w^2)$

**16.** $\dfrac{6z}{2w}$

**17.** $\dfrac{2x - 4z}{y + (-z)}$

**18.** $\dfrac{3x - y}{w - x}$

**19.** $\dfrac{x(y^2 - z^2)}{(y + z)(y - z)}$

**20.** $\dfrac{y(x - w)^2}{x^2 - 2xw + w^2}$

**11.3** *List the terms of the expressions.*

**21.** $4a^3 - 3a^2$

**22.** $5x^2 - 7x + 3$

*Circle like terms.*

**23.** $5m^2, -3m, -4m^2, 5m^3, m^2$

**24.** $4ab^2, 3b^2, -5a, ab^2, 7a^2, -3ab^2, 4a^2b$

*Combine like terms.*

**25.** $5c + 7c$

**26.** $2x + 5x$

**27.** $4a - 2a$

**28.** $6c - 3c$

**29.** $9xy - 6xy$

**30.** $5ab^2 + 2ab^2$

**31.** $7a + 3b + 12a - 2b$

**32.** $6x - 2x + 5y - 3x$

**33.** $5x^3 + 17x^2 - 2x^3 - 8x^2$

**34.** $3a^3 + 5a^2 + 4a - 2a^3 - 3a^2 - a$

**35.** Subtract $4a^3$ from the sum of $2a^3$ and $12a^3$.

**36.** Subtract the sum of $3x^2$ and $5x^2$ from $15x^2$.

**11.4** *Tell whether the number shown in parentheses is a solution for the given equation.*

**37.** $7x + 2 = 16$ (2)

**38.** $5x - 8 = 3x + 2$ (4)

**39.** $7x - 2 = 2x + 8$ (2)

**40.** $4x + 3 = 2x - 11$ (−7)

**41.** $x + 5 + 3x = 2 + x + 23$ (6)

**42.** $\frac{2}{3}x - 2 = 10$ (21)

*Solve the following equations and check your results.*

**43.** $x + 5 = 7$

**44.** $x - 9 = 3$

**45.** $5x = 4x - 5$

**46.** $3x - 9 = 2x$

**47.** $5x - 3 = 4x + 2$

**48.** $9x + 2 = 8x - 7$

**49.** $7x - 5 = 6x + (-4)$

**50.** $3 + 4x - 1 = x - 7 + 2x$

**51.** $4(2x + 3) = 7x + 5$

**52.** $5(5x - 3) = 6(4x + 1)$

**11.5–11.6** *Solve the following equations and check your results.*

**53.** $5x = 35$

**54.** $7x = -28$

**55.** $-6x = 24$

**56.** $-9x = -63$

**57.** $\frac{x}{4} = 8$

**58.** $-\frac{x}{5} = -3$

**59.** $\frac{2}{3}x = 18$

**60.** $\frac{3}{4}x = 24$

**61.** $5x - 3 = 12$

**62.** $4x + 3 = -13$

**63.** $7x + 8 = 3x$

**64.** $3 - 5x = -17$

**65.** $3x - 7 = x$

**66.** $2 - 4x = 5$

**67.** $\frac{x}{3} - 5 = 1$

**68.** $\frac{3}{4}x - 2 = 7$

**69.** $6x - 5 = 3x + 13$

**70.** $3x + 7 = x - 9$

**71.** $7x + 4 = 2x + 6$

**72.** $9x - 8 = 7x - 3$

73. $2x + 7 = 4x - 5$

74. $3x - 15 = 7x - 10$

75. $\frac{10}{3}x - 5 = \frac{4}{3}x + 7$

76. $\frac{11}{4}x - 15 = 5 - \frac{5}{4}x$

77. $3.7x + 8 = 1.7x + 16$

78. $5.4x - 3 = 8.4x + 9$

79. $3x - 2 + 5x = 7 + 2x + 21$

80. $8x + 3 - 2x + 5 = 3 - 4x$

81. $5(3x - 1) - 6x = 3x - 2$

82. $5x + 2(3x - 4) = 14x + 7$

# self-test 11

Name ____________

Section ________ Date ________

The purpose of this chapter test is to help you check your progress so that you can find sections and concepts that you need to review before the next exam. Allow yourself about an hour to take this test. At the end of that hour, check your answers against those given in the back of this text. If you missed any, note the section reference that accompanies the answer. Go back to that section and reread the examples until you have mastered that particular concept.

*Write, using symbols.*

1. 5 less than $a$

2. The product of 6 and $m$

3. 4 times the sum of $m$ and $n$

4. The quotient when the sum of $a$ and $b$ is divided by 3

*Evaluate the expressions if $a = -2$, $b = 6$, and $c = -4$.*

5. $4a - c$

6. $5c^2$

7. $6(2b - 3c)$

8. $\dfrac{3a - 4b}{a + c}$

*Combine like terms.*

9. $8a + 7a$

10. $8x^2y + (-5x^2y)$

11. $10x + 8y + 9x - 3y$

12. Subtract $9a^2$ from the sum of $12a^2$ and $5a^2$.

*Tell whether the number shown in parentheses is a solution for the given equation.*

13. $7x - 3 = 25$ (5)

14. $8x - 3 = 5x + 9$ (4)

*Solve the following equations and check your results.*

15. $x - 7 = 4$

16. $7x - 12 = 6x$

17. $9x - 2 = 8x + 5$

## Answers

1. ____________
2. ____________
3. ____________
4. ____________
5. ____________
6. ____________
7. ____________
8. ____________
9. ____________
10. ____________
11. ____________
12. ____________
13. ____________
14. ____________
15. ____________
16. ____________
17. ____________

## Answers

18. ______

19. ______

20. ______

21. ______

22. ______

23. ______

24. ______

*Solve the following equations and check your results.*

**18.** $7x = 49$

**19.** $\frac{1}{4}x = -3$

**20.** $\frac{4}{5}x = 20$

*Solve the following equations and check your results.*

**21.** $7x - 5 = 16$

**22.** $10 - 3x = -2$

**23.** $7x - 3 = 4x - 5$

**24.** $2x - 7 = 5x + 8$

# cumulative review chapters 1-11

Name ______

Section ______ Date ______

The following exercises are presented to help you review concepts from earlier chapters. This is meant as review material and not as a comprehensive exam. The answers are presented in the back of the text. Beside each answer is a section reference for the concept. If you have difficulty with any of these exercises, be certain to at least read through the summary related to that section.

*Name the property that is illustrated.*

**1.** $(7 + 3) + 8 = 7 + (3 + 8)$

**2.** $6 \times 7 = 7 \times 6$

**3.** $5 \cdot (2 + 4) = 5 \cdot 2 + 5 \cdot 4$

*Round the numbers to the indicated place value.*

**4.** 5,873 to the nearest hundred

**5.** 953,150 to the nearest ten-thousand

**6.** Evaluate: $2 + 8 \times 3 \div 4$

**7.** Write the prime factorization of 264.

**8.** Find the least common multiple (LCM) of 6, 15, and 45.

**9.** Convert to a mixed number: $\frac{22}{7}$

**10.** Convert to an improper fraction: $6\frac{5}{8}$

*Perform the indicated operations.*

**11.** $\frac{2}{3} \times 1\frac{4}{5} \times \frac{5}{8}$

**12.** $2\frac{2}{7} \div 1\frac{11}{21}$

**13.** $4\frac{7}{8} + 3\frac{1}{6}$

**14.** $9 + \left(-5\frac{3}{8}\right)$

**Answers**

1. ______
2. ______
3. ______
4. ______
5. ______
6. ______
7. ______
8. ______
9. ______
10. ______
11. ______
12. ______
13. ______
14. ______

Answers

15. ____________

16. ____________

17. ____________

18. ____________

19. ____________

20. ____________

21. ____________

22. ____________

23. ____________

24. ____________

25. ____________

26. ____________

27. ____________

28. ____________

15. **Construction** A $6\frac{1}{2}$-in. bolt is placed through a wall that is $5\frac{7}{8}$ in. thick. How far does the bolt extend beyond the wall?

16. **Business and Finance** You pay for purchases of $13.99, $18.75, $9.20, and $5 with a $50 bill. How much cash will you have left?

17. **Geometry** Find the area of a circle whose diameter is 3.2 ft. Use 3.14 for $\pi$ and round the result to the nearest hundredth.

18. **Construction** A 14-acre piece of land is being developed into home lots. If 2.8 acres of land will be used for roads, and each home site is to be 0.35 acre, how many lots can be formed?

19. Write the decimal equivalent of $\frac{8}{11}$. Use bar notation.

20. Solve for the unknown: $\frac{5}{m} = \frac{0.4}{9}$

21. **Business and Finance** You are using a photocopy machine to reduce an advertisement that is 14 in. wide by 21 in. long. If the new width is to be 8 in., what will the new length be?

*In exercises 22 to 24, write as percents.*

22. 0.003

23. $\frac{5}{8}$

24. $3\frac{1}{2}$

25. 120% of what number is 180?

26. 72 is 12% of what number?

27. **Business and Finance** Luisa works on an 8% commission basis. If she wishes to earn $2,200 in commissions in 1 month, how much must she sell during that period?

28. Complete the statement: 300 mg = ________ g

Answers

29. ______
30. ______
31. ______
32. ______
33. ______
34. ______
35. ______
36. ______
37. ______
38. ______
39. ______
40. ______
41. ______
42. ______
43. ______

29. **Business and Finance** According to the line graph, what is the difference in benefits between 2000 and 2002?

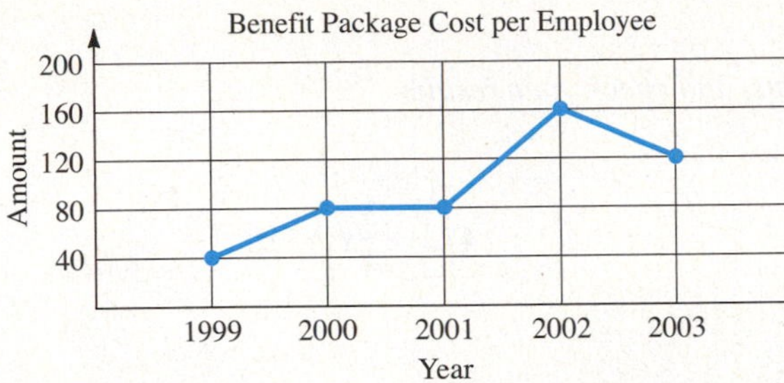


30. The opposite of 8 is ________.

31. The absolute value of $-20$ is ________.

*In exercises 32 to 37, evaluate.*

32. $-(-12)$

33. $|-5|$

34. $-12 + (-6)$

35. $-8 - (-4)$

36. $(-6)(15)$

37. $48 \div (-12)$

*Write, using symbols.*

38. 3 times the sum of $x$ and $y$

39. The quotient when 5 less than $n$ is divided by 3

*In exercises 40 to 43, evaluate the expressions if* $a = 5$, $b = -3$, $c = 4$, *and* $d = -2$.

40. $6ad$

41. $3b^2$

42. $3(c - 2d)$

43. $\dfrac{2a - 7d}{a - b}$

**Answers**

44. ____

45. ____

46. ____

47. ____

48. ____

49. ____

*Combine like terms.*

**44.** $10a^2 + 5a + 2a^2 - 2a$

**45.** $5x - 3y + 2x + y - 7x$

*Solve the following equations, and check your results.*

**46.** $9x - 5 = 8x$

**47.** $-\frac{3}{4}x = 18$

**48.** $2x + 3 = 7x + 5$

**49.** $\frac{4}{3}x - 6 = 4 - \frac{2}{3}x$

# Answers to Pretests, Reading Your Text, Summary Exercises, Self-Tests, and Cumulative Reviews

## Pretest Chapter 1

**1.** One hundred seven thousand, nine hundred forty-five
**2.** Associative property of addition **3.** 27,614
**4.** Commutative property of multiplication **5.** 19,992
**6.** 56,100 **7.** 462 **8.** 1,024 r 3 **9.** 81 r 131
**10.** (a) 26; (b) 56 **11.** 93 **12.** \$151 **13.** 6
**14.** $P = 14$ yd; $A = 10$ yd$^2$

## Reading Your Text

**Section 1.1** (a) decimal; (b) digit; (c) comma; (d) million
**Section 1.2** (a) natural; (b) addition; (c) order; (d) read
**Section 1.3** (a) difference; (b) less; (c) read; (d) borrowing
**Section 1.4** (a) rounding; (b) place value; (c) rounding; (d) less
**Section 1.5** (a) reasonable; (b) product; (c) powers; (d) rectangle
**Section 1.6** (a) quotient; (b) remainder; (c) zero; (d) subtraction
**Section 1.7** (a) first; (b) exponent; (c) inside; (d) one

## Summary Exercises for Chapter 1

**1.** Hundreds
**3.** Twenty-seven thousand, four hundred twenty-eight
**5.** 37,583 **7.** Commutative property of addition **9.** 1,416
**11.** 4,801 **13.** (a) 27; (b) 11; (c) 15 **15.** \$540
**17.** 18,800 **19.** 2,574 **21.** \$536 **23.** 7,000
**25.** 550,000 **27.** $>$ **29.** 22 in.
**31.** Distributive property of multiplication over addition
**33.** 1,856 **35.** 154,602 **37.** 24 ft$^2$ **39.** 30,960 **41.** 0
**43.** 308 r 5 **45.** 497 r 1 **47.** 28 mi/gal **49.** 21
**51.** 1,000 **53.** 4 **55.** 36 **57.** 33 **59.** 0

## Self-Test for Chapter 1

**[1.1] 1.** Hundred thousands
**2.** Three hundred two thousand, five hundred twenty-five
**3.** 2,430,000 **[1.2] 4.** Commutative property of addition
**5.** Associative property of addition **6.** 1,918 **7.** 13,103
**8.** 21,696 **9.** 55,978 **[1.3] 10.** 235 **11.** 12,220
**12.** 30,770 **13.** 40,555 **14.** 72 lb **15.** 7,700
**[1.4] 16.** $>$ **17.** $<$ **18.** 12 in.
**[1.5] 19.** Associative property of multiplication
**20.** Distributive property of multiplication over addition
**21.** 4,984 **22.** 55,414 **23.** \$308,750 **24.** 12 in.$^2$
**[1.6] 25.** 3,041 r 2 **26.** 76 r 7 **27.** 209 r 145
**28.** \$223 **[1.7] 29.** 15 **30.** 39

## Pretest Chapter 2

**1.** 1, 2, 3, 6, 7, 14, 21, 42
**2.** Prime: 2, 3, 7, 17, 23; composite: 6, 9, 18, 21
**3.** 2, 3, and 5 **4.** $2 \times 5 \times 5 \times 7$ **5.** 4 **6.** $9\frac{1}{4}$
**7.** $\frac{46}{7}$ **8.** Yes **9.** $\frac{3}{5}$ **10.** $\frac{5}{6}, \frac{3}{8}, \frac{20}{21}, \frac{5}{11}$
**11.** $\frac{8}{7}, \frac{13}{9}, \frac{15}{8}, \frac{9}{9}, \frac{16}{5}$ **12.** $2\frac{3}{5}, 7\frac{2}{9}, 3\frac{2}{7}$ **13.** $\frac{21}{5}$
**14.** $\frac{2}{3}$ **15.** $\frac{63}{5}$ **16.** $\frac{4}{5}$ **17.** $\frac{2}{3}$ **18.** $\frac{3}{4}$ **19.** 14 blocks

## Reading Your Text

**Section 2.1** (a) One (1); (b) prime; (c) composite; (d) even
**Section 2.2** (a) commutative; (b) prime; (c) common; (d) greatest
**Section 2.3** (a) denominator; (b) proper; (c) mixed; (d) numerator
**Section 2.4** (a) cross; (b) equivalent; (c) Fundamental; (d) common
**Section 2.5** (a) simplest; (b) simplify; (c) reasonableness; (d) reasonable
**Section 2.6** (a) reciprocal; (b) complex; (c) divisor; (d) units

## Summary Exercises for Chapter 2

**1.** 1, 2, 4, 13, 26, 52 **3.** Prime: 2, 5, 7, 11, 17, 23, 43; composite: 14, 21, 27, 39 **5.** None **7.** $2 \times 2 \times 3 \times 5 \times 7$
**9.** $2 \times 3 \times 3 \times 5 \times 5 \times 5$ **11.** 1 **13.** 13 **15.** 11
**17.** Numerator: 17; denominator: 23
**19.** Fraction: $\frac{5}{6}$; numerator: 5; denominator: 6 **21.** $6\frac{5}{6}$
**23.** $7\frac{2}{3}$ **25.** $\frac{61}{8}$ **27.** $\frac{37}{7}$ **29.** No
**31.** $\frac{2}{3}$ **33.** $\frac{7}{9}$ **35.** Yes **37.** $\frac{1}{9}$ **39.** $1\frac{1}{2}$ **41.** $9\frac{3}{5}$
**43.** 8 **45.** \$204 **47.** \$45 **49.** 408 mi **51.** $\frac{5}{8}$
**53.** $\frac{2}{3}$ **55.** $\frac{3}{16}$ **57.** $\frac{3}{7}$ **59.** $\frac{3}{4}$ ft **61.** 56 mi/h
**63.** 48 lots

## Self-Test for Chapter 2

**[2.1] 1.** Prime: 5, 13, 17, 31; composite: 9, 22, 27, 45
**2.** 2 and 3 **3.** $2 \times 2 \times 2 \times 3 \times 11$ **[2.2] 4.** 12
**5.** 8 **[2.3] 6.** $\frac{5}{6}$; numerator: 5; denominator: 6
**7.** $\frac{5}{8}$; numerator: 5; denominator: 8
**8.** $\frac{3}{5}$; numerator: 3; denominator: 5
**9.** Proper: $\frac{10}{11}, \frac{1}{8}$; improper: $\frac{9}{5}, \frac{7}{7}, \frac{8}{1}$; mixed number: $2\frac{3}{5}$
**10.** $4\frac{1}{4}$ **11.** $4\frac{1}{4}$ **12.** $9\frac{1}{4}$ **13.** 3 **14.** 15 **15.** $\frac{37}{7}$
**16.** $\frac{35}{8}$ **17.** $\frac{74}{9}$ **[2.4] 18.** Yes **19.** Yes **20.** No
**21.** $\frac{7}{9}$ **22.** $\frac{3}{7}$ **23.** $\frac{8}{23}$ **24.** Yes **25.** No **26.** Yes

**[2.5] 27.** $\frac{10}{21}$ **28.** $\frac{9}{16}$ **29.** $3\frac{3}{7}$ **30.** $9\frac{1}{5}$ **31.** $\frac{4}{15}$
**32.** 4 **33.** $1.32 **34.** 20 $\text{yd}^2$ **35.** 190 mi
**[2.6] 36.** $1\frac{1}{7}$ **37.** $\frac{5}{8}$ **38.** $1\frac{3}{11}$ **39.** $2\frac{2}{3}$
**40.** 47 homes **41.** 48 books

## Cumulative Review Chapters 1–2

**[1.1] 1.** Hundred thousands
**2.** Three hundred two thousand, five hundred twenty-five
**3.** 2,430,000 **[1.2] 4.** Commutative property of addition
**5.** Additive identity **6.** Associative property of addition
**7.** 966 **8.** 23,351 **[1.4] 9.** 5,900 **10.** 950,000
**11.** 7,700 **12.** $>$ **13.** $<$ **[1.3] 14.** 3,861 **15.** 17,465
**[1.2] 16.** 905 **17.** $17,579
**[1.5] 18.** Associative property of multiplication
**19.** Commutative property of multiplication
**20.** Distributive property **21.** 378,214 **22.** 686,000
**23.** $1,008 **[1.6] 24.** 67 r43 **25.** 103 r176
**[1.7] 26.** 38 **27.** 56 **28.** 36 **29.** 8 **30.** $58
**[2.1] 31.** Prime: 5, 13, 17, 31; composite: 9, 22, 27, 45
**32.** 2 and 3 **[2.2] 33.** $2 \times 2 \times 2 \times 3 \times 11$ **34.** 12 **35.** 8
**[2.3] 36.** Proper: $\frac{7}{12}, \frac{3}{7}$; improper: $\frac{10}{8}, \frac{9}{9}, \frac{7}{1}$; mixed numbers: $3\frac{1}{5}, 2\frac{2}{3}$
**[2.4] 37.** $2\frac{4}{5}$ **38.** 4 **39.** $\frac{13}{3}$ **40.** $\frac{63}{8}$ **41.** Yes
**42.** No **43.** $\frac{2}{3}$ **44.** $\frac{3}{8}$ **[2.5] 45.** $\frac{8}{27}$ **46.** $\frac{4}{15}$
**47.** $5\frac{2}{5}$ **48.** $22\frac{2}{3}$ **49.** $\frac{3}{4}$ **[2.6] 50.** $1\frac{1}{3}$ **51.** $4\frac{1}{2}$
**52.** $\frac{5}{6}$ **53.** $1\frac{1}{2}$ **54.** $540 **55.** 88 sheets

## Pretest Chapter 3

**1. (a)** $\frac{5}{7}$; **(b)** $\frac{5}{9}$ **2.** 48 **3. (a)** 200; **(b)** 120 **4.** $\frac{17}{30}$
**5.** $2\frac{1}{8}$ **6.** $\frac{5}{72}$ **7.** $6\frac{17}{24}$ **8.** $1\frac{1}{36}$ **9.** $\frac{13}{50}$ **10.** $\frac{19}{36}$
**11.** $18\frac{1}{4}$ $\text{yd}^2$ **12.** $4\frac{3}{8}$ points

## Reading Your Text

**Section 3.1** **(a)** like; **(b)** numerators; **(c)** simplify; **(d)** difference
**Section 3.2** **(a)** multiples; **(b)** smallest; **(c)** denominator; **(d)** greater
**Section 3.3** **(a)** LCM; **(b)** unlike; **(c)** equivalent; **(d)** numerator
**Section 3.4** **(a)** LCD; **(b)** regrouping; **(c)** like; **(d)** smaller
**Section 3.5** **(a)** grouping; **(b)** exponents; **(c)** invert; **(d)** improper
**Section 3.6** **(a)** dual; **(b)** estimation; **(c)** approximations; **(d)** estimation

## Summary Exercises for Chapter 3

**1.** $\frac{2}{3}$ **3.** $1\frac{2}{13}$ **5.** $1\frac{1}{3}$ **7.** $1\frac{2}{9}$ **9.** 12 **11.** 72 **13.** 60
**15.** 72 **17.** $\frac{7}{12}, \frac{5}{8}$ **19.** $>$ **21.** $<$ **23.** $\frac{36}{120}, \frac{75}{120}, \frac{70}{120}$
**25.** 36 **27.** 200 **29.** 132 **31.** 24 **33.** $\frac{19}{24}$ **35.** $\frac{7}{12}$
**37.** $1\frac{17}{90}$ **39.** $\frac{7}{8}$ **41.** $1\frac{13}{72}$ **43.** $\frac{5}{9}$ **45.** $\frac{1}{2}$ **47.** $\frac{5}{24}$
**49.** $\frac{7}{18}$ **51.** $\frac{11}{24}$ **53.** $\frac{13}{42}$ **55.** $\frac{1}{3}$ **57.** $\frac{1}{4}$ **59.** $\frac{7}{60}$
**61.** $\frac{35}{72}$ **63.** $9\frac{37}{60}$ **65.** $4\frac{1}{3}$ **67.** $2\frac{19}{24}$ **69.** $\frac{5}{12}$ cup
**71.** $19\frac{9}{16}$ in. **73.** $53\frac{9}{16}$ in. **75.** $3\frac{7}{12}$ gal **77.** $1\frac{3}{4}$ yd
**79.** $\frac{37}{96}$ **81.** $\frac{1}{48}$ **83.** $1\frac{1}{2}$ **85.** $2,500 **87.** 24

## Self-Test for Chapter 3

**[3.1] 1.** $\frac{9}{10}$ **2.** $\frac{2}{3}$ **3.** 72 **[3.2] 4.** 60 **5.** 36
**[3.3] 6.** $\frac{4}{5}$ **7.** $\frac{25}{42}$ **8.** $\frac{19}{24}$ **9.** $1\frac{11}{60}$ **10.** $1\frac{23}{40}$
**11.** $\frac{7}{12}$ **12.** $1\frac{5}{12}$ cups **13.** $\frac{1}{3}$ **14.** $\frac{1}{9}$ **15.** $\frac{23}{30}$
**16.** $\frac{1}{4}$ h **[3.4] 17.** $7\frac{7}{10}$ **18.** $10\frac{1}{4}$ **19.** $7\frac{11}{12}$ **20.** $12\frac{3}{40}$
**21.** $1\frac{3}{4}$ **22.** $1\frac{11}{18}$ **23.** $3\frac{23}{24}$ **24.** $1\frac{8}{15}$ **25.** $9\frac{1}{7}$
**26.** $13\frac{11}{20}$ **27.** $5\frac{3}{4}$ h **[3.5] 28.** $\frac{1}{12}$ **29.** $12\frac{11}{24}$
**[3.6] 30.** 39,000 cups

## Cumulative Review Chapters 1–3

**[1.1–1.2] 1.** 7,173 **2.** 1,918 **3.** 2,731 **4.** 13,103
**[1.3] 5.** 235 **6.** 12,220 **7.** 429 **8.** 3,239
**[1.5] 9.** 174 **10.** 1,911 **11.** 4,984 **12.** 55,414
**[1.6] 13.** 24 r191 **14.** 22 r21 **15.** 209 r145
**[1.7] 16.** 5 **17.** 7 **18.** 3 **19.** 16 **20.** 20 **21.** 3
**[2.3] 22.** Proper: $\frac{5}{7}, \frac{2}{5}$; improper: $\frac{15}{9}, \frac{8}{8}, \frac{11}{1}$; mixed numbers: $4\frac{5}{6}, 3\frac{5}{6}$
**23.** $1\frac{7}{9}$ **24.** $7\frac{1}{5}$ **25.** $\frac{23}{4}$ **26.** $\frac{55}{9}$ **[2.4] 27.** Yes
**28.** No **[2.5] 29.** $\frac{1}{9}$ **30.** $\frac{1}{6}$ **31.** $1\frac{1}{2}$ **32.** $2\frac{1}{8}$
**33.** $9\frac{3}{5}$ **34.** $11\frac{1}{3}$ **35.** 8 **[2.6] 36.** $\frac{2}{3}$ **37.** $\frac{5}{6}$
**38.** $\frac{3}{16}$ **[3.1] 39.** $\frac{4}{5}$ **40.** $\frac{61}{75}$ **41.** $1\frac{31}{40}$ **[3.3] 42.** $\frac{1}{2}$
**43.** $\frac{5}{36}$ **44.** $\frac{5}{9}$ **[3.4] 45.** $6\frac{2}{7}$ **46.** $8\frac{1}{24}$ **47.** $4\frac{5}{9}$
**48.** $4\frac{1}{24}$ **49.** $3\frac{5}{8}$ **50.** $3\frac{13}{24}$ **51.** $14\frac{19}{30}$ h **52.** $\frac{5}{8}$ in.
**53.** $1\frac{11}{12}$ h

## Pretest Chapter 4

**1.** Thousandths **2.** 2.371; two and three hundred seventy-one thousandths **3. (a)** 63.29; **(b)** 2.375 **4.** $2.36
**5. (a)** 0.86037; **(b)** 536.2 **6.** 2.36 **7.** $28.05
**8.** 25.12 yd **9.** 4.25 **10.** 2.435 **11.** $7.60
**12.** 0.0534 **13. (a)** 0.375; **(b)** 0.29 **14.** $\frac{5}{8}$

### Reading Your Text

**Section 4.1** **(a)** decimal; **(b)** places; **(c)** exact; **(d)** ten-thousandths
**Section 4.2** **(a)** divide; **(b)** bar; **(c)** terminating; **(d)** places
**Section 4.3** **(a)** decimal points; **(b)** value; **(c)** Perimeter; **(d)** following
**Section 4.4** **(a)** add; **(b)** product; **(c)** zeros; **(d)** right
**Section 4.5** **(a)** above; **(b)** past; **(c)** whole number; **(d)** left

### Summary Exercises for Chapter 4

**1.** Hundredths **3.** 0.37 **5.** Seventy-one thousandths **7.** 4.5
**9.** > **11.** < **13.** 5.84 **15.** 4.876 **17.** $21\frac{857}{1{,}000}$
**19.** 0.429 **21.** 3.75 **23.** $\frac{21}{250}$ **25.** 3.47
**27.** 37.728 **29.** 23.32 **31.** 1.075 **33.** 28.02 cm
**35.** 6.15 cm **37.** $18.93 **39.** 0.000261 **41.** 0.0012275
**43.** 450 **45.** $287.50 **47.** $5,742 **49.** 4.65 **51.** 2.664
**53.** 1.273 **55.** 0.76 **57.** 0.0457 **59.** 39.3 mi/gal
**61.** 29.8 mi/gal

### Self-Test for Chapter 4

**[4.1] 1.** Ten-thousandths **2.** 0.049
**3.** Two and fifty-three hundredths **4.** 12.017 **5.** < **6.** >
**[4.3] 7.** 16.64 **8.** 47.253 **9.** 12.803 **10.** 50.2 gal
**11.** 10.54 **12.** 24.375 **13.** 3.888 **14.** $3.06
**[4.4] 15.** 17.437 **16.** 0.02793 **17.** 1.4575 **18.** 7.525 $\text{in.}^2$
**19.** 735 **20.** 12,570 **21.** $543 **[4.1] 22.** 0.598
**23.** 23.57 **24.** 36,000 **[4.5] 25.** 0.465 **26.** 2.35
**27.** 0.051 **28.** 2.55 **29.** 2.385 **30.** 7.35 **31.** 0.067
**32.** 32 lots **33.** 0.004983 **34.** 0.00523 **35.** $573.40
**[4.2] 36.** 0.4375 **37.** 0.429 **38.** $0.\overline{63}$ **39.** $\frac{9}{125}$
**40.** $4\frac{11}{25}$ **41.** > **42.** $\frac{229}{500}$

### Cumulative Review Chapters 1–4

**[1.1] 1.** Two hundred eighty-six thousand, five hundred forty-three
**2.** Hundreds **[1.2] 3.** 34,594 **[1.3] 4.** 48,888
**[1.5] 5.** 5,063 **6.** 70,455 **[1.6] 7.** 17 **8.** 35 r11
**[1.7] 9.** 29 **[1.4] 10.** 4,000 **[1.8] 11.** *P*: 24 ft; *A*: 35 $\text{ft}^2$
**[2.4] 12.** $\frac{5}{17}$ **[2.5] 13.** $\frac{3}{4}$ **14.** $2\frac{6}{7}$ **[2.7] 15.** $\frac{9}{17}$
**[3.1] 16.** $\frac{5}{7}$ **[3.3] 17.** $\frac{7}{30}$ **[3.4] 18.** $3\frac{9}{10}$
**[4.3] 19.** 12.468 **[4.5] 20.** 3.9 **21.** 0.005238
**[4.4] 22.** 1.1385 **23.** 15,300 **[4.2] 24.** $\frac{43}{100}$
**25. (a)** 0.625; **(b)** 0.39 **[4.5] 26.** 0.429 **[4.4] 27.** 17.21
**[4.5] 28.** 39.829

### Pretest Chapter 5

**1.** $\frac{7}{10}$ **2.** $\frac{4}{3}$ **3.** $34.5\frac{\text{mi}}{\text{h}}$ **4.** 4 min **5.** Yes **6.** No
**7.** 10 **8.** 12 **9.** $6.30 **10.** 11 gal

### Reading Your Text

**Section 5.1** **(a)** fraction; **(b)** like; **(c)** simplest; **(d)** mixed
**Section 5.2** **(a)** like; **(b)** rate; **(c)** Mixed; **(d)** Unit
**Section 5.3** **(a)** equal; **(b)** variable; **(c)** proportion; **(d)** rates
**Section 5.4** **(a)** equation; **(b)** coefficient; **(c)** read; **(d)** proportional

### Summary Exercises for Chapter 5

**1.** $\frac{4}{17}$ **3.** $\frac{5}{8}$ **5.** $\frac{4}{9}$ **7.** $\frac{7}{36}$ **9.** $100\frac{\text{mi}}{\text{h}}$ **11.** $50\frac{\text{cal}}{\text{oz}}$
**13.** $200\frac{\text{ft}}{\text{s}}$ **15.** $6\frac{1}{2}\frac{\text{hits}}{\text{game}}$ **17.** Marisa **19.** $\frac{9¢}{\text{oz}}$
**21.** $\frac{9.5¢}{\text{oz}}$ **23.** $\frac{\$14.95}{\text{CD}}$ **25.** $\frac{4}{9} = \frac{20}{45}$
**27.** $\frac{110\text{ mi}}{2\text{ h}} = \frac{385\text{ mi}}{7\text{ h}}$ **29.** No **31.** Yes **33.** Yes
**35.** Yes **37.** $m = 2$ **39.** $t = 4$ **41.** $w = 180$
**43.** $x = 100$ **45.** $135 **47.** 15 in. **49.** 28 parts
**51.** 140 g

### Self-Test for Chapter 5

**[5.1] 1.** $\frac{7}{19}$ **2.** $\frac{5}{3}$ **3.** $\frac{2}{3}$ **4.** $\frac{1}{12}$ **5.** $\frac{26}{33}; \frac{26}{7}$
**[5.2] 6.** $4.8\frac{\text{mi}}{\text{gal}}$ **7.** $8.25\frac{\text{dollars}}{\text{h}}$ **8.** $\frac{\$2.56}{\text{gal}}$ **9.** Yes
**10.** No **11.** Yes **12.** No **[5.4] 13.** $x = 20$
**14.** $a = \frac{216}{13}$ **15.** $p = 3$ **16.** $m = 16$ **17.** $2.28
**18.** 576 mi **19.** 600 mufflers **20.** 24 tsp

### Cumulative Review Chapters 1–5

**[1.1] 1.** Forty-five thousand, seven hundred eighty-nine
**2.** Ten thousands **[1.2] 3.** 26,304 **[1.3] 4.** 47,806
**[1.5] 5.** 4,408 **[1.6] 6.** 78 r67 **[1.3] 7.** $568
**[1.7] 8.** 3 **[1.8] 9.** *P*: 16 ft; *A*: 12 $\text{ft}^2$ **10.** $1,104
**[2.2] 11.** $2 \times 2 \times 3 \times 7 \times 11$ **12.** 14 **[2.4] 13.** $\frac{1}{4}$
**[2.5] 14.** $1\frac{1}{5}$ **15.** 10 **[2.7] 16.** $4\frac{1}{2}$ **17.** $1\frac{9}{13}$
**[3.3] 18.** $\frac{25}{44}$ **[3.4] 19.** $7\frac{7}{12}$ **20.** $4\frac{1}{2}$ **[3.2] 21.** 180
**[2.6] 22.** 176 mi **[2.7] 23.** $48\frac{\text{mi}}{\text{h}}$ **[4.2] 24.** 7.828
**[4.5] 25.** 1.23 **[4.3] 26.** 6.6015 **[4.7] 27.** $\frac{9}{25}$
**[4.6] 28.** 0.32 **[4.2] 29.** 14.06 m **[4.4] 30.** 50.24 $\text{cm}^2$
**[5.1] 31.** $\frac{6}{13}$ **32.** $\frac{10}{3}$ **33.** $\frac{4}{5}$ **[5.3] 34.** Yes **35.** No
**[5.4] 36.** $x = 2$ **37.** $x = 15$ **[5.2] 38.** $\frac{24.4¢}{\text{oz}}$
**[5.4] 39.** 600 km **40.** 50

### Pretest Chapter 6

**1.** $\frac{7}{100}$ **2.** 0.23 **3.** 3.5% or $3\frac{1}{2}\%$ **4.** 80%
**5.** Rate: 32%; base: 240; amount: 76.8

**6.** Rate: unknown; base: 580; amount: 18 **7.** 63 **8.** 9%
**9.** \$560 **10.** \$1,200

## Reading Your Text

**Section 6.1** **(a)** hundred; **(b)** fraction; **(c)** left; **(d)** greater
**Section 6.2** **(a)** right; **(b)** decimal; **(c)** percent; **(d)** zeros
**Section 6.3** **(a)** base; **(b)** rate; **(c)** rate; **(d)** wholesale
**Section 6.4** **(a)** amount; **(b)** rate; **(c)** less; **(d)** principal

## Summary Exercises for Chapter 6

**1.** 75% **3.** $\frac{1}{5}$ **5.** $1\frac{1}{2}$ **7.** 3 **9.** 0.04 **11.** 0.135
**13.** 2.25 **15.** 37.5% **17.** 700% **19.** 0.5% **21.** 70%
**23.** 125% **25.** 27.3% **27.** 140% **29.** 12.5% **31.** 75
**33.** 75 **35.** 175% **37.** \$1,800 **39.** 7.5% **41.** \$102
**43.** 720 students **45.** \$3,157.50 **47.** 500 s (8 min 20 s)
**49.** \$114.50

## Self-Test for Chapter 6

**[6.1] 1.** 80% **2.** $\frac{7}{100}$ **3.** $\frac{18}{25}$ **4.** 0.42 **5.** 0.06
**6.** 1.6 **[6.2] 7.** 3% **8.** 4.2% **9.** 40% **10.** 62.5%
**[6.3] 11.** *A*: 50; *R*: 25%; *B*: 200 **12.** *A*: unknown; *R*: 8%; *B*: 500
**13.** *R*: 6%; *A*: \$30; *B*: unknown **[6.4] 14.** 11.25 **15.** 500
**16.** 750 **17.** 20% **18.** 7.5% **19.** 175% **20.** 800
**21.** 300 **22.** \$4.96 **23.** 60 questions **24.** \$70.20
**25.** 12% **26.** 24% **27.** 8% **28.** \$18,000
**29.** 6,400 students **30.** \$18,500

## Cumulative Review Chapters 1–6

**[1.1] 1.** Thousands **[1.5] 2.** 11,368 **[1.6] 3.** 89
**[1.7] 4.** 5 **5.** 9 **6.** 42 **[2.1] 7.** 53, 59, 61, 67
**[2.2] 8.** $2 \times 2 \times 5 \times 13$ **9.** 28 **[3.2] 10.** 180
**[2.5] 11.** $8\frac{1}{2}$ **[2.7] 12.** $1\frac{1}{3}$ **[3.4] 13.** $8\frac{7}{12}$ **14.** $4\frac{19}{24}$
**[2.6] 15.** \$286 **[2.7] 16.** $54\frac{\text{mi}}{\text{h}}$ **[3.4] 17.** $41\frac{1}{4}$ in.
**[4.1] 18.** Hundredths **19.** Ten-thousandths **20.** $<$
**21.** $=$ **[4.7] 22.** $\frac{9}{25}$ **23.** $5\frac{1}{8}$ **[4.3] 24.** 11.284
**[4.5] 25.** 17.04 **[4.3] 26.** \$108.05 **[4.4] 27.** 75.36 ft
**28.** 452.16 ft$^2$ **[5.1] 29.** $\frac{2}{3}$ **30.** $\frac{17}{12}$ **[5.4] 31.** $x = 18\frac{2}{3}$
**32.** $y = 0.4$ **33.** 350 mi **34.** \$140 **[6.1] 35.** 0.34; $\frac{17}{50}$
**[6.2] 36.** 0.55; 55% **[6.4] 37.** 45 **38.** 500
**39.** 125 employees **40.** 8.5%

## Pretest Chapter 7

**1.** 108 **2.** 700 **3.** 6 min 30 s **4.** 13 gal 3 qt
**5.** 2 lb 14 oz **6.** **(c)** **7.** 8,000 **8.** m **9.** mL
**10.** 25 **11.** 59

## Reading Your Text

**Section 7.1** **(a)** unit; **(b)** Denominate; **(c)** like; **(d)** largest
**Section 7.2** **(a)** metric; **(b)** yard; **(c)** centi-; **(d)** kilometer
**Section 7.3** **(a)** gram; **(b)** milligram; **(c)** liter; **(d)** millimeter
**Section 7.4** **(a)** calculator; **(b)** kilometers; **(c)** Celsius; **(d)** hot

## Summary Exercises for Chapter 7

**1.** 132 **3.** 24 **5.** 64 **7.** 4 **9.** 4 ft 11 in.
**11.** 9 lb 3 oz **13.** 5 ft 7 in. **15.** 4 h 15 min **17.** 33 h 35 min
**19.** **(a)** **21.** **(c)** **23.** mm **25.** cm **27.** 30
**29.** 8,000 **31.** 0.008 **33.** **(b)** **35.** **(b)** **37.** kg
**39.** 5,000 **41.** 5,000 **43.** **(c)** **45.** **(b)** **47.** L
**49.** mL or cm$^3$ **51.** 5,000 **53.** 9,000 **55.** 326.77
**57.** 6.75 **59.** 5.51 **61.** 62.6 **63.** 37 **65.** 15
**67.** 41

## Self-Test for Chapter 7

**[7.1] 1.** 96 **2.** 48 **3.** 6 ft 9 in. **4.** 11 ft 5 in.
**5.** 2 lb 9 oz **6.** 15 h 20 min **7.** 4 lb 6 oz **8.** \$1,050
**[7.2] 9.** **(b)** **10.** **(b)** **[7.3] 11.** **(b)** **12.** **(b)**
**[7.2] 13.** 5,000 **[7.3] 14.** 3,000 **15.** 3 **[7.4] 16.** 101.6
**17.** 3.2 **18.** 55.1 **19.** 67.5 **20.** 30.5
**21.** 14.4 **22.** 75.2

## Cumulative Review Chapters 1–7

**[1.5] 1.** \$896 **[1.6] 2.** \$62 **[1.7] 3.** 16
**[2.2] 4.** $2 \times 2 \times 2 \times 3 \times 7$ **5.** 4 **[2.4] 6.** $\frac{3}{5}, \frac{5}{8}, \frac{2}{3}$
**[2.5] 7.** $\frac{3}{4}$ **[2.7] 8.** $\frac{5}{12}$ **[3.2] 9.** 60 **[3.3] 10.** $1\frac{1}{30}$
**[3.4] 11.** $3\frac{13}{24}$ **[4.2] 12.** \$18.30 **[4.3] 13.** 27.84 cm$^2$
**[4.2] 14.** 21.5 cm **[4.6] 15.** 0.5625 **16.** 0.538
**[5.4] 17.** $x = 24$ **[5.5] 18.** 400 mi **19.** 450 mi
**[6.2] 20.** 37.5% **[6.1] 21.** $\frac{1}{8}$ **[6.4] 22.** 3,526
**23.** 225% **24.** 150 **[6.5] 25.** \$142,500 **[7.1] 26.** 120
**27.** 1 min 35 s **28.** 21 lb 11 oz **29.** 2 ft 9 in.
**30.** 21 h 20 min **[7.2] 31.** 0.43 **[7.3] 32.** 62,000
**[7.2] 33.** 74 **[7.3] 34.** 14,000 **35.** 0.5 **[7.4] 36.** 13.3
**37.** 149.6 **38.** 29.4 **39.** 48.2

## Pretest Chapter 8

**1.** 25.12 yd **2.** **(a)** 153.86 in.$^2$; **(b)** 7.5 in.$^2$; **(c)** 6 ft$^2$
**3.** Neither **4.** 60°; acute **5.** 100° **6.** Isosceles
**7.** Obtuse **8.** **(a)** and **(c)** **9.** 40° **10.** 5 **11.** 15
**12.** 55 m **13.** 10

## Reading Your Text

**Section 8.1** **(a)** circumference; **(b)** radius; **(c)** pi; **(d)** acre
**Section 8.2** **(a)** earth; **(b)** perpendicular; **(c)** obtuse;
**(d)** complementary

Section 8.3 (a) equilateral; (b) isosceles; (c) similar; (d) corresponding
Section 8.4 (a) radical; (b) hypotenuse; (c) Pythagorean; (d) whole

## Summary Exercises for Chapter 8

**1.** 37.7 ft **3.** 314 ft$^2$ **5.** 400 in.$^2$ **7.** $787.50
**9.** $\angle COD$; right; 90° **11.** $\angle RST$; obtuse; 140°
**13.** $\angle BAC$; acute; 60° **15.** 252° **17.** 98° **19.** 42°
**21.** 112° **23.** 71° **25.** 30°; isosceles **27.** 60°; scalene
**29.** 15.8 **31.** 18 **33.** 13.75 **35.** 55

## Self-Test for Chapter 8

**[8.1] 1.** 10.05 ft
**[8.2] 2. (a)** Parallel; **(b)** neither; **(c)** perpendicular; **(d)** neither
**3. (a)** Straight; **(b)** obtuse; **(c)** right **4.** 50° **5.** 135°
**6.** 300° **7.** 137° **8.** 66° **[8.3] 9.** Obtuse **10.** Right
**11.** Scalene **12.** Isosceles **13. (b)** and **(c)** **14.** 67°
**15.** 53.9 **[8.4] 16.** 21 **17.** 65 m **18.** 112 mm

## Cumulative Review Chapters 1–8

**[1.1] 1.** Ten-thousands **[1.7] 2.** 64 **[2.1] 3.** $2 \cdot 3 \cdot 3 \cdot 5 \cdot 7$
**4.** 4 **5.** 120 **[2.5] 6.** $\frac{3}{4}$ **[2.6] 7.** $4\frac{1}{2}$ **8.** $578
**9.** $60 \frac{\text{mi}}{\text{h}}$ **[3.3] 10.** $1\frac{13}{30}$ **[3.4] 11.** $3\frac{13}{24}$ **12.** $6\frac{7}{12}$ mi
**[4.1] 13.** Hundredths **14.** Ten–thousandths
**[4.4] 15.** $P = 24.2$ ft; $A = 35.04$ ft$^2$ **[4.2] 16.** $\frac{1}{8}$
**17.** 0.3125 **[5.2] 18.** $20,250/yr **[5.4] 19.** $w = 4$
**20.** 1,125 ft$^2$ **[6.1] 21.** 0.085 **22.** $\frac{3}{8}$ **[6.2] 23.** 67.5%
**[6.4] 24.** 51.2 **25.** 1,500 **26.** 1,350
**[7.1–7.4] 27.** 5,280 **28.** 0.25 **29.** 5,800
**[8.1] 30.** 49.6 ft **31.** 342 cm$^2$ **32.** 1,256 m$^2$
**33.** 60.63 ft$^2$ **[8.2] 34.** 45° **[8.3] 35.** 73°; isosceles
**[8.4] 36.** $x = 40$ **37.** 17 ft

## Pretest Chapter 9

**1.** Mean: 27; median: 24; mode: 42 **2.** 35 million pounds
**3.** 91 million pounds **4.** Approximately 25,000
**5.** A decrease of 20,000 **6.** 1980–1985; 20,000
**7.** $17,200,000,000 **8.** $22,800,000,000
**9.** Min: $1.74; $Q_1$: $1.78; median: $1.83; $Q_3$: $1.86; max: $1.92
**10.**

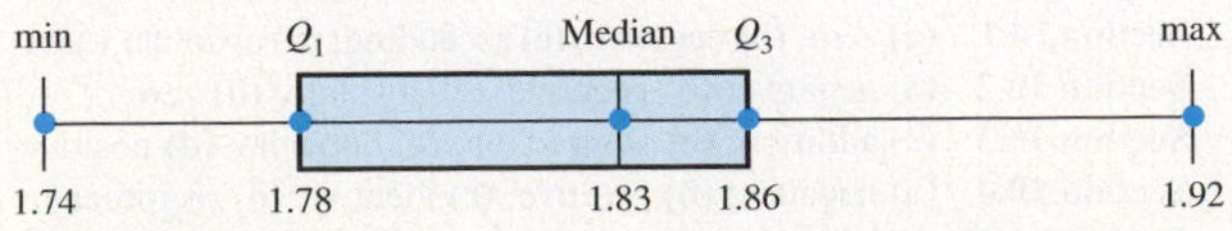


## Reading Your Text

Section 9.1 (a) mean; (b) median; (c) mode; (d) bimodal
Section 9.2 (a) parallel; (b) cell; (c) pictograph; (d) legend
Section 9.3 (a) line; (b) future; (c) extrapolation; (d) predict
Section 9.4 (a) table; (b) approximation; (c) pie; (d) circle
Section 9.5 (a) quartiles; (b) three; (c) summary; (d) box-and-whisker

## Summary Exercises for Chapter 9

**1.** 6 **3.** 120 **5.** 85 **7.** 18.6; 20 **9.** 29.4; 28 **11.** 92
**13.** 32,000 vehicles; 10,145,000 vehicles **15.** 60%
**17.** 17.7% **19.** 60%
**21.** 1 house = $100,000

**23.** 5,000 students **25.** 250,000 computers
**27.** About 400,000 computers
**29.**

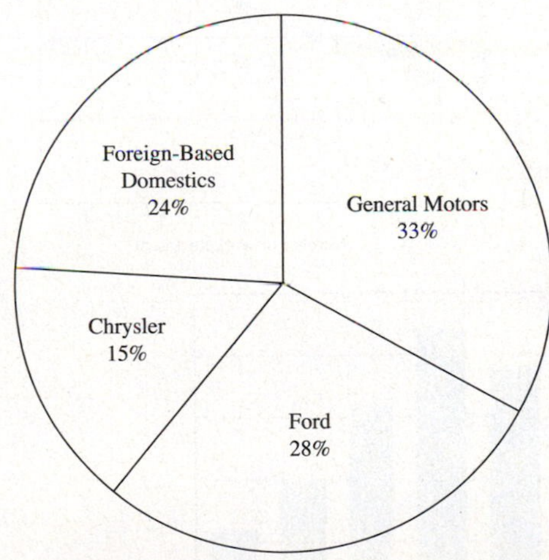


**31.** 47% **33.** 21, 24, 28, 30, 35 **35.** The data do not exhibit any striking features (the data set is mildly skewed right).

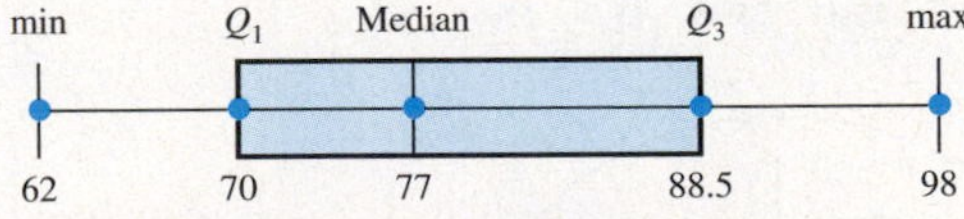


The data do not exhibit any striking features. A statistician might say that the data are mildly *skewed to the right.*

## Self-Test for Chapter 9

**[9.1] 1.** 15 **2.** 8 **3.** 11 **4.** 6 **5.** 204 **6.** 93
**7.** Brown **[9.2] 8.** 31,162 **9.** 41,378 **10.** 324%
**11.** 124%

**12.**

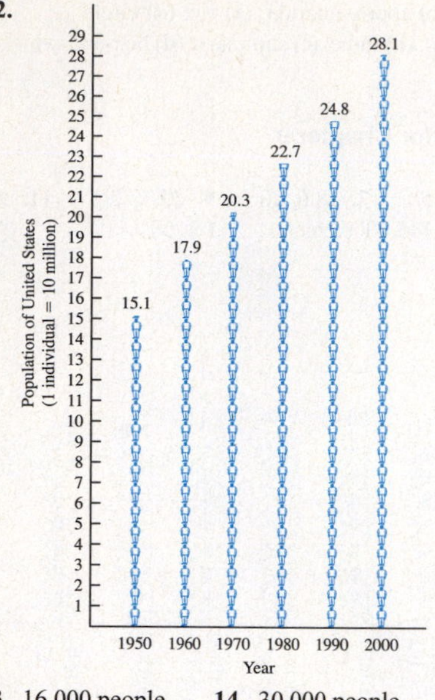


**13.** 16,000 people **14.** 30,000 people **15.** 10,000 people
**16.** 16,000 people **17.** 2000 to 2001 **[9.3] 18.** December
**19.** August and September
**20.** about 21 defects

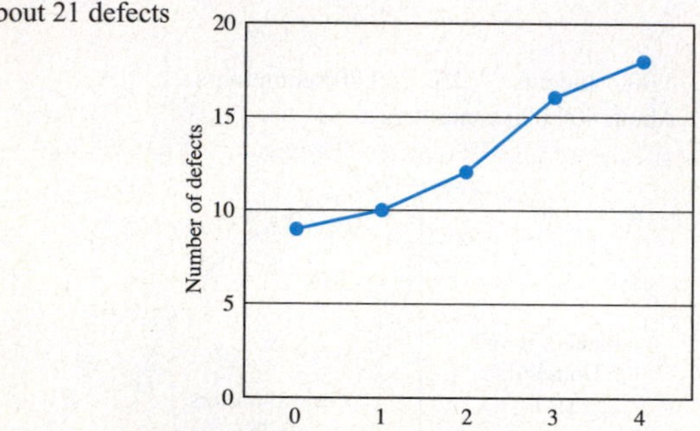


**[9.4] 21.**

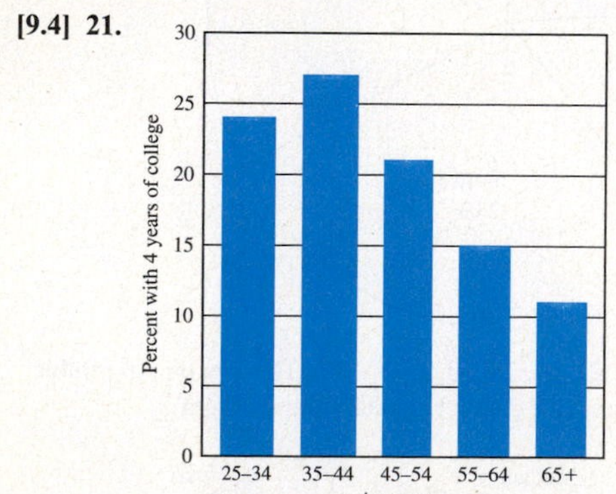


**22.**

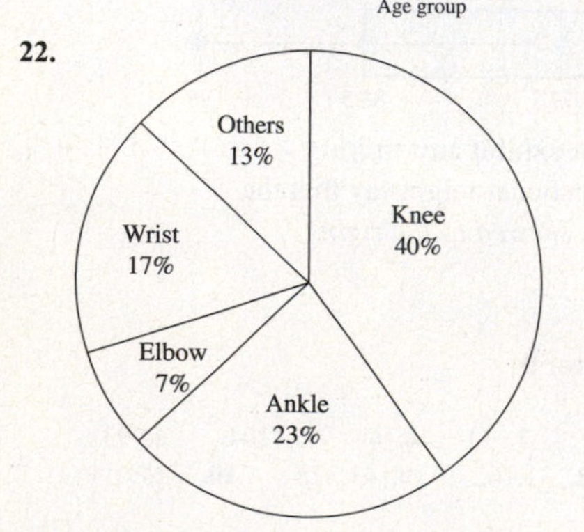


**23.** 45% **24.** 85%
**[9.5] 25.** min $Q_1$ Median $Q_3$ max
0.00 0.35 0.56 1.03 1.83

## Cumulative Review Chapters 1–9

**[1.1] 1.** Thousands **[1.2] 2.** 32,278 **[1.3] 3.** 39,288
**[1.5] 4.** 26,230 **[1.6] 5.** 308 r 6 **[1.3] 6.** 48,588
**[4.3] 7.** 75.215 **[2.5] 8.** $\frac{2}{3}$ **[2.7] 9.** $\frac{6}{11}$
**[3.4] 10.** $7\frac{1}{12}$ **[5.4] 11.** $x = 14$ **12.** $x = 9$
**[6.1] 13.** 0.18; $\frac{9}{50}$ **14.** 0.425; 42.5% **[7.1] 15.** 11 lb 5 oz
**16.** 1 min 35 s **[7.2] 17.** 8,000 **[7.3] 18.** 3
**19.** 5 **20.** 250 **[9.3] 21.** 2002 and 2003
**[9.4] 22.**

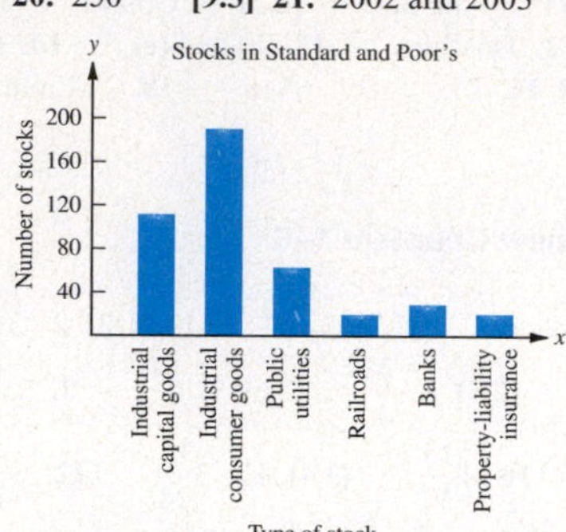


**[9.1] 23.** Mean: 9; median: 9; mode: 11 **[5.2] 24.** 70 gal
**25.** $88\ \frac{\text{ft}}{\text{s}}$ **26.** $19\ \frac{\text{pitches}}{\text{inning}}$ **[1.8] 27.** $408
**28.** $48\ \frac{\text{mi}}{\text{h}}$ **29.** $28\frac{3}{5}$ cm **30.** $55\frac{1}{3}$ ft **[6.4] 31.** 133
**32.** 0.2% **33.** 185 **[2.5] 34.** $4\frac{1}{3}$
**[6.5] 35.** 2,800 students

## Pretest Chapter 10

**1.** $-7, -3, -\frac{3}{4}, \frac{1}{2}, 1, 4, 9$ **2.** Max: 7; min: −5 **3.** 4
**4.** −6 **5.** **(a)** 7; **(b)** −12 **6.** **(a)** −22; **(b)** 17
**7.** **(a)** −36; **(b)** 54 **8.** **(a)** 3; **(b)** −5 **9.** 24 **10.** 20

## Reading Your Text

**Section 10.1** **(a)** zero; **(b)** negative; **(c)** ascending; **(d)** absolute value
**Section 10.2** **(a)** negative; **(b)** negative; **(c)** absolute; **(d)** zero
**Section 10.3** **(a)** addition; **(b)** subtraction; **(c)** opposite; **(d)** positive
**Section 10.4** **(a)** negative; **(b)** positive; **(c)** identity; **(d)** reciprocal
**Section 10.5** **(a)** negative; **(b)** positive; **(c)** Division; **(d)** positive

## Summary Exercises for Chapter 10

**1.**

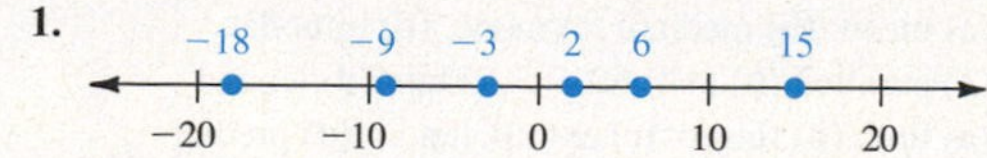

3. $-\frac{4}{5}, -\frac{2}{3}, -\frac{1}{2}, \frac{3}{5}, \frac{7}{10}, \frac{5}{6}$ **5.** Max: 8; min: $-9$ **7.** 9
**9.** $-9$ **11.** $-4$ **13.** $-4$ **15.** 6 **17.** $-32$ **19.** $-1$
**21.** $-14$ **23.** $-17$ **25.** $-4$ **27.** 2 **29.** 16
**31.** $-4$ **33.** $-4$ **35.** $-2$ **37.** $-1$ **39.** $-70$
**41.** 45 **43.** 0 **45.** $-\frac{3}{2}$ **47.** 80 **49.** 10 **51.** $-2$
**53.** $-60$ **55.** $-9$ **57.** 0 **59.** Undefined **61.** $-1$

## Self-Test for Chapter 10

**[10.1] 1.**

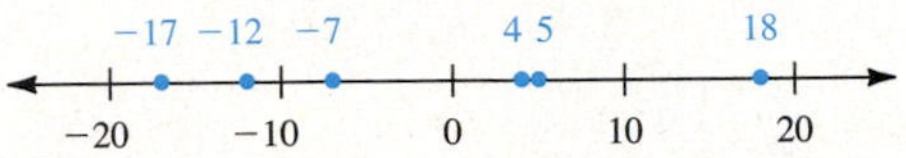


**2.** $-6, -3, -2, 0, \frac{1}{2}, \frac{3}{4}, 2, 4, 5$ **3.** Max: 6; min: $-5$ **4.** 7
**5.** 7 **6.** 11 **7.** 11 **[10.2] 8.** $-13$ **9.** $-3$
**10.** $-21$ **11.** 1 **[10.3] 12.** $-6$ **13.** $-24$
**14.** 9 **15.** 0 **[10.4] 16.** $-40$ **17.** 63 **18.** $-27$
**19.** $-24$ **[10.5] 20.** $-25$ **21.** 3 **22.** $-5$
**23.** Undefined **24.** $-24$ **25.** $-10$

## Cumulative Review Chapters 1–10

**[1.1] 1.** Ten thousands **[1.2] 2.** 142,231 **[1.3] 3.** 29,573
**[1.5] 4.** 53,445 **[1.6] 5.** 402 r 28 **[4.2] 6.** 13.687
**[4.3] 7.** 1,837.353 **[2.5] 8.** $\frac{3}{7}$ **[2.7] 9.** 1
**[3.4] 10.** $8\frac{19}{54}$ **[5.4] 11.** $x = 4$ **[6.1] 12.** 0.58; $\frac{29}{50}$
**[6.2] 13.** 0.48; 48% **[7.1] 14.** 8 ft 10 in. **15.** 9 lb 4 oz
**16.** 12 ft 6 in. **17.** 2 lb 14 oz **18.** 10 h 30 min
**19.** 2 min 9 s **20.** 14 ft 8 in.
**[5.2] 21.** The three smaller bottles **[4.4] 22.** 49.02 $cm^2$
**23.** $72\frac{1}{4}$ $in.^2$ **24.** 25.7 ft **[6.4] 25.** 80 **[6.5] 26.** 7.5%
**[7.3] 27.** 0.017 **[7.2] 28.** 820 **[7.5] 29.** 160°
**[9.2] 30.** 5,000 students **[9.1] 31.** Mean: 17; median: 17; mode: 17 **[10.2] 32.** 4 **[10.3] 33.** 20 **34.** $-32$
**35.** $-15$ **36.** 31 **[10.4] 37.** 108 **[10.5] 38.** 4
**39.** 41 **40.** 72

## Pretest Chapter 11

**1.** **(a)** $y - 1$; **(b)** $2(a + b)$ **2.** **(a)** 22; **(b)** 45 **3.** $11m^2 - 2m$
**4.** $2a + 4$ **5.** $x = -4$ **6.** $x = 10$ **7.** $x = 6$
**8.** $x = -20$ **9.** $x = 3$ **10.** $x = -9$

## Reading Your Text

**Section 11.1** **(a)** variables; **(b)** difference; **(c)** multiplication; **(d)** letter
**Section 11.2** **(a)** variable; **(b)** fraction; **(c)** operations; **(d)** positive
**Section 11.3** **(a)** term; **(b)** number; **(c)** like; **(d)** distributive
**Section 11.4** **(a)** equation; **(b)** true; **(c)** solution(s); **(d)** sentence
**Section 11.5** **(a)** equivalent; **(b)** multiplication; **(c)** dividing; **(d)** reciprocal
**Section 11.6** **(a)** isolate; **(b)** addition; **(c)** original; **(d)** solved

## Summary Exercises for Chapter 11

**1.** $y + 5$ **3.** $8a$ **5.** $5mn$ **7.** $17x + 3$ **9.** $-7$
**11.** 15 **13.** 19 **15.** 25 **17.** 1 **19.** $-3$
**21.** $4a^3, -3a^2$ **23.** $5m^2, -4m^2, m^2$ **25.** $12c$ **27.** $2a$
**29.** $3xy$ **31.** $19a + b$ **33.** $3x^3 + 9x^2$ **35.** $10a^3$
**37.** Yes **39.** Yes **41.** No **43.** $x = 2$ **45.** $x = -5$
**47.** $x = 5$ **49.** $x = 1$ **51.** $x = -7$ **53.** $x = 7$
**55.** $x = -4$ **57.** $x = 32$ **59.** $x = 27$ **61.** $x = 3$
**63.** $x = -2$ **65.** $x = \frac{7}{2}$ **67.** $x = 18$ **69.** $x = 6$
**71.** $x = \frac{2}{5}$ **73.** $x = 6$ **75.** $x = 6$ **77.** $x = 4$
**79.** $x = 5$ **81.** $x = \frac{1}{2}$

## Self-Test for Chapter 11

**[11.1] 1.** $a - 5$ **2.** $6m$ **3.** $4(m + n)$ **4.** $\frac{a + b}{3}$
**[11.2] 5.** $-4$ **6.** 80 **7.** 144 **8.** 5 **[11.3] 9.** $15a$
**10.** $3x^2y$ **11.** $19x + 5y$ **12.** $8a^2$ **[11.4] 13.** No
**14.** Yes **15.** $x = 11$ **16.** $x = 12$ **17.** $x = 7$
**[11.5] 18.** $x = 7$ **19.** $x = -12$ **20.** $x = 25$
**[11.6] 21.** $x = 3$ **22.** $x = 4$ **23.** $x = -\frac{2}{3}$
**24.** $x = -5$

## Cumulative Review Chapters 1–11

**[1.2] 1.** Associative property of addition
**[1.5] 2.** Commutative property of multiplication
**3.** Distributive property **[1.4] 4.** 5,900 **5.** 950,000
**[1.7] 6.** 8 **[2.2] 7.** $2 \times 2 \times 2 \times 3 \times 11$ **[3.2] 8.** 90
**[2.3] 9.** $3\frac{1}{7}$ **10.** $\frac{53}{8}$ **[2.5] 11.** $\frac{3}{4}$ **[2.6] 12.** $1\frac{1}{2}$
**[3.4] 13.** $8\frac{1}{24}$ **14.** $3\frac{5}{8}$ **15.** $\frac{5}{8}$ in. **[4.3] 16.** \$3.06
**[8.1] 17.** 8.04 $ft^2$ **[4.5] 18.** 32 **[4.2] 19.** $0.\overline{72}$
**[5.4] 20.** $m = 112.5$ **[5.5] 21.** 12 in. **[6.2] 22.** 0.3%
**23.** 62.5% **24.** 350% **[6.4] 25.** 150 **26.** 600
**27.** \$27,500 **[7.3] 28.** 0.3 **[9.3] 29.** 80
**[10.2] 30.** $-8$ **[10.1] 31.** 20 **32.** 12 **33.** 5
**[10.2] 34.** $-18$ **[10.3] 35.** $-4$ **[10.4] 36.** $-90$
**37.** $-4$ **[11.1] 38.** $3(x + y)$ **39.** $\frac{n - 5}{3}$
**[11.2] 40.** $-60$ **41.** 27 **42.** 24 **43.** 3
**[11.3] 44.** $12a^2 + 3a$ **45.** $-2y$ **[11.4] 46.** $x = 5$
**[11.5] 47.** $x = -24$ **[11.6] 48.** $x = -\frac{2}{5}$ **49.** $x = 5$